Transcriptional Regulation in Eukaryotes

Concepts, Strategies, and Techniques

Transcriptional Regulation in Eukaryotes

Concepts, Strategies, and Techniques

Michael Carey
*University of California,
Los Angeles*

Stephen T. Smale
*Howard Hughes Medical Institute and
University of California, Los Angeles*

CSHL
PRESS

COLD SPRING HARBOR LABORATORY PRESS
COLD SPRING HARBOR, NEW YORK

Transcriptional Regulation in Eukaryotes
Concepts, Strategies, and Techniques

Developmental Editor: Judy Cuddihy
Assistant Developmental Editor: Birgit Woelker
Project Coordinator: Maryliz M. Dickerson
Production Editor: Patricia Barker

Desktop Editors: Danny deBruin, Susan Schaefer
Interior Book Design: Denise Weiss
Cover Design: Tony Urgo
Cover art rendered by Michael Haykinson

Cover illustration: The cover schematically illustrates the structure of the RNA polymerase II transcription complex emerging from a black box. It is a composite illustration of the TFIIA-TBP-TATA (Geiger et al. *Science* 272: 830–836 [1996]; Tan et al. *Nature* 381: 127–151 [1996]), and TFIIB-TBP-TATA (Nikolov et al. *Nature* 377: 119–128 [1995]) crystal structures rendered by Michael Haykinson (UCLA) using the Molecular Graphics structure modeling computer program Insight II.

Library of Congress Cataloging-in-Publication Data

Carey, Michael (Michael F.)
    Transcriptional regulation in eukaryotes: concepts, strategies, and techniques/Michael
Carey, Stephen T. Smale.
      p. cm.
    Includes bibliographical references and index.
    ISBN 0-87969-537-4 (alk.paper)
    1. Genetic transcription--Regulation. 2. Transcription factors. 3. Genetic
transcription--Regulation--Research--Methodology. I. Smale T. II. Title.

QH450.2.C375 1999
572.8′845--dc21

99-049636

10 9 8 7 6 5 4 3 2 1

Contents

Preface

Since the advent of recombinant DNA technology three decades ago, thousands of eukaryotic genes have been isolated. The differential expression of these genes is critical for both normal cellular processes and abnormal processes associated with disease. To understand these processes, a growing number of investigators from diverse fields of biology have begun to study the molecular mechanisms regulating gene transcription. Furthermore, the genome projects under way throughout the world have led to the identification of the entire gene complements of *Saccharomyces cerevisiae, Caenorhabditis elegans,* and numerous archaeal and eubacterial organisms. Within the next few years, the approximately 100,000 genes within the human genome will have been identified. After this goal is realized, the need to dissect mammalian transcriptional control regions and regulatory mechanisms rigorously will increase dramatically.

Despite the global interest in elucidating mechanisms of transcriptional regulation, a comprehensive source of strategic, conceptual, and technical information has not been available for those entering the field for the first time. Although protocols for numerous techniques have been published, the strategic decisions necessary to carry out a step-by-step analysis have not been outlined. This deficiency became apparent to us while we were serving as instructors for the Eukaryotic Gene Expression course held each summer at Cold Spring Harbor Laboratory. This laboratory course was designed for physician-scientists interested in understanding the regulation of a specific disease-related gene, Ph.D. scientists trained in other fields who became interested in the regulatory mechanisms for a gene involved in a particular biological process, and graduate students or postdoctoral fellows who were initiating transcriptional regulation projects. This book is targeted toward this same diverse group of scientists who have developed an interest in transcriptional regulation.

In writing this book, we have focused on issues that the average investigator faces when undertaking a transcriptional regulation analysis, and we have outlined recommended strategies for completing the analysis. One risk of describing a prescribed step-by-step approach is that it may suppress creativity and may not be applicable to all regulatory scenarios. To the contrary, our hope is that our recommendations will enhance creativity by allowing it to evolve from an informed perspective.

We thank the many participants in the Eukaryotic Gene Expression Course from 1994 through 1998 for providing the inspiration and motivation for this book. We also acknowledge our colleagues at UCLA, the members of our laboratories, and our co-instructors for the Eukaryotic Gene Expression course, including Marc Learned, Ken Burtis, Grace Gill, David Gilmour, and Jim Goodrich, for many valuable discussions. We are deeply indebted to a number of colleagues for specific contributions and reading of sections, including Doug Black, Mike Haykinson, Leila Hebshi, Reid Johnson, Ranjan Sen, and Amy

Weinmann. We are particularly grateful to our editor Judy Cuddihy and the book's reviewers, Grace Gill, Bill Tansey, and Steve Hahn, whose generous contribution of time and ideas made the undertaking intellectually rewarding and personally enjoyable. The book was greatly improved by the work of Birgit Woelker and Maryliz Dickerson at Cold Spring Harbor Laboratory Press, as well as Jan Argentine, Pat Barker, and Denise Weiss. Finally, we acknowledge Cold Spring Harbor Laboratory Press Director John Inglis, whose encouragement was essential for the completion of this novel project.

M.C. and S.T.S.

Overview

The goal of this book is to provide a detailed description of the approaches to be employed and issues to be considered when undertaking a molecular analysis of the transcriptional regulatory mechanisms for a newly isolated gene, or a biochemical analysis of a new transcription factor. Our emphasis is on mammalian transcription, which is complicated by the combinatorial nature of regulation and the lack of facile genetics. We refer periodically to studies in yeast, *Drosophila*, and other organisms where more tractable genetic approaches have led to a detailed understanding of particular mechanistic issues. The topics covered in the book extend from the determination of whether a gene is in fact regulated at the level of transcription initiation to advanced strategies for characterizing the biochemical mechanism underlying its combinatorial regulation by activators. Although numerous specialized and detailed techniques are included, the unique characteristics of this book are its strategic and conceptual emphasis on analysis of individual genes and the transcription factors that regulate them.

Chapter 1 reviews the current state of the RNA polymerase II transcription field. This chapter provides an investigator entering the field with a comprehensive introduction into areas of active research and the types of regulatory strategies that will be confronted. We have defined the general properties of known regulatory regions (i.e., enhancers, promoters, silencers), components of the transcriptional machinery (mediator components and the general transcription factors), activators, and repressors. Select review articles and online information sources are included for the novice interested in additional details on the various topics. Emphasis is placed on the role of macromolecular complexes in regulation.

Chapters 2–9 were conceived as a step-by-step guide for an investigator who wants to pursue the regulatory mechanisms for a new gene that has been identified. Chapter 2 presents general strategic issues to consider before the analysis is initiated. First and foremost is a discussion of the goals of the analysis. This topic was included because it has become apparent that many investigators enter the transcription field with unrealistic expectations. Presumably, these expectations arise because a preliminary analysis of a control region, using basic reporter assays and electrophoretic mobility shift assays, is relatively straightforward. To the contrary, a substantial amount of effort is usually required to make meaningful progress toward an understanding of a gene's regulatory mechanisms. Chapter 2 also contains a discussion of the feasibility of achieving the goals. The feasibility is largely dependent on the availability of particular tools, including appropriate cell lines for functional and biochemical studies, and an appropriate functional assay. The chapter concludes with a discussion of whether to begin the analysis by studying the promoter or, alternatively, distant control regions, with a brief description of the initial steps required for each starting point. In this book, the phrase "distant control regions" is used in reference to any control region that is distinct from the promoter, such as enhancers, locus control regions, and silencers.

One issue that will become apparent in Chapter 2 and in all subsequent chapters is that specific protocols are not included for many of the methods described. Instead, references are given to standard methods manuals, in particular Sambrook et al. (1989) and Ausubel et al. (1994). The intention was to avoid duplication of the valuable information provided in pre-existing manuals and to instead focus on strategic advice. Although the book could have been written without any protocols, since they all can be found in the literature, we chose to include selected protocols for three reasons. First, some of the protocols were chosen because we felt that the reader would benefit from a detailed explanation of the specific steps and history of the methodology, information generally not found in other manuals. Second, in some instances we felt it necessary to provide the reader with a sense of the mechanics of a technique while reading the book. Finally, several protocols were included because of their special nature (e.g., permanganate footprinting, TFIID binding studies) and the fact that no general source exists for such methods.

Chapter 3 continues the step-by-step guide by describing how to determine the mode of regulation for a new gene. At the outset, this chapter emphasizes the fact that the regulation of a biochemical activity does not necessarily mean that the gene encoding the protein is subject to regulation. Alternative possibilities are the regulation of protein synthesis or degradation, or posttranslational regulation of the biochemical activity itself. Furthermore, if the gene is found to be regulated, it is not necessarily regulated at the level of transcription initiation. Rather, it may be regulated at the level of transcription elongation, mRNA stability, pre-mRNA splicing, polyadenylation, or mRNA transport. Because regulation at the level of transcription initiation is most difficult to distinguish from regulation of mRNA stability and transcription elongation, the basic principles of these latter modes of regulation are discussed. Furthermore, strategies for distinguishing between the various modes of regulation are presented, along with a detailed protocol for one important technique, the nuclear run-on.

As stated above, one critical decision discussed in Chapter 2 is whether to begin an analysis of transcriptional regulation by studying the promoter or, alternatively, the distant control regions for the gene. If the investigator opts to study the promoter, the approaches detailed in Chapters 4 and 5 should be followed if the gene is found to be regulated at the level of transcription initiation. Chapter 4 describes methods for determining the location of the transcription start site, an essential first step in every promoter analysis. Four methods for start-site mapping are described, including the primer extension, RNase protection, S1 nuclease, and RACE methods. The advantages and limitations of each method are discussed, and detailed protocols are included for the first three.

Chapter 5 considers the development of a functional assay for a promoter; in other words, the development of an assay that can be used to identify, by mutagenesis (see Chapter 7), the individual control elements required for promoter activity. Transient and stable transfection assays are discussed in detail, including an overview of transfection procedures, reporter genes, vectors, and assays, and the initial design and interpretation of experiments. Alternative functional assays, including in vitro transcription and transgenic assays, are also briefly mentioned, along with their advantages and disadvantages. Chapter 5 is the first of several chapters where the text becomes strongly focused toward a discussion of transcriptional activation, with very little discussion of transcriptional repression. The intention was not to minimize the importance of repression mechanisms for transcriptional regulation; however, a discussion of each point from the perspective of both activation and repression would have been unmanageable. In most cases, it therefore is left to the reader to determine how the principles discussed can be applied to a repression analysis.

If an investigator chooses to pursue distant control regions instead of, or in addition to, the promoter, Chapters 5 and 6 are designed to follow Chapter 3. Chapter 5, as described above, contains basic information regarding the design of functional assays. This information is applicable to both promoters and distant regions. Chapter 6 describes approaches for identifying distal control regions, including the recommended starting point of performing DNase I hypersensitivity experiments. Chapter 6 also describes special strategies not discussed in Chapter 5 for developing functional assays to analyze distant control regions.

After a functional assay is developed for a promoter (Chapters 4 and 5) or distant control region (Chapters 5 and 6), the next step is to dissect the individual DNA elements constituting the region. These procedures, which usually involve a systematic mutant analysis, are described in Chapter 7. This chapter stresses the benefits of a mutant analysis, but also describes other strategies that may lead to the identification of important DNA elements within a control region.

After the DNA elements are identified, the proteins that bind to them must be identified and their genes cloned. These procedures are described in Chapter 8, beginning with the development of EMSA and DNase I footprinting assays for use with crude nuclear extracts. These assays are discussed in greater detail in Chapter 13 from the perspective of an analysis of a pure recombinant protein. An attempt was made to minimize the duplication of information between these two chapters. However, to maintain the logical progression of the book, some redundancy was unavoidable. Various strategies that can be used to clone the gene encoding a DNA-binding protein are then described, including protein purification, the yeast one-hybrid screen, in vitro expression library screen, mammalian expression cloning, degenerate PCR, and database approaches.

Chapter 9 completes the step-by-step outline of the characterization of a new gene by focusing on a crucial issue: After a factor that binds an important DNA element in vitro is identified, how can one determine whether that factor is indeed responsible for the function of the control element in vivo? Although no experiment is available that can provide conclusive evidence that the protein is functionally relevant, twelve experimental strategies are described that can be used to test the hypothesis. As with all science, the strength of the hypothesis will correspond to the number of rigorous tests to which it has been subjected.

The analysis of a control region, using the strategies described in Chapters 2–9, relies on the use of artificial assays, such as transfection assays and in vitro DNA-binding assays. To complement these approaches, it can be helpful to study the properties of the endogenous control region within its natural environment. Chapter 10 describes experimental strategies for such a characterization, beginning with genomic footprinting and in vivo crosslinking/immunoprecipitation strategies for visualizing specific protein–DNA interactions at the endogenous locus. Chromatin structure is also known to be an important contributor to gene regulation and is best studied in the context of the endogenous locus. Therefore, strategies are included for determining nucleosome positioning and remodeling. Strategies for analyzing DNA methylation status and subnuclear localization of a gene are also briefly discussed.

From a biochemical point of view, an understanding of the mechanism of gene regulation involves recreating regulated transcription in vitro and delineating the precise protein–protein and protein–DNA interactions involved in the process. Chapters 11–15 describe approaches for recreating and studying gene regulation in vitro using purified and reconstituted biochemical systems.

The initial starting point in a biochemical analysis of any regulatory protein is to synthesize the protein and its derivatives in recombinant form. Chapter 11 provides a list of approaches for expressing proteins, and guides the investigator through the strategic and technical decisions encountered in choosing an appropriate system for diverse applications. The chapter outlines the fundamentals of using *E. coli* to generate small regulatory molecules (e.g., DNA-binding domains of activators and repressors) and baculovirus and retroviral systems to generate multi-protein complexes.

Typically, as an investigator proceeds through different stages of an analysis, it becomes imperative to delineate the protein domains engaged in interactions with other regulatory proteins and with the transcriptional machinery. This information is essential for completing a biochemical analysis of mechanism. The approach employed to gain such insights is termed "structure–function" analysis. This is not a trivial task, and the approach and decision-making are often based on the particular type of regulatory protein being studied. Chapter 12 discusses structure–function analysis from several perspectives. Approaches for studying protein interactions are described briefly to permit the investigator to design specific assays for analyzing the relevant domains. Simple deletion analysis is discussed as a means to delineate how different regions of a regulatory protein contribute to different aspects of DNA binding and transcriptional regulation. This discussion serves as a springboard to more advanced approaches, including domain swapping, a straightforward means to ascribe precise functions to portions of proteins. Most importantly, however, a molecular understanding of transcription is often derived from knowledge of the specific amino acid residues mediating the relevant contacts. Particular emphasis is placed on guiding the investigator through different conceptual approaches to generating site-directed mutants, how such mutants are modeled, and case studies in which mutagenesis is compared with the results of crystal structures. Finally, the chapter discusses the exciting and emerging concept that structural information can be employed to generate novel "altered specificity" genetic systems for analyzing transcriptional mechanisms.

DNA recognition by combinations of proteins is the major contributor to the cell and developmental specificity of a transcriptional response. The mechanisms employed by proteins to bind a promoter or enhancer, both alone and cooperatively with other proteins, are key areas of study in the transcription field. As new transcription factors are identified from the genome project, even more focus will be placed on understanding DNA-binding cooperativity and combinatorial interactions. Chapter 13 describes the fundamentals of equilibrium binding. It introduces the concepts of DNA recognition, describes the chemistry of DNA–protein interactions to the novice, and finally, discusses how chemical and nuclease probes can be employed to generate detailed models for DNA binding. Furthermore, the chapter outlines case studies where models derived from chemical probing are compared with the results of crystal structures of DNA–protein co-complexes. Finally, but most importantly, the chapter provides a basic introduction to the concept and study of nucleoprotein complexes called enhanceosomes, an emerging area of research that underlies the combinatorial action of transcription factors.

Ultimately, the investigator may wish to understand the detailed biochemical steps affected by activators. This goal involves two undertakings: First, development of a robust in vitro transcription system that recreates the regulatory phenomenon in vitro and, second, design of mechanistic experiments with highly specialized reagents including purified transcription factors and chromatin templates. Chapter 14 guides the investigator through the logistical decisions and reagents necessary to design the appropriate reporter templates and to develop active transcription systems. The chapter discusses how in vitro transcription reactions are measured and optimized, including G-less cassettes and primer extension, while expanding on the nuances of in vitro systems presented originally in Chapter 8. Descriptions of the available methods for generating reconstituted systems with crude or pure general factors and Pol II and the development of systems for analyzing chromatin templates are also presented.

Once activators are shown to stimulate transcription in vitro, the investigator may wish to further pursue the biochemical mechanism of activated transcription using purified transcription reagents. This is a rapidly evolving area in terms of both new concepts and specialized reagents. Chapter 15 presents a historical overview of how different methods were originally applied for understanding basal and activated transcription. The chapter then outlines numerous strategies employed to study specific steps in activated transcription using crude and pure reagents. These include approaches for analyzing transcription complex assembly including sarkosyl sensitivity, the immobilized template approach, permanganate probing, and others. The emphasis is on assay development and data interpretation. The chapter also attempts to provide an up-to-date tabulation of sources for specialized reagents including systems for expressing and purifying recombinant transcription factors and multi-component complexes such as the human holoenzyme, chromatin remodeling machines, human mediator, and TFIID.

REFERENCES

Ausubel F.M., Brent R.E., Kingston E., Moore D.D., Seidman J.G., Smith J.A., and Struhl K. 1994. *Current protocols in molecular biology*. John Wiley and Sons, New York.

Sambrook J., Fritsch E.F., and Maniatis T. 1989. *Molecular cloning: A laboratory manual*. Cold Spring Harbor Laboratory Press, Cold Spring Harbor, New York.

Abbreviations and Acronyms

In addition to standard abbreviations for metric measurements (e.g., ml) and chemical symbols (e.g., HCl), the abbreviations and acronyms below are used throughout this manual.

A, adenine
AcPNV, *Autographa californica* polyhedrosis virus
AdMLP, adenovirus major late promoter
AMV, avian myeloblastosis virus
AR, androgen receptor
ARC, activator-recruited co-factor
ARS, autonomous replication sequence
AOX1, alcohol oxidase
ARE, AU-rich response element
ATP, adenosine triphosphate
att site, attachment site

BAC, bacterial artificial chromosome
BEAF, boundary element-associated factor
bHLH, basic helix-loop-helix
BrdU, bromodeoxyuridine
BRE, TFIIB recognition element
BSA, bovine serum albumin
bZIP, basic leucine zipper

C, cytosine
CAP, catabolite activator protein
CAT, chloramphenicol acetyltransferase
CBP, CREB-binding protein
cDNA, complementary DNA
C/EBP, CCAAT enhancer-binding protein
CHD, chromodomain SWI/SNF-like helicase/ATPase domain and DNA-binding domain
CITE, cap-independent translational enhancers
CMV, cytomegalovirus
CREB, cAMP receptor element binding protein
cRNA, complementary RNA
cs, cold sensitive
CTD, carboxy-terminal domain
CTP, cytosine triphosphate

DAN, deadenylating nuclease
dATP, deoxyadenosine triphosphate
dCTP, deoxycytidine triphosphate
DEPC, diethyl pyrocarbonate
dGTP, deoxyguanosine triphosphate
DHFR, dihydrofolate reductase
DMP, dimethyl pimelidate dihydrochloride
DMS, dimethyl sulfate
DMSO, dimethyl sulfoxide
DPE, downstream core promoter element
DR, direct repeat
DRB, 5,6-dichloro-1-β-D-ribofuranosylbenzimidazole
DTT, dithiothreitol

ECMV, encephalomyocarditis virus
EBV, Epstein-Barr virus
EDTA, ethylenediaminetetraacetic acid
EKLF, erythroid Kruppel-like factor
EMCV, encephalomyocarditis virus
EMSA, electrophoretic mobility shift assay (gel shift)
ES, embryonic stem (cells)
EST, expressed sequence tag
ETL, early-to-late promoter
*Exo*III, exonuclease III

FACS, fluorescence-activated cell sorting
FISH, fluorescence in situ hybridization
FBS, fetal bovine serum

β-gal, β-galactosidase
G, guanine
GFP, green fluorescent protein
GTFs, general transcription factors
gpt, guanine phosphoribosyltransferase
GST, glutathione-S-transferase
GTP, guanosine triphosphate

H$_2$O$_2$, hydrogen peroxide
HA, hemagglutinin
HAT, histone acetyltransferase
HCF, host cell factor
HEBS, HEPES-buffered saline
HEPES, N-2-hydroxyethylpiperazine-N′-2-ethanesulfonic acid
HisD, histidinol dehydrogenase
HIV, human immunodeficiency virus
HIV-1, human immunodeficiency virus type 1
HKLM, heat-killed *Listeria monocytogenes*
HLH, helix-loop-helix
HMBA, hexamethylene bisacetamide

HMG, high mobility group
HMK, heart muscle kinase
hpi, hours post induction
HPLC, high-performance liquid chromatography
HS, hypersensitive
hsp70, heat shock protein 70
HSTF, heat shock transcription factor
HSV, herpes simplex virus
HSV-1, herpes simplex virus type 1
HSV-TK, herpes simplex virus thymidine kinase

IFN-β, interferon-β
Ig, immunoglobulin
IgM, immunoglobulin heavy-chain protein
IL-2, interleukin-2
IL-12, interleukin-12
Inr, initiator elements
int, integrase
IPTG, isopropyl-β-D-thiogalactoside
IRE, iron-responsive element
IRP, iron-regulating protein
ISWI, imitation SWI

LCR, locus control region
LIS, lithium diiodosalicylate
LM-PCR, ligation-mediated PCR
LPS, lipopolysaccharide
LTR, long terminal repeat

M, molar
MAR, matrix attachment region
MBP, maltose binding protein
MEL, mouse erythroleukemia (cells)
MMLV, Moloney murine leukemia virus
MMTV, mouse mammary tumor virus
MNase, micrococcal nuclease
moi, multiplicity of infection
MOPS, 3-(*N*-morpholino) propanesulfonic acid
MPE, methidium propyl EDTA
mRNA, messenger RNA
MTX, methotrexate

NAT, negative activator of transcription
NER, nucleotide excision repair
neo, aminoglycoside phosphotransferase
NHP, nonhistone proteins
Ni-NTA, nickel-nitriloacetic acid
NMR, nuclear magnetic resonance
NP-40, Nonidet P-40

NTP(s), nucleotide triphosphate(s)
NURF, nucleosome remodeling factor
$\mathbf{O_L}$, leftward operator
$\mathbf{O_R}$, rightward operator
OH-radical, hydroxyl-radical
ONPG, O-nitrophenyl-β-D-galactopyranoside
OP-Cu, Cu-phenanthroline
ORC, origin recognition complex
ori, origin of replication

$\mathbf{P_R}$, promoter in rightward direction
$\mathbf{P_{RM}}$, promoter for repressor maintenance
PAGE, polyacrylamide gel electrophoresis
PAN, poly(A) nuclease
PBS, phosphate-buffered saline
pc, positive control
PCR, polymerase chain reaction
PCV, packed cell volume
PEG, polyethylene glycol
PEI, polyethylenimine
PIC, preinitiation complex
PIPES, piperazine-N,N'-bis(2-ethanesulfonic acid)
PMSF, phenylmethylsulfonyl fluoride
PNK, polynucleotide kinase
PPAR-γ, peroxisome proliferator-activated receptor-γ

RACE, rapid amplification of cDNA ends
RAR, retinoic acid receptor
Rb, retinoblastoma
rbs, ribosome binding site
RDA, representative difference analysis
RNA, ribonucleic acid
RNAP, RNA polymerase
RRE, Rev-responsive element
RSV, Rous sarcoma virus
RT, reverse transcriptase
RT-PCR, reverse transcription polymerase chain reaction

SAAB, selected and amplified binding site analysis
SAGA, SPT-ADA-GCN acetyltransferase
SAR, scaffold-associated region
SCAP, SREBP cleavage-activating protein
SDS, sodium dodecyl sulfate
SDS-PAGE, sodium dodecyl sulfate polyacrylamide gel electrophoresis
SEAP, secreted alkaline phosphatase
SMCC, SRB MED co-activator complex
sn, small nuclear
SPR, surface plasmon resonance

SRB, suppressor of RNA polymerase B
SREBP-1, sterol response element binding protein
SSC, standard saline citrate
Su(Hw), suppressor of hairy wing

Tac, Trp-Lac (promoter)
TAE, Tris-acetate-EDTA
TAF, TBP-associated factor
TAg, T antigen
TBE, Tris-borate-EDTA
TBP, TATA-box binding protein
TCR-α, T-cell receptor-α
TdT, terminal transferase
TE, Tris/EDTA buffer
TES, N-Tris[hydroxymethyl]methyl-2-amino ethane sulfonic acid
Tet, tetracycline
TetR, Tet repressor
TFII, transcription factor for Pol II
TICS, TAF_{II}- and initiator-dependent cofactors
TK, thymidine kinase
TLC, thin-layer chromatography
TR, thryoid receptor
TRAP, TR-associated protein
TRF, TFIID-related factor
tRNA, transfer RNA
TRRD, transcription regulatory region database
TSA, trichostatin A

U, uracil
UAS, upstream activating sequence
UAS$_G$, galactose upstream activating sequence
USA, upstream stimulatory activity
UTL, untranslated leader
UTP, uridine triphosphate

VAF, virus-inducible transcription activator complex
VHL, von Hippel-Lindau
VSV, vesicular stomatitis virus

WCE, whole cell extract

Xis, excision protein

YAC, yeast artificial chromosome

CHAPTER 1

A Primer on Transcriptional Regulation in Mammalian Cells

INTRODUCTION

One of the central goals of the gene expression field is understanding how a mammalian organism regulates transcription of approximately 100,000 genes in the proper spatial and temporal patterns. Knowledge of how transcription factors function during this "differential" gene expression can be applied to fundamental issues in the fields of biology and medicine. To decipher these mechanisms, we need to understand the numerous processes influencing transcription and develop technical and strategic approaches for addressing them. This chapter provides an introduction to basic aspects of RNA polymerase II transcription. The goal is to prepare the novice for the issues raised in subsequent chapters and to provide a general overview of the field as of this writing. However, this field is evolving rapidly and the reader is encouraged to consult recent reviews in the literature. *Current Opinion in Cell Biology* and *Current Opinion in Genetics and Development* publish such reviews in the June and April issues, respectively. Some of the topics are quite advanced, although we have cited numerous review articles to allow the novice to explore unfamiliar areas. Almost all of the topics are covered in subsequent chapters and may help clarify concepts discussed briefly in this chapter.

A General Model for Regulation of a Gene

In eukaryotes, DNA is assembled into chromatin, which maintains genes in an inactive state by restricting access to RNA polymerase and its accessory factors. Chromatin is composed of histones, which form a structure called a nucleosome. Histones can be modified posttranslationally to decrease the ability of the nucleosome to inhibit transcription factor binding. Nucleosomes themselves are assembled into higher-order structures with different properties depending on the regulatory context. During the process of development, genes are turned on and off in a pre-programmed fashion, a process that eventually generates cell specificity. This developmental program is orchestrated by transcription factors, which bind to specific DNA sites near genes they control. A single transcription factor is not dedicated to each regulatory event. Instead, a mechanism called combinatorial control is employed. In combinatorial control, different combinations of ubiquitous and cell-type-specific regulatory proteins are used to turn genes on and off in different regulatory contexts (Britten and Davidson 1969). The ability of an organism to employ small numbers of regulatory proteins to elicit a larger number of regulatory decisions is based on the principles of cooperativity and synergy, issues we discuss later in the chapter.

Activating a Gene

To provide a framework for the issues involved in transcription regulation, consider a model for how a typical gene is turned on (Fig. 1.1) and then off again. In a typical gene, a DNA sequence called the core promoter is located immediately adjacent to and upstream of the gene. The core promoter binds RNA polymerase II (Pol II) and its accessory factors ("the general transcription machinery") and directs the Pol II to begin transcribing at the correct start site. In vivo, in the absence of regulatory proteins, the core promoter is generally inactive and fails to interact with the general machinery. A caveat is that some core promoters such as the heat-shock promoter in *Drosophila* and the Cyc-1 promoter in yeast appear to contain partial complements of general factors (i.e., RNA Pol and TATA box-binding protein [TBP], respectively) when inactive, but these factors are insufficient for transcription in the absence of regulatory proteins. Immediately upstream of the core promoter is a regulatory promoter, and farther away either upstream or downstream are enhancer sequences (Fig. 1.1A). Regulatory promoters and enhancers bind proteins called activators, which "turn on" or activate transcription of the gene. Activation generally occurs by recruitment of the general machinery to the core promoter via interactions between the activator bound to promoter DNA and the general machinery in solution. Some activators are ubiquitously expressed, whereas others are restricted to certain cell types, regulating genes necessary for a particular cell's function.

To activate a gene, the chromatin encompassing that gene and its control regions must be altered or "remodeled" to permit transcription. There are different levels of modification needed at different levels and stages of the transcription process. Higher-order chromatin structures comprising networks of attached nucleosomes must be decondensed, specific nucleosomes over gene-specific enhancers and promoters must be made accessible to cell-specific activators, and, finally, nucleosomes within the gene itself must be remodeled to permit passage of the transcribing RNA polymerases (Fig. 1.1B). There are different types of enzymes involved in chromatin remodeling and these must be directed, perhaps by a limited set of activators or other sequence-specific DNA-binding proteins, to the "target" genes. These enzymes fall into two broad classes: ATP-dependent remodeling enzymes and histone acetyltransferases (or simply histone acetylases). Once they bind near a gene, these enzymes remodel the chromatin so that other activators and the general machinery can bind. The mechanisms of remodeling are unclear, but they involve changes in the structure of chromatin and in modification of histones that somehow increase accessibility to transcription factors. Remodeling achieved at a local level affects only the chromatin close to a gene. In some instances, however, a single gene or locus of related genes might spread over 100 kb or more. In these cases, genes might be under control of not simply specific enhancers and regulatory promoters but also of locus control regions

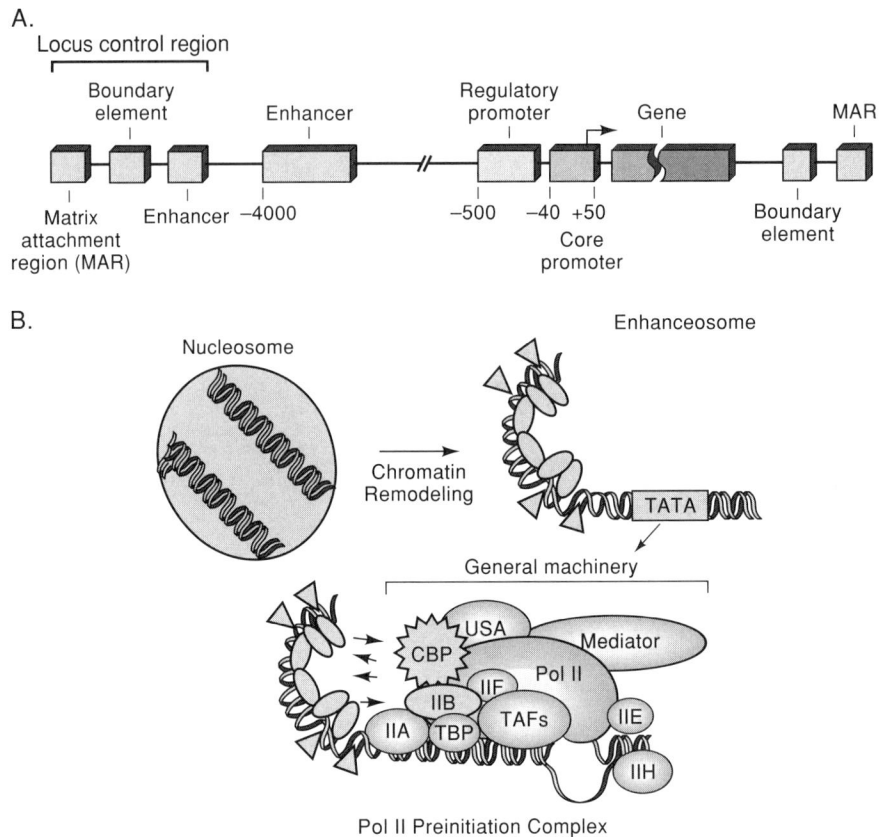

A.

Locus control region

Boundary element

Enhancer

Regulatory promoter

Gene

MAR

Matrix attachment region (MAR)

Enhancer −4000

−500 −40 +50

Core promoter

Boundary element

B.

Nucleosome

Enhanceosome

Chromatin Remodeling

TATA

General machinery

USA

CBP

Mediator

Pol II

IIF

IIB

IIA TBP TAFs

IIE

IIH

Pol II Preinitiation Complex

FIGURE 1.1. (*A*) Model of typical gene and components involved in gene activation and inactivation. (*B*) Acitvation of a gene and assembly of the Pol II preinitiation complex. (Redrawn, with permission, from Carey 1998 [copyright Cell Press].)

(LCRs), which remodel chromatin and control global access of activators over an extended region. Once enhancers are accessible, they can stimulate transcription of a gene. However, because enhancers are known to activate transcription when they are positioned large distances from a gene, they could inadvertently activate other nearby genes in the absence of appropriate regulation. To focus the action of the enhancer or LCR on the appropriate gene or set of genes, the gene and its regulatory regions are thought to be assembled into a domain. Domain formation appears to involve boundary elements and matrix attachment regions (MARs). Boundary or insulator elements are thought to flank both sides of an individual gene or gene locus. These elements bind proteins that prevent the enhancer from communicating with genes on the opposite side of the insulator. MARs also flank some active genes and tether them as loops to the nuclear periphery or matrix although a gene function for MARs has not been established.

The current view is that once the enhancer and promoter are accessible they bind to combinations of activators. Binding of activators is generally cooperative, where one protein binds weakly, but multiple activators engage in protein–protein interactions that increase each of their affinities for the regulatory region. The nucleoprotein structures comprising these combinatorial arrays of activators are called enhanceosomes (Fig. 1.1B). The enhanceosome interacts with the general transcription machinery and recruits it to a core promoter to form "the preinitiation complex." The enhanceosome, the general machinery, and the core promoter form a complicated network of protein–protein and protein–DNA interactions that dictate the frequency of transcription initiation. The interactions between the enhanceosome and components of the general machinery are rarely direct but are bridged or linked by proteins called coactivators. It is important to note that the term "coactivator" has several definitions depending on the regulatory context: In some cases, coactivators are part of the general machinery and in other cases they are not. The term will be defined on a case-by-case basis in this chapter.

Inactivating a Gene

In many instances, genes are activated transiently and then later turned off. In these cases, the hypothetical sequence of events would include inactivation of the preinitiation complex and establishment of a repressive chromatin environment over the gene and its regulatory regions. Establishment of an inactive chromatin environment involves ATP-dependent remodeling and histone deacetylases. However, higher-order interactions with the nuclear periphery may also occur to form domains of inactive "heterochromatin." The mechanisms for inactivating a gene vary, but generally they involve the binding of sequence-specific repressors to silencer elements. Genes are often methylated to maintain the inactive state. Methylation also leads to recruitment of histone deacetylases.

Although the sequence of events described above provides a framework for gene activation and inactivation, the regulatory strategies employed vary considerably. We attempt in the following sections to evaluate different aspects of this simple model and to alert the reader to alternate regulatory strategies.

Overview

In Section I, we summarize the basic mechanics of the transcription process, including an overview of core promoter structure and the composition of the general machinery. The general machinery consists of general transcription factors, or GTFs, and Pol II, which are necessary for the catalytic process of transcription. However, the machinery also comprises coactivators and corepressors, which allow activators and repressors to communicate with the GTFs and chromatin. In Section II, we discuss regulatory DNA sequences, including enhancers and silencers, and regulatory proteins, including activators and repressors. In Section III, we consider the structure of chromatin and the enzymes involved in "remodeling" it. There is an emphasis on the roles of remodeling enzymes in establishing the active and repressed states of genes. Finally, we end with Section IV, entitled "the enhanceosome," where we discuss the concepts of enhancer complexes and the basis for combinatorial control. Note that some of the topics are covered in greater detail in the ensuing chapters and we have abbreviated our description of these here to prevent unnecessary redundancy.

CONCEPTS AND STRATEGIES: I. PROMOTERS AND THE GENERAL TRANSCRIPTION MACHINERY

A typical core promoter encompasses DNA sequences between approximately –40 and +50 relative to a transcription start site (Smale 1994). Core promoter DNA elements (1) bind to and control assembly of the preinitiation complex containing Pol II, the general transcription factor, and coactivators; (2) position the transcription start site and control the directionality of transcription; and (3) respond to nearby or distal activators and repressors in a cell. In most cases, the core promoter elements do not play a direct role in regulated transcription. The core promoter alone is generally inactive in vivo, but in vitro it can bind to

TABLE 1.1. *Composition of coactivator/mediator complexes*

Factor	Yeast gene(s)	Mass (kD)	Essential	Characteristics	Human gene(s)	Mass (kD)
RNA Polymerase II	RPB1	192	yes	heptapeptide repeat		220
	RPB2	139	yes			150
	RPB3	35	yes			44
	RPB4	25	no			32
	RPB5	25	yes	shared with Pol I, II, III		27
	RPB6	18	yes	shared with Pol I, II, III		23
	RPB7	19	yes			16
	RPB8	17	yes	shared with Pol I, II, III		14.5
	RPB9	14	no			12.6
	RPB10	8	yes	shared with Pol I, II, III		10 (β)
	RPB11	14	yes			12.5
	RPB12	8	yes	shared with Pol I, II, III		10 (α)
TBP	SPT15	27	yes	binds TATA element; nucleates PIC assembly; recruits TFIIB	TBP	38
TFIIA	TOA1	32	yes	required for activation	TFIIA	37 (α)
	TOA2	13.5	yes	required for activation		19 (β)
						13 (γ)
TFIIB (factor e)	SUA7	38	yes	stabilizes TATA-TBP interaction; recruits RNA Pol II-TFIIF; affects start site selection; zinc ribbon	TFIIB	35
TFIIF (factor g)	TFG1, SSU71	82	yes	facilitates RNA Pol II- promoter targeting; stimulates elongation; functional interaction with TFIIB	RAP74	58
	TFG2	47	yes	σ factor homology; destabilizes nonspecific RNA Pol II–DNA interactions	RAP30	26
	TFG3, ANC1, SWP29, TAF30	27	no	common subunit of TFIID, TFIIF, and the SWI/SNF complex	AF-9, ENL	20
TFIIE (factor a)	TFA1	66	yes	recruits TFIIH; stimulates TFIIH catalytic activities; functions in promoter melting and clearance; zinc-binding domain	TFIIE-α	56
	TFA2	43	yes	—	TFIIE-β	34
TFIIH (factor b)	SSL2, RAD25	95	yes	functions in promoter melting and clearance; ATP-dependent DNA helicase (3´-5´); DNA-dependent ATPase; ATPase/helicase required for both transcription and NER	XPB, ERCC3	89

Group	Gene	Size		Description	Mammalian homolog	Size
	RAD3	85	yes	ATP-dependent DNA helicase (5´-3´; DNA-dependent ATPase; ATPase/helicase required for NER but not transcription	XPD, ERCC2	80
	TFB1	73	yes	required for NER	p62	62
	TFB2	59	yes	required for NER	p52	52
	SSL1	50	yes	required for NER; zinc binding	hSSL1	44
	CC1	47, 45	yes	TFIIK subcomplex with Kin 28	Cyclin H	37
	TFB4	37	yes		p34	34
	TFB3	32	yes	zinc RING finger; links core-TFIIH with TFIIK; unlike Mat1, not a subunit of kinase/cyclin subcomplex	Mat1	32
Srbs	KIN28	33	yes	TFIIK subcomplex with Ccl1	MO15, Cdk7	40
	SRB2	23	no	interacts with GAL4		
	SRB4	78	yes			
	SRB5	34	no			
	SRB6	14	yes			
	SRB7	16	yes			SRB7 (p20)
	SRB8	166	no			
	SRB9	160	no			
	SRB10	63	no	cyclin/kinase pair with Srb11; repr. of glucose reg. genes	Srb10, Cdk8, (p53)	
	SRB11	38	no	see Srb10	Srb11 (p32), cyclin C	
Meds	MED1	64	no	interacts with Med2, neg. regulated by Srb10/11	p70	
	MED2	48	no	see Med1; required for Gal4-activation		
	MED4	32	yes			
	MED6	33	yes		MED6 (p33)	
	MED7	26	yes		MED7 (p33)	
	MED8	25	yes			
Gal11	GAL11	120	no			
Rgr1	RGR1	123	yes	required for repression of glucose-regulated genes	TRAP 170	
Sin4	SIN4	111	no	in subcomplex with Gal11, Rgr1, Pgd1, Med2		
Pgd1	PGD1	47	no			
Rox3	ROX3	25	yes	inv. in glucose-regulated transcription		

the general machinery and support low or "basal" levels of transcription. The amount of basal transcription is dictated by the DNA sequences in the core promoter. Activators great-ly stimulate transcription levels, and the effect is called activated transcription.

The preinitiation complex that binds the core promoter comprises two classes of factors (see Table 1.1): (1) the GTFs including Pol II, TFIIA, TFIIB, TFIID, TFIIE, TFIIF, and TFIIH (Orphanides et al. 1996); and (2) coactivators and corepressors that mediate response to regulatory signals (see Hampsey 1998; Myer and Young 1998). In mammalian cells, coacti-vator complexes are heterogeneous and occasionally can be purified as discrete entities or as part of a larger Pol II holoenzyme, a point on which we elaborate below. This section describes the properties of core promoters and the general transcription machinery.

Core Promoter Architecture

TFIID is the only general transcription factor capable of binding core promoter DNA both independently and specifically. TFIID is a multisubunit protein containing TBP and 10 or more TBP-associated factors or $TAF_{II}s$ (described in more detail below). The TFIID DNase I footprint extends from –40 to +50 and encompasses most of the DNA constituting the core promoter. TFIID does not contact all of the bases in the footprint, and other general factors bind in a sequence-specific fashion within these "open" regions. The other general factors bind DNA weakly on their own, however, and it is the network of cooperative pro-tein–protein interactions with TFIID and each other that allows them to form stable, spe-cific DNA interactions. The role of cooperativity in assembly of transcription complexes is a recurring theme in gene regulation, one that we return to throughout this book. The amount of basal transcription and the ability to respond to activators are likely related to the affinities of the GTFs and TFIID for the core promoter (discussed in Lehman et al. 1998). A typical core promoter contains the following DNA sequence elements (Fig. 1.2):

1. *The TATA motif.* This sequence element, with the consensus TATAAA, was originally dis-covered by David Hogness and is called the Hogness box in the older literature. It is locat-ed 25–30 bp upstream of the transcription start site. The TATA box is capable of inde-pendently directing a low level of transcription by Pol II on naked DNA templates in vitro or transfected DNA templates in vivo. The TATA box is sufficient for directing activated transcription when an activator protein binds to a nearby regulatory element. The TBP subunit (Hernandez 1993; Burley and Roeder 1996) of TFIID (Table 1.1) makes direct contact with the TATA motif. The binding of TFIID to the TATA box nucleates the bind-ing of the remaining general transcription factors, currently thought to be present in the form of a multifactor "holoenzyme," an issue we discuss below (for review, see Myer and Young 1998). We discuss TBP-binding mechanisms in Chapters 13 and 15.

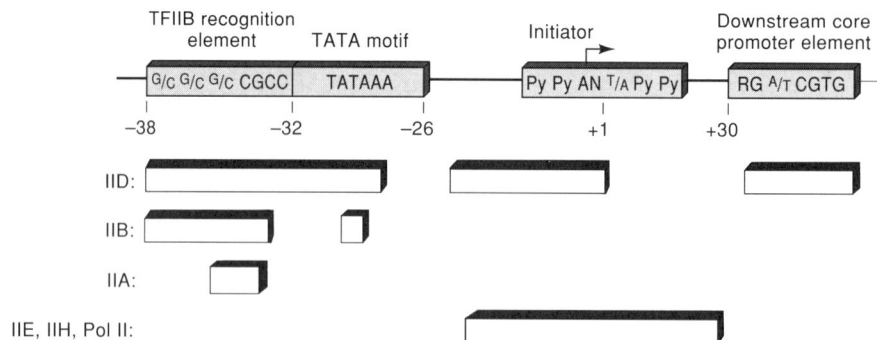

FIGURE 1.2. Sequence elements in a typical core promoter.

2. *The initiator element.* A second type of core promoter element that appears to be functionally analogous to the TATA box is the initiator (Inr; Smale 1994). Although this element carries out the same functions as TATA by directing the formation of a preinitiation complex, determining the location of the start site, and mediating the action of upstream activator proteins, it directly overlaps the transcription start site. Functional Inr activity depends on a loose consensus sequence of approximately PyPyA+1NT/APyPy. The basal Inr appears to be recognized by two independent proteins: a TAF_{II} and Pol II (for review, see Smale 1997). The TAF_{II} that binds the Inr has not been firmly established but may be $TAF_{II}250$, with binding further stabilized by $TAF_{II}150$ (Kaufmann et al. 1998). TFIID binding to the Inr appears to be influenced by both TFIIA and cofactors called TICs (TAF_{II}- and initiator-dependent cofactors) (Emami et al. 1997; Martinez et al. 1998). A plausible model for the initiation of transcription from a TATA-less promoter containing an Inr is as follows: The TFIID complex recognizes the Inr, possibly with the assistance of TFIIA and TIC-1. At some promoters, this recognition event directs the TBP subunit of TFIID to associate with the −30 region of the promoter in a TATA sequence-independent manner, although at some promoters TBP binding may be unnecessary. Following the stable binding of TFIID to the core promoter, the remaining steps leading to formation of a functional preinitiation complex and transcription initiation proceed by a similar mechanism and require a similar set of general transcription factors as TATA-containing promoters. The specific interactions between Pol II and the Inr may become important at later steps in preinitiation complex formation.

3. *The downstream core promoter element.* The downstream core promoter element (DPE) is a 7-nucleotide sequence first identified in *Drosophila*. The DPE bears the consensus sequence $RG^A/_TCGTG$ and is centered approximately 30 bp downstream of the initiation site. It is found in many, but not all, *Drosophila* promoters and most likely many mammalian promoters (Burke and Kadonaga 1996). In *Drosophila*, where the DPE element has been studied in the greatest detail, the DPE is found in TATA-less promoters and acts in conjunction with the Inr element to direct specific initiation of transcription. Crosslinking and DNA-binding studies suggest that the DPE recognizes *Drosophila* $TAF_{II}60$ and perhaps $TAF_{II}40$ directly (Burke and Kadonaga 1997).

4. *The TFIIB recognition element.* The TFIIB recognition element (BRE) was discovered by Ebright and colleagues (Lagrange et al. 1998), who recognized the potential for specific DNA binding by TFIIB based on the position of TFIIB relative to the major groove in the crystal structure of the TBP–TFIIB–TATA ternary complex (for review, see Burley and Roeder 1996). Binding-site-selection experiments (discussed in Chapter 13) revealed that TFIIB bound specifically to a sequence with the consensus $^G/_C{}^G/_C{}^G/_ACGCC$ located from −32 to −38, just upstream of the TATA box. The BRE is found in a substantial number of eukaryotic promoters.

It is likely that other general transcription factors display a limited degree of sequence-specific recognition. TFIIA, TFIIF, and Pol II all interact with the major groove as revealed in the crystal structures of TFIIA–TBP–TATA and in photocrosslinking experiments with Pol II and TFIIF (Kim et al. 1997; Robert et al. 1998). Additional TAF_{II}s may also bind DNA specifically, based on photocrosslinking data (Oelgeschlager et al. 1996). Further research is needed to understand the significance of the interactions occurring on the core promoter. However, as discussed above, preliminary indications are that the ability of the core promoter to respond to activators and direct high levels of transcription is dependent on cooperative binding of multiple general factors.

At least one class of promoter appears to lack both TATA and Inr elements but instead contains several transcription initiation sites, a high G/C content, and multiple binding sites for the ubiquitous mammalian transcription factor Sp1 (see Smale 1994). On these promoters, which often are associated with "housekeeping genes," Sp1 directs the formation of preinitiation complexes to a region 40 to 100 bp downstream of its binding sites. Within that window, TFIID may direct preinitiation complex formation at the DNA sequences that most closely resemble TATA or Inr elements.

One issue that remains unresolved is why core promoters have evolved to contain widely varying sequence organizations or architectures, particularly because the mechanisms of initiation appear to be similar on different promoters. An explanation for core promoter heterogeneity will likely emerge from studies that have revealed a requirement for specific core promoter structures during transcriptional regulation. For example, the activity of the lymphocyte-specific terminal transferase promoter depends on its Inr element, because the promoter does not function efficiently if a TATA box is inserted and the Inr eliminated (Garraway et al. 1996). Thus, the specific core promoter structure found in a given gene is likely to play a role in transcriptional regulation, not by interacting with cell-type-specific proteins in most cases, but by providing an appropriate context for efficient activation or repression.

The General Transcription Machinery

Mammalian gene regulation involves a complicated interplay between activators, repressors, the general transcription machinery, and chromatin. The general transcription machinery consists of Pol II, the GTFs TFIIA, TFIIB, TFIID, TFIIE, TFIIF, and TFIIH (for review, see Orphanides et al. 1996), and a complex of coactivators termed the mediator. Pol II is a large multisubunit enzyme. One important feature of Pol II is the heptapeptide repeat constituting the carboxyl terminus of the largest subunit. This carboxy-terminal domain is phosphorylated extensively by different kinases involved in transcription regulation. Biochemical studies show that the GTFs support basal transcription and carry out many of the catalytic functions required for initiation. Coactivators with the mediator are believed to bridge the activators and GTFs. Analysis of genome-wide transcription using DNA microarray technology suggests that coactivators are only required for transcription from subsets of genes (see Holstege et al. 1998). Coactivators and many of the GTFs are both part of a complex termed the holoenzyme. We define coactivators that reside in the holoenzyme (e.g., SRBs and CBP/p300; see below) as components of the general machinery.

The reader should be aware of two caveats. First, there is one report of a cell- and possibly promoter-specific general factor called TRF. TRF is a TBP-related factor from *Drosophila* that can substitute for the TBP subunit of TFIID in basal transcription; it is found in the central nervous system and gonads of flies, where it is believed to mediate cell-specific gene expression (for review, see Hori and Carey 1998). Reports of other factors like TRF will probably appear in the literature in the next few years. Second, a complete set of GTFs may not be required at all promoters. In vitro experiments in cell-free systems and genetic studies in yeast support the view that some GTFs are required for transcription of only subsets of genes (see Holstege et al. 1998). We will use the term "general machinery" throughout this chapter with the understanding that the reader is aware of the caveats of usage.

The subunits constituting all of the GTFs have been cloned (Table 1.1), and a rudimentary knowledge of their function and mechanism has emerged from studies in both mammalian cells and yeast. Genetic studies in yeast have provided insightful data on general factor mechanism and have been essential for evaluating the validity and ramifications of biochemical and functional studies performed in mammalian systems (for review, see Hampsey 1998). Here we discuss (1) how the general factors alone assemble into transcription complexes and (2) their association with coactivators and mediators in the form of the holoenzyme.

Basal Transcription Complex Assembly

Purified GTFs and Pol II mediate basal transcription on a core promoter in vitro but cannot support activated transcription in the absence of coactivators (see below). The study of the basal transcription reaction has revealed important insights into the catalytic events necessary for initiation. In the original studies on transcription complex assembly, TBP was used in place of TFIID because TBP was small, it could support basal transcription in the presence of the other GTFs, and, finally, it could be subjected to electrophoretic mobility shift analysis (EMSA) for detailed structure–function analyses. Furthermore, TFIID was not sufficiently purified at that time to analyze basal complexes containing it.

These early studies indicated that purified GTFs and TBP assembled into a transcription preinitiation complex on the DNA in a stepwise fashion. The complex is nucleated by the binding of TBP to the TATA box, a reaction aided by TFIIA or TFIIB, which can bind in any order (for review, see Orphanides et al. 1996). The crystal structures of TBP and the TBP–TFIIB and TBP–TFIIA complexes with DNA have been solved, revealing insights into the process of promoter recognition (for review, see Burley and Roeder 1996; Tan and Richmond 1998); both TFIIA and TFIIB contact DNA and TBP, increasing the stability of TBP binding. After TFIIB binds to TBP, a complex of TFIIF in association with Pol II is recruited, followed by sequential binding of TFIIE and TFIIH.

Conformational Changes during Transcription Complex Assembly

The transcription complex undergoes several important conformational changes during the course of its assembly. These transitions have been studied by crystallography, DNase I footprinting, and chemical crosslinking. The transitions are characterized by dramatic changes in the path or structure of the promoter DNA. For example, the crystal structure of TBP with the TATA box reveals that, upon binding, the DNA is dramatically bent and unwound, a point discussed further in Chapter 15 (for review, see Burley and Roeder 1996). Crosslinking and DNase I footprinting have shown that $TAF_{II}s$ in conjunction with TFIIA induce conformational changes in the complex leading to wrapping of the core promoter around TFIID (for review, see Hoffmann et al. 1997). Other studies have revealed extensive changes in the path of DNA upon binding of Pol II, possibly indicative of wrapping of the promoter around Pol II in the intact basal complex (Kim et al. 1997; Robert et al. 1998); in one study TFIIF greatly stimulated the wrapping (Robert et al. 1998), an issue we discuss in Chapter 15. Finally, several lines of evidence suggest that TFIIE and TFIIH assist Pol II in melting the start site in an ATP-dependent fashion immediately prior to initiation (Holstege et al. 1997 and references therein). The energetic melting step is mediated by the ERCC3/XPB helicase activity of TFIIH (see Tirode et al. 1999); TFIIE binds to and stabilizes the melted DNA. Although many of the conformational transitions discussed above may simply represent unregulated mechanistic steps in the process of initiation, the changes in TFIID conformation are regulated by activators (Chi and Carey 1996).

TFIIH is an unusual basal factor in that it has a dual role: It participates in transcription and in nucleotide excison repair (NER) (see Table 1.1). During transcription, TFIIH probably melts the DNA to form the open complex using a helicase subunit called ERCC3/XPB. In NER, a process in which damaged nucleotides are removed and repaired by a repairosome complex containing TFIIH, both the ERCC3/XPB and ERCC2/XPD helicases are required. In this case, the helicases melt the DNA around the lesion to allow access by the repair machinery. TFIIH can be divided into several subcomplexes, one of which is termed the CAK complex. CAK contains a cyclin-dependent kinase/cyclin pair implicated in phosphorylation of the carboxy-terminal domain (CTD) of the largest Pol II subunit. CTD phosphorylation by CAK (Kin28-Ccl1 in yeast) can be influenced by many factors, including general factors such as TFIIE and activators like the HIV Tat protein (for review, see Yankulov and Bentley 1998). In simple basal systems, phosphorylation of the CTD correlates with early stages of initiation and may be necessary to dissociate Pol II from the other GTFs (for review, see Orphanides et al. 1996). TFIIH has also been shown to participate in promoter escape, the stage at which the polymerase releases from the general machinery and begins productive elongation. We discuss some of these studies in Chapter 15. It is likely that CAK's role in transcription is complicated.

TAF$_{II}$s

There are at least three classes of TAFs; each class associates with TBP and is required for transcription by a different nuclear polymerase (for review, see Hernandez 1993). RNA Pol II requires the TAF$_{II}$s present in TFIID. Pol I employs TAF$_I$s associated with a factor termed SL-1. Pol III employs TAF$_{III}$s present in the TFIIIB fraction. SNAPc represents a unique TBP–TAF complex that mediates transcription by both Pol II and III of small nuclear (sn) RNA genes. (Note, however, that it is not entirely clear whether TBP is a stable component of the SNAPc complex, an issue that will be resolved in the near future.) TAF$_{II}$s also play roles outside of the TBP complexes. They are found in chromatin-remodeling enzymes in yeast and humans.

We will focus on TAF$_{II}$s required for TFIID function and transcription by Pol II of protein-coding genes. The TAF$_{II}$s are somewhat conserved from yeast to humans (Table 1.2). Although the TAF$_{II}$s appear to be ubiquitous, at least one example of a cell-specific TAF$_{II}$ has been reported (Dikstein et al. 1996). The association of this TAF$_{II}$, TAF$_{II}$105, with TFIID appears to be restricted to B lymphocytes, although paradoxically the mRNA is ubiquitously expressed. TAF$_{II}$s are involved in core promoter recognition, interaction with other GTFs, activated transcription, and chromatin remodeling. We have already discussed their roles as core promoter recognition factors and discuss here their role in activated transcription. In addition, certain TAF$_{II}$s share homology with histones, and one contains histone acetylase activity (Table 1.2). It has been speculated that these features permit TAF$_{II}$s to recapitulate nucleosome-like functions, a point we discuss below.

The precise roles and physiological function of TAF$_{II}$s have been the subject of much study. One of the earliest proposals was that TAF$_{II}$s were required for transcriptional activation. Indeed, biochemical and genetic studies in mammalian cells and *Drosophila* revealed that TAF$_{II}$s directly interacted with activators and mediated the action of those activators in vitro (Fig. 1.3) (Goodrich and Tjian 1994). Furthermore, numerous functional studies on transcription complex assembly in mammalian, *Drosophila*, and yeast biochemical systems showed that activators recruited complexes containing TFIID and TFIIA to a core promoter, and this effect required TAF$_{II}$s (see Chi et al. 1995). The biochemistry was buttressed by genetic experiments in *Drosophila* pointing to a role for TAF$_{II}$s in activated transcription during development (Zhou et al. 1998). Genetic studies in yeast showed that the TAF$_{II}$s were essential genes and that transient inactivation of the histone-like TAF$_{II}$s by temperature-sensitive mutants abolished activated transcription (for review, see Hahn 1998). Inactivation of other TAF$_{II}$s (e.g., yeast TAF$_{II}$145) in growing yeast cells by

TABLE 1.2. *Comparison of TAF-containing complexes from different species*

| Functions | TAF$_{II}$-containing complexes | | | | | |
| | containing TBP | | | lacking TBP | | |
	yTFIID	dTFIID	hTFIID	ySAGA	hTFTC	hPCAF/GCN5
Protein kinase; Hat activity; required for cell cycle progression	145(130)	230	250	—	—	—
Interacts with promoter DNA; required for initiator function	TSM1	150	150	—	ND	ND
Interacts with Sp1, E1A, and CREB; coactivator for RAR, TR, and VDR	—	110	135	—	135	—
Contains WD40 repeats; interacts with TFIIFβ	90	80	100	90	100	(PAF65β)
Coactivator for p53, histone H4 motif, and forms histone-like pair with hTAF$_{II}$32	60	62	80(70)	60	80	(PAF65α)
Interacts with several activators, Sp1, VDR, and TR	67(68)	55	55	—	55	ND
Histone H3 motif and forms histone-like pair with hTAF$_{II}$80; interacts with TFIIB and the VP16 AAD	17(20)	42	31(32)	17(20)	31(32)	31(32)
Interacts with the AF-2a domain of the ER Required for ER activity in vitro	25	ND	30	25	30	30
Coactivator for ER, VDR, RXR, and Tax; forms histone pair with hTAF$_{II}$18	40	30β	28	—	—	ND
Contains histone H2B-like motif	68(61)	30α	20(15)	68(61)	20(15)	20(15)
Contains novel type of histone fold; forms histone-like pair with hTAF$_{II}$28	19(FUN81)	ND	18	—	—	ND
Binds to promoter TATA element	yTBP	dTBP	hTBP	—	—	—
HAT	—	—	—	yGcn5	hGcn5-L	PCAF/Gcn5-s

(ND) Not determined. Reprinted, with permission, from Björklund et al. 1999 (Copyright 1999 Cell Press).

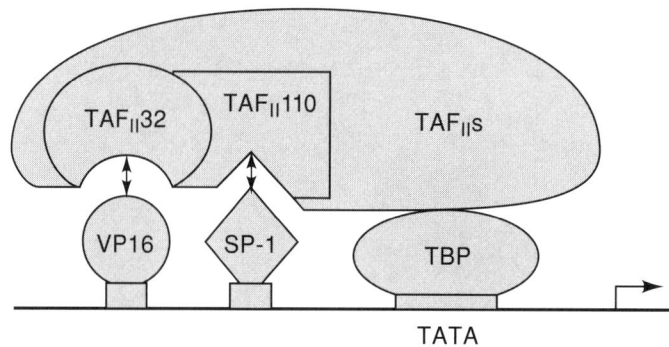

FIGURE 1.3. Interactions between activators and specific TAF$_{II}$s recruit TFIID to the core promoter.

directed proteolysis or temperature-sensitive mutants had a much smaller effect on activated transcription (for review, see Hahn 1998).

The significant homology between TAF$_{II}$s and histones (Table 1.2) has led to the hypothesis that TFIID may mimic nucleosome function, perhaps as a means of stabilizing its binding to DNA or as a way to displace nucleosomes during transcription complex assembly. Indeed, chemical crosslinking and topoisomerase relaxation studies suggest that the DNA may be wrapped around TFIID, perhaps like DNA is wrapped around a nucleosome (see below and Fig. 1.5) (for review, see Oelgeschlager et al. 1996). However, examination of the histone surfaces that interact within the nucleosome crystal structure suggests that residues mediating the critical contacts (largely arginines with the minor groove and phosphate backbone; Luger et al. 1997) are not well conserved in the TAF$_{II}$s (for review, see Oelgeschlager et al. 1996 and Hahn 1998). Another view is that TAF$_{II}$s allow internucleosomal contacts as a means of allowing TFIID to dock with a nearby nucleosome. TFIID could then modify the surrounding chromatin via the histone acetylase activity of TAF$_{II}$250, for example (see section III below). Alternatively, the TAF$_{II}$s may simply share the histone fold as a structural core.

The Holoenzyme and Mediators

The concept that the general factors were assembled into transcription complexes in a stepwise manner was attractive from the standpoint that different steps could, in principle, be regulated by activators and repressors. Such a mechanism would help to explain the diversity in gene expression patterns. The significance of the finding, however, was dependent on GTFs being differentially limiting at promoters. Although there are examples where GTFs have different affinities for core promoters (e.g., TBP binding to consensus and nonconsensus TATAs; TFIIB binding to a consensus vs. a degenerate BRE; TFIID binding to an Inr-containing vs. Inr-less promoter), differential binding of GTFs has not yet emerged as a major regulatory theme. Instead, research has focused on recruitment of a single large GTF-containing complex called the holoenzyme. In contrast to the complexity of the stepwise pathway, the holoenzyme provides a single target through which activators bound to an enhancer or promoter can recruit the general machinery in a concerted manner (for review, see Myer and Young 1998).

Discovery of the Pol II Holoenzyme

The holoenzyme was discovered initially in yeast through a combination of biochemistry and genetics (for review, see Myer and Young 1998). Partial deletions in the CTD of the Pol

II large subunit elicited a cold-sensitive (*cs*) growth phenotype in *Saccharomyces cerevisiae*. The CTD in yeast is composed of 26 copies of a heptapeptide repeat (Tyr-Ser-Pro-Ser-Thr-Pro-Ser) rich in amino acids that can be phosphorylated by serine-threonine and tyrosine kinases. Genetic suppressors of the *cs* mutants were identified and called SRBs (*s*uppressor of *R*NA polymerase *B*). The SRB-encoded proteins were initially shown to co-purify in a stable complex with Pol II.

Later studies revealed that the SRB complex could support activated transcription. Reactions containing the SRB complex, TBP, and TFIIE elicited basal levels of transcription and responded to the activators in vitro. This observation led to the finding that the SRB complex contained a subset of the yeast general transcription factors (for review, see Hampsey 1998; Myer and Young 1998), including TFIIB, TFIIF, and TFIIH. The multifactor composition of the SRB complex inspired the term "RNA Pol II holoenzyme." (The term holoenzyme is somewhat ambiguous because, unlike its prokaryotic counterpart, the yeast holoenzyme does not contain DNA recognition capabilities in vitro; i.e., it lacks TBP.) Genetic studies revealed that several of the SRB genes were essential, and inactivation, in growing cells, of temperature-sensitive SRB4 mutants led to an immediate cessation of Pol II transcription. These data, combined with biochemical studies demonstrating that most SRBs were associated with the holoenzyme, led to the proposal that the holoenzyme is the functional form of Pol II in vivo.

Interestingly, genome-wide analysis has revealed that different SRBs are required for transcription of different subsets of genes. This observation suggests that different SRBs may interact with different activators (Holstege et al. 1998). Mammalian holoenzymes have been studied in less detail than the yeast complex but appear to contain many orthologous subunits (for review, see Parvin and Young 1998). In the section below, we focus on the composition of the holoenzyme complexes and briefly discuss how activators stimulate transcription in the context of the holoenzyme.

Composition of the Yeast Holoenzyme

The yeast holoenzyme is composed of five different classes of factors: (1) Pol II, a 12-subunit enzyme in yeast, whose roles and subunit interactions have been defined genetically by several labs. The CTD of the largest subunit has been most extensively studied, since its phosphorylation is associated with different phases of initiation and elongation control (for review, see Myer and Young 1998). (2) The TFIIB, TFIIH, and TFIIF GTFs (for review, see Hampsey 1998). (3) A core mediator complex originally proposed by Kornberg and colleagues to be the direct target of activators (for review, see Li et al. 1996). The mediator is believed to be tethered directly to the CTD of Pol II. (4) An SRB10-11 complex that contains subunits which act as corepressors in certain yeast regulatory circuits (for review, see Myer and Young 1998). (5) The SWI/SNF complex involved in chromatin remodeling (for review, see Workman and Kingston 1998).

We have already discussed Pol II and the GTFs. The yeast core mediator subcomplex confers on GTFs the ability to respond to activators in vitro. The composition of the mediator is currently believed to include the following components: SRB2, 4, 5, 6, and 7; MED1, 2, 3, 4, 6, 7, and 8; and GAL11, PGD1, RGR1, ROX3, and SIN4. The discovery that all of these proteins exist in a complex, which mediates activation and repression, clarified a wide body of genetic data implicating the various components in positive and negative gene regulation (for review, see Li et al. 1996; Myer and Young 1998).

The SRB10–11 subcomplex comprises SRB8, 9, 10, and 11. SRB10 and 11 form a CDK-cyclin pair that, like the CAK subunits of TFIIH, regulates Pol II activity by CTD phosphorylation. Genetic data suggest that this subcomplex plays a negative regulatory role in transcription, possibly acting as a corepressor by mediating the action of certain DNA-bound repressors (see below). Phosphorylation of the CTD in solution by SRB10 and 11 prior to phosphorylation on promoter DNA by the CDK-cyclin subunits of TFIIH inactivates holoenzyme activity (for review, see Myer and Young 1998).

The SWI/SNF chromatin-remodeling complex has also been found associated with the yeast holoenzyme, and it is plausible that other chromatin-remodeling enzymes will be found associated as the characterization continues (for review, see Workman and Kingston 1998). SWI/SNF contains several subunits identified in genetic screens as important for chromatin remodeling. We describe the isolation and characterization of SWI/SNF below.

The regulatory implications of the holoenzyme are fairly straightforward. Because the complex is largely intact, a DNA-bound activator should recruit the holoenzyme to DNA by stably contacting any subunit or surface (for review, see Carey 1998 and Ptashne and Gann 1998). Different surfaces of the holoenzyme might be designed to interact with different activators or combinations of activators (i.e., enhanceosomes) on a promoter.

Mammalian Holoenzymes

Shortly after the discovery of the yeast holoenzyme, several labs reported the isolation of mammalian holoenzymes (Table 1.3). These complexes varied considerably in their composition (for review, see Greenblatt 1997 and Parvin and Young 1998). The examples ranged from holoenzyme complexes containing all of the GTFs to those lacking most GTFs but containing coactivators and mammalian SWI/SNF homologs. The variation in holoenzyme composition is likely a reflection of purification differences but may also belie the existence of multiple complexes that are differentially employed in gene regulation (for review, see Hampsey and Reinberg 1999).

TABLE 1.3. *Mammalian holoenzyme components identified using different purification strategies*

Factor	Neish[a]	Maldonado[b]	Pan[c]	Ossipow[d]
Pol II	+	+	+	+
SRB/Mediator	+	+	+	+
TFIIF	+	+	+	+
TFIIE	+	+	+	+
TFIIH	+	+	+	+
TFIIB	–	–	+	+
TBP	–	–	+	+
TAFs	–	ND	+	+
TFIIA	ND	ND	–	ND
CBP	+	ND	–	ND
p300	+	ND	ND	ND
BRCA1	+	ND	ND	ND
RHA	+	ND	ND	ND
Recomb, repair	–	+	–	ND
YY 1	–	+	ND	ND
SWI/SNF	+	ND	ND	ND
Poly A factors	ND	ND	+	ND
Rb	–	ND	ND	ND
SRC1/PCAF	–	ND	ND	ND

(ND), not determined; (+), present; (–), absent. Reprinted, with permission, from Parvin and Young 1998 (Copyright 1998 Elsevier Science). Strategies: [a] Data from Scully et al. 1997; Nakajima et al. 1997a,b; Neish et al. 1998; J. Parvin, unpubl. [b] Data from Maldonado et al. 1996; Cho et al. 1997. [c] Data from McCracken et al. 1997; Pan et al. 1997. [d]Data from Ossipow et al. 1995; U. Schibler, unpubl.

A hallmark of mammalian holoenzyme preparations is the presence of coactivators including p300/CBP and mediator components such as CDK8. p300 and CBP are highly related proteins (see, e.g., Eckner 1996; Torchia et al. 1998). p300 was identified as a protein interacting with E1A in coimmunoprecipitation experiments. CBP or CREB-binding protein was shown to be a coactivator of the cAMP receptor element binding protein (CREB). Phosphorylation of CREB by protein kinase A leads to recruitment of CBP. Because CBP is a component of the holoenzyme, recruitment of CBP by CREB could, in principle, lead to holoenzyme recruitment. CBP and p300 have been shown by affinity chromatography experiments to interact with numerous activators. CDK8 is a mammalian SRB10 homolog. It appears to be present in many but not all of the mammalian versions of the mediator complex.

Several mammalian mediator complexes that contain homologs of the yeast SRB and MED proteins have been isolated. These mediators are at best distantly related to their yeast counterparts but may be functionally analogous. The current list of mediators includes entities called SMCC (also called TRAP or DRIP complex) negative activator of transcription (NAT), the human mediator complex, CRSP, and activator-recruited cofactor (ARC). Some of these probably represent the same complex, although subtle differences may exist. It is not yet clear whether the mammalian mediator, like yeast, exists in a stable complex with Pol II, although substoichiometric amounts of Pol II do copurify.

The SMCC (SRB MED coactivator complex), isolated by Roeder and colleagues, contains human homologs of SRB7, 10 (CDK8), 11 (Cyclin C), MED6, and RGR1. SMCC also contains proteins called TRAPs or DRIPs, which associate with nuclear receptors (thyroid receptor, TRAPs; vitamin D receptor, DRIPs; for review, see Freedman 1999). Interestingly, mammalian homologs of yeast proteins were identified in SMCC that had not previously been reported as part of the yeast holoenzyme (see Gu et al. 1999 and Ito et al. 1999 and references therein). The ability of SMCC to mediate activation was demonstrated in a reconstituted system.

Another complex called ARC was isolated by Tjian and colleagues (Näär et al. 1999). ARC was isolated on the basis of its requirement for synergistic gene activation in vitro by the sterol response element binding protein (SREBP-1a) and Sp1. The complex shares many subunit similarities with NAT, SMCC (TRAP, DRIP), and human mediator. The complex has a high affinity for several activators in affinity chromatography experiments, including SREBP1a, VP16, and the p65 subunit of NF-κB. It apparently requires chromatin templates to manifest its coactivator function in vitro.

A very similar complex called NAT functions negatively in vitro, although the differences between it and ARC and SMCC are not entirely clear. The differences may lie simply in the assays used to study transcriptional activity (Sun et al. 1998).

Two other complexes required for Sp-1 and E1A-mediated transcription, called CRSP and human mediator, have also been isolated. CRSP shares some subunits with NAT and SMCC, including hRGR-1 and hMED7 homologs (Ryu et al. 1999). Although CRSP may be a subcomplex of ARC, it contains unique subunits (Ryu and Tjian 1999).

The human mediator complex, on the other hand, appears to be very similar to SMCC (Boyer et al. 1999). It was isolated by virtue of the interaction between the adenovirus E1A activator and human Sur2/DRIP130, the homolog of a *Caenorhabditis elegans* protein implicated in the MAP kinase signal transduction pathway. As additional genes encoding proteins from the various complexes are cloned and analyzed, the functional similarities and differences will become more apparent.

The existence of large coactivator complexes in yeast and humans raises important issues. Why are such complexes needed for regulation? How do activators and repressors interact with the complexes? What are the consequences of such interactions on transcription? A clue as to the function of the complexes derives from genetic and biochemical experiments in yeast, where activators have been shown to interact directly with components of the mediator complex (for review, see Hampsey 1998 and Myer and Young 1998). Yeast GAL4, for example, interacts directly with the SRB4 subunit of the holoenzyme in affinity experiments. Furthermore, mutations in SRB4 suppress mutations in the GAL4 activation domain.

The current view is that interaction of activators with any surface of the holoenzyme, be it a GTF or one of the mediator components, allows recruitment of the holoenzyme to DNA. By this model, GAL4 would recruit the holoenzyme through interaction with SRB4. This view is supported by several independent experiments, including the observation that fusion of a LexA or GAL4 DNA-binding domain to almost any component of the general machinery can lead to activated transcription (for review, see Carey 1995). These findings suggest two possibilities: (1) that recruiting even a subset of the general machinery leads to complex assembly because the factors are all associated in the form of a holoenzyme in solution; (2) that the remaining general factors can nucleate around a recruited GTF in any order; i.e., recruiting TFIIB via a LexA fusion would lead to recruitment of TBP, the mediator, and the other GTFs because of the cooperative protein–protein and protein–DNA interactions at the promoter. Both are examples of what is termed a "concerted" recruitment reaction.

The notion of concerted recruitment of the general machinery is attractive, since any surface of the general machinery, given the appropriate affinity for an activator, could lead to transcription complex assembly. The model is further supported by in vivo crosslinking experiments where inactivation of numerous components of the general machinery by temperature-sensitive mutants (i.e., TFIIB or SRB4) leads to a complete block to transcription complex assembly (Li et al. 1999).

Similar mechanisms would be applicable to the human mediator complexes. When the activator-binding data from studies on different human mediator complexes are taken together, they suggest that the mediator is a global coactivator that directly interacts with numerous activators and allows them to communicate with the GTFs to stimulate transcription. For example, in the human mediator, hSUR2 and TRAP220/DRIP205 have been shown to be direct targets of E1A and the TR (or VDR) nuclear receptor, respectively. Furthermore, the VP16 and p53 activation domains interact with the TRAP80 subunit (Ito et al. 1999) (Fig. 1.4). The observation that different subunits of the mediator interact with different activators raises the possibility that multiple activators, possibly in the form of an enhanceosome, simultaneously interact with the mediator complex. It is these types of interactions that probably underlie the process of transcriptional synergy (see Carey et al. 1990) and combinatorial control (discussed in Ito et al. 1999). Moreover, because the human homologs of SRB10 and 11 (CDK8 and cyclin C) are also present in the human mediators, it suggests that the complex could also serve as a corepressor (see below). Although we have emphasized the concept of recruitment and synergy, the mediator and its machinations are likely to become far more complicated as our understanding of the mechanism progresses.

CONCEPTS AND STRATEGIES: II. ACTIVATORS AND REPRESSORS

Regulatory Promoters and Enhancers

Transcriptional regulation is controlled by the binding of sequence-specific DNA-binding proteins to regulatory promoters and enhancers (for review, see Blackwood and Kadonaga

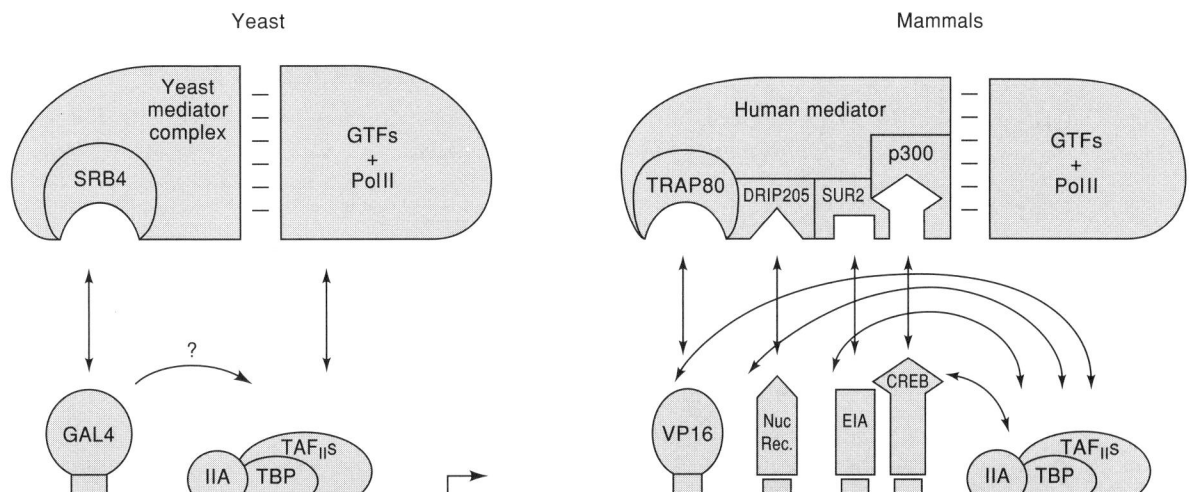

FIGURE 1.4. Recruitment of the general machinery by interaction with mediators and TAF$_{II}$s. Activators interact with distinct subunits of mediator and distinct TAF$_{II}$s. These interactions lead to recruitment of TFIID and the mediator. Individual interactions between mediator and GTFs and TFIID stabilize the complex.

1998). This section summarizes the properties of activators/repressors and enhancers/silencers. For the purposes of this chapter, the regulatory promoter is defined as the region surrounding the core promoter and within a few hundred base pairs of the transcription start site. An enhancer is defined as a control region found at a greater distance from the transcription start site, either upstream or downstream of the gene or within an intron. The distinctions between regulatory promoters and enhancers have even become blurred over time because control elements found in an enhancer can often function in the context of a promoter. Conversely, individual promoter elements can often impart enhancer activity if multimers of that element are inserted at a more distant location. A current compilation of promoters, where the transcription start sites have been mapped, is available from the eukaryotic promoter database (http://www.epd.isb-sib.ch/).

Many of the concepts and approaches employed today to understand cell-specific enhancers were derived from studying viral enhancers (see, e.g., Khoury and Gruss 1983; Ondek et al. 1988; Schaffner et al. 1988). The SV40 enhancer contains two 72-bp repeated sequences positioned upstream of three 21-bp repeats (largely Sp1 sites) and a core promoter driving T antigen (TAg) expression; TAg is expressed early in the viral lytic cycle and controls lytic DNA replication. Studies on the SV40 enhancer revealed many insights into how enhancers were organized and functioned. These principles include (1) the use of activator combinations to regulate expression; (2) the use of binding sites for regulatory proteins that permitted the enhancer to function in different cell types; (3) the ability of an enhancer to function at a distance irrespective of orientation; (4) the concept that an enhancer was composed of modules that could be multimerized to augment activity; and (5) the concept that these modules themselves consisted of multiple binding sites, previously called enhansons, whose spatial relationship was sometimes fixed—often an indication of cooperative binding.

Currently, the specific properties that permit an enhancer to function from a great distance have not been determined. Current models for enhancer action have been reviewed recently (Kadonaga 1998). Two of the original ideas were (1) the enhancer transmits changes in DNA structure to the core promoter or (2) the enhancer serves as a bidirectional binding site for Pol II, which then tracks along the DNA to the core promoter. These hypotheses were disproven by studies demonstrating that enhancers could be linked via a protein bridge to a core promoter and still retain functionality (Mueller-Storm et al. 1989), a result inconsistent with the aforementioned models.

The current view is that enhancers bind activators and other sequence-specific proteins that are involved in chromatin remodeling. Once bound, these activators loop out the intervening DNA to interact with proteins bound to the regulatory and core promoters (i.e., other activators and the general machinery). These interactions are believed to stabilize transcription complex assembly. The looping model is attractive for two reasons. First, the energetics of DNA looping have been studied extensively in model systems by ligation of large DNA molecules and by cooperativity among distant proteins (for review, see Wang and Giaever 1988; Rippe et al. 1995). These studies have borne out the feasibility of looping as a mechanism for facilitating distal protein–protein interactions. Second, in a looping model, chromatin could play a positive architectural role by compacting the DNA that is between the enhancer and promoter, facilitating long-range interactions.

Only in recent years have enhancers been studied in sufficient detail to allow a mechanistic understanding of how activators bind to them and turn on a gene. This delay has largely been due to the time necessary to isolate and clone the numerous factors that bind to any given cellular enhancer. Studies on several enhancers, including Igμ, T-cell receptor-α(TCR-α), and interferon-β (IFN-β) have revealed the details of how combinations of activators bind cooperatively and how the cooperativity contributes to specific gene expression patterns (for review, see Grosschedl et al. 1994; Ernst and Smale 1995; Carey 1998). The concept of enhanceosomes and how they confer specificity is discussed in a later section.

Transcriptional Activators

Modular Activators

Activators are modular proteins with distinct domains for DNA binding and transcriptional activation (Johnson and McKnight 1989; Triezenberg 1995). The current view is that the DNA-binding domain targets the activator to a specific site, perhaps in conjunction with cooperativity domains that allow combinatorial interactions with other activators. The activation domain, on the other hand, interacts with the general machinery to recruit it to the promoter. In some instances, these domains are part of the same polypeptide (i.e., the yeast GAL4 and GCN4 proteins), whereas in others, the domains are located on separate subunits of a multiprotein complex. This multisubunit organization provides additional opportunities for combinatorial control and regulatory diversity.

The Oct-1 complex, for example, employs multisubunit modularity as a means of combinatorial control. In mammalian cells, Oct-1 is a ubiquitous DNA-binding protein containing a homeodomain and a separate POU domain. On its own, Oct-1 binds weakly to 8-bp DNA sites termed octamers. Oct-1 interacts with OCA-B, a B-cell-specific factor, and VP16, a herpes simplex virus factor. Additional specificity in binding is imparted by VP16 and OCA-B, which interact with different surfaces of the Oct-1 POU domain (part of the DNA-binding surface of Oct-1). VP16 and OCA-B by themselves bind DNA very weakly but, when combined with Oct-1, target the complexes to different promoters (see Babb et al. 1997). Therefore, Oct-1 is an example in which a single activator can combine with different coactivators to elicit unique regulatory effects. However, VP16 and OCA-B are considered coactivators because they bridge the DNA-binding domain (Oct-1) to the general machinery (see Luo et al. 1998 and references therein). However, VP16 is not a coactivator in the sense of a mediator complex component, but it does serve as a bridge between the Oct-1 and the mediator or TAF$_{II}$s. We briefly discuss the DNA-binding and activation domains here, although a more thorough review and strategies for studying these domains are presented in Chapter 12.

DNA-binding Domains

Regulatory proteins are often grouped into families according to the sequence and structure of their DNA-binding domains. The targeting function of the DNA-binding domain defines the site of activator action and the contribution of an activator to differential gene regulation. For this reason, understanding how activators bind specific sites and distinguish between related sites has become a major focus of investigation in the gene expression field. Dozens of classes of DNA-binding domains have been described in higher eukaryotes (Pabo and Sauer 1992). Some common motifs whose structures have been solved include the homeodomain, a variety of zinc-nucleated DNA-binding domains with different structures (i.e., Zif268, GAL4, GR), the basic leucine zipper (bZIP), basic helix-loop-helix (bHLH), Rel homology, Ets homology, Myb homology, high-mobility-group (HMG) domain, and others (for review, see Luisi 1995). Some DNA-binding proteins do not fit into any of the defined families, whereas in others these broad classes have been further subdivided. Not surprisingly, members of some protein families bind to similar DNA sequences. In other families, however, there is little similarity between recognition sites for the different family members, largely because the key recognition amino acids are highly variable among family members.

The list of DNA-binding domains has become too large for a comprehensive overview. However, updates are reviewed yearly in several journals and volumes including *Current Opinion in Structural Biology* and *Annual Reviews of Biochemistry and Biophysics.* More importantly, a structural database of DNA-binding domains (http://ndbserver.rutgers.edu/NDB/structure-finder/dnabind/index.html) exists that allows one to search for structures of various crystallized DNA-binding proteins and then see the structures using shareware rendering programs (i.e., RASMOL; http://www. umass.edu/microbiol/rasmol). If a DNA-binding domain fits into a defined family, the crystal structure of related family members can provide a starting point for a structure–function study (see, e.g., Hanes and Brent 1989). A complete listing of DNA-binding domains and family members is present as part of the TRANSFAC database (http://transfac.gbf.de/TRANSFAC/cl/cl.html).

TRANSFAC, TRRD (Transcription Regulatory Region Database), and COMPEL are databases that contain information on the factors and DNA elements involved in transcription. Sequence-specific transcription factors and their recognition sites are located in TRANSFAC, the regulatory elements of entire genes are located in TRRD, and properties of composite DNA elements are located in COMPEL (for review, see Heinemeyer et al. 1998). The databases are accessible via the World Wide Web (http://transfac.gbf.de/index.html).

Activation Domains

The term "transcriptional activation domain" has been used to refer loosely to a wide variety of protein domains that interact either directly with components of the general transcription machinery or with coactivators. The popular definition of activation domain is a region of protein that stimulates transcription when attached to a heterologous DNA-binding domain in a so-called domain swap experiment (see Chapter 12). However, although many activators are modular in structure, there are examples where residues important for activation are interdigitated with the DNA-binding domain. Two model cases are MyoD and the glucocorticoid receptor (Schena et al. 1989; Davis et al. 1990).

Most previous classification schemes have employed amino acid composition (i.e., acidic, glutamine-rich) to define activation domains. These schemes are somewhat dated, and it is not clear that they signify mechanistic differences (for review, see Triezenberg 1995). As the field evolves, functional differences will be employed to distinguish and classify activation domains more precisely. Chapter 12 describes methods for delineating activation domains, and Chapter 13 describes methods for determining whether and how proteins bind DNA.

Many, if not all, activators contain multiple activation domains. GAL4, for example, contains one domain on the amino terminus adjacent to the DNA-binding domain and another on the carboxyl terminus (see Gill and Ptashne 1987 and references therein). There appears to be considerable flexibility in organization within a domain. Deletion analysis of GCN4, for example, reveals a functional redundancy, where deletion of one or the other segment elicits a negligible effect on activation; deletion of the entire domain is necessary to abrogate activity (Hope et al. 1988). Many studies show that activation domains within a regulatory protein act additively or synergistically on activation potential, e.g., as with the two activation domains of the glucocorticoid receptor (Hollenberg and Evans 1988). Artificial multimerization of large or small pieces of activation domains can also lead to synergistic transcription (see, e.g., Hollenberg and Evans 1988; Emami and Carey 1992).

The rationale for an activator to possess multiple activation domains is twofold. First, multiple domains can increase the potency of an activator by increasing the probability that any individual domain will interact with a target, e.g., a subunit of the mediator complex. Alternatively, different domains of an activator may contact different targets. The NF-κB subunit p65 contains at least two domains, one that interacts with the coactivator p300 and is necessary for the synergistic effect of p65 in the context of the enhanceosome and another that contacts a different subunit of the general machinery (Merika et al. 1998).

In studying transcriptional activators, considerable attention has been focused not only on their mechanisms of action, but also on how their functions are modulated. Although many activators are regulated at the transcriptional level, many other activators are regulated by posttranslational mechanisms. The subcellular location of the NF-κB and NFAT complexes can be modulated by signaling pathways, resulting in sequestration from target genes (Liou and Baltimore 1993; Israel 1994; Ghosh et al. 1998). Furthermore, phosphorylation can affect the activity of a transcription factor either positively or negatively. Phosphorylation of CREB, for example, leads to recruitment of CBP (Parker et al. 1996; for review, see Torchia et al. 1998). CBP, like many other coactivators, also modifies chromatin and other transcription factors through acetylation. Another interesting example of posttranslational modification of a transcription factor is the sterol-response element binding protein, which controls genes involved in cholesterol metabolism. The transcription factors SREBP-1 and -2 are tethered to the endoplasmic reticulum. Low cholesterol initiates a signaling cascade that leads to release of SREBP and subsequent nuclear translocation to activate genes involved in cholesterol uptake and metabolism (for review, see Brown and Goldstein 1997).

Structural Aspects of Activation Domains

Very few activation domains have been characterized at the structural level. However, ultimately the cocrystal structures of activation domains with their targets are necessary to understand fully how activation domains function. One of the paradigms for understanding activation domain structure has been VP16. The carboxy-terminal 90 amino acids of

VP16 represent one of the most potent activation domains studied to date. VP16 has been subjected to numerous structure–function studies by Triezenberg and colleagues (for review, see Triezenberg 1995). Although VP16 is classified as an acidic activation domain, mutagenesis studies suggest that bulky hydrophobic amino acids apparently mediate direct protein–protein interactions with VP16's targets. VP16 binds to TFIIB and TBP and has been proposed to recruit TFIIB into the TBP or TFIID-promoter complex (Stringer et al. 1990; Roberts et al. 1993). VP16 also interacts with TFIIH (Xiao et al. 1994), a TAF_{II} (Goodrich et al. 1993), and the mediator complex (Näär et al. 1999). It is not clear which if any of these are the physiological targets of VP16. The structure of a portion of the VP16 activation domain has been solved by nuclear magnetic resonance (NMR) (Uesugi et al. 1997), revealing it to be an α helix. The α helix is somewhat disordered in solution and becomes structured upon interacting with human $TAF_{II}32$.

The crystal structures of other activation domains alone and complexed with their targets have been solved. These include several nuclear receptor ligand-binding domains, which contain an activation domain referred to as AF-2 (see Torchia et al. 1998). The ligand-binding domains of nuclear receptors share a conserved 12-helical structure. AF-2 is apparently located in helix 12 and is an amphipathic helix. The crystal structure of the peroxisome proliferator-activated receptor-γ (PPAR-γ), a ligand-dependent nuclear receptor involved in adipocyte differentiation, was recently solved alone and in a complex with the coactivator histone acetylase SRC-1 (discussed below). The study revealed that the ligand dramatically repositions AF-2 (for a discussion of the effect of ligand on LBD structure, see Moras and Gronemeyer 1998). Glutamate and lysine residues, conserved in nuclear receptor ligand-binding domains, then form a charge clamp that contacts backbone atoms of the LXXLL helices of SRC-1, permitting SRC-1 recruitment to the receptor in a ligand-dependent fashion (Nolte et al. 1998); LXXLL motifs tether nuclear receptor coactivators to the ligand-binding domain. The p53 activation domain bound to its inhibitor MDM1 has also been solved. It reveals that the activation domain is a short amphipathic helix (Kussie et al. 1996). As more domains appear in the literature, structural themes are bound to emerge. Furthermore, with the cloning and characterization of the human mediator complexes, the list of possible targets in the general machinery is hopefully coming to an end. A systematic effort can now be undertaken to classify activation domains based on function.

Repressors and Corepressors

General Mechanisms

Gene expression is often regulated by repressors and corepressors. Repression mechanisms are, however, less well understood than activation mechanisms. Several older reviews discuss transcriptional repression in eukaryotes and propose different conceptual and mechanistic classes (Herschbach and Johnson 1993; Hanna-Rose and Hansen 1996). In general, transcriptional repression can be divided into three broad categories. First, repression can occur by inactivation of an activator, which can be accomplished by several distinct mechanisms: (1) posttranslational modification of the activator (e.g., acetylation of HMG-I inactivates its activity on the IFN-β enhanceosome; Munshi et al. 1998), (2) dimerization of the activator with a nonfunctional partner (e.g., MyoD's interaction with I_d inactivates its activity; Benezra et al. 1990), (3) competition for the activator's binding site, or a direct repressor–activator interaction that results in masking of the activator's function (e.g., p53 interacts with MDM2 in mammalian cells; Momand et al. 1992). Second, repression can be mediated by proteins that associate tightly with GTFs and thereby inhibit the formation of a preinitiation complex. Examples include interaction of MOT1 and Dr2 with TBP (for review, see Orphanides et al. 1996). These "global" repressors may play an integral role in the transcriptional activation process, in that some activators may stimulate the formation of a preinitiation complex by displacing a global repressor. The third category of repression is mediated by a specific DNA element and DNA-binding protein, which act dominantly to repress both activated and basal transcription of a given gene. Recent studies suggest that in these cases interactions with the general machinery (Hanna-Rose and Hansen 1996) or chromatin can lead to gene inactivation.

Sequence-specific Repressors

Repression through a sequence-specific DNA-binding protein has been studied in great detail in *S. cerevisiae*, which exists as two haploid cell types, **a** and α. In **a** cells, the cell-type identity is maintained by the repression of **a**-specific genes by a protein called α2 (or MATα2; see Herschbach and Johnson 1993). The α2 protein, which is a member of the homeodomain family, binds cooperatively with an activator, MCM1, to a control element found upstream of the TATA box in a variety of coordinately repressed genes. Genetic data have revealed that the repression by α2 depends on the activities of two corepressors, TUP1 and SSN6.

Corepressors can be defined as proteins that mediate repression when recruited to a promoter by protein–protein interactions. TUP1 appears to control repression directly, because it can repress transcription when artificially brought to a promoter via fusion to a heterologous DNA-binding domain (Tzamarias and Struhl 1994). Biochemical studies have shown that TUP1 and SSN6 associate with each other and that TUP1 links the complex to the α2 protein (see Komachi et al. 1994; Tzamarias and Struhl 1994). The precise mechanism of repression by this complex is not known, but the targets of repression appear to be a general transcription factor and/or a chromatin component (Herschbach and Johnson 1993; Cooper et al. 1994).

One possibility for the α2 target is SRB10, a component of the holoenzyme. SRB10 was identified in yeast genetic studies as a gene required for α2-mediated repression (Wahi and Johnson 1995). SRB10 acts negatively by phosphorylating the CTD of Pol II prior to initiation (for review, see Myer and Young 1998). Mutations in SRB10 and 11 appear to diminish the ability of α2 to repress transcription (Kuchin and Carlson 1998). One might imagine that α2 directly stimulates SRB10 CTD kinase activity to inactivate the holoenzyme. An alternative theory is that α2 interacts with chromatin via direct interaction between the corepressor TUP1 and the histone tails of H3 and H4. Mutations in the basic tails, for example, appear to cause derepression of genes controlled by α2 (Huang et al. 1997).

Another class of repressor action is exemplified by UME6. UME6 is a sequence-dependent repressor that interacts with a corepressor complex containing the yeast negative regulator SIN3 and a histone deacetylase. The interaction of repressor proteins with histone deacetylases appears to be an emerging mechanism for repression, an issue we discuss in more detail below. The key point is, however, that acetylation favors chromatin remodeling and gene activation, whereas deacetylation promotes repression by establishing a transcriptionally repressive chromatin environment (for review, see Struhl 1998).

The properties of corepressors combined with the properties of coactivators discussed earlier are interesting to consider with regard to combinatorial gene regulation. By modulating the expression patterns or activities of corepressors and coactivators, as well as of the DNA-binding proteins that interact with them, a limited set of proteins may contribute to diverse expression patterns. Furthermore, if differentially expressed coactivators and corepressors interact with the same DNA-binding protein, a single control element could mediate both activation and repression of a given gene.

CONCEPTS AND STRATEGIES: III. CHROMATIN AND GENE REGULATION

Genetic and biochemical studies have begun to reveal the critical role that chromatin plays in modulating transcription. In this section, we focus initially on important aspects of chromatin structure and then elaborate on how chromatin's structural integrity is regulated by remodeling enzymes during gene activation and repression.

Chromatin

Structure and Organization

In a mammalian interphase nucleus, DNA is incorporated into a 10-nm-diameter nucleosomal fiber, with each nucleosome containing a core histone octamer (containing two molecules of H2A, H2B, H3, and H4) and one linker histone H1 or H5 (van Holde 1989). The 10-nm nucleosome fiber is assembled into a higher-order structure, referred to as the 30-nm filament (Fig. 1.5A). In addition to the histones, several abundant non-histone proteins (NHP), including various HMG proteins, are commonly associated with chromatin. Correlative evidence strongly suggests that genes within the "condensed" 30-nm chromatin filament must be "decondensed" to be competent for transcription. The status of decondensed DNA associated with nucleosomes is less clear (for review, see Workman and Kingston 1998).

The crystal structure of the nucleosome (Fig. 1.5B) has recently been solved (Luger et al. 1997), providing insight into how DNA is packaged (for review, see Workman and Kingston 1998). A typical nucleosome contains 146 bp of DNA wrapped 1.65 times, in a toroidal supercoil, around the histone octamer. The structure reveals that the nucleosome has a disk-like shape. The octamer can be subdivided into two heterodimers of H3/H4 and H2A/H2B. Two H3/H4 heterodimers stably associate to form an H3/H4 tetramer.

Each histone contains a characteristic motif termed the histone fold. A histone fold comprises a long α helix linked at either end to two shorter α helices via short β turns. The histones insert arginines into all 14 of the minor grooves facing the surface of the octamer in a mononucleosome. These contacts provide most of the energy driving the DNA-octamer interactions.

In addition to interactions of the arginines with the minor groove, other sources of nucleosome stability are the lysine-rich amino-terminal tails of all four histones. The tails are believed to stabilize intra- and internucleosomal DNA interactions. Despite the fact that the tails are somewhat disordered in the crystal, some interesting features have emerged. The H4 tail, for example, clearly interacts with the H2A/H2B interface of an adjacent nucleosome (Luger et al. 1997). Such interactions may facilitate nucleosome compaction and transcriptional repression in vivo. The tails of H4 in particular are known to be acetylated during gene activation. Acetylation causes neutralization of the lysine basic charge, which in turn is speculated to loosen contacts between the histone tails and other histones or DNA. Biochemical experiments have shown that octamers bearing trypsinized or acetylated tails allow greater access to transcription factors in vitro (for review, see Workman and Kingston 1998).

A.

B.

☐ H2A
▨ H2B

☐ H3
■ H4

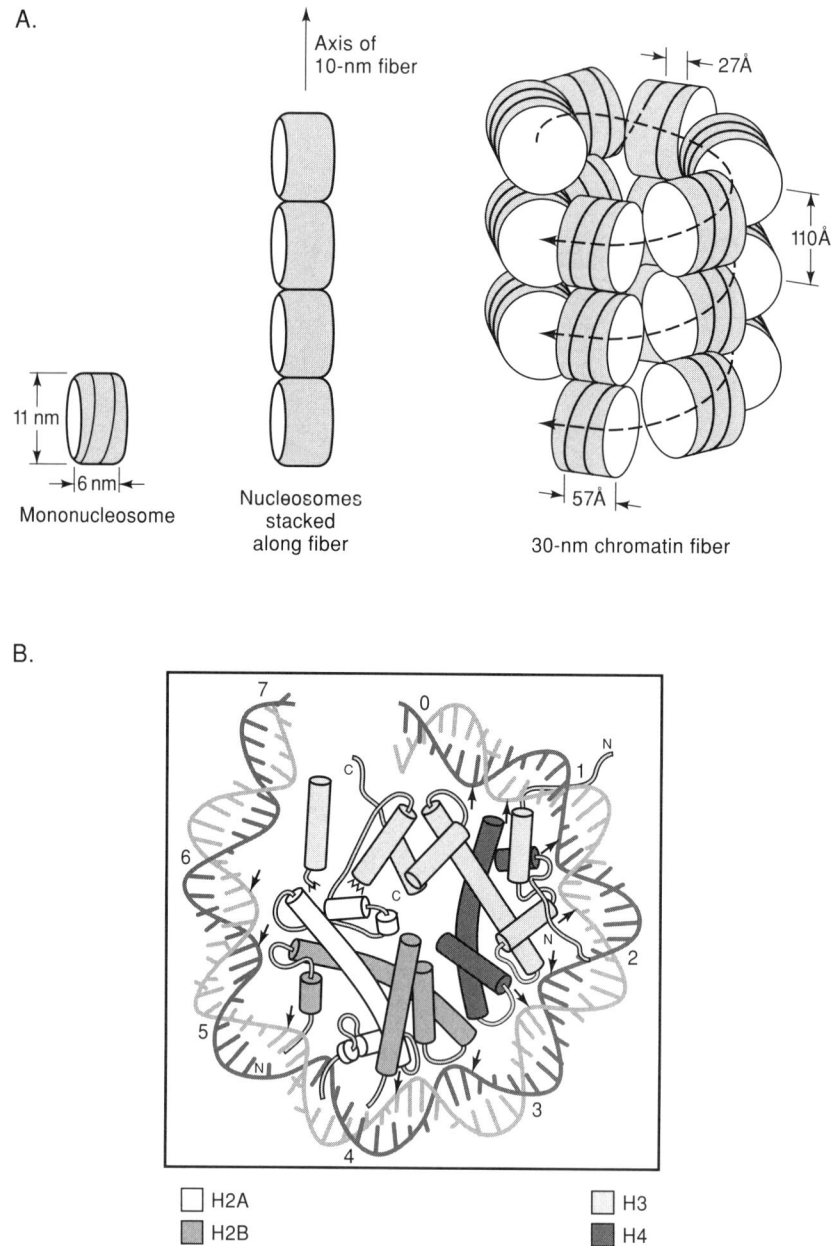

FIGURE 1.5. DNA and nucleosomal structure. (*A*) Mononucleosome (*left*), 10-nm-diameter nucleo-somal fiber (*middle*), and 30-nm filament (*right*). (Reproduced, with permission, from Lewin 1990 [Copyright Oxford University Press].) (*B*) Crystal structure of the nucleosome to a resolution of 2.8 Å. Small arrows indicate the points of protein–DNA contacts. Only one turn of the DNA around the nucleosome is shown. (Reproduced, with permission, from *Nature* [Luger et al. 1997 and Rhodes 1997] copyright 1997 Macmillan Magazines Ltd.)

Binding of Transcription Factors to Chromatin

To recruit acetylases or other chromatin-remodeling components to specific genes, tran-scription factors must devise a mechanism to gain localized access to chromatin. Transcription factors can bind to nucleosomes in vitro but do so weakly and tend to bind at the edges, where biophysical studies have shown the interactions between DNA and the octamer are weakest. On multimerized binding sites, however, once one transcription factor molecule is bound to the edge, the other molecules are able to bind inward toward the nucleosome axis in a cooperative fashion (for review, see Workman and Kingston 1998). Such a mechanism may allow local entry of a limited set of factors to specific sites in vivo; these factors would then recruit enzymes that remodel the chromatin more extensively and permit subsequent recruitment of other activators and the general machinery.

In the absence of cooperative binding, localized access of individual factors may still be possible. Nucleosomes appear in some cases to be positioned along the DNA strand so that the same DNA sequences consistently face either toward or away from the nucleosomal core. In some cases, the DNA recognition sequence for an individual factor may therefore be accessible. Positioning of nucleosomes is influenced by two variables: sequence-specific factors that bind to and fix the location of the DNA helix relative to the octamer (such as the winged helix family of factors) and the DNA sequences that have inherent positioning capabilities; statistically, the minor groove of AT-rich sequences tends to face toward and contact the octamer. Both natural and artificial nucleosome positioning sequences have been isolated (see, e.g., Widlund et al. 1997 and references therein). Such positioning sequences also fix the position of one nucleosome relative to another, in some cases leaving the linker sequences between nucleosomes accessible. In some cases, these linker sequences can serve as binding sites for transcriptional activators (see, e.g., review by Lohr 1997).

Genetic Links between Gene Activation and Chromatin

Studies in *S. cerevisiae* have provided the most convincing evidence for a dynamic role of chromatin in gene regulation. First, nucleosome depletion in vivo has been found to result in transcriptional activation in a general manner (for review, see Grunstein 1990), suggesting that removal of nucleosomes is coupled to gene activation. Evidence for how the coupling might occur was derived from mutations of a gene called SWI2 (or SNF2). Mutations in this gene led to down-regulation both of the mating type switching gene, HO, and of a gene involved in sucrose fermentation called SUC2. The resulting mutant was called SWI2 by one group and SNF2 by the other, hence the term SWI2/SNF2 or simply SWI/SNF. Remarkably, the mutations in SWI2 could be suppressed by mutations in histone H4. Taken together, the data suggested that SWI2/SNF2 played a role in gene expression by modulating chromatin structure. It was later found that the SWI2/SNF2-encoded protein was a component of a larger multiprotein complex that could disrupt or reconfigure nucleosomes and stabilize the binding of transcription factors to nucleosomal DNA in vitro. The resulting complex was termed the SWI/SNF complex (Imbalzano et al. 1994; Kwon et al. 1994).

Independent genetic studies in yeast have revealed that the amino termini of histones H3 and H4 are important for transcriptional activation of a subset of genes (Grunstein 1990). Furthermore, a strong correlation has been established between transcriptionally active genes and acetylation of amino-terminal lysine residues in histone H4 (Hebbes et al. 1988; Paranjape et al. 1994). Recently, several histone acetylase complexes have been isolated from yeast and mammalian cells. Some of these complexes have been shown to be associated with activators, providing a mechanism for targeting the acetylases to certain regions of DNA, a point we discuss below.

ATP-dependent Remodeling Complexes

SWI/SNF Complexes

SWI2/SNF2 was found to be the prototype of a large family of protein complexes characterized biochemically in yeast, *Drosophila*, and humans. Many such complexes have been

TABLE 1.4. *ATP-dependent nucleosome remodeling machines*

Organism	Complex	Name	Molecular mass	No. of subunits	ATPase
Saccharomyces cerevisiae	SWI/SNF	switching mating type sucrose-fermenting	2 MD	11	SWI2/SNF2
	RSC	remodels the structure of chromatin	1 MD	15	STH1
Homo sapiens	BAF (mammalian SWI/SNF)	BRG1- or hbrm-associated factors	2 MD	9–12	BRG1
			2 MD	9–12	hbrm
	NuRD	nucleosome remodeling histone deacetylase complex	1 MD	7	Mi-2
	NRD	nucleosome remodeling and deacetylase	ND	6	CHD3/4
	NURD	nucleosome remodeling and histone deacetylase	1.5 MD	18	CHD4
	RSF	remodeling and spacing factor	500 kD	2	hSNF2h (homolog of *Drosophila* ISWI)
Xenopus laevis	Mi-2 complex		1–1.5 MD	6	Mi-2 (Snf2 superfamily member)
Drosophila melanogaster	Brahma complex		2 MD	ND	BRM
	NURF	nucleosome remodeling factor	500 kD	4	ISWI
	CHRAC	chromatin accessibility complex	670 kD	5	ISWI
	ACF	ATP-dependent chromatin assembly and remodeling factor	220 kD	4	ISWI

Mi-2, NuRD, NRD, and NURD all have reported ATP-dependent remodeling and histone deacetylase activity. Several complexes have common subunits. Reprinted, with permission, from Björklund et al. 1999 (Copyright 1999 Cell Press).

identified (RSC, CHRAC, NURF, and ACF) and the list is growing (see Table 1.4). All of these complexes contain a subunit with an evolutionarily conserved ATPase motif. ATP is required to remodel nucleosomal arrays in vitro and enhance accessibility to transcription factors and nucleases used to probe nucleosome structure and DNA accessibility. Biochemical and genetic studies suggest that these complexes remodel chromatin in unique ways. The complexes differ in their effects on remodeling artificial arrays of nucleosomes in vitro, accessibility of transcription factors in vitro, and whether or not they are essential for viability of an organism (i.e., yeast) after gene disruption.

The human SWI/SNF complex contains two proteins called BRG1 and hBRM. These appear to be homologous to the SWI2 gene found in yeast SWI/SNF and the STH gene found in the RSC complex. The two proteins are part of distinct mammalian SWI/SNF complexes. The three *Drosophila* complexes, ACF, CHRAC, and NURF, contain a subunit called imitation SWI or ISWI. The ISWI complexes are smaller than RSC or SWI/SNF, typically comprising three or four subunits. The ISWI and BRG1 subunits alone appear to have some limited remodeling capabilities (Corona et al. 1999). CHRAC contains topoisomerase II, suggesting that alterations in DNA topology are linked to its function. The presence of the multiple *Drosophila* complexes suggests that other mammalian complexes will be discovered. General aspects of the different complexes are reviewed by Varga-Weisz and Becker (1998).

Mechanisms and Targeting

The three central issues regarding the SWI/SNF complex are (1) what is the mechanism of chromatin remodeling, (2) are the complexes functionally equivalent or do they play different roles in vivo (i.e., remodeling of euchromatic vs. heterochromatic structures), and (3) how are the enzymes targeted to DNA during activation and repression? For the most part, the mechanisms of ATP-dependent remodeling are unknown.

SWI/SNF models for ATP-dependent remodeling should account for the following observations: First, SWI/SNF disrupts the characteristic 10-bp period repeat of DNase I cleavage within a positioned nucleosome in vitro and enhances access of transcription factors to the DNA. Furthermore, the remodeled state appears to be stable after the remodeling enzyme is removed. Second, SWI/SNF can form loops on long linear and circular DNA molecules, forming isolated domains where it could conceivably act catalytically on several intradomain nucleosomes (Bazett-Jones et al. 1999). Finally, careful quantitation of the DNA path and length in nucleosomal arrays in the presence and absence of SWI/SNF by high-resolution microscopy techniques shows that nucleosome disruption is accompanied by a decrease in the amount of DNA packaged into the nucleosome, a result consistent with partial unwrapping of the DNA from the nucleosome. Such an unwrapping mechanism would explain in part how the SWI/SNF complex enhances transcription factor binding in vitro and in vivo. The ability to form domains would further explain how the SWI/SNF complex might remodel multiple nucleosomes over a promoter and downstream gene.

Interestingly, inactive mutants of SWI2/SNF2 can be suppressed by particular mutations in histone H4. One such mutation is at an arginine known to interact with the minor groove of the nucleosomes in the crystal structure. Presumably, weakening this contact of the histone with DNA can substitute for loss of SWI/SNF activity. These data imply that SWI/SNF normally weakens the octamer–DNA interactions as part of its mechanism, consistent with the unwinding studies cited above (for review, see Workman and Kingston 1998). Note, however, that other mutations at histone dimer interfaces also suppress SWI/SNF defects, suggesting that the nucleosome protein–protein contacts may also be affected by the SWI/SNF complex.

Studies on the related RSC complex show that it catalyzes transfer of the histone octamer from a nucleosome onto a naked DNA particle (Lorch et al. 1999). The data suggest that RSC in an ATP-dependent manner can lead to loop formation within a single nucleosome allowing transfer of the octamer in *cis* from one stretch of the DNA to another (for review, see Kornberg and Lorch 1999). Again, the mechanism suggests unwinding of the DNA from the core. While SWI/SNF and RSC appear to promote unwinding of the DNA around the nucleosome, the ISWI-containing complexes may function by a different mechanism. Both CHRAC and NURF facilitate sliding of nucleosome monomers along the DNA without disruption of the nucleosome transfer to competing DNA. Sliding would allow transcription factor access by moving the nucleosome off of a particular DNA site (Hamiche et al. 1999; Längst et al. 1999).

The issue of how the complexes are targeted to a specific location is still unclear. SWI/SNF has been reported to be a component of the yeast holoenzyme. It is plausible that an activator could gain limited access to chromatinized DNA and recruit the holoenzyme, which would then remodel the chromatin via the SWI/SNF subunits to permit stable binding of the general machinery (Fig. 1.6). The general machinery in turn might reciprocally stabilize activator binding both directly through contacts with the activators and indirectly by keeping the promoter nucleosome-free. Some versions of the mammalian holoenzyme have been shown to contain the mammalian homologs of SWI/SNF, providing an evolutionarily conserved mechanism for chromatin remodeling (Parvin and Young 1998). Conversely, it has been proposed that activator may directly interact with SWI/SNF complexes independently of holoenzyme (for review, see Workman and Kingston 1998). By such a mechanism, chromatin remodeling by SWI/SNF would precede holoenzyme recruitment. For example, Nasmyth and colleagues recently showed using chromatin immunoprecipitation assays that the activator SWI5 recruits SWI/SNF to the HO promoter and that this event leads to recruitment of the histone acetylase complex SAGA, which then leads to recruitment of another transcription factor called SBF. SBF was then proposed to recruit the general machinery, although this was not shown directly (Cosma et al. 1999).

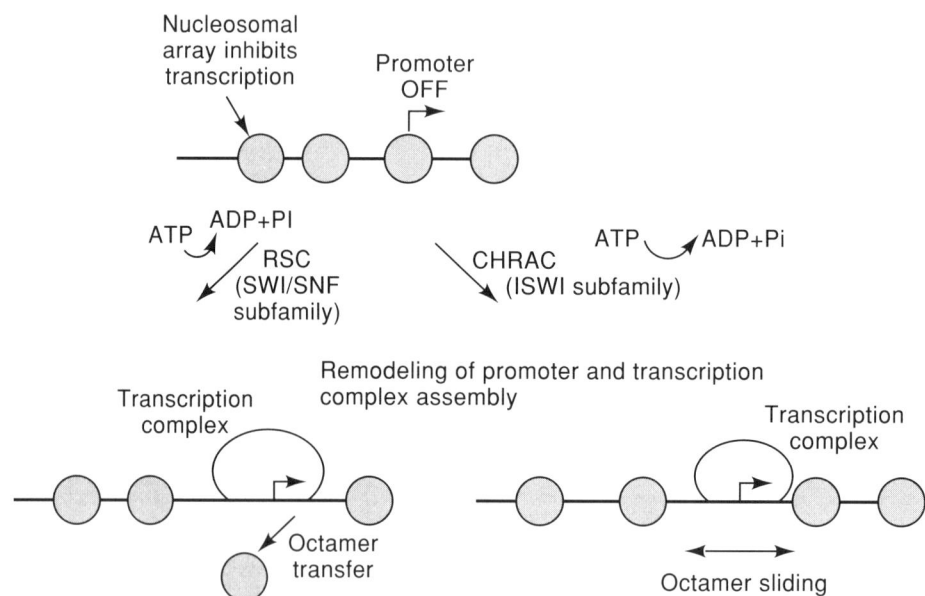

FIGURE 1.6. Models for ATP-dependent chromatin remodeling. (Adapted, with permission, from Hamiche et al. 1999 [Copyright 1999 Cell Press].)

It should also be emphasized that SWI/SNF may be involved in repression as well as activation. First, analysis of genes affected by mutations of SWI/SNF in yeast show that numerous genes are both down- and up-regulated when SWI/SNF is inactivated (Holstege et al. 1998). Second, other ATP-dependent chromatin-remodeling complexes are associated with complexes known to be involved in transcription repression, an issue we address below.

In addition to providing a block to transcription factor binding, chromatin also hinders transcriptional elongation. Studies with pure nucleosomes and RNA polymerase III show that although RNA polymerase can transcribe through a nucleosome, the process is slow. Remarkably, however, the nucleosome does not leave the template, and as the polymerase elongates, the octamer appears to transfer directly to the DNA behind it (Studitsky et al. 1997; for review, see Kornberg and Lorch 1999). This process is likely to be facilitated in vivo by factors that directly assist in chromatin remodeling. The SWI/SNF complex, for example, can facilitate heat shock transcription factor-mediated elongation in vitro through a nucleosome-imposed pause on the HSP70 gene (Brown and Kingston 1997). Additional complexes that facilitate elongation in vitro have been isolated. Using a system containing pure general transcription factors, an energy-dependent factor called RSF, containing the human ISWI protein, was shown to be required for initiation but was insufficient for elongation on nucleosomal templates. However, an energy-independent factor called FACT was required for elongation. The mechanism of FACT action is currently being investigated (LeRoy et al. 1998).

Acetylation of Chromatin

As discussed above, acetylation and deacetylation of the histone tails represents a major regulatory mechanism during gene activation and repression. Histone hyperacetylation has been associated with histone deposition during replication and repair, and with transcriptionally active chromatin. Actively transcribed regions of the genome tend to be hyperacetylated, whereas inactive regions are hypoacetylated. There are two classes of histone acetylases (HATs) called type A (nuclear HATs) and type B (cytoplasmic HATs). The cytoplasmic HATs acetylate H3 and H4 posttranslationally. The acetylation is believed to be important so that chromatin assembly factor CAF-1 can deposit these histones onto chromatin during DNA replication and repair (for review, see Workman and Kingston 1998).

The first direct link between the nuclear HATs and gene activation came with the discovery by Allis and colleagues that the tetrahymena type A HAT, p55, was homologous to the GCN5 (Brownell et al. 1996). The yeast GCN5 protein had previously been implicated in GCN4-mediated transcription of amino acid biosynthesis genes. GCN5 was later shown to be present in a 0.8-MD complex with two other proteins called ADA2 and 3, previously found to be involved in gene activation in yeast. The GCN5 and ADA proteins were also present in a larger complex containing SPT proteins. The SPT proteins were originally isolated as suppressors of defects in gene activation caused by promoter insertions of the Ty transposon. The SPT-containing complex was called SAGA (SPT-ADA-GCN acetyltransferase). A direct correlation between acetylation and activation was established when it was found that GCN5-regulated genes are hyperacetylated in vivo during activated transcription and that mutation of GCN5 significantly diminishes the levels of acetylation (Kuo et al. 1998). These discoveries initiated a flurry of studies in yeast and mammalian systems to examine the mechanism by which acetylases could be targeted to specific genes.

It is likely, on the basis of studies in yeast and mammalian systems, that activators can directly recruit the acetylases. GAL4 derivatives were shown to facilitate SAGA-mediated acetylation of mononucleosomes containing GAL4 sites in vitro, and numerous studies have shown direct interactions of activators with acetylase complexes in mammalian systems (Utley et al. 1998).

Mammalian Acetylases

The mammalian protein p300 and its related homolog CBP, for example, were shown to possess histone acetylase activity. p300 and CBP interact with the activation domains of numerous signal-dependent activators and the ligand-dependent nuclear receptors (for review, see Torchia et al. 1998). These interactions provide a direct means of targeting the acetylation to a surrounding regulatory region. However, p300 and CBP proteins not only acetylate histones, but also acetylate GTFs (i.e., TFIIE), activator/repressors (i.e., p53), and architectural proteins (i.e., HMG I (Y)). Thus, the direct targets of the acetylases are not entirely clear. p300 also interacts with other acetylases called PCAF and ACTR forming large multifunctional histone–transcription factor acetylase complexes. Promoters and enhancers are likely to be bound to multiple histone acetylases (including $TAF_{II}250$) during gene activation. Despite the correlation of acetylation with activation, the precise mechanism by which the acetylated chromatin permits access to transcription factors is unknown. It is intriguing, however, that the mammalian homologs of SWI/SNF and p300/CBP are both components of the mammalian holoenzyme. The ability of activators to target these chromatin-remodeling components as part of a holoenzyme complex would provide an economical method of remodeling chromatin and assembling the preinitiation complex in a single concerted step. However, as shown in the study by Nasmyth and colleagues, these need not be concerted steps, since SWI/SNF recruitment by SWI5 appears to preclude SAGA recruitment (Cosma et al. 1999).

TAFs and Chromatin Remodeling

An intriguing link between the general machinery and acetylation complexes is the observation that some complexes contain TAF_{II}s as integral components. The yeast SAGA complex contains several TAFs ($TAF_{II}90$, –68/61, –60, –25/23, and –20/17). Genetic experiments showed that $TAF_{II}68$ is essential for SAGA activity (Grant et al. 1998). Similarly, the PCAF acetylase can be found in a large 20-subunit complex in mammalian cells containing several TAF_{II}s and TAF-like molecules. The histone H3- and H2B-like subunits within PCAF complex are identical to TFIID TAFs, $hTAF_{II}31$ and $hTAF_{II}20/15$. PCAF also contain a histone H4-like subunit homologous to $hTAF_{II}80$ (Ogryzko et al. 1998). Again, the relevance of the histone-like TAFs in the acetylation complexes is unknown but may be linked to the way the complexes interact with chromatin during acetylation. Recall that the TFIID complex contains the histone-like TAFs in a complex with the $TAF_{II}250$ histone acetylase. In some respects, the functional organization of TFIID is similar to that of PCAF, suggesting that, whatever the mechanism, the two complexes may share related functions in chromatin remodeling.

Histone Deacetylation, Transcriptional Repression, and Silencing

The view that histone acetylation is associated with activation would lead to the inference that deacetylation is associated with repression. Although this view appears to be largely correct, the details are complicated. Histones are deacetylated by large complexes containing histone deacetylase activity (for review, see Pazin and Kadonaga 1997). Yeast cells contain at least two such complexes, called histone deacetylase A and B. The yeast protein RPD3 is a catalytic component of complex B, whereas a related protein called HDA1 is the catalytic component of complex A. Studies in yeast show that disrupting the genes encoding HDA1 or RPD3 leads to hyperacetylation of histones H3 and H4 in vivo. Although such mutants can apparently enhance transcription from reporter genes, they paradoxically increase repression at telomeric gene loci in yeast. Furthermore, mutations in *Drosophila* RPD3, like yeast, enhance position-effect variegation (a point we discuss below) and lead to spreading of inactive heterochromatin. Therefore, a simple correlation does not exist between repression and deacetylases. Nevertheless, the evidence in favor of deacetylation being involved in repression is compelling. First, as discussed above, chromatin immuno-precipitation experiments have revealed that inactive genes are hypoacetylated. Second, deacetylase complexes are associated with repressors in vivo. Finally, the deacetylase activity is necessary to mediate repression based on genetics and trichostatin A (TSA) and trapoxin inhibitor studies (for review, see Workman and Kingston 1998).

Repression and Deacetylases

One typical example of deacetylase targeting is the UME6 repressor discussed briefly above. UME6 is a sequence-specific repressor that binds to sequences upstream of several yeast genes. RPD3 and SIN3 are present in the same complex in vivo (the HDA B complex) and have been implicated in repression genetically. The SIN3/RPD3 complex can be directly targeted to promoters via interaction with UME6. Chromatin immunoprecipitation shows that the deacetylation around the UME6 site is localized to only two nucleosomes encompassing the promoter and is somewhat specific for histones H3 and H4 and lysines 5 and 12 (Kadosh and Struhl 1998; Rundlett et al. 1998).

In mammalian cells, the mammalian Sin3 homologs, mSIN3A and B, have also been shown to associate with two mammalian RPD3 homologs, histone deacetylases 1 and 2 (HDAC1 and 2), in large complexes. Mammalian repressor proteins including Mad-Max, the unliganded thyroid receptor, and RXR-RAR nuclear receptors, YY1, and others have all been shown to interact directly with HDAC complexes. In most of these cases, adapter proteins bind the HDAC to the repressors. For example, the nuclear receptor corepressor N-CoR or SMRT is believed to bind the ligand-binding domain of the receptor and recruit the mSIN3-HDAC-containing complexes to the promoter (for review, see Laherty et al. 1998).

The nuclear receptors are an exciting example of regulatory proteins that switch from repressor to activator modes during signaling, i.e., in the presence of ligand (for review, see Xu et al. 1999). Thus, the unliganded receptor binds NcoR, recruits deacetylases, and represses target genes. In the presence of ligand, however, the receptor binds acetylases such as NCoA and SRC-1, which modify chromatin, and other coactivators such as the TRAP and DRIP complex, which appear to be more directly involved in recruitment of the general transcription machinery. A two-step mechanism has been proposed where the nuclear receptor binds and first recruits the acetylases, which remodels the chromatin. Following this step, the acetylase is released and the activator then binds other coactivators such as SMCC, which makes direct contacts with the GTFs, perhaps as part of a holoenzyme (for review, see Freedman 1999).

Linking Deacetylation and ATP-remodeling Machines

An intriguing link between histone deacetylases and ATP-dependent chromatin remodeling came with the recent isolation of chromatin-remodeling complexes containing both activities. All three reported complexes appear to be similar in composition and contain HDAC1 and 2 and CHD4 (also called Mi-2). CHD proteins contain a *c*hromodomain SWI/SNF-like *h*elicase/ATPase domain and a *D*NA- binding motif, hence the name. The complexes also contain the retinoblastoma protein (Rb)-associated protein RbAp48/46, which is known to interact with chromatin. The HDAC1/2 and RbAp48/46 also appear in the corepressor complexes with SIN3 (discussed above), but SIN3A is not present in the CHD4-containing complexes. Biochemical assays of these so-called nucleosome-remodeling and deacetylation complexes called NRD or NURD show that they contain ATP-dependent remodeling and deacetylation capabilities in vitro. Currently, it is not known why a single complex contains both activities. However, one model is that ATP-dependent remodeling is necessary for making the nucleosome accessible to the deacetylase (Tong et al. 1998; Xue et al. 1998; Zhang et al. 1998).

The mechanism of targeting such dual complexes to specific genetic loci is still murky. Recent studies have, however, revealed intriguing connections between hunchback, a protein involved in polycomb-dependent repression of HOX genes in *Drosophila* and *Drosophila* Mi-2 (Kehle et al. 1998). A similar connection has been found in mammals between NURD and Ikaros, a sequence-specific DNA-binding protein (Kim et al. 1999). Ikaros is a zinc-finger transcriptional regulator that, based on genetic disruptions in mice, plays a major role in B- and T-cell differentiation. Ikaros has many isoforms, and those proteins work in concert with related proteins called Aiolos and Helios. These proteins localize to heterochromatin in cytogenetic experiments (Brown et al. 1997). Remarkably, in T-cell extracts the majority of Ikaros is found associated with the NURD complex. It remains to be shown, however, that Ikaros can directly target the NURD complex in vitro and in vivo, although the Ikaros–NURD cocomplex displays general remodeling and deacetylation functions (Kim et al. 1999). Therefore, much as the SWI/SNF and acetylase activities are linked as part of the holoenzyme or preinitiation complex targeted by activators, SWI/SNF-like activities and histone deacetylases are components of other complexes targeted by repressors.

Methylation and Repression

A link between methylation and deacetylation has also recently emerged. In mammalian somatic cells, the CpG dinucleotides found in tissue-specific genes often contain methylated cytosine bases when transcriptionally inactive and unmethylated cytosines when transcriptionally active. This correlation has suggested that methylation and demethylation may play critical roles in the regulation of tissue-specific genes, with methylation inhibiting transcription either through the binding of methyl-CpG-binding proteins or through the inhibition of specific transcriptional activators (for review, see Ng and Bird 1999). Studies showed that in vitro DNA methylation of various genes can result in stable repression and that demethylation, following addition of 5-azacytidine, can result in transcriptional activation.

A methylation-specific transcriptional repressor called MeCP2 was isolated that bound to methylated DNA. MeCP2 was found to copurify in a complex with SIN3 and histone deacetylases, providing a mechanism linking methylation-specific repression and histone deacetylation (Jones et al. 1998; Nan et al. 1998). Demethylation also involves specific demethylases. At least two such activities have been characterized, although their role in gene regulation is currently a source of investigation (Weiss and Cedar 1997; Bhattacharya et al. 1999).

Transcriptional Silencing

Silencers are DNA elements that modulate transcription by influencing chromatin structure through an extended DNA region (for review, see Ogbourne and Antalis 1998). The best-characterized examples of silencing are found in *S. cerevisiae* (see Brand et al. 1985; Laurenson and Rine 1992) and *Drosophila* (Paro 1993). *S. cerevisiae* strains typically possess copies of the two mating-type genes, **a** and α, in repressed (silenced) loci termed HML**a** and HMRα (for review, see Laurenson and Rine 1992). A mating-type gene is activated only if the entire gene, including sequences required for transcriptional activation, is transposed to the MAT locus during mating-type switching. The DNA sequence elements required for silencing flank the HML and HMR loci and contain an autonomous replication sequence (ARS) and binding sites for the origin recognition complex (ORC) and RAP1. Silencing of the HML and HMR loci (and also of telomeres) depends on multiple *trans*-acting factors, including SIR2, SIR3, and SIR4.

Several studies have pointed to a model in which silencer binding proteins and SIR proteins direct the formation and maintenance of an inaccessible heterochromatin configuration (for review, see Grunstein 1997). Moreover, the amino termini of histones H3 and H4 are needed for silencing and have been shown to associate with specific SIR proteins. Silencing also has been strictly correlated with deacetylation of core histones (Braunstein et al. 1993). Finally, in cells and in isolated nuclei, the silenced loci are relatively inaccessible to DNA methylation and to nucleases. Together, these findings suggest that RAP1 recruits SIR proteins to the loci, where the SIR proteins directly interact with histones. SIR–histone complexes may then be propagated throughout the loci, leading to an inaccessible chromatin configuration (for review, see Grunstein 1997). Recent studies show that perinuclear localization of chromatin facilitates transcriptional silencing, although the significance of this observation needs to be further established (Andrulis et al. 1998; for review, see Cockell and Gasser 1999).

Silencer activities have been detected in many tissue-specific genes (for review, see Chapter 6 and Ogbourne and Antalis 1998). Further studies are needed to determine whether the silencing found in mammalian cells occurs by a mechanism similar to that found in yeast or by a more direct repression mechanism involving methylation and/or histone deacetylase complexes. Recent studies do indicate association of silencing with proteins such as the polycomb group, which controls position-specific repression of homeotic genes during *Drosophila* development (see Busturia et al. 1997 and references therein).

Locus Control Regions, Insulators, and Matrix Attachment Regions

Locus Control Regions

LCRs are regulatory regions that modulate transcription by influencing chromatin structure through an extended DNA region and appear to induce and maintain accessibility of that region to transcription factors as measured by nuclease sensitivity (for review, see Fraser and Grosveld 1998; Grosveld 1999) (see also Chapter 6). The actual mechanism of LCR function remains poorly understood, although studies in the last few years have revealed some insights into function (Fig. 1.7). LCRs were identified during studies of the β-globin locus in transgenic mice. β-Globin transgenes typically were expressed at low levels, and the expression level was strongly influenced by the site of insertion into the chromosome. However, when the transgene contained a specific DNA fragment from the dis-

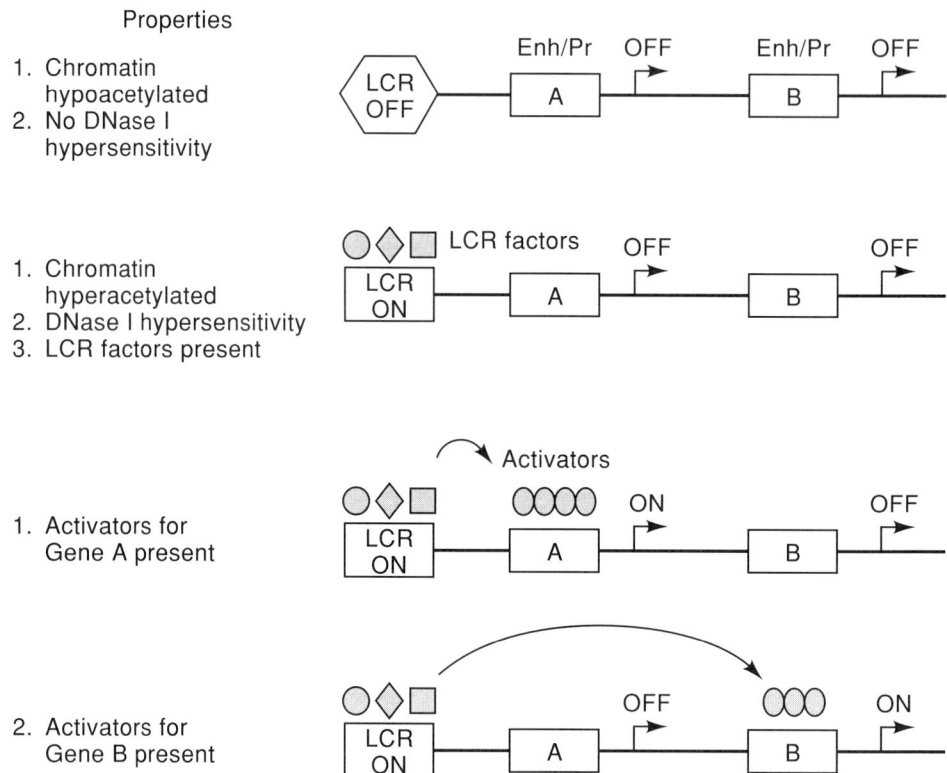

FIGURE 1.7. Models for LCR function. The LCR is inactive in inappropriate cells. Cell-type LCR factors, including activators, boundary element factors, and proteins that recruit acetylase complexes, then bind in appropriate cells. The locus is still inactive transcriptionally but the LCR contains hypersensitive sites and hyperacetylated chromatin. Activators for gene A or B bind to the enhancer/promoter (Enh/Pr) and communicate with activators at the LCR to elicit specific gene activation. (Adapted, with permission, from Blackwood and Kadonaga 1998 [Copyright 1998 American Association for the Advancement of Science].)

tal end of the globin locus, which is now known to encompass the LCR, high levels of position-independent (or integration site-independent) expression were observed. The β-globin LCR is unusual in that it is quite compact and contains five closely spaced DNase I hypersensitive (HS) sites, each containing 150–300 bp of DNA; other LCRs appear to be more spread out among the genes in a locus. Each DNase I HS site may represent a distinct functional module, because in the chicken locus one site can act as a classic enhancer whereas others do not have enhancer properties, although one other site appears to contain insulator activity (see below). Because of its compact nature, the β-globin LCR in mice, humans, and chickens has become the paradigm for understanding the function of other LCRs.

The LCR has several defining characteristics: (1) the ability to confer high levels of induced expression on linked genes in transgene experiments; (2) expression levels directly linked to copy number; (3) the ability to enhance transcription in a chromatin setting; (4) the ability to open chromatin over long distances (the β-globin LCR affects DNase I sensitivity over 100 kb), a point that is currently being debated (for discussion see Grosveld 1999); and (5) the ability to confer position-independent expression even in the presence of centromeric heterochromatin, which generally inactivates integrated transgenes. Note, however, that copy-number-dependent, position-independent expression is considered by many to be the minimal defining feature of an LCR.

In general, LCRs share properties with cell-specific enhancers, in that they coincide with cell-specific DNase I hypersensitive sites and bind to typical transcription factors. Furthermore, they stimulate hyperacetylation of surrounding chromatin. Despite these similarities, LCRs are clearly distinct from enhancers, because, in many instances, they enhance transcription only when integrated into a chromosome. Moreover, classic enhancers alone do not impart high levels of position-independent expression in transgenic mice.

One current view is that an LCR loops out and interacts with the regulatory promoter and enhancer of linked genes. This view is based on cytological studies and in situ hybridization, which have established that only one gene in the β-globin locus is active at any given time. The frequency and duration of contact between an LCR and linked gene are believed to control the period of time during which the gene is transcriptionally active. Deletions in the LCR still retain minimal function, although they can diminish the activity and decrease position independence. However, there is disagreement in the field as to the precise mechanism, and further studies will be required to elucidate LCR mechanism.

Boundary Elements

As described above, control regions like enhancers, LCRs, and silencers are capable of influencing gene expression and/or chromatin structure over long distances. One issue that must be considered is how these control regions are prevented from influencing transcription at adjacent loci. Two different classes of regulatory regions that may be involved in this function have been described: (1) Boundary elements or insulators (for review, see Geyer 1997; Bell and Felsenfeld 1999) (Fig. 1.8) and (2) matrix (or scaffold) attachment regions (MARs or SARs) (for review, see Hart and Laemmli 1998).

Boundary elements have been best described in *Drosophila*, although several have been identified in mammalian cells, chicken, and *Xenopus* (for a current compilation, see Bell and Felsenfeld 1999; see also Chapter 6). In the *Drosophila* heat shock protein 70 (hsp70) locus, the insulator regions are located at a considerable distance both upstream and downstream (these are termed scs and scs´) of the locus, with each insulator containing a pair of nuclease-hypersensitive sites that surround a 250- to 350-bp nuclease-resistant core regulatory sequence. The scs´ binds a protein called boundary element-associated factor (BEAF) 32. BEAF32 appears to bind cooperatively to sites within the boundary elements. Although no natural regulatory mutations in BEAF have been isolated, the protein appears to bind a wide range of boundary elements or insulators on polytene chromosomes. Another class of *Drosophila* insulators are characterized by the gypsy retroposon. The gypsy insulator binds the suppressor of hairy wing (su[Hw]) proteins (in one case the insulator comprises 12 su[Hw] binding sites). Unlike BEAF32, mutations in su(Hw) inactivate insulator activity.

Most boundary elements were identified on the basis of their location. The boundary elements in general appear to be in the vicinity of junctions between decondensed and condensed chromatin, which presumably correspond to junctions between active and inactive loci. Like LCRs, most boundary elements were found to impart position-independent expression to a heterologous gene. However, boundary elements appear to be distinct from LCRs because they do not enhance transcription and, in the heat shock locus, insulators are needed both upstream and downstream of the locus for their activity. Furthermore, boundary elements can prevent transcription when placed between an enhancer and a promoter. Thus, the insulator appears to provide a functional boundary for both accessible and inaccessible chromatin.

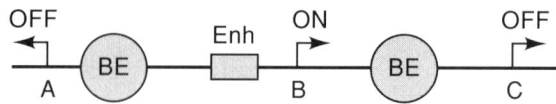

FIGURE 1.8. Boundary elements (BE) prevent the enhancer from activating genes A and C, thereby focussing the enhancer on gene B. (Adapted, with permission, from Blackwood and Kadonaga 1998 [Copyright 1998 American Association for the Advancement of Science].)

Several models have been proposed to explain insulator function. In one model the insulator acts as a decoy, by trapping enhancer activity and possibly disrupting the action of proteins necessary to mediate enhancer promoter looping (see Chapter 6 and Bell and Felsenfeld 1999). However, the boundary elements are directional and they only prevent an enhancer from communicating with a promoter on the other side of the element (Fig. 1.8). The enhancer can still function on a promoter lacking the intervening boundary element. Although the decoy idea is attractive, it suggests that the insulator might possess silencing capabilities, but it generally does not. Another model proposes that the insulator simply serves as a strong block to impeding chromatin. Because some insulators must flank the gene, they could presumably form a loop, which protects the enclosed domain. However, the mechanism is somewhat unclear because scs-insulators can work in *trans* when the insulator and the transgene are on catenated plasmids. Perhaps the simplest view is that they serve as a neutral boundary to neighboring regulatory elements. Further study will be required to elucidate their action.

MARs

The study of MARs initiated with the idea that, within a eukaryotic nucleus, chromosomes are incorporated into higher-order looped structures, with the loops fastened to the intranuclear matrix (Gasser and Laemmli 1987). These loops have been proposed to influence gene expression by separating a chromosome into individual regulatory domains. Numerous MARs have been isolated as DNA fragments capable of associating with the nuclear structures that remain following a stringent extraction and wash with high salt or detergent (Phi-Van and Strätling 1990). MARs are A/T-rich sequences that often are located at the boundaries of transcription units or in the vicinity of transcriptional enhancers. Several unique proteins associate with MARs (for review, see Hart and Laemmli 1998). However, because MARs are defined by physical rather than functional characteristics, their actual functions may be heterogeneous. Indeed, in some cases, MARs have been found to function similarly to insulators or LCRs (McKnight et al. 1992 and references therein). However, MARs are not often capable of insulating a gene from the action of an upstream enhancer. Additional studies are needed to establish the precise relationship between MARs, insulators, and LCRs.

CONCEPTS AND STRATEGIES: IV. THE ENHANCEOSOME

Combinatorial Control, Cooperativity, and Synergy

As we stated in the introduction, combinatorial transcription control in eukaryotes employs the principles of cooperativity and transcriptional synergy. Cooperativity in the context of DNA–protein interactions is a phenomenon whereby two proteins, which bind to adjacent sites on DNA, engage in protein–protein interactions that reciprocally stabilize each other's binding. This stabilization allows simultaneous binding of the two proteins to adjacent sites. The proteins bind with much higher affinity than they would bind independently if they did not interact. An RNA Pol II enhancer and regulatory promoter employ cooperativity to generate specific responses to combinations of activators. An enhancer designed to respond to any given signal might contain multiple binding sites for a given activator organized in a manner that promotes direct interaction and cooperative binding to DNA. Alternately, an enhancer that responds only to a combination of signals would organize binding sites for activators that respond to those signals in a manner that allows cooperative binding of the activator combination. The cooperativity plays two roles. First, it permits binding of activator combinations only when the appropriate set of activators is present in the cell. If only one activator were present, the absence of the cooperative interactions with its partner would prohibit its binding to DNA. A second effect of cooperativity is that it ensures the activation and inactivation of genes in response to subtle changes in activator concentration (or activity, e.g., phosphorylated vs. unphosphorylated), an issue we discuss in more detail in Chapter 13. Combinatorial control would be impossible if the Pol II general machinery could respond to a single molecule of activator. Therefore, the Pol II transcriptional machinery is designed to respond in a greater-than-additive or synergistic fashion to multiple activators. This requirement for multiple activators for efficient transcription permits combinations of activators to be employed to turn on genes. By the combinatorial model, a single factor might join up with unique subsets of factors to activate different genes in different regulatory contexts (Carey 1998; Ptashne and Gann 1998).

Although combinatorial regulation is often characterized by cooperative binding, synergistic activation need not be congruent with cooperativity. Saturating the DNA with multiple activators via cooperative or noncooperative binding mechanisms still generates synergistic transcription because of the energetic effects multiple activators versus a single activator have on recruiting the general machinery. Furthermore, even in cases where cooperative binding occurs, it may not require protein–protein interactions; for example, cooperative binding of activators to chromatin in vitro (for review, see Workman and Kingston 1998).

The Enhanceosome Theory

Recent studies have established that cooperative binding of activators to enhancers and regulatory promoters leads to assembly of nucleoprotein structures termed "enhanceosomes"(Thanos and Maniatis 1995). The enhanceosome displays two layers of specificity necessary for gene activation. In one, the context-dependent activator–activator interactions promote cooperative assembly of the enhanceosome on naked DNA or chromatin (e.g., see Thanos and Maniatis 1995). In the other, the enhanceosome displays a specific activation surface that is complementary to a "target" surface within the Pol II transcriptional machinery, recruiting it to the promoter to generate synergistic transcription (Fig. 1.9) (see Kim and Maniatis 1997).

Cooperative binding of the activators to DNA in enhanceosomes often, but not always, depends on the precise arrangement of activator recognition sites and the correct combinations of bound activators, which together generate a network of protein–protein and protein–DNA interactions unique to a given enhancer. Concentrations of activators below a key threshold and altered positioning of the activator binding sites prevent the necessary cooperative interactions and inhibit enhanceosome assembly (see Chapters 9 and 12). Some of the caveats to this model are discussed in subsequent chapters. Thus, studying binding sites out of context may yield aberrant results.

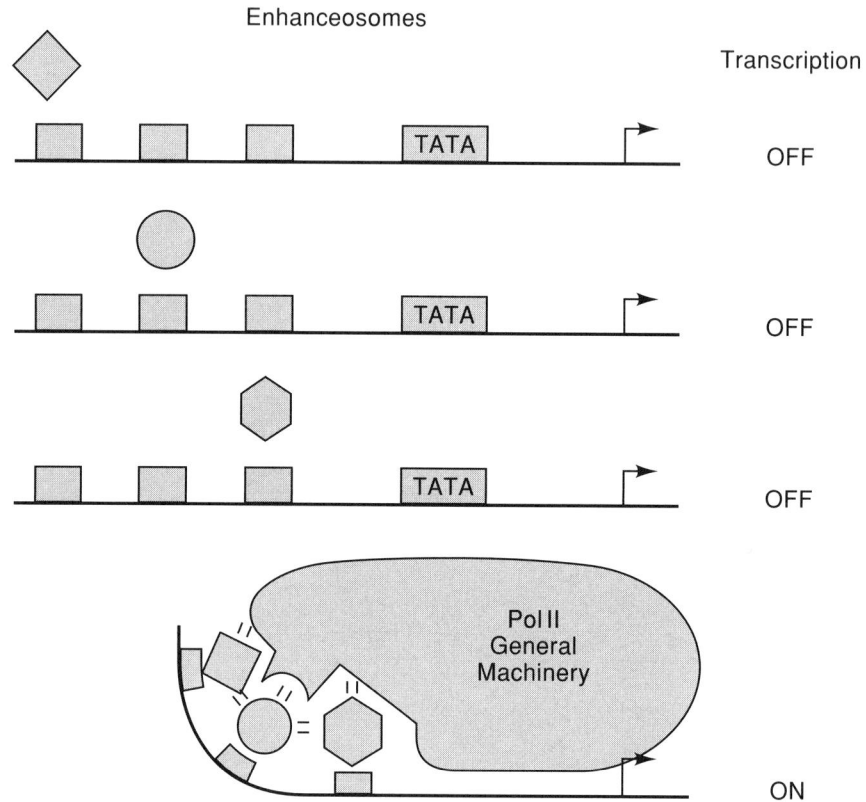

FIGURE 1.9. Enhanceosome theory. Individual activators bind DNA weakly on their own. Together, they bind cooperatively, often with the help of architectural proteins that bend DNA. Once bound, they recruit the general machinery to generate synergistic transcription.

The IFN-β Enhanceosome

We discuss the concept of enhanceosome using the IFN-β enhancer as a paradigm. Important issues include: How do enhanceosomes contribute to specificity, how are they assembled, and how do they function in vitro and in vivo? We attempt to bring together some of the principles described in earlier sections to provide a comprehensive view of how a single regulatory unit functions in mammalian cells.

The IFN-β enhanceosome comprises a 57-bp DNA sequence consisting of four previously mapped regulatory sequences called PRD I–IV. Virus infection induces transcription from this control region. Cell transfection studies, in vitro transcription, and DNase I footprinting experiments demonstrated that ATF2/c-jun, IRF-1, and NF-κB bound cooperatively to these sites with the architectural protein HMG I (Y). Later studies showed that IRF-1 was not required, and a complex between IRF-3, IRF-7, and p300 was isolated and shown to bind the enhancer (Wathelet et al. 1998). This complex, called virus-inducible transcriptional activator complex or VAF, also bound cooperatively with NF-κB and ATF-2/c-jun. Most of the biochemical analyses have, however, focused on IRF-1, and in the studies discussed below IRF-1 was employed. In the case of the IFN-β enhancer, artificial promoters bearing tandem copies of each activator binding site (i.e., ATF) responded moderately to a range of signals. In contrast, when the sites were assembled into the enhancer, transcriptional induction was only observed in response to viral infection.

One potential explanation is that the activators are modified by kinase-dependent signaling systems, and the phosphorylation increases their affinities for each other and for coactivators like p300. Due to cooperative binding, the simultaneous, albeit modest, increase in the affinities of multiple interacting activators would partially explain the highly synergistic response observed in vivo (see Chapter 13 for a discussion of the theory). At this stage, one might argue that the synergy was due to cooperative binding of the activators at limiting concentrations.

The cooperativity alone, however, is not sufficient, and the enhanceosome also assembles a specific surface that recruits the holoenzyme. Kim and Maniatis (1997) were able to reproduce the combinatorial regulation of IFN-β in cell-free extracts. After depleting mammalian nuclear extracts of endogenous IFN-β-binding proteins by DNA and immunoaffinity chromatography, the authors were able to supplement the extract with limiting concentrations of the individual recombinant proteins and recapitulate the highly synergistic transcriptional response observed in vivo. The ability to manipulate activator concentration allowed the authors to test whether synergistic transcription would be obtained under conditions in which the template sites were occupied and cooperative binding was bypassed. Saturating concentrations of NF-κB, IRF-1, ATF-2, and c-Jun could circumvent the requirement for HMG I. However, synergy was abolished by omitting any of the activators (i.e., ATF-2) or repositioning them, even under conditions where the site was saturated and cooperativity was not a factor. These observations provided circumstantial evidence that the activators in IFN-β collectively create a stereospecific interface for docking with and recruitment of the transcriptional machinery.

Biochemical Mechanism of Activation

Two models have been proposed to explain synergistic effects. In one, interaction of activators with multiple targets influences different kinetically limiting steps, which results in synergistic activation (Herschlag and Johnson 1993). Another model proposes that the activators form an enhanceosome, where different activation domains interact with different portions of a common surface of the general machinery, leading to cooperative recruitment and assembly of the preinitiation complex and synergistic gene activation (for review, see Carey 1998). The models are not mutually exclusive, and there is evidence for both in the literature.

Thanos and colleagues provided compelling biochemical evidence for the latter model in the IFN-β case. CBP/p300 was shown by EMSA to bind to the preassembled enhanceosome. Removing the activation domains from any of the activators or a newly discovered synergy domain from p65 subunit of NF-κB abolished the synergy and recruitment. Because CBP/p300 is a component of many of the reported mammalian holoenzymes, its ability to serve as an interface with the surface of the enhanceosome provides a means for the enhanceosome to recruit the general machinery to the core promoter.

Studies by Kim and Maniatis suggested that the enhanceosome assembled in two phases. First, complexes containing TFIID, TFIIA, the USA coactivator fraction, and TFIIB were stabilized by the enhanceosome (Kim and Maniatis 1997). Second, later studies showed that the enhanceosome directly recruited a CBP/p300-containing holoenzyme to the TFIID-TFIIA-TFIIB-USA complex (Kim et al. 1998). More details on how these experiments were performed are presented in Chapter 15.

The significance of p300 recruitment in vitro was reinforced by in vivo studies showing virus-inducible hyperacetylation of histones H3 and H4 within the IFN-β enhancer. A cell line expressing a dominant inhibitor version of IRF-3 that bound DNA but not p300 prevented virus induction in vivo (Parekh and Maniatis 1999). Thus, the enhancer recruitment of p300 appears to correlate with acetylation, supporting the importance of the biochemical acetylation activity ascribed to p300.

These studies raise many interesting but relevant questions: Is the enhancer in a nucleosome before virus induction? Does the nucleosome need to be remodeled partially before enhanceosome assembly? That is, do certain activators bind first and then recruit remodeling enzymes, which then allow other activators to bind as in activation of the HO promoter in yeast (Cosma et al. 1999)? Alternatively, do the factors bind in a concerted reaction as they do in vitro?

Perspective

The information presented in this chapter reveals that multiple types of control regions and controlling events may contribute to the precise regulation of a given gene. The multiplicity of regulatory strategies is consistent with and, in fact, predicted by, the combinatorial theory of gene regulation. It is important to note, however, that most of these regulatory strategies (e.g., regulation by a silencer, LCR, MAR, coactivator, or DNA methylation) have been studied in depth for very few genes. At this point, it is unlikely that any mammalian gene has been studied in sufficient detail to reveal every mode of regulation for that gene. Thus, the number of regulatory strategies involved in the expression of a gene and the precise order of events leading from a heterochromatin-associated locus to an actively transcribed gene remain to be established.

REFERENCES

Andrulis E.D., Neiman A.M., Zappulla D.C., and Sternglanz R. 1998. Perinuclear localization of chromatin facilitates transcriptional silencing. *Nature* **394:** 592–595.

Babb R., Cleary M.A., and Herr W. 1997. OCA-B is a functional analog of VP16 but targets a separate surface of the Oct-1 POU domain. *Mol. Cell. Biol.* **17:** 7295–7305.

Bazett-Jones D.P., Côté J., Landel C.C., Peterson C.L., and Workman J.L. 1999. The SWI/SNF complex creates loop domains in DNA and polynucleosome arrays and can disrupt DNA-histone contacts within these domains. *Mol. Cell. Biol.* **19:** 1470–1478.

Bell A.C. and Felsenfeld G. 1999. Stopped at the border: Boundaries and insulators. *Curr. Opin. Genet. Dev* **9:** 191–198.

Benezra R., Davis R.L., Lockshon D., Turner D.L., and Weintraub H. 1990. The protein Id: A negative regulator of helix-loop-helix DNA binding proteins. *Cell* **61:** 49–59.

Bhattacharya S.K., Ramchandani S., Cervoni N., and Szyf M. 1999. A mammalian protein with specific demethylase activity for mCpG DNA. *Nature.* **397:** 579–583.

Björklund S., Almouzni G., Davidson I., Nightingale K.P., and Weiss K. 1999. Global transcription regulators of eukaryotes. *Cell* **96:** 759-767.

Blackwood E.M. and Kadonaga J.T. 1998. Going the distance: A current view of enhancer action. *Science* **281:** 61–63.

Boyer T.G., Martin M.E.D., Lees E., Ricciardi R.P., and Berk A.J. 1999. Mammalian Srb/Mediator complex is targeted by adenovirus E1A protein. *Nature* **399:** 276–279.

Brand A.H., Breeden L., Abraham J., Sternglanz R., and Nasmyth K. 1985. Characterization of a "silencer" in yeast: A DNA sequence with properties opposite to those of a transcriptional enhancer.*Cell* **41:** 41–48.

Braunstein M., Rose A.B., Holmes S.G., Allis C.D., and Broach J.R. 1993. Transcriptional silencing in yeast is associated with reduced nucleosome acetylation. *Genes. Dev.* **7:** 592–604.

Britten R.J. and Davidson E.H. 1969. Gene regulation for higher cells: A theory. *Science* **165:** 349–357.

Brown K.E., Guest S.S., Smale S.T., Hahm K., Merkenschlager M., and Fisher A.G. 1997. Association of transcriptionally silent genes with Ikaros complexes at centromeric heterochromatin. *Cell* **91:** 845–854.

Brown M.S. and Goldstein J.L. 1997. The SREBP pathway: Regulation of cholesterol metabolism by proteolysis of a membrane-bound transcription factor. *Cell* **89:** 331–340.

Brown S.A. and Kingston R.E. 1997. Disruption of downstream chromatin directed by a transcriptional activator. *Genes Dev.* **11:** 3116–31121.

Brownell J.E., Zhou J., Ranalli T., Kobayashi R., Edmondson D.G., Roth S.Y., and Allis C.D. 1996. *Tetrahymena* histone acetyltransferase A: A homolog to yeast Gcn5p linking histone acetylation to gene activation. *Cell* **84:** 843–851.

Burke T.W. and Kadonaga J.T. 1996. *Drosophila* TFIID binds to a conserved downstream basal promoter element that is present in many TATA-box-deficient promoters. *Genes Dev.* **10:** 711–724.

———.1997. The downstream core promoter element, DPE, is conserved from *Drosophila* to humans and is recognized by TAF$_{II}$60 of *Drosophila*. *Genes Dev.* **11:** 3020–3031.

Burley S.K. and Roeder R.G. 1996. Biochemistry and structural biology of transcription factor IID (TFIID). *Annu. Rev. Biochem.* **65:** 769–799.

Busturia A., Wightman C.D., and Sakonju S. 1997. A silencer is required for maintenance of transcriptional repression throughout *Drosophila* development. *Development* **124:** 4343–4350.

Carey M.F. 1995. Transcriptional activation—A holistic view of the complex. *Curr. Biol.* **5:** 1003–1005.

———.1998. The enhanceosome and transcriptional synergy. *Cell* **92:** 5–8.

Carey M., Lin Y.S., Green M.R., and Ptashne M. 1990. A mechanism for synergistic activation of a mammalian gene by GAL4 derivatives. *Nature* **345:** 361–364.

Chi T. and Carey M. 1996. Assembly of the isomerized TFIIA–TFIID–TATA ternary complex is necessary and sufficient for gene activation. *Genes Dev.* **10:** 2540–2550.

Chi T., Lieberman P., Ellwood K., and Carey M. 1995. A general mechanism for transcriptional synergy by eukaryotic activators. *Nature* **377:** 254–257.

Cho H., Maldonado E., and Reinberg D. 1997. Affinity purification of a human RNA polymerase II complex using monoclonal antibodies against transcription factor IIF. *J. Biol. Chem.* **272:** 11495–11502.

Cockell M. and Gasser S.M. 1999. Nuclear compartments and gene regulation. *Curr. Opin. Genet. Dev.* **9:** 199–205.

Cooper J.P., Roth S.Y., and Simpson R.T. 1994. The global transcriptional regulators, Ssn6 and Tup1, play distinct roles in the establishment of a repressive chromatin structure. *Genes Dev.* **8:** 1400–1410.

Corona D.F., Langst G., Clapier C.R., Bonte E.J., Ferrari S., Tamkun J.W., and Becker P.B. 1999. ISWI is an ATP-dependent nucleosome remodeling factor. *Mol. Cell* **3:** 239–245.

Cosma M.P., Tanaka T., and Nasmyth K. 1999. Ordered recruitment of transcription and chromatin remodeling factors to a cell cycle- and developmentally regulated promoter. *Cell* **97:** 299–311.

Coulombe B. and Burton Z.F. 1999. DNA bending and wrapping around RNA polymerase: A "revolutionary" model to describe transcriptional mechanisms. *Microbiol. Mol. Biol. Rev.* **63:** 457–478.

Current Opinion in Cell Biology. ISSN 0955-0674. Published bimonthly by Current Biology Ltd., 34-42 Cleveland St., London, W1P 6LB, United Kingdom.

Current Opinion in Genetics & Development. ISSN 0959-437X. Published bimonthly by Current Biology Ltd., 34-42 Cleveland St., London, W1P 6LB, United Kingdom.

Davis R.L., Cheng P.F., Lassar A.B., and Weintraub H. 1990. The MyoD DNA binding domain contains a recognition code for muscle-specific gene activation. *Cell* **60:** 733–746.

Dikstein R., Zhou S., and Tjian R. 1996. Human TAF$_{II}$105 is a cell type-specific TFIID subunit related to hTAFII130. *Cell* **87:** 137–146.

Eckner R. 1996. p300 and CBP as transcriptional regulators and targets of oncogenic events. *Biol.*

Chem. **377:** 685–688.

Emami K.H. and Carey M. 1992. A synergistic increase in potency of a multimerized VP16 transcriptional activation domain. *EMBO J.* **11:** 5005–5012.

Emami K.H., Jain A., and Smale S.T. 1997. Mechanism of synergy between TATA and initiator: Synergistic binding of TFIID following a putative TFIIA-induced isomerization. *Genes Dev.* **11:** 3007–3019.

Ernst P. and Smale S.T. 1995. Combinatorial regulation of transcription II: The immunoglobulin μ heavy chain gene. *Immunity* **2:** 427–438.

Fraser P. and Grosveld F. 1998. Locus control regions, chromatin activation and transcription. *Curr. Opin. Cell Biol.* **10:** 361–365.

Freedman L.P. 1999. Increasing the complexity of coactivation in nuclear receptor signaling. *Cell* **97:** 5–8.

Garraway I.P., Semple K., and Smale S.T. 1996. Transcription of the lymphocyte-specific terminal deoxynucleotidyltransferase gene requires a specific core promoter structure. *Proc. Natl. Acad. Sci.* **93:** 4336–4341.

Gasser S.M. and Laemmli U.K. 1987. A glimpse at chromosomal order. *Trends Genet.* **3:** 16–22.

Geyer PK. 1997. The role of insulator elements in defining domains of gene expression. *Curr. Opin. Genet. Dev.* **7:** 242–248.

Ghosh S., May M.J., and Kopp E.B. 1998. NF-κB and REL proteins: Evolutionarily conserved mediation of immune responses. *Annu. Rev. Immunol.* **16:** 225–260.

Gill G. and Ptashne M. 1987. Mutants of GAL4 protein altered in an activation function. *Cell* **51:** 121–126.

Goodrich J.A. and Tjian R. 1994. TBP-TAF complexes: Selectivity factors for eukaryotic transcription. *Curr. Opin. Cell Biol.* **6:** 403–409.

Goodrich J.A., Hoey T., Thut C., Admon A., and Tjian R. 1993. *Drosophila* TAF$_{II}$40 with both a VP16 activation domain and basal transcription factor TFIIB. *Cell* **75:** 519–530.

Grant P.A., Schieltz D., Pray-Grant M.G., Steger D.J., Reese J.C., Yates J.R.III, and Workman J.L. 1998. A subset of TAF(II)s are integral components of the SAGA complex required for nucleosome acetylation and transcriptional stimulation [see comments]. *Cell* **94:** 45–53.

Greenblatt J. 1997. RNA polymerase II holoenzyme and transcriptional regulation. *Curr. Opin. Cell Biol.* **9:** 310–319.

Grosschedl R., Giese K., and Pagel J. 1994. HMG domain proteins: Architectural elements in the assembly of nucleoprotein structures. *Trends Genet.* **10:** 94–100.

Grosveld F. 1999. Activation by locus control regions? *Curr. Opin. Genet. Dev.* **9:** 152–157.

Grunstein M. 1990. Histone function in transcription. *Annu. Rev. Cell. Biol.* **6:** 643–678.

———.1997. Molecular model for telomeric heterochromatin in yeast. *Curr. Opin. Cell Biol.* **9:** 383–387.

Gu W., Malik S., Ito M., Yuan C.X., Fondell J.D., Zhang X., Martinez E., Qin J., and Roeder R.G. 1999. A novel human SRB/MED-containing cofactor complex, SMCC, involved in transcription regulation. *Mol. Cell* **3:** 97–108.

Hahn S. 1998. The role of TAFs in RNA polymerase II transcription. *Cell* **95:** 579–582.

Hamiche A., Sandaltzopoulos R., Gdula D.A., and Wu. C. 1999. ATP-dependent histone octamer sliding mediated by the chromatin remodeling complex NURF. *Cell* **97:** 833–842.

Hampsey M. 1998. Molecular genetics of the RNA polymerase II general transcriptional machinery. *Microbiol. Mol. Biol. Rev.* **62:** 465–503.

Hampsey M. and Reinberg D. 1999. RNA polymerase II as a control panel for multiple coactivator complexes. *Curr. Opin. Genet. Dev.* **9:** 132–139.

Hanes S.D. and Brent R. 1989. DNA specificity of the bicoid activator protein is determined by homeodomain recognition helix residue 9. *Cell* **57:** 1275–1283.

Hanna-Rose W. and Hansen U. 1996. Active repression mechanisms of eukaryotic transcription repressors. *Trends Genet.* **12:** 229–234.

Hart C.M. and Laemmli U.K. 1998. Facilitation of chromatin dynamics by SARs. *Curr. Opin. Genet. Dev.* **8:** 519–525.

Hebbes T.R., Thorne A.W., and Crane-Robinson C. 1988. A direct link between core histone acetylation and transcriptionally active chromatin. *EMBO J.* **7:** 1395–1402.

Heinemeyer T., Wingender E., Reuter I., Hermjakob H., Kel A.E., Kel O.V., Ignatieva E.V., Ananko E.A., Podkolodnaya O.A., Kolpakov F.A., Podkolodny N.L., and Kolchanov N.A. 1998. Databases on transcriptional regulation: TRANSFAC, TRRD and COMPEL. *Nucleic Acids Res.* **26:** 362–367.

Hernandez N. 1993. TBP, a universal eukaryotic transcription factor? *Genes Dev.* **7:** 1291–1308.

Herschbach B.M. and Johnson A.D. 1993. Transcriptional repression in eukaryotes. *Annu. Rev. Cell. Biol.* **9:** 479–509.

Herschlag D. and Johnson F.B. 1993. Synergism in transcriptional activation: A kinetic view. *Genes Dev.* **7:** 173–179.

Hoffmann A., Oelgeschlager T., and Roeder R.G. 1997. Considerations of transcriptional control mechanisms: Do TFIID-core promoter complexes recapitulate nucleosome-like functions? *Proc. Natl. Acad. Sci.* **94:** 8928–8935.

Hollenberg S.M. and Evans R.M. 1988. Multiple and cooperative trans-activation domains of the human glucocorticoid receptor. *Cell* **55:** 899–906.

Holstege F.C., Fiedler U., and Timmers H.T. 1997. Three transitions in the RNA polymerase II transcription complex during initiation. *EMBO J.* **16:** 7468–7480.

Holstege F.C., Jennings E.G., Wyrick J.J., Lee T.I., Hengartner C.J., Green M.R., Golub T.R., Lander E.S., and Young R.A. 1998. Dissecting the regulatory circuitry of a eukaryotic genome. *Cell* **95:** 717–728.

Hope I.A., Mahadevan S., and Struhl K. 1988. Structural and functional characterization of the short acidic transcriptional activation region of yeast GCN4 protein. *Nature* **333:** 635–640.

Hori R. and Carey M. 1998. Transcription: TRF walks the walk but can it talk the talk? *Curr. Biol.* **8:** R124–127.

Huang L., Zhang W., and Roth S.Y. 1997. Amino termini of histones H3 and H4 are required for a1-α repression in yeast. *Mol. Cell. Biol.* **17:** 555–562.

Imbalzano A., Kwon H., Green M.R, and Kingston R.E. 1994. Facilitated binding of TATA-binding protein to nucleosomeal DNA. *Nature* **370:** 481–485.

Israel A. 1994. Immunology. NF-AT comes under control. *Nature* **369:** 443–444.

Ito M., Yuan C.-X., Malik S., Gu W., Fondell J.D., Yamamura S., Fu Z.-Y., Zhang X., Qin J., and Roeder R.G. 1999. Identity between TRAP and SMCC complexes indicates novel pathways for the function of nuclear receptors and diverse mammalian activators. *Mol. Cell* **3:** 361–370.

Johnson P.F. and McKnight S.L. 1989. Eukaryotic transcriptional regulatory proteins. *Annu. Rev. Biochem.* **58:** 799–839.

Jones P.L., Veenstra G.J., Wade P.A., Vermaak D., Kass S.U., Landsberger N., Strouboulis J., and Wolffe A.P. 1998. Methylated DNA and MeCP2 recruit histone deacetylase to repress transcription. *Nature Genet.* **19:** 187–191.

Kadonaga JT. 1998. Eukaryotic transcription: An interlaced network of transcription factors and chromatin-modifying machines. *Cell* **92:** 307–313.

Kadosh D. and Struhl K. 1998. Targeted recruitment of the Sin3-Rpd3 histone deacetylase complex generates a highly localized domain of repressed chromatin in vivo. *Mol. Cell. Biol.* **18:** 5121–5127.

Kaufmann J., Ahrens K., Koop R., Smale S.T., and Muller R. 1998. CIF150, a human cofactor for transcription factor IID-dependent initiator function. *Mol. Cell. Biol.* **18:** 233–239.

Kehle J., Beuchle D., Treuheit S., Christen B., Kennison J.A., Bienz M., and Muller J. 1998. dMi-2, a hunchback-interacting protein that functions in polycomb repression. *Science* **282:** 1897–1900.

Kim J., Said S., Jones B., Jackson A., Koipally J., Heller E.,Winanady S., Viel A., Sawyer A., Ikeda T., Kingston R., and Georgopoulos K. 1999. Ikaros DNA binding proteins direct formation of chromatin remodeling complexes in lymphocytes. *Immunity* **10:** 345–355.

Kim T.K. and Maniatis T. 1997. The mechanism of transcriptional synergy of an in vitro assembled interferon-β enhanceosome. *Mol. Cell* **1:** 119–129.

Kim T.K., Kim T.H., and Maniatis T. 1998. Efficient recruitment of TFIIB and CBP-RNA polymerase

II holoenzyme by an interferon-enhanceosome in vitro. *Proc. Natl. Acad. Sci.* **95:** 12191–12196.

Kim T.K., Lagrange T., Wang Y.H., Griffith J.D., Reinberg D., and Ebright R.H. 1997. Trajectory of DNA in the RNA polymerase II transcription preinitiation complex. *Proc. Natl. Acad. Sci.* **94:** 12268–12273.

Khoury G. and Gruss P. 1983. Enhancer elements. *Cell* **33:** 313–314.

Komachi K., Redd M.J., and Johnson A.D. 1994. The WD repeats of Tup1 interact with the homeo domain protein a2. *Genes Dev.* **8:** 2857–2867.

Kornberg R.D. and Lorch Y. 1999. Chromatin-modifying and -remodeling complexes. *Curr. Opin. Genet. Dev.* **9:** 148–151.

Kuchin S. and Carlson M. 1998. Functional relationships of Srb10-Srb11 kinase, carboxy-terminal domain kinase CTDK-I, and transcriptional corepressor Ssn6-Tup1.*Mol. Cell. Biol.* **18:** 1163–1171.

Kuo M.H., Zhou J., Jambeck P., Churchill M.E., and Allis C.D. 1998. Histone acetyltransferase activity of yeast Gcn5p is required for the activation of target genes in vivo. *Genes Dev.* **12:** 627–639.

Kussie P.H., Gorina S., Marechal V., Elenbaas B., Moreau J., Levine A.J., and Pavletich N.P. 1996. Structure of the MDM2 oncoprotein bound to the p53 tumor suppressor transactivation domain [comment]. *Science* **274:** 948–953.

Kwon H., Imbalzano A., Khavari P.A., Kingston R.E., and Green M.R. 1994. Nucleosome disruption and enhancement of activator binding by a human SWI/SNF complex. *Nature* **370:** 477–481.

Lagrange T., Kapanidis A.N., Tang H., Reinberg D., and Ebright R.H. 1998. New core promoter element in RNA polymerase II-dependent transcription: Sequence-specific DNA binding by transcription factor IIB. *Genes Dev.* **12:** 34–44.

Laherty C.D., Billin A.N., Lavinsky R.M., Yochum G.S., Bush A.C., Sun J.M., Mullen T.M., Davie J.R., Rose D.W., Glass C.K., Rosenfeld M.G., Ayer D.E., and Eisenman R.N. 1998. SAP30, a component of the mSin3 corepressor complex involved in N-CoR-mediated repression by specific transcription factors. *Mol. Cell* **2:** 33–42.

Längst G., Bonte E.J., Corona D.F.V., and Becker P.B. 1999. Nucleosome movement by CHRAC and ISWI without disruption or trans-displacement of the histone octamer. *Cell* **97:** 843–852.

Laurenson P. and Rine J. 1992. Silencers, silencing, and heritable transcriptional states. *Microbiol. Rev.* **56:** 543–560.

Lehman A.M., Ellwood K.B., Middleton B.E., and Carey M. 1998. Compensatory energetic relationships between upstream activators and the RNA polymerase II general transcription machinery. *J. Biol. Chem.* **273:** 932–939.

LeRoy G., Orphanides G., Lane W.S., and Reinberg D. 1998. Requirement of RSF and FACT for transcription of chromatin templates in vitro [see comments]. *Science* **282:** 1900–1904.

Lewin B. 1990. *Genes IV.* John Wiley & Sons/Oxford University Press, New York.

Li X-Y., Virbasius A., Zhu X., and Green M.R 1999. Enhancement of TBP binding by activators and general transcription factors. *Nature* **399:** 605–609.

Li Y., Bjorklund S., Kim Y.J., and Kornberg R.D. 1996. Yeast RNA polymerase II holoenzyme. *Methods Enzymol.* **273:** 172–175.

Liou H.C. and Baltimore D. 1993. Regulation of the NF-κB/rel transcription factor and IκB inhibitor system. *Curr. Opin. Cell Biol.* **5:** 477–487.

Lohr D. 1997. Nucleosome transactions on the promoters of the yeast GAL and PHO genes. *J. Biol. Chem.* **272:**26795–26798.

Lorch Y., Zhang M., and Kornberg R.D. 1999. Histone octamer transfer by a chromatin-remodeling complex. *Cell* **96:** 389–392.

Luger K., Mäder A.W., Richmond R.K., Sargent D.F., and Richmond T.J. 1997. Crystal structure of the nucleosome core particle at 2.8 Å resolution. *Nature* **389:** 251–260.

Luisi B. 1995. DNA-protein interaction at high resolution. In *DNA-protein: Structural interactions* (ed. Lilleys D.M.J.), vol. 7, p.1. Oxford University Press, New York and IRL Press at Oxford University Press, United Kingdom.

Luo Y., Ge H., Stevens S., Xiao H., and Roeder R.G. 1998. Coactivation by OCA-B: Definition of critical regions and synergism with general cofactors. *Mol. Cell. Biol.* **18:** 3803–3810.

Maldonado E., Shiekhattar R., Sheldon M., Cho H., Drapkin R., Rickert P., Lees E., Anderson C.W., Linn S., and Reinberg D. 1996. A human RNA polymerase II complex associated with SRB and DNA-repair proteins [published erratum appears in *Nature* 1996 Nov 28;384(6607):384]. *Nature* **2**: 86–89.

Martinez E., Ge H., Tao Y., Yuan C.X., Palhan V., and Roeder R.G. 1998. Novel cofactors and TFIIA mediate functional core promoter selectivity by the human TAFII150-containing TFIID complex. *Mol. Cell. Biol.* **18**: 6571–6583.

McCracken S., Fong N., Yankulov K., Ballantyne S., Pan G., Greenblatt J., Patterson S.D., Wickens M., and Bentley D.L. 1997. The C-terminal domain of RNA polymerase II couples mRNA processing to transcription. *Nature* **385**: 357–361.

McKnight R.A., Shamay A., Sankaran L., Wall R.J., and Hennighausen L. 1992. Matrix-attachment regions can impart position-independent regulation of a tissue-specific gene in transgenic mice. *Proc. Natl. Acad. Sci.* **89**: 6943–6947.

Merika M., Williams A.J., Chen G., Collins T., and Thanos D. 1998. Recruitment of CBP/p300 by the IFN β enhanceosome is required for synergistic activation of transcription. *Mol. Cell* **1**: 277–287.

Momand J., Zambetti G.P., Olson D.C., George D., and Levine A.J. 1992. The mdm-2 oncogene product forms a complex with the p53 protein and inhibits p53-mediated transactivation. *Cell* **69**: 1237–1245.

Moras D. and Gronemeyer H. 1998. The nuclear receptor ligand binding domain: Structure and function. *Curr. Opin. Cell Biol.* **10**: 384–391.

Mueller-Storm H.P., Sogo J.M., and Schaffner W. 1989. An enhancer stimulates transcription in trans when attached to the promoter via a protein bridge [published erratum appears in *Cell* 1989 Oct 20;59(2):405]. *Cell* **58**: 767–777.

Munshi N., Merika M., Yie J., Senger K., Chen G., and Thanos D. 1998. Acetylation of HMG I(Y) by CBP turns off IFN β expression by disrupting the enhanceosome. *Mol. Cell* **2**: 457–467.

Myer V.E. and Young R.A. 1998. RNA polymerase II holoenzymes and subcomplexes. *J. Biol. Chem.* **273**: 27757–27760.

Näär A.M., Beaurang P.A., Zhou S, Abraham S., Solomon W., and Tjian R. 1999. Composite coactivator ARC mdiates chromatin-directed transcriptional activation. *Nature* **398**: 828–832.

Nakajima T., Uchida C., Anderson S.F., Parvin J.D., and Montminy M. 1997a. Analysis of a cAMP-responsive activator reveals a two-component mechanism for transcriptional induction via signal-dependent factors. *Genes Dev.* **11**: 736–747.

Nakajima T., Uchida C., Anderson S.F., Lee C.G., Hurwitz J., Parvin J.D., and Montminy M. 1997b. RNA helicase A mediates association of CBP with RNA polymerase II. *Cell* **90**: 1107–1112.

Nan X., Ng H.H., Johnson C.A., Laherty C.D., Turner B.M., Eisenman R.N., and Bird A. 1998. Transcriptional repression by the methyl-CpG-binding protein MeCP2 involves a histone deacetylase complex [see comments]. *Nature* **393**: 386–389.

Neish A.S., Anderson S.F., Schiegal B.P., Wei W., and Parvin J.D. 1998. Factors associated with the mammalian RNA polymerase II holoenzyme. *Nucleic Acids Res.* **26**: 847–863.

Ng H.H. and Bird A. 1999. DNA methylation and chromatin modification. *Curr. Opin. Genet. Dev.* **9**: 158–163.

Nolte R.T., Wisely G.B., Westin S., Cobb J.E., Lambert M.H., Kurokawa R., Rosenfeld M.G., Willson T.M., Glass C.K., and Milburn M.V. 1998. Ligand binding and coactivator assembly of the peroxisome proliferator-activated receptor-γ. *Nature* **395**: 137–143.

Oelgeschlager T., Chiang C.M., and Roeder R.G. 1996. Topology and reorganization of a human TFIID-promoter complex. *Nature* **382**: 735–738.

Ogbourne S. and Antalis T.M. 1998. Transcriptional control and the role of silencers in transcriptional regulation in eukaryotes. *Biochem. J.* **331**: 1–14.

Ogryzko V.V., Kotani T., Zhang X., Schiltz R.L., Howard T., Yang X.J., Howard B.H., Qin J., and Nakatani Y. 1998. Histone-like TAFs within the PCAF histone acetylase complex [see comments]. *Cell* **94**: 35–44.

Ondek B., Gloss L., and Herr W. 1988. The SV40 enhancer contains two distinct levels of organization. *Nature* **333**: 40–45.

Orphanides G., Lagrange T., and Reinberg D. 1996. The general transcription factors of RNA polymerase II. *Genes Dev.* **10:** 2657–2683.

Ossipow V., Tassan J.P., Nigg E.A., and Schibler U. 1995. A mammalian RNA polymerase II holoenzyme containing all components required for promoter-specific transcription initiation. *Cell* **83:** 137–146.

Pabo C.O. and Sauer R.T. 1992. Transcription factors: Structural families and principles of DNA recognition. *Annu. Rev. Biochem.* **61:** 1053–1095.

Pan G., Aso T., and Greenblatt J. 1997. Interaction of elongation factors TFIIS and elongin A with a human RNA polymerase II holoenzyme capable of promoter-specific initiation and responsive to transcriptional activators. *J. Biol. Chem.* **272:** 24563–24571.

Paranjape S.M., Kamakaka R.T., and Kadonaga J.T. 1994. Role of chromatin structure in the regulation of transcription by RNA polymerase II. *Annu. Rev. Biochem.* **63:** 265–297.

Parekh B.S. and Maniatis T. 1999. Virus infection leads to localized hyperacetylation of histones H3 and H4 at the IFN-β promoter. *Mol. Cell* **3:** 125–129.

Parker D., Ferreri K., Nakajima T., LaMorte V.J., Evans R., Koerber S.C., Hoeger C., and Montminy M.R. 1996. Phosphorylation of CREB at Ser-133 induces complex formation with CREB-binding protein via a direct mechanism. *Mol. Cell. Biol.* **16:** 694–703.

Paro R. 1993. Mechanisms of heritable gene repression during development of *Drosophila. Curr. Opin. Cell Biol.* 5, 999–1005.

Parvin J.D. and Young R.A. 1998. Regulatory targets in the RNA polymerase II holoenzyme. *Curr. Opin. Genet. Dev.* **8:** 565–570.

Pazin M.J. and Kadonaga J.T. 1997. What's up and down with histone deacetylation and transcription? *Cell* **89:** 325–328.

Phi-Van L. and Strätling W.H. 1990. Association of DNA with nuclear matrix. *Prog. Mol. Subcell. Biol.* **11:** 1–11.

Ptashne M. and Gann A. 1998. Imposing specificity by localization: Mechanism and evolvability. *Curr. Biol.* **8:** R812–22.

Rhodes D. 1997. The nucleosome core all wrapped up. *Nature* **389:** 231–233.

Rippe K., von Hippel P.H., and Langowski J. 1995. Action at a distance: DNA-looping and initiation of transcription. *Trends Biochem. Sci.* **20:** 500–506.

Robert F., Douziech M., Forget D., Egly J.M., Greenblatt J., Burton Z.F., and Coulombe B. 1998. Wrapping of promoter DNA around the RNA polymerase II initiation complex induced by TFIIF. *Mol. Cell* **2:** 341–351.

Roberts S.G.E., Ha I., Maldonado E., Reinberg D., and Green M.R. 1993. Interaction between an acidic activator and transcription factor IIB is required for transcriptional activation. *Nature* **363:** 741–744.

Rundlett S.E., Carmen A.A., Suka N., Turner B.M., and Grunstein M. 1998. Transcriptional repression by UME6 involves deacetylation of lysine 5 of histone H4 by RPD3. *Nature* **392:** 831–835.

Ryu S. and Tjian R. 1999. Purification of transcription cofactor complex CRSP. *Proc. Natl. Acad. Sci.* **96:** 7137–7142.

Ryu S., Zhou S., Ladurner A.G., and Tjian R. 1999. The transcriptional cofactor complex CRSP is required for activity of the enhancer-binding protein Sp1. *Nature* **397:** 446–450.

Schaffner G., Schirm S., Muller-Baden B., Weber F., and Schaffner W. 1988. Redundancy of information in enhancers as a principle of mammalian transcription control. *J. Mol. Biol.* **201:** 81–90.

Schena M., Freedman L.P., and Yamamoto K.R. 1989. Mutations in the glucocorticoid receptor zinc finger region that distinguish interdigitated DNA binding and transcriptional enhancement activities. *Genes Dev.* **3:** 1590–1601.

Scully R., Anderson S.F., Chao D.M., Wei W., Ye L., Young R.A., Livingston D.M., and Parvin J.D. 1997. BRCA1 is a component of the RNA polymerase II holoenzyme. *Proc. Natl. Acad. Sci.* **94:** 5605–5610.

Smale S.T. 1994. Core promoter architecture for eukaryotic protein-coding genes. In *Transcription: Mechanisms and regulation* (ed. Conaway R.C. and Conaway J.W.), pp. 63–81. Raven Press, New York.

————.1997. Transcription initiation from TATA-less promoters within eukaryotic protein-coding genes. *Biochim. Biophys. Acta* **1351:** 73–88.

Stringer K.F., Ingles C.J., and Greenblatt J. 1990. Direct and selective binding of an acidic transcriptional activation to the TATA-box factor TFIID. *Nature* **345:** 783–786.

Struhl K. 1998. Histone acetylation and transcriptional regulatory mechanisms. *Genes Dev.* **12:** 599–606.

Studitsky V.M., Kassavetis G.A., Geiduschek E.P., and Felsenfeld G. 1997. Mechanism of transcription through the nucleosome by eukaryotic RNA polymerase [see comments]. *Science* **278:** 1960–1663.

Sun X., Zhang Y., Cho H., Rickert P., Lees E., Lane W., and Reinberg D. 1998. NAT, a human complex containing Srb polypeptides that functions as a negative regulator of activated transcription. *Mol.Cell* **2(2):** 213–222.

Tan S. and Richmond T.J. 1998. Eukaryotic transcription factors. *Curr. Opin. Struct. Biol.* **8:** 41–48.

Thanos D. and Maniatis T. 1995. Virus induction of human IFN β gene expression requires the assembly of an enhanceosome. *Cell* **29:** 1091–1100.

Tirode F., Busso D., Coin F., and Egly J.M. 1999. Reconstitution of the transcription factor TFIIH: Assignment of functions for the three enzymatic subunits, XPB, XPD, and cdk7. *Mol. Cell* **3:** 87–95.

Torchia J., Glass C., and Rosenfeld M.G. 1998. coactivators and corepressors in the integration of transcriptional responses. *Curr. Opin. Cell Biol.* **10:** 373–383.

Tong J.K., Hassig C.A., Schnitzler G.R., Kingston R.E., and Schreiber S.L. 1998. Chromatin deacetylation by an ATP-dependent nucleosome remodelling complex. *Nature* **395:** 917–921.

Triezenberg S.J. 1995. Structure and function of transcriptional activation domains. *Curr. Opin. Genet. Dev.* **5:** 190–196.

Tzamarias D. and Struhl K. 1994. Functional dissection of the yeast Cyc8-Tup1 transcriptional corepressor complex. *Nature* **369:** 758–761.

Uesugi M., Nyanguile O., Lu H., Levine A.J., and Verdine G.L. 1997. Induced α helix in the VP16 activation domain upon binding to a human TAF. *Science* **277:** 1310–1313.

Utley R.T., Ikeda K., Grant P.A., Cote J., Steger D.J., Eberharter A., John S., and Workman J.L. 1998. Transcriptional activators direct histone acetyltransferase complexes to nucleosomes. *Nature* **394:** 498–502.

van Holde K.E. 1989. *Chromatin.* Springer-Verlag, New York.

Varga-Weisz P.D. and Becker P.B. 1998. Chromatin-remodeling factors: Machines that regulate? *Curr. Opin. Cell Biol.* **10:** 346–353.

Wahi M. and Johnson A.D. 1995. Identification of genes required for α2 repression in *Saccharomyces cerevisiae. Genetics* **140:** 79-90.

Wathelet M.G., Lin C.H., Parekh B.S., Ronco L.V., Howley P.M., and Maniatis T. 1998. Virus infection induces the assembly of coordinately activated transcription factors on the IFN-β enhancer in vivo. *Mol. Cell* **1:** 507–518.

Wang J.C. and Giaever G.N. 1988. Action at a distance along a DNA. *Science* **240:** 300–304.

Weiss A. and Cedar H. 1997. The role of DNA demethylation during development. *Genes Cells* **2:** 481–486.

Widlund H.R., Cao H., Simonsson S., Magnusson E., Simonsson T., Nielsen P.E., Kahn J.D., Crothers D.M., and Kubista M. 1997. Identification and characterization of genomic nucleosome-positioning sequences. *J. Mol. Biol.* **267:** 807–817.

Workman J.L. and Kingston R.E. 1998. Alteration of nucleosome structure as a mechanism of transcriptional regulation. *Annu. Rev. Biochem.* **67:** 545–579.

Xiao H., Pearson A., Coulombe B., Truant R., Zhang S., Regier J.L., Triezenberg S.J., Reinberg D., Flores O., and Ingles C.J. 1994. Binding of basal transcription factor TFIIH to the acidic activation domains of VP16 and p53. *Mol. Cell. Biol.* **14:** 7013–7024.

Xue Y., Wong J., Moreno G.T., Young M.K., Côté J., and Wang W. 1998. NURD, a novel complex with both ATP-dependent chromatin-remodeling and histone deacetylase activities. *Mol. Cell* **2:** 851-861.

Xu L., Glass C.K., and Rosenfeld M.G. 1999. Coactivator and corepressor complexes in nuclear receptor function. *Curr. Opin. Genet. Dev.* **9:** 140–147.

Yankulov K. and Bentley D. 1998. Transcriptional control: Tat cofactors and transcriptional elongation. *Curr. Biol.* **8:** R447–449.

Zhang Y., LeRoy G., Seelig H.P., Lane W.S., and Reinberg D. 1998. The dermatomyositis-specific autoantigen Mi2 is a component of a complex containing histone deacetylase and nucleosome remodeling activities. *Cell* **95:** 279–289.

Zhou J., Zwicker J., Szymanski P., Levine M., and Tjian R. 1998. TAFII mutations disrupt Dorsal activation in the *Drosophila* embryo. *Proc. Natl. Acad. Sci.* **95:** 13483–13488.

WWW RESOURCES

http://www.ncbi.nlm.nih.gov/Structure/ The Structure Group at NCBI developed an up-to-date database of 3-D macromolecular structures along with tools for comparing and visualizing structures and associating these data with other information sources such as taxonomy and literature databases. The Molecular Modelling Database, MMDB, contains experimentally determined biopolymer 3D structures obtained from the Protein Data Bank (PDB). The structures contained in MMDB are stored in ASN.1 format, and make explicit representation of many structural features usually left only implicit by the PDB specifications. By making this information explicit, the extent of redundant and sometimes trivial computation required of structure comparison algorithms and molecular modeling packages is often greatly reduced. Cn3D is NCBI's 3D structure viewerspecifically developed to read the structure files stored in MMDB. MMDB is fully integrated into Entrez, and in addition to the Cn3D structure viewer, provides users with a powerful search tool (VAST) for rapidly identifying similar 3D substructures.

http://genome.nhgri.nih.gov/histones/ The Histone Sequence Database is an annotated and searchable collection of all available histone and histone fold sequences and structures. Particular emphasis has been placed on documenting conflicts between similar sequence entries from a number of source databases, conflicts that are not necessarily documented in the source databases themselves. New additions to the database include compilations of posttranslational modifications for each of the core and linker histones, as well as genomic information in the form of map loci for the human histone gene complement, with the genetic loci linked to Online Mendelian Inheritance in Man (OMIM)

Rasmol homepage: http://www.umass.edu/microbio/rasmol/ RasMol is a molecular graphics program intended for the visualization of proteins, nucleic acids, and small molecules. The program was developed by Roger Sayle and is aimed at display, teaching, and generation of publication-quality images. The program has been developed at the University of Edinburgh's Biocomputing Research Unit and the Biomolecular Structures Group at Glaxo Research and Development, Greenford, UK.

http://www.nhgri.nih.gov:80/Data/ The NCBI database of links to genome research.

http://www.epd.isb-sib.ch/ The Eukaryotic Promoter Database (EDP), is developed and maintained by members of the Bioinformatics Group of the Swiss Institute for Experimental Cancer Research (ISREC), a founding member of the Swiss Institute for Bioinformatics (SIB).

http://ndbserver.rutgers.edu/NDB/structure-finder/dnabind/ The Nucleic Acid Database Project (NDB) at Rutgers, The State University of New Jersey, maintains this server, which has several interfaces for searching the DNA-Binding Protein Database.

http://transfac.gbf.de/index.html/ TRANSFAC (for sequence-specific transcription factors and their recognition sites), TRRD (for regulatory elements of entire genes), and COMPEL (for properties of composite DNA elements) databases are accessible from this website.

CHAPTER 2

Initial Strategic Issues

Important issues

- *The first step in a gene regulation analysis is to determine the mode of regulation.*
- *The time commitment and resources required for a rigorous gene regulation analysis are often underestimated.*
- *The feasibility of an analysis should be assessed before it is initiated.*
- *Transcriptional regulation mechanisms can be pursued by studying the promoter, distant control regions, or both.*

INTRODUCTION

After a new gene is isolated, the most important objectives are to elucidate the structure, biochemical activities, and biological functions of the encoded protein. A related goal is to study the mechanisms used to regulate the protein's biochemical activities within a cell and in different cell types and tissues. A biochemical activity can be regulated by posttranslational modifications, by interactions with other proteins or ligands, or by modulating protein stability. Furthermore, an activity can be regulated at the level of gene expression by modulating (1) the transcription initiation rate, (2) the transcription elongation rate, (3) the pre-mRNA splicing pattern, (4) mRNA stability, (5) mRNA transport, or (6) the translation rate.

A gene's regulatory mechanisms may be of interest for a wide variety of reasons. The gene may be expressed with an interesting temporal or spatial pattern during the development of a cell or organism, suggesting that the gene's regulatory factors play an important developmental role. Alternatively, aberrant regulation of the gene may contribute to a particular disease, or the gene may be specifically expressed in a cell type associated with a disease. In these instances, the regulatory factors may contribute to an understanding of disease pathogenesis or may provide targets for therapeutic intervention.

For the above reasons, many investigators have chosen to divert significant resources toward studies of gene regulation. This is an admirable goal because it will contribute to our knowledge of the numerous regulatory circuits that control important cellular processes. Nevertheless, prior to initiating such an analysis, there is a critical need to evaluate the time commitment, resources, and tools required to achieve the specific objectives. A flowchart of the experimental and conceptual progression of an initial transcriptional regulation analysis is presented in Figure 2.1. This chapter contains a candid discussion of these initial steps and considerations. Although potential difficulties are raised in this discussion, most researchers will find that the analysis being contemplated is feasible and likely to yield exciting insight into mechanisms of transcriptional regulation.

CONCEPTS AND STRATEGIES

The Initial Steps in a Gene Regulation Analysis

Before beginning an analysis, an essential step is to determine whether the underlying mode of regulation is truly at the level of the gene (Fig. 2.1). Perhaps it is only known that the biochemical activity of a protein is detectable in one cell type, but not in another. This could be due to modulation of the protein's activity, its steady-state abundance, or the steady-state abundance of its mRNA. Three scenarios are possible:

1. If the protein's activity and abundance, as well as its mRNA abundance, differ in the two cell types, the activity could be modulated at least in part at the mRNA (i.e., gene expression) level.

2. If the protein's activity and steady-state abundance, but not mRNA abundance, differ in the two cell types, the activity would be modulated primarily at the protein level, most likely through differential translation or protein degradation.

3. If the protein's activity, but not protein or mRNA abundance, differs in the two cell types, the activity would most likely be regulated by posttranslational modifications or physical interactions with other proteins or ligands.

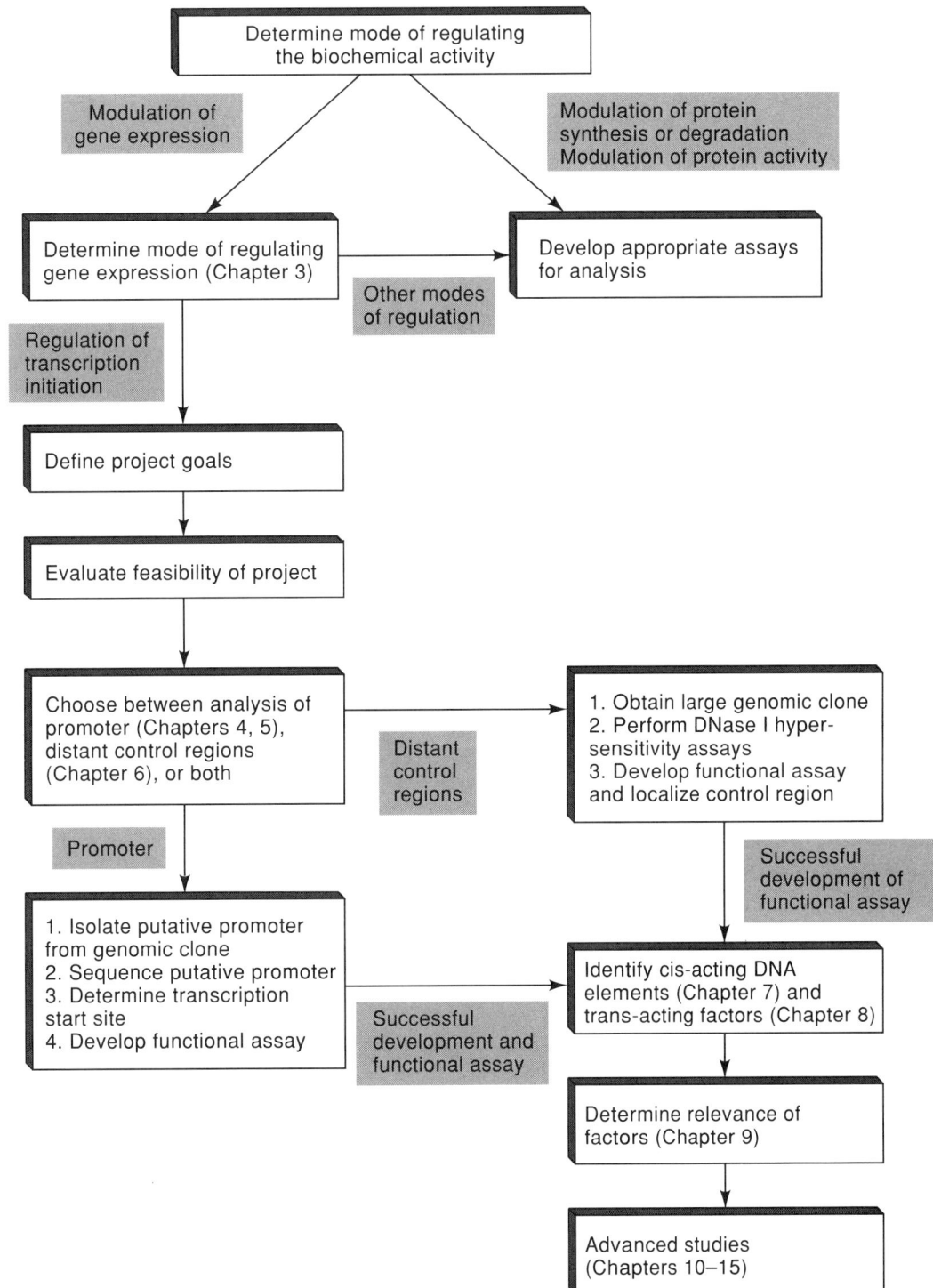

FIGURE 2.1. Analysis of a transcriptional regulation mechanism.

To distinguish among the above three scenarios, a simple comparison in the two cell types is needed of the protein's biochemical activity, the steady-state protein amount, and the steady-state mRNA amount. Steady-state protein levels can be measured by a variety of methods, including standard immunoblot or immunoprecipitation assays. Steady-state mRNA levels can be determined by Northern blot, primer extension, RNase protection, S1 nuclease, or quantitative RT-PCR analyses.

We focus exclusively on the first scenario, in which regulation is primarily at the level of steady-state mRNA abundance or gene expression. Nevertheless, the biochemical activities of an enormous number of proteins are regulated directly, and a large number of proteins appear to be subject to the regulation of translation or degradation (Hershey 1991; Pickert 1997; Varshavsky 1997). If a protein's activity is regulated directly, biochemical and structural studies of the protein can be initiated to identify relevant posttranslational modifications, interacting proteins, or other interacting molecules. If regulation is at the level of steady-state protein abundance, studies can be initiated to distinguish between regulation of translation and protein degradation, followed by experiments to dissect the relevant pathways.

If regulation is primarily at the level of steady-state mRNA abundance, it is important to remember that it does not necessarily involve modulation of transcription initiation. Differences in steady-state mRNA abundance can be regulated at the level of transcription initiation, elongation (attenuation), differential mRNA degradation, mRNA transport, or pre-mRNA splicing. The strategies for addressing these possibilities are described in detail in Chapter 3. For most genes, mRNA abundance is regulated, at least in part, at the level of transcription initiation. The following discussion highlights issues to consider before pursuing an analysis of control regions (e.g., promoters and enhancers) that regulate transcription initiation. Many of the same issues need to be considered for analyses of other modes of gene regulation.

Consider the Time Commitment and Resources Needed to Reach a Defined Goal

Before beginning an analysis of transcriptional regulation (i.e., the regulation of transcription initiation), two issues that must be considered are the amount of effort and the resources that are required. The effort and resources are often underestimated because of the ease with which the preliminary steps of an analysis can be completed.

Two General Strategies That Provide Preliminary Albeit Superficial Insight into Transcriptional Regulation Mechanisms

The recommended strategy for a rigorous, yet basic, analysis of transcriptional regulation begins with the identification of an important control region, such as a promoter, enhancer, silencer, or LCR (see Chapter 1; the latter three types of regions are often referred to as distant control regions). The identification of a control region is followed by (1) the delineation of relevant DNA sequence elements using a comprehensive mutant analysis, (2) the identification of factors that interact with the elements, (3) the cloning of the genes encoding the factors, and (4) a demonstration that the factors truly regulate the gene. The specific steps involved are described in Chapters 4–9. However, to emphasize the effort and resources that are required, two related strategies that are commonly followed for an initial promoter analysis can be considered. These strategies are valid, but are not ideal (see Chapter 7).

One basic strategy is to determine the promoter sequence and identify potential regulatory elements using a computer algorithm that reveals homologies with known control elements (see Chapter 1). If potential control elements are identified, their relevance can be assessed by testing specific substitution mutations in a standard transient transfection assay (see Chapters 5 and 7). If the mutations decrease promoter activity, the ability of the predicted proteins to bind the elements can be confirmed by electrophoretic mobility shift analysis (EMSA) (see Chapters 8 and 13), using either recombinant proteins or a crude nuclear extract. The identities of proteins detected with a crude extract can be confirmed using specific antibodies, which will either abolish or alter (i.e., supershift) the EMSA complex.

In numerous cases, this approach has provided insight into a gene's regulatory mechanisms, but for the most part, the information obtained must be considered preliminary. In the absence of an unbiased mutant analysis to identify important control elements, it is difficult to determine whether the element identified by the computer program plays a major or minor role relative to the roles played by unidentified control elements in contributing to cell-specific promoter activity. Moreover, the protein found to bind an element often represents a large family of DNA-binding proteins, whose members possess similar site-recognition specificities. A considerable amount of effort is required to determine unambiguously which family member is responsible for regulating the gene of interest. As described in Chapters 1, 9, and 13, protein binding to a site within an intact control region is thought to be context-specific and dependent on interactions with proteins bound to nearby sites; thus, the protein responsible for an abundant EMSA complex on an isolated site is not necessarily the functional protein in vivo. A final limitation of this basic approach is that it is strongly biased toward known rather than novel transcription factors. Overall, this strategy provides a valid starting point, but much more effort is required to rigorously define critical DNA elements and regulatory proteins for a gene.

A second strategy that provides rapid yet preliminary insight into a gene's regulatory mechanisms involves the development of an appropriate functional assay, such as a transient transfection assay (Chapter 5), followed by the identification of an important control region (e.g., a promoter or a distant control region) and a systematic deletion analysis of the region (Chapter 7). The deletion analysis may define a short DNA fragment (e.g., 50 bp) that contains important control elements, and EMSA or DNase I footprinting (Chapters 8 and 13) can be used to identify proteins that bind that region. This strategy is less biased than the computer-based strategy and has greater potential for the identification of novel DNA-binding proteins. Nevertheless, the identification of a binding activity that interacts with a crudely defined control region requires further validation. Additional experiments are needed to demonstrate that the protein identified is the physiological regulator, because many proteins bind to crudely defined control regions (see Chapter 9). Furthermore, if the protein appears to be novel, its identification is only a starting point toward its characterization and the isolation of its gene, both of which will require additional effort (Chapter 8).

An Example of a Rigorous, Yet Incomplete, Gene Regulation Analysis: The Immunoglobulin μ Heavy-chain Gene

To illustrate more clearly the effort required to define regulatory mechanisms in organisms that are not amenable to classical genetics, it may be useful to consider the murine immunoglobulin (Ig) μ heavy-chain gene. The transcriptional regulatory mechanisms for Ig μ have been studied since 1983, longer than any other developmentally regulated gene (Calame and Ghosh 1995; Ernst and Smale 1995). In fact, the first cellular transcriptional enhancer was identified in the Ig μ locus (Banerji et al. 1983; Gilles et al. 1983), and the first cell-specific DNA-binding protein was identified during studies of the Ig μ promoter (Staudt et al. 1986). Since these initial studies, numerous laboratories have contributed to the dissection of Ig μ promoters and enhancers.

Despite these intensive efforts, the Ig μ transcriptional regulatory mechanisms remain largely undefined. Many different control elements have been implicated in Ig μ regulation, yet there is considerable uncertainty regarding precisely which DNA-binding proteins mediate promoter and enhancer activity (Fig. 2.2) (Calame and Ghosh 1995; Ernst and Smale 1995). Indeed, the first tissue-specific regulatory factor found to interact with the Ig μ promoter, Oct-2, is now known to be a member of a large family of DNA-binding proteins (Rosenfeld 1991); the identity of the physiologically relevant family member remains unresolved. Furthermore, one of the most important control elements within the enhancer, μA, was not reported until 1993, 10 years after the discovery of the enhancer itself (Libermann and Baltimore 1993; Nelsen et al. 1993; Rivera et al. 1993). Another element within the enhancer, μE3, was thought for the past few years to function through its interaction with TFE3, a basic helix–loop–helix (HLH) leucine zipper protein. However, recent data have suggested that TFE3 may be the wrong protein and may not even be a member of the correct family (Erman et al. 1998).

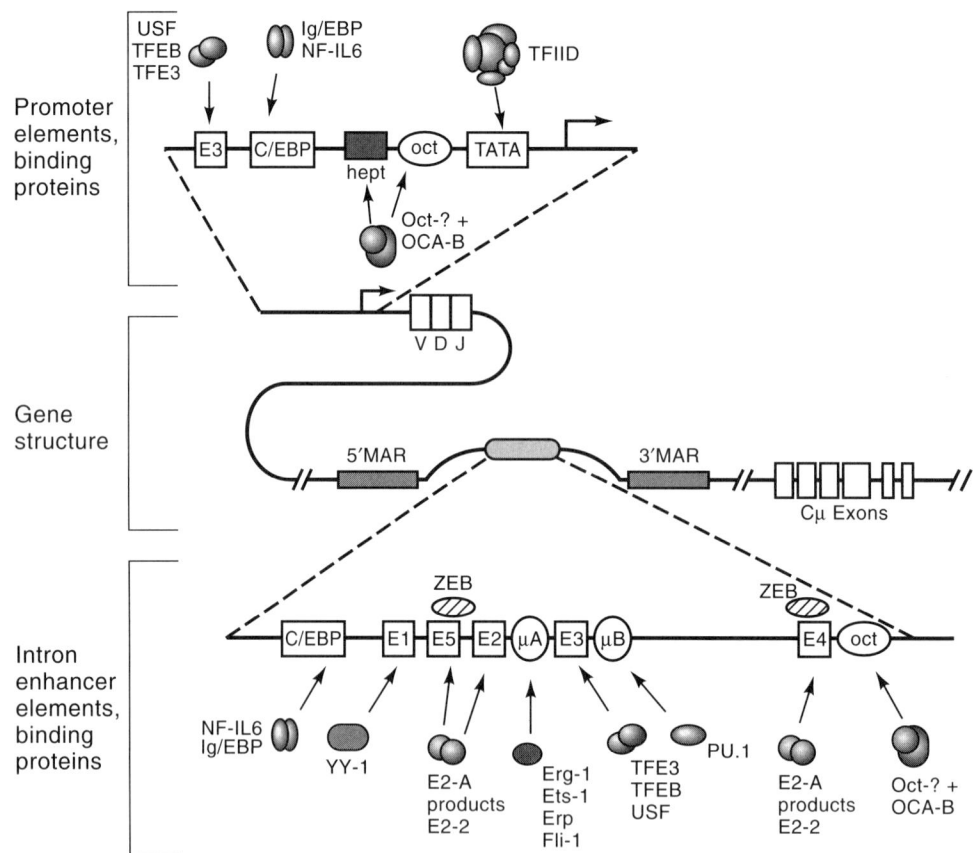

FIGURE 2.2. The murine immunoglobulin μ heavy-chain gene and control elements. Also shown are proteins capable of binding each element. In most cases, the functionally relevant binding protein remains unknown. (Redrawn, with permission, from Ernst and Smale 1995 [copyright Cell Press].)

Two reasons for the apparently slow pace of this analysis are the large number of regulatory elements and factors, and the fact that each factor is a member of a large multiprotein family. It is noteworthy that, after more than 15 years of analysis, it remains unclear how the B-cell specificity of Ig μ expression is achieved. Most likely, it results from the combined action of several proteins, none of which is strictly B-cell-specific, but all of which have properties that, when combined, result in the desired expression pattern.

Defining the Project Goals

The above discussion serves to focus attention on the importance of defining the project goals and understanding the commitment and resources that are necessary to achieve the goals. The effort and resources will vary tremendously with the objectives. At one extreme, the primary objective may simply be to identify a control region that can direct transcription in a specific tissue for the purpose of expressing proteins in transgenic mice. This objective will often (but not always) involve a relatively straightforward series of experiments to identify control regions and test for functional activity. At the other extreme, the objectives may be to define novel transcription factors, isolate their genes, and confirm that they are indeed important regulators of the gene of interest. Although technology is improving at a rapid pace, a few, and more likely several, years will be required to achieve these objectives. If, after defining the regulatory factors, the objective is to elucidate the detailed mechanism by which the various factors act in concert with each other to regulate transcription initiation by RNA polymerase II, an entire career will be required, because this objective has not yet been realized for any gene in a higher eukaryote.

It should be emphasized that many important advances can be made during the course of a comprehensive analysis. For example, the basic delineation of the control elements and potential regulatory proteins for a control region represents a significant step. Also significant are (1) the identification and cloning of a novel DNA-binding protein, (2) an analysis that rigorously establishes the relevance of a DNA-binding protein for the regulation of a gene, (3) a gene disruption experiment that provides insight into the biological function of a DNA-binding protein, and (4) an analysis of the molecular mechanisms and signal transduction pathways that regulate the activity of a DNA-binding protein. Thus, although a complete understanding of a gene's regulatory mechanism is difficult to achieve, the pursuit of that goal is likely to result in a number of fundamentally important advances.

Evaluate the Feasibility of the Analysis

A second issue to consider is the feasibility of the analysis, based on the available tools. Among the most important issues are the source of cells for functional and biochemical studies and the success in developing an appropriate functional assay.

Appropriate Source of Cells for Functional Studies

A principal objective during the initial stage of an analysis is to delineate the critical control elements within a promoter or distant control region. These elements are identified by comparing the function of a wild-type control region to that of mutant control regions. Because it currently is not reasonable to carry out a systematic mutant analysis of an endogenous gene within a chromosome, artificial assays must be employed (see Chapter 5). Assays that are most useful for such an analysis involve transient or stable transfection of cells with plasmids containing wild-type or mutant control regions that regulate transcription of an appropriate "reporter" gene. These types of assays are referred to as "functional" assays because they provide a semi-quantitative measure of the stimulatory function of the control region.

To determine the feasibility of studying a given gene, one of the key considerations is whether or not an appropriate cell source is available for the functional assays. For mammalian genes, such assays inevitably depend on the availability of a cell line or population that can be grown in culture, and that can take up DNA upon transfection. In most cases, transformed (immortal) cell lines that express the gene of interest are the most suitable. Transformed cell lines are generally easy to grow in culture and to maintain in sufficient quantities. Furthermore, they usually are more amenable to transfection than primary cells and, for many cell lines, optimal transfection procedures and conditions have already been reported.

The disadvantage of using transformed cell lines is that the transcription factors regulating the gene of interest may be altered relative to primary cells. Factors that regulate the gene in primary cells might not be expressed in transformed cell lines. Moreover, there are numerous examples of transcription factors expressed in transformed cells that are not expressed in untransformed cells, including many oncoproteins. Nevertheless, if the gene of interest is expressed in a transformed cell line, it should be possible to define the factors responsible for its transcription in that line. After the factors and regulatory mechanisms are elucidated, subsequent experiments in primary cells or animals can be designed to assess the importance of the factors identified and the validity of proposed regulatory mechanisms. For example, if studies in transformed cells lead to the hypothesis that a particular transcription factor is required for the expression of a target gene, the gene encoding the transcription factor can be disrupted by homologous recombination to determine if it is indeed essential for target gene expression in primary cells or animals.

If the objective is to elucidate the mechanism by which a gene is induced or repressed during a specific developmental process, or in response to a particular extracellular stimulus, it would be beneficial at the initial stages of the analysis to employ a cell line that can mimic the induction or repression event. The activity of the control region before and after induction or repression can be compared to define relevant DNA elements and transcription factors. A less attractive, but valid, alternative is to compare transformed cell lines that express and do not express the gene of interest. To study the mechanisms responsible for the cell-specific activation of the Ig μ gene, for example, the activity of the Ig μ enhancer in B-lymphocyte cell lines was compared to its activity in common nonlymphoid cell lines (Banerji et al. 1983; Gilles et al. 1983). This approach is less desirable than an inducible system because the different cell lines are likely to possess many different properties that may yield misleading results. Nevertheless, it is often an essential and valuable approach for studying cell-specific gene regulation.

The absence of an appropriate immortalized cell line for an analysis presents a serious obstacle. In some cases, the studies can be performed in primary cells that can be grown in culture for at least a few days, if an efficient transfection procedure can be developed. A second alternative is to carry out functional studies in transgenic mice. Transgenic mice are useful for identifying control regions and can be used for a limited mutant analysis to identify discrete control elements. This alternative is not feasible for an in-depth analysis of a promoter or enhancer, however, because the cost and effort would be prohibitive. A final solution is to make use of an in vitro transcription assay. This assay is sometimes feasible if an abundant source of cells expressing the gene of interest can be obtained. These alternative methods are discussed in greater detail in Chapters 5 and 14.

To summarize, an important requirement for a gene regulation analysis is the availability of an appropriate cell source for the functional dissection of *cis*-acting control elements. If a transformed cell line expressing the endogenous gene of interest is available and transfectable by standard techniques, it may be possible to gain considerable insight into the regulatory mechanisms. If an appropriate cell line is not available, other more challenging alternatives need to be considered.

Source of Cells for Protein Extract Preparation

After the DNA sequence elements responsible for regulating a gene have been defined, the interacting proteins must be identified and characterized. The cell lines employed for the functional studies are often useful as sources of cell-free extracts for biochemical analyses. The primary advantages of transformed cell lines are that they are (1) homogeneous, (2) capable of being grown in large quantities, and (3) suitable for a direct comparison between the functional and biochemical studies (if the same cell lines are used for both purposes). A disadvantage is that the factors present may not provide an accurate representation of those present in primary cells.

One issue that was not discussed in the previous section is the quantity of cells that can be obtained. This issue is often more important for biochemical studies than for functional studies, in particular if a protein needs to be purified (see Chapter 8). Transformed cell lines are useful for this purpose, primarily if they are nonadherent. Nonadherent cells can often be grown in large quantities in spinner culture (usually in specially designed flasks with either a stir bar or a suspended stirring arm) or in roller bottles (with the bottles 90% filled with growth medium). Adherent cells are difficult to obtain in the large quantities needed for protein purification, but in some cases, adherent cells can be adapted to nonadherent growth. The HeLa cell line that is most commonly used for studies of ubiquitously expressed proteins provides an example of this adaptation process, because it originally was an adherent cell line that has been adapted for growth in spinner culture (see Chapter 14).

Although transformed cell lines are frequently used for protein purification, the cost and effort required to grow cultured cells in large quantities make it worthwhile to consider the use of fresh or frozen animal tissues. Much success has been achieved using animal tissues for the purification and characterization of transcription factors. NF-κB provides a key example of a transcription factor that was purified for peptide sequence analysis from a primary tissue, specifically rabbit lung (Ghosh et al. 1990). If a tissue source can be identified, it may provide an inexpensive and less time-consuming method of preparing cells for protein purification. It will be necessary, however, to characterize carefully the extracts prepared because some tissues are unusually susceptible to protein degradation. For this reason, it is important to perform pilot studies comparing various fresh and frozen tissues.

If an abundant source of cells is unavailable, alternative routes can be considered for isolating the gene encoding a relevant transcription factor. For example, a cDNA library can be prepared from far fewer cells than are necessary to obtain sufficient quantities of extract for protein isolation. There are several strategies for using cDNAs and cDNA libraries to clone genes encoding DNA-binding proteins that circumvent the need for protein purification. These approaches are described in Chapter 8.

Success in Developing an Appropriate Functional Assay

If an appropriate source of cells for protein and functional studies is available, it usually will be possible to make significant progress toward dissecting a regulatory mechanism.

However, the success of the project depends on the development of a viable functional assay for a promoter or a distant control region. Several difficulties have been encountered during assay development that are unrelated to the availability of appropriate cell lines for transfection experiments.

With some promoters, for reasons that remain obscure, it has been very difficult to detect promoter activity using a standard transient transfection assay. The terminal transferase (TdT) promoter, for example, functions poorly in transient transfection assays in pre-B and pre-T cells, where the endogenous gene is normally expressed (Lo et al. 1991). The promoter may simply be weak in the absence of distant enhancers, or an unknown property of the promoter may prevent it from functioning in artificial assays. In other cases, promoter activity is readily detected, but does not exhibit the cell specificity of the endogenous gene. The promoter for the RAG-1 gene, which is also expressed in pre-B and pre-T cells, provides an example of this phenotype (Brown et al. 1997). It is not yet clear whether promoters with this phenotype simply do not contribute to the cell specificity of their respective genes or whether the cell-specific factors do not function properly in artificial assays. The altered chromatin structure and large copy number of transiently transfected plasmids may lead to aberrant function (see Chapter 5).

If the primary objective is to identify and characterize distant control regions for a particular gene (e.g., an enhancer, silencer, or LCR, any of which can be quite far from the promoter), the feasibility of the study may be limited by the investigator's ability to identify the relevant regions. Control regions have been identified within introns or dozens of kilobase pairs upstream or downstream of genes. The strategies described in Chapter 6 may lead to the identification of the distant regions, but in some cases these approaches will be unsuccessful.

The chances of encountering these types of difficulties can sometimes be evaluated before beginning the analysis, because some types of genes have proven to be more amenable to transcriptional analyses than other types. For example, many rapidly induced genes, including early response and inducible cytokine genes (e.g., the c-*fos* and interleukin-2 [IL-2] genes; Treisman 1985; Siebenlist et al. 1986), are amenable to transcriptional analyses because rapid gene induction is usually mediated by promoter sequences, rather than enhancer sequences (the enhancers for these genes usually appear to be more important for dictating cell-type specificity). The promoters for these genes also tend to be highly active and inducible in transient transfection assays. In contrast, the promoters for some developmentally regulated genes, such as the TdT and RAG-1 genes mentioned above, function poorly in standard transient transfection assays and sometimes fail to reflect the proper regulation of the endogenous gene. Thus, insight into the potential success of the study can be gained by surveying the literature to determine the level of success achieved by other investigators studying similar types of genes.

In summary, the feasibility of a transcriptional regulation analysis is dependent on (1) the availability of an appropriate source of cells for functional and protein studies, and (2) the success in developing an appropriate functional assay for analysis of the control region of interest. Before experiments are performed, a preliminary evaluation of the feasibility of the project can be carried out. If appropriate cell sources are available, the project can easily be initiated but will need to be reevaluated following the initial attempts to develop a functional assay. If the control region of interest does not exhibit detectable or properly regulated activity in standard transient transfection assays, more sophisticated assays can be attempted (e.g., stable transfection or transgenic assays), or the objectives of the project may need to be reconsidered. If the desired activity is observed in a functional assay, it should be possible to elucidate the key regulatory mechanisms for the gene.

Initiate an Analysis of Transcriptional Regulation

Beginning with the Promoter or Distant Control Regions

After demonstrating that the gene is regulated at the level of transcription initiation and that the goals are justified and feasible, a key decision must be made: whether to initiate an analysis of the promoter region, or to begin a comprehensive search for distant control elements, such as enhancers, silencers, and LCRs. A third alternative is to begin both studies simultaneously. Although a complete understanding of the regulatory mechanisms will require an understanding of every control region, the necessary resources might not be available to approach the project from both directions. The decision of whether to begin with an analysis of the promoter, distant control regions, or both must be based on multiple criteria, the most important of which is the interests of the investigator.

For most genes, there is a good chance that important regulatory events will occur both at the promoter and at distant control regions. For example, both the promoter and intronic enhancer for the Ig μ gene contribute to B-cell-specific expression (Calame and Ghosh 1995; Ernst and Smale 1995). Some investigators consider distant control regions to be of higher priority because these regions are often powerful and may control the accessibility of the entire locus. In other words, distant control regions may alter the chromatin structure throughout the locus, making it possible for transcription factors to bind the promoter. Thus, distant control regions may be considered to be at a higher position in the regulatory hierarchy. A disadvantage of focusing on distant control regions is that they may be difficult to find. A second disadvantage of studying some distant control regions, such as LCRs, is that they often do not function in a standard transient transfection assay; instead, more difficult and time-consuming assays are needed, such as stable transfection or transgenic assays (see Chapter 6).

The major advantage of studying the promoter is that, by definition, it is located in the immediate vicinity of the transcription start site. Thus, the promoter can be localized definitively and relatively easily by mapping the transcription start site. Second, there is a greater chance that the biochemical mechanism of promoter activity can be reproduced in an in vitro assay, which can facilitate detailed mechanistic studies at an advanced stage of the analysis. The primary disadvantage of beginning a transcriptional regulation study with an analysis of a promoter region is that, in the regulatory hierarchy, it may be downstream of the distant control regions.

If sufficient time and resources are available, the most desirable strategy may be to initiate studies of the promoter and distant control regions simultaneously. After the first steps in each analysis are completed, the choice of whether to pursue one or both of the directions could become clearer. For example, in the functional assays that are developed, the promoter may mimic, at least in part, the regulation of the endogenous gene, whereas attempts to isolate distant control regions may fail. This scenario may lead to a valid decision to study the promoter to define some of the events that lead to appropriate regulation. In contrast, the initial studies may result in the identification of a powerful and appropriately regulated enhancer or silencer, but the initial promoter analysis may reveal that its activity is ubiquitous in the functional assays that were employed. In this instance, the valid decision may be to focus on the regulatory contributions of the enhancer or silencer, and to evaluate the contributions of the promoter at a later date.

The above strategy for choosing between the promoter and distant control regions is useful for many genes, but for specific classes of genes, the starting point may be more obvious. As mentioned above, genes that are rapidly induced following cell stimulation, such as cytokine genes within the immune system and early response genes in general, seem to concentrate major determinants of induction at the promoter. These genes are likely to have distant control regions that contribute to regulation, but the distant regions may primarily be responsible for regulating the accessibility of the locus in a specific cell type, so that the gene is poised for rapid activation by proteins binding with the promoter. Thus, for this class of genes, distant control regions need to be analyzed to elucidate the cell-type specificity of transcription, but the promoter must usually be the primary focus to elucidate the mechanism of rapid induction.

Initiating an Analysis of a Promoter

The strategies for initiating a promoter analysis are described in Chapters 4 and 5. The initial objectives are to isolate the putative promoter region from a genomic clone, localize the transcription start site(s), sequence the region surrounding the start site(s), and develop an appropriate functional assay for identifying relevant control elements. A genomic clone for isolating the putative promoter is usually obtained by screening a genomic library with a probe derived from the 5′ end of a full-length cDNA.

Initiating an Analysis of Distant Control Regions

The strategies for initiating an analysis of distant control regions are described in Chapter 6. The first objective is to isolate genomic clones containing the exons and introns for the gene of interest and extending several kilobase pairs on either side of the coding region. Because it is not unusual for distant control regions to be located several kilobases upstream or downstream of the coding region, it could be beneficial to isolate cosmid, P1, bacterial artificial chromosome (BAC), or yeast artificial chromosome (YAC) clones, which contain larger inserts than λ phage clones. These clones can be used to perform DNase I hypersensitivity studies, or for functional studies to identify the distant control regions, as described in Chapter 6.

SUMMARY

An analysis of transcriptional regulation must begin with experiments to determine whether a protein's activity is truly regulated at the level of gene expression, and whether gene expression is regulated at the level of transcription initiation. If these experiments demonstrate that transcription initiation is an important mode of regulation, the goals of the project should be evaluated along with a careful consideration of the time commitment and resources that will be required to achieve the goals. The feasibility of the project also must be evaluated with respect to the available tools (e.g., cell lines). Next, one must decide whether to initiate an analysis of the promoter, distant control regions, or both. A control region must then be isolated and a functional assay developed for its analysis.

With an appropriate functional assay, important advances toward the elucidation of transcriptional regulatory mechanisms are well within reach. Although eukaryotic gene regulation mechanisms have been studied in considerable detail for more than two decades, many fundamental questions remain to be answered. An individual laboratory can make important strides toward elucidating the regulatory circuitry that governs the development and function of a particular cell type, or the abnormal circuitry that contributes to disease pathogenesis. Perhaps even more intriguing is the possibility that from the collective studies performed in many laboratories, an underlying order to gene regulation circuitry will begin to emerge.

REFERENCES

Banerji J., Olson L., and Schaffner W. 1983. A lymphocyte-specific cellular enhancer is located downstream of the joining region in immunoglobulin heavy chain genes. *Cell* **33**: 729–740.

Brown S.T., Miranda G.A., Galic Z., Hartman I.Z., Lyon C.J., and Aguilera R.J. 1997. Regulation of the RAG-1 promoter by the NF-Y transcription factor. *J. Immunol.* **158**: 5071–5074.

Calame K. and Ghosh S. 1995. Factors regulating immunoglobulin transcription. In *Immunoglobulin genes* (eds. Honjo T., Alt F.W., and Rabbits T.H.), pp. 397–422. Academic Press, London, United Kingdom.

Erman B., Cortes M., Nikolajczyk B.S., Speck N.A., and Sen R. 1998. ETS-core binding factor: A common composite motif in antigen receptor gene enhancers. *Mol. Cell. Biol.* **18**: 1322–1330.

Ernst P. and Smale S.T. 1995. Combinatorial regulation of transcription II: The immunoglobulin μ heavy chain gene. *Immunity* **2**: 427–438.

Ghosh S., Gifford A.M., Riviere L.R., Tempst P., Nolan G.P., and Baltimore D. 1990. Cloning of the p50 DNA binding subunit of NF-kappa B: Homology to rel and dorsal. *Cell* **62**: 1019–1029.

Gillies S.D., Morrison S.L., Oi V.T., and Tonegawa S. 1983. A tissue-specific transcription enhancer element is located in the major intron of a rearranged immunoglobulin heavy chain gene. *Cell* **33**: 717–728.

Hershey J.W. 1991. Translational control in mammalian cells. *Annu. Rev. Biochem.* **60**: 717–755.

Libermann T.A. and Baltimore D. 1993. Pi, a pre-B-cell-specific enhancer element in the immunoglobulin heavy-chain enhancer. *Mol. Cell. Biol.* **13**: 5957–5969.

Lo K., Landau N.R., and Smale S.T. 1991. LyF-1, a transcriptional regulator that interacts with a novel class of promoters for lymphocyte-specific genes. *Mol. Cell. Biol.* **11**: 5229–5243.

Nelsen B., Tian G., Erman B., Gregoire J., Maki R., Graves B., and Sen R. 1993. Regulation of lymphoid-specific immunoglobulin μ heavy chain gene enhancer by ETS-domain proteins. *Science* **261**: 82–86.

Pickart C.M. 1997. Targeting of substrates to the 26S proteasome. *FASEB J.* **11**: 1055–1066.

Rivera R.R., Stuiver M.H., Steenbergen R., and Murre C. 1993. Ets proteins: New factors that regulate immunoglobulin heavy-chain gene expression. *Mol. Cell. Biol.* **13**: 7163–7169.

Rosenfeld M.G. 1991. POU-domain transcription factors: Pou-er-ful developmental regulators. *Genes Dev.* **5**: 897–907.

Siebenlist U., Durand D.B., Bressler P., Holbrook N.J., Norris C.A., Kamoun M., Kant J.A., and Crabtree G.R. 1986. Promoter region of interleukin-2 gene undergoes chromatin structure changes and confers inducibility on chloramphenicol acetyltransferase gene during activation of T cells. *Mol. Cell. Biol.* **6**: 3042–3049.

Staudt L.M., Singh H., Sen R., Wirth T., Sharp P.A., and Baltimore D. 1986. A lymphoid-specific protein binding to the octamer motif of immunoglobulin genes. *Nature* **323**: 640–643.

Treisman R. 1985. Transient accumulation of c-fos RNA following serum stimulation requires a conserved 5′ element and c-fos 3′ sequences. *Cell* **42**: 889–902.

Varshavsky A. 1997. The ubiquitin system. *Trends Biochem. Sci.* **22**: 383–387.

CHAPTER 3

Modes of Regulating mRNA Abundance

Important issues

- *Many genes are regulated at the level of transcription initiation, but other common modes of regulating gene expression exist.*

- *mRNA transcripts are degraded by distinct mechanisms that confer a defined half-life.*

- *The half-life of a specific transcript can be modulated and contribute to gene regulation.*

- *Assays are available for determining whether a gene is regulated at the level of transcription initiation or mRNA stability.*

- *Mechanisms of regulating transcription elongation have been elucidated in prokaryotes and eukaryotes.*

- *Methods are available for determining whether a gene is regulated at the level of transcription elongation.*

INTRODUCTION

It could be argued that the most economical method for modulating the expression of a gene is to control the rate of transcription initiation. If the gene product is not needed in a particular cell type, then little energy will be expended if the gene is not transcribed and is instead maintained in an inaccessible chromatin configuration.

Given this consideration, it is not surprising that transcription initiation is a principal mode of regulating many, if not most, genes. Nevertheless, genes can be regulated at several other points in the expression pathway. These alternative modes of regulation include transcription elongation, mRNA stability, pre-mRNA splicing, mRNA transport, polyadenylation, and translation. There are distinct advantages of using these mechanisms to regulate certain classes of genes. Thus, some genes are regulated solely by one of these alternative mechanisms, and many other genes are regulated by a combination of mechanisms.

It is noteworthy that many of the above modes result in modulation of a gene's steady-state mRNA abundance. Therefore, if mRNA abundance is regulated, one cannot simply assume that regulation is mediated solely at the level of transcription initiation. Transcription initiation might instead play a partial role or no role in regulation. In this chapter, the strategies and methods for determining the mode of regulating mRNA abundance are discussed.

CONCEPTS AND STRATEGIES

Transcription Initiation Versus mRNA Stability

The steady-state abundance of every mRNA is dependent on its rate of synthesis and rate of decay (Ross 1995; Caponigro and Parker 1996; Jacobson and Peltz 1996). It follows that the intrinsic stability of an mRNA can have a profound influence on the abundance of the mRNA and the abundance of the encoded polypeptide. The stabilities of mRNAs in mammalian cells vary widely, with approximate half-lives ranging from 15 minutes (e.g., c-*fos*) to 10 hours (e.g., β-actin) (Ross 1995). Moreover, mRNA stabilities can be tightly regulated, and in some cases are regulated coordinately with transcription initiation.

This section contains a brief overview of mRNA degradation pathways and the interrelationship between mRNA stability and transcription initiation. In addition, common methods which can be used to confirm that a gene is regulated at the level of transcription initiation and/or mRNA stability are described, along with a strategic discussion of when these techniques are necessary and practical.

Basic mRNA Degradation Pathways

The degradation pathways for typical mRNAs within a eukaryotic cell have not been defined in detail, nor is it known how many different pathways can lead to degradation. However, recent studies performed primarily in *Saccharomyces cerevisiae* have provided considerable insight into one basic degradation pathway, called the deadenylation-dependent pathway (Caponigro and Parker 1996; Jacobson and Peltz 1996) (Fig. 3.1). This pathway appears to be conserved in mammalian cells and may be responsible for the passive and regulated degradation of many or most mRNAs. Other mRNAs are degraded by deadenylation-independent pathways, which may be quite similar to the deadenylation-dependent pathway following the event that initiates degradation (see Fig. 3.1). As discussed below, the degradation of transferrin receptor mRNA may be an example of a deadenylation-independent mechanism.

One of the earliest steps in the deadenylation-dependent pathway is the shortening of the poly(A) tail by 3′ to 5′ ribonucleases (Fig. 3.1, Deadenylation). Nucleases that may be involved in this step include the poly(A) nuclease (PAN) in yeast (Jacobson and Peltz 1996) and the deadenylating nuclease (DAN) in mammalian cells (Korner and Wahle 1997). Poly(A) shortening appears to take place in the cytoplasm. This event has been documented in both yeast and mammalian cells by monitoring the properties of specific mRNAs over time. From studies of mRNAs containing sequence determinants that regulate mRNA stability, poly(A) shortening has been shown to precede mRNA degradation, and the rate of poly(A) shortening correlates with the degradation rate. Proteins that bind poly(A) tails have been identified and may be responsible for linking poly(A) tail length to mRNA degradation (Jacobson and Peltz 1996).

After the poly(A) tail is shortened, the next step appears to be removal of the 7-methyl-guanosine cap structure at the 5′ end of the mRNA by a cap-specific pyrophosphatase (Fig. 3.1, Decapping). The removal of the cap makes the mRNA susceptible to cytoplasmic 5′ to

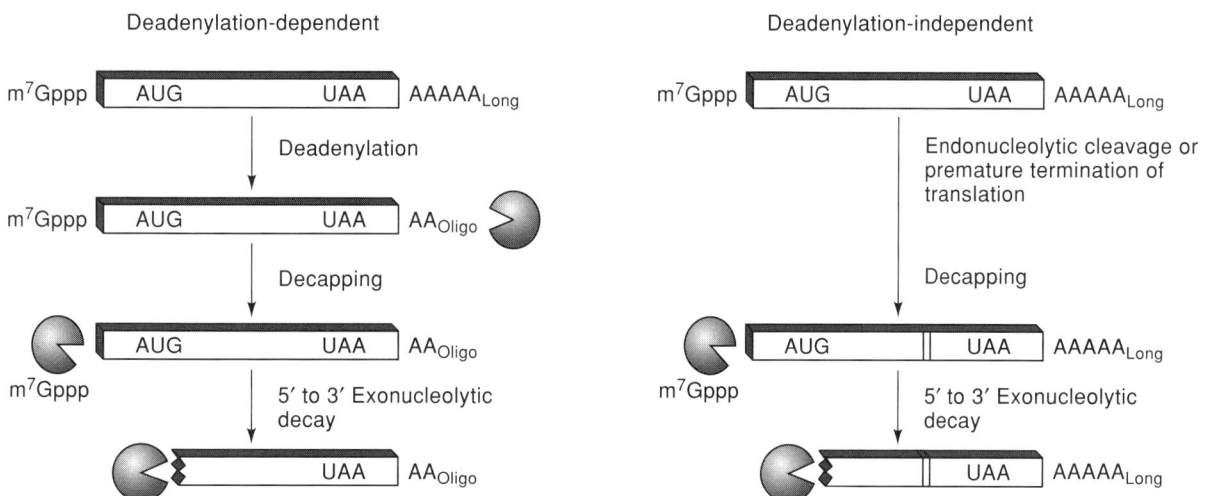

FIGURE 3.1. mRNA degradation pathways. (Updated, with permission, from Caponigro and Parker 1996 [Copyright American Society for Microbiology].)

3′ exonucleases, which complete the degradation process (Fig. 3.1, 5′ to 3′ Exonucleolytic decay). Endonucleolytic cleavage of mRNA or premature termination of transcription may promote degradation via a deadenylation-independent pathway similar to that diagramed in Figure 3.1 (right).

It has been speculated that the determinants of intrinsic mRNA stability include primary mRNA sequences, mRNA secondary structures, and mRNA-binding proteins that alter the rate of poly(A) shortening, cap removal, or the 5′ to 3′ exonuclease activity (Ross 1995; Caponigro and Parker 1996; Jacobson and Peltz 1996; Rajagopalan and Malter 1997). One common determinant is an element called an AU-rich response element (ARE), which contributes to the short half-lives of a number of early-response mRNAs (Chen and Shyu 1995; Ross 1995; Jacobson and Peltz 1996; Wilson and Brewer 1999). AREs, often containing the sequence AUUUA, are typically found in multiple copies in the 3′-untranslated regions of relevant mRNAs. These elements clearly contribute to mRNA instability, as they are capable of destabilizing heterologous mRNAs when fused at their 3′ ends. The precise nature of an ARE has not been established, and it is not known whether all AREs function in a similar manner. Moreover, the proteins that functionally interact with AREs have not been clearly defined, although several ARE-binding proteins have been reported. Nevertheless, at least a subset of the AREs appear to enhance the rate of mRNA degradation by stimulating poly(A) shortening.

c-Fos is one well-studied mammalian gene containing an ARE (Wilson and Treisman 1988). A careful analysis of mRNA degradation mediated by this element revealed a close link to poly(A) shortening (Shyu et al. 1991). Degradation was consistently preceded by shortening, and every mutation in the ARE that reduced the degradation rate also reduced the rate of shortening. These results suggest that some or all AREs alter mRNA stability by directly influencing the deadenylation-dependent pathway described above.

Regulation of mRNA Stability and Degradation

Because mRNA stabilities vary, it is not surprising that the degradation process can be tightly regulated and contribute to gene regulation. The stabilities of many mRNAs are known to be regulated, but only a few have been studied in sufficient detail to yield insight into the underlying mechanism.

Perhaps the best-characterized regulatory mechanisms are associated with genes whose expression levels depend on intracellular iron availability (Hentze and Kühn 1996). The mRNAs for several iron-regulated genes contain iron-responsive elements (IREs) that interact with either of two cytoplasmic RNA-binding proteins, iron-regulatory proteins 1 and 2 (IRP-1 and IRP-2). IREs consist of a stem-loop structure, with the IRP recognition sequence located primarily in the 6-nucleotide loop (Fig. 3.2).

The 3′-untranslated region of the transferrin receptor mRNA contains five IREs, which bind IRPs in the presence of low concentrations of iron. IRP binding stabilizes the mRNAs, enhancing transferrin receptor gene expression. When the intracellular concentration of iron is increased, it binds the IRPs and catalyzes their dissociation from the mRNA molecules. IRP release enhances the rate of mRNA degradation. The precise mechanism of degradation is not known, but IRP release appears to expose a sequence within the 3′-untranslated region that is recognized and cleaved by an endonuclease. The endonucleolytic cleavage may promote degradation by catalyzing 5′ cap removal, similar to the mechanism by which deadenylation promotes degradation.

Although IRP binding stabilizes transferrin receptor mRNAs, it destabilizes other mRNAs, such as ferritin mRNA, by binding to their 5′ ends (Hentze and Kühn 1996).

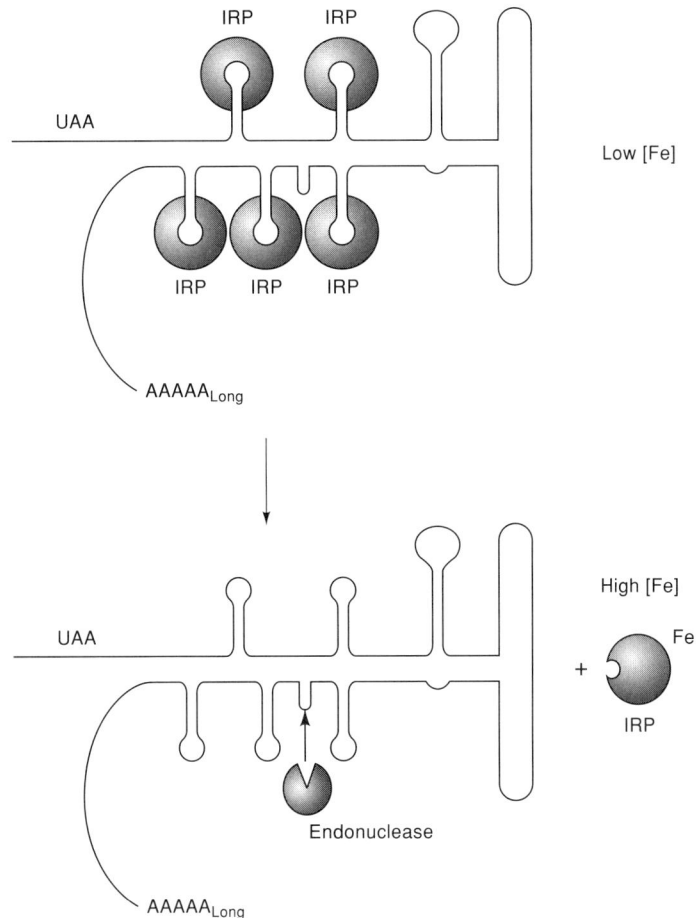

FIGURE 3.2. Regulated degradation of transferrin receptor mRNA. (Adapted, with permission, from Hentze and Kühn 1996 [Copyright 1996 National Academy of Sciences, U.S.A.].)

Destabilization is linked to translation and may result from an inhibition of ribosome elongation by IRP binding. Thus, the iron regulatory system makes use of two very different mechanisms of regulating mRNA stability, using the same mRNA-binding proteins.

The regulation of ferritin expression represents one of several examples in which mRNA stability and translation are linked. Another classic example is the regulation of α- and β-tubulin mRNA stability by a mechanism dependent on the first few amino acids of the nascent peptide (Theodorakis and Cleveland 1993; Ross 1995). A current hypothesis is that a cofactor, possibly unassembled tubulin subunits, binds the nascent peptide during translation, leading to the recruitment of a nuclease that degrades the mRNA. Rapid mRNA degradation also occurs in mammalian cells when the translation machinery encounters a nonsense codon at an inappropriate location (Hentze and Kulozik 1999). Nonsense-codon-mediated degradation, which has been studied in considerable detail, is thought to prevent the synthesis of truncated proteins that can have dominant-negative effects.

Despite the considerable progress toward understanding the regulation of mRNA stability, much remains to be learned. Perhaps the most important unanswered question is whether most or all events that promote mRNA degradation feed into one pathway, or whether several parallel pathways exist (Wilson and Brewer 1999). Clearly, there are several mechanisms for initiating degradation, including deadenylation, endonucleolytic cleavage, and premature termination of translation (see Fig. 3.1). Further studies are needed to determine whether this diversity of initiating events is due to the existence of several distinct pathways for mRNA degradation, or whether the various initiating events converge on one or a few general pathways. In light of the current data, the possible convergence of pathways is an intriguing concept because it suggests extraordinary interplay between mRNA degradation, the poly(A) tail, the 5′ cap, and the translation machinery.

Interrelationship between mRNA Stability and Transcription Initiation

Differential mRNA stability is the primary mode of regulating a large number of genes, including a subset of those described above. Other genes are regulated solely by modulating the rate of transcription initiation. In many instances, however, transcription initiation and mRNA stability must be considered simultaneously (see also Ross 1995).

First, it is important to consider the influence of intrinsic mRNA stability on the expression of a gene that is regulated primarily at the level of transcription initiation. This issue is exemplified most clearly by considering early-response genes, whose mRNAs are often needed at maximum concentrations soon after induction, and then must be rapidly down-regulated (e.g., the c-*fos* gene; Greenberg and Ziff 1984). These mRNAs often possess and benefit from intrinsically short half-lives. The short half-lives are of obvious benefit for rapid down-regulation because the mRNA rapidly degrades after transcription ceases. The benefit derived from a short half-life is illustrated by the observation that stabilization of the c-*fos* mRNA, by removal of its 3′ ARE, can result in cell transformation (Ross 1995). A second benefit is that the short half-life allows the mRNA to reach its steady-state abundance in a relatively short time period following the induction of transcription (see Fig. 3.3). Regardless of the rate of transcription initiation, an mRNA with a long half-life would not reach its steady-state abundance for a relatively long time.

In contrast to the benefits of short half-lives for early-response genes, long half-lives are beneficial for ubiquitous or tissue-specific genes that do not require rapid induction or repression (e.g., the β-globin gene; Shyu et al. 1989). A long mRNA half-life minimizes the energy needed to express the gene. These considerations suggest that the intrinsic short or long half-life of an mRNA does not necessarily indicate that the gene's expression is controlled by regulation of mRNA stability (e.g., by an increase in stability when the gene is active or induced). Instead, the gene may be regulated solely at the level of transcription initiation, with the intrinsic stability of the mRNA contributing to its expression pattern and abundance.

According to the above scenario, a gene that is regulated solely at the level of transcription initiation may encode an mRNA with a short half-life simply because it must rapidly achieve its maximum steady-state mRNA abundance. This leads to a second and more complex scenario linking mRNA stability and transcription initiation: If transient expression of a gene is desired, an even more rapid method of achieving the maximal steady-state mRNA abundance is to induce transcription initiation transiently and simultaneously to stabilize the mRNA in a transient manner. An example of this type of regulation is found in T cells, where several cytokine genes are transiently induced by enhancement of both transcription initiation and mRNA stability (Lindstein et al. 1989). The cytokine mRNAs rapidly accumulate when both processes are induced, and then rapidly decay when transcription ceases and mRNA stability is reduced.

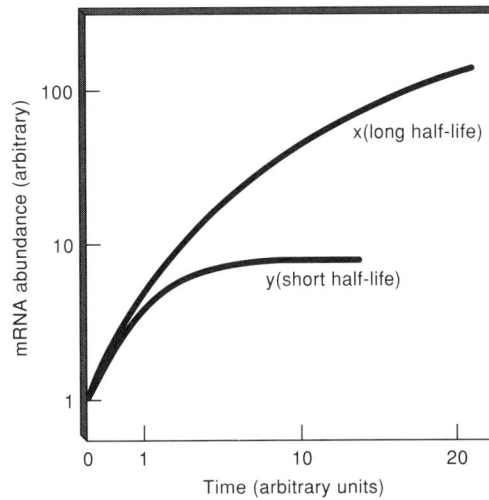

FIGURE 3.3. mRNAs with short half-lives reach a steady-state abundance more rapidly than mRNAs with long half-lives. (Adapted, with permission, from Ross 1995 [Copyright American Society for Microbiology].)

Confirming That the Rate of Transcription Initiation Contributes to Gene Regulation

The primary purpose of the above discussion is to emphasize the important role played by mRNA stability in regulating the expression of some genes. Gene expression can be influenced by the intrinsic stability of an mRNA, but mRNA stability itself can be regulated. Moreover, mRNA stability may be the sole mode of regulation, or it may play no role. For the most effective regulation, transcription initiation and mRNA stability may both be involved.

To determine if a gene is regulated at the level of transcription initiation or mRNA stability, it is necessary to address each potential mode of regulation separately. In other words, there is no single experiment that can definitively distinguish between these two modes of regulation or determine their relative contributions.

Nuclear run-ons. Two methods can be used to address the contribution of transcription initiation to the regulation of a given gene: the nuclear run-on assay (Box 3.1) and standard transfection procedures (see below and Chapter 5). The most desirable method, but also the most difficult, is the nuclear run-on (see Protocol 3.1, p. 87). In general, this method provides an indirect measure of the in vivo rate of transcription initiation from a gene using experimental conditions that greatly reduce or eliminate potential contributions from regulation of mRNA stability.

Nuclear run-ons yield results that are not generally influenced by mRNA degradation rates for two principal reasons. First, most mRNA degradation takes place in the cytoplasm, yet the radionucleotide pulse is performed with isolated nuclei. Second, even if some degradation occurs in the nuclei, the short incubation time with the radiolabeled nucleotides reduces the contribution of mRNA stability, unless the relevant nuclease rapidly attacks the nascent transcript. Given this latter caveat, the nuclear run-on is not conclusive, but it provides a strong indication as to whether or not an mRNA is regulated at the level of transcription initiation.

Box 3.1

Nuclear Run-on Transcription Assay

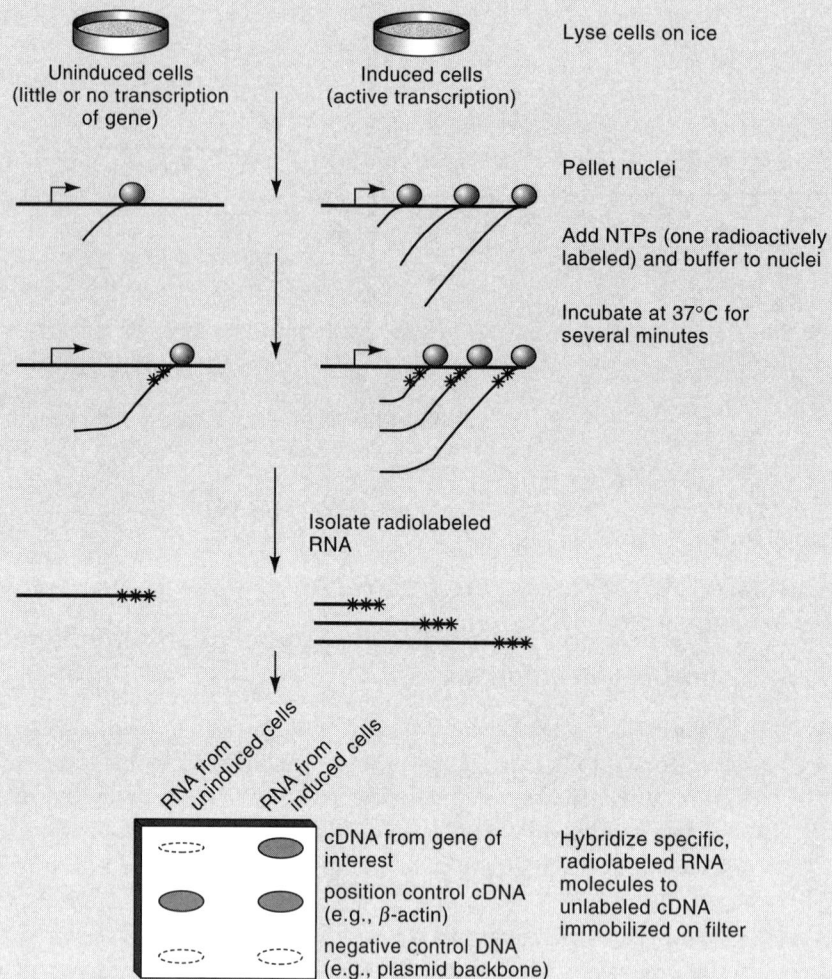

FIGURE 3.4. Nuclear run-on transcription assay.

The nuclear run-on technique (Derman et al. 1981; Greenberg and Ziff 1984; Powell et al. 1984; Ucker and Yamamoto 1984) is typically performed with cell populations that express the gene of interest at different levels. When studying an inducible gene, for example, cell populations prior to induction and following induction are used. For a tissue-specific gene, cells expressing the gene are compared to cells that lack expression.

The first step in the procedure is to lyse the cell membranes on ice without disrupting the nuclear envelope (Fig. 3.4). At the time of lysis and chilling, elongating polymerase molecules become "frozen" on the genes that are being actively transcribed. The nuclei are then pelleted by centrifugation and washed. The resulting nuclei are then suspended in a reaction buffer containing one ^{32}P-labeled nucleoside triphosphate (NTP) and three unlabeled NTPs. The mixture is incubated at 37°C for a short time period (5–15 minutes) to allow the DNA-bound polymerase molecules to resume elongation. De novo transcription initiation does not occur in isolated nuclei. The radiolabeled pre-mRNA is isolated and hybridized to a membrane containing immobilized, unlabeled DNA sequences from the gene of interest. The specific radiolabeled RNA is quantified by autoradiography and densitometry or by phosphorimager analysis.

The key theoretical concept is that the amount of specific radiolabeled RNA observed by autoradiography is approximately proportional to the number of polymerase molecules associated with the gene when the cells were lysed and chilled. The number of polymerase molecules, sometimes referred to as the "polymerase density," should, in turn, be proportional to the rate of transcription initiation from the gene. Thus, if the gene is not transcribed prior to induction, but is actively transcribed after induction, the amount of specific radiolabeled RNA will be much higher after induction than before induction, reflecting the change in steady-state mRNA abundance. In contrast, if the rate of transcription initiation does not change following induction, the amount of specific radiolabeled RNA will be similar before and after induction. If the steady-state mRNA abundance changes during induction, this latter nuclear run-on result would suggest that mRNA abundance is regulated by a different mechanism, such as mRNA stability.

Transfection approaches. A second method that provides evidence of regulation at the level of transcription initiation is to test the activity of the gene's transcriptional control regions using transient or stable transfection procedures (see Chapter 5). The promoter for the gene can be inserted upstream of a standard reporter gene, such as a luciferase or chloramphenicol acetyltransferase (CAT) gene, and the resulting plasmid can be transfected into appropriate cell lines. If reporter activity is regulated in a manner that reflects the regulation of the endogenous gene, the gene is likely to be regulated, at least in part, at the level of transcription initiation. The transfection experiments would of course need to be performed with appropriate controls, as described in Chapter 5.

One disadvantage of the transfection approach is that it involves an analysis of the regulatory sequences in an artificial context, whereas the nuclear run-on procedure provides a measure of transcription initiation from the endogenous gene. A second disadvantage of the transfection approach is that the control regions responsible for appropriate transcriptional regulation may be difficult to locate. Thus, the gene may be regulated at the level of transcription initiation, but the control regions analyzed with the transfection assay (e.g., the promoter) may not reflect the regulation.

In summary, if clear regulation of promoter or enhancer activity is observed in a transfection assay, the endogenous gene is likely to be regulated, in part, at the level of transcription initiation. On the other hand, if regulation is not observed with the transfection assay, a conclusion is not possible because of the considerable chance that the wrong region was tested or that the artificial assay abolished regulation.

Measuring mRNA Stabilities

The above experiments can provide insight into the possibility that a gene is regulated by modulation of transcription initiation. If the results suggest that transcription initiation is not of primary importance for gene regulation, an alternative possibility is that the gene is regulated at the level of mRNA stability. Even if transcription initiation is found to play an important role in regulation, it may be worthwhile determining whether mRNA stability makes a contribution.

Inhibitor studies. For most genes in higher eukaryotes, straightforward methods can provide approximate mRNA half-lives, although accurate half-lives are difficult to determine. The simplest methods make use of transcriptional inhibitors, such as actinomycin D,

5,6-dichloro-1-β-D-ribofuranosylbenzimidazole (DRB), and α-amanitin (Ross 1995; Rajagopalan and Malter 1997). Actinomycin D readily enters most cells, intercalates into DNA, and rapidly blocks transcription and DNA replication. DRB apparently inhibits a kinase necessary for efficient transcription elongation, and α-amanitin inhibits transcription by binding a subunit of RNA polymerase II. The amount of intact mRNA remaining at various time points after addition of an inhibitor can be monitored by Northern blot, primer extension, S1 nuclease, or RNase protection analyses. The amount of time required for 50% of the mRNA to degrade represents an approximate half-life for the mRNA.

Over the years, many groups have incorporated inhibitor analyses of mRNA stability into their studies of gene regulation. The schematic in Figure 3.5 is loosely adapted from a study of the Igγ2b heavy-chain gene in two different myeloma cell lines (Harrold et al. 1991). Following addition of the inhibitor (both actinomycin D and DRB were tested in the study cited), mRNA abundance was monitored at different time points in each cell line by Northern analysis. The results revealed that the half-life of the mRNA was approximately 70% shorter in one cell line than in the other.

Transcriptional inhibitors are simple to use and can provide valuable information, but each inhibitor possesses one or more limitations. α-Amanitin, for example, does not enter all cells, whereas actinomycin D can enhance mRNA stability, leading to inaccurate half-life measurements (Ross 1995; Rajagopalan and Malter 1997). In addition to the specific limitations of each inhibitor, all of the inhibitors are of limited utility for determining mRNA half-lives exceeding 1 or 2 hours. By blocking transcription, the drugs induce cell death beginning after only a few hours of treatment. If the half-life of an mRNA is 10 hours, for example, it would be desirable to measure the half-life by monitoring mRNA abundance 10, 20, 40, and 80 hours following addition of the drug. Unfortunately, cells are generally unhealthy by the 10-hour time point, when mRNA abundance would be decreased by only twofold.

Despite these caveats, transcriptional inhibitors remain useful for measuring approximate mRNA half-lives and for determining whether mRNA stability contributes to gene regulation. Although the half-life measurements may not be precise, the method can indicate whether the half-life is reasonably short (i.e., 1 hour or less) or long. Moreover, the experiment can provide an approximate comparison of mRNA half-lives in two or more cell populations.

If a precise half-life is needed, multiple methods should be used until a consensus is reached. In the study of the Igγ2b mRNAs cited above, four different methods were employed for determination of mRNA half-lives (Harrold et al. 1991). All of the methods showed the mRNA half-life in one cell line to be 70% less than the half-life in the other line. However, the absolute half-lives estimated by each method varied considerably. For example, in one line, the half-life of the mRNA was found to be 6.4 hours by DRB treatment, 5.7 hours by steady-state accumulation (see Harrold et al. 1991; Ross 1995), 2.9 hours by actinomycin D treatment, and 3.8 hours by pulse chase (see below) (Harrold et al. 1991; Ross 1995).

A final limitation of using inhibitors to compare mRNA stabilities is that they cannot be used to measure the half-life of an mRNA for which the steady-state level cannot be detected. As an example of this point, consider an inducible gene, such as the c-*fos* gene, whose steady-state mRNA is undetectable by Northern blot, etc., prior to induction, but is easily detected after induction (Greenberg and Ziff 1984). If the transcripts are undetectable prior to induction, it is impossible to use inhibitors to determine their half-life. The half-life can only be determined after induction when the transcripts can be detected.

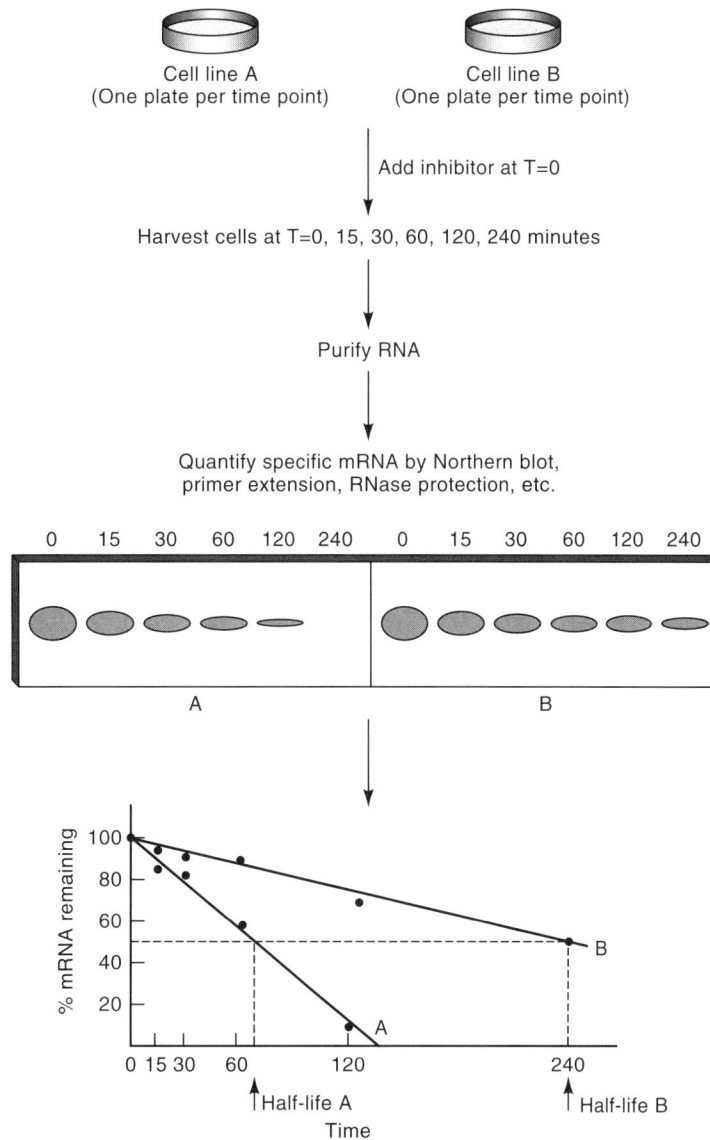

FIGURE 3.5. A typical inhibitor analysis of mRNA stability. Actinomycin D is added to cells at $t = 0$, and cells are harvested for mRNA at various time points. Regression analysis is performed on the quantitated levels of mRNA to determine the half-life of uninduced versus induced transcripts. Results will determine whether or not the gene of interest is posttranscriptionally regulated by differential mRNA stability.

Although this severely limits the usefulness of transcriptional inhibitors, the stability measurement after induction may nevertheless provide insight into the possibility that mRNA stabilization contributes to induction. If the half-lives of the transcripts after induction are found to be extremely short, they are unlikely to have been significantly stabilized relative to the uninduced transcripts.

Basic transfection approaches. To circumvent the limitations of inhibitors, alternative strategies for measuring mRNA half-lives and relative stabilities can be used. These strategies make use of recombinant DNA and transfection techniques. One approach is to measure the stability of an mRNA expressed from a heterologous promoter that is equally active in the two cell populations being compared (e.g., in cells before and after induction). There are several different variations on this theme. The simplest is to insert the cDNA downstream of a strong promoter that is constitutively active, such as the strong viral promoters from cytomegalovirus (CMV) or Rous sarcoma virus (RSV) (Fig. 3.6). After transfecting the two cell populations with the plasmid, steady-state mRNA levels are determined by Northern blot, primer extension, and so forth. The rates of mRNA synthesis in the two populations are likely to be similar because the viral promoter is likely to be equally active. Thus, the relative mRNA levels will be dictated by the relative mRNA stabilities. In other words, if the steady-state mRNA abundances in the two cell populations are different, the likely cause is different mRNA degradation rates. To confirm that mRNA synthesis driven by the viral promoter is similar in the two cell populations, a control should be performed with the same promoter driving transcription of a standard reporter gene, such as luciferase, CAT, or β-globin.

Because the *cis*-acting determinants of mRNA stability are often found in 3′-untranslated regions, a variation of the above strategy is to fuse the 3′ region from the gene of interest to a standard reporter, such as luciferase or CAT. If mRNA stability is mediated through the 3′-untranslated region, this region may be sufficient for conferring regulation on the heterologous transcript.

Pulse chase. The above procedures may provide information about relative mRNA stabilities, but the half-lives obtained are often inaccurate. To obtain more accurate measurements of mRNA half-lives, pulse-chase experiments are preferred (see Ross 1995; Rajagopalan and Malter 1997). This technique, used for studies in prokaryotes and some eukaryotes, involves incubation of cells with a radiolabeled nucleoside for several minutes (the pulse). The radioactivity is then diluted with cold nucleotides (the chase), such that only the nascent RNA is labeled. The rate of degradation of the specific radiolabeled mRNA can then be monitored by hybridization to an unlabeled cDNA. Unfortunately, this procedure is rarely useful in mammalian cells or other higher eukaryotes because the high concentration of intracellular nucleotides prevents specific mRNAs from being labeled to a high specific activity and prevents the radioactivity from being rapidly diluted during the chase period. Because of these limitations, pulse-chase experiments can only be used to study very abundant transcripts with long half-lives.

Pulse-chase by transfection. Fortunately, several methods related to the pulse chase can provide valuable information about mRNA half-lives in mammalian cells (Ross 1995; Rajagopalan and Malter 1997). One relatively straightforward method makes use of promoters that can be induced rapidly and transiently. The c-*fos* promoter is useful for this procedure because it is transiently induced by serum in some cell types following incubation at low serum concentrations (Kabnick and Housman 1988; Shyu et al. 1989). The *Drosophila* hsp70 promoter has also been used for this purpose because it can be transiently induced at high temperatures (Helms and Rottman 1990). The promoter is fused to the cDNA of interest and the resulting plasmid is transfected into cultured cells. Promoter induction leads to synthesis of a population of nascent mRNA, whose steady-state abundance reaches a peak that is dependent on the kinetics of promoter induction and inactivation. From this peak, an approximate mRNA half-life can be determined by measuring mRNA abundance at various time points by Northern blot, for example. This strategy is imprecise because the pulse of mRNA synthesis is likely to cover a longer time period than an authentic pulse-chase experiment. Nevertheless, it is likely to provide more accurate half-lives than those obtained using transcriptional inhibitors.

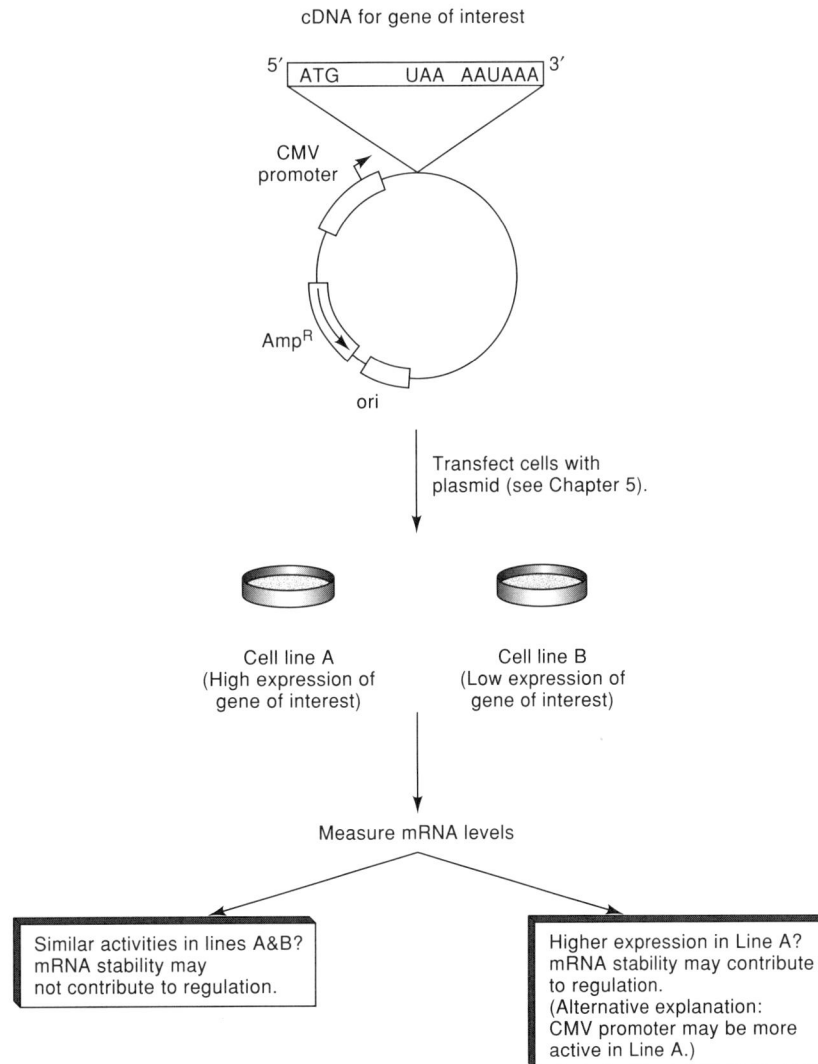

cDNA for gene of interest

CMV promoter

Amp^R

ori

Transfect cells with plasmid (see Chapter 5).

Cell line A
(High expression of gene of interest)

Cell line B
(Low expression of gene of interest)

Measure mRNA levels

Similar activities in lines A&B?
mRNA stability may
not contribute to regulation.

Higher expression in Line A?
mRNA stability may contribute
to regulation.
(Alternative explanation:
CMV promoter may be more
active in Line A.)

FIGURE 3.6. A simple transfection method for testing whether mRNA stability contributes to gene regulation.

Recommended Approach for Demonstrating Regulation of Transcription Initiation and mRNA Stability

On the basis of the above information, the following approach is recommended for exploring the possiblity that transcription initiation and/or mRNA stability are responsible for the expression of a gene of interest. If the properties of the gene suggest that transcription initiation is likely to play a major role in regulation (i.e., if other genes of the same type are regulated at the level of initiation), this possiblity should be addressed first using either the nuclear run-on assay or a functional assay (for a promoter or distant control region, as described in Chapters 5 and 6). If a functional assay clearly reveals that a promoter or distant control region recapitulates the regulation observed with the endogenous gene, the nuclear run-on assay is usually not necessary. Instead, one could proceed directly to a functional dissection of the control region (Chapter 7).

If, on the other hand, functional analyses of a promoter or distant control region fail to demonstrate regulation of transcription initiation (or if the results are ambiguous), nuclear run-ons should be performed to determine whether the gene is in fact regulated at that level. If the nuclear run-on results clearly demonstrate regulation of transcription initiation, more sophisticated functional assays may need to be pursued until a control region and assay are identified that recapitulate the regulation. If the nuclear run-on results suggest that transcription initiation is not of primary importance for regulation, experiments to examine the possibility of regulation at the level of mRNA stability can be performed. A simple starting point would be to perform inhibitor experiments and, if necessary, one or more of the other approaches described above. If those assays reveal regulation of mRNA stability, appropriate functional assays to study the mechanism of regulating stability can be developed.

Even if the results of nuclear run-on and functional assays suggest that a gene is regulated at the level of transcription initiation, it still may be worthwhile investigating the possibility that mRNA stability contributes to the regulation. An analysis of this possibility may reveal that mRNA stability makes a significant contribution, which may inspire a broader approach to the analysis of gene regulation than simply dissecting the transcriptional control regions. The c-*fos* gene provides an excellent example of this concept, as the gene was first found to be regulated at the level of transcription initiation (Greenberg and Ziff 1984); subsequently, the gene proved to be an excellent model for studying mRNA stability, even though the mRNA half-life is not modulated during gene induction.

If a gene is subject to regulation of both transcription initiation and mRNA stability, a decision will need to be made as to which mechanism to pursue. This judgment may depend on the estimated relative contributions of the two mechanisms. Another major factor will be the personal interests of the investigator. For those interested in defining fundamental mechanisms of gene regulation, an analysis of mRNA stability could be particularly attractive, because much less is known about these regulatory events, despite their obvious importance. Recent review articles can be consulted to determine how to proceed with such an analysis, which is beyond the scope of this manual (Ross 1995; Caponigro and Parker 1996; Jacobson and Peltz 1996; Rajagopalan and Malter 1997; Wilson and Brewer 1999).

Transcription Elongation

Basic Mechanism of Elongation

Prokaryotic and eukaryotic RNA polymerases catalyze the elongation of transcription in a highly processive manner following initiation and promoter escape (Uptain et al. 1997). Although transcription is usually regulated at the initiation event, a large number of genes are now known to be regulated during elongation (Greenblatt et al. 1993; Bentley 1995; Reines et al. 1996; Shilatifard et al. 1997; Uptain et al. 1997). Promoter escape is an equally attractive target for regulation, but in eukaryotes, this event cannot be measured unambiguously in vivo or separated from initiation and elongation.

Elongation involves stable ternary complexes containing the polymerase, DNA template, and nascent transcript (Uptain et al. 1997; Nudler 1999). Polymerization occurs within a DNA "bubble" that is formed by unwinding and reannealing of DNA as the polymerase moves along the gene. The current model to explain polymerase movement, based largely on studies of *E. coli* RNA polymerase, is referred to as the sliding clamp model (Nudler 1999). According to this model, the polymerase contacts nucleic acid at three sites. One contact, which involves strong, nonionic interactions, is with the double-stranded DNA that precedes the transcription bubble. The second contact, which is relatively weak, is with the RNA–DNA hybrid immediately adjacent to the site of ribonucleotide addition.

The third contact is with the nascent transcript upstream of the hybrid binding site. Each binding site contributes to the stability of the complex and restricts movement in the vicinity of the other binding sites. At the same time, the presence of three sites allows each to be intrinsically flexible, permitting movement following ribonucleotide addition. Thus, this structure provides the stability and flexibility that are essential for processive and efficient elongation.

The precise determinants of the rate of transcription elongation have not been established, but numerous studies in prokaryotes and eukaryotes have revealed that elongation is a complicated and dynamic process that can be influenced by the components of the ternary complex, by the conformation of the complex by nucleosomes and proteins bound to the template or the nascent transcript, or by the primary and secondary structures of the template and transcript. All of these determinants have the potential to increase or decrease the rate of elongation, most likely by enhancing or suppressing polymerase pausing, arrest, or termination. According to Uptain et al. (1997), pausing is defined as a reversible cessation of elongation for a defined period of time. Arrest is similar to pausing, but arrested polymerase molecules cannot continue unless supplementary factors are added. It often is difficult to distinguish between pausing, arrest, and termination of elongation, leading to the use of a more generic term, an elongation block.

Regulation of Transcription Elongation in Prokaryotes

The numerous determinants of elongation rate have suggested many different mechanisms of regulating gene expression at the level of elongation. Several mechanisms in prokaryotes have been dissected in considerable detail. One classic example occurs at the *Escherichia coli* trp operon, which is efficiently transcribed in the absence of tryptophan, but is transcribed less frequently as the concentration of tryptophan increases. Transcription of this operon is regulated in part at the level of transcription initiation. However, transcription is also regulated at the level of elongation, by a process called attenuation (Yanofsky 1988).

In the presence of tryptophan, elongating polymerase molecules frequently pause on the trp operon after synthesizing a 130-nucleotide leader transcript, rather than the entire 7000-nucleotide operon. The leader RNA consists of four intercomplementary regions and encodes a short peptide rich in tryptophans. Transcription termination is signaled by the formation of a hairpin between two of the complementary regions, regions 3 and 4. The 3:4 hairpin forms in the presence of tryptophan because the ribosome efficiently synthesizes the leader peptide. By following closely behind the RNA polymerase, the ribosome promotes formation of the 3:4 hairpin by sterically preventing region 3 from instead forming a hairpin with region 2.

When tryptophan is absent or present in low concentrations, translation is less efficient. The reduced rate of translation allows region 3 to pair with region 2, rather than 4, preventing the transcriptional arrest and thereby promoting transcription of the entire operon.

A second classic example of the regulation of elongation involves the bacteriophage λ Q protein, which is required for efficient elongation through the λ late genes (Greenblatt et al. 1993; Uptain et al. 1997). Regulation by Q exhibits several similarities to the eukaryotic examples described below and may therefore be particularly relevant to this discussion. In the absence of Q, the *E. coli* RNA polymerase pauses after transcribing only 16 or 17 nucleotides of the late gene. Even if the polymerase is released from this pause site, it is susceptible to pausing at several other locations within the gene. The pause appears to result from a sequence-specific interaction between the polymerase-associated sigma factor and the nontemplate strand of DNA in the transcription bubble. The Q protein relieves the pause and facilitates elongation by physically associating with the polymerase and presumably altering its conformation. The Q–polymerase complex is resistant, not only to the initial pause site, but also to the other pause sites within the gene. Interestingly, Q binds the polymerase only after the sigma factor promotes pausing at the promoter proximal site. Thus, the activity of the initiation factor (i.e., the sigma factor) is directly linked to the elongation competence of the polymerase.

Regulation of Transcription Elongation in Eukaryotes

Much of our knowledge of the regulation of elongation in eukaryotes has emerged from studies of the heat-shock, c-*myc,* and HIV genes (Greenblatt et al. 1993; Lis and Wu; Bentley 1995; Jones 1997; Uptain et al. 1997). Two characteristics of these genes are strikingly similar and therefore may be generalized to many eukaryotic genes controlled at the level of elongation. First, they all are subject to rapid induction of expression. This common feature suggests that many genes of this type are regulated in part at the level of elongation. Second, they all appear to promote pausing of RNA polymerase II within 20–60 bp of the transcription start site, similar to the location of the pause site controlled by the λ Q protein. The locations of the eukaryotic pause sites suggest that regulation may involve a mechanism similar to the Q mechanism. Furthermore, this location makes it feasible for elongation to be influenced by regulatory proteins bound to the nearby promoter.

It is not yet clear whether Q protein homologs exist in eukaryotic cells, but several studies have revealed a link between the relief of pausing and control elements within the promoter. The *Drosophila* hsp70 gene provides a prime example of this link (Lis and Wu 1993; Uptain et al. 1997). Analyses of promoter mutants revealed that, prior to heat shock and the induction of hsp70 expression, multiple binding sites for the GAGA transcription fac-

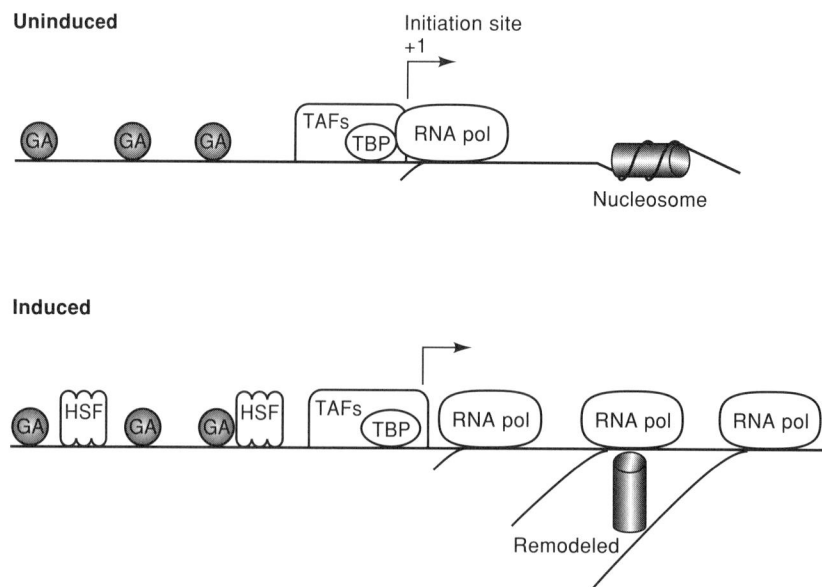

FIGURE 3.7. The in vivo architecture of the hsp70 promoter before and after heat shock. (Adapted, with permission, from Lis and Wu 1993 [Copyright 1993 Cell Press].)

tor stimulate initiation from the core promoter by RNA polymerase II (Fig. 3.7). Full-length transcripts are not produced, but rather the polymerase pauses after synthesizing a transcript of only 20–40 nucleotides. Nuclear run-on experiments were used to demonstrate that transcription of the 5′ end of the gene occurs in the absence of heat shock. In vivo UV crosslinking experiments demonstrated that the polymerase is paused near the 5′ end of the gene. Potassium permanganate experiments provided further evidence that the polymerase is paused. This chemical selectively modifies thymine residues in single-stranded DNA, promoting subsequent cleavage by piperidine. Strong cleavage was observed at the 5′ end of the hsp70 transcription unit in *Drosophila* cells prior to heat shock, presumably due to the presence of a transcription bubble (Lis and Wu 1993).

Upon heat shock, the heat shock factor (HSF) is induced to bind the promoter and to facilitate the release of the paused polymerase, allowing the polymerase molecule to complete the synthesis of the full-length primary transcript (Lis and Wu 1993). HSF also stimulates further rounds of initiation, at least in part by stimulating the assembly of a functional preinitiation complex. Thus, HSF plays an important role in stimulating both initiation and elongation, resulting in a rapid and potent induction of hsp70 expression.

The HIV-1 and c-*myc* genes appear to be regulated in a similar fashion, but through the activities of different transcription factors (Greenblatt et al. 1993; Bentley 1995; Jones 1997; Uptain et al. 1997). During HIV-1 transcription, most polymerase molecules that initiate transcription pause after synthesizing transcripts of 55–60 nucleotides. In fact, transcription often terminates, with the short transcripts released. When the infected cells are activated, the viral Tat protein leads to an increase in the synthesis of full-length transcripts, and in the overall rate of initiation from the LTR. The Tat protein functions by binding the TAR region at the 5′ end of the nascent transcript in association with a set of host proteins that includes the TAK/P-TEFb kinase complex (Zhou and Sharp 1996; Zhu et al. 1997; Wei et al. 1998). This complex appears to increase both the rate of transcription and the rate of elongation by phosphorylating the carboxy-terminal domain of RNA polymerase II. The general transcription factor TFIIH may also be important for Tat function (Parada and Roeder 1996; Cujec et al. 1997; Garcia-Martinez et al. 1997).

The precise mechanisms by which HSF and Tat enhance the efficiency of transcription elongation are not known, but one model is that they convert RNA polymerase II from a relatively nonprocessive form to a processive form. This conversion may depend on a ligand-induced conformational change, a posttranslational modification, and/or the stable association of one or more proteins with the polymerase. Particularly noteworthy are recent studies suggesting that the modulation of elongation efficiency is not restricted to a small set of activators like Tat and HSF, but that a wide range of transcriptional activators are capable of enhancing elongation (Blau et al. 1996).

General elongation stimulatory factors may also contribute to the regulation of elongation. Several factors of this type have been described, including SII, P-TEFb, TFIIF, elongin, ELL, and TFIIH (Reines et al. 1996; Shilatifard et al. 1997; Uptain et al. 1997). The precise mechanisms by which these proteins stimulate elongation remain to be elucidated, but biochemical studies have revealed that they can (1) help an arrested polymerase resume elongation by stimulating the 3–5′ nuclease inherent to Pol II (SII), (2) suppress polymerase pausing (ELL, elongin), and (3) modify the polymerase by phosphorylation (TFIIH, P-TEFb). Although some regulatory factors may enhance the intrinsic rate of elongation, most appear to influence the polymerase's ability to bypass impediments. Further studies will be needed to determine whether any of the general elongation factors contribute to gene-specific regulation of elongation.

Strategies for Distinguishing between Regulation of Elongation and Regulation of Initiation

In some respects, it is more difficult to distinguish between regulation of initiation and elongation than between initiation and RNA stability. At the same time, it may be less important to distinguish between initiation and elongation at the early stages of a gene regulation analysis. Both of these assertions derive from the fact that regulation of both initiation and elongation may depend on the function of the promoter. If the promoter is responsible for regulating both processes, it becomes difficult to distinguish between the two. Nevertheless, it may not be necessary to know which process is regulated until the critical regulatory elements within the promoter have been identified and the proteins that interact with those elements isolated.

The technique most frequently used for distinguishing between regulation of initiation and elongation is the nuclear run-on (see, e.g., Blau et al. 1996). The nuclear run-on was described above (see Box 3.1, p. 72), but the procedure must be modified to achieve the current goal (Fig. 3.8). To gain evidence that regulation is mediated at the level of tran-

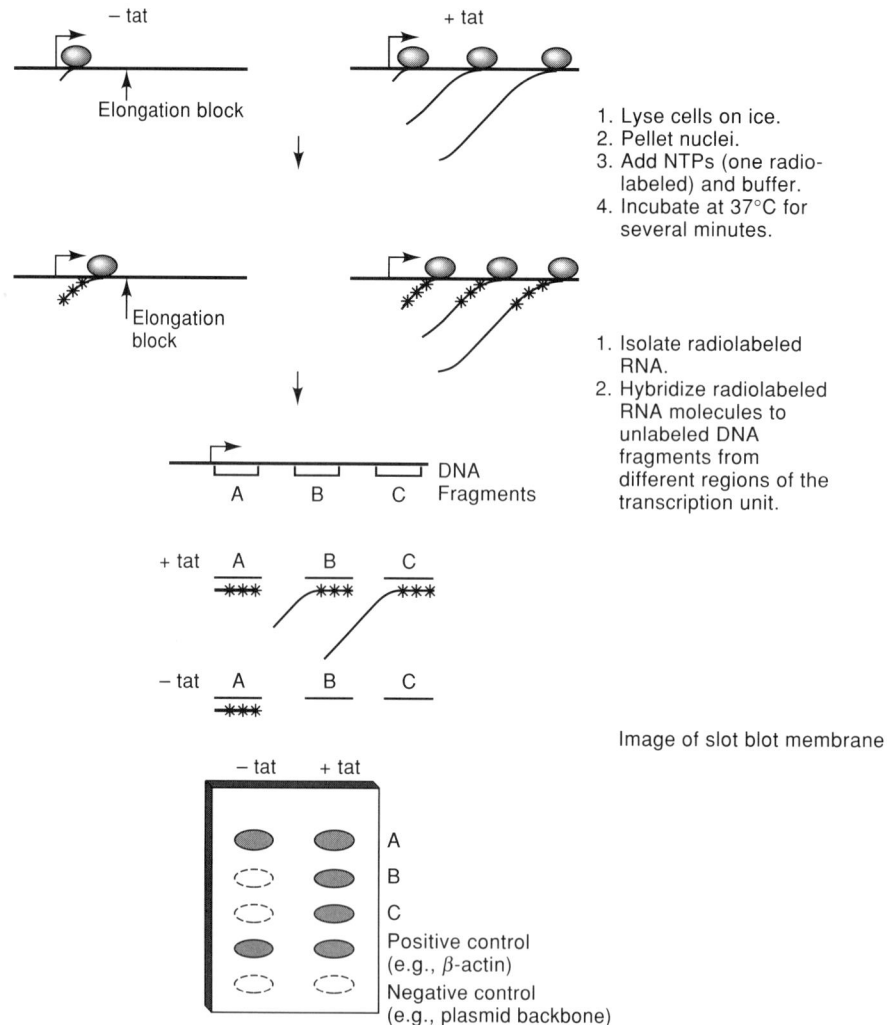

FIGURE 3.8. Nuclear run-on assay for assessing regulation of elongation using HIV as an example.

scription initiation, radiolabeled nascent RNA is hybridized to unlabeled DNA from the gene of interest. If cells prior to induction of gene expression are compared to cells after induction, a nuclear run-on signal is expected only in the induced sample if expression is regulated at the level of transcription initiation (see Box 3.1). Equivalent signals are expected in both samples if expression is regulated primarily at the level of mRNA stability.

If elongation contributes to regulation, the nuclear run-on signal obtained with uninduced cells following hybridization to unlabeled DNA from the 5′ end of the gene is likely to be much stronger than the signal obtained following hybridization to unlabeled DNA from the 3′ end. To understand this concept, consider the HIV-1 gene (Kao et al. 1987; Jones 1997). In the absence of Tat, polymerase molecules can be found associated with the 5′ but not the 3′ end of the gene. When nuclear run-on experiments are performed, the polymerase molecules elongate only through the first 55–60 nucleotides of the gene and then pause (Kao et al. 1987). Thus, the radiolabeled transcripts extend only through this short region. In contrast, when nuclear run-on experiments are performed in the presence of Tat, polymerase molecules are distributed throughout the gene and radiolabeled nucleotides are incorporated uniformly. A comparison of the results obtained with the two different unlabeled probes (Fig. 3.8) demonstrates that expression is regulated at the level of elongation rather than initiation.

This strategy can be generally applied for distinguishing between regulation at the level of initiation and elongation. Furthermore, it is important to consider this strategy when using the nuclear run-on assay to distinguish between regulation of transcription initiation and mRNA stability, as described earlier in this chapter. To achieve the highest level of confidence that initiation is being measured, the unlabeled DNA fragment should be from the 5′ end of the gene. If the only DNA fragment used is from the 3′ end, a positive result could be indicative of regulation at the level of either initiation or elongation.

A final point is that the strength of a run-on signal obtained following hybridization to one DNA fragment cannot be compared directly to that obtained following hybridization to another DNA fragment. For example, the signal obtained with a DNA fragment from the 5′ end of a gene cannot be directly compared to the signal obtained with a fragment from the 3′ end. Direct comparisons are meaningless because the intrinsic hybridization efficiencies of the two fragments may be different (see Blau et al. 1996). The signals obtained with the 5′ fragment must initially be compared in the two radiolabeled RNA samples (e.g., the samples derived from cells before and after induction or minus and plus tat). Similarly, the signals obtained with the 3′ probe must be compared in the two RNA samples. If the 5′ signal is enhanced by 2-fold upon induction and the 3′ probe is enhanced by 20-fold, the results would suggest that the gene is regulated primarily at the level of elongation, with a 2-fold contribution from regulation at the level of initiation. On the other hand, if both the 5′ and 3′ signals are enhanced by 5-fold upon induction, the results would suggest that regulation is primarily at the level of initiation. To gain additional confidence in the results, the relative hybridization efficiencies of the 5′ and 3′ probes can be normalized by performing a control experiment using radiolabeled RNA synthesized in vitro by T7, T3, or Sp6 RNA polymerases.

Recommended Approach for Demonstrating Regulation of Transcription Initiation or Elongation

The recommended strategy for assessing regulation of transcription initiation and elongation is as follows. If a functional assay can be developed for a promoter or distant control region that recapitulates the regulation observed with the endogenous gene (Chapters 5 and 6), one can proceed directly to an analysis of the DNA sequence elements (Chapter 7) and transcription factors (Chapters 8 and 9) that are required for the function of the control region. The possibility that elongation contributes to regulation can usually be pursued when the analysis reaches an advanced stage. For example, when dissecting the functional domains of an important transcription factor, it would be interesting to determine whether the factor influences initiation and/or elongation.

If, on the other hand, the initial functional studies fail to identify a control region that recapitulates regulation, a nuclear run-on assay should be developed. One reason for developing the nuclear run-on is to distinguish between regulation of transcription and mRNA stability as described earlier in this chapter. Because the nuclear run-on is also useful for distinguishing between the regulation of initiation and elongation, it would be worthwhile developing the assay with this possibility in mind. In other words, unlabeled DNA probes from both the 5′ and 3′ ends of the gene should be used. By designing the nuclear run-on in this way, the maximum amount of insight into the regulatory strategies employed by the gene can be obtained. This insight can help determine whether future efforts should be directed (1) toward more sophisticated functional assays for identifying control regions that regulate transcription initiation, (2) toward the possibility that mRNA stability is regulated, or (3) toward the possibility that elongation is regulated through a mechanism that does not rely on DNA sequence elements within the promoter or distant control regions.

Extending an Analysis of Elongation Regulation

If the nuclear run-on experiments suggest that a gene is regulated primarily at the level of elongation, a variety of different strategies can be followed to dissect the mechanism further. An initial goal may be to determine whether regulation is mediated by control elements within the promoter or by sequences within the transcription unit itself.

To determine whether the promoter mediates the regulation of elongation, the promoter region could be fused to a standard reporter gene and analyzed by transient or stable transfection. Regulation of reporter activity would implicate promoter sequences in the regulation. A nuclear run-on assay can be used to confirm that the promoter–reporter fusion is indeed regulated at the level of elongation. This experiment is similar to that described above for an endogenous gene: Radiolabeled RNA from cells transfected with the reporter plasmid is hybridized to unlabeled DNA from both the 5′ and 3′ ends of the reporter gene. In cells that do not express the reporter, one would expect to observe hybridization primarily to the 5′ fragment if regulation is at the level of elongation. In cells that express the reporter, efficient hybridization to both the 5′ and 3′ fragments would be expected. One common reporter gene that has been used for this purpose is the luciferase gene (Blau et al. 1996). Other laboratories have preferred to use a reporter gene that contains a strong, previously described pause site, such as that found in the c-*myc* gene (Krumm et al. 1993).

To determine whether a sequence within the transcription unit mediates the regulation of elongation (e.g., by directly blocking elongation), the transcribed region of the gene could be fused to a constitutively active promoter. Synthetic promoters driven by ectopically expressed activators would be suitable for this purpose. If the transcription unit mediates regulation, one would expect higher steady-state mRNA levels in cells that express the gene of interest than in cells that lack expression. As a control, the same synthetic promoter should direct equivalent expression of a standard reporter gene in both cell populations. Once again, a nuclear run-on assay should be performed to confirm that transcription from the transfected plasmid is indeed regulated at the level of elongation.

If the promoter is primarily responsible for stimulating elongation, the next step would likely be to dissect the promoter region using a standard mutant analysis (see Chapter 7). After the control elements are localized and the functional DNA-binding proteins identified, more sophisticated studies could be performed to pursue the precise biochemical mechanism by which these proteins influence elongation. The specific strategies to follow would be similar to those followed during the analysis of hsp regulation, in which specific promoter sequences influence elongation (Lis and Wu 1993).

If the transcription unit itself is essential for the regulation of elongation, the DNA sequences that impart regulation can be identified. Nuclear run-on experiments using a series of probes hybridizing to different regions of the gene can be used to localize the site of arrest. These experiments must be approached with caution, however, because they may be misleading (see Krumm et al. 1993). Early studies of the c-*myc* gene, for example, revealed a strong pause site within the first intron. This pause site was thought to be the key to understanding the regulation of c-*myc* elongation, until recent studies suggested that regulation is primarily mediated through the promoter. Pausing within the first intron appears to occur when an elongation-incompetent polymerase transcribes the gene.

Differential Pre-mRNA Splicing, mRNA Transport, and Polyadenylation

Basic Principles

Differential pre-mRNA splicing (Wang and Manley 1997) and polyadenylation (Colgan and Manley 1997) most likely evolved as regulatory mechanisms for generating related but functionally distinct protein products from the same gene and precursor transcript. These mechanisms, in addition to nucleocytoplasmic transport of mRNAs (Gorlich and Mattaj 1996; Mattaj and Englmeir 1998), contribute to the regulation of many genes. They are discussed less extensively than mRNA stability and transcription elongation, however, because they are much easier to distinguish from regulation of transcription initiation. As described below, Northern blots are usually sufficient to suggest that splicing, transport, or polyadenylation contributes to gene expression.

Of these three modes of regulation, differential pre-mRNA splicing appears to be by far the most common. In fact, in metazoans, the number of genes regulated by splicing may be comparable to the number regulated by transcription initiation. Despite their obvious importance, alternative splicing mechanisms remain poorly understood, although a few systems have recently been dissected in moderate detail (Wang and Manley 1997).

Far fewer examples of differential transport and polyadenylation have been reported, but key paradigms have been studied in sufficient detail to establish the validity and importance of these processes. A well-established example of differential polyadenylation occurs at the immunoglobulin (Ig) heavy-chain locus (Colgan and Manley 1997). The Ig heavy-chain protein, IgM, is initially produced in a form, μm, that binds the plasma membrane. However, when mature B cells are induced to differentiate into plasma cells, an alternative carboxy-terminal domain results in the production of a secreted form of the protein, μs. The μs carboxyl terminus is encoded by an exon that resides upstream of the exons encoding the transmembrane domain of μm. In mature B cells, the exon encoding the carboxyl terminus of μs is excised by pre-mRNA splicing, with polyadenylation occurring downstream of the μs exons. In plasma cells, a second polyadenylation signal between the μs and μm exons is activated, leading to the generation of polyadenylated mRNAs containing the μs exon and lacking the μm exons. Recent studies suggest that the μs polyadenylation signal is relatively weak and is activated only in plasma cells, where a general polyadenylation factor, CstF-64, is present at a relatively high concentration (Colgan and Manley 1997).

An established example of regulation of mRNA transport occurs with HIV-1 transcripts (Cullen 1992). At an early stage of the viral life cycle, primary HIV-1 transcripts are efficiently spliced, so that the desired proteins can be encoded by the mature, processed mRNAs. Later in the viral life cycle, however, proteins are needed that can only be encoded by the unspliced transcript. A virus-encoded protein, Rev, facilitates the transport of the unspliced transcripts to the cytoplasm. Rev accomplishes this goal by binding a specific control element, the Rev-responsive element, near one of the splice acceptor sites of the HIV-1 pre-mRNA. The protein–RNA interaction interferes with splicing and promotes cytoplasmic transport of the precursor transcript. Recent studies suggest that Rev contains a leucine-rich nuclear export signal that redirects the Rev–RNA complexes away from the typical mRNA processing pathway and toward a nuclear export pathway that involves an export receptor called CRM1 (Fornerod et al. 1997). It remains to be determined whether any RNAs for cellular protein-coding genes are regulated by a similar mechanism.

Identifying Regulation of Pre-mRNA Splicing, Transport, and Polyadenylation

Among the possible modes of regulating the steady-state abundance of a specific mRNA, the contributions from alternative splicing, transport, and polyadenylation are generally the easiest to establish. If alternative splicing or polyadenylation regulates an mRNA's abundance, another mRNA will usually be generated that is partially homologous. If differential transport regulates an mRNA's abundance, its relative abundance in the nucleus and cytoplasm will vary.

A simple hypothetical scenario to consider is one in which a small fragment of a cDNA has been used to probe a Northern blot, revealing a single mRNA product that is present in one cell type, but not in another. At first glance, this result appears to suggest that the gene is differentially transcribed in the two cell lines and, indeed, differential transcription is a likely explanation for the result. Nevertheless, this same result is consistent with the possibility that the gene is subject to alternative pre-mRNA splicing or alternative polyadenylation, with the portion of the transcript complementary to the probe included in the mature mRNA only in one cell line. In other words, this portion of the transcript could be derived from an exon that is deleted during splicing in one cell type, but not in another. An obvious method to rule out this possibility is a Northern blot with probes derived from other portions of the cDNA or with the full-length cDNA. If the abundance of the original mRNA product is regulated by alternative splicing, other mRNA products should be detected with the new probes. If not, one can conclude with considerable confidence that mRNA abundance is not regulated solely by alternative splicing.

Relatively straightforward approaches can also be employed to determine whether a gene is regulated at the level of alternative polyadenylation or differential mRNA transport, although alternative polyadenylation is often difficult to distinguish from alternative splicing. Alternative polyadenylation is addressed using a similar strategy to that described above, whereas differential mRNA transport is studied by comparing RNA abundance in cytoplasmic and nuclear RNA fractions.

TECHNIQUES

PROTOCOL 3.1

Nuclear Run-on Assay

SUMMARY

The nuclear run-on assay was developed as a method for establishing that the transcription initiation rate contributes to the regulated expression of mammalian genes (Derman et al. 1981; Greenberg and Ziff 1984; Powell et al. 1984; Ucker and Yamamoto 1984). The difference between monitoring gene expression by the nuclear run-on assay versus other assays (e.g., Northern blot, primer extension, S1 nuclease, or RNase protection) is that the nuclear run-on assay provides a measure of the frequency of transcription initiation and is largely independent of the effects of RNA stability (see chapter text and Derman et al. 1981). In contrast, the other assays measure steady-state RNA abundance. The nuclear run-on assay can also be used to determine whether polymerase pausing or attenuation contributes to gene regulation.

Briefly, the nuclear run-on assay begins with samples of cells that contain different steady-state amounts of the mRNA or protein of interest (see Box 3.1, p. 72). The cells are chilled and the plasmid membranes are permeabilized or lysed. These steps result in polymerase pausing. The nuclei are then incubated at 37°C for a short time in the presence of NTPs and radiolabeled UTP. New transcripts are not initiated during this incubation, but the radiolabeled nucleotide becomes incorporated into transcripts that were being synthesized when the cells were first chilled and lysed. The number of nascent transcripts on the gene at the time of chilling is thought to be proportional to the frequency of transcription initiation. To determine the relative number of nascent transcripts in each sample, the radiolabeled RNA is purified and hybridized to a membrane containing immobilized DNA from the gene of interest. The amount of radioactivity that hybridizes to the membrane is approximately proportional to the number of nascent transcripts.

Multiple variations of the nuclear run-on protocol have been developed. The following protocol is adapted from the protocols of Greenberg and Ziff (1984) and Linial et al. (1985). A similar protocol with additional considerations can be found in Ausubel et al. (1994, Unit 4.10).

TIME LINE AND ORGANIZATION

Before beginning the nuclear run-on procedure, the unlabeled DNA, which will be immobilized on the membrane for hybridization to the radiolabeled RNA, must be prepared. The following should be considered when preparing DNA:

1. For the basic nuclear run-on procedure, either double-stranded or single-stranded DNA from the gene of interest can be used. Longer genomic DNA fragments will yield stronger signals than shorter fragments, but shorter fragments may be desirable to monitor the number of nascent transcripts associated with a particular region of the locus.

2. To increase the chances that the assay will measure the frequency of transcription initiation independent of effects of polymerase pausing or attenuation, a fragment from the 5′ end of the transcription unit is preferred.

3. Single-stranded DNA fragments complementary to the nascent transcript are preferred over double-stranded fragments because the background signal due to antisense transcription (which is surprisingly common in mammalian cells) will be reduced. Single-stranded DNA molecules can be generated by inserting the gene fragment of interest into M13 or phagemid vectors (Sambrook et al. 1989, Chapter 4; Ausubel et al. 1994, Units 1.14 and 1.15).

4. In addition to the DNA corresponding to the gene of interest, positive and negative control DNAs are also strongly recommended. A negative control is usually a plasmid vector lacking a gene insert. An appropriate positive control would be a DNA fragment from a constitutively expressed gene, such as β-actin.

With the purified single-stranded or double-stranded DNA in hand, the nuclear run-on procedure can be performed in 2 days. In most instances, the unlabeled DNA is first immobilized on the membrane using a dot-blot or slot-blot apparatus, which can be purchased from a variety of companies, including Schleicher & Schuell (cat. # 447-850) and Bio-Rad (cat. # 170-3938). Nuclei are then prepared from the cells of interest by lysis of the plasma membranes. The nuclei are incubated in the presence of the radiolabeled NTP, and the radiolabeled transcripts are then purified and hybridized to the membrane strips containing the unlabeled DNA. Following an overnight hybridization, the membrane is often treated with RNases to digest unannealed RNA, washed, and analyzed by autoradiography or phosphorimager analysis. Long exposures of 2–3 weeks (by autoradiography) may be required for analysis of weakly expressed genes.

Days 1 and 2: Nuclear run-on assay
Day 3: Data analysis

OUTLINE

Nuclear run-on assay

Step 1: Prepare buffers (1 hour)
Step 2: Immobilize unlabeled DNA on membrane (3 hours)
Step 3: Harvest cells and prepare nuclei (1 hour)
Step 4: Radiolabel nascent transcripts and begin hybridization (4 hours and overnight)
Step 5: Wash membrane (4 hours)

PROCEDURE

CAUTIONS: *Ammonium acetate, CaCl$_2$, Chloroform, Glycerol, HCl, KCl, MgCl$_2$, NaOH, Phenol, SDS. See Appendix I.*

Step 1: Prepare buffers

NP-40 lysis buffer:
10 mM Tris-HCl (pH 7.4)
10 mM NaCl
3 mM MgCl$_2$
0.5% NP-40

Nuclear freezing buffer:
> 50 mM Tris-HCl (pH 8.3)
> 40% v/v glycerol
> 5 mM MgCl$_2$
> 0.1 mM EDTA

5X Run-on buffer:
> 25 mM Tris-HCl (pH 8.0)
> 12.5 mM MgCl$_2$
> 750 mM KCl
> 1.25 mM ATP
> 1.25 mM GTP
> 1.25 mM CTP

Proteinase K buffer:
> 10% SDS
> 50 mM EDTA
> 10 mM Tris-HCl (pH 7.4)
> 3 mg/ml Proteinase K (add fresh)

TE (pH 8):
> 10 mM Tris-HCl (pH 8)
> 1 mM EDTA

MgCl$_2$/CaCl$_2$ buffer:
> 10 mM MgCl$_2$
> 5 mM CaCl$_2$

CaCl$_2$ solution:
> 10 mM CaCl$_2$

NaOH solutions:
> 1 N NaOH
> 3 N NaOH

Ammonium acetate solution:
> 5 M ammonium acetate

HEPES:
> 1 M HEPES (free acid)

6x and 2x SSC:
> 20x stock:
> 3 M NaCl
> 0.3 M sodium citrate x 2H$_2$O
>> Adjust pH to 7 with 1 M HCl.

TES/NaCl solution:
> 10 mM TES (pH 7.4)
> 10 mM EDTA
> 0.2% SDS
> 0.6 M NaCl

Step 2: Immobilize unlabeled DNA on membrane

1. Use approximately 10 µg of double-stranded plasmid DNA or single-stranded DNA for each slot or dot of the slot-blot or dot-blot apparatus. (Use a similar amount of DNA for each positive and negative control sample.) If using double-stranded DNA, first cleave with a restriction endonuclease to linearize. Extract with phenol:chloroform and precipitate with ethanol in the presence of 0.3 M sodium acetate.

2. Dilute the DNA sample for each slot or dot into 400 µl of TE buffer.

3. Add 40 µl of 3 M NaOH to denature and incubate for 30 minutes at room temperature.

4. Add 300 µl of 2.5 M ammonium acetate (pH 7) to neutralize.

5. Apply the sample to a 0.45-µm nitrocellulose membrane on the slot-blot or dot-blot apparatus. Apply vacuum according to manufacturer's instructions. Rinse each slot with 6x SSC (1 ml).

   Note: Hybridization membranes other than nitrocellulose may yield successful results, but be aware that some membranes that are useful for Northern or Southern blots may yield very high background when used for nuclear run-ons.

6. Air-dry and then bake the membrane for 2 hours at 80°C in a vacuum oven.

7. Wash the filter with 2x SSC.

Step 3: Prepare nuclei

Begin with 10^7–10^8 cells for each sample:

1. For cultured cells, pellet the cells at 1500 rpm for 10 minutes in a clinical centrifuge, using disposable conical tubes.

2. Discard the supernatant. Wash the cell pellet twice with cold (4°C) PBS by suspending the cells and pelleting by centrifugation at 1500 rpm for 10 minutes. The chilling of the cells will lead to polymerase pausing on actively transcribed genes.

3. Resuspend the cell pellet in 5 ml of cold NP-40 lysis buffer. Incubate for 5 minutes on ice to lyse the plasma membranes. (A few microliters of the sample can be analyzed with a phase-contrast microscope to confirm that the majority of the cells have lysed.)

4. Pellet the nuclei in a clinical centrifuge at 1000 rpm for 10 minutes at 4°C. Discard the supernatant.

5. Resuspend the nuclear pellet in 500 µl of nuclear freezing buffer and store at –70°C until needed. Nuclei stored at –70°C are stable for a few years.

Step 4: Radiolabel nascent transcripts and begin hybridization

1. Thaw the frozen nuclei on ice and mix the contents of the tube by tapping.

2. Add 225 µl of the mixture containing nuclei to 60 µl of 5× run-on buffer.

3. Add 15 µl (150 µCi) of 3000 Ci/mmole (10 mCi/ml) [α-^{32}P]UTP to each tube and incubate at 37°C for 15 minutes. In this step the transcripts are elongated and radioactively labeled. Initiation does not generally occur.

   Note: Ideally, multiple time points (e.g., 0, 10, 20, and 30 minutes at 37°C) should be performed to confirm that the incorporation of radioactivity is approximately linear during the time range used for the experiment. However, given the large amount of radioactivity needed for each sample, this is not always desirable and is probably only necessary for genes whose transcription rates are thought to vary by only a few fold or less among the cell samples being analyzed.

4. Add 20 µl of 10 mM $CaCl_2$ and 10 µl of 1 mg/ml RNase-free DNase. Incubate at 30°C for 5 minutes to digest the genomic DNA and reduce its viscosity.

5. Add 35 µl of proteinase K buffer.

6. Incubate at 37°C for 45 minutes to degrade the cellular proteins.

7. Extract twice with 400 µl of phenol:chloroform and once with 400 µl of chloroform. (Be very careful not to transfer any of the precipitate at the interphase along with the aqueous phase.) Remove any precipitate in the aqueous solution following extraction by centrifugation.

8. Dilute the aqueous phase 1:1 with 5 M ammonium acetate (~ 350 µl). Add an equal volume of isopropanol to the sample and precipitate at –20°C for 30 minutes.

9. Centrifuge for 10 minutes at 14,000 rpm in a microfuge. Discard the supernatant, which should contain most of the unincorporated radionucleotides, in the radioactive waste.

10. Suspend the pellet in 100 µl of TE (pH 8.0).

    Note: At this point, the solution contains the radiolabeled nascent transcripts. Most of the unincorporated radionucleotides should have been removed by the ethanol precipitation in the presence of ammonium acetate. The radiolabeled RNA is therefore theoretically ready for hybridization to the nitrocellulose filters. For abundant transcripts, it may be possible to proceed directly to the hybridization step. However, in most instances, further purification of the labeled RNA may be beneficial to reduce background radioactivity on the filters that appears to be trapped by cellular DNA and proteins that were not removed by the preceding steps. The following steps result in further purification of the RNA and a reduction in background. Treatment of the resulting RNA with NaOH cleaves the transcripts into smaller fragments, which also reduce background. An alternative procedure for purifying the RNA, which makes use of more dilute DNase and proteinase K digestions, followed by TCA precipitation of the RNA, can be found in Ausubel et al. (1994, Unit 4.10).

11. Add 100 µl of a solution containing 10 mM $MgCl_2$ and 5 mM $CaCl_2$.

12. Add 10 µl of 1 mg/ml RNase-free DNase. Incubate 5 minutes at 37°C.

13. Chill on ice for 5 minutes.

14. Add 50 µl of 1 N NaOH to degrade transcripts into smaller fragments. Incubate 2 minutes on ice.

15. Immediately add 77 µl of 1 M HEPES (free acid) to neutralize. Also add 340 µl of 5 M ammonium acetate and 700 µl of isopropanol.

16. Chill at –20°C for 1 hour. Pellet the RNA by centrifugation in a microfuge for 10 minutes at 14,000 rpm. Remove the supernatant.

17. Dry the pellet on the bench top at room temperature, and then resuspend the pellet in 100 μl of TE.

18. Spot 1 μl of solution onto filter paper and count in a scintillation counter (Cerenkov counting). A typical incorporation of 3×10^5 cpm/μl should be obtained.

19. Wash the nitrocellulose filter strips in 2x SSC.

20. Hybridize an aliquot of radiolabeled RNA to the appropriate filter strips in plastic bags or a scintillation vial. Filter strips used for each hybridization should contain a slot or dot with the cDNA of interest and with appropriate positive and negative control DNAs. For each hybridization, try approximately 6×10^6 cpm and 1 ml of TES/NaCl hybridization solution. Hybridize for at least 24 hours at 65°C.

    Note: The hybridization buffer and conditions suggested here were derived from Ausubel et al. (1994). However, hybridization buffers and conditions may need to be optimized. In general, any hybridization conditions that are useful for Northern blots, including formamide buffers and inclusion of dextran sulfate to enhance hybridization, may be useful for nuclear run-on hybridizations.

Step 5: Wash membrane

1. Wash the nitrocellulose in 2x SSC at 65°C for 1 hour. Repeat. Monitor membranes with a Geiger counter. No radioactivity should be detected hybridizing to negative control DNAs.

    Note: Additional washes may be necessary to reduce the background, possibly with 2x SSC containing 0.1% SDS. Background can also be reduced by including RNase A (~ 10 μg) in one of the 2x SSC washes (30 minutes at 37°C), followed by an additional wash in the absence of RNase (to remove the RNase). The RNase will digest the regions of the radiolabeled transcripts that are not annealed to the unlabeled cDNAs on the filters. In addition to reducing the background, the RNase treatment may be important if the experiment is being performed to quantify the nascent transcripts associated with a specific region of the gene of interest, because only the portion of the transcript that is directly annealed to the DNA on the membrane will be resistant to RNase digestion.

2. Wrap the membranes in plastic wrap and expose to film or phosphorimager screen. If the gel image reveals that the background is too high, it may be possible to reduce background further with additional washes of the membranes.

ADDITIONAL CONSIDERATIONS

1. The nuclear run-on assay is reasonably straightforward for analysis of transcripts from highly active genes. However, even with highly active genes, a large amount of radioactivity is required and numerous manipulations following the addition of the radioactivity are involved. For less active genes, the assay can be difficult, with the most difficult step being generation of a convincing signal-to-background ratio: Weakly active genes result in weak or undetectable signals, and the assay is unusually susceptible to high background. Because of these considerations, several attempts may be required before the assay succeeds.

2. The NP-40 lysis method for preparation of nuclei described in Step 3 (p. 90) has been found to yield successful results with several different types of cultured cells. However,

alternative methods of preparing nuclei have also been described and may yield much better results (i.e., greater incorporation of the radiolabeled nucleotides into nascent transcripts). Methods for isolating nuclei for nuclear run-on by Dounce homogenization and by sucrose gradient centrifugation are described in Ausubel et al. (1994, Unit 4.10). A method for permeabilization of cell membranes using digitonin is described in Ucker and Yamamoto (1984).

3. When immobilizing unlabeled double-stranded DNA, instead of unlabeled single-stranded DNA, to the nitrocellulose, it is important to be aware that the resulting radioactive signals may be due to antisense transcription. For example, consider the analysis of an inducible gene using an immobilized, double-stranded gene fragment. If a signal is detected only with the induced sample, one can conclude with considerable confidence that the gene is induced (at least in part) at the level of transcription initiation. However, if signals with similar intensities are obtained with both the uninduced and induced samples, the results must be interpreted with caution. Perhaps, as suggested by the results, the gene is not regulated at the level of transcription initiation. Alternatively, antisense transcription through the locus occurs in both uninduced and induced cells, obscuring the detection of sense transcription, which may be induced. To obtain more conclusive results, single-stranded DNA should be immobilized on the nitrocellulose membrane.

4. Unlabeled RNA generated by bacteriophage T7, T3, or SP6 RNA polymerases can be immobilized to the nitrocellulose filters instead of unlabeled DNA, but the susceptibility to background problems is greatly increased. This strategy is therefore not recommended.

TROUBLESHOOTING

Poor incorporation of the radiolabeled nucleotide into nascent transcripts

Possible cause: Inadequate method for preparation of nuclei.
Solutions: Confirm lysis of cell membranes by microscopy following lysis. Try other methods for preparing nuclei or for permeabilizing cells (see above, Additional Considerations, number 2).

Possible cause: RNA may have been lost during the purification procedure.
Solutions: Repeat. Test incorporation of label into TCA precipitable material following initial DNase I and proteinase K digestions.

High background radioactivity retained on washed filters

Possible cause: Stringency of hybridization or washing too low.
Solutions: Try different hybridization and washing conditions, following basic parameters for hybridization and washing of Northern blots (Sambrook et al. 1989, Chapter 7; Ausubel et al. 1994, Unit 4.9). Increase extent of digestion with RNase A during washes.

Possible cause: Inadequate purification of radiolabeled RNA, allowing residual viscous DNA or proteins to trap radiolabeled RNA on filters.
Solutions: Use fresh batches of DNase I and proteinase K when purifying radiolabeled RNA. Increase digestion times with DNase I and proteinase K. Try a different method of purifying radiolabeled RNA (e.g., TCA precipitation as described in Ausubel et al. 1994, Unit 4.10).

Weak or undetectable signal obtained on nitrocellulose membranes

Possible causes: Inefficient hybridization or rare transcripts.

Solutions: Try different hybridization and washing conditions, following basic parameters for hybridization and washing of Northern blots (Sambrook et al. 1989, Chapter 7; Ausubel et al. 1994, Unit 4.9). Increase amount of unlabeled DNA on membrane. Increase amount of radiolabeled RNA added to hybridization. Decrease total volume of hybridization. Increase time of hybridization. Decrease stringency of washing. Increase concentration of radiolabeled nucleotide when labeling nascent transcripts.

REFERENCES

Ausubel F.M., Brent R.E., Kingston E., Moore D.D., Seidman J.G., Smith J.A., and Struhl K. 1994. *Current protocols in molecular biology.* John Wiley and Sons, New York.

Bentley D.L. 1995. Regulation of transcriptional elongation by RNA polymerase II. *Curr. Op. Genet. Dev.* **5:** 210–216.

Blau J., Xiao H., McCracken S., O'Hare P., Greenblatt J., and Bentley D. 1996. Three functional classes of transcriptional activation domains. *Mol. Cell. Biol.* **16:** 2044–2055.

Caponigro G. and Parker R. 1996. Mechanisms and control of mRNA turnover in *Saccharomyces cerevisiae. Microbiol. Rev.* **60:** 233–249.

Chen C.-Y.A. and Shyu A.-B. 1995. AU-rich elements: Characterization and importance in mRNA degradation. *Trends Biochem. Sci.* **20:** 465–470.

Colgan D.F. and Manley J.L. 1997. Mechanism and regulation of mRNA polyadenylation. *Genes Dev.* **11:** 2755–2766.

Cujec T.P., Okamoto H., Fujinaga K., Meyer J., Chamberlin H., Morgan D.O., and Peterlin B.M. 1997. The HIV transactivator TAT binds to the CDK-activating kinase and activates the phosphorylation of the carboxy-terminal domain of RNA polymerase II. *Genes Dev.* **11:** 2645–2657.

Cullen B.R. 1992. Mechanism of action of regulatory proteins encoded by complex retroviruses. *Microbiol. Rev.* **56:** 375–394.

Derman E., Krauter K., Walling L., Weinberger C., Ray M., and Darnell J.E., Jr. 1981. Transcriptional control in the production of liver-specific mRNAs. *Cell* **23:** 731–739.

Fornerod M., Ohno M., Yoshida M., and Mattaj I.W. 1997. CRM1 is an export receptor for leucine-rich nuclear export signals. *Cell* **90:** 1051–1060.

Garcia-Martinez L.F., Mavankal G., Neveu J.M., Lane W.S., Ivanov D., and Gaynor R.B. 1997. Purification of a Tat-associated kinase reveals a TFIIH complex that modulates HIV-1 transcription. *EMBO J.* **16:** 2836–2850.

Gorlich D. and Mattaj I.W. 1996. Nucleocytoplasmic transport. *Science* **271:** 1513–1518.

Greenberg M.E. and Ziff E.B. 1984. Stimulation of 3T3 cells induces transcription of the c-fos proto-oncogene. *Nature* **311:** 433–438.

Greenblatt J., Nodwell J.R., and Mason S.W. 1993. Transcriptional antitermination. *Nature* **364:** 401–406.

Harrold S., Genovese C., Kobrin B., Morrison S.L., and Milcarek C. 1991. A comparison of apparent mRNA half-life using kinetic labeling techniques vs decay following administration of transcriptional inhibitors. *Anal. Biochem.* **198:** 19–29.

Helms S.R. and Rottman F.M. 1990. Characterization of an inducible promoter system to investigate decay of stable mRNA molecules. *Nucleic Acids Res.* **18:** 255–259.

Hentze M.W. and Kühn L.C. 1996. Molecular control of vertebrate iron metabolism: mRNA-based regulatory circuits operated by iron, nitric oxide, and oxidative stress. *Proc. Natl. Acad. Sci.* **93:** 8175–8182.

Hentze M.W. and Kulozik A.E. 1999. A perfect message: RNA surveillance and nonsense-mediated decay. *Cell* **96:** 307–310.

Jacobson A. and Peltz S.W. 1996. Interrelationships of the pathways of mRNA decay and translation in eukaryotic cells. *Annu. Rev. Biochem.* **65:** 693–739.

Jones K.A. 1997. Taking a new TAK on Tat transactivation. *Genes Dev.* **11:** 2593–2599.

Kabnick K.S. and Housman D.E. 1988. Determinants that contribute to cytoplasmic stability of human f-fos and β-globin mRNAs are located at several sites in each mRNA. *Mol. Cell. Biol.* **8:** 3244–3250.

Kao S.Y., Calman A.F., Luciw P.A., and Peterlin B.M. 1987. Anti-termination of transcription within the long terminal repeat of HIV-1 by tat gene product. *Nature* **330:** 489–493.

Korner C.G. and Wahle E. 1997. Poly(A) tail shortening by a mammalian poly(A)-specific 3′-exoribonuclease. *J. Biol. Chem.* **272:** 10448–10456.

Krumm A., Meulia T., and Groudine M. 1993. Common mechanisms for the control of eukaryotic transcriptional elongation. *BioEssays* **15:** 659–665.

Lindstein T., June C.H., Ledbetter J.A., Stella G., and Thompson C.B. 1989. Regulation of lymphokine messenger RNA stability by a surface-mediated T cell activation pathway. *Science* **244:** 339–343.

Linial M., Gunderson N., and Groudine M. 1985. Enhanced transcription of c-myc in bursal lymphoma cells requires continuous protein synthesis. *Science* **230:** 1126–1132.

Lis J. and Wu C. 1993. Protein traffic on the heat shock promoter: Parking, stalling, and trucking along. *Cell* **74:** 1–4.

Mattaj I.W. and Englmeier L. 1998. Nucleocytoplasmic transport: The soluble phase. *Annu. Rev. Biochem.* **67:** 265–306.

Nudler E. 1999. Transcription elongation: Structural basis and mechanisms. *J. Mol. Biol.* **288:** 1–12.

Parada C.A. and Roeder R.G. 1996. Enhanced processivity of RNA polymerase II triggered by Tat-induced phosphorylation of its carboxy-terminal domain. *Nature* **384:** 375–378.

Powell D.J., Friedman J.M., Oulette A.J., Krauter K.S., and Darnell J.E., Jr. 1984. Transcriptional and post-transcriptional control of specific messenger RNAs in adult and embryonic liver. *J. Mol. Biol.* **179:** 21–35.

Rajagopalan L.E. and Malter J.S. 1997. Regulation of eukaryotic messenger RNA turnover. *Progr. Nucleic Acids Res.* **56:** 257–286.

Reines D., Conaway J.W., and Conaway R.C. 1996. The RNA polymerase II general elongation factors. *Trends Biochem. Sci.* **21:** 351–355.

Ross J. 1995. mRNA stability in mammalian cells. *Microbiol. Rev.* **59:** 423–450.

Sambrook J., Fritsch E.F., and Maniatis T. 1989. *Molecular cloning: A laboratory manual.* Cold Spring Harbor Laboratory Press, Cold Spring Harbor, New York.

Shilatifard A., Conaway J.W., and Conaway R.C. 1997. Mechanism and regulation of transcriptional elongation and termination by RNA polymerase II. *Curr. Op. Genet. Dev.* **7:** 199–204.

Shyu A.-B., Belasco J.G., and Greenberg M.E. 1991. Two distinct destabilizing elements in the c-fos message trigger deadenylation as a first step in rapid mRNA decay. *Genes Dev.* **5:** 221–231.

Shyu A.-B., Greenberg M.E., and Belasco J.G. 1989. The c-fos transcript is targeted for rapid decay by two distinct mRNA degradation pathways. *Genes Dev.* **3:** 60–72.

Theodorakis N.G. and Cleveland D.W. 1993. Translationally coupled degradation of tubulin mRNA. In *Control of messenger RNA stability.* (ed. J. Belasco and G. Brawermans), pp. 219–328. Academic Press, San Diego, California.

Ucker D.S. and Yamamoto K.R. 1984. Early events in the stimulation of mammary tumor virus RNA synthesis by glucocortocoids. Novel assays of transcription rates. *J. Biol. Chem.* **259:** 7416–7420.

Uptain S.M., Kane C.M., and Chamberlin M.J. 1997. Basic mechanisms of transcript elongation and its regulation. *Annu. Rev. Biochem.* **66:** 117–172.0

Wang J. and Manley J.L. 1997. Regulation of pre-mRNA splicing in metazoa. *Curr. Op. Genet. Dev.* **7:** 205–211.

Wei P., Garber M.E., Fang S.M., Fischer W.H., and Jones K.A. 1998. A novel CDK9-associated C-type cycling interacts directly with HIV-1 Tat and mediates its high-affinity, loop-specific binding to

TAR RNA. *Cell* **92:** 451–462.

Wilson G.M. and Brewer G. 1999. The search for trans-acting factors controlling messenger RNA decay. *Prog. Nucleic Acids Res.* **62:** 257–291.

Wilson T. and Treisman R. 1988. Removal of poly(A) and consequent degradation of c-fos mRNA facilitated by 3′ AU-rich sequences. *Nature* **336:** 396–399.

Yanofsky C. 1988. Transcription attenuation. *J. Biol. Chem.* **263:** 609–612.

Zhou Q. and Sharp P.A. 1996. Tat-SF1: Cofactor for stimulation of transcriptional elongation by HIV-1 Tat. *Science* **274:** 605–610.

Zhu Y., Pe'ery T., Peng J., Ramanathan Y., Marshall N., Marshall T., Amendt B., Mathews M.B., and Price D.H. 1997. Transcription elongation factor P-TEFb is required for HIV-1 tat transactivation in vitro. *Genes Dev.* **11:** 2622–2632.

CHAPTER 4

Transcription Initiation Site Mapping

Important issues

- *It is important to determine the location of a transcription initiation site before beginning a promoter analysis.*
- *Four methods are available for mapping a transcription initiation site.*
- *Each method possesses distinct advantages and disadvantages.*

INTRODUCTION

A common starting point for analyzing a gene regulated at the level of transcription initiation is to identify its promoter and characterize the *cis*-acting sequence elements and *trans*-acting proteins responsible for promoter activity. As described in Chapter 1, the promoter includes the DNA sequence elements in the vicinity of the initiation site (start site) that direct activation or repression of transcription. On the basis of this definition, the location of the promoter region can be identified simply by mapping the transcription initiation site.

Accurate mapping of the initiation site is extremely important for the success of the subsequent promoter analysis. In more than one instance, investigators have studied a putative promoter in considerable detail, only to find that the region under investigation was quite far from the start site and played no role in gene regulation. In some cases, the error occurred because the start site was not determined at the beginning of the analysis. In others, the start site was localized to the wrong position.

The transcription start site for a gene is determined by identifying the 5′ end of the encoded mRNA. It generally is assumed that the sequence at the 5′ end of an mRNA corresponds to the DNA sequence at which transcription initiates. This is not always an accurate assumption, because mRNAs can degrade or be cleaved, which may expose a 5′ end that does not correspond to the authentic start site. A more accurate method would include a demonstration that the 5′ nucleotide contains a cap structure, which is added to the 5′ end of transcripts synthesized by RNA polymerase II. The indirect methods that are commonly used today do not provide evidence that a cap structure exists. Instead, confirmation that the start site has been mapped correctly is provided by future experiments demonstrating that the surrounding DNA contains a functional promoter with discrete sequence elements that control initiation from the site identified.

The first methods developed for identifying the 5′ ends of eukaryotic mRNAs were time-consuming and cumbersome. These techniques were also insensitive and could therefore be used only on the most abundant cellular and viral transcripts. To identify the 5′ ends of rabbit α- and β-globin mRNAs, for example, Lockard and RajBhandary (1976) first purified the specific mRNAs from reticulocyte polysomes by oligo-dT chromatography and SDS sucrose density gradient centrifugation. The 5′ caps were then removed by periodate oxidation, allowing the 5′ ends to be radiolabeled with [γ-^{32}P]ATP and T4 polynucleotide kinase. The end-labeled mRNAs were purified by gel electrophoresis and were digested with nuclease T1. The nucleotide sequence at the 5′ end was then determined by enzymatic analysis of the radiolabeled oligonucleotide generated by T1 digestion.

Shortly after the above technique and a few other cumbersome techniques were described, two facile and sensitive RNA analysis methods were developed. Both methods were indirect, meaning that radiolabeled cDNAs were detected rather than the RNA transcripts and ribonucleotides themselves. The first method, called the S1 nuclease method (see Protocol 4.3, p. 130), was developed in 1977 by Arnold Berk and Philip Sharp (Berk and Sharp 1977). The second method, now called primer extension (see Protocol 4.1, p. 116), was first described in 1978 by P.K. Ghosh and S.M. Weissman (Ghosh et al. 1978). In the 1980s, two additional indirect techniques, RNase protection (see Protocol 4.2, p. 124; Melton et al. 1984) and RACE (rapid amplification of cDNA ends, Frohman et al. 1988) were developed. These four techniques remain the most common for determining transcription start sites.

The concept underlying each of the four techniques is described below, along with the major advantages and disadvantages. For most genes, more than one technique should be tested because it is difficult to predict which one will yield convincing results. Another reason to use more than one technique is that the data obtained with any single method may be relatively inconclusive. Figure 4.1 shows the results obtained in parallel RNase protection, primer extension, and S1 nuclease experiments performed to map the transcription start site for the murine terminal transferase (TdT) gene (Smale and Baltimore 1989; Lo et al. 1991). This example will be referred to throughout this chapter.

CONCEPTS AND STRATEGIES

Initial Considerations

Reagents Needed before Proceeding

Before determining the location of a transcription start site, it is highly desirable to isolate a cDNA containing as much of the 5′ untranslated leader region as possible. A full-length cDNA can often be isolated from a high-quality cDNA library prepared from mRNA by oligo(dT) or random-nucleotide priming (Sambrook et al. 1989, Chapter 8). Alternatively, specifically primed cDNA can be amplified by PCR using the 5′ RACE method described in this chapter.

In addition to obtaining a cDNA that is as close to full-length as possible, it is desirable, and often necessary (e.g., for the RNase protection and S1 nuclease procedures), to obtain a genomic clone spanning the 5′ end of the first exon. A genomic clone can be isolated by screening a genomic library with a radiolabeled probe corresponding to the 5′ end of the cDNA. Alternatively, CLONTECH Laboratories offers a PCR-based GenomeWalker kit (cat. # K-1803-1) that allows a genomic DNA fragment to be isolated using a single specific primer derived from the 5′ end of the cDNA. After a genomic clone is obtained, the DNA sequence extending at least a few hundred base pairs upstream of the first exon should be determined.

Information Provided by the DNA Sequence

cDNA and upstream genomic DNA sequences are necessary for designing primers and probes for the start-site mapping techniques. Furthermore, the sequences may provide evidence that the start site is nearby or, alternatively, that the clones lack the start site and perhaps the entire first exon. A comparison between the cDNA sequence and the genomic DNA sequence will provide initial insight into this issue. If the 5′ end of the cDNA matches the genomic DNA, the genomic DNA may indeed contain the first exon and transcription start site. In contrast, if the 5′-most sequence of the cDNA does not match the genomic sequence, with the two sequences homologous only through a region slightly downstream,

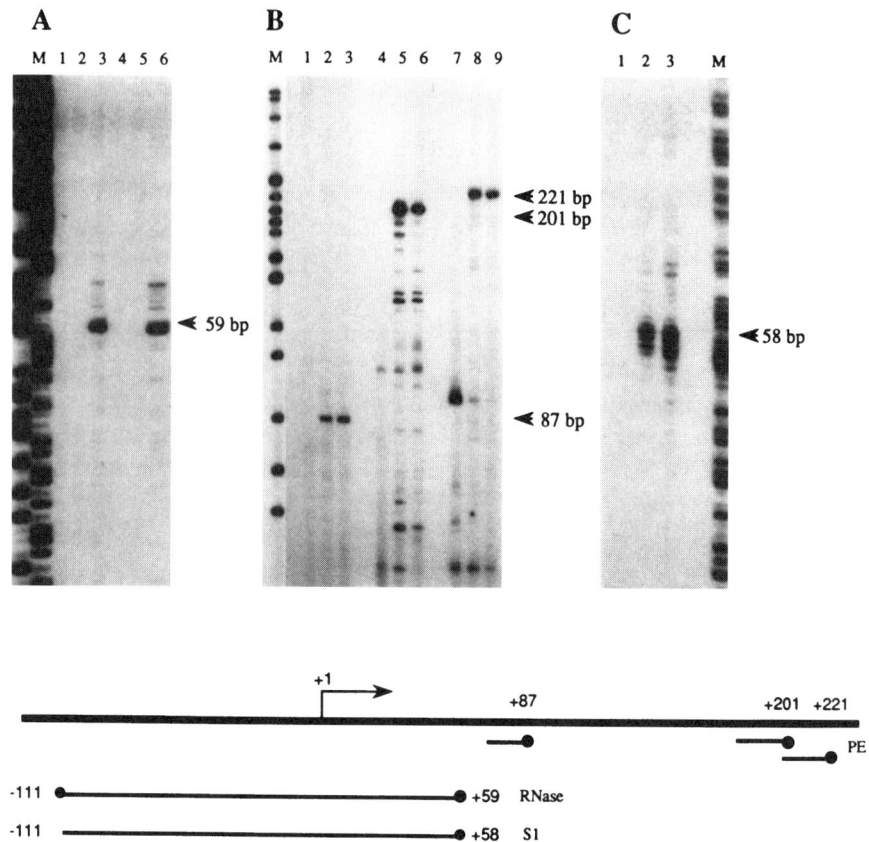

FIGURE 4.1. Results from RNase protection, primer extension, and S1 nuclease experiments to map the transcription initiation site for the murine terminal transferase (TdT) gene. (*A*) The RNase protection assay was performed with cytoplasmic RNA from two cell lines that express the TdT gene (lanes *3* and *6*) and from four cell lines that lack expression (lanes *1*, *2*, *4*, and *5*). The labeled probe extended from nucleotide +59 to –111. The expected product is 59 nucleotides. Sequencing markers are also shown (M). (Reprinted, with permission, from Lo et al. 1991.) (*B*) The primer extension assay was performed with cytoplasmic RNA from two cell lines that express the TdT gene (lanes *2*, *3*, *5*, *6*, *8*, and *9*) and from one line that lacks expression (lanes *1*, *4*, and *7*). Three different primers were used whose 5′ ends hybridize 87 (lanes *1–3*), 201 (lanes *4–6*), and 221 (lanes *7–9*) bp downstream of the anticipated start site. Markers (M) are 5′-end-labeled *Mbo*I restriction fragments from the plasmid pBR322. (*C*) The S1 nuclease assay was performed with cytoplasmic RNA from two cell lines that express the TdT gene (lanes *2* and *3*) and from one line that lacks expression (lane *1*). The labeled probe extended from nucleotide +58 to nucleotide –111. The expected product is 58 nucleotides. Sequencing markers are shown (M). (Reprinted, with permission, from Smale and Baltimore 1989 [Copyright 1989 Cell Press].)

the genomic DNA most likely contains the junction between the first intron and second exon. The first exon may be farther upstream in the genomic clone or may be missing from the clone.

The cDNA sequence may also reveal the presence of a downstream translation initiation codon (ATG) surrounded by a Kozak consensus sequence (GCCACC<u>ATG</u>G; Kozak 1996) and followed by an open reading frame. The presence of a translation initiation codon suggests that the transcription start site may be nearby. In vertebrate genes, the distance from the transcription start site to the ATG is usually 20–100 nucleotides, although 5′ noncoding regions of more than 100 nucleotides are found in approximately 25% of vertebrate genes (Kozak 1987). Furthermore, translation usually initiates within the first exon, although in several genes translation initiates within other exons (e.g., RAG-1 [Schatz et al. 1989] and IL-12 p40 [Murphy et al. 1995]).

Additional information provided by the DNA sequence includes the presence or absence of consensus splice acceptor sites. If a splice acceptor is identified near the 5′ end of the full-length cDNA, the DNA fragment isolated may not contain the first exon. The RNase protection, S1 nuclease, and RACE experiments described below can be used to determine whether the sequence identified functions as a splice acceptor.

If the cDNA and genomic clones are found to lack the first exon, additional attempts to isolate a full-length cDNA may be required. Alternatively, one could continue to sequence the genomic DNA farther upstream, with the hope that the intron is short. This route is risky because the intron may be several kilobase pairs in length.

A computer analysis of the genomic DNA sequence (see Chapter 1) may reveal sequences that match known control elements and that therefore may be involved in regulating transcription of the gene. Of particular relevance for determining the location of the transcription start site are sequences similar to consensus TATA or initiator (Inr) elements (see Chapter 1). The transcription start site may be located 25–30 bp downstream of a sequence that matches a consensus TATA box, or within a sequence that matches a consensus Inr element. The absence of a TATA or Inr sequence does not rule out the possibility that the start site is nearby, however, as many genes do not contain these elements. Furthermore, the mere presence of a consensus TATA or Inr element upstream of the 5′ end of a cDNA does not necessarily indicate the location of a transcription start site. The reason for this is that only a fraction of consensus TATA and Inr elements within the genome are functional, in particular those flanked by an appropriate array of activator and repressor elements.

The genomic DNA sequence surrounding the 5′ end of a cDNA might also reveal a high G/C content combined with a large number of CpG dinucleotides (Slansky and Farnham 1996). Several promoters have been described that exhibit this feature, which is most commonly found associated with so-called "housekeeping" genes. (Housekeeping genes generally exhibit little cell-type specificity and little developmental regulation.) The high G/C content within these promoters usually correlates with a large number of binding sites for the ubiquitous transcription factor Sp1, which recognizes the G/C-rich sequence, CCCGCC. Indeed, functional studies of several G/C-rich promoters suggest that Sp1 and other Sp1 family members are key regulators of promoter activity (e.g., Slansky and Farnham 1996).

Regarding the current topic of transcription start-site mapping, G/C-rich promoters can be difficult to work with for two reasons. First, these promoters often contain multiple transcription start sites that can span hundreds of base pairs. Second, the highly-G/C-rich sequence can form unusual secondary structures in vitro, which can hinder use of the common methods for start-site mapping.

Finally, a computer analysis of the genomic sequence may reveal homologies with known control elements (see Chapter 1), suggesting the existence of important regulatory elements. As discussed in Chapter 7, the presence of consensus binding sites for known transcription does not necessarily provide strong evidence that the promoter has been identified. A computational analysis of virtually any DNA sequence will reveal putative binding sites for known transcription factors. As described above for TATA and Inr elements, the sites that are functionally relevant are those that are appropriately oriented relative to the binding sites for a constellation of other transcription factors, resulting in a control region that is competent for nucleosome remodeling and for stimulating the formation of a preinitiation complex containing RNA polymerase II.

Primer Extension

Advantages and Disadvantages

Primer extension (Box 4.1) is recommended as the first method to try when attempting to identify the initiation site for a new gene. A principal advantage of primer extension is that it is the easiest and fastest method to perform. It can be completed in approximately 5 hours, and the only specific reagents required are an mRNA sample and a radiolabeled oligonucleotide primer complementary to a sequence within the gene. With a primer that efficiently hybridizes to the mRNA, this method can be extremely sensitive.

A disadvantage of the primer extension protocol is that it can be difficult to find a primer that works well for a new gene. Moreover, background bands often appear that can make it difficult to determine unambiguously the location of the start site. Furthermore, for some mRNAs, including those with G/C-rich 5′ ends, and thus stable secondary structure, the reverse transcriptase may be incapable of efficient extension. Because of these limitations, it usually is important to confirm the results using one of the other methods described in this chapter.

Design of Oligonucleotide Primers

Oligonucleotide primers of many different lengths can work well for primer extension. Before synthetic oligonucleotides became widely available, single-stranded primers derived from restriction fragments of 100 nucleotides or more were employed successfully. Primers of this length are rarely used today because synthetic oligonucleotides are now relatively inexpensive and easy to obtain. Synthetic oligonucleotides of 9 nucleotides are sufficient for hybridization and extension, but are not used for this method because the hybridization lacks specificity and efficiency. Oligonucleotides of 20–25 bp typically yield excellent results. Several software programs are available for designing primers with length and melting temperature (T_m) in mind. Operon Technologies, Inc. (see http://www.operon.com) has several online programs or tool kits for primer design in the custom DNA Synthesis link. Commercially available programs include the GeneRunner Sequence Analysis Program from Hastings Software, Inc.; PrimerSelect, bundled with the Lasergene Biocomputing Software, from DNASTAR, Inc. (see http://www.dnastar.com/lasergen); and Oligo Primer Analysis Software from Molecular Biology Insights (MBInsights; see http://www.mbinsights.com).

Although computer programs are recommended for the design of primers, the practical success obtained with different primers can be variable. If less-than-ideal results are obtained with an initial primer, one or even several additional primers complementary to other sequences near the 5′ end of the mRNA should be tested, because they may yield better results. It is not known why the quality of the data obtained with different primers possessing similar T_m values can be widely variable, but differences in primer labeling efficiency or mRNA secondary structure might be responsible.

To localize a transcription start site accurately, it is best for the 5′ end of the primer to anneal to a position on the mRNA that is expected to be about 50–150 nucleotides from the mRNA start site. This length allows the resulting cDNA products to be measured accurately on a denaturing polyacrylamide gel, if electrophoresed adjacent to a DNA sequencing ladder. cDNA products shorter than 50 bp are undesirable because reverse transcriptase reactions often terminate or pause after extending a very short distance, resulting in strong background bands that are 10–20 bp longer than the primer. Additionally, reverse

Box 4.1

Primer Extension

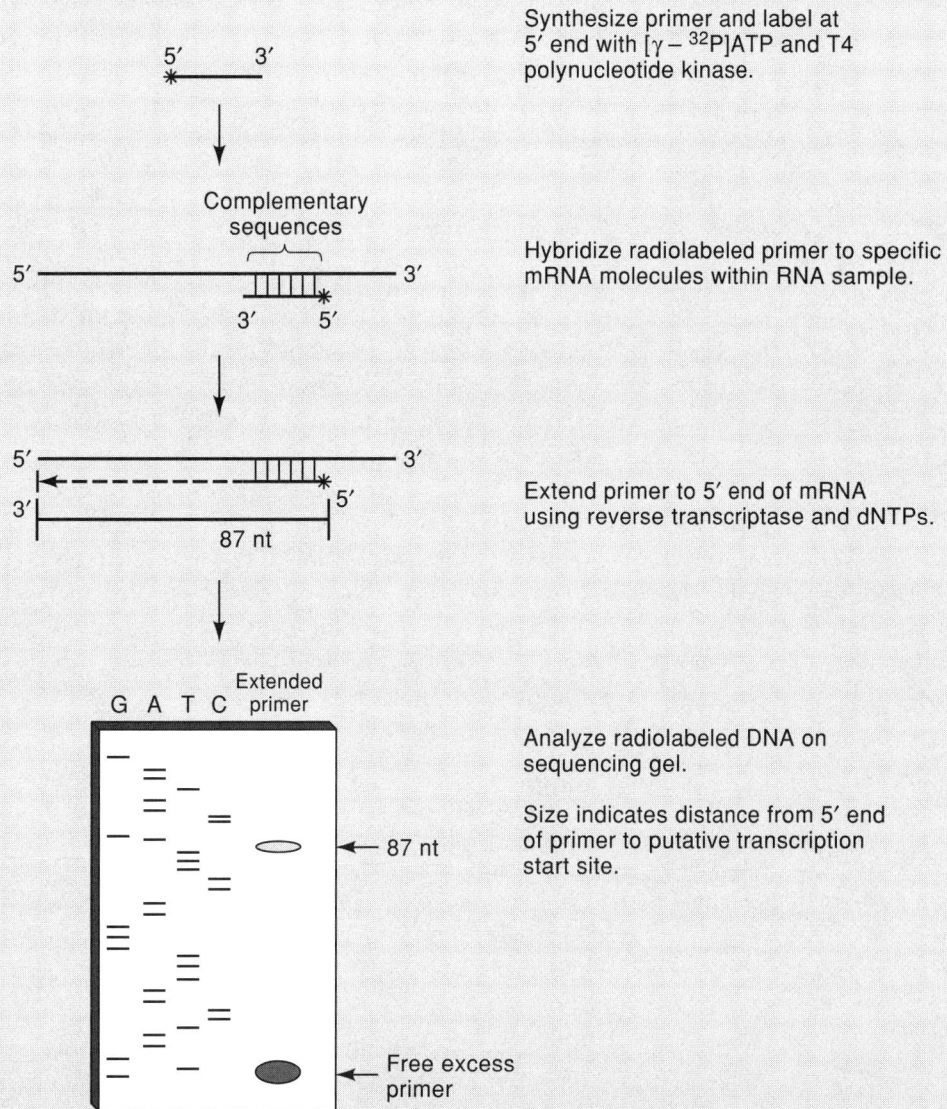

5′   3′

Synthesize primer and label at 5′ end with [γ – ^{32}P]ATP and T4 polynucleotide kinase.

Complementary
sequences

5′ ——————————————— 3′

3′   5′

Hybridize radiolabeled primer to specific mRNA molecules within RNA sample.

5′ ——————————————— 3′

3′   5′

87 nt

Extend primer to 5′ end of mRNA using reverse transcriptase and dNTPs.

Extended
G A T C primer

← 87 nt

← Free excess
primer

Analyze radiolabeled DNA on sequencing gel.

Size indicates distance from 5′ end of primer to putative transcription start site.

FIGURE 4.2. Primer extension.

The primer extension protocol (Fig. 4.2) begins with a 5′ ^{32}P-labeled primer, usually a synthetic oligonucleotide, that is complementary to an mRNA sequence approximately 50–150 nucleotides downstream of the anticipated 5′ end. A large molar excess of the radiolabeled primer is annealed under empirically determined reaction conditions to the specific RNA molecules within total RNA or oligo-dT purified mRNA isolated from the cells of interest. Reverse transcriptase, deoxyribonucleoside triphosphates, and appropriate buffer components are added to the primer–mRNA hybrids to catalyze elongation of the primer to the 5′ end of the mRNA. The resulting radiolabeled cDNA products are analyzed by denaturing polyacrylamide gel electrophoresis. The sizes of the bands detected on the gel, as compared to an adjacent sequencing ladder or molecular weight standards, provide a measure of the distance from the 5′ end of the synthetic oligonucleotide to the beginning of the mRNA transcripts. If the labeled cDNA products are within the resolution range of the gel, the transcription start site can be determined with an accuracy of plus or minus one nucleotide.

transcriptase will sometimes synthesize a copy of the primer, leading to the so-called "primer dimer" artifact. If the cDNA synthesized is at least 50 nucleotides (e.g., a 20-nucleotide primer plus a 30-nucleotide extension), it will usually be well-separated from the majority of these background bands. cDNA products longer than 150 bp can sometimes be obtained, but the efficiency of reverse transcriptase extension decreases with distance because of pausing and premature termination. Furthermore, the longer the extension product, the more difficult it will be to determine an accurate size, due to the decreased resolution of large fragments on a denaturing polyacrylamide sequencing gel.

Primer Annealing and Reverse Transcription

For each primer tested, the annealing conditions must be determined empirically, because the temperatures predicted on the basis of nucleotide content are not always accurate. For the primer extension procedure listed at the end of this chapter, the annealing reaction is performed in 250 mM KCl. At this salt concentration, suggested temperatures to test are 37, 45, 60, and 68°C. One can also heat the annealing reaction to 68°C in a temperature block and then allow it to cool slowly to room temperature.

The extension reaction in the protocol listed employs Moloney murine leukemia virus (MMLV) reverse transcriptase that had been cloned and expressed in *E. coli*. MMLV reverse transcriptase contains an RNase H activity that can lead to enhanced background, but modified reverse transcriptases lacking RNase H activity are now sold by Life Technologies, Inc. (SuperScript RT, GIBCO/BRL cat. # 18053-017). Extension reactions can be carried out using other reverse transcriptases, including avian reverse transcriptase, but the reaction conditions are different and more sensitive to variations in pH.

Analysis of Example Data

For the terminal transferase (TdT) primer extension example shown in Figure 4.1, three different primers were tested (Smale and Baltimore 1989). The 5′ end of one primer hybridizes at a location that is 87 bp downstream of the transcription start site, whereas the 5′ ends of the other two primers hybridize 201 and 221 bp downstream of the start site. The large excess of free primer, which migrates very rapidly through the gel, was removed from the bottom of the autoradiograph shown in this figure. To confirm that the primer was in considerable excess, it is beneficial to retain the free primer on the bottom of the gel. The largest extension products observed with each primer map to a single transcription start site. The second and third primers hybridize somewhat farther from the transcription start site than was recommended above (201 and 221 bp rather than 50–150 bp). Nevertheless, these primers were efficiently extended to the 5′ end of the mRNA in this experiment. In other experiments in which extension efficiency was lower, the full-length extension products were less abundant and accompanied by shorter, partial extension products (data not shown). The signal intensities obtained with the three different primers, although comparable on this gel, were variable, with the amounts of reaction product normalized to generate bands of similar intensity on the autoradiograph. Variability between primers can result from differences in the efficiency of labeling or annealing.

For each primer, total cellular mRNA was tested from two murine cell lines that express the TdT gene and from one murine cell line that does not express the gene. The use of a negative control from the same species can be invaluable. Background bands are often observed that result from hybridization of the primer to unrelated RNAs. The presence of these bands in the negative control lane reveals that they are unrelated to the tissue-specific transcript of interest. Two examples of this type of background band are observed with the second primer and two with the third primer, with one example in each case in the middle section of the autoradiograph and one at the bottom. The third primer yielded an unusually strong background band in the negative control lane that was much less abundant in the positive lanes and was not detected with the other primers. The absence of this band with the second primer suggests that it was caused by an unrelated transcript to which the third primer fortuitously hybridized. Apparently, this transcript was strongly expressed only in the negative control cell line.

Other background bands that are evident (primarily with the second primer) are likely to have resulted from reverse transcriptase pausing during extension on full-length mRNAs. Background bands can also be caused by severe mRNA degradation, hybridization to unrelated transcripts as described above, or the presence of multiple transcription start sites. A primary limitation of the primer extension procedure is that background bands like those observed with the second primer are quite common; it is very difficult to determine which bands correspond to authentic transcription start sites and which are experimental artifacts. For the TdT gene, the use of multiple primer extension primers and subsequent confirmation by RNase protection and S1 nuclease analysis allowed a precise determination of the authentic mRNA start site.

RNase Protection

Advantages and Disadvantages

The RNase protection procedure (Box 4.2) is the second most common method for mapping start sites and is recommended as an attractive alternative to primer extension. Because primer extension and RNase protection rely on different principles, our strongest recommendation is to use both methods. Such an approach produces a high degree of confidence in the results if the two methods map the start site to the same position. RNase protection is more time-consuming than primer extension, but it often is the most sensitive of the quantitative methods for start-site localization. RNase protection is sensitive for two reasons. First, the probe is usually synthesized with high-specific-activity $[\alpha\text{-}^{32}P]UTP$, such that ~10% of the uracil residues within the resulting probe can be radiolabeled; in contrast, primer extension primers contain an average of less than one radiolabeled phosphate atom per molecule. Second, the RNA–RNA hybrids formed during RNase protection are thermodynamically stable, much more so than the RNA–DNA hybrids formed during S1 nuclease analysis. This stability increases sensitivity by diminishing the frequency of "breathing" (i.e., transient dissociation) of the hybrids during the nuclease digestion step.

A primary disadvantage of the RNase protection procedure is that the assay can be difficult to establish for a new gene because the hybridization and digestion conditions must be determined empirically. Furthermore, the radiolabeled probe is susceptible to formation of stable secondary structures that can generate unwanted RNase-resistant products. Finally, the high-specific-activity radiolabeled probes can be used for only a few days because they tend to undergo radiolysis rapidly. In contrast, labeled oligonucletoide primers for primer extension can often be used for up to a month.

Probe Preparation

The challenges associated with construction of a plasmid for probe preparation are often minimal. The simplest method is to amplify a genomic sequence spanning the transcription start

Box 4.2

RNase Protection

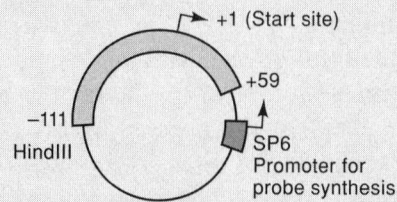

Plasmid includes region of gene containing putative transcription start site and 59 nucleotides downstream of this start site, fused to the SP6 promoter and flanked by convenient restriction sites.

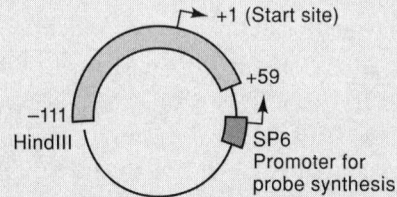

Cut with HindIII to linearize plasmid.

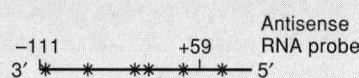

Add transcription buffer, SP6 RNA polymerase, NTPs plus [α-^{32}P]UTP (asterisks) to generate antisense probe.

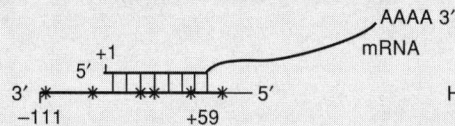

Hybridize probe to isolated mRNA.

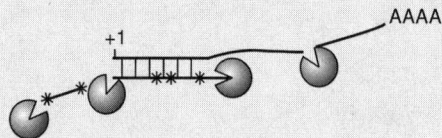

Digest with RNase T$_1$ and RNase A (cleaves single-stranded RNAs).

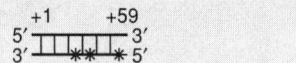

Creates a 59-nucleotide RNA-RNA hybrid.

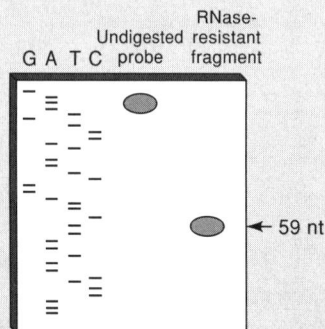

Denature and analyze by denaturing gel electrophoresis.

FIGURE 4.3. RNase protection.

The RNase protection procedure (Fig. 4.3) begins with an RNA probe that is uniformly labeled by incorporation of one [α-^{32}P]NTP, usually [α-^{32}P]UTP. The RNA probe is synthesized by bacteriophage RNA polymerase (SP6, T7, or T3), which initiates transcription from specific phage promoters that have been engineered into a number of common plasmid vectors. For start-site mapping, the plasmid template contains a genomic DNA fragment spanning the region thought to contain the transcription start site for the gene of interest. This genomic fragment is subcloned into a plasmid downstream of the phage promoter in the antisense orientation, so that a portion of the 5′ end of the resulting RNA probe will be complementary to the mRNA of interest. DNA templates for probe preparation can also be prepared by PCR, by including the bacteriophage polymerase promoter sequence into one of the PCR primers and then amplifying the genomic DNA sequence. The radiolabeled probe is annealed to cytoplasmic or total cellular mRNA purified from the cells of interest, with the hybridization reaction proceeding for several hours or overnight, usually in a formamide buffer. RNase A and/or RNase T1 is then added to the hybridization reactions. These nucleases digest the single-stranded overhang regions of RNA molecules, but RNA–RNA hybrids will be resistant to cleavage. This resistance forms the conceptual basis for the procedure, because the region of the probe that anneals to the specific mRNA will be resistant to digestion. The length of the resistant region of the probe will correspond to the distance from the 5′ end of the probe to the transcription start site. Because the probe is radiolabeled, the size of the resistant fragment can be visualized by electrophoresis on a high-resolution, denaturing polyacrylamide gel.

site by PCR, using a genomic clone or genomic DNA as a template and two PCR primers, one of which contains the bacteriophage promoter sequence (see Box 4.2). The bacteriophage promoter should be included in the downstream primer, so that the genomic sequence ends up in an antisense orientation. The PCR product can be used directly for probe preparation.

If the same probe will be prepared frequently, it may be beneficial to prepare a plasmid containing the bacteriophage promoter and antisense genomic DNA fragment. This can be accomplished by PCR amplification of a genomic fragment using a primer containing a bacteriophage promoter, followed by insertion into any plasmid vector. Alternatively, a genomic DNA fragment can be cloned in an antisense orientation into a vector that already contains bacteriophage promoters, such as the Bluescript (Stratagene) or pSP (Promega) vectors. Plasmid construction is straightforward, but if PCR is used to amplify the genomic DNA prior to insertion into the vector, the resulting insert must be sequenced to confirm that no mutations have been introduced; mutations would disrupt the integrity of the RNA–RNA duplex, resulting in unwanted RNase cleavages. As with all procedures, a prediction regarding the likely location of the transcription start site must be made to determine the boundaries of the genomic DNA fragment used for the probe. If the prediction proves to be incorrect, other probes spanning different regions of the locus will be needed.

One critical question that must be addressed is how long the probe should be. In general, the longer the probe extends into the transcribed region, the greater the sensitivity of the assay. The reason is that longer probe molecules incorporate a greater number of radioactive phosphate atoms. Longer probes have a greater chance of containing secondary structure, however, which may lead to background signals or inefficient hybridization. The amount of secondary structure present in different regions can be calculated using Michael Zuker's mfold algorithm. This and other secondary structure prediction programs can be accessed through the Zuker lab RNA page of the Institute for Biomedical Computing at Washington University in St. Louis, Missouri website (see http:// www.ibc.wustl.edu/~zuker/rna). Longer probe molecules may also contain DNA regions through which the phage polymerase cannot efficiently extend. Because a large number of incomplete transcripts are obtained, it usually is necessary to gel-purify longer probes before they can be used. There is a better chance that short-probe molecules (e.g., <300 nucleotides) can be used without gel purification. To determine whether purification is necessary, the approximate percentage of full-length probe molecules can be determined by visualizing a small aliquot of the radioactive reaction product on a gel.

Probe Annealing and RNase Digestion

As mentioned above, annealing and RNase digestion conditions must be determined empirically. The annealing temperature can vary considerably from probe to probe and is only partially dependent on the calculated T_m of the RNA–RNA duplex. The precise concentration of RNases needed to degrade single-stranded products can also vary from probe to probe.

To simplify the development of an RNase protection assay for a new gene, commercially available kits can be used (e.g., Ambion Inc.'s RPA kits, cat. # 1410, 1412, 1420). These kits contain RNase solutions that have been titered for their effectiveness. The kits also contain a template for preparation of a control actin probe that is useful for confirming that the reagents are functioning properly. Because carefully titered RNase stocks and a control probe are used, the variables that need to be considered when setting up and troubleshooting an RNase protection assay are more limited.

The development of an assay can also benefit from a comparison of RNA samples from cell lines of the same species that express and do not express the gene of interest. Relevant bands will of course not be detected in reactions performed with mRNA from cells that do not express the gene, allowing one to identify some of the irrelevant bands easily.

Analysis of Example Data

In the example shown in Figure 4.1A, an RNase protection probe to analyze the TdT transcription start site was prepared from genomic sequences extending from nucleotides –111 to +59 (Lo et al. 1991). This fragment was inserted in an antisense orientation downstream of an SP6 promoter in pSP72 (Promega Corp.; cat. # P2191). Prior to probe preparation, the plasmid was cleaved with a restriction enzyme at a site adjacent to nucleotide –111, so that the SP6 RNA polymerase would synthesize a 180-nucleotide radiolabeled RNA product before running off the cleaved template. Because the probe was relatively short and contained no strong termination or pause sites for SP6 polymerase, it did not require gel purification. The radiolabeled probe was hybridized to RNA samples from six different murine cell lines, only two of which express the TdT gene (lanes 3 and 6). The samples were subsequently cleaved with RNases A and T1 to digest the single-stranded RNA overhangs of the hybrids. The most abundant resistant product, when analyzed by denaturing gel electrophoresis, migrated at approximately 59 nucleotides, as determined by comparison to the DNA sequence markers. Based on the size of this product, the 5′ end of the mRNA maps to the same location as that mapped by primer extension.

In the two lanes from cell lines that express TdT, weaker bands were detected that are slightly longer or shorter than the major band. If RNase protection were the only assay used to localize the transcription start site, it would be difficult to determine whether these bands represented minor transcription start sites. Because these weak bands were not detected by primer extension at the same locations, they are likely to be present for other reasons. Most likely, the larger bands are caused by the fact that RNases A and T1, both of which are endonucleases rather than exonucleases, did not always cleave at the last phosphodiester bond separating the double-stranded RNA from the single-stranded RNA, and instead cleaved at sites that left short overhangs. The bands that are slightly shorter than the major product most likely resulted from breathing at the end of the RNA–RNA duplex, with the RNases occasionally cleaving internal phosphodiester bonds. These results reveal the utility of using two distinct assays to distinguish between bands representing authentic 5′ ends and those representing experimental artifacts of each assay.

S1 Nuclease Analysis

Advantages and Disadvantages

Today, the S1 nuclease procedure (Box 4.3) is used much less frequently than the primer extension or RNase protection procedures, but it offers a valuable alternative if other methods prove to be troublesome for a particular gene. When using 5′-end-labeled probes, which are quite common for S1 nuclease analysis, the primary advantage relative to RNase protection is that the background signals are often diminished, presumably because aberrant cleavage products are detectable only if they include the radiolabeled 5′ nucleotide. The primary disadvantage when using a 5′- end-labeled probe is that the procedure is less sensitive because the probe molecules possess an average of less than 1 atom of radioactive phosphorus. Another potential disadvantage is that the RNA–DNA duplexes can be more susceptible to breathing and subsequent S1 nuclease cleavage of the single-stranded regions. Breathing can be particularly troublesome if the hybrid contains long A-T-rich stretches, which form relatively unstable base pairs.

Uniformly labeled probes can also be used for S1 nuclease analysis, making it nearly as sensitive as RNase protection. Nevertheless, the hybrids remain susceptible to breathing, which can cause considerable background. An additional disadvantage is that the generation of uniformly labeled probes often involves the preparation of single-stranded M13 plasmids or phagemids. The techniques involved are quite simple, but more complicated and variable than those used to prepare RNase protection probes.

Probe Preparation

Radiolabeled probes for S1 nuclease analysis can be generated by several methods, and the probe can be either 5′-end-labeled or internally labeled. Single-stranded, 5′-end-labeled DNA probes can be prepared from long, synthetic oligonucleotides that span the anticipated start site, using [γ-^{32}P]ATP and T4 polynucleotide kinase (Ausubel et al. 1994, Unit 4.6). Alternatively, a double-stranded restriction fragment can be 5′-end-labeled with polynucleotide kinase and the single-stranded probe then separated from its complement on a native polyacrylamide, strand-separation gel (Sambrook et al. 1989, p. 6.49). The double-stranded DNA sample to be applied to the strand-separation gel is denatured by boiling and then chilled rapidly. The rapid chilling allows each strand to form a unique tertiary structure, which prevents the two strands from annealing to one another. The two strands usually migrate with different mobilities because of nuances of secondary structure. This

---

Box 4.3

S1 Nuclease Protection

The conceptual basis of the S1 nuclease procedure (Fig. 4.4) is similar to that of the RNase protection procedure, except the radiolabeled probe is usually a single-stranded or double-stranded DNA molecule rather than an RNA molecule. The DNA probe can be prepared by a variety of procedures. One procedure involves the extension of a primer annealed to a single-stranded M13-based plasmid, using Klenow and radiolabeled dNTPs (see text). The plasmid should contain a genomic DNA fragment spanning the anticipated transcription start site. After restriction enzyme cleavage and gel purification of the radiolabeled antisense probe, it is hybridized to specific mRNA molecules within an RNA sample. S1 nuclease is then added to digest single-stranded regions of the DNA probe and annealed mRNA. The sizes of the radiolabeled, S1-resistant products are determined by denaturing polyacrylamide gel electrophoresis, followed by autoradiography or phosphorimager analysis. The length of the resistant product should correspond to the distance from the 5′ end of the probe to the transcription start site.

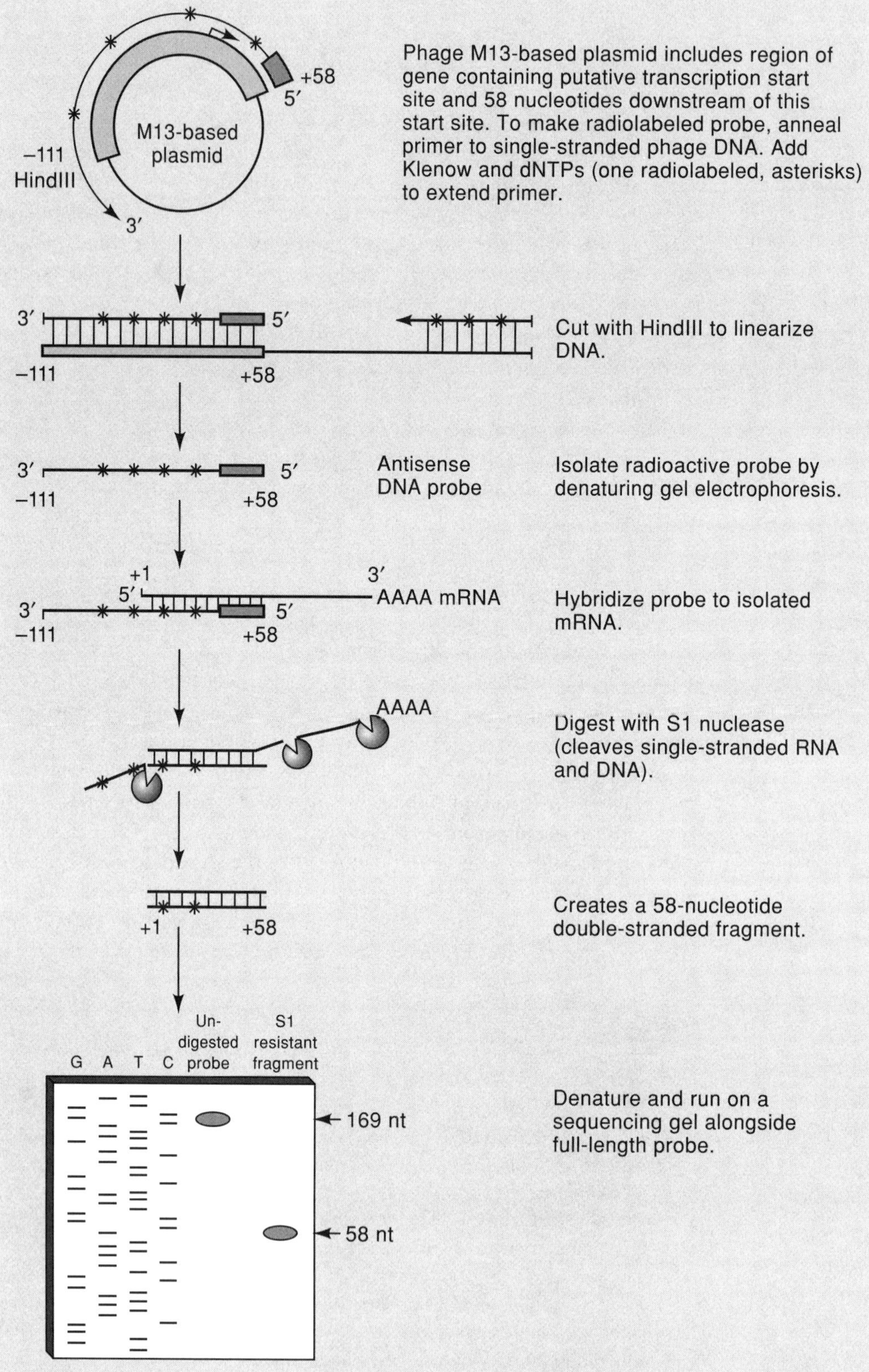
+58
5′
M13-based plasmid
−111
HindIII
3′
Phage M13-based plasmid includes region of gene containing putative transcription start site and 58 nucleotides downstream of this start site. To make radiolabeled probe, anneal primer to single-stranded phage DNA. Add Klenow and dNTPs (one radiolabeled, asterisks) to extend primer.
3′
5′
−111
+58
Cut with HindIII to linearize DNA.
3′
5′
−111
+58
Antisense DNA probe
Isolate radioactive probe by denaturing gel electrophoresis.
+1
5′
3′
AAAA mRNA
3′
5′
−111
+58
Hybridize probe to isolated mRNA.
AAAA
Digest with S1 nuclease (cleaves single-stranded RNA and DNA).
+1
+58
Creates a 58-nucleotide double-stranded fragment.
G A T C
Un-digested probe
S1 resistant fragment
169 nt
58 nt
Denature and run on a sequencing gel alongside full-length probe.

FIGURE 4.4. S1 nuclease protection.

method has been used for many studies, but because the strand-separation gels are laborious and inconsistent, it is rarely used today.

Internally labeled probes can be generated from a single-stranded DNA template, such as M13, containing a genomic DNA fragment spanning the anticipated start site. An oligonucleotide primer is annealed to the M13 plasmid and extended with Klenow fragment in the presence of an $[\alpha\text{-}^{32}P]dNTP$ (see Fig. 4.4 and Protocol 4.3) (Sambrook et al. 1989, pp. 10.18–10.26; Ausubel et al. 1994, Unit 4.6). The radiolabeled DNA fragment will have a consistent and discrete starting point dictated by the oligonucleotide, but will have variable 3′ ends, depending on the efficiency of elongation by the polymerase. To create a discrete probe, the double-stranded hybrids are cleaved with a restriction enzyme (Fig. 4.4). The single-stranded antisense probe is then isolated on a denaturing polyacrylamide gel. This strategy can also be used to generate 5′-end-labeled probes by annealing a kinased oligonucleotide to the M13 template and extending in the presence of unlabeled dNTPs (Sambrook et al. 1989, pp. 10.18–10.26; Ausubel et al. 1994, Unit 4.6)

Several other methods for preparing single-stranded DNA probes, including asymmetric PCR (Kaltenboeck et al. 1992; Ausubel et al. 1994, Unit 15.2) and a biotinylation method (Armes and Fried 1995) can be envisioned and have been employed. In addition, S1 nuclease analyses have been performed frequently and successfully using denatured, double-stranded DNA probes that are labeled only on one strand, by annealing the probe to the mRNA under conditions that favor RNA–DNA duplexes as opposed to DNA–DNA duplexes (Berk and Sharp 1977; Weaver and Weissmann 1979; Sambrook et al. 1989, pp. 7.58–5.65). Double-stranded probes are usually easier to prepare than single-stranded probes, but because DNA–DNA duplexes must be minimized, the optimization of hybridization conditions can be more difficult and the signals obtained are likely to be weaker. Double-stranded probes are therefore recommended only when the transcripts being analyzed are abundant.

Analysis of Example Data

In the example shown in Figure 4.1C, S1 nuclease experiments were performed with a single-stranded genomic DNA probe extending from nucleotide –111 to nucleotide +58 relative to the TdT transcription start site (+1). The probe was prepared from a plasmid containing a genomic DNA fragment. The plasmid was first cleaved at +58 with *Bam*HI, dephosphorylated with calf-intestine alkaline phosphatase, radiolabeled with $[\gamma\text{-}^{32}P]ATP$ and T4 polynucleotide kinase, cleaved at –111 with *Sac*I, and then isolated as a single-stranded radiolabeled probe on a native polyacrylamide strand separation gel. The short radioactive band representative of the 5′-end-labeled probe was excised, eluted from the gel, purified, quantitated by scintillation counting, and then hybridized to the mRNA samples.

The results shown in Figure 4.1C were derived from reactions performed with mRNA samples from two cell lines that express the TdT gene (lanes 2 and 3) and one sample that does not express the gene (lane 1). It is important to note that this panel of the figure was derived from a much longer autoradiographic exposure than the other panels, reflecting the decreased sensitivity of an S1 nuclease assay using a 5′-end-labeled probe. The results, like the RNase protection data, are very clean throughout most of the gel. However, in the region representing the start site, there are four or five intense bands, unlike the single intense bands observed by primer extension or RNase protection. The intense band with the slowest mobility maps to the 5′ end of the TdT cDNA, as determined by the primer extension and RNase protection results. The other bands presumably arise because of breathing at the ends of the RNA–DNA duplex, allowing relatively efficient cleavage by S1 nuclease. This exemplifies the point made earlier, that the RNA–DNA duplex is less stable than the RNA–RNA duplex, making the S1 nuclease procedure slightly less precise than RNase protection.

Rapid Amplification of cDNA Ends

Advantages and Disadvantages

The clear advantage of rapid amplification of cDNA ends (RACE) (Box 4.4) is its tremendous sensitivity, because it is the only method that relies on PCR amplification. This sensitivity provides a unique opportunity to localize transcription start sites for genes that are inefficiently transcribed, or for genes that have proven to be refractory to other methods. An important disadvantage is that PCR can preferentially amplify products that do not represent the authentic 5′ ends of the transcripts. For example, during the initial reverse transcriptase step, the reverse transcriptase is likely to have paused or terminated with low frequency at specific nucleotides before reaching the 5′ ends of the transcripts. Because these products are shorter than the full-length cDNA products, they are likely to be preferentially amplified, making them appear to represent the major transcription start site. A similar problem arises if transcription of a gene begins at multiple sites. The site that yields the shortest cDNA product might be preferentially amplified and appear to represent the major transcription start site, even if it represents a minor start site.

A related disadvantage of the RACE procedure is that it is far less quantitative than primer extension, RNase protection, or S1 nuclease analysis. The latter three methods are sufficiently quantitative to allow them to be useful, not only for qualitatively determining the location of a transcription start site, but also for quantitating the amount of steady-state mRNA derived from a specific gene. These assays are frequently used in place of Northern blot analysis to monitor mRNA abundance in various cell lines and tissues. The RACE procedure is not as useful for this purpose.

Data Analysis

To minimize the limitations described above, RACE must be performed carefully and the results analyzed with caution. Sequencing of a large number (>30) of individual clones generated during the procedure is strongly recommended. Special attention should be paid

---

Box 4.4

RACE Procedure

The RACE procedure (Fig. 4.5) (Frohman et al. 1988; Frohman 1995) begins with a primer extension step: An oligonucleotide primer is annealed to a sequence within the first 100 or 200 nucleotides of the mRNA of interest and is extended to the 5′ end of the mRNA using reverse transcriptase. In this case, however, the primer is not radiolabeled. The subsequent objective is to use PCR to amplify the cDNA products, whose 3′ ends correspond to the transcription start site. Because the 3′ end is undefined, an additional step must be added to the PCR procedure to facilitate amplification. The new step involves extension of the 3′ end of the cDNA to yield a short region of known sequence that can serve as a primer-binding site during PCR. The 3′ end can be extended by ligation of a single-stranded oligonucleotide using RNA ligase, or by addition of a homopolymeric sequence using terminal transferase. Using a primer complementary to the extended sequence and a second primer slightly internal to the primer used for the original cDNA synthesis step, the cDNA fragment is amplified. The amplified products are inserted into a vector, several individual plasmid clones are isolated, and the sequences of each clone are determined. By sequencing many clones derived from the PCR mixture, the locations of the putative RNA start sites can be deduced. An alternative method is to perform the final PCR assays with a 5′-end-labeled primer and then to determine the size of the products on a denaturing polyacrylamide gel. By comparing the size of the products to appropriate markers, the location of the transcription start site can be predicted. CLONTECH Laboratories, Inc., Roche Molecular Biochemicals Division, and Life Technologies, Inc. (GIBCO/BRL) sell RACE kits that are accompanied by detailed protocols and troubleshooting guides. CLONTECH's kit is called the Marathon cDNA Amplification Kit (cat.# K1802-1). Roche Molecular Biochemical's product is the 5′/3′ RACE kit (cat. # 1734792); Life Technologies sells separate 5′ and 3′ amplification kits (cat. # 18374-025 and 18373-019, respectively).

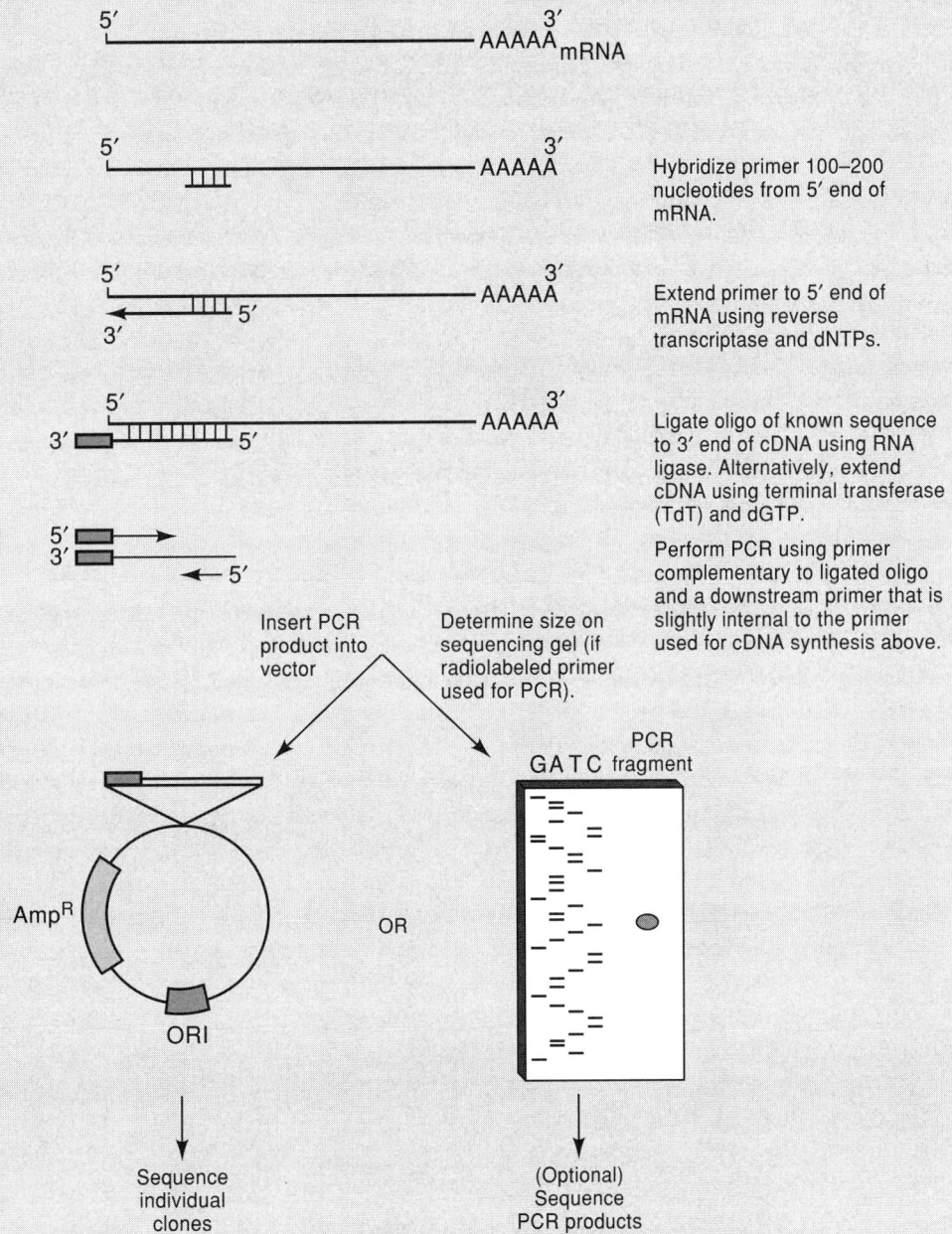

FIGURE 4.5. 5′ RACE procedure.

to the longest products obtained, rather than the most abundant products, unless independent data suggest that the shorter products represent a major start site. If the location of the major start site cannot be confirmed by another method, more advanced studies of the promoter may provide information that helps to determine the importance of the various initiation sites identified by this method.

Box 4.5

Effect of Introns on Interpretation of Start-site Mapping Results

A
Primer hybridization to Exon 1

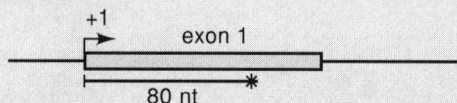

Both structures will lead to 80-nucleotide cDNA products. If primer hybridizes to exon 2 (bottom) but is thought to hybridize to exon 1 (top), start site will be incorrectly assigned to a location within first intron (bottom). By sequencing clones obtained by RACE and comparing to genomic sequence, this error will be eliminated.

Primer hybridization to Exon 2

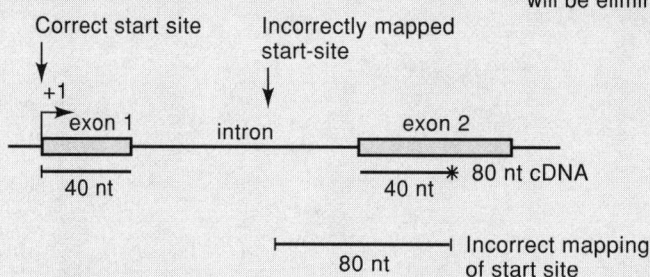

B
5′ End of probe within Exon 1

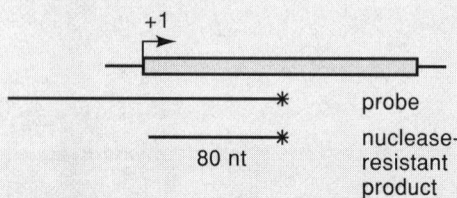

Both structures will lead to 80-nucleotide nuclease-resistant products. If probe spans boundary between first intron and second exon (bottom), rather than transcription start site (top), results will indicate distance from probe to exon-intron junction (bottom), not to transcription start site.

5′ End of probe within Exon 2

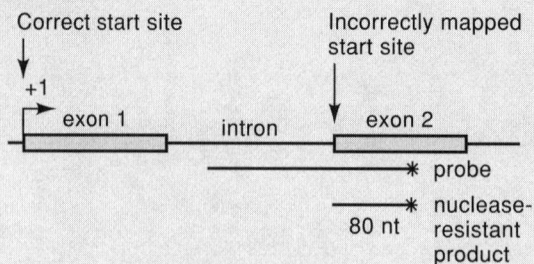

FIGURE 4.6. Influence of exon–intron structure on start-site mapping by primer extension and RACE (*A*) and RNase protection and S1 nuclease analysis (*B*).

An important consideration in developing any start-site mapping technique is the effect of introns on the interpretation of the results (see Fig. 4.6). When performing primer extension or RACE, the presence of an intron between the primer-binding site and the transcription start site can yield results that are misleading. Similarly, the presence of a splice donor or acceptor within the region spanned by an RNase protection or S1 nuclease probe has the potential to lead to an incorrect interpretation of the results. The potential impact of an intron must be emphasized because investigators have occasionally spent months studying a DNA region that was assumed to contain a promoter based on start-site mapping data, only to find that the location of the start site had been miscalculated by hundreds or thousands of base pairs because an intron influenced the outcome of the experiments.

For primer extension and RACE, the sizes of the cDNAs correspond to the distance from the 5′ end of the primer to the 5′ end of the mature mRNA. If the primer is complementary to a sequence in the first exon of the gene, the transcription start site can easily be determined if the sequence of the primer, the sequence of the genomic DNA, and the size of the cDNA product are known. If the primer is complementary to a sequence in the second exon, the start site can still be determined quite easily, but only if the location and sequence of the first exon are known, as well as the precise splice donor and acceptor sites. Most importantly, one must be careful not to assume that the primer hybridizes to the first exon simply because it is near the 5′ end of the isolated cDNA. If a primer is believed to hybridize to the first exon, but actually hybridizes to the second exon, the primer extension procedure will suggest, albeit incorrectly, that the transcription start site resides within the first intron. This concept is illustrated in Figure 4.6A. The RACE procedure provides an advantage in helping to prevent misinterpretation of the data. When the individual clones derived from RACE are sequenced, the sequences will not be colinear with the genomic DNA sequence if the original primer hybridized to a sequence within the second exon. Rather, the sequences of the RACE clones will diverge from the genomic sequence at the boundary between the second exon and the first intron, revealing the presence of the intron.

For RNase protection and S1 nuclease methods, the size of the probe fragment that is resistant to cleavage by RNases or S1 will correspond to the distance from the 5′ end of the labeled probe to the 5′ end of the mRNA, if the 5′ end of the probe hybridizes to a sequence within the first exon. However, if the 5′ end of the probe hybridizes to a sequence within the second exon (or the first intron), the resistant products will not correspond to the desired distance. The products obtained may correspond to any of a variety of distances, depending on where the probe actually hybridizes and whether the probe is internally labeled or 5′-end-labeled. An internally labeled probe has the potential to yield a band corresponding to any fragment that hybridizes to a portion of the mRNA. In contrast, a 5′-end-labeled probe will yield a band corresponding to the distance from the labeled nucleotide to the exon–intron junction (if the labeled nucleotide hybridizes within the second exon). These potential problems are illustrated in Figure 4.6B.

TECHNIQUES

PROTOCOL 4.1

Primer Extension Assay

SUMMARY

The general methodology for using retroviral reverse transcriptase and DNA primers to prepare cDNAs of specific mRNAs emerged during the early 1970s, during the development of basic molecular biology techniques and following the discovery of reverse transcriptase. This technology was adopted a few years later for the mapping of transcription start sites and the determination of relative mRNA concentrations (e.g., Ghosh et al. 1978; see Box 4.1). The start-site mapping technique using reverse trancriptase is commonly referred to as the primer extension assay.

In short, the primer extension assay requires a specific DNA primer, typically a synthetic oligonucleotide of about 20 residues, that is complementary to a sequence near the 5′ end of the mRNA of interest. The primer is 5′-end-labeled using [γ-^{32}P]ATP and T4 polynucleotide kinase and is annealed to the specific mRNA molecules within an RNA sample. The annealed primer molecules are then extended to the 5′ end of the mRNAs using reverse transcriptase and unlabeled nucleoside triphosphates. The sizes of the radiolabeled cDNA products are determined by denaturing polyacrylamide gel electrophoresis followed by autoradiography. In theory, the 3′ end of the cDNA will coincide with the 5′ end of the mRNA. Thus, the size of the radiolabeled cDNAs should represent the distance from the labeled 5′ end of the primer to the 5′ end of the mRNA (i.e., the 3′ end of the cDNA; see Box 4.1). In other words, if the cDNA product on the autoradiograph migrates at a size of 50 nucleotides when compared to adjacent DNA markers, the transcription start site may be located 50 nucleotides from the labeled 5′ nucleotide of the primer.

The primer extension assay can be performed with kits purchased from commercial sources (e.g., Promega Corp.; cat. # E3030) or by following any of a variety of protocols (e.g., Sambrook et al. 1989, pp. 7.79–7.83) that make use of different reverse transcriptases expressed in *E. coli* or purified from retrovirus-infected cells. A basic protocol that employs MMLV reverse transcriptase follows (adapted from McKnight and Kingsbury 1982). Unlike the RNase protection assay, the primer extension assay gains very few benefits from the use of commercially available kits.

TIME LINE AND ORGANIZATION

Before beginning the primer extension assay, an oligonucleotide primer complementary to a sequence within the mRNA of interest is needed. Purified, synthetic oligonucleotides are usually employed and are designed as described in the text. In addition, cytoplasmic RNA, total cellular RNA, or polyadenylated mRNA must be prepared from appropriate cell lines or tissues using any of a variety of kits or published methods (see Sambrook et al. 1989, pp. 7.1–7.36). With the purified oligonucleotide and RNA samples in hand, all steps of the primer extension assay can be completed in 1 day.

Day 1: Primer extension assay
Day 2: Data analysis

OUTLINE

Primer extension assay (time commitment: 1 day)

Step 1: Kinase the purified oligonucleotide primer (2 hours)
Step 2: Remove unincorporated [γ-^{32}P]ATP (30 minutes)
Step 3: Make buffers (30 minutes)
Step 4: Primer extension (3–4 hours)
Step 5: Polyacrylamide electrophoresis (2–3 hours)

PROCEDURE

CAUTIONS: *Actinomycin D, Bromophenol blue, DTT, Formamide, MgCl$_2$, NaOH, KCl, Radioactive substances. See Appendix I.*

Step 1: Kinase the purified oligonucleotide primer

Note: To enhance the efficiency of the kinasing reaction, contaminants should be removed from the synthetic oligonucleotide preparation by gel or column purification (see, e.g. Sambrook et al. 1989, pp. 11.21–11.30). The concentration can then be determined from the OD$_{260}$ of a diluted sample (see Sambrook et al. 1989, p. 11.30). Protocols for determining the precise extinction coefficient are available on the Operon Technologies, Inc. website (http://www.operon.com), among others.

1. Mix:

10x T4 polynucleotide kinase buffer	5 μl
gel-purified oligonucleotide	2.5 picomoles
high specific activity [γ-^{32}P]ATP	
(>5000 Ci/mmol; 10 mCi/ml)	5 μl (~10 pmoles)
dH$_2$O	to 49 μl
T4 polynucleotide kinase	10 Units (1 μl)
Total	50 μl

   10x Bacteriophage T4 polynucleotide kinase buffer:

   0.5 M Tris-Cl (pH 7.6)
   0.1 M MgCl$_2$
   50 mM DTT
   1 mM spermidine HCl
   1 mM EDTA (pH 8.0)

2. Incubate at 37°C for 45 minutes to 1 hour.

3. Heat-inactivate the kinase for 10 minutes at 68°C.

Step 2: Remove unincorporated [γ-^{32}P] ATP

Note: Unincorporated [γ-^{32}P]ATP molecules should be removed from the primer preparation to prevent them from interfering with the primer extension reactions and, in particular, with gel electrophoresis and autoradiography. Unincorporated nucleotides can be removed by differential ethanol precipitation in the presence of ammonium acetate or by column chromatography (see Sambrook et al. 1989, pp. 11.33–11.39). The rapid chromatography procedure described here makes use of a Stratagene NucTrap Probe Purification Column (cat. # 400701), which contains a gel filtration resin. The sample is applied to the top of the column, and the column is then attached to an extended disposable 10-cm³ syringe. The sample is forced through the resin by pushing down on the syringe plunger. To prevent slippage of the column during the procedure, the Push Column Beta Shield Device is strongly recommended (Stratagene, cat. # 400700).

1. Add 1 μl of 5 M NaCl to the heat-inactivated reaction sample.

2. Prewash a Stratagene NucTrap column by applying 70 μl of STE buffer.

   STE buffer:
   100 mm NaCl
   20 mm Tris-HCl (pH 7.5)
   10 mm EDTA

3. Extend the plunger of a 10-cm³ B-D syringe with a Luer-Lok tip and screw the syringe onto the column to form a seal between the column and the syringe according to the manufacturer's protocol. Do not screw the syringe on too tightly.

4. Force the buffer down the length of the column until a small drop exits at the end. Use the prewetted column within 5–10 minutes.

5. Attach the column to the Beta Shield according to the manufacturer's instructions.

6. Apply the radioactive solution to the column, attach the extended syringe, and slowly (25–35 seconds) push the sample through the column with the plunger. Collect the flowthrough in a microfuge tube. The unincorporated nucleotides will remain in the resin and the radiolabeled primer will pass through the column into the tube.

7. Apply 70 μl of STE solution to the column, push through with the syringe, and collect the flowthrough in the same tube.

8. Dilute the primer solution with an additional 140 μl of STE buffer.

9. Quantitate the radioactivity by scintillation counting (Cerenkov counting; i.e., in the absence of scintillation fluid). Cerenkov measurements vary from machine to machine, but roughly 50,000–100,000 cpm per μl should be obtained in the final primer solution.

10. Store the primer at –20°C. It can usually be used for up to 1 month.

Step 3: Make buffers for primer annealing and extension

5x PE (primer extension) hybridization buffer:
   1.25 M KCl
   50 mM Tris-HCl (pH 7.5)
   5 mM EDTA

RT (reverse transcription) buffer (store at –20°C):
   25 mM KCl
   50 mM Tris-HCl (7.5)
   10 mM DTT
   3.5 mM MgCl$_2$
   0.5 mM dCTP, dGTP, dATP, dTTP
   100 μg/ml BSA

Step 4: Primer extension

Note: The optimal hybridization temperature should be determined empirically for each primer. A good starting point is to test temperatures of 37, 45, 60, and 68°C. Some primers work well if the hybridization reactions are first heated to 68°C, then allowed to cool slowly to room temperature in a metal block.

Negative control RNAs should be included whenever possible and are essential for mapping unknown start sites. A tRNA or yeast RNA control yields less information, but can be used as an additional control. Ideally, negative control RNA should be prepared from a cell line or tissue that does not express the RNA being measured and, ideally, this cell line should be from the same species as the experimental RNA. If possible, RNAs prepared from multiple positive and negative cell lines or tissues should be tested. Without appropriate negative controls, mapping of a novel start site will be less compelling.

1. Add 1 μl of radiolabeled primer (approximately 50,000 – 100,000 cpm as measured by Cerenkov counting) to 10–60 μg of total RNA (or about 2–5 μg of poly A+ mRNA). The optimal amount of RNA to use for a primer extension reaction will vary and may need to be determined empirically. In general, the success of the procedure is less dependent on the amount of RNA than on other variables, such as the specific activity of the primer, the annealing temperature, and the presence or absence of RNA secondary structures. During initial experiments, 30 μg of total cellular RNA or total cytoplasmic RNA is an appropriate amount to use.

2. Add 3 M sodium acetate (pH 5.2) to a concentration of 0.3 M, followed by 2.5 volumes of ethanol. Incubate in dry ice for 10 minutes or at –20°C for 30 minutes to precipitate the RNA and primer.

4. Pellet the RNA and primer by centrifugation at 14,000*g* for 10 minutes in a microfuge.

5. Carefully remove the supernatant and dry the pellet in a SpeedVac or by evaporation on the bench top. Virtually all of the radioactivity should be present in the pellet, with trace amounts in the supernatant.

6. Suspend the dried pellet in 8 μl of TE buffer. The pellet will not suspend in this small volume of buffer by vortexing because the buffer will not remain at the bottom of the tube. The preferred method for pellet suspension is to "flick" the bottom of the tube repeatedly with one's finger. Alternatively, the pellet can be dissolved by repetitive pipetting of the solution with a pipetman.

7. Add 2 μl of 5× PE buffer to the tube. Mix and centrifuge briefly.

   Note: The 5× PE buffer should not be added until the RNA has dissolved in the TE because the RNA will dissolve much more slowly in the high-salt PE buffer.

8. Anneal the primer to the RNA by incubating 90 minutes at the empirically determined temperature. During the incubation, centrifuge the samples briefly every 20 minutes, so that the solution does not condense at the top of the tube.

9. Prepare a solution containing enough RT buffer and Moloney murine leukemia virus (MMLV) reverse transcriptase for all reactions (39.5 μl of RT buffer and 0.5 μl of reverse transcriptase). Since pipetmen are not perfectly accurate, it is best to make up a larger amount of the solution than is needed.

   Note: We find that the cloned MMLV SuperscriptI RT from Life Technologies, Inc. (cat. # 18053-017) works well, but excellent results can be obtained with enzymes from other sources.

10. Add 40 μl of RT mix to each reaction tube. Mix, but do not vortex. Centrifuge briefly.

11. Incubate for 60 minutes at 37°C. Higher temperatures (e.g., 42, 45, 55°C) can be used in an attempt to diminish unwanted effects of the RNA secondary structure.

12. Stop the reaction and precipitate the nucleic acids by adding 6 μl of 3 M sodium acetate and 150 μl of ethanol. Vortex well. Chill on dry ice for 10 minutes or at –20°C for 30 minutes.

13. Centrifuge for 10 minutes. A small white pellet should be visible on the bottom of the tube. Remove the ethanol. Blot the top of the tube on a Kimwipe to remove excess liquid. Most of the radioactivity should remain in the pellet.

14. Dry the pellet well in a SpeedVac roto-evaporator.

Step 5: Denaturing polyacrylamide gel electrophoresis

1. Suspend the pellet in 4 μl of formamide loading dye and 2 μl of 0.1 M NaOH.

   Note: At this point, the amount of radioactivity in each tube should still be similar to the amount added to the reactions in Step 4.1: If the amount is less, the precipitations probably did not work well or the pellets were lost.

2. Boil for 2 minutes to denature the DNA/RNA hybrids. Centrifuge briefly.

3. Load 3 μl of each sample onto an 8% denaturing polyacrylamide sequencing gel, after rinsing the wells of the gel.

4. Run the gel until the bromophenol blue dye is near the bottom. (The goal is to have the excess labeled primer migrate near the bottom of the gel.)

Step 6: Product analysis

1. Dry the gel and expose it to XAR-5 film in the presence of an intensifying screen, or analyze on a phosphorimager. Strong signals on a film should be detectable with an overnight exposure.

ADDITIONAL CONSIDERATIONS

1. The above protocol is designed for MMLV reverse transcriptase. Other reverse transcriptases, such as avian myeloblastosis virus (AMV) reverse transcriptase are commonly used, but require different reaction buffers. A protocol using AMV reverse transcriptase can be found in Chapter 14, Protocol 14.2, coupled to the in vitro transcription protocol.

2. Nuclease-free buffers and reagents, and RNase inhibitors, can be employed for RNA preparation and the primer extension reactions, but they generally are not necessary. Extensive degradation would be needed to affect the results of a primer extension assay noticeably, because only the 5′ end of each mRNA molecule needs to be intact for the primer extension assay to succeed.

3. Actinomycin D inhibits the synthesis of double-stranded DNA by reverse transcriptases and is sometimes included in the reverse transcriptase reactions to prevent the synthesis of hairpin molecules. Hairpin molecules are rarely a problem with short synthetic oligonucleotide primers, unless the expected extension products are also very

short. If desired, actinomycin D can be added to the reverse transcription buffer at a concentration of 50 µg/ml. Stock solutions of actinomycin D are prepared at 5 mg/ml in ethanol, stored in the dark at –20°C, and added to the buffer immediately before use. Alternatively, solutions of mannitol-solubilized actinomycin D (Sigma cat. # A-515) can be prepared and stored as above.

4. Synthetic oligonucleotides are typically labeled efficiently by T4 polynucleotide kinase. Rigorous quantitation of the incorporation of radioactivity therefore is not often necessary. A final Cerenkov count of the radiolabeled primer is usually sufficient as described in Step 2: Removing unincorporated $[\gamma\text{-}^{32}P]$ATP, number 10. If concerns arise about the efficiency of kinasing, more rigorous methods of quantitating incorporation can be employed, as described in Sambrook et al. (1989, p. 11.32).

TROUBLESHOOTING

The primer extension assay is difficult to troubleshoot because each problem has many potential causes and each potential cause has several possible solutions. The most common problems are weak signals and multiple bands, making it difficult to establish which bands correspond to the correct start site(s). If a reasonable amount of effort fails to yield a clear and convincing transcription start site, it will be necessary to employ at least one additional assay (i.e., RNase protection, S1 nuclease analysis, RACE) to enhance confidence in the location of the start site. Even with multiple assays, it sometimes is difficult to determine the location of a gene's transcription start sites with confidence.

Poor yield of radiolabeled primer

Possible cause: Reagents (e.g., kinase may be inactive, radiolabeled ATP may be hydrolyzed, kinase buffer may be improperly prepared, or DTT may be oxidized).
Solutions: Test new reagents. Use more rigorous method of quantitating incorporation of ^{32}P (Sambrook et al. 1989, p. 11.32).

Possible cause: Oligonucleotide may contain excess contaminants or quantitation may be inaccurate.
Solutions: Purify a new aliquot of oligonucleotide or confirm concentration. Compare yields with the yields obtained using another oligonucleotide. Use a different method of purifying oligonucleotide. Use a more rigorous method of quantitating incorporation (Sambrook et al. 1989, p. 11.32).

Possible cause: Method of removing unincorporated ATP may result in loss of radiolabeled primer.
Solutions: Repeat procedure. Try a different method of removing unincorporated nucleotides.

Most radioactivity in supernatant

At beginning of Step 4:5, most radioactivity remains in supernatant following ethanol precipitation.

Possible cause: Unincorporated ATP may have been inefficiently removed during Step 2:6.
Solutions: Repeat Step 2:6 with remaining preparation of radiolabeled primer. Try a different method of removing unincorporated nucleotides.

Possible cause: Ethanol precipitation of radiolabeled primer may have been inefficient.
Solutions: Mix ethanol/primer/RNA solution more thoroughly before chilling. Chill on dry ice for a longer amount of time. Add another 0.5 volume of ethanol. Add another 0.1 M sodium acetate.

Possible cause: Pellet may have been poured off into supernatant.
Solution: Centrifuge supernatant again and try to detect the pellet when removing the supernatant. Remove the supernatant with a drawn-out pasteur pipet.

Blank gel image, no primer signal, and no cDNA products

Possible cause: Yield of radiolabeled primer may have been low. Radioactivity in primer preparation may have been due primarily to unincorporated nucleotides.
Solutions: Run an aliquot of radiolabeled primer on the gel and expose the gel to film to determine if preparation contains radiolabeled primer. If not, proceed as in Troubleshooting: Poor yield of radiolabeled primer.

Possible cause: Radiolabeled primer may have been lost during an ethanol precipitation step.
Solutions: Carefully monitor amount of radioactivity in supernatants and pellets during each precipitation.

Weak extension products

The gel image shows a strong primer signal at the bottom of the gel, but no extension products or only very short extension products that are also present in negative controls.

Possible cause: Inefficient annealing to specific mRNA.
Solutions: Try different annealing temperatures. Try different annealing buffer components, such as formamide (see Sambrook et al. 1989, pp. 7.79–7.83). Try different primers that anneal to slightly distal or proximal sequences. In some cases, several different primers need to be tested to find one that anneals efficiently. The reasons for inefficient annealing of some primers are not known.

Possible cause: Inefficient extension by reverse transcriptase.
Solutions: Use a higher concentration or a different batch of reverse transcriptase. Prepare a new batch of reverse transcription buffer. Perform reverse transcription reactions at a higher temperature to reduce mRNA secondary structures. Check the pH of reaction buffer (in particular for the avian enzyme, which is sensitive to pH).

Possible cause: Concentration of specific mRNAs too low to detect.
Solutions: Try using more mRNA for experiment. Try purified polyadenylated mRNA. Try a method of mapping the transcription start site that is potentially more sensitive, such as the RNase protection method or RACE.

Too many background bands on gel image

Difficult to distinguish specific cDNA products from background.
Possible cause: Primer hybridizing to other RNAs in sample.
Solutions: Try higher annealing and/or extension temperatures. Try a different primer complementary to a slightly proximal or distal sequence. Confirm start site using a different

method, such as RNase protection, S1 nuclease analysis, or RACE. Try using less reverse transcriptase.

Possible cause: Inefficient extension by reverse transcriptase.

Solutions: Try higher extension temperatures to reduce RNA secondary structure. Add actinomycin D to the reverse transcription buffer. Add more reverse transcriptase or try a new preparation of reverse transcription buffer. Try a different RNA preparation, which may contain fewer contaminants that inhibit the reverse transcriptase.

PROTOCOL 4.2

RNase Protection Assay

SUMMARY

The RNase protection assay was developed in the early 1980s as a sensitive means of quantitating mRNA transcripts initiated at a specific nucleotide (Zinn et al. 1983). The method subsequently proved to be useful for defining transcription start sites and can also be used for the delineation of RNA 3′ ends and splice junctions.

Briefly, to define a transcription initiation site, a radiolabeled RNA probe spanning the anticipated start site must be prepared. This is accomplished by inserting a genomic DNA fragment in an antisense orientation downstream of a bacteriophage T7, T3, or SP6 promoter, in one of several commercially available vectors. Alternatively, PCR can be performed of the relevant genomic DNA fragment, with a bacteriophage promoter sequence included in one of the primers. Antisense transcripts from the genomic fragment are prepared using the bacteriophage RNA polymerase and are radiolabeled throughout the transcript at high specific activity by including [α-^{32}P]NTPs. The labeled probes are isolated, sometimes gel purified, and annealed to specific mRNA molecules within an RNA sample. Unannealed (single-stranded) regions of RNA are then digested with RNases. In theory, the sizes of the undigested, radiolabeled RNA fragments correspond to the distance from the transcription start site for the mRNA of interest to the end of the genomic fragment contained within the probe (see Box 4.2). As discussed in the text, the use of a commercial kit for the initial setup of an RNase protection assay is strongly recommended. The kits typically contain carefully titrated reagents and useful control templates. Kits that yield successful results are sold by Ambion, Inc. (cat. # 1410, 1412, 1420). The manufacturer's procedure included with the kit should be followed. An alternative procedure, adapted from that of Zinn et al. (1983) and Sambrook et al. (1989, pp. 7.71–7.78 and 10.29–10.37) follows.

TIME LINE AND ORGANIZATION

Before the RNase protection assay is begun, a genomic DNA fragment spanning the anticipated transcription start site must be cloned in an antisense orientation into an appropriate vector containing a bacteriophage T7, T3, or SP6 promoter (available from Promega, Stratagene, or Ambion), or amplified by PCR using a primer containing a bacteriophage promoter sequence. Before an RNA probe is prepared from a plasmid, it must be cleaved with an appropriate restriction enzyme, which will result in the termination of bacteriophage transcription at a specific point, yielding radiolabeled probe molecules of a discrete size (see Box 4.2). The restriction enzyme should yield a blunt end or a 5′ overhang, and the cleaved plasmid DNA (or PCR product) should be extracted with phenol:chloroform and precipitated with ethanol (see Sambrook et al. 1989, p. 10.31). Also needed prior to performing the RNase protection assay is purified cytoplasmic RNA, total cellular RNA, or polyadenylated mRNA from the appropriate cell lines or tissues. RNA can be purified using any of a variety of kits or published methods (see Sambrook et al. 1989, pp. 7.1–7.36). With the linearized plasmid template (or PCR-derived template) and cellular RNA in hand, the RNase protection assay can be performed in approximately 2 days.

Day 1: Prepare radiolabeled RNA probe and begin annealing to specific mRNAs
Day 2: RNase digestion and gel electrophoresis
Day 3: Data analysis

OUTLINE

Prepare radiolabeled RNA probe and begin annealing to specific RNAs (time commitment: 1–2 days)

Step 1: Make buffers (1 hour)
Step 2: Prepare probe (2 hours)
Step 3: Purify probe via PAGE (optional, 2–3 hours followed by overnight incubation)
Step 4: Hybridize probe and RNA (1 hour followed by overnight incubation)

RNase digestion and gel electrophoresis (time commitment: 1 day)

Step 5: Digest unannealed RNA with RNase (1 hour 30 minutes)
Step 6: Polyacrylamide gel electrophoresis (2–4 hours)

PROCEDURE

CAUTIONS: *Ammonium acetate, Bromophenol blue, Chloroform, DEPC, DTT, Formamide, Glycerol, KC1, MgCl$_2$, Phenol, Radioactive substances, SDS, Xylene. See Appendix I.*

Step 1: Make buffers

Note: To diminish RNase contamination, water treated with DEPC can be used to prepare the following buffers.

5x Transcription buffer:
> 200 mM Tris (pH 7.5)
> 30 mM MgCl$_2$
> 10 mM spermidine
> 50 mM NaCl

Elution buffer (needed if probe required gel purification):
> 0.5 M ammonium acetate
> 1 mM EDTA
> 0.1% SDS

Hybridization buffer:
> 80% formamide
> 40 mM PIPES (pH 6.4)
> 400 mM NaCl
> 1 mM EDTA

RNase digestion buffer:
> 10 mM Tris-HCl (pH 7.5)
> 5 mM EDTA
> 300 mM NaCl
>
> Add RNases just before use:
> 40 µg/ml RNase A
> 2 µg/ml RNase T1

Step 2: Prepare probe

1. Mix the following components in a microfuge tube. Mix at room temperature in the order shown:

dH$_2$O	2 μl
linearized template DNA (1 mg/ml)	1 μl
10 mM ATP	1 μl
10 mM CTP	1 μl
10 mM GTP	1 μl
100 mM DTT	2 μl
5x transcription buffer	4 μl
placental RNase inhibitor (10 Units/ μl)	1 μl
[α-^{32}P]UTP (10 mCi/ml; 800 Ci/mmol)	5 μl
SP6, T7, or T3 RNA polymerase (5 Units/ μl)	2 μl
Total	20 μl

2. Mix the contents by tapping the outside of the tube (do not vortex) and then briefly centrifuge.

3. Incubate the reaction mix for 1 hour at 37°C.

4. Digest the template DNA by adding 1 μl of RNase-free DNase I (2 units/μl) to the reaction. Mix by tapping and centrifuge briefly.

5. Incubate the reaction for an additional 15 minutes at 37°C.

   Note: Some probes, particularly longer probes (> 100 nucleotides), may need to be gel-purified to remove shorter RNA products that can increase the number of background products observed in the RNase protection assay. The radiolabeled probe can be applied to a gel for purification at this point (see Step 3: Purify probe via PAGE, below).

6. Add 80 μl of 2.5 M ammonium acetate and 5 μg of yeast RNA (or tRNA) to the reaction.

7. Extract with 100 μl of phenol:chloroform, followed by 100 μl of chloroform.

8. Add 300 μl of ethanol. Chill for 10 minutes on dry ice or 30 minutes at –20°C.

9. Pellet the RNA in a microcentrifuge for 10 minutes at 4°C.

10. Resuspend the RNA in 100 μl of 0.3 M sodium acetate. Precipitate as above following addition of 300 μl of ethanol.

11. Dry the RNA pellet.

12. Suspend the pellet in 50 μl of hybridization buffer.

13. Determine the activity of the probe by counting 1 μl in a scintillation counter.

14. The probe may be stored at –20°C for a few days, but optimal results are obtained when a newly prepared probe is used for the subsequent steps.

Step 3: Purify probe via PAGE

1. Following the DNase digestion in Step 2:5, add an equal volume of formamide gel loading buffer (21 μl; 95% formamide, 20 mM EDTA, 0.05% bromophenol blue, 0.05% xylene cyanol) to the reaction mix and heat for 3–5 minutes at 85–95°C.

2. Load the entire reaction on an 8% denaturing acrylamide gel.

3. Excise the full-length fragment from the gel and elute overnight in 350 µl of elution buffer.

   To excise a radiolabeled fragment from a polyacrylamide gel, mark the corners of the gel with fluorescent dye, cover the wet gel with Saran Wrap, and expose to X-ray film. A short exposure of <1 minute should be sufficient to detect the radiolabeled RNA. Use the fluorescent markers to align the film and the covered gel. Excise the polyacrylamide fragment containing the radiolabeled RNA through the Saran Wrap using a scalpel or razor blade. Slice the fragment into several pieces and transfer to a microfuge tube. Alternatively, the fragment can be crushed in the microfuge tube using a glass rod. Add the elution buffer.

4. Separate the gel fragment from the probe by centrifugation followed by transfer of the aqueous solution containing the radioactive RNA to a new microfuge tube. A second centrifugation and transfer of the aqueous solution can be used to extract the remaining probe from the gel fragments.

5. Determine the specific activity of the probe by counting 1 µl in a scintillation counter.

6. The probe may be stored in elution buffer at –20°C for 3–4 days.

Step 4: Hybridize probe and RNA

1. Mix the labeled probe (2–8×10^4 cpm) with 10–30 µg of each RNA sample.

2. Add 0.1 volume of 3 M sodium acetate (pH 5.2) and 2.5 volumes of ethanol. Mix thoroughly and place on dry ice for 10 minutes.

3. Pellet the probe and RNA by centrifugation for 15 minutes in a microfuge at 14,000 rpm at 4°C.

4. Carefully remove the ethanol from each tube and dry the pellet on the bench top at room temperature.

5. Resuspend each pellet in 30 µl of hybridization buffer by vortexing and then briefly centrifuging the tubes.

6. Heat the reaction tubes at 90°C for 3–4 minutes, then vortex and centrifuge briefly.

7. Immediately place tubes in a water bath set at the empirically determined annealing temperature (often 45°C).

8. Allow hybridization to proceed overnight (or 8 hours).

Step 5: Digest hybridized probe with RNase

1. Add 300 µl of RNase digestion buffer (including RNases) to each reaction. Mix and centrifuge briefly in a microfuge.

2. Incubate the tubes for 60 minutes at 30°C.

3. Stop the RNase digestion by adding proteinase K (10 µl of a 10 mg/ml solution) and SDS (10 µl of a 20% solution). Mix and centrifuge briefly. Incubate for 15 minutes at 37°C.

4. Extract with 350 µl of phenol:chloroform.

5. Add 2 µg of tRNA or yeast RNA and 1 ml of ethanol. Chill on dry ice for 10 minutes or at –20°C for 30 minutes.

6. Pellet the RNA by centrifugation in a microfuge for 15 minutes at 14,000 rpm and 4°C. Carefully remove all supernatant from each tube. Residual supernatant will cause aberrant migration of bands in the gel. Dry the pellets on the bench top at room temperature.

7. Dissolve the pellets in 8 μl of formamide gel loading buffer (95% formamide, 20 mM EDTA, 0.05% bromophenol blue, 0.05% xylene cyanol).

Step 6: Gel electrophoresis and product analysis

1. Heat the tubes for 3–4 minutes at 90°C.

2. Vortex and centrifuge briefly before loading the samples on an 8% denaturing polyacrylamide gel. Run the gel until bromophenol blue is at the bottom of the gel.

3. Dry the gel and expose to X-ray film with an intensifying screen or analyze on a phosphorimager.

ADDITIONAL CONSIDERATIONS

1. An ideal length for an RNase protection probe is difficult to determine. Longer probes (250 nucleotides or more) possess the advantage of higher specific activities, as a larger number of radiolabeled nucleotides can be incorporated into each probe molecule. Short probes (less than 250 nucleotides) can be advantageous because the efficiency of full-length probe production by the bacteriophage RNA polymerase is increased and gel purification of the probe is often not necessary.

2. The above protocol for probe preparation makes use of radiolabeled UTP. Other radiolabeled nucleotides can be used instead of UTP. GTP has a lower K_m for some bacteriophage polymerases, allowing higher-specific-activity radionucleotides to be used during probe preparation, resulting in higher specific activity probes. ATP has a particularly high K_m and is rarely used for probe preparation.

3. As with the primer extension assay, the best negative control for the RNase protection assay is RNA from a cell line or tissue that is not expected to contain the gene of interest. If possible, the RNA should be derived from the same species as the RNAs used for the experiment.

TROUBLESHOOTING

Low yield of probe

Possible cause: Inefficient RNA synthesis due to polymerase termination at fortuitous sites.
Solutions: Increase concentrations of bacteriophage RNA polymerase and/or radiolabeled nucleotide, and gel-purify the resulting full-length probe. (Although most transcripts may still terminate prematurely, sufficient quantities of full-length probe may be obtained for the subsequent experiments.) Insert the same genomic fragment downstream of the promoter for a different bacteriophage polymerase, or try a slightly shorter genomic DNA fragment that still spans the anticipated transcription start site.

Possible cause: Contamination of reagents or tubes with RNases.
Solutions: Follow standard procedures for elimination and inhibition of RNases (Sambrook et al. 1989, pp. 7.3–7.5). Use highly purified plasmid DNA.

Possible cause: Inactive or improperly prepared reagents.
Solutions: Test new batches of reagents, in particular the radiolabeled nucleotide, bacteriophage RNA polymerase, and the template DNA. Use highly purified plasmid DNA to reduce contamination by RNases.

No RNase-resistant products on gel image

Possible cause: Overdigestion of probe/RNA hybrids by RNases.
Solutions: Reduce the concentration of RNases or the time of digestion. Try different concentrations of both RNase A and RNase T1. Try RNase T1 in the absence of RNase A. As a control, add no RNases to the probe/RNA hybrids; with this control, a full-length probe should be observed.

Possible cause: Inefficient annealing of probe and specific RNAs.
Solutions: In formamide hybridization buffers, most probes should anneal to the specific mRNAs at 45°C. However, the optimal hybridization temperature for some probes may be different.

Possible cause: Contamination of reagents or tubes with RNases.
Solutions: Follow standard procedures for elimination and inhibition of RNases (Sambrook et al. 1989, pp. 7.3–7.5). Use highly purified plasmid DNA.

Possible cause: Low abundance or absence of specific mRNA.
Solutions: Use higher concentrations of cellular RNA for the experiment. Use polyadenylated mRNA selected by oligo-dT chromatography. Prepare a longer probe to increase specific activity.

Multiple RNase-resistant products on gel image

Possible cause: Underdigestion of probe/RNA hybrids by RNases.
Solutions: Increase concentration of RNases or time of digestion. Try different concentrations of both RNase A and RNase T1.

Possible cause: Self-annealing of regions of the probe RNA, annealing to the probe molecules generated from the opposite strand of the template, or annealing to other RNAs in the cellular RNA sample.
Solutions: Perform controls to test for these possibilities. Self-annealing or annealing to probe molecules from the opposite strand will result in RNase-resistant products when no mRNA is included in the reaction. Annealing to other RNAs in the cellular RNA sample is likely to result in RNase-resistant products when RNA from a cell line that does not express the gene of interest is used for the experiment. The RNA should be derived from the same species. By including proper controls, it may not be necessary to eliminate unwanted bands because the bands that do not result from specific hybridization can be identified. Nevertheless, a shorter probe, a higher hybridization temperature, or an increased concentration of RNases may help to eliminate these unwanted bands.

PROTOCOL 4.3

S1 Nuclease Assay

SUMMARY

The S1 nuclease assay was developed by Berk and Sharp (1977) as a method for mapping of 5′ ends of transcripts derived from the adenovirus genome. The method proved to be generally useful for this purpose as well as for determining exon–intron junctions of mRNAs. Today, the S1 assay is used less frequently than the RNase protection and primer extension assays, which often are more sensitive (both assays) and can be performed in a shorter amount of time (primer extension). Nevertheless, the S1 nuclease assay remains a valid alternative for identifying or confirming the location of a transcription start site.

Briefly, to measure a transcription start site using the S1 assay, a radiolabeled single- or double-stranded DNA probe spanning the anticipated start site must be prepared. This can be accomplished by any of several methods (see text and below). The radiolabeled DNA probe is annealed to the specific mRNA molecules within an RNA sample. Unannealed regions of the probe are then digested with the S1 endonuclease, leaving resistant fragments of a specific size that allow the putative location of the transcription start site to be determined. In theory, the sizes of the S1-resistant, radiolabeled DNA fragments correspond to the distance from the transcription start site for the mRNA of interest to the end of the genomic fragment contained within the probe (see Box 4.3).

An S1 nuclease assay kit is available from Ambion, Inc. (cat. #1425). Several protocols for performing the S1 assay have been published (e.g., Sambrook et al. 1989, pp. 7.58–7.70; Ausubel et al. 1994, Unit 4.6). The published protocols vary primarily in the type of probe (single-stranded DNA, double-stranded DNA, or synthetic oligonucleotide) used for the experiment and the corresponding modifications in the hybridization and digestion conditions that are needed for each type of probe. The example described below makes use of a single-stranded DNA probe and was derived from the method of Berk and Sharp (1977) as modified by Henschel et al. (1980).

TIME LINE AND ORGANIZATION

Before the S1 nuclease assay is performed, a single-stranded, radiolabeled DNA probe spanning the anticipated transcription start site must be prepared. If a double-stranded probe is preferred, Sambrook et al. (1989, pp. 7.58–7.65) should be consulted for the probe preparation method and hybridization conditions (which must promote RNA–DNA hybrids over DNA–DNA hybrids). A protocol for preparing single-stranded probes is not included here because of the abundance of available methods, as described in the text. The simplest method for preparing a single-stranded probe is to phosphorylate the 5′ end of a long, synthetic oligonucleotide spanning the anticipated transcription start site. This type of probe can be prepared using the method described in Protocol 4.1 for preparation of a primer extension primer. Ausubel et al. (1994, Unit 4.6) should also be consulted for more detailed information. An important limitation of an oligonucleotide probe for S1 analysis is that the location of the transcription start site must already be known with considerable precision because the probe length is restricted by the current oligonucleotide synthesis technology. Other single-stranded probe preparation methods that should be considered involve the use of M13-based plasmids (see text and Fig. 4.3; Sambrook et al. 1989, pp. 10.18–10.26; Ausubel et al. 1994, Unit 4.6) or asymmetric PCR (Kaltenbroeck et al. 1992; Ausubel et al. 1994, Unit 15.2).

In addition to the single-stranded probe, purified cytoplasmic RNA, total cellular RNA, or polyadenylated mRNA from the appropriate cell lines or tissues must be obtained for the S1 assay. RNA can be purified using any of a variety of kits or published methods (see Sambrook et al. 1989, pp. 7.3–7.36).

With the probe and RNA in hand, the S1 nuclease assay can be performed in 2 days. The probe and RNA are mixed and hybridized to one another to completion, usually overnight. The samples are then digested with nuclease S1. The S1-resistant products are analyzed on a denaturing polyacrylamide gel, followed by autoradiography.

Day 1: Begin hybridization
Day 2: S1 digestion and gel electrophoresis
Day 3: Data analysis

OUTLINE

Step 1: Prepare buffers (1 hour)
Step 2: Hybridize RNA to radioactive probe (2 hours, incubate overnight)
Step 3: S1 nuclease digestion (2 hours)
Step 4: Polyacrylamide gel electrophoresis (3 hours)
Step 5: Data analysis

PROCEDURE

CAUTIONS: *Bromophenol blue, Formamide, MOPS, Radioactive substances, Xylene. See Appendix I.*

Step 1: Prepare buffers

Hybridization buffer:
　　8 µl of formamide
　　2 µl of 5x hybridization mix

5x Hybridization mix:
　　2 M NaCl
　　200 mM MOPS (pH 6.7)
　　5 mM EDTA

S1 buffer:
　　300 mM NaCl
　　30 mM sodium acetate (pH 4.5)
　　3 mM $ZnSO_4$

　　Note: A 1 M sodium acetate pH 4.5 stock solution can be prepared by mixing 1 M acetic acid and 1 M sodium acetate until the desired pH is obtained. The pH of the solution may change when diluted to 30 mM and may require readjustment.

Stop buffer:
　　400 mM EDTA
　　5 µg/µl tRNA or yeast RNA

Step 2: Hybridize RNA to radioactive probe

1. Add 5,000–10,000 cpm of the single-stranded probe to 30 µg of RNA.

2. Precipitate probe and RNA by adding 3 M sodium acetate to a final concentration of 0.3 M and 2.5 volumes of ethanol. Chill on dry ice for 10 minutes. Pellet probe and RNA by microcentrifugation at 14,000 rpm for 10 minutes.

3. Carefully remove the supernatant and discard in the radioactive waste. Wash the pellet with 80% ethanol and centrifuge for 5 minutes at 14,000 rpm. Discard the supernatant. Dry the pellet at room temperature on the bench top.

4. Dissolve the dried pellet in 10 µl of hybridization buffer.

5. Place at 90°C for 5 minutes to denature probe and RNA.

   Note: If the S1 nuclease analysis is being performed to analyze RNA expressed from transiently transfected cells or RNA synthesized in an in vitro transcription assay, the hybridization mixture should not be heated to denature the probe and RNA. Heating to 90°C would denature the plasmid DNA that usually copurifies with the RNA. The denatured plasmid DNA would then hybridize to many of the specific mRNA molecules and also to some of the probe molecules, resulting in increased background and a weaker signal. In most cases, efficient hybridization will occur in the absence of the 90°C heating step. Two alternative methods for eliminating the effects of plasmid DNA are to (1) purify the RNA from the plasmid DNA by oligo(dT) chromatography, or (2) degrade the plasmid DNA by DNase I digestion, followed by phenol:chloroform extraction.

6. Immediately transfer the reaction tubes into a water bath or temperature block for hybridization. The optimum hybridization temperature must be determined empirically. An appropriate starting point would be to try 37°C, 42°C, 50°C, and 60°C.

7. Incubate overnight (or at least 6 hours) to ensure complete hybridization.

Step 3: S1 nuclease digestion

1. Add 190 µl of ice-cold S1 buffer to the hybridization reaction.

2. Add 20–50 units of S1 nuclease.

   Note: S1 nuclease amounts needed for digestion must be determined empirically. Usually, 20–50 units provide an appropriate starting point. The amounts can be varied after the results from the initial experiments are observed following autoradiography.

3. Incubate for 1 hour at 37°C. In this step, S1 nuclease digests the unannealed single-stranded DNA probe.

4. Stop the reaction by adding 5 µl of Stop buffer.

5. Add 2.5 volumes of ethanol.

6. Chill on dry ice for 15 minutes to precipitate the nucleic acid.

7. Centrifuge for 15 minutes in a microfuge at 14,000 rpm. Before centrifugation, warm the reaction tubes to room temperature to prevent excess salt from pelleting.

8. Discard the supernatant and wash the pellet with 80% ethanol. Centrifuge at 14,000 rpm for 5 minutes. Discard the supernatant and dry the pellet.

9. Dissolve the pellets in 8 µl of formamide gel loading buffer (95% formamide, 20 mM EDTA, 0.05% bromophenol blue, 0.05% xylene cyanol).

Step 4: Polyacrylamide gel electrophoresis

1. Heat the tubes for 3–4 minutes at 90°C.

2. Vortex and centrifuge briefly before loading the samples on an 8% denaturing polyacrylamide gel. Include the products of the DNA sequencing reactions as markers. Also, load 1,000 cpm of probe preparation as a marker for the full-length probe.

Step 5: Data analysis

1. Dry the gel under vacuum onto 3MM Whatman paper.

2. Perform autoradiography with an intensifying screen or phosphorimager analysis.

ADDITIONAL CONSIDERATIONS

1. The ideal length for the radiolabeled probe is difficult to determine. To allow the size of the S1 nuclease-resistant product to be mapped accurately, it should be 150 nucleotides or less. Therefore, the 5′ end of the probe should be no more than 150 nucleotides from the anticipated transcription start site. The extent to which the probe extends past the transcription start site in the opposite direction is less relevant.

2. As with the primer extension and RNase protection assays, the best negative control is RNA from a cell line or tissue that does not express the gene of interest. If possible, the negative control RNA should be derived from the same species as the RNAs used for the experiment.

3. When using the S1 nuclease assay to measure RNAs transcribed from transiently transfected cells or RNAs synthesized in an in vitro transcription assay, S1 nuclease-resistant products that correspond to the full-length genomic sequence within the probe are frequently observed, in addition to the S1 nuclease-resistant product that represents accurately initiated transcripts. The full-length resistant products are obtained because transcripts initiate at numerous cryptic sites within any transiently transfected plasmid and within any plasmid used in an in vitro transcription assay. Therefore, transcripts that span the entire probe molecule will be present and will lead to full-length S1-resistant products.

TROUBLESHOOTING

Weak signal or no signal observed on gel image

Possible cause: Suboptimal hybridization or S1 digestion conditions.
Solutions: Perform several reactions with different concentrations of the S1 nuclease. Perform several reactions with different hybridization temperatures. Try heating hybridization reactions to 90°C in temperature block and then allowing them to cool slowly to room temperature.

Possible cause: Low abundance of specific mRNA.
Solutions: Use more RNA for each reaction. Try an internally labeled probe to increase sensitivity. Try RNase protection or primer extension assays, which are usually more sensitive. Optimize hybridization temperature and S1 nuclease digestion to enhance specific signals.

Several bands observed on gel image

Note: The presence of numerous bands on a gel image is a common result and makes the S1 assay (or the RNase protection and primer extension assays) extremely difficult to interpret. Specifically, it is difficult to determine if the bands observed correspond to multiple transcription start sites, to background bands, or to a combination of the two.

Possible cause: Suboptimal S1 digestion or hybridization conditions.

Solutions: Optimize conditions as described above for problem 1. The presence of multiple bands suggests that the probe is hybridizing to some mRNA molecules within the cell. However, optimization of the hybridization conditions may result in an enhancement of the specific hybrid and a decrease in nonspecific hybrids. If a body-labeled probe is being used, try using an end-labeled probe to decrease background.

Possible cause: Multiple authentic transcription start sites.

Solutions: Confirm using another assay, such as RNase protection, primer extension, or RACE.

REFERENCES

Armes N. and Fried M. 1995. The genomic organization of the region containing the *Drosophila melanogaster* rpL7a (Surf-3) gene differs from those of the mammalian and avian surfeit loci. *Mol. Cell. Biol.* **15:** 2367–2373.

Ausubel F.M., Brent R.E., Kingston E., Moore D.D., Seidman J.G., Smith J.A., and Struhl K. 1994. *Current protocols in molecular biology.* John Wiley and Sons, Inc., New York.

Berk A.J. and Sharp P.A. 1977. Sizing and mapping of early adenovirus mRNAs by gel electrophoresis of S1 endonuclease-digested hybrids. *Cell* **12:** 721–732.

Frohman M.A. 1995. Rapid amplification of cDNA ends. In *PCR primer: A laboratory manual* (ed. C.W. Dieffenbach and G.S. Dveksler), pp. 381–409. Cold Spring Harbor Laboratory Press, Cold Spring Harbor, New York.

Frohman M.A., Dush M.K., and Martin G.R. 1988. Rapid production of full-length cDNAs from rare transcripts: Amplification using a single gene-specific oligonucleotide primer. *Proc. Natl. Acad. Sci.* **85:** 8998–9002.

Ghosh P.K., Reddy V.B., Swinscoe J., Lebowitz P., and Weissman S.M. 1978. Heterogeneity and 5′-terminal structures of the late RNAs of simian virus 40. *J. Mol. Biol.* **126:** 813–846.

Hentschel C., Probst E., and Birnstiel M.L. 1980. Transcriptional fidelity of histone genes injected into *Xenopus* oocyte nuclei. *Nature* **288:** 100–102.

Kaltenboeck B., Spatafora J.W., Zhang X., Kousoulas K.G., Blackwell M., and Storz J. 1992. Efficient production of single-stranded DNA as long as 2 kb for sequencing of PCR-amplified DNA. *Biotechniques* **12:** 168–171.

Kozak M. 1987. An analysis of 5′-noncoding sequences from 699 vertebrate messenger RNAs. *Nucleic Acids Res.* **15:** 8125–8148.

———1996. Interpreting cDNA sequences: Some insights from studies on translation. *Mamm. Genome.* **7:** 563–574.

Lo K., Landau N.R., and Smale S.T. 1991. LyF-1, a transcriptional regulator that interacts with a novel class of promoters for lymphocyte-specific genes. *Mol. Cell. Biol.* **11:** 5229–5243.

Lockard R.E. and RajBhandary U.L. 1976. Nucleotide sequences at the 5′ termini of rabbit alpha and beta globin mRNA. *Cell* **9:** 747–760.

McKnight S.L. and Kingsbury R. 1982. Transcriptional control signals of a eukaryotic protein-coding gene. *Science* **217:** 316–324.

Melton D.A., Krieg P.A., Rebagliati M.R., Maniatis T., Zinn K., and Green M.R. 1984. Efficient in vitro synthesis of biologically active RNA and RNA hybridization probes from plasmids containing a bacteriophage SP6 promoter. *Nucleic Acids Res.* **12:** 7035–7056.

Murphy T.L., Cleveland M.G., Kulesza P., Magram J., and Murphy K.M. 1995. Regulation of interleukin 12 p40 expression through an NF-kappa B half-site. *Mol. Cell. Biol.* 15: 5258–5267.

Sambrook J., Fritsch E.F., and Maniatis T. 1989. *Molecular cloning: A laboratory manual.* Cold Spring Harbor Laboratory Press, Cold Spring Harbor, New York.

Schatz D.G., Oettinger M.A., and Baltimore D. 1989. The V(D)J recomination activating gene, RAG-1. *Cell* 59: 1035–1048.

Slansky J.E. and Farnham P.J. 1996. Transcriptional regulation of the dihydrofolate reductase gene. *Bioessays* **18:** 55–62.

Smale S.T. and Baltimore D. 1989. The "initiator" as a transcription control element. *Cell* **57:** 103–113.

Weaver R.F. and Weissmann C. 1979. Mapping of RNA by a modification of the Berk-Sharp procedure: The 5′ termini of 15 S beta-globin mRNA precursor and mature 10 S beta-globin mRNA have identical map coordinates. *Nucleic Acid Res.* **7:**1175—1193.

Zinn K., DiMaio D., and Maniatis T. 1983. Identification of two distinct regulatory regions adjacent to the human beta-interferon gene. *Cell* **34:** 865–879.

WWW RESOURCES

http://www.ibc.wustl.edu/~zuker/rna Dr. Michael Zuker's laboratory's RNA webpage at Institute for Biomedical Computing, Washington University, 700 S. Euclid Ave., St. Louis, MO 63110. This site contains much information on, as well as access to, RNA secondary structure prediction programs and documentation.

http://www.operon.com Operon Technologies, Inc., 1000 Atlantic Ave., Ste. 108, Alameda, CA 94501 (phone 510-865-8644, toll-free 800-688-2248; fax 510-865-5255). Follow the DNA Synthesis link to tool kit for determining the precise extinction coefficient.

CHAPTER 5

Functional Assays for Promoter Analysis

Important issues

- *The development of a functional assay is an essential early step in the analysis of a promoter.*

- *Transient and stable transfection assays, in vitro transcription assays, and transgenic and homologous recombination assays possess distinct advantages and disadvantages.*

- *Several issues must be considered during the development of transient and stable transfection assays.*

INTRODUCTION

A critical goal when initiating an analysis of a gene's transcriptional regulatory mechanisms is the development of an assay for measuring the activity of relevant *cis*-acting control regions (see Chapter 2). With an appropriate functional assay, it is possible to assess a control region's ability to mimic accurately the expression pattern of the endogenous gene. Mutations can then be introduced into the control region to identify important regulatory elements and, ultimately, important transcription factors. In the absence of a functional assay, it is very difficult, if not impossible, to identify and assess the relevance of DNA sequence elements and proteins that may contribute to gene regulation.

Several types of functional assays have been used to study transcriptional regulation. Most common is the transient transfection assay, in which plasmids containing the control region of interest are introduced by one of several transfection procedures into cells maintained in culture (Fig. 5.1). Typically, the control region regulates transcription of a so-called "reporter gene," a known gene whose mRNA or protein level can be measured easily and accurately. If the regulatory region of interest is a promoter, it is placed immediately upstream of the reporter gene, such that the promoter will drive reporter gene transcription. If the control region of interest is an enhancer or other control region that appears to function at a distance from the promoter, a well-characterized promoter is usually placed upstream of the reporter gene, with the enhancer inserted upstream of the promoter or downstream of the reporter gene. The downstream location is generally preferred because placement of an enhancer in close proximity to a promoter may alter its properties, as individual elements may cooperate in an aberrant manner with the nearby promoter elements.

At a specific time point following transfection of cultured cells with the resulting plasmid, the activity of the control region is assessed by measuring mRNA or protein synthesis from the reporter gene. This assay is considered to be transient because the plasmids remain episomal and rarely integrate into the host genome. As such, mRNA or protein production must be measured within a short time period, ranging from 1 to 3 days; otherwise, the plasmids degrade or are diluted as the cells grow and divide. Despite a number of limitations (see below), the transient assay is usually employed for an initial analysis of the *cis*-acting DNA sequences and *trans*-acting factors that regulate gene expression.

For some control regions, alternative assays are essential or more desirable than transient assays. Alternative assays may be necessary, for example, for control regions that do not exhibit the expected activity in a transient assay, or for control regions that depend on a specific chromatin structure (see Chapter 1 and below). One alternative is the stable transfection assay, in which the plasmid containing the reporter gene and control region of interest becomes stably incorporated into the genome (Fig. 5.2). A drug-resistance gene

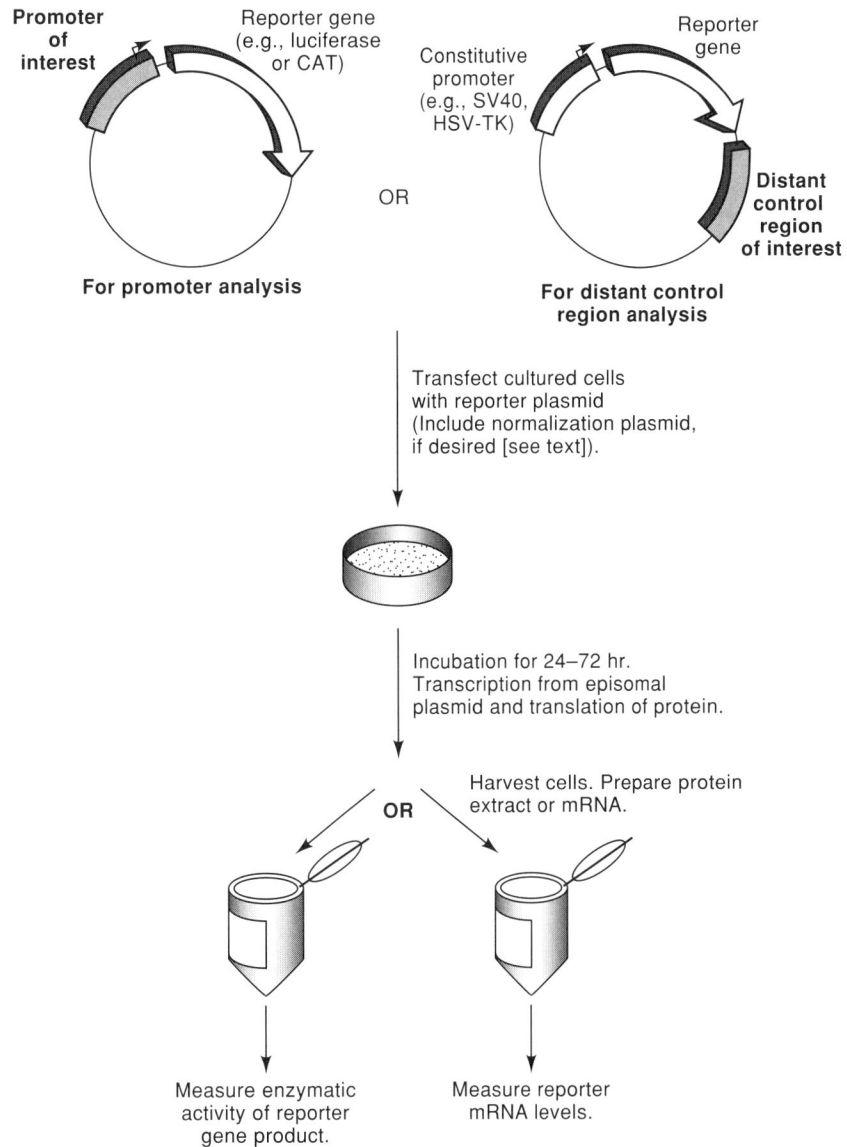

FIGURE 5.1. Transient transfection assay.

(i.e., dominant selectable marker gene) under the control of a constitutively active promoter is also needed, either within the same plasmid as the reporter gene or within a separate plasmid.

The plasmid or plasmids containing the reporter and drug-resistance genes are transfected into cultured cells. Cells that have stably integrated the plasmid into a chromosome are selected by supplementing the growth medium with the drug, which kills the cells that do not stably express the drug-resistance gene. In most cases, several plasmid molecules become ligated to one another within the cell, and each multimer integrates at a fairly random location within the genome. Each stably transfected cell therefore possesses a unique integration site. The activity of the control region of interest is then determined by measuring reporter gene activity in pools of stably transfected cells. Alternatively, several individual cell clones can be isolated and expanded, so that all cells in each clonal culture have the plasmid integrated into the same chromosomal location (see below).

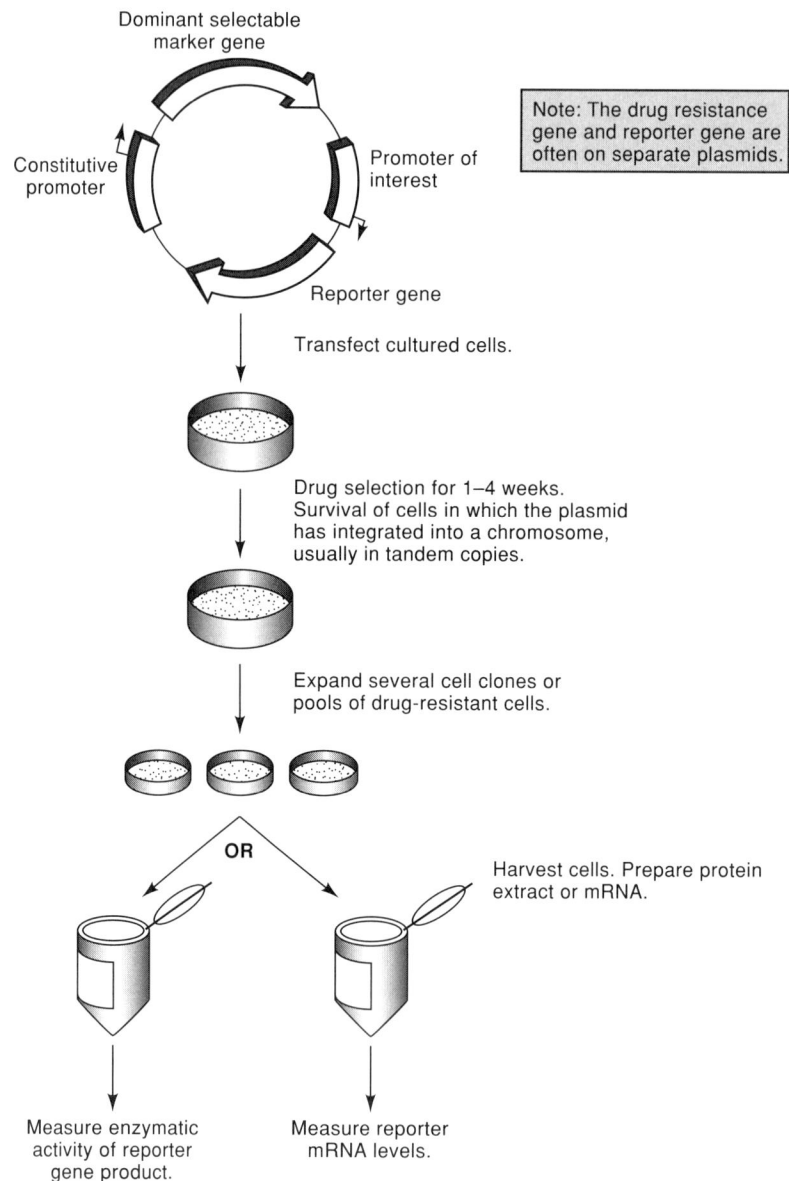

Note: The drug resistance gene and reporter gene are often on separate plasmids.

Dominant selectable marker gene

Constitutive promoter

Promoter of interest

Reporter gene

Transfect cultured cells.

Drug selection for 1–4 weeks. Survival of cells in which the plasmid has integrated into a chromosome, usually in tandem copies.

Expand several cell clones or pools of drug-resistant cells.

OR

Harvest cells. Prepare protein extract or mRNA.

Measure enzymatic activity of reporter gene product.

Measure reporter mRNA levels.

FIGURE 5.2. Stable transfection assay.

A variation of the stable transfection assay is to use episomal reporter plasmids that are stably maintained in a constant copy number via cell-cycle regulation of replication (see Sambrook et al. 1989, pp. 16.23–16.27). Replication origins from appropriate viruses, along with a drug-resistance gene under the control of a constitutively active promoter, must be included in the reporter plasmid to allow maintenance of the episome. Following transfection, the drug is added to the medium to select the transfected cells, and reporter gene activity is measured to determine the activity of the control region of interest.

In addition to the transfection methods, three other functional assays can be considered. One is the in vitro transcription assay, in which the function of a control region is measured in a cell-free extract (Fig. 5.3). Another assay that has been used to study the regulation of transcription is a transgenic assay in which the reporter gene and control region of interest are stably integrated at fairly random locations into the genome of an animal,

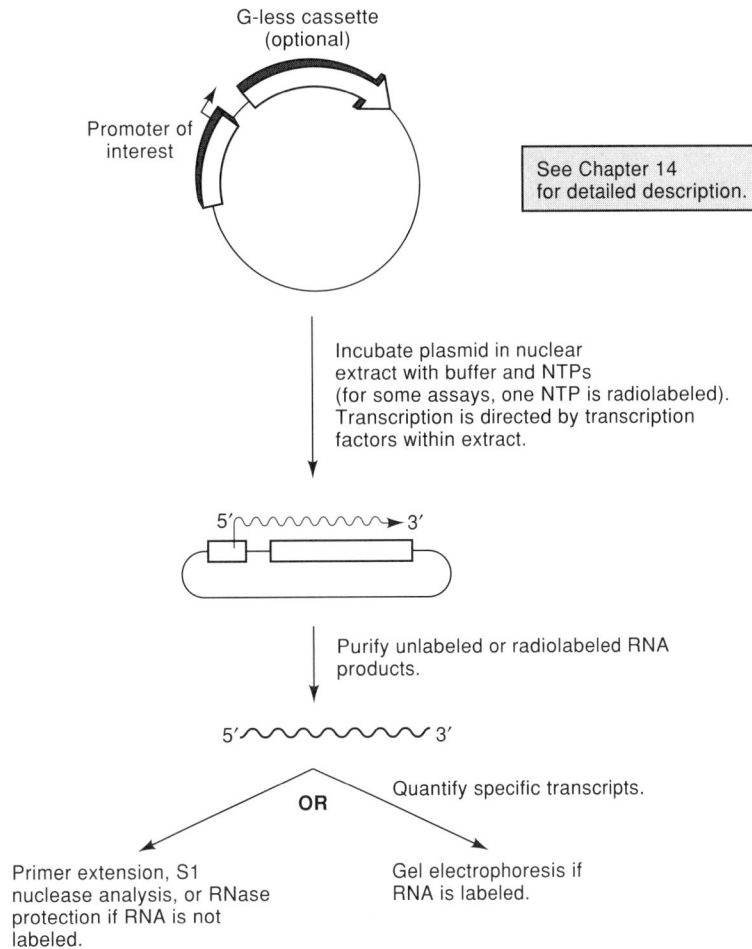

FIGURE 5.3. In vitro transcription assay.

to measure the precise activities of the control region in a more natural context (Fig. 5.4). Although rarely used, a final assay for delineating functional control regions and control elements is a homologous recombination assay in which the endogenous gene is manipulated within the genome of a cultured cell or animal (Fig. 5.5).

The Concepts and Strategies section below describes the issues that must be considered when choosing an appropriate functional assay. Strategic issues and methods are then discussed in detail for the transient and stable transfection assays. In vitro transcription assays are discussed separately in Chapter 14. Specific strategies and techniques are not presented for animal assays or homologous recombination, as they are beyond the scope of this book. Stable transfection experiments using plasmids maintained as episomes are also not discussed in detail.

CONCEPTS AND STRATEGIES

Choosing an Assay: Advantages and Disadvantages of Each Assay

The functional assays available for analyzing a control region often determine the level of success of the analysis. Considerable thought must therefore be applied when choosing an assay. Unfortunately, none is ideal and each has important limitations.

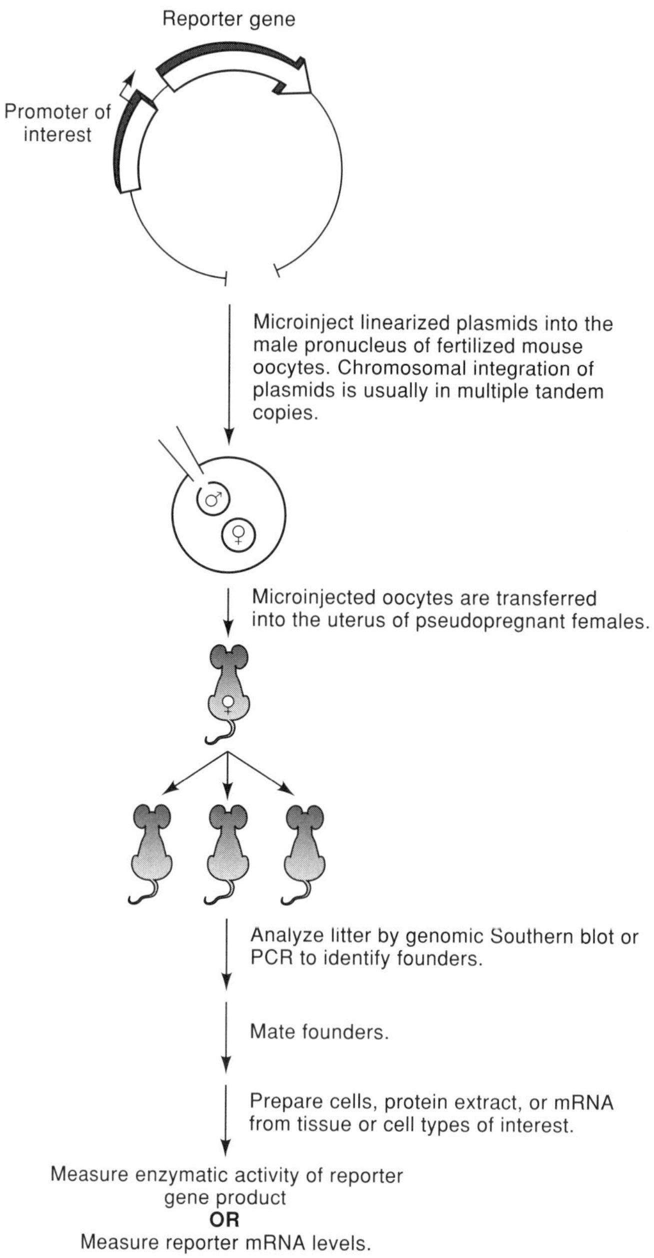

FIGURE 5.4. Transgenic assay. (Adapted, with permission, from Janeway and Travers 1996 [Copyright Garland Publishing].)

Transient Transfection Assay

For most genes, transient transfection offers a number of advantages that make it the assay of choice, at least at the initial stages of an analysis. The transient assay is rapid and simple to perform, and the results obtained are easy to quantitate. The importance of these advantages becomes apparent when considering the limitations of the alternative assays (see below).

There are two primary limitations of the transient assay. First, within a transfected cell, the plasmids exist in an artificial configuration and copy number that may lead to inactiv-

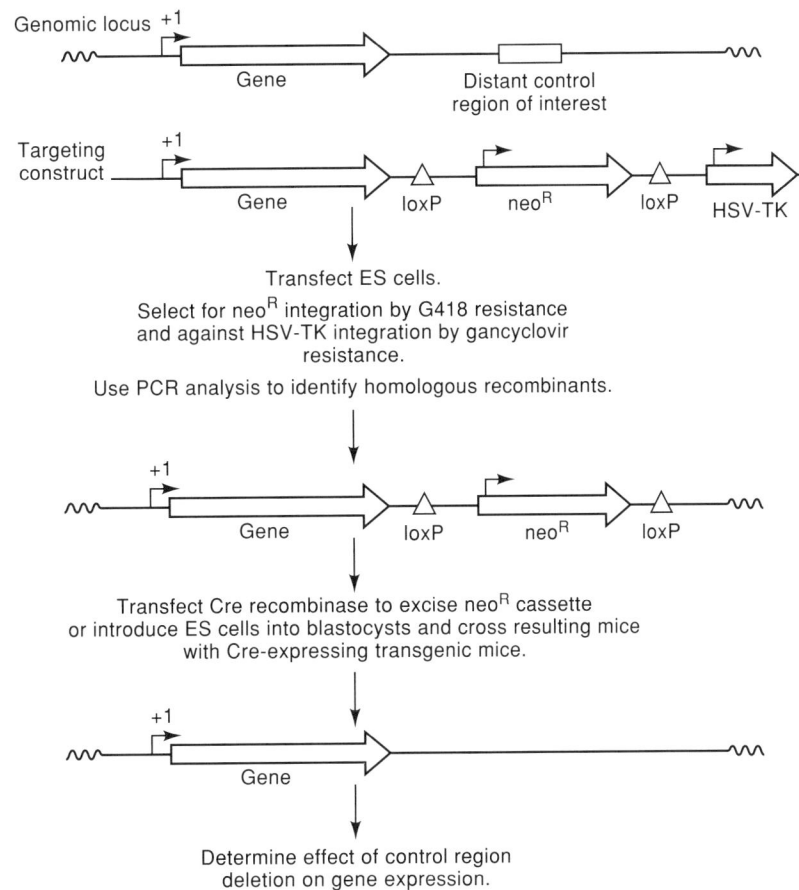

FIGURE 5.5. Homologous recombination assay.

ity or aberrant function of specific control elements (Mercola et al. 1985; Smith and Hager 1997). This limitation cannot be overemphasized. Hundreds or thousands of plasmid molecules can enter a transfected cell; since essential transcription factors may be present in limited quantities, very few plasmids may become associated with the full complement of proteins needed for the proper function of the control region. Indeed, it has been documented that large numbers of transfected reporter plasmids can suppress the activities of specific control elements and DNA-binding proteins (see, e.g., Mercola et al. 1985).

The episomal and nonreplicating nature of the plasmids is another artificial feature that may cause a control region or element to function aberrantly. In this case, the aberrant function appears to result from the fact that the plasmids are not in an appropriate chromatin configuration (Smith and Hager 1997). Locus control regions (LCRs) and silencers are two types of control regions that often do not function in transient assays (see Chapter 1 and below). Some promoters and enhancers also rely on specific chromatin structures and therefore are not regulated properly in transient assays.

An additional limitation of the transient assay is that it cannot be used to measure the activity of a region that requires an induction or differentiation period exceeding the time limitations of the transient assay (48–72 hours). Mouse erythroleukemia cells (MEL), for example, can be induced to undergo an erythroid differentiation process that often requires several days. A transient transfection assay is therefore not useful for monitoring the activity of a control region during and after MEL differentiation (see, e.g., Wright et al. 1983; Forrester et al. 1989).

Stable Transfection Assay by Integration into Host Chromosome

The most common form of stable transfection assay involves integration of a reporter plasmid into the host chromosome. A primary advantage of this assay is that the control region/reporter gene being analyzed is usually in a more natural chromatin configuration and at a more natural copy number than when analyzed by transient transfection (see Smith and Hager 1997 and references therein). Since these features allow a control region to mimic its normal function more accurately, the stable assay may be useful when studying a control region that is inactive or does not exhibit its expected activity in a transient assay.

The selection of cells that have integrated the transfected DNA provides an additional advantage for cell types that are relatively resistant to transfection; the number of cells that have taken up DNA may be too small to analyze by transient transfection, but in the stable assay the transfected cells are selectively expanded, resulting in enhanced transcription signals.

The murine terminal transferase (TdT) promoter provides an example of these advantages. This promoter does not direct significant reporter gene activity in a standard transient transfection assay, even in the presence of a heterologous enhancer and even when the most sensitive reporter assays are employed (Lo et al. 1991). In contrast, transcripts from stably transfected promoter/reporter plasmids are readily detected (P. Ernst and S.T. Smale, unpubl.). Presumably, the transient assay fails because of the poor transfection efficiency of the cell lines that must be used and/or the high copy number of the episomal plasmids.

The activity of the c-*jun* promoter in F9 embryonal carcinoma cells provides another example (Smith and Hager 1997). This promoter is highly active following transient transfection, but proper regulation is observed only following stable integration. A stable assay may be particularly useful when analyzing control regions such as LCRs and silencers, which often function only when integrated into a chromosome. The LCR that regulates the β-globin locus provides an example, as it and most of its subdomains are active in stable transfection and transgenic mouse assays, but not in transient transfection assays (Blom van Assendelft et al. 1989; Talbot and Grosveld 1991).

A final advantage of the stable transfection assay is that there are no time limits for the subsequent transcriptional analysis. Therefore, the stable assay is useful for studying differentiation and induction events that require relatively long incubation periods (see, e.g., Wright et al. 1983; Forrester et al. 1989). Stable integration of the reporter gene is also beneficial when measuring the induction or repression of reporter gene activity by a variety of external agents. With the transient assay, the cells need to be transfected prior to each experiment, but with a stable assay, the inducers or repressors can be added to the established cell line.

A primary limitation of the stable transfection assay involving chromosomal integration is that it is more difficult and time-consuming than the transient assay. Thus, instead of the 48–60 hours needed to complete a transient transfection assay, a stable transfection assay often requires several weeks, with most of the time required for drug selection and cell expansion.

A second limitation is that the transcriptional activity and regulation of a reporter gene can be strongly influenced by inherent "accessibility" of the site into which it integrates. The term "accessibility" is poorly defined, but presumably corresponds to differences in chromatin structure in the vicinity of active versus inactive genes. It is thought that the activity of an integrated reporter gene is strong only if it integrates into an accessible chromatin structure in the vicinity of an actively transcribed endogenous gene. Some control regions, such as LCRs and insulators, are capable of overcoming differences in integration site accessibility (see Chapters 1 and 6 and below). Because of these integration site variations, it usually is necessary to characterize multiple clones derived from each stable transfection experiment and to interpret with caution the results obtained. Stable transfection assays are often used, for example, for analysis of promoter and enhancer mutants, to delineate physiologically relevant control elements. Unfortunately, it is virtually impossible to analyze mutants that alter activity by only two- to threefold, because the clone-to-clone variation will be more than that amount. Thus, the stable assay may be useful for determining the effects of mutants that severely alter the activity of a control region, but not of mutants that have relatively minor effects.

Stable Transfection of Episomally Maintained Plasmids

An alternative stable transfection assay makes use of reporter plasmids containing replication origins from viruses that are subject to cell-cycle and copy number regulation, such as Epstein-Barr virus (EBV) and bovine papillomavirus (Kushner et al. 1982; Zinn et al. 1982; Yates et al. 1985; Ohe et al. 1995). This assay possesses many of the same advantages and disadvantages as the more common, genomic-integration form of stable transfection assay. However, it possesses its own set of advantages and disadvantages as well. A primary advantage relative to chromosomal integration is that gene transcription is not subject to variations caused by chromosomal integration sites. Thus, the results obtained from clone to clone should be more comparable. Furthermore, although the transfected plasmid remains episomal, its ability to replicate allows it to assemble into a chromatin configuration that resembles the endogenous control region of interest (Beato 1996; Smith and Hager 1997 and references therein).

An important limitation of this assay is that episomally maintained plasmids are often unstable and undergo recombination events as the cells are propagated (see, e.g., Chittenden et al. 1989). If the new episomal species generated by recombination confers a growth advantage to the cells, it will eventually become the predominant episomal plasmid in the culture. Alternatively, if the rearrangement inactivates the replication origin, the resulting plasmid may integrate into the genome and confer drug resistance from its genomic location. Because of these limitations, stable cell lines generated with episomal plasmids must be characterized frequently and extensively at the molecular level to ensure that the plasmids remain episomal and unrearranged. For this reason, the episomal stable transfection assay is recommended only for specialized studies of the influence of chromatin structure on gene activation. A prominent example of this use is the analysis of nucleosomes regulating the mouse mammary tumor virus promoter (Bartsch et al. 1996; Beato 1996). This assay will not be discussed further in this manual, but the published manuscripts from the Beato laboratory can be consulted for further information on the relevant methods and strategies. Finally, it is worth noting that Invitrogen offers a series of EBV-based vectors (pREP4 [cat. #V004-50], pREP8 [cat. #V008-50], and pREP9 [cat. #V009-50]) that can be adapted for studies of transcriptional regulation.

In Vitro Transcription Assay

An in vitro transcription assay may be useful for cell types that are particularly resistant to transfection, for promoters that do not yield detectable activity in a transfection assay, or for advanced studies of gene regulation. A significant advantage of this assay is that efficient transfection conditions are not needed. In addition, antibodies can be added to the in vitro reactions and transcription factors can be added or depleted, allowing the function of a given transcription factor to be assessed (see Chapter 14). It also is noteworthy that, among the assays described here, the in vitro assay is the most rapid, in that an entire experiment can be performed in a single day.

The primary limitation of the in vitro assay is that factors involved in transcriptional regulation are much more likely than with in vivo assays to be inactive, weakly active, or to function aberrantly. Many transcription factors fail to retain their full activities during extract preparation or simply fail to stimulate transcription in vitro. One likely explanation for the poor function is that the factors are much less concentrated in a nuclear extract than in a nucleus. Furthermore, some factors depend on the presence of nucleosomes or may require other chromatin-associated proteins. Nucleosomes can also enhance responsiveness to a factor by suppressing basal transcription. Recent efforts to obtain regulated transcription in vitro with templates assembled into nucleosomes have met with greater success than studies using naked templates (Barton et al. 1993; Pazin et al. 1996; Kraus and Kadonaga 1998).

Despite the above limitations, in vitro assays have been used to recapitulate the regulated or tissue-specific activity of several promoters, including the immunoglobulin (Ig) (Pierani et al. 1990), human immunodeficiency virus type 1 (HIV-1) (Pazin et al. 1996), interleukin-2 (IL-2) (Flanagan and Crabtree 1992), and TdT (Garraway et al. 1996) promoters. In contrast to the success obtained with promoters, distant control regions rarely function in vitro (see Carey et al. 1990). To detect the activity of a distant enhancer, it often is necessary to insert it immediately upstream of the promoter, which may lead to aberrant function by allowing some of the enhancer elements to synergize improperly with nearby promoter elements. Future studies of distant enhancers in vitro will most likely involve nucleosome-reconstituted templates, which appear to facilitate enhancer function from a distance. This strategy has been employed in only a small number of studies, however (Kadonaga 1991; Laybourn and Kadonaga 1992; Barton et al. 1993, 1997), and the extent to which it will be generally useful for the analysis of enhancer function remains to be established.

Given the above considerations, it is recommended that the in vitro assay be used to complement an in vivo assay, or to study the properties of specific control elements and proteins after their functions and properties have been characterized in vivo (see Chapters 7 and 9). The in vitro assay is recommended as an initial or primary assay only if attempts to develop an in vivo functional assay have been unsuccessful.

Transgenic Assays

Because this manual deals primarily with mammalian systems, the transgenic mouse is discussed as the prototypical animal assay. In general, transgenic mice have been used for transcriptional regulation studies to address two specific issues. First, this assay is useful for confirming that a particular control region or element identified in a transfection assay is actually important in the animal. Transgenic studies of the Ig μ enhancer provide an example (Ernst and Smale 1995). This enhancer was discovered using transient and stable transfection assays and was predicted to confer B-cell specificity to Ig μ transcription (Banerji et al. 1983; Gilles et al. 1983). However, subsequent transgenic studies were needed to determine the precise tissue specificity of the enhancer (Jenuwein and Grosschedl 1991). Transgenic mice were also generated with constructs containing mutations in individual Ig μ control elements originally identified in transient transfection experiments (Jenuwein and Grosschedl 1991; Annweiler et al. 1992). These transgenic experiments confirmed the importance of the control elements.

Transgenic assays are also useful for identifying LCRs, which often do not function in transient or stable transfection assays. An LCR stimulates high levels of integration-site-independent transcription (see Chapters 1 and 6). Although a stable transfection assay can provide preliminary evidence of integration-site independence, the results obtained are less conclusive because a cointegrated drug-resistance gene is needed. Assays for studying LCRs are discussed in greater detail in Chapter 6.

Except for the above two uses, the transgenic assay is limited in its usefulness. In particular, because of the time commitment and cost that would be required, a transgenic assay has not been used for a detailed mutant analysis of a control region (see Chapter 7).

Homologous Recombination Assay

Mutation of an endogenous control region or control element by homologous recombination (Capecchi 1989), in either a cell line or animal, provides the most conclusive evidence that it is important for a gene's function. Unfortunately, this assay is rarely employed for this purpose because the methodology is relatively difficult. When homologous recombination is performed in a transformed cell line, the difficulty stems from the need to manipulate large DNA fragments to generate the recombination substrate and from the challenge of selecting the rare cell clones that have integrated the substrate by homologous recombination. When homologous recombination is performed in a mouse, the difficulty is compounded by the need to maintain healthy embryonic stem cells and to obtain germ-line transmission of the recombined allele. To date, the assay has been used primarily to determine the importance of specific enhancer regions for the expression of an endogenous gene. Two examples are the T-cell receptor-β enhancer (Bories et al. 1996) and the Ig κ intronic enhancer (Xu et al. 1996).

A second limitation is that the homologous recombination technique usually depends on integration of a drug-resistance gene at the site of the mutation. The promoter and enhancer associated with the drug-resistance gene may contribute to the mutant phenotype either positively or negatively. The enhancer may stimulate transcription from the endogenous gene of interest. Alternatively, the promoter associated with the drug-resistance gene may compete with the natural promoter for the stimulatory effect of surrounding regulatory regions. Fortunately, this limitation can be largely eliminated by methods, such as the Cre-loxP method, for deleting the drug-resistance cassette (Sternberg and Hamilton 1981; Gu et al. 1993).

Homologous recombination has rarely, if ever, been used to analyze the importance of individual control elements, despite the fact that it would provide highly relevant information. The primary reason, once again, is the amount of effort required. In the future, as methods for performing homologous recombination in mammalian cells become easier and less time-consuming, this approach is likely to become an important tool for assessing the function of individual DNA sequence elements.

Transient Transfection Assays

To develop a transient transfection assay to analyze a gene's promoter, the following steps must be completed: (1) Identify a source of cultured cells; (2) establish a transfection procedure for introducing DNA into the cells; (3) choose a reporter gene and corresponding reporter assay; (4) identify a putative promoter region for the gene of interest (see Chapter 4); and (5) insert the promoter upstream of the reporter gene in an appropriate vector.

After completing these initial steps, transient transfection experiments can be attempted to determine whether promoter-dependent reporter gene activity is detectable.

Cells

The cells used for transient transfection should fulfill three criteria: (1) They should express the endogenous gene from which the promoter of interest is derived; (2) it should be possible to maintain consistent quantities of cells in culture for the 2–3 days needed to complete the experiment; and (3) a procedure must be available for transfecting plasmid DNA into the cells.

Transformed (immortal) cell lines are usually the best choice for a transient transfection analysis because lines derived from numerous cell types are readily available from commercial (e.g., American Type Culture Collection [ATCC, #1-800-638-6597], National Institute of General Medical Sciences cell repository) and academic sources. In addition, transformed cell lines are relatively easy to maintain and, for many cell lines, efficient transfection procedures have been established and published. In contrast, primary cell populations grown in culture are often difficult to isolate and maintain and are resistant to common transfection procedures. If a transformed cell line that expresses the endogenous gene of interest is not available, it may be necessary to determine the feasibility of employing a primary cell population, or to establish an immortal cell line by transformation with an oncogene.

Transfection Procedures

If a cell line is available in which the endogenous gene of interest is expressed, but for which transfection conditions have not been established, optimal transfection conditions must be determined. Several transfection methods have been employed for transient promoter analyses, including calcium phosphate, DEAE-dextran, electroporation, lipofection, transfection with other cationic reagents, and protoplast fusion. Box 5.1 contains an introduction to the most common of these procedures. More detailed information about transfection procedures can be obtained from the specific references cited, as well as from Ausubel et al. (1994, pp. 9.1–9.5), Sambrook et al. (1989, pp. 16.30–16.55), and Spector et al. (1998, pp. 86.1–86.6).

Box 5.1

Common Transfection Methods

- *Calcium phosphate:* DNA is mixed with calcium chloride in a phosphate buffer, resulting in the formation of a DNA–calcium phosphate precipitate, which is layered on the cells (Graham and van der Eb 1973). The precipitate is taken up by endocytosis. This procedure is widely used for both transient and stable transfection experiments in fibroblasts and other adherent cell types. It can be performed quite easily with reagent-grade chemicals (see Protocols section). Alternatively, reagents and kits that are specifically designed for calcium phosphate transfections can be purchased from commercial sources (e.g., Stratagene Calcium Phosphate [cat. #200285], Invitrogen Calcium Phosphate Transfection Kit [cat. #K2780-01], Promega ProFection Kit [cat. # E1200], or CLONTECH Calcium Phosphate Maximizer and Transfection Kit [cat. #K2051-1]). Stratagene also markets a related calcium phosphate-like transfection kit (MBS Mammalian Transfection Kit, cat. #200388).

- *DEAE-dextran:* The negatively charged DNA binds to the cationic DEAE-dextran (McCutchan and Pagano 1968). The soluble complex is added to the cells, which take it up by endocytosis. This procedure is used for transient transfection of both adherent and nonadherent cells. However, it rarely is used for stable transfection experiments because of toxicity that prevents long-term maintenance of the transfected cells. Like the calcium phosphate method, DEAE-dextran transfections can easily be performed with reagent-grade chemicals, although commercial kits are available (e.g., Promega ProFection [cat. #E1200] and Stratagene DEAE-dextran [cat. #200386]).

- *Electroporation:* The DNA and cells are mixed in an empirically defined buffer. The mixture is then subjected to a high-voltage electroshock, which induces or stabilizes pores in the plasma membrane through which the DNA enters the cell (Neumann et al. 1982; Zimmermann 1982). This method is rapid and can be used for many types of adherent and nonadherent cells.

- *Lipofection:* The negatively charged DNA binds the positively charged end of cationic lipid compounds (Felgner et al. 1987; Mannino and Gould-Fogerite 1988). The DNA-bound lipids associate with the cell membrane, leading to DNA internalization. With some cells, lipofection yields exceptionally high transfection efficiencies, but with others, the efficiencies are comparable or inferior to those obtained with the other methods. Proprietary lipid compounds are marketed by several different companies, including CLONfectin from CLONTECH (cat. #8020-1), Tfx and Transfectam from Promega (cat. #E1811 and E1231), LipoTAXI from Stratagene (cat. #204110), PerFect Lipid from Invitrogen (cat.# K925-01), and DOSPER and DOTAP from Roche Molecular Biochemicals (cat. # 18811169 and 1811177). These compounds are slightly different from one another; therefore, the transfection efficiencies observed can be quite variable. For a new cell line, it is difficult to predict which one works best. One disadvantage of lipofection is its cost relative to methods like calcium phosphate and DEAE-dextran.

- *Polycations:* Several polycations, including Polybrene (Chaney et al. 1986) and polyethyleneimine (Boussif et al. 1995), have been shown to promote efficient transfer of plasmid DNA into some mammalian cell lines. Proprietary polycations, such as the SuperFect reagent from QIAGEN (cat. #301305), have recently been marketed. These reagents are easy to use and result in efficient transfection of some cell lines. Most likely, these chemicals are similar to DEAE-dextran in their mechanism of action.

To determine which transfection procedures are most likely to succeed for a given cell line, a literature search should first be performed to determine which methods have been used successfully for that line or for related cell lines (i.e., cell lines derived from a similar tissue). Most lymphocyte lines, for example, have been successfully transfected by DEAE-dextran, electroporation, or lipofection methods, but rarely with calcium phosphate. Therefore, calcium phosphate would be a poor choice when establishing conditions for a new lymphocyte line. It is also important to note that the transfection strategy should be chosen with the gene of interest in mind. For example, it may be difficult to study calcium-inducible genes using a calcium phosphate transfection strategy.

When developing a functional assay for a new cell line, the transfection procedures should be tested and optimized using a standard reporter plasmid containing a strong viral promoter and enhancer fused to a reporter gene for which a sensitive assay is available (see below). The pGL3-Control (cat. #E1741), pCAT3-Control (cat. #E1851), and pSV-β-Galactosidase Control (cat. #E1081) vectors from Promega, which contain reporter genes under the control of the SV40 promoter and enhancer, are useful for optimizing transfection conditions using the luciferase, CAT, and β-galactosidase assays, respectively (see description of reporter assays below). Plasmids containing reporter genes under the control of other strong promoters and enhancers are available (e.g., pRL-CMV from Promega). Viral promoters and enhancers are useful for optimizing transfection conditions because they are stronger than control regions from most cellular genes and therefore will confer greater sensitivity during optimization attempts. Furthermore, viral promoters and enhancers contain binding sites for an unusually large number of transcription factors, with considerable redundancy so that they are strongly active in most cell types (Schmidt et al. 1990).

After conditions are established that result in detectable activity with a control reporter plasmid, various transfection parameters should be systematically varied to optimize reporter activity. With most transfection procedures, the optimal signal depends on a balance between transfection efficiency and cell death. Because harsh conditions often allow more DNA to enter the cells, the ideal conditions for DNA uptake usually cause many or most cells to die. In contrast, conditions that result in 100% cell survival are often too mild for DNA uptake. Optimal reporter gene activity will usually be obtained with conditions that promote some cell death, but that leave enough healthy cells to drive expression of the reporter gene. The parameters that are typically varied to optimize transfection conditions are indicated in the Techniques section.

If a cell line that yields strong reporter activity cannot be identified, a few additional transient transfection strategies can be attempted. First, if a cell line is available that is poorly transfected, a method can be employed to isolate and concentrate the small number of transfected cells, allowing reporter gene expression from those cells to be monitored (see Box 5.5). Second, replication-competent plasmids can be used to boost the plasmid copy number and the resulting reporter activity in the cells (see Box 5.6). Finally, depending on the nature of the regulatory region of interest, it may be reasonable to test a closely related cell type that does not express the relevant endogenous gene. A related cell type may contain many of the transcription factors needed for gene activation. Because the mechanism that represses the endogenous gene in that cell type may not function in the artificial transient assay, promoter activity may be detectable. This tactic is far from ideal, however, and generally should be used as a last resort.

Reporter Genes, Vectors, and Assays

In addition to identifying an appropriate source of cells for transient transfection, an appropriate reporter gene must be chosen. The reporter gene's mRNA, protein, or enzyme levels will be used as a measure of promoter activity. One advantage of using a reporter gene, rather than the physiological gene regulated by the control region, is that the reporter gene eliminates difficulties in distinguishing transcripts derived from the transfected plasmid and the endogenous gene.

Common reporter genes. The reporter genes that are most commonly used are derived from insects (luciferase, Box 5.2) or prokaryotes (CAT [Box 5.3] and β-galactosidase [Box 5.4]) and encode enzymatic activities not typically found in the cells of most eukaryotes. The promoter region of interest is inserted upstream of the reporter gene's coding sequence. Because the reporter vectors usually contain a Kozak consensus sequence (i.e., the sequence encompassing the eukaryotic translation initiation codon) and a downstream polyadenylation signal, the mRNA produced by RNA polymerase II is competent for efficient translation into protein. Expression of these reporter genes is usually monitored by measuring their enzymatic activity, which should be roughly proportional to the transcription initiation frequency from the promoter, thereby providing a reasonably accurate assessment of promoter activity.

Boxes 5.2 through 5.4 describe the most commonly used reporter genes and assays, and list sources of vectors and assay kits. Protocols for luciferase, CAT, and β-galactosidase assays are presented at the end of this chapter. Detailed information and protocols can also be found in the specific references cited, as well as in Ausubel et al. (1994, Unit 9.6) and Sambrook et al. (1989, pp. 16.56–16.66). Ausubel et al. (1994) contains a table with detailed information about a large number of reporter assays (Unit 9.6.8). Furthermore, many of the companies that market reporter vectors and assays provide information; one source of detailed information is Promega's *Protocols and Applications Guide*.

Box 5.2

Luciferase Reporter Gene Assay

Perhaps the most commonly used reporter gene is the luciferase gene from the firefly *Photinus pyralis* (deWet et al. 1987). This gene encodes a 61-kD enzyme that oxidizes D-luciferin in the presence of ATP, oxygen, and Mg^{++}, yielding a fluorescent product that can be quantified by measuring the released light (see Promega *Protocols and Applications Guide*). Inclusion of coenzyme A in the reaction enhances the sensitivity of the assay and provides a longer, sustained light reaction. The luciferase assay is extremely rapid, simple, relatively inexpensive, sensitive, and possesses a broad linear range. Promega states that the luciferase assay is 30–1000 times more sensitive than the CAT assay and provides a linear response over eight orders of magnitude.

Promega offers a variety of luciferase reporter vectors. The pGL2-Basic (cat. #E1641) and pGL3-Basic (cat. #E1751) vectors contain a luciferase gene downstream of a multiple cloning site, into which the promoter of interest is inserted. To reduce luciferase background activity due to transcription initiating within the vector, a polyadenylation signal is located upstream of the multiple cloning site. A few unique restriction enzyme sites are present downstream of the luciferase gene for insertion of enhancers or other distant control regions (see Chapter 6). The pGL2-Enhancer (cat. #E1621) and pGL3-Enhancer (cat. #E1771) vectors contain an SV40 enhancer downstream of the luciferase gene to boost transcription driven by weak promoters. The pGL2-Promoter (cat. # E1631) and pGL3-Promoter (cat. # E1761) vectors contain an SV40 promoter upstream of the luciferase gene; these vectors are used for the identification and analysis of enhancer regions (see Chapter 6). The pGL2-Control (cat. #E1611) and pGL3-Control (cat. #E1741) vectors contain both the SV40 promoter and enhancer. Other luciferase vectors are offered by Promega (e.g., pGEM-*luc* [cat. #E1541] and pSP-*luc*+[cat, #E1781]) for those interested in excising the luciferase gene and inserting it into their own plasmids. The Promega literature should be consulted for additional information. Promega recently began offering a new set of luciferase reporters containing the gene from the sea pansy *Renilla reniformis* (Matthews et al. 1977; Liu et al. 1997). The encoded enzyme has a different substrate specificity from the *Photinus* enzyme, generating light in the presence of coelenterazine instead of D-luciferin. The different specificity is the basis of Promega's Dual-Luciferase Reporter Assay System (Promega, cat. # E1910). This system provides a rapid means of comparing the activity of a promoter of interest to the activity of a control promoter (for the purpose of normalization). The promoter of interest can be inserted into the pGL3-Basic vector, for example, and the resulting plasmid cotransfected into cells with the pRL-CMV vector, which contains the *Renilla* luciferase gene under the control of the CMV promoter/enhancer. Extracts of the transfected cells can first be supplemented with D-luciferin to measure the *Photinus* luciferase activity. A second reagent can then be added to quench the *Photinus* enzyme and activate the *Renilla* enzyme.

In addition to the Dual-Luciferase Reporter Assay System, Promega and other companies (e.g., Roche Molecular Biochemicals [Luciferase Reporter Gene Assay, cat. #1 669 893 and 1 814 036]) offer kits for the traditional measurement of *Photinus* luciferase activity (Promega, cat. #E1500). These kits generally include a cell lysis buffer, assay substrate, and assay buffer. Luciferase activities are usually measured in a luminometer (available from Wallach, Promega, Packard Instruments, and others) but can also be measured in a scintillation counter (see Techniques section).

One caution about the Promega luciferase vectors is that the vectors containing the SV40 promoter include the SV40 origin of replication. The origin allows plasmid replication to high copy number in cell lines, such as COS-7 and 293T, which express the SV40 large-T-antigen gene. Another caution about all luciferase reporters is that they yield experimental artifacts in at least a few inducible cell lines (see, e.g., Plevy et al. 1997). The luciferase coding region apparently contains an element that confers induction to a variety of stimuli, regardless of the promoter or enhancer driving transcription. The basis of this promoter-independent induction has not been determined, but may involve a nonspecific stabilization of luciferase mRNA in response to signal transduction pathways (Plevy et al. 1997). This nonspecific effect will not interfere with an analysis unless inducible activation or repression of transcription is being monitored.

Box 5.3

CAT Reporter Gene Assay

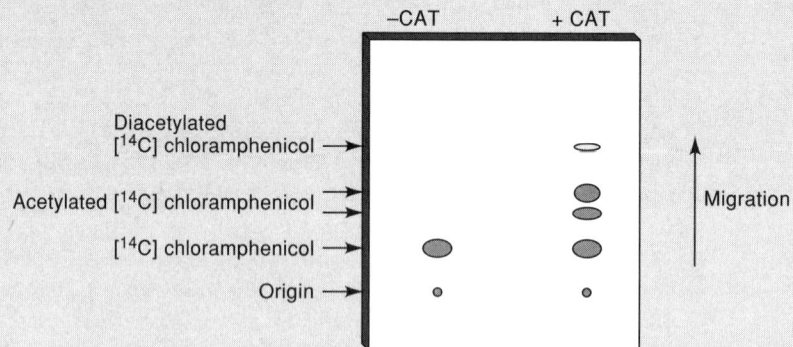

FIGURE 5.6. CAT assay using thin-layer chromatography.

A second commonly used reporter gene is the *E. coli* transposon CAT gene (Gorman et al. 1982). The CAT enzyme catalyzes the acetylation of chloramphenicol, with the acetyl group donated by acetyl CoA or, in some commercial assays, *n*-butyryl CoA. The acetylated chloramphenicol can be monitored by a variety of methods. Most commonly, [^{14}C]chloramphenicol is used as the reaction substrate, with acetylation monitored by autoradiography following thin-layer chromatography (TLC) to separate the acetylated from the unacetylated forms (Gorman et al. 1982; Fig. 5.6). The percent conversion of [^{14}C]chloramphenicol to acetyl-[^{14}C]chloramphenicol can be measured by phosphorimager analysis of the TLC plate, by excising the radioactive spots from the TLC plate and counting in a scintillation counter, or by densitometry analysis of an autoradiograph. The acetylated ^{14}C-labeled product can also be quantified without TLC, by organic extraction and scintillation counting. Reagent-grade chemicals purchased separately can be used for these assays (from Promega, Roche Molecular Biochemicals, Sigma, etc). Alternatively, a kit containing most of the reagents is available from Promega (CAT Enzyme Assay System [cat. #E1000]). Non-radioactive methods for measuring CAT protein include an ELISA method (Roche Molecular Biochemicals CAT ELISA [cat. #1363727]) and a TLC-based method using a fluorescently labeled chloramphenicol substrate (Stratagene FLASH CAT Nonradioactive CAT Assay Kit [cat. # 200394 and 200396]).

Although this assay is less sensitive and more expensive than the luciferase assay, it was developed several years before the luciferase assay system, and therefore is still commonly used. It is strongly recommended for analysis of inducible genes that might be subject to experimental artifacts with the luciferase assay.

Promega offers the pCAT and pCAT3 series of vectors (cat. # E1861, E1871, E1881) that are similar to the pGL2 and pGL3 luciferase vectors described in Box 5.2. Each series includes a vector with no promoter or enhancer (pCAT-Basic [cat. # E1041] and pCAT3-Basic [cat. #1871]), a vector containing the SV40 promoter (pCAT-Promoter [cat. # E1031] and pCAT3-Promoter [cat. #1861]), a vector containing the SV40 enhancer (pCAT-Enhancer [cat. # E1021] and pCAT3-Enhancer [cat. # 1881]), and a vector containing both the promoter and enhancer (pCAT-Control [cat. # E1011] and pCAT3-Control [cat. #1851]).

Box 5.4

LacZ, SEAP, and GFP Reporter Gene Assays

Although luciferase and CAT are the most commonly used reporter genes, several other reporters are available. The *E. coli* lacZ gene, encoding β-galactosidase (β-gal), is one of the more popular alternatives (Hall et al. 1983). β-gal catalyzes the hydrolysis of β-galactoside sugars. The substrates used in the standard β-gal assay, including 5-bromo-4-chloro-3-indolyl-β-D-galactopyranoside (X-gal) and o-nitrophenyl-β-D-galactopyranoside (ONPG), generate reaction products that can be quantitated spectrophotometrically. The LacZ reporter is often used as an internal control for transfection studies using CAT or luciferase reporters (see text). One limitation of the β-gal assay is background caused by endogenous mammalian enzymes. This background can be reduced by a heat treatment that selectively inactivates the mammalian enzymes (Young et al. 1993). β-gal reporter vectors are available from CLONTECH (cat. # 6044-1), with β-gal control vectors available from Promega (cat. # E1081). Reagents for spectrophotometric detection of enzymatic activity can be purchased separately or as kits from Promega, Stratagene, and Invitrogen. Roche Molecular Biochemicals (cat. # 1758241) and CLONTECH (cat. # K2047-1 and K2048-1) sell sensitive chemiluminescence kits for monitoring β-gal enzymatic activity, and Roche Molecular Biochemicals sells an ELISA kit for measuring β-gal protein (cat. # 1539426).

Another useful reporter is the secreted form of human placental alkaline phosphatase (SEAP), which is expressed from reporter plasmids marketed by CLONTECH (cat. # 6049-1, 6050-1, and 6051-1). SEAP protein can be quantified by either fluorescent or chemiluminescent detection, using CLONTECH's kits. An important advantage of this reporter for some purposes is that the reporter protein is secreted into the culture medium, allowing reporter detection without cell lysis. Because the cells can be maintained in culture as SEAP expression is measured, the kinetics of expression can be monitored and the transfected cells can be harvested for other purposes at a later time.

One recent addition to the selection of reporter genes is green fluorescent protein (GFP), a 28-kD protein from the jellyfish *Aequorea victoria* (Prasher et al. 1992; Chalfie et al. 1994). GFP contains an intrinsic peptide chromophore that emits green light (with an emission peak wavelength of 509 nm) following oxidation and excitation with ultraviolet or blue light. The emitted light can usually be measured in intact cells, resulting in a simple assay for reporter gene expression. Plasmid vectors for using GFP as a reporter are available from CLONTECH (cat. # U17997) and Packard Bioscience (CytoGem). The GFPs expressed from the commercial plasmids contain specific amino acid substitutions that enhance fluorescent emission and sometimes alter the emission wavelength and stability of the protein. GFP plasmids are routinely used for monitoring gene expression by fluorescent microscopy and flow cytometry. In addition, GFP can be used as a substitute for CAT or luciferase reporters, by measuring fluorescence in a fluorometer. GFP has not yet been used as a reporter for the detailed dissection of a control region, however. One reason is historical: The GFP gene was cloned quite recently. In addition, the available data suggest that GFP's linear range and signal-to-background ratios are more limited than either the luciferase or CAT reporters. At this time, GFP cannot be recommended as a substitute for the more common reporters for the dissection of a control region (as described in Chapter 7). Detailed information about the properties and uses of GFP expression plasmids can be found in the CLONTECH and Packard Bioscience literature.

Advantages of indirect reporter assays. The reporter assays mentioned above are often termed *indirect* assays because enzyme levels are used as an indirect measure of promoter activity. In contrast, an assay that measures mRNA levels (primer extension, S1 nuclease, or RNase protection) is considered a direct assay for promoter activity (although it does not involve direct visualization of the mRNA). The three principal advantages of the indirect assays are simplicity, speed, and sensitivity. For example, the luciferase assay, which is the most sensitive of the indirect assays, is roughly 1000-fold more sensitive than the primer extension and RNase protection assays. A luciferase assay can also be performed in about 15 minutes with only two solutions, far less than the 5–24 hours needed for primer extension and RNase protection, which involve radiolabeled probes and several buffers. Thus, it is not surprising that these indirect assays are used far more frequently than the direct assays.

Limitations of indirect reporter assays. One limitation of indirect reporter assays is that the proteins are generally more stable than their mRNAs. Therefore, it is especially important to perform appropriate kinetics to ensure that enzymatic activity is being measured in a linear range.

A second limitation is that, during an analysis of promoter function, the results obtained can be misleading because the assay does not distinguish transcripts that initiate at the authentic transcription start site from those that initiate aberrantly within the vector or promoter. These "cryptic" transcripts are not uncommon; in fact, during a transient assay, a low level of cryptic transcription always initiates at numerous positions within the plasmid. A subset of these transcripts will be translated into functional protein, which will contribute to the total signal obtained in the enzymatic assay. Much of the background can be alleviated by using plasmid vectors that have been designed to block aberrant transcription through the reporter gene, and by including appropriate controls in the initial transfection experiments (see below).

Minimizing background transcription and confirming start site. A useful strategy for minimizing the background transcription has been to insert a polyadenylation signal immediately upstream of the promoter, such that aberrant transcripts initiating within the vector will be cleaved and polyadenylated. Almost all of the reporter vectors available from Promega contain an upstream polyadenylation signal. With this strategy, the only stable mRNA molecules produced should be those that initiate downstream of the polyadenylation signal (i.e., within the promoter region). It should be noted that this procedure will not eliminate background transcripts that may initiate at aberrant sites within the inserted promoter.

If ambiguous results are obtained using an indirect assay, it may be necessary to confirm the results with a direct assay, such as a primer extension or RNase protection assay (see Chapter 4), which measures transcripts initiating only at a specific location. If strong reporter activities are obtained by transient transfection, it may be relatively easy to monitor the location of the transcription start site. The primer extension assay is the first assay that should be considered because primer sequences that yield successful results have been determined for at least two of the common reporter genes, luciferase and CAT (e.g., 5'-ACCAACAGTACCGGAATGCC-3' for luciferase [see Plevy et al. 1997]; 5'-TATATCAACG-GTGGTATATCCAGTG-3' for CAT [see Liston and Johnston 1999]). Because the sequences to which these primers hybridize are within the reporter genes themselves, they should be useful for most luciferase and CAT plasmids. The RNase and S1 nuclease assays may be more sensitive than the primer extension assay in some instances, but a different probe will be needed for each promoter/reporter fusion.

The small number of reporter transcripts in a transiently transfected population often makes direct mRNA analysis difficult. To enhance sensitivity, methods to separate the transfected cells within a population from the untransfected cells can be considered (see Box 5.5). Alternatively, an enhancer can be placed downstream of the reporter gene, or replication-competent reporter plasmids can be tested (see Box 5.6, below). At first glance, it might seem desirable to use an mRNA detection assay that is more sensitive than primer extension or RNase protection, such as RT-PCR. Unfortunately, RT-PCR assays, as well as Northern blots, usually cannot be used to measure specific transcripts synthesized in a transient transfection assay. The reason is that these assays measure all RNA molecules complementary to the probe or primer sequences, regardless of whether the RNAs initiated at the correct start site. As stated above, a substantial number of transcripts initiate at aberrant vector and promoter sequences during a transient assay. RT-PCR primers or Northern blot probes will hybridize to the aberrant transcripts as efficiently as to the correctly initiated transcripts. These assays would therefore provide no distinction between transcripts initiated at the correct site relative to those that initiated at cryptic sites.

5' RACE is the only assay that might be considered as a more sensitive alternative to the primer extension, RNase protection, and S1 nuclease assays (see Chapter 4). This PCR-based method can provide some evidence that transcripts initiate at the correct transcription start site. However, the fact that RACE is less quantitative than primer extension, RNase protection, etc. (see Chapter 4) is a clear disadvantage.

Plasmid Construction

To analyze the activity of a promoter, the putative promoter sequence must be inserted upstream of the reporter gene. To choose the appropriate DNA fragment for insertion into the reporter vector, it is important first to determine the locations of the translation initiation codon and transcription start site(s). For mammalian genes, the transcription start site is usually 50–200 bp upstream of the translation initiation codon. However, as described in Chapter 4, it is inappropriate simply to assume that the promoter will be directly upstream of the ATG codon, even if a potential TATA box and other potential regulatory elements are identified by sequence homology. It may be possible to detect reporter activity in the presence of the sequence flanking the ATG for the gene of interest. However, unless the transcription start site has been mapped, it will remain unclear whether the activity detected is driven by the authentic promoter or by cryptic control elements present throughout the genome that appear capable of driving low levels of transcription when tested with a sensitive reporter assay. Since promoter regions always appear to surround transcription start sites, the confidence in the validity of the results is greatly increased if the transcription start sites have been clearly defined.

Downstream boundary of inserted fragment. In general, the downstream boundary of the promoter fragment chosen for the initial studies should be between the transcription start site and the ATG. Functional promoter elements have been reported within this transcribed leader region, and thus it is useful to include as much of the untranslated region as possible. (It unfortunately has also been found that some transcribed leader sequences inhibit promoter activity in a transient assay. Therefore, if promoter activity is not detected with a plasmid containing the transcribed leader, a related plasmid containing leader sequences extending only about 10 bp downstream of the transcription start site can be tested.) If a reporter assay is being employed that depends on the detection of reporter protein or enzymatic activity, the ATG and any splice donor sequences should be omitted from the inserted fragment, as they may interfere with the processing and translation of the reporter gene. Alternatively, the ATG from the inserted promoter can be fused directly to the ATG of the reporter gene (i.e., in place of the reporter gene ATG).

Upstream boundary of inserted fragment. The appropriate upstream boundary of the promoter fragment inserted into the reporter vector is more difficult to predict. Some promoters are strongly active and appropriately regulated with only 200–300 bp upstream of the transcription start site. Other promoters are not regulated or fail to yield detectable activity unless accompanied by a distant control region, such as an enhancer. Because one common location for an enhancer is upstream of the promoter, the probability of detecting appropriately regulated activity increases with a larger fragment. It should be noted, however, that constitutively active negative regulatory regions have been detected upstream of several functional promoter regions (e.g., Malone et al. 1997). In most cases, the biological significance of these regions for the regulation of the endogenous gene is not known, but they can interfere with detection of promoter activity.

To gain insight into the amount of sequence that may be required for strong, regulated reporter activity, it is worthwhile searching the literature for precedents with related genes, as similarities in regulatory mechanisms appear to exist among the genes within a given class. As discussed in Chapter 2, early response genes and inducible cytokine genes usually depend on sequences within a few hundred base pairs of the start site for regulated transcription (see, e.g., Goodbourn et al. 1985; Treisman 1986; Plevy et al. 1997). Although transcription of this class of genes is likely to be influenced by more distant control regions, much can be learned about their regulation by studying the first few hundred base pairs. In contrast, promoters for some genes involved in early development of lymphocytes, many of which lack TATA boxes, exhibit very low or undetectable activity in transient transfection assays (see, e.g., Lo et al. 1991). Promoters for other developmentally regulated genes, such as the β-globin gene, are reasonably active in transient transfections, but poorly regulated (Charnay et al. 1985). Thus, the initial reporter plasmids for this class of genes should probably include a relatively large amount of upstream DNA.

Use of enhancer-containing vectors. When preparing the initial constructs, it is recommended that the promoter fragments be introduced into at least two different vectors: one containing only the promoter-less reporter gene and the other containing a constitutively active enhancer downstream of the reporter gene (e.g., the SV40 enhancer; see above). Insertion into the first vector will reveal the strength and properties of the isolated promoter. The enhancer added in the second vector should help to amplify the signal if the isolated promoter is weak (see, e.g., Wirth et al. 1987). The viral enhancers included in standard reporter vectors stimulate transcription from most but not all promoters (Treisman et al. 1983; Robbins et al. 1986), and usually will not mask promoter regulation, although this possibility should be addressed experimentally for each promoter.

Plasmid purification. The plasmids to be used in the transfection experiments can be purified using either a CsCl gradient method (Sambrook et al. 1989, pp.1.42–1.48) or a standard column chromatography method (e.g., QIAGEN, cat. #12263). For some cell

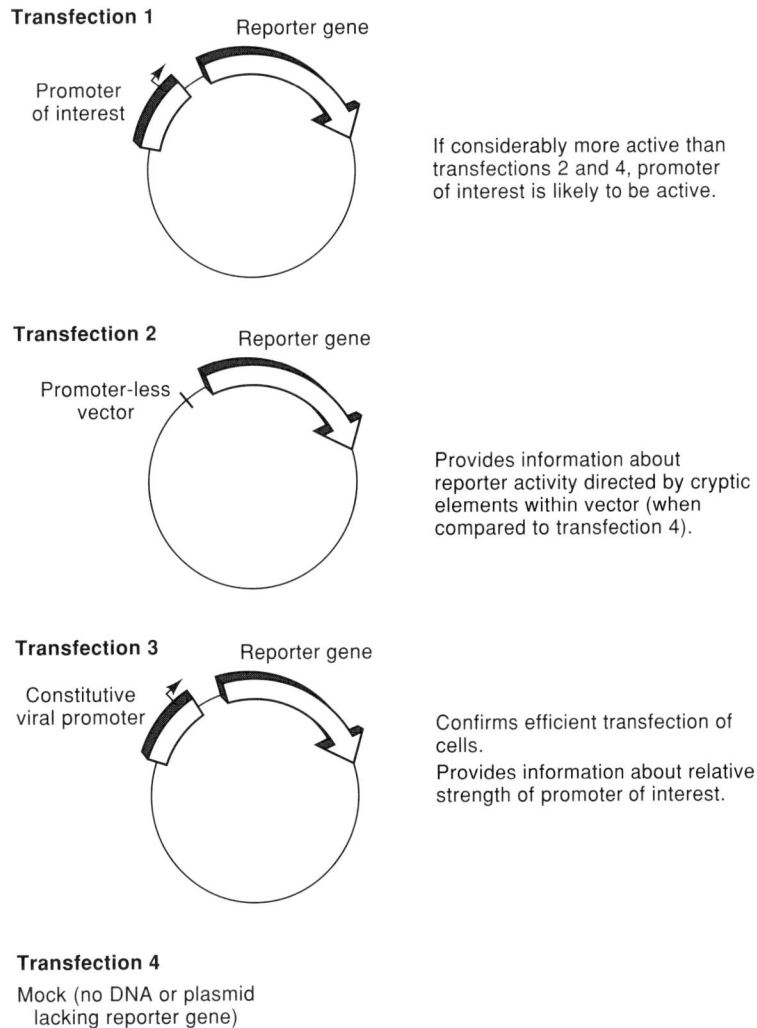

Transfection 1

Reporter gene

Promoter of interest

If considerably more active than transfections 2 and 4, promoter of interest is likely to be active.

Transfection 2

Reporter gene

Promoter-less vector

Provides information about reporter activity directed by cryptic elements within vector (when compared to transfection 4).

Transfection 3

Reporter gene

Constitutive viral promoter

Confirms efficient transfection of cells.
Provides information about relative strength of promoter of interest.

Transfection 4

Mock (no DNA or plasmid lacking reporter gene)

FIGURE 5.7. Initial transfection experiments.

lines, transfection efficiency and reporter gene activity can be influenced by the quality and purity of the DNA. Therefore, to optimize or troubleshoot an experiment, different DNA preparation methods can be tested. The results obtained with some cell lines (see, e.g., Plevy et al. 1997) are improved if the plasmids are prepared using a method that removes endotoxin (LPS) from the DNA preparation. QIAGEN offers a plasmid purification kit that includes a procedure for endotoxin removal (cat. #12362).

Initial Transfection Experiments

Control transfections. For an initial transfection experiment with a reporter plasmid containing the promoter of interest and a standard reporter gene (e.g., CAT or luciferase), three primary control transfections should be included (Fig. 5.7). One control is a transfection with the promoter-less vector into which the promoter of interest was inserted. A second control transfection should be performed with a plasmid containing a strong viral promoter fused to the reporter gene. Finally, a mock transfection should be performed with no plasmid or with a plasmid that lacks the reporter gene.

The signal detected with the strong control promoter will confirm that the cells were efficiently transfected and will provide information about the relative strength of the promoter of interest (Fig. 5.7). The signal detected with the promoter-less vector relative to the mock transfection provides an indication of reporter gene activity directed by cryptic control elements within the vector. Every vector will direct transcription of the reporter gene in the absence of an inserted promoter, with a small subset of these transcripts competent for translation. For this reason, a background signal is often detected with a sensitive reporter assay.

If the signal detected with the plasmid containing the promoter of interest is much stronger than that detected with the promoter-less vector, the difference is likely to correspond to authentic promoter activity. If the signal is only slightly or moderately above that obtained with the promoter-less vector, it may correspond to weak, authentic promoter activity, or it may result from aberrant transcription initiating within the inserted fragment. In either case, to confirm that the activity detected with an enzymatic assay corresponds to specific promoter activity, it is helpful to perform a primer extension, S1 nuclease, or RNase protection assay to demonstrate that transcripts initiate from the correct transcription start site. Unfortunately, if the promoter activity is weak, this confirmation may be difficult or impossible. It therefore may be difficult to confirm that the activity corresponds to authentic promoter activity until a much more extensive analysis is carried out. This extensive analysis would include a demonstration that the promoter activity is appropriately regulated and depends on a set of discrete, functional control elements (as revealed by a detailed mutant analysis [see Chapter 7]).

Internal controls. When measuring reporter gene expression with an indirect enzymatic assay, an internal control plasmid is usually included to normalize all of the transfection experiments performed at that time. This control is a plasmid containing a standard promoter fused to a different reporter gene, whose activity can be measured with a different protein or enzymatic assay. For example, if the luciferase gene is used for the experimental plasmids, a plasmid containing a CAT gene or β-gal gene driven by a strong viral promoter/enhancer can be included in each transfection as a control. This control is used to normalize the transfection efficiency obtained with each sample within a given experiment. Furthermore, if the same cell extract is used for the experimental and control enzyme assays (e.g., for both CAT and β-gal), the control plasmid provides normalization for the consistency of extract preparation,

An internal control of this sort can be particularly useful when comparing a series of promoter mutants. However, it is important to note that internal controls do not diminish the need to rely primarily on experimental repetition for establishing the validity of the results. Measurement errors can be made in aliquoting DNA prior to transfection, as well as in the transfection procedure itself, raising the possibility of inaccurate normalization. The viral promoters/enhancers used in the internal control plasmids also have the potential to compete for limiting transcription factors (Farr and Roman 1992), making it essential to rely on experimental repetition in some instances.

Controls when performing direct RNA analysis. When using an assay that involves a more direct measure of mRNA abundance (e.g., primer extension, S1 nuclease, or RNase protection), the initial transfection experiments are performed with similar controls to those described above. An advantage of mRNA measurement is that, if a clear specific signal is obtained, there is much more confidence that it corresponds to correctly initiated transcription. Nevertheless, during the initial experiments, it is important to include a mock transfection and a transfection with a promoter-less vector to demonstrate that the expected band is obtained only when the promoter is included, since background bands are often detected with each of the direct assays. A control plasmid to normalize transfection efficiencies and, perhaps, mRNA preparation can be included as described above for the enzymatic assays. However, since the mRNA detection assays involve numerous manipulations that can result in variability from sample to sample, the validity of the results obtained must be based primarily on several repetitions of the experiment.

Assessing Appropriate Promoter Regulation

The final step in an initial transient transfection analysis is to determine whether the promoter being studied is appropriately regulated in the artificial assay. For genes that are cell-type specific, "appropriate regulation" refers to the degree to which the control region is active in the same cell types as the endogenous gene. For genes that are inducible, it refers to the degree to which the control region is induced with similar kinetics and by similar agents as the endogenous gene.

Inducible genes. For inducible genes, an assessment of the degree to which the transient transfection assay mimics the regulation of the endogenous gene is relatively straightforward. Aliquots of the appropriate cells are transfected with the promoter/reporter plasmid and the cells are allowed to recover from the transfection procedure for a period of time, which will vary depending on the particular situation. Then, a subset of the aliquots are induced with an appropriate agent and the other aliquots are mock-induced, but maintained in culture so that both populations can be harvested at the same time. If the promoter is properly regulated in the transient assay, its activity should be selectively induced, relative to the activity of a control promoter.

Cell-type-specific genes. As a crude test of cell-type specificity of a promoter, the promoter activity in the expressing cell line can be compared to the activity in nonexpressing cell lines. Although weak or negligible activity in one or two cell lines does not conclusively demonstrate that the promoter activity is tightly or properly regulated, it provides preliminary evidence of regulation.

In most instances, two difficulties arise when comparing the activity of a promoter in multiple cell types. First, most promoters analyzed in a transient assay are not as tightly regulated as the endogenous gene. The transient activity is usually detected in some nonexpressing cell types and may not be completely absent in any cell type. This could be due to the fact that chromatin and other regulatory regions (e.g., silencers) may be needed to keep the endogenous gene inactive in nonexpressing cells. In addition, the presence of multiple copies of the plasmid can cause leakiness, because a subset of the positive and negative regulatory proteins may be in limiting quantities.

The second difficulty encountered when comparing promoter activity in multiple cell types is that transfection efficiencies vary from cell line to cell line, such that the absolute activities are not directly comparable. Transfection efficiencies are usually normalized using a reporter plasmid under the control of a strong viral promoter/enhancer that is thought to function similarly in all cell types, such as the CMV, RSV, or SV40 promoters and enhancers. Although this method is commonly used for normalization, it is far from ideal because it is difficult to establish that the viral control region is indeed similarly active in the cell types examined. An alternative method for comparing transfection efficiencies is to perform immunofluorescence to determine the number of cells transfected. This is even less desirable, because transfection outcomes can vary, not only with regard to the number of cells transfected, but also with regard to the number of molecules introduced into each cell. A third possibility is to isolate the plasmid DNA from the transfected cells to determine the quantities of DNA. This is also misleading, because plasmid DNAs adhere to the cell surface, such that the DNA isolated contains DNA within the nucleus combined with a substantial amount of DNA attached to the surface.

Although no ideal method exists for comparing promoter activities from cell line to cell line in a transient assay, perhaps the preferred method is to normalize the promoter activities using two or more control plasmids. Thus, if one of the viral promoter plasmids has an aberrant activity in one of the cell lines, the activity of the other control plasmid may reveal the disparity. Unfortunately, if a disparity is observed, it will be difficult to determine which viral promoter is most appropriate for normalizing signals obtained with the various cell lines. Ultimately, more advanced experiments will be needed to determine whether the control region being studied actually contains control elements that contribute to cell-type-specific regulation. The goal is to demonstrate that a substitution mutation in a specific control element affects promoter activity in transient transfection experiments performed in one set of cell lines (presumably those that express the endogenous gene of interest), but not in another set of cell lines (those that do not express the endogenous gene). Differential effects of promoter mutants provide significant evidence that the promoter is indeed differentially regulated in the various cell lines, leading to more advanced experiments to confirm the differential regulation and determine its mechanism.

Stable Transfection Assays by Chromosomal Integration

To develop a stable transfection assay for the analysis of a promoter or other control region, many of the issues that must be considered are similar to those discussed above for the transient assay. Namely, a source of cells, transfection procedure, and reporter gene must be chosen. Then, a reporter plasmid must be constructed and the initial experiments performed with appropriate controls to determine whether the control region is functioning properly. One additional decision that is unique to stable transfection assays is the drug selection method to be used to isolate cells that have stably integrated the reporter plasmid.

The above issues and a general outline of the experimental approach are discussed below. Unfortunately, we are unaware of published articles that provide more detailed information about the methods involved in studying gene regulation using a stable transfection assay. Ausubel et al. (1994, Unit 9.5) and Spector et al. (1998, pp. 86.4–86.6) can be consulted for in-depth information about stable transfections from the perspective of an investigator interested in overexpressing a protein by stable integration of its gene. Even with this information, however, it is strongly recommended that a researcher performing stable transfection experiments for the first time rely on personal advice and assistance from a nearby investigator with stable transfection experience. As discussed below, a few of the steps, including the initial growth and maintenance of drug-resistant cell clones or populations, can be quite tedious and difficult for the beginner.

General Strategies

The basic stable transfection assay involving chromosomal integration can be performed using two general strategies. The most common and generally preferred method is to transfect cells with two genes, one containing the promoter of interest driving a standard reporter gene, and the other containing the drug-resistance gene (i.e., dominant selectable marker gene) usually under the control of a viral promoter/enhancer. The two genes can be on the same plasmid or on separate plasmids. After transfection, the cells are incubated for a short period of time (typically 1–2 days) and then are subjected to drug selection. After drug-resistant cell clones or populations are obtained (see below), promoter activity is quantified by measuring the production of mRNA, protein, or enzymatic activity of the reporter gene or protein.

Box 5.5

Isolation of Transiently Transfected Cells

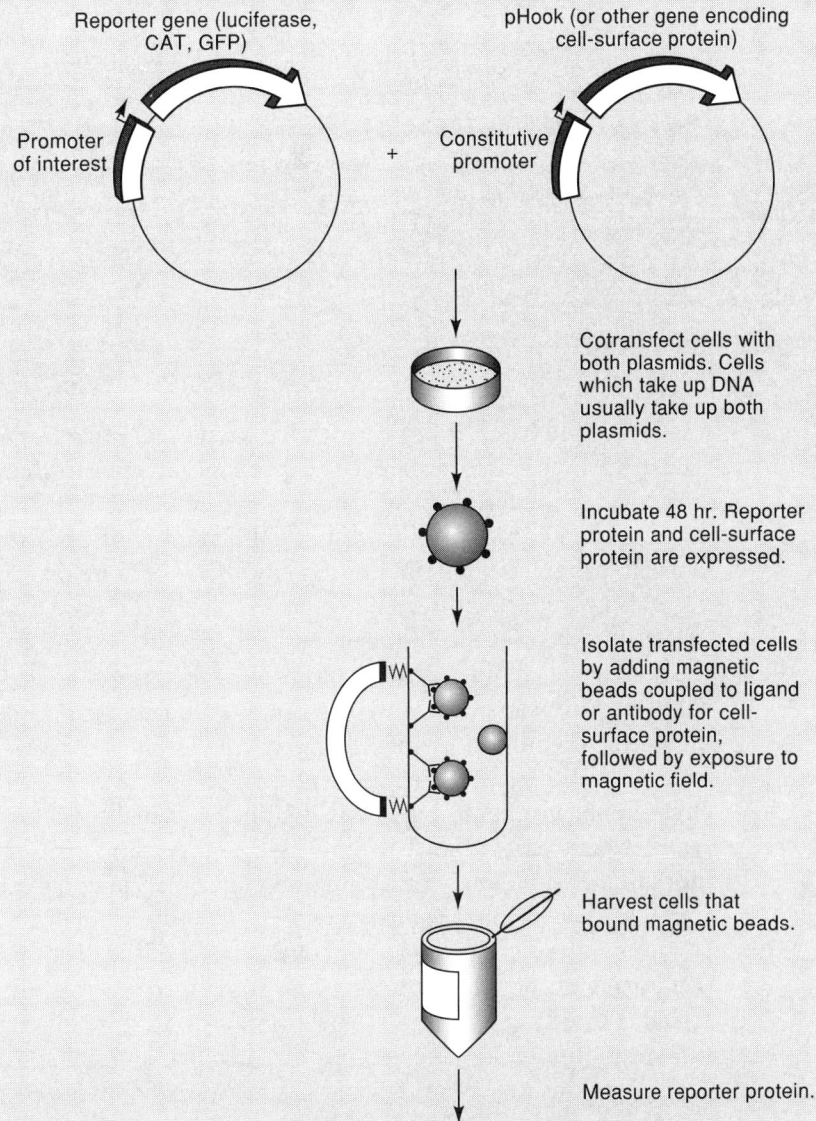

FIGURE 5.8. Isolating transiently transfected cells. (Adapted, with permission, from Invitrogen Corp. [Copyright Invitrogen Corp. All rights reserved.].)

One key challenge in establishing a transient transfection assay is the development of a procedure for transfecting cells with an efficiency that is sufficient to detect reporter gene activity. If efficient transfection is not obtained with the available techniques, one strategy for enhancing the detection of activity is to isolate the transfected cells from the untransfected cells. It should be easier to detect reporter gene activity from a pool of transfected cells that have been separated from the much larger number of untransfected cells. Alternatively, the isolated transfected cells can be analyzed individually. Multiple variations of this strategy have been used to improve reporter gene detection.

One approach is to cotransfect cells with the reporter plasmid and with a second plasmid containing a gene encoding a cell-surface protein under the control of a viral promoter/enhancer (Fig. 5.8). Cells that take up DNA usually take up both plasmids. The cells that express the cell-surface protein can be isolated by a variety of methods, including binding to magnetic beads linked to an appropriate ligand or antibody, or fluorescence-activated cell sorting (FACS). Extracts can be prepared from the isolated cells and reporter gene activity can be measured.

The Invitrogen Capture-Tec System is one example of this strategy (Chesnut et al. 1996; [Invitrogen cat.# K650-01]). This kit contains a plasmid, called pHook (Fig. 5.8), that expresses a single-chain antibody on the surface of transfected cells. The antibody gene is under the control of a CMV promoter/enhancer and is fused to the coding region for a transmembrane domain, facilitating membrane anchoring. Cells are cotransfected with this plasmid and the reporter plasmid, and the transfected cells are isolated using magnetic beads linked to the hapten recognized by the antibody. The cells that bind the beads are separated from the untransfected cells by exposure to a magnetic field.

The second basic strategy, which has been used much less frequently, involves placing a drug-resistance gene, or oncogene capable of cell transformation, directly under the control of the promoter of interest (see, e.g., Fromm and Berg 1982; Zhong and Krangel 1997; see also Chapter 6). With this strategy, the number of colonies obtained appears to correlate with promoter strength. Presumably, this is because drug resistance or transformation depends on a specific threshold concentration of the gene product—the stronger the promoter, the greater the likelihood that the threshold concentration will be achieved. Because the number of colonies is related to promoter strength, the approximate activities of promoter mutants can be measured by simply counting the number of drug-resistant colonies obtained. In addition to the relative simplicity of this assay, the advantage of placing a drug-resistance gene or oncogene under the control of the promoter of interest is that it circumvents a limitation of the first strategy. Namely, when a drug-resistance gene and a reporter gene are cotransfected into a cell, they almost always cointegrate into the same location in the cell, even if they are transfected on separate plasmids. The result of the cointegration event is that the reporter gene is almost always in close proximity to the promoter and/or enhancer that is driving transcription of the drug-resistance gene, preventing the control region of interest from being studied independently of another nearby control region. By placing a drug-resistance gene or oncogene directly under the control of the promoter of interest, there is no requirement for another control region.

Despite this important advantage, this strategy possesses two severe limitations. First, no colonies will be obtained if the promoter is too weak to allow production of sufficient gene product. Second, the number of colonies obtained is unlikely to be directly proportional to the promoter strength, although it may be related. Because these disadvantages severely limit the usefulness of this second alternative, the remaining discussion focuses on the first strategy. However, if the advantages of this second strategy outweigh its disadvantages for the particular control region being studied, it can provide valuable information regarding gene regulation mechanisms.

Cells and Transfection Procedures

The cell and transfection procedure considerations for a stable assay are similar to those for the transient assay. Namely, the cells must express the gene of interest, it must be possible to maintain the cells in culture through the course of the assay, and conditions must be available for transfecting DNA into the cells. Optimal transfection conditions are usually

Box 5.6

Replication-competent Vectors

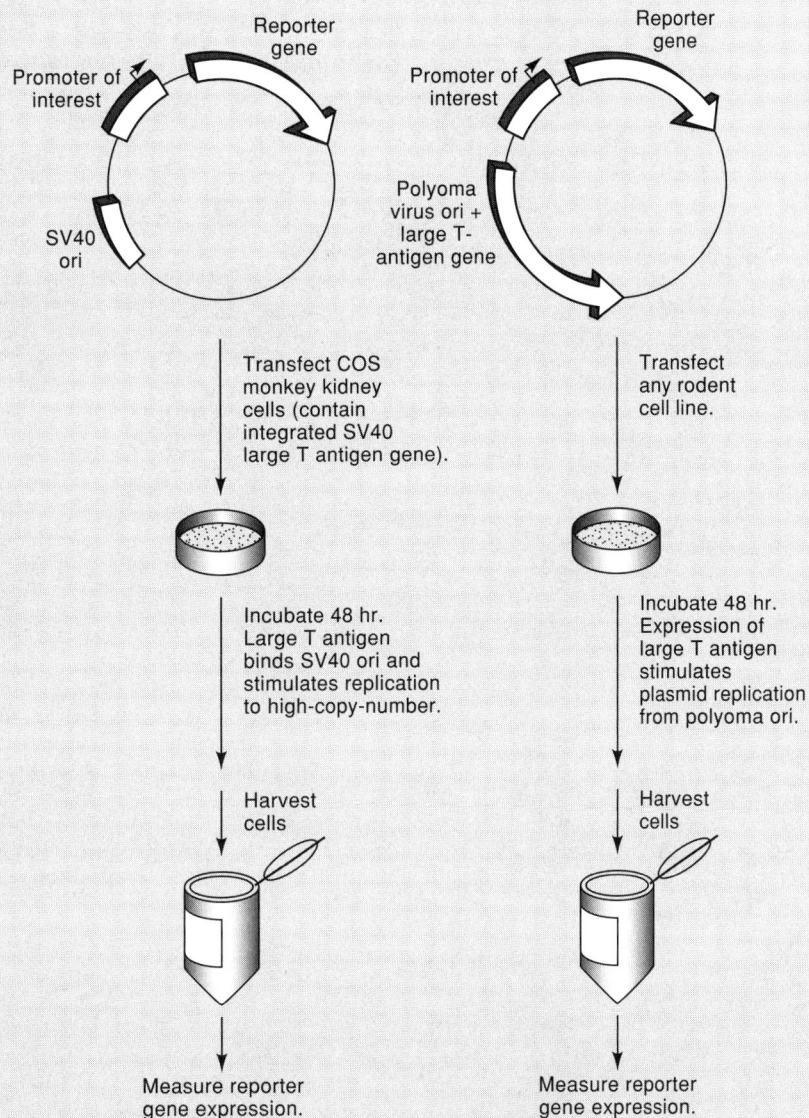

FIGURE 5.9. High-copy replication systems; SV40 and polyomavirus.

An alternative transient transfection strategy employs vectors containing viral replication origins that promote plasmid replication to high copy number (Fig. 5.9). This strategy may permit the detection of activity from promoters whose activities cannot be detected in a conventional transient transfection assay. Moreover, the signals obtained may be stronger because of the increased copy number, which may make it easier to monitor and confirm transcription start sites. The primary disadvantage of this approach is that the high plasmid copy number increases the chances that the promoter will be regulated aberrantly. Of greatest concern is the possibility that important transcription factors will be limiting within the nucleus. With this scenario, transcription might be regulated primarily by the most abundant factors capable of interacting with the promoter, rather than by the factors that are most important for regulating the endogenous gene. A second limitation is that the function of the viral replication origin usually is somewhat dependent on the presence of the adjacent viral enhancer. Almost all plasmids used for this assay therefore contain viral enhancers, which may act on the promoter of interest and make it difficult to study promoter activity independent of a nearby enhancer. Thus, although this assay may be useful for gaining some insight into control regions and control elements that contribute to the regulation of a gene, it is far from ideal.

The two high-copy replication systems that have been used most commonly in mammalian cells rely on the SV40 and polyomavirus origins of replication. Vectors containing SV40 origins only replicate to high copy number in primate cells in the presence of the SV40 large-T antigen. A well-known African green monkey cell line called COS-7 contains a stably integrated SV40 fragment encoding the large-T-antigen gene (Gluzman 1981; see also Sambrook et al. 1989, pp. 16.17–16.21). Transfection of COS-7 cells with plasmids containing the SV40 replication origin often results in plasmid replication to a copy number of more than 50,000 molecules per cell. Vectors containing the polyomavirus origin of replication only replicate in rodent cell lines expressing the polyomavirus large-T-antigen gene. The polyomavirus replication system has been used to study transcriptional regulation in a variety of cell types, including lymphocytes (Grosschedl and Baltimore 1985; Lo et al. 1991). To accomplish this objective, plasmids were prepared that contain both the polyomavirus origin of replication and the coding sequence for the polyomavirus large-T-antigen gene (Grosschedl and Baltimore 1985). The promoter of interest and a linked reporter gene can be inserted into this same plasmid. The plasmid can then be transfected into any mouse cell line. The large-T antigen will be expressed and will then bind to the replication origin and stimulate replication to high copy number. These assays are not discussed further in this manual, but the appropriate vectors and procedures can be found in the references cited above.

established using a transient assay, by optimizing the conditions that lead to efficient transient expression of a standard reporter gene (e.g., CAT or luciferase). It is important to be aware, however, that some transfection procedures, such as DEAE-dextran, are useful for transient transfection experiments, but not for stable transfections because of long-term toxicity problems (see Box 5.1).

One difference in the criteria for choosing cells for the stable assay as opposed to the transient assay is that it is necessary to maintain the cells in culture for a longer time period. The cells must also be amenable to drug selection. Not all drug selection procedures are effective with all cell lines, and the optimal drug concentrations for killing and selection can vary over a wide range from cell line to cell line (see below).

When identifying an appropriate transfection procedure, it should be noted that the stable assay can often succeed with transfection efficiencies that are too low for a successful transient assay with a control region of interest. For a stable assay to succeed, it often is necessary for only a few cells within a population to take up DNA. As long as the drug selection procedures are adequate, those cells can be selected to grow, resulting in cell clones containing the transfected plasmid. With a transient transfection procedure, selection is generally not involved. In practical terms, when optimizing transfection conditions by transient transfection of a viral promoter–CAT or viral promoter–luciferase reporter plasmid, it may not be necessary to identify transfection conditions that yield strong promoter activity. If the optimal conditions yield only low or modest levels of reporter activity, it may suffice for the stable transfection assay.

Reporter Genes and Assays

The selection of an appropriate reporter gene for a stable transfection assay involves similar considerations to the selection of a reporter gene for a transient assay. Any of the standard reporter genes that can be monitored by enzymatic assay or by mRNA analysis are appropriate. The one new consideration is that the use of a stable transfection assay instead of a transient assay usually increases the likelihood that specific transcripts can be effectively monitored by mRNA analysis in addition to or instead of an indirect enzymatic assay. As described earlier, the use of an assay that provides information about the amount of transcription that initiates from a specific transcription initiation site can greatly enhance one's confidence in the validity of the results.

Drug-resistance Genes and Vectors

Several selectable genes have been used for the generation of stable transfectants since the general method was developed in the late 1970s. In the earliest studies, the selectable genes restored enzymatic deficiencies in metabolic pathways. For example, thymidine kinase mutant cells were transfected with a herpes simplex virus thymidine kinase (HSV-TK) gene (Wigler et al. 1977; Mantei et al. 1979). Cells that stably integrated the HSV-TK gene were selected in HAT medium, which prevents de novo thymidine synthesis from deoxyuridine, leading to the requirement for an active thymidine kinase.

Beginning in the early 1980s, the need for mutant cell lines was eliminated by the development of an array of dominant selectable marker genes. The *E. coli* xanthine-guanine phosphoribosyltransferase (*gpt*, Mulligan and Berg 1981) and aminoglycoside phosphotransferase (*neo*, Southern and Berg 1982) genes were the first used for this purpose. Both remain in common use. The *E. coli* GPT enzyme catalyzes the synthesis of GMP from xanthine via a salvage pathway. When the growth medium is supplemented with aminopterin and mycophenolic acid to block de novo synthesis of GMP by mammalian enzymes, cell survival depends on *gpt* integration and expression. The *E. coli* *neo* gene confers resistance to the antibiotic G418. When G418 is added to the growth medium, cell survival depends on stable integration and expression of the *neo* gene. Some of the other dominant selectable marker genes used for stable transfection experiments are listed in Table 5.1.

Determining which strategy to employ depends on several factors. One is the availability of appropriate vectors encoding the drug-resistance gene, because vectors for some of the genes are not yet available from commercial sources (see below). Another important consideration can be the cost of the drug; drug costs vary and, if numerous experiments are planned, the total cost can be substantial. The cost can be strongly influenced by the amount of drug needed to kill the cells being used, which can vary widely among cell lines. A third factor that can influence the selection of a drug-resistance strategy is the speed with which the drugs kill cells and allow stably transfected cells to be selected. Invitrogen claims that Blasticidin kills cells much more rapidly than other drugs and allows the selection and growth of stably transfected cells much more rapidly. A final consideration is the proximity of researchers who have prior experience with a particular drug-resistance protocol, because some protocols exhibit idiosyncrasies that are understood only by those with experience using the protocol.

Vectors containing drug-resistance genes are available from several academic and commercial sources. CLONTECH markets vectors containing neo (pSV2neo, cat. #U02434), pac (pPUR, cat. #6156-1), and hyg (pTK-Hyg, cat. #U40398). Invitrogen sells vectors con-

TABLE 5.1. *Dominant selectable markers for use in stable transfections into mammalian cells*

Dominant selectable marker	Gene encodes	Selection*	Reference	Notes
NeoR	E.coli aminoglycoside phosphotransferase, which confers resistance to neomycin or the neomycin analog G418.	Selection in G418, an aminoglycoside that blocks protein synthesis.	Southern and Berg 1982	High cost of G418 is a disadvantage.
HisD	Salmonella typhimurium histidinol dehydrogenase (HisD), which catalyzes the synthesis of histidine from histidinol.	Selection in histidinol, which is toxic unless oxidized to histidine by HisD. A more stringent selection uses histidinol in histidine-free medium; cell survival depends on the synthesis of histidine catalyzed by HisD.	Hartman and Mulligan 1988	Use of histidine-free medium can improve chances of success, but selection often succeeds in complete medium.
PAC	Streptomyces alboniger N-acetyl transferase, which confers resistance to puromycin.	Selection in puromycin, which blocks protein synthesis unless detoxified by acetylation.	Vara et al. 1986	Often preferred because of consistency, speed of selection, and low cost of puromycin.
HPH	E.coli hygromycin B aminoglycoside phosphotransferase, which confers resistance to the antibiotic hygromycin B.	Selection in hygromycin B, which blocks protein synthesis.	Blochlinger and Diggelmann 1984	HPH selection has been commonly used in conjunction with Neo selection when two separate selections are needed.
DHFR	Murine dihydrofolate reductase (DHFR) containing a single amino acid mutation that confers resistance to methotrexate (MTX). DHFR is involved in purine biosynthesis.	Selection in MTX, which is a competitive inhibitor of wild-type DHFR. MTX does not bind mutant DHFR. Therefore, cells expressing mutant DHFR are resistant to MTX.	Simonsen and Levinson 1983	Wild-type DHFR can also be used as a selectable marker, but cells must be grown in dialyzed medium (to remove purines). In addition, many cells express endogenous DHFR, reducing the success of this selection strategy. Wild-type DHFR is typically used to amplify transfected genes.

Ble (ZeoR)	*Streptoalloteichus hindustanos* (Sh) ble gene, which confers resistance to bleomycin, phleomycin, and Zeocin.	Selection in the glycopeptidic antibiotics bleomycin, phleomycin, or Zeocin, a phleomycin derivative and trademark of CAYLA (licensed to Invitrogen). All three antibiotics cleave DNA and are neutralized by stoichiometric binding to the ble gene product.	Mulsant et al. 1988	Additional genes that confer resistance to these antibiotics have been isolated from Tn5, a resistant strain of *Actinomycetes*, and from *Streptomyces verticillus*.
GPT (XGPRT)	*E.coli* xanthine-guanine phosphoribosyl-transferase, which catalyzes the synthesis of GMP from xanthine in guanine-free medium.	Selection in aminopterin and mycophenolic acid to block de novo GMP synthesis. Xanthine and hypoxanthine must also be provided as substrates for GMP synthesis catalyzed by XGPRT. To prevent synthesis of GMP from guanine in the growth medium, it must be dialyzed and then supplemented with thymidine and glutamine.	Mulligan and Berg 1981	GPT was the first dominant selectable marker discovered and is useful in virtually all mammalian cell types.
BSD	*Asperigillus terreus* bsd gene, which confers resistance to blasticidin.	Selection in blasticidin, an antibiotic isolated from *Streptomyces griseochromogenes* which inhibits protein synthesis.	Izumi et al. 1991	Recently marketed by Invitrogen as an unusually rapid selection strategy.
OuabainR	A mutant version of the rat Na,K-ATPase α1 subunit which confers resistance to ouabain.	Selection in ouabain, a cardiac glycoside which inhibits the Na,K-ATPase.	Emanuel et al. 1988	Marketed by PharMingen and recommended primarily for primate cells.

taining neo (pcDNA3.1, cat. #V790-20), hyg (pcDNA3.1/Hygro, cat. #V870-20), zeo (pcDNA3.1/Zeo, cat. #V860-20; pCMV/Zeo, cat. #501-20; pSV40/Zeo, cat. #V502-20), his (pREP8, cat. #V008-50), and bsd (pcDNA6/V5-His). Unfortunately, none of the commercially available vectors was designed specifically for analyzing promoters or enhancers by stable transfection. Thus, none of the vectors contains the ideal components, which would be a drug-resistance gene under the control of a viral promoter, and a common reporter gene downstream of a multiple cloning site. The vectors typically contain a drug-resistance gene downstream of a strong viral promoter and enhancer. These vectors can be cotransfected into cells along with a reporter plasmid containing the promoter of interest. Alternatively, the drug-resistance cassette can be excised and inserted into a standard reporter plasmid.

To establish selection conditions, the drug-resistance strategy must be chosen. Then, serial dilutions of the drug must be tested to define the drug concentration needed for killing cells that lack the integrated drug-resistance gene. This test is often performed in 24-well or 96-well plates. The concentration chosen for the experiment is usually slightly above the minimum concentration needed to kill the cells. A concentration that is too high decreases the chances of obtaining stable transfectants. After the optimal concentration is determined, the initial transfection experiments can be performed.

Plasmid Construction

Choosing between cotransfecting reporter and drug-resistance plasmids, or including both genes on the same plasmid. As described above, a stable transfection assay using a reporter gene and a separate drug-resistance gene can be performed using either of two strategies: The reporter gene and drug-resistance gene can be on the same plasmid, or the two genes can be on separate plasmids and cotransfected with one other.

The primary advantage of including the reporter and drug-resistance genes on the same plasmid is that one's knowledge about the relative locations of the integrated control region of interest and the viral promoter/enhancer used to drive transcription of the drug-resistance gene will be somewhat greater. Because the viral promoter/enhancer may affect the activity of the promoter of interest, it is helpful to have it at a consistent location. This advantage is diminished by the fact that most cells will integrate multiple tandem copies of the plasmid, presumably in both head-to-head and head-to-tail orientations, leading to considerable variability in the location of the viral promoter/enhancer relative to the promoter of interest, unless the inserts are carefully mapped to determine their number and orientations. A secondary advantage of placing the reporter gene and drug-resistance gene on the same plasmid is that the frequency of cointegration of the reporter gene and drug-resistance gene will be enhanced; very few drug-resistant cells will be isolated that lack an integrated reporter gene.

The primary advantage of having the reporter and drug-resistance genes on different plasmids is the following: If appropriate reporter genes are already available on a separate plasmid, the cotransfection experiment can be performed immediately. Despite the advantages of the first strategy, the second strategy usually works quite well.

Plasmid construction considerations. If the strategy chosen is to cotransfect the reporter plasmid with a separate drug-resistance plasmid, the considerations for insertion of the control region of interest will be the same as those described above for construction of plasmids for transient transfection experiments, and the same vectors can be used. The only exception is that reporter vectors containing viral replication origins cannot be used in cells in which the origin will be active. An active replication origin will prevent stable integration, and the high plasmid copy number that results from replication will promote cell death.

If the chosen strategy is to construct a plasmid containing both the reporter gene and the drug-resistance gene, it will be necessary to insert the control region of interest into a vector that already contains both the reporter gene and the drug-resistance gene. Unfortunately, very few vectors that are designed for simple insertion of a control region are available. In most cases, it will be necessary to generate one's own vectors.

Regardless of which strategy is chosen, the integration site variability observed during stable transfection experiments may be reduced by placing insulator fragments on either side of the reporter gene cassette (see Chapters 1 and 6) (Chung et al. 1997; Zhong and Krangel 1997). The insulators may isolate the reporter cassette from the enhancer that regulates the drug-resistance gene, as well as from other control regions in the vicinity of the integration site.

Drug Selection

With optimized transfection conditions, an optimized drug concentration for cell killing, and an appropriate plasmid, the initial experiments can be performed (see also Ausubel et al. 1994, Unit 9.5; Spector et al. 1998, pp. 86.4–86.6).

Basic transfection and selection procedure. For each plasmid, an aliquot of cells is transfected according to the established procedure. Following transfection, the cells are transferred to tissue culture dishes or multiwell plates, depending on whether pools of drug-resistant cells or individual cell clones will be isolated and analyzed (see below). If the decision is to isolate pools of drug-resistant cells, the cells can be transferred to standard tissue culture dishes. If individual clones will be isolated, the transfected cells are usually aliquoted immediately onto 96-well plates. (An alternative, which is sometimes used for adherent cells, is to add the cells to a standard tissue culture dish, and then to isolate individual colonies using a cloning ring or limiting dilution; see Spector et al. 1998, pp. 86.5–86.6.) The number of cells to add to each well will vary depending on the cell type, drug selection scheme, and transfection efficiency, and will need to be determined empirically. Ideally, most wells should have no more than one drug-resistant cell. After the cells are allowed to recover from the transfection procedure, usually for 1 or 2 days, drug is added to begin selection. At this point, the cells must be carefully monitored to maintain the health of the cells to permit the growth of drug-resistant clones.

Maintenance of cells during drug selection. The most difficult step of the stable transfection procedure is often the maintenance of the cells at the initial stages of drug selection (Ausubel et al. 1994, Unit 9.5; Spector et al. 1998, pp. 86.4–86.6). The drug-resistant cells are faced with the formidable challenge of remaining sufficiently healthy to grow when surrounded by cell debris and a vast number of dying cells. At the same time, they must express sufficient quantities of the protein produced by the integrated drug-resistance gene to avoid their own death. The goal of the researcher is to passage the cells and change the cell growth medium at appropriate times to aid in the growth of resistant clones. The timing of these events and the general monitoring of cell growth and death will vary from cell line to cell line. In many instances, the most important factor in determining the success of a stable transfection experiment is the level of experience of the researcher who must make the subjective judgments. For this reason, it is strongly recommended that those developing a stable transfection assay for the first time obtain the assistance and advice of a nearby researcher who has experience with stable transfection procedures and can visually inspect the transfected cells during drug selection.

Choosing between analysis of cell clones or pools. The recommended strategy for analyzing the activity of a control region is to isolate and assay several individual clones obtained with each reporter plasmid, to determine the range of activities. This is more desirable than analyzing pools of clones (as described below) because it provides more information about the range of activities of the control region of interest in different chromosomal integration sites. As described earlier in this chapter, the activity of a control region that lacks an LCR or insulator often varies considerably from stable cell line to stable cell line because the nature of the integration site can result in enhancement or suppression of the activity of the control region. The copy number of the integrated plasmid also varies and influences activity.

By analyzing several independent clones that are obtained with a given reporter plasmid, one can gain knowledge about the range of activities and mean activity that are obtained. This information is invaluable when attempting to compare the activity of the control region in different cell lines or when attempting to compare a wild-type control region to mutant control regions. The variability observed from clone to clone makes these comparisons difficult, but a lack of knowledge about the degree of variability is much more problematic. If the goal is to monitor the inducible activity of a particular control region in the stably transfected cells, it is equally important to isolate and analyze individual cell clones, because the integration site can influence the degree of induction.

Analyzing cell clones. Although an analysis of several different clones is worthwhile, in most instances it is not necessary to determine rigorously that each putative clone is actually clonal. If a small percentage of wells within the 96-well plate contains drug-resistant colonies, each is likely to be clonal. For some experiments, it may be important to establish definitively that each population is clonal (by further subcloning of the cells by limiting dilution and by Southern blot analysis to determine integration sites). However, because several cell clones will be compared and considerable variability is expected due to integration site differences, the rigorous determination of clonality is probably not necessary. A rigorous determination of copy number (by quantitative Southern blot) also may not be necessary, unless there is a need to determine precisely how much of the variability observed with individual clones is due to variability in copy number versus variability in integration site. This information will be useful in some instances, such as for determining whether a control element conforms to the definition of an LCR. However, for most gene regulation studies, this information will be of little value.

Analyzing pools of cells. An alternative to the isolation of several individual colonies is to compare several independent pools of cells. The potential advantage of this strategy is that variations in each pool due to integration site and copy number differences can be averaged, reducing variability from pool to pool. A disadvantage, however, is that the relative abundance of each clone within the pool can change dramatically over time, if one or a few clones exhibit a growth advantage (due to integration site or level of expression of the reporter gene or drug-resistance gene). Thus, the results obtained with a pool of cells at one time might differ from the results obtained after the cells are passaged for an additional few weeks. This limits the long-term usefulness of the cells.

With regard to the selective outgrowth of cells, it also is important to note that the use of cell pools does not reduce the number of samples that need to be tested with each reporter plasmid, relative to the number tested when individual clones are isolated. Although the use of pools theoretically leads to an average reporter gene signal derived from a large number of individual clones, the potential for selective outgrowth of one or a few clones makes it essential for several pools to be tested.

Controls and Interpretation of Results

Controls are needed with the stable transfection assays to confirm that the promoter is properly regulated. If the promoter is expected to be cell-type-specific, a determination can be made by performing stable transfection experiments with different cell lines. As with transient transfection experiments, it is beneficial to normalize the results obtained with the various cell lines by performing parallel experiments with a reporter driven by a viral promoter/enhancer. The range of activities obtained with the control promoter/enhancer provides information about intrinsic differences between cell lines with regard to the average copy number and integration site accessibility. An alternative method for normalizing among a variety of cell lines is to develop an assay to monitor transcription of the drug-resistance gene itself.

If the promoter is inducible, the control to show that induction is specific to the promoter of interest is to perform a similar experiment with the reporter gene under the control of a different promoter. If transcription is induced only from the promoter of interest, this result confirms that specific control elements within the promoter are responsible.

After convincing evidence is obtained that the stable assay, or any of the other assays mentioned in this chapter, mimics at least in part the proper regulation of the promoter of interest, the next step is to identify the *cis*-acting DNA sequences that are important for promoter function. The strategies for approaching this next goal are outlined in Chapter 7.

TECHNIQUES

COMMON TRANSFECTION METHODS FOR MAMMALIAN CELLS

SUMMARY

The first methods for inducing cultured mammalian cells to take up exogenous DNA were developed in the late 1960s and early 1970s (McCutchan and Pagano 1968; Graham and van der Eb 1973). These original cell transfection methods, the DEAE-dextran and calcium phosphate methods, were developed as a means of introducing wild-type and mutant genomes from DNA tumor viruses into cultured cells, which could then express the viral proteins and produce infectious virions. The transfection methods were soon adopted for use with plasmids containing specific genes or control regions of interest.

Two general types of transfection experiments are in common use: transient transfections and stable transfections. In a transient transfection experiment, the plasmid is maintained for only one or two cell generations, either because the plasmid cannot replicate or because it replicates to very high copy number and is therefore toxic (see Boxes 5.5 and 5.6). Gene expression occurs from the episomal plasmid and must be monitored during a relatively short time period of up to 72 hours. A sensitive assay is usually needed to monitor gene expression because, for most cell lines, only a small fraction of the cells take up the exogenous DNA.

In stable transfection experiments, the exogenous DNA is maintained for many cell generations. Maintenance is usually accomplished by selecting for expression of a dominant selectable marker gene. Stable expression of the marker gene occurs when the exogenous DNA integrates into a chromosome. Alternatively, some viral replication origins permit long-term maintenance of episomes. In both instances, the selection process leads to a cell population in which all of the cells contain the exogenous DNA.

Transfection experiments are most commonly used for two purposes. First, a transcriptional control region of interest can be inserted into a plasmid and used to regulate the expression of a reporter gene, allowing the activity of the control region to be monitored. Second, the cDNA encoding a protein of interest can be placed under the control of a promoter and enhancer, leading to expression or overexpression of the protein. Transfection experiments are used to study many other fundamental processes, including DNA replication, recombination, and RNA processing.

Numerous methods have been used to transfect mammalian cells. The underlying basis of each method is briefly summarized in Box 5.1. The three most common methods are the calcium phosphate, DEAE-dextran, and electroporation methods. A basic example of each of these methods is provided below. The parameters of each protocol were specifically designed for the cell line or type indicated and are intended only as a general introduction to the protocol. A calcium phosphate method is described for murine 3T3 cells, a DEAE-dextran method for B and T lymphocytes, and an electroporation method for macrophages. Transfection parameters vary from cell line to cell line. Therefore, before experiments are initiated with a line of interest, the published literature should be surveyed to identify conditions that have been used successfully.

The protocols provided here are intended as a basic introduction to the steps involved in transfection experiments for investigators with little or no exposure to these techniques.

Much more detailed information about the optimization and performance of the calcium phosphate, DEAE-dextran, and electroporation methods can be found in Sambrook et al. (1989, Chapter 16), Ausubel et al. (1994, Chapter 9), and Spector et al. (1998, Chapter 86). For those who are unfamiliar with transfection methodology, these references are strongly recommended.

PROTOCOL 5.1

Calcium Phosphate Transfection of 3T3 Fibroblasts

TIME LINE AND ORGANIZATION

The calcium phosphate protocol described here for transfection of murine 3T3 fibroblasts is derived from Graham and van der Eb (1973) as modified by Parker and Stark (1979). The cells are usually grown on 60- or 100-mm tissue culture dishes and typically passaged 1 day prior to transfection; they are about 50% confluent at the time of transfection. This degree of confluence allows the cells to be transfected when most are not in direct contact with other cells. In addition, transfection at 50% confluence allows the cells to reach 100% confluence about 48 hours post-transfection, at the time most transiently transfected cells are harvested.

The procedure begins with the formation of a precipitate containing DNA and calcium phosphate. The precipitate is formed by gradually mixing a phosphate-containing buffer (HEBS) with a solution of DNA and calcium chloride. The mixture is then layered onto a plate containing the adherent cultured cells and growth medium. The DNA–calcium phosphate precipitate adheres to the cells and is taken up by endocytosis. After the cells have been exposed to the precipitate for a defined period of time, the medium containing the precipitate is removed and fresh medium is added to prevent toxicity.

Day 1: Formation of DNA–calcium phosphate precipitate and addition to cells
Day 1 or 2: Removal of precipitate and addition of fresh growth medium
Day 2 or 3: For transient transfections, harvest cells. For stable transfections, add drug to begin selection

PROCEDURE

CAUTIONS: $CaCl_2$, KCl, Na_2HPO_4. See Appendix I.

Step 1: Prepare buffers

2 × HEPES buffered saline (HEBS):

280 mM NaCl
10 mM KCl
1.5 mm $Na_2HPO_4 \cdot 2\ H_2O$
12 mM dextrose
50 mM HEPES
To pH 7.05 with 10 N NaOH (accurate pH is critical for efficient transfection)
Sterilize by filtration through 0.45 μm nitrocellulose filter.
Store at –20°C.

1 M $CaCl_2$:

Sterilize by filtration through 0.45 μm nitrocellulose filter. Store at –20°C.

Step 2: DNA precipitation and transfection

Note: The amounts indicated below are for transfection of a 60-mm plate of cells. For a 100-mm plate, the solution amounts should be doubled or tripled.

1. Add in order (Solution A)

   H_2O (volume calculated to achieve a total volume of 250 μl)
   20 μg of total DNA
   62.5 μl of 1 M $CaCl_2$
   _____

   250 μl total

   Prepare this solution in 6-ml transparent, sterile tubes, 250 μl per transfection.

   Notes: The indicated amount of total DNA for a 60-mm plate is 20 μg, although some researchers prefer more or less. For many experiments, it is best to use 20 μg of the plasmid of interest. However, a smaller amount of the plasmid can be combined with sufficient "filler" DNA, usually an empty plasmid vector, to achieve the 20-μg total. In other experiments, the 20 μg will include a specific amount of a reporter plasmid combined with an expression plasmid for a transcription factor.

   To sterilize the DNA, it can be precipitated with ethanol and then resuspended in sterile water. However, sterilization is often unnecessary for transient transfection experiments.

2. Add 250 μl of 2x HEBS (Solution B) dropwise to the tube. After each drop, flick the tube vigorously or vortex. Allow the precipitate to develop at room temperature for 20–30 minutes. The solution should turn uniformly cloudy and the fine precipitate should be visible under a microscope.

3. Add the precipitate to a 60-mm dish of 50% confluent 3T3 cells containing 5 ml of growth medium (often DMEM/10% FBS).

   Notes: The ratio of precipitate volume to medium volume should be approximately 1:10.

   An alternative method for adding the precipitate is first to aspirate the medium from the dish. Then, add the precipitate, tilt the dish to ensure complete coverage, and incubate for 30 minutes at room temperature. Add 5 ml of growth medium. Some investigators find that this method significantly increases the transfection efficiency.

4. Incubate the cells for 6–8 hours or overnight (up to 12 hours) at 37°C.

   Note: Each cell line has a different tolerance to the precipitate. For most cell lines, an overnight incubation is acceptable, but the precipitate should never be left on longer than 12 hours.

5. Replace the medium by aspirating gently and adding back 5 ml of fresh medium.

6. For stable transfections, add drug 1–2 days after the initial addition of the precipitate. For transient transfections, harvest the cells approximately 48 hours after addition of the precipitate.

PROTOCOL 5.2

DEAE-dextran Transfection of Lymphocyte Cell Lines

TIME LINE AND ORGANIZATION

The DEAE-dextran transfection method described here for B and T lymphocyte cell lines is derived from Luthman and Magnusson (1983) and Lenardo et al. (1987). For each transfection experiment, approximately 10^7 exponentially growing cells are used. The procedure described is for cells like lymphocytes, which are nonadherent and grow in suspension. Several DEAE-dextran protocols for adherent cells can be found in Sambrook et al. (1989) and Ausubel et al. (1994).

DEAE-dextran is frequently used for transient transfections but is rarely useful for stable transfections because of toxicity. The DEAE-dextran method described here involves an initial incubation with a DNA/DEAE-dextran mixture followed by a second incubation in growth medium supplemented with chloroquine. The negatively charged DNA binds the positively charged DEAE moieties of the DEAE-dextran polymer. The cells then take up this complex by endocytosis. Chloroquine is thought to enhance transfection efficiencies by neutralizing lysosomal hydrolases, which can degrade the DNA. One notable advantage of the DEAE-dextran method is that a relatively small amount of DNA is needed for each experiment.

Day 1: Treat cells with DNA/DEAE-dextran mixture and then with chloroquine.
Day 3: Harvest cells.

PROCEDURE

CAUTIONS: *CaCl$_2$, Chloroquine, DEAE-dextran, KCl, MgCl$_2$, Na$_2$HPO$_4$. See Appendix I.*

Step 1: Prepare buffers

TD:

> For 1 liter:
> 8 g of NaCl
> 0.38 g of KCl
> 0.1 g of Na$_2$HPO$_4$·7H$_2$O
> 3 g of Tris
> Adjust pH to 7.4.
> Sterilize by autoclaving.

MgCl$_2$/CaCl$_2$ stock solution:

> 10 mg/ml MgCl$_2$
> 10 mg/ml CaCl$_2$
> Sterilize by filtration through 0.45 μm nitrocellulose filter.
> Store at 4°C.

DEAE-dextran stock solution:

> 10 mg/ml in TD
> Sterilize by filtration through 0.45 μm nitrocellulose filter.
> Store at 4°C.

Chloroquine stock solution (100x):

> 10 mg/ml in H_2O
> Sterilize by filtration through 0.45 μm nitrocellulose filter.
> Store at −20°C in 5-ml aliquots.

Step 2: Transfection

1. Prepare fresh TS buffer by mixing 100 ml of TD with 1 ml of $MgCl_2$/$CaCl_2$ stock solution.

2. Prepare DNA/DEAE-dextran mixture:

   950 μl of TS
   2 μg of plasmid DNA
   50 μl of DEAE-dextran stock (500 μg/ml final conc.)

3. Pellet 10^7 cells (usually about 10 ml of cultured cells at 10^6/ml) by centrifugation for 5 minutes at 2000 rpm in a tabletop centrifuge. Use 15-ml sterile, disposable conical tubes. Decant medium.

4. Wash cells by suspending in 5 ml of TS. Pellet cells by centrifugation as above and remove supernatant.

5. Suspend the cells in the TS/DNA/DEAE-dextran mixture (1 ml).

6. Incubate in a conical tube at room temperature for 15 minutes.

7. Add 10 ml of medium + serum (usually RPMI and 10% FBS for lymphocytes) supplemented with 100 μM chloroquine (1:100 dilution from 10 mM stock).

8. Incubate in conical tube for 30 minutes at 37°C. Place tube on side in incubator.

9. Pellet cells by centrifugation at 2000 rpm for 5 minutes. Remove supernatant.

10. Wash cells with 5 ml of growth medium plus serum.

11. Suspend cells in 5 ml of growth medium plus serum and transfer to a T flask (T75) containing another 25 ml of growth medium plus serum.

12. Incubate at 37°C and harvest after approximately 48 hours.

Transfection by Electroporation of RAW264.7 Macrophages

TIME LINE AND ORGANIZATION

The electroporation method of introducing DNA into cultured eukaryotic cells was originally developed by Neumann et al. (1982). The method described here for RAW264.7 macrophages is from Plevy et al. (1997).

Electroporation apparatuses can be obtained from several companies and fall into two basic categories (Potter 1988). Most devices provide an exponentially decaying pulse of current, which is passed through the DNA sample. The current is supplied by a capacitor that is charged to a defined voltage prior to discharge. The size of the capacitor and the voltage to which it is charged are varied to optimize the transfection efficiency for each cell line. (Increasing the size of the capacitor increases the amount of current stored at a given voltage, which increases the pulse time.) A third important variable is the resistance of the buffer containing the cells and DNA. The most popular electroporation apparatus, the Bio-Rad Gene Pulser (cat. #165-2105), is of this type. The second category of electroporation device provides a square wave of current and the potential for repetitive pulses. Although square wave devices have been reported to yield higher transfection efficiencies with some cell lines, they are more expensive and are used less frequently than the exponential decay devices.

The procedure begins by placing the DNA and cells to be transfected in an appropriate buffer, which establishes the resistance during the electric shock. The solution is then placed in a cuvette that is compatible with the electroporation device being used. The cells are then subjected to a brief electric shock. For a short time before and after the shock, the cuvette is incubated at room temperature (for some cell lines, the incubation is done on ice). Finally, the cells are transferred to a tissue culture plate or T flask along with growth medium and serum. One attractive feature of the electroporation method is that it is more rapid and involves fewer manipulations than the other methods.

Day 1: Electric shock cells in the presence of DNA.
Day 2 or 3: For transient transfections, harvest cells. For stable transfections, add drug to begin selection.

PROCEDURE

Step 1: Prepare buffers

Step 2: Transfection

1. Grow the RAW264.7 macrophages on 100-mm or 150-mm petri dishes in DMEM supplemented with 10% low-endotoxin fetal bovine serum (FBS). Harvest for transfection when cells are between 50% and 80% confluent. To harvest, scrape the cells off the plate and transfer along with the growth medium to a 50-ml disposable conical tube.

2. Determine cell density using a hemacytometer.

3. Pellet the cells in the 50-ml tube at 4°C, 1500 rpm for 10 minutes in a clinical centrifuge. Remove the supernatant.

4. Resuspend the cells at a concentration of 3.75 ×10^7/ml in DMEM/10% low-endotoxin FBS.

5. Add 200 μl of cells and 20 μg of plasmid DNA (in a 50-μl volume in 1× phosphate-buffered saline (PBS) to an electroporation cuvette (0.4-cm electrode gap, Bio-Rad cat. # 165-2088).

6. Incubate at room temperature for 10 minutes.

7. Apply a 250-V shock to each cuvette with a Bio-Rad Gene Pulser, using a 960 μF capacitor.

    Note: The optimal voltage and capacitance should be determined intermittently, as it can vary with the health and age of the cells.

8. Incubate the cuvette at room temperature for 10 minutes.

9. Transfer the cells to a 15-ml disposable conical tube containing 5 ml of 1× PBS to wash.

10. Pellet the cells at 4°C for 5 minutes and 1500 rpm. Aspirate the supernatant.

11. Resuspend the cells in 5 ml of DMEM/10% low-endotoxin FBS and transfer to a 60-mm petri dish. Incubate at 37°C with humidity.

12. For stable transfections, add drug 1–2 days post-transfection. For transient transfections, harvest the cells approximately 48 hours post-transfection.

ADDITIONAL CONSIDERATIONS FOR PROTOCOLS 5.1–5.3

1. A chloroquine treatment was included in the DEAE-dextran protocol described above. Chloroquine enhances DEAE-dextran transfection efficiencies for some, but not all, cell lines by neutralizing lysosomal hydrolases, which can degrade the DNA. Another treatment that can enhance the efficiency of either DEAE-dextran or calcium phosphate transfections involves a brief "shock" with DMSO or glycerol. These chemicals are thought to act by increasing permeability of the plasma membrane (see Sambrook et al. 1989, pp. 16.54–16.55, or Ausubel et al. 1994, Unit 9.3).

2. In addition to the optional treatments described above, a variety of parameters should be varied to optimize transfection efficiencies. Even when parameters have already been reported in the literature, optimization in a new laboratory is worthwhile. For the calcium phosphate procedure, the key parameters to optimize are the amount of DNA, the length of exposure to the calcium phosphate precipitate, and the optional treatment with DMSO or glycerol. For the DEAE-dextran method, the amount of DNA, concentration of DEAE-dextran, and the length of treatment with DEAE-dextran should be optimized. In addition, the DEAE-dextran method should be tested in the presence and absence of chloroquine, and with and without a glycerol or DMSO shock. The length of the chloroquine treatment and the glycerol/DMSO shock can also be varied. For some cell lines, the optimal conditions may include a relatively short treatment with DEAE-dextran and a long treatment with chloroquine. Since both chemicals are toxic to cells, a long treatment with one chemical can be tolerated if the treatment with the other is minimized. Optimization of the electroporation method can be difficult because an endless variety of buffers have been reported in the literature. The buffer solution is a critical parameter for electroporation because it defines the resistance during the electric shock. Ausubel et al. (1994) suggest four standard buffer solutions that can be tested during optimization of a procedure. For each buffer, the voltage and capacitance must be optimized.

3. Although the three transfection methods described above are the most popular, several other methods can be attempted for cell lines that are refractory to these methods. A variety of methods are briefly described in Box 5.1. For most cell lines that are refractory to the three basic procedures, testing of a few of the commercially available lipofection reagents is recommended. Most of these proprietary reagents are similar to one another, but each possesses unique structural features that may contribute to success with particular cell types. Unfortunately, it is difficult to predict which reagent is most likely to work with a given cell line.

COMMON REPORTER ENZYME ASSAYS

SUMMARY

When a transient or stable transfection assay is developed for a promoter, a primary objective is to quantitate promoter strength. Promoter strength is generally considered to be proportional to the number of specific mRNA transcripts synthesized by the RNA polymerase. To distinguish transcripts derived from the transfected promoter from those derived from the endogenous gene, the promoter is usually fused to the coding sequence for a heterologous gene. Direct methods for quantitating transcription of the heterologous gene include measurements of de novo RNA synthesis using the nuclear run-on assay (Chapter 3) or measurements of steady-state mRNA levels by primer extension, RNase protection, S1 nuclease analysis, or Northern blot (see Chapter 4). Unfortunately, mRNA can be difficult to detect by these methods during transient transfection experiments if the transfection efficiency of the cells is low. RNA can also be difficult to detect in stable transfection assays if the promoter is weak. Because of these detection problems, promoters are commonly fused to heterologous reporter genes, which encode enzymes that can be quantified using highly sensitive assays. The enzymatic activity within a transfected cell population is roughly proportional to the amount of enzyme, which is roughly proportional to the steady-state mRNA level. Additional advantages of reporter enzyme assays are their simplicity and speed.

As described in Boxes 5.2–5.4, several different reporter genes and reporter assays have been used to quantify transcription. The three most common are the luciferase (Box 5.2), CAT (Box 5.3), and β-galactosidase (Box 5.4) genes and assays. Brief protocols for these three assays are provided below. Additional information about these and other reporter assays can be found in Sambrook et al. (1989, Chapter 16), Ausubel et al. (1994, Unit 9), and the Promega *Protocols and Applications Guide.*

PROTOCOL 5.4

Luciferase Assay

TIME LINE AND ORGANIZATION

The basic firefly luciferase assay described here is a modified version of the methods described in de Wet et al. (1987) and the Promega *Protocols and Applications Guide.* The procedure makes use of solutions marketed by Promega, the leading supplier of luciferase products. Although the assay can be performed effectively with solutions prepared in the laboratory (see Ausubel et al. 1994, Unit 9), Promega's solutions are convenient, relatively inexpensive, and include proprietary modifications that yield stronger signals. Promega also states that its solutions prolong the fluorescent signal by enhancing enzymatic turnover.

The first step in the assay is lysis of cells transfected with a luciferase reporter plasmid using a detergent-containing buffer. Alternatively, the cells can be lysed by freeze-thawing (see Protocol 5.5). The cell debris is then removed by microcentrifugation. In most cases, the luciferase activity is measured using a luminometer. Some luminometers directly inject the solution containing the D-luciferin substrate, ATP, and coenzyme A into the cell lysate, and then measure the resulting fluorescence. Automation allows the signal to be measured at a precise time following injection, which can increase the consistency of the results because the fluorescence decreases fairly rapidly after mixing. For manual luminometers, the substrate solution is mixed by hand with the cell lysate, with the fluorescence read for a defined time following mixing (often 10 seconds). Luciferase activity can also be measured in a scintillation counter (see Nguyen et al. 1988 and the Promega *Protocols and Applications Guide*).

PROCEDURE

1. Remove growth medium from transfected cells and wash with PBS. Adherent cells can be washed directly on the culture plate. Nonadherent cells can be transferred to a 15-ml disposable conical tube for washing.

2. Add 1× Promega reporter lysis buffer (cat. # E3971) to the cells. Use 400 μl for adherent cells on a 60-mm dish or 100 μl for each 10^6 nonadherent cells. The cells are lysed by detergents within the lysis buffers.

   Notes: Lysis buffer can be added directly to plates containing adherent cells. The cells can then be scraped off the plate and transferred to a microfuge tube.

   For nonadherent cells, lysis can be performed in the 15-ml tubes in which the cells were washed. After adding lysis buffer, cells should be resuspended using a pipetman, which will facilitate the disruption of cell clumps. Lysed cells can then be transferred to a microfuge tube.

   Cells can also be lysed by freeze-thawing (see Protocol 5.5) or using a Triton X-100-containing buffer prepared in the laboratory (see Ausubel et al.1994, Unit 9).

3. Incubate cells in lysis buffer on ice for 10–20 minutes to allow complete lysis.

4. Pellet cell debris in a microcentrifuge at full speed for 5–10 minutes.

   Note: Inadequate centrifugation will lead to inconsistent luminometer readings.

5. Perform a Bradford assay to determine overall protein concentration of each lysate.

6. Within 1 hour, use 20 μl of supernatant to measure luciferase activities with a luminometer. Mix sample briefly with 100 μl of Promega luciferase assay reagent (cat. #

E1483), containing D-luciferin, ATP, and coenzyme A. For automated luminometers, a tube connected to the automatic injector will be placed in the assay reagent. Measure fluorescence according to the recommendations of the manufacturer.

Notes: Both the sample and assay reagent should be at room temperature at the time of the assay.

Duplicate assays are recommended to ensure validity of results.

If the fluorescent signal exceeds the linear range of the luminometer, which can be determined empirically, the sample should be diluted with lysis buffer.

Assay reagent can also be prepared in the laboratory (see Ausubel et al. 1994, Unit 9).

PROTOCOL 5.5

Chloramphenicol Acetyltransferase Assay

TIME LINE AND ORGANIZATION

The basic CAT assay protocol described here is a modified version of the protocol of Gorman et al. (1982). This protocol monitors the acetylation of [^{14}C]chloramphenicol by the *E. coli* CAT enzyme, with acetylation detected by thin-layer chromatography (TLC). As described in Box 5.3, several other methods of monitoring CAT activity have been developed. Additional information about the TLC method and the other methods can be found in Sambrook et al. (1989, Chapter 16), Ausubel et al. (1994, Unit 9), and the literature of specific companies that market CAT assay reagents and kits.

The TLC method begins with the lysis of cells transfected with a CAT reporter plasmid. Lysis is often accomplished by freeze-thawing. After removing cell debris by centrifugation, the lysate is incubated with a solution containing [^{14}C]chloramphenicol and acetyl-coenzyme A. The acetylated chloramphenicol products and the unmodified reactants are separated from the aqueous solution by organic extraction with ethyl acetate. The ethyl acetate is then removed by lyophilization, and the chloramphenicol derivatives are resuspended in a small volume of ethyl acetate and spotted onto a TLC plate. After resolution, the plate is exposed to X-ray film or analyzed by phosphorimager analysis.

PROCEDURE

CAUTIONS: *Chloramphenicol, Chloroform, KCl, KH$_2$PO$_4$, Methanol, Na$_2$HPO$_4$, Radioactive substances. See Appendix I.*

Step 1: Prepare buffers

Phosphate-buffered saline (PBS):

For 10x stock solution, 1 liter:
80 g of NaCl
2 g of KCl
11.5 g of Na$_2$HPO$_4$•7 H$_2$O
2 g of KH$_2$PO$_4$

Working solution (pH 7.3)
137 mM NaCl
2.7 mM KCl
4.3 mM Na$_2$HPO$_4$•7 H$_2$O
1.4 mM KH$_2$PO$_4$

0.25 M Tris-HCl, pH 7.8

5 mg/ml acetyl coenzyme A

Methanol/chloroform solution:

19 parts chloroform: 1 part methanol

Step 2: Cell lysis and CAT assay

1. Remove growth medium from transfected cells and wash with PBS. Adherent cells can be washed directly on the culture plate. Nonadherent cells can be transferred to a 15-ml disposable conical tube for washing (pellet cells after washing in a tabletop centrifuge at 1500 rpm for 5 minutes).

2. Transfer cells to a 1.5-ml microcentrifuge tube. Adherent cells should be scraped off the dish with a small volume of PBS and transferred using a pipetman. Nonadherent cells can be suspended in a small volume of PBS for transfer.

3. Pellet cells in a microcentrifuge at low speed (8000 rpm) for 1 minute. Remove supernatant and wash with 1 ml of PBS. Pellet cells as above and remove supernatant.

4. Suspend the cell pellet in 2 volumes of 0.25 M Tris-HCl, pH 8. Vortex.

5. Freeze the solution in a dry ice–ethanol bath. Thaw in a 37°C water bath. Repeat the freeze-thaw cycle three times to ensure complete cell lysis.

   Note: Cells can also be lysed using detergent-containing buffers. Promega's reporter lysis buffer (see Protocol 5.4) was designed for compatibility with luciferase, CAT, and β-galactosidase assays.

6. Heat suspensions to 65°C for 10 minutes.

   Note: If a plasmid encoding β-galactosidase is cotransfected for the purpose of normalizing, this heat treatment must be omitted because it will inactivate the β-galactosidase. Alternatively, an aliquot of the extract can be removed for the β-galactosidase assay prior to heat treatments.

7. Cool on ice and then pellet cell debris in a microcentrifuge for 5 minutes at 14,000 rpm. Transfer the supernatant (containing the soluble CAT protein) to a new tube. The supernatant can be frozen at this point at –70°C.

8. Perform a Bradford assay to determine overall protein concentration of each lysate. Use 75 μg of each lysate for the CAT assay.

9. To each reaction, add

   5 μl of Acetyl CoA. 5 mg/ml
   2 μl of ^{14}C-labeled chloramphenicol (54 mCi/mmole, 25 μCi/ml)
   75 μg of cell lysate
   0.25 M Tris-HCl (pH 7.8) to 125 μl

10. Incubate at 37°C for 4 hours.

    Note: The reaction time can be varied according to the strength of the signal. For experiments in which the lysates contain very little CAT enzyme, the reactions can proceed overnight. For experiments in which the lysates contain abundant CAT enzyme, the reactions can be terminated after 1 hour.

11. Add 500 μl of ethyl acetate to each reaction. Vortex at least 10 seconds.

12. Separate phases by centrifugation in a microfuge at room temperature at 14,000 rpm for 15 seconds.

13. Transfer the top (organic) layer to a fresh tube and lyophilize in a SpeedVac concentrator (with heat). At this step, the pellet can be stored at –70°C.

14. Resuspend the pellets in 15 μl of ethyl acetate and spot onto a silica-gel TLC plate (Baker, cat. #4462-04). Spot slowly and do not spot the entire 15 μl all at once. Allow the spot to evaporate before adding the rest. Spot 1 cm away from the bottom edge of the plate.

    Note: A blow-dryer can facilitate the rapid evaporation of the solvent during spotting.

15. Place the plate in an enclosed glass container with a 19:1 chloroform:methanol solution. The container can be lined with Whatman 3MM filter paper in contact with the eluant to prevent evaporation of the solvent during resolution.

16. Resolve until the solvent front covers two-thirds of the plate. Remove the plate from the container, dry face up in a fume hood, and wrap in Saran Wrap.

17. Expose the plate to film or place on a phosphorimager overnight.

    Notes: Four spots may be apparent above the spotting origin (see Fig. 5.6). The lowest spot is the nonacetylated chloramphenicol, the next two are the monoacetylated products, and the top spot is the diacetylated product, which will only be observed when a large amount of activity is present (and when the assay is out of the linear range).

    To quantitate the CAT activity in the absence of a phosphorimager, the spots can be excised with a razor blade (after aligning the TLC plate and the film) and counted in a scintillation counter in the presence of scintillation fluid. Data are usually presented as percent conversion, corresponding to the counts obtained in the acetylated spots divided by the total counts (acetylated plus nonacetylated spots).

PROTOCOL 5.6

β-Galactosidase Assay

TIME LINE AND ORGANIZATION

Although the *E. coli lacZ* gene, encoding β-galactosidase, can be used like the CAT and luciferase genes as a standard reporter for monitoring the strength of a promoter or enhancer in a transient or stable transfection assay, it is predominantly used as an internal control during transient transfection experiments (see Box 5.3 and the Internal Controls section of this chapter). The basic colorimetric assay described here is the simplest and least expensive assay for quantitating β-galactosidase activity. This method is economical and convenient when a β-galactosidase reporter plasmid is employed as an internal control (with the *lacZ* gene under the control of a strong viral promoter/enhancer). However, it is less sensitive than either the luciferase or CAT assays and therefore is not useful for monitoring the activity of a weak or moderately active promoter. Nevertheless, the sensitivity of the β-galactosidase reporter can be greatly enhanced by using the commercially available chemiluminescent and ELISA assays (see Box 5.4 and Ausubel et al. 1994, Unit 9).

The first step in the assay described here is lysis of the cells transfected with a β-galactosidase reporter plasmid. When used as an internal control, the cells will have been transfected with the control plasmid, containing a ubiquitously active viral promoter fused to the *E. coli lacZ* gene, and an experimental plasmid containing another reporter gene (e.g,. luciferase or CAT) under the control of the promoter or enhancer of interest. The cells can be lysed using either the freeze/thaw method described in Protocol 5.5 (and below) or the detergent method described in Protocol 5.4. After determining the total protein concentration in the extracts (usually by Bradford assay), an aliquot of the extract is mixed with the reaction substrate, o-nitrophenyl-β-D-galactopyranoside (ONPG), in a buffer containing sodium phosphate and magnesium chloride. When the yellow product becomes visible, the reaction is stopped with sodium carbonate and the optical densities of the samples are determined at a wavelength of 410 nm.

PROCEDURE

CAUTIONS: *β-Mercaptoethanol, NaH_2PO_4, Na_2HPO_4. See Appendix I.*

Step 1: Prepare buffers

Mg^{++} buffer:

1 M $MgCl_2$	100 μl
β-mercaptoethanol (14 M)	350 μl
H_2O	550 μl
	1.0 ml

1x O-nitrophenyl β-D-galactopyranoside (ONPG):

4.0 g/l = 4.0 mg/ml
Dissolve in 0.1 M phosphate buffer (pH 7.5)

0.8 M Phosphate buffer (dilute 1:8 to make ONPG):
> 4.41 g of $NaH_2PO_4 \times H_2O$ (fw=120)
> 23.86 g of Na_2HPO_4 (fw=141.96)
> Add H_2O to 250 ml

Step 2: Cell lysis and β-galactosidase assay

1. Follow steps 1–8 from Protocol 5.5 for harvesting of transfected cells, lysis by the freeze/thaw method, and determination of protein concentration.

2. Set up reaction tubes for each cell lysate (duplicate samples are highly recommended).

Mg^{++} buffer	3 μl
1× ONPG	66 μl
cell lysate	25 μg

   0.1 M phosphate buffer (pH 7.5) to 300 μl total volume

   Note: Since many mammalian cells contain endogenous β-galactosidase activities, it is critical to include a negative control sample (lysate from untransfected cells).

3. Incubate at 37°C until a faint yellow color becomes apparent, at least in the tubes expected to contain the most activity. The incubation time usually is between 0.5 and 2 hours.

4. Read absorbance at 420 nm.

REFERENCES

Annweiler A., Muller U., and Wirth T. 1992. Functional analysis of defined mutations in the immunoglobulin heavy-chain enhancer in transgenic mice. *Nucleic Acids Res.* **20:** 1503–1509.

Ausubel F.M., Brent R.E., Kingston E., Moore D.D., Seidman J.G., Smith J.A., and Struhl K. 1994. *Current protocols in molecular biology.* John Wiley and Sons, New York.

Banerji J., Olson J., and Schaffner W. 1983. A lymphocyte-specific cellular enhancer is located downstream of the joining region in immunoglobulin heavy chain genes. *Cell* **33:** 729–740.

Barton M.C., Madani N., and Emerson B.M. 1993. The erythroid protein GATA-1 functions with a stage-specific factor to activate transcription of chromatin-assembled beta-globin genes. *Genes Dev.* **7:** 1796–1809.

———. 1997. Distal enhancer regulation by promoter derepression in topologically constrained DNA in vitro. *Proc. Natl. Acad. Sci.* **94:** 7257–7262.

Bartsch J., Truss M., Bode J., and Beato M. 1996. Moderate increase in histone acetylation activates the mouse mammary tumor virus promoter and remodels its nucleosome structure. *Proc. Natl. Acad. Sci.* **93:** 10741–10746.

Beato M. 1996. Chromatin structure and the regulation of gene expression: Remodeling at the MMTV promoter. *J. Mol. Med.* **74:** 711–724.

Blochlinger K. and Diggelmann H. 1984. Hygromycin B phosphotransferase as a selectable marker for DNA transfer experiments with higher eucaryotic cells. *Mol. Cell. Biol.* **4:** 2929–2931.

Blom van Assendelft G., Hanscombe O., Grosveld F., and Greaves D.R.. 1989. The β-globin dominant control region activates homologous and heterologous promoters in a tissue-specific manner. *Cell* **56:** 969–977.

Bories J.C., Demengeot J., Davidson L., and Alt F.W. 1996. Gene-targeted deletion and replacement mutations of the T-cell receptor β-chain enhancer: The role of enhancer elements in controlling V(D)J recombination accessibility. *Proc. Natl. Acad. Sci.* **93:** 7871–7876.

Boussif O., Lezoualc'h F., Zanta M.A., Mergny M.D., Scherman D., Demeneix B., and Behr J.P. 1995. A versatile vector for gene and oligonucleotide transfer into cells in culture and in vivo: Polyethylenimine. *Proc. Natl. Acad. Sci.* **92:** 7297–7301.

Capecchi M.R. 1989. Altering the genome by homologous recombination. *Science* **244:** 1288–1292.

Carey M., Leatherwood J., and Ptashne M.. 1990. A potent GAL4 derivative activates transcription at a distance in vitro. *Science* **247:** 710–712.

Chalfie M., Tu Y., Euskirchen G., Ward W.W., and Prasher D.C. 1994. Green fluorescent protein as a marker for gene expression. *Science* **263:** 802–805.

Chaney W.G., Howard D.R., Pollard J.W., Sallustio S., and Stanley P. 1986. High-frequency transfection of CHO cells using polybrene. *Somat. Cell Mol. Genet.* **12:** 237–244.

Charnay P., Mellon P., and Maniatis T. 1985. Linker scanning mutagenesis of the 5'-flanking region of the mouse β-major-globin gene: Sequence requirements for transcription in erythroid and nonerythroid cells. *Mol. Cell. Biol.* **5:** 1498–1511.

Chesnut J.D., Baytan A.R., Russell M., Chang M.P., Bernard A., Maxwell I.H., and Hoeffler J.P. 1996. Selective isolation of transiently transfected cells from a mammalian cell population with vectors expressing a membrane anchored single-chain antibody. *J. Immunol. Methods* **193:** 17–27.

Chittenden T., Lupton S., and Levine A. J. 1989. Functional limits of oriP, the Epstein-Barr virus plasmid origin of replication. *J. Virol.* **63:** 3016–3025.

Chung J.H., Bell A.C., and Felsenfeld G. 1997. Characterization of the chicken β-globin insulator. *Proc. Natl. Acad. Sci.* **94:** 575–580.

de Wet J.R., Wood K.V., DeLuca M., Helinski D.R., and Subramani S. 1987. Firefly luciferase gene: Structure and expression in mammalian cells. *Mol. Cell. Biol.* **7:** 725–737.

Emanuel J.R., Schulz X.M., and Zhou E.A. 1988. Expression an an ouabain-resistant Na, K-ATPase in CV-1 cells after transfection with a cDNA encoding the rat Na , K-ATPase α1 subunit. *J. Biol. Chem.* **263:** 7726–7733.

Ernst P. and Smale S.T. 1995. Combinatorial regulation of transcription II: The immunoglobulin μ heavy chain gene. *Immunity* **2:** 427–438.

Functional Assays for Promoter Analysis ■ 189

Farr A. and Roman A. 1992. A pitfall of using a second plasmid to determine transfection efficiency. *Nucleic Acids Res.* **20:** 920.

Felgner P.L., Gadek T.R., Holm M., Roman R., Chan H.W., Wenz M., Northrop J.P., Ringold G.M., and Danielsen M. 1987. Lipofection: A highly efficient lipid-mediated DNA transfection procedure. *Proc. Natl. Acad. Sci.* **84:** 7413–7417.

Flanagan W.M. and Crabtree G.R. 1992. In vitro transcription faithfully reflecting T-cell activation requirements. *J. Biol. Chem.* **267:** 399–406.

Forrester W.C., Novak U., Gelinas R., and Groudine M. 1989. Molecular analysis of the human β-globin locus activation region. *Proc. Natl. Acad. Sci.* **86:** 5439–5443.

Fromm M. and Berg P. 1982. Deletion mapping of DNA regions required for SV40 early region promoter function in vivo. *J. Mol. Appl. Genet.* **1:** 457–481.

Garraway I.P., Semple K., and Smale S.T. 1996. Transcription of the lymphocyte-specific terminal deoxynucleotidyltransferase gene requires a specific core promoter structure. *Proc. Natl. Acad. Sci.* **93:** 4336–4341.

Gillies S.D., Morrison S.L., Oi V.T., and Tonegawa S. 1983. A tissue-specific transcription enhancer element is located in the major intron of a rearranged immunoglobulin heavy chain gene. *Cell* **33:** 717–728.

Gluzman Y. 1981. SV40-transformed simian cells support the replication of early SV40 mutants. *Cell* **23:** 175–182.

Goodbourn S., Zinn K., and Maniatis T. 1985. Human β-interferon gene expression is regulated by an inducible enhancer element. *Cell* **41:** 509–520.

Gorman C.M., Moffat L.F., and Howard B.H. 1982. Recombinant genomes which express chloramphenicol acetyltransferase in mammalian cells. *Mol. Cell. Biol.* **2:** 1044–1051.

Graham F.L. and Van der Eb A.J. 1973. A new technique for the assay of infectivity of human adenovirus 5 DNA. *Virology* **52:** 456–467.

Grosschedl R. and Baltimore D. 1985. Cell-type specificity of immunoglobulin gene expression is regulated by at least three DNA sequence elements. *Cell* **41:** 885–897.

Gu H., Zou Y.R., and Rajewsky K. 1993. Independent control of immunoglobulin switch recombination at individual switch regions evidenced through Cre-loxP-mediated gene targeting. *Cell* **73:**1155–1164.

Hall C.V., Jacob P.E., Ringold G.M., and Lee F. 1983. Expression and regulation of *Escherichia coli* lacZ gene fusions in mammalian cells. *J. Mol. Appl. Genet.* **2:** 101–109.

Hartman S.C. and Mulligan R.C. 1988. Two dominant-acting selectable markers for gene transfer studies in mammalian cells. *Proc. Natl. Acad. Sci.* **85:** 8047–8051.

Izumi M., Miyazawa H., Kamakura T., Yamaguchi I., Endo T., and Hanaoka F. 1991. Blasticidin S-resistance gene (bsr): A novel selectable marker for mammalian cells. *Exp. Cell Res.* **197:** 229–233.

Janeway C.A. and Travers P. 1996. *Immunobiology: The immune system in health and disease,* 2nd ed. Current Biology, London, United Kingdom and San Francisco, California; Garland, New York.

Jenuwein T. and Grosschedl R. 1991. Complex pattern of immunoglobulin μ gene expression in normal and transgenic mice: Nonoverlapping regulatory sequences govern distinct tissue specificities. *Genes Dev.* **5:** 932–943.

Kadonaga J.T. 1991. Purification of sequence-specific binding proteins by DNA affinity chromatography. *Methods Enzymol.* **208:** 10–23.

Kraus W.L. and Kadonaga J.T. 1998. p300 and estrogen receptor cooperatively activate transcription via differential enhancement of initiation and reinitiation. *Genes Dev.* **12:** 331–342.

Kushner P.J., Levinson B.B., and Goodman H.M. 1982. A plasmid that replicates in both mouse and *E. coli* cells. *J. Mol. Appl. Genet.* **1:** 527–538.

Laybourn P.J. and Kadonaga J.T. 1992. Threshold phenomena and long-distance activation of transcription by RNA polymerase II. *Science* **257:** 1682–1685.

Lenardo M., Pierce J.W., and Baltimore D. 1987. Protein-binding sites in Ig gene enhancers determine transcriptional activity and inducibility. *Science* **36:** 1573-1577.

Liston D.R. and Johnson P.J. 1999. Analysis of a ubiquitous promoter element in a primitive eukaryote: Early evolution of the initiator element. *Mol. Cell. Biol.* **19:** 2380–2388.

Liu J., O'Kane D.J., and Escher A. 1997. Secretion of functional *Renilla reniformis* luciferase by mammalian cells. *Gene* **203:** 141–148.

Lo K., Landau N.R., and Smale S.T. 1991. LyF-1, a transcriptional regulator that interacts with a novel class of promoters for lymphocyte-specific genes. *Mol. Cell. Biol.* **11:** 5229–5243.

Luthman H. and G. Magnusson. 1983. High efficiency polyoma DNA transfection of chloroquine-treated cells. *Nucleic Acids Res.* **11:** 1295–1308.

Malone C.S., Omori S.A., and Wall R. 1997. Silencer elements controlling the B29 (Igβ) promoter are neither promoter- nor cell-type-specific. *Proc. Natl. Acad. Sci.* **94:** 12314–12319.

Mannino R.J. and Gould-Fogerite S. 1988. Liposome mediated gene transfer. *Biotechniques* **6:** 682–690.

Mantei N., Boll W., and Weissmann C. 1979. Rabbit β-globin mRNA production in mouse L cells transformed with cloned rabbit β-globin chromosomal DNA. *Nature* **281:** 40–46.

Matthews J.C., Hori K., and Cormier M.J. 1977. Purification and properties of *Renilla reniformis* luciferase. *Biochemistry* **16:** 85–91.

McCutchan J.H. and Pagano J.S. 1968. Enhancement of the infectivity of simian virus 40 deoxyribonucleic acid with diethylaminoethyl dextran. *J. Natl. Cancer Inst.* **41:** 351–357.

Mercola M., Goverman J., Mirell C., and Calame K. 1985. Immunoglobulin heavy-chain enhancer requires one or more tissue-specific factors. *Science* **227:** 266–270.

Mulligan R.C. and Berg P. 1981. Selection for animal cells that express the *Escherichia coli* gene coding for xanthine-guanine phosphoribosyltransferase. *Proc. Natl. Acad. Sci.* **78:** 2072–2076.

Mulsant P., Gatignol A., Dalens M., and Tiraby G. 1988. Phleomycin resistance as a dominant selectable marker in CHO cells. *Somatic Cell Mol. Genet.* **14:** 243–252.

Neumann E., Schaefer-Ridder M., Wang Y., and Hofschneider P.H. 1982. Gene transfer into mouse lyoma cells by electroporation in high electric fields. *EMBO J.* **1:** 841–845.

Nguyen V.T., Morange M., and Bensaude O. 1988. Firefly luciferase luminescence assays using scintillation counters for quantitation in transfected mammalian cells. *Anal. Biochem.* **171:** 404–408.

Ohe Y., Zhao D., Saijo N., and Podack E.R. 1995. Construction of a novel bovine papillomavirus vector without detectable transforming activity suitable for gene transfer. *Hum. Gene Ther.* **6:** 325–333.

Parker B.A. and Stark G.R. 1979. Regulation of simian virus 40 transcription: Sensitive analysis of the RNA species present early in infections by virus or viral DNA. *J. Virol.* **31:** 360–369.

Pazin M.J., Sheridan P.L., Cannon K., Cao Z., Keck J.G., Kadonaga J.T., and Jones K.A. 1996. NF-κB-mediated chromatin reconfiguration and transcriptional activation of the HIV-1 enhancer in vitro. *Genes Dev.* **10:** 37–49.

Pierani A., Heguy A., Fujii H., and Roeder R.G. 1990. Activation of octamer-containing promoters by either octamer binding transcription factor 1 (OTF-1) or OTF-2 and requirement of an additional B-cell-specific component for optimal transcription of immunoglobulin promoters. *Mol. Cell. Biol.* **10:** 6204–6215.

Plevy S.E., Gemberling J.H.M., Hsu S., Dorner A.J., and Smale S.T. 1997. Multiple control elements mediate activation of the murine and human interleukin 12 p40 promoters: Evidence of functional synergy between C/EBP and Rel proteins. *Mol. Cell. Biol.* **17:** 4572–4588.

Potter H. 1988. Electroporation in biology: Methods, applications, and instrumentation. *Anal. Biochem.* **174:** 361–373.

Prasher D.C., Eckenrode V.K., Ward W.W., Prendergast F.G., and Cormier M.J. 1992. Primary structure of the *Aequorea victoria* green-fluorescent protein. *Gene* **111:** 229–233.

Promega Corporation. 1996. *Protocols and applications guide*, 3rd ed. Promega, Madison, WI.

Robbins P.D., Rio D.C., and Botchan M.R. 1986. *trans* Activation of the simian virus 40 enhancer. *Mol. Cell. Biol.* **6:** 1283–1285.

Sambrook J., Fritsch E.F., and Maniatis T. 1989. *Molecular cloning: A laboratory manual*, 2nd ed. Cold Spring Harbor Laboratory, Cold Spring Harbor, New York.

Schmidt E.V., Christoph G., Zeller R., and Leder P. 1990. The cytomegalovirus enhancer: A pan-active control element in transgenic mice. *Mol. Cell. Biol.* **10:** 4406–4411.

Simonsen C.C. and Levinson A.D. 1983. Isolation and expression of an altered mouse dihydrofolate reductase cDNA. *Proc. Natl. Acad. Sci.* **80:** 2495–2499.

Smith C.L. and Hager G.L. 1997. Transcriptional regulation of mammalian genes in vivo: A tale of two templates. *J. Biol. Chem.* **272:** 27493–27496.

Southern, P.J. and Berg P. 1982. Transformation of mammalian cells to antibiotic resistance with a bacterial gene under control of the SV40 early region promoter. *J. Mol. Appl. Genet.* **1:** 327–341.

Spector D.L., Goldman R.D., and Leinwand L.A. 1998. *Cells: A laboratory manual,* vol. 2. Light microscopy and cell structure, pp. 86.1–86.6. Cold Spring Harbor Laboratory Press, Cold Spring Harbor, New York.

Sternberg N. and D. Hamilton. 1981. Bacteriophage P1 site-specific recombination: Recombination between loxP sites. *J. Mol. Biol.* **150:** 467–486.

Talbot D. and F. Grosveld. 1991. The 5′HS2 of the globin locus control region enhances transcription through the interaction of a multimeric complex binding at two functionally distinct NF-E2 binding sites. *EMBO J.* **10:** 1391–1398.

Treisman R. 1986. Identification of a protein-binding site that mediates transcriptional response of the c-fos gene to serum factors. *Cell* **46:** 567–574.

Treisman R., Green M.R., and Maniatis T. 1983. *cis* and *trans* activation of globin gene transcription in transient assays. *Proc. Natl. Acad. Sci.* **83:** 7428–7432.

Vara J.A., Portela A., Ortin J., and Jimenez A. 1986. Expression in mammalian cells of a gene from *Streptomyces alboniger* conferring puromycin resistance. *Nucleic Acids Res.* **14:** 4617–4624.

Wigler M., Silverstein S., Lee L.S., Pellicer A., Cheng Y.C., and Axel R. 1977. Transfer of purified herpes virus thymidine kinase gene to cultured mouse cells. *Cell* **11:** 223–232.

Wirth T., Staudt L., and Baltimore D. 1987. An octamer oligonucleotide upstream of a TATA motif is sufficient for lymphoid-specific promoter activity. *Nature* **329:** 174–178.

Wright S., deBoer E., Grosveld F.G., and Flavell R.A. 1983. Regulated expression of the human β-globin gene family in murine erythroleukaemia cells. *Nature* **305:** 333–336.

Xu Y., Davidson L., Alt F.W., and Baltimore D. 1996. Deletion of the Ig κ light chain intronic enhancer/matrix attachment region impairs but does not abolish VκJκ rearrangement. *Immunity* **4:** 377–385.

Yates J.L., Warren N., and Sugden B. 1985. Stable replication of plasmids derived from Epstein-Barr virus in various mammalian cells. *Nature* **313:** 812–815.

Young D.C., Kingsley S.D., Ryan K.A., and Dutko F.J. 1993. Selective inactivation of eukaryotic β-galactosidase in assays for inhibitors of HIV-1 TAT using bacterial β-galactosidase as a reporter enzyme. *Anal. Biochem.* **215:** 24–30.

Zhong X.-P. and Krangel M.S. 1997. An enhancer blocking element between α and δ gene segments within the T cell receptor α/δ locus. *Proc. Natl. Acad. Sci.* **94:** 5219–5224.

Zimmermann U. 1982. Electric field-mediated fusion and related electrical phenomena. *Biochim. Biophys. Acta* **694:** 227–277.

Zinn K., Mellon P., Ptashne M., and Maniatis T. Regulated expression of an extrachromosomal human β-interferon gene in mouse cells. *Proc. Natl. Acad. Sci.* **79:** 4897–4901.

CHAPTER 6

Identification and Analysis of Distant Control Regions

Important issues

- *Distant control regions are usually, if not always, required for proper gene regulation.*
- *Distant control regions can be difficult to find.*
- *DNase I hypersensitive sites and matrix attachment regions can facilitate the identification of distant control regions.*
- *A functional assay is needed to confirm that a distant control region has been identified.*

INTRODUCTION

As described in Chapter 2, an analysis of transcriptional regulation can begin with a focus on either the promoter or distant control regions. Both strategies are likely to provide insight into the mechanisms of regulated transcription. The promoter often provides an easier starting point because its identification relies solely on the determination of the transcription start site. Distant control regions are usually more difficult to identify, but they are considered by many researchers to be the primary determinants of transcriptional regulation. Current models propose that higher-order chromatin structures must be altered before the promoter can be bound by transcription factors and RNA polymerase. Distant control regions are likely to be involved in regulating the accessibility of a locus by influencing the chromatin structure (see Chapter 1 and Blackwood and Kadonaga 1998). Therefore, distant control regions are likely to be upstream of the promoter in the regulatory heirarchy.

Many types of distant control regions have been identified, including enhancers, silencers, locus control regions (LCRs), insulators, and matrix attachment regions (MARs) (see Chapter 1). Every protein-coding gene in metazoans is likely to contain one or more of these regions, in addition to the promoter. The fraction of genes that are regulated by each type of distant region has not been established, however.

As described in Chapter 1 and below, one challenge in studying distant control regions is that their defining characteristics are imprecise. Some LCRs, for example, function only in the context of a chromosome, whereas others function on episomes in transient assays. Furthermore, some insulators provide no enhancer function, whereas others exhibit both insulator and enhancer activities. MARs appear to exhibit a wide range of different functions. The molecular mechanisms responsible for these variable properties are unknown. Despite the dramatic progress that has been made in understanding the basic mechanisms of transcriptional activation by DNA-binding proteins, and the role of chromatin structure in gene regulation, the general mechanisms by which distant control regions regulate transcription remain poorly understood.

Like the analysis of promoters (see Chapter 5), the analysis of distant control regions is strongly dependent on the development of a functional assay. In many cases, the functional assays are similar to those used to study and dissect promoters, such as transient and stable transfection assays. However, more difficult and time-consuming assays, in particular transgenic assays, are often needed.

Another fundamental difference between promoters and distant control regions concerns the strategies used for their initial identification and localization. To identify a promoter, it simply is necessary to localize the transcription start site, because the promoter and start site, by definition, are closely linked (see Chapter 1). The methods used to identify distant control regions are quite different. The recommended approach begins with the identification of DNA regions in the vicinity of the gene that are hypersensitive to digestion with nucleases, primarily DNase I. Nuclease hypersensitivity indicates that the region is nucleosome-free or is assembled into nucleosomes with altered (or remodeled) structures, as would be expected for distant control regions that are associated with an array of sequence-specific DNA-binding proteins. Because of this property, nuclease hypersensitivity has proven to be a relatively rapid method for localizing distant control regions. After a nuclease hypersensitive region is identified, an assay needs to be developed to establish the functional relevance of the region.

Other strategies, in addition to the recommended strategy, can be employed to identify distant control regions. For example, DNA fragments surrounding a gene can be tested in a functional assay, such as a transient or stable transfection assay, in absence of prior nuclease hypersensitivity studies. Alternatively, MARs can be identified on the basis of their physical association with the nuclear matrix. MARs sometimes possess functional activity themselves or are closely linked to important control regions. The identification of a MAR may therefore lead to the identification of a distant control region.

In this chapter, the DNase I hypersensitivity, matrix attachment, and functional strategies for identifying distant regulatory regions are described. Then, the development of an appropriate functional assay for the analysis and dissection of a distant control region is discussed.

CONCEPTS AND STRATEGIES

DNase I Hypersensitivity

Basic Principles of DNase I Sensitivity and Hypersensitivity

Since the early 1970s, it has been known that genomic DNA is assembled into nucleosomes, which can then form higher-order chromatin structures (for a review of early advances, see Elgin and Weintraub 1975). One question that arose from this knowledge was whether chromatin on active genes would be the same as, or different from, that on inactive genes. In 1976, Harold Weintraub and Mark Groudine addressed this issue by asking whether the chromatin associated with active and inactive genes was equally accessible to cleavage when pancreatic DNase I was added to isolated nuclei. Their results revealed that the coding regions of the chicken globin genes were more sensitive to DNase I digestion in isolated nuclei from chicken erythrocytes than in nuclei from other cell types that did not express globin (Weintraub and Groudine 1976).

This "DNase I sensitivity assay" (Fig. 6.1) involved the treatment of aliquots of isolated nuclei with increasing concentrations of DNase I. In the initial study (Weintraub and Groudine 1976), the extent of digestion of the globin locus and of control loci was determined using a solution hybridization method. However, this method was soon replaced by a genomic Southern blot (Stalder et al. 1980). For this method of analysis, the DNase-digested genomic DNA was first purified and cleaved with a restriction enzyme that yielded fragments within the globin locus of known size. The DNA fragments were then separated by agarose gel electrophoresis and transferred to a nitrocellulose membrane. Specific restriction fragments were detected by hybridization to radiolabeled probes from globin or control genes.

In samples derived from erythrocyte nuclei, the DNA sequences within the globin locus were efficiently digested by DNase I (Fig. 6.1, left). The efficient digestion was apparent from the loss of the corresponding restriction fragment on the genomic Southern blot. In contrast, in cell types that did not express globin, the globin DNA sequences were resistant to DNase I (Fig. 6.1, right). Furthermore, in the erythrocyte samples, genes that were transcriptionally inactive were resistant to DNase I. From these data, Weintraub and Groudine concluded that the chromatin structure of the entire globin locus was altered when the genes were actively transcribed (Weintraub and Groudine 1976; Stalder et al. 1980). The precise alteration that renders a locus sensitive to DNase digestion remains unknown to this day but is likely to result from the decompaction of the locus, despite the fact that the locus remains largely assembled into nucleosomes.

In 1979, further investigation of the DNase I sensitivity phenomenon led to the discovery of DNase I hypersensitivity, which is more relevant to the topic of this chapter (i.e., identifying transcription control regions). When performing DNase sensitivity experiments with *Drosophila* heat-shock genes, Carl Wu, Sarah Elgin, and colleagues noticed the

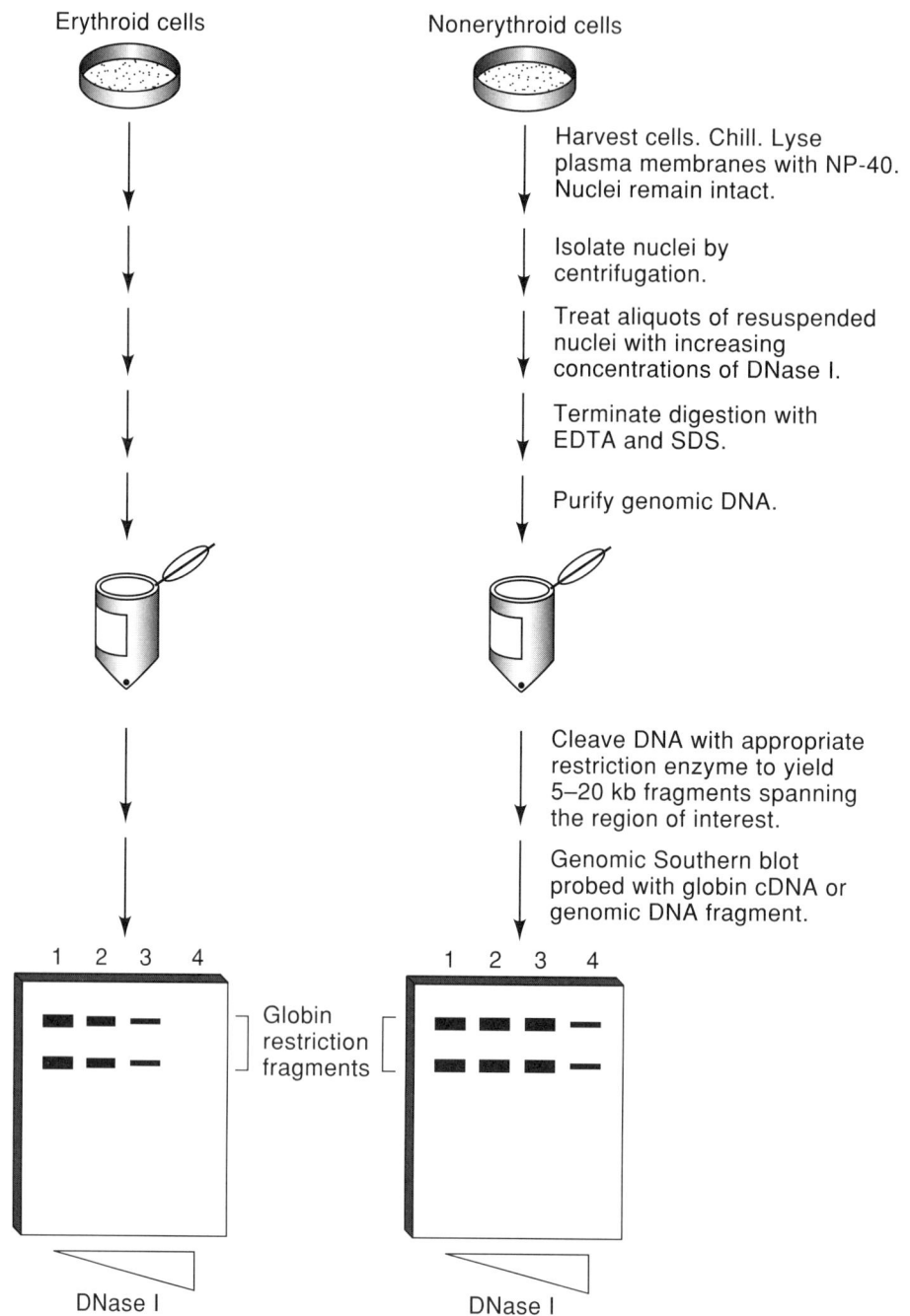

FIGURE 6.1. DNase sensitivity.

following: As a restriction fragment associated with an active gene decreased in intensity on a Southern blot autoradiograph following digestion of nuclei with DNase I, new "sub-bands" appeared that were smaller than the original restriction fragments (Wu et al. 1979a,b). By probing the Southern blot with different radiolabeled DNA fragments and by cleaving the DNase-digested genomic DNA with different restriction enzymes prior to Southern analysis, the subbands, now referred to as DNase I hypersensitive sites, were localized to specific regions flanking the heat-shock genes (Wu 1980; see Box 6.1).

When the above experiments were performed in the late 1970s, distant control regions had not yet been demonstrated using functional assays. Nevertheless, it was proposed from these data, as well as from complementary data from the globin system, that the hypersensitive sites might correspond to regions containing particularly severe alterations in chromatin structure, which might coincide with transcriptional control regions (Wu et al. 1979a,b; Stalder et al. 1980; McGinnis et al. 1983). It is important to note that the correlation between DNase hypersensitivity and putative control regions was quite different from the DNase sensitivity of an entire transcriptionally active locus.

Proof of the hypothesis that DNase I hypersensitive regions can correspond to important distant control regions was soon provided by a number of studies. One of the first studies published identified a functional enhancer in the intron of the immunoglobulin kappa locus that coincided with a previously described DNase I hypersensitive site (Picard and Schaffner 1984). DNase I hypersensitive regions have been found within other types of distant control regions, including LCRs (Talbot et al. 1989), silencers (Sawada et al. 1994; Siu et al. 1994), insulators (Udvardy et al. 1985), and MARs (Levy-Wilson and Fortier 1989). As established by the early studies of Wu (Wu et al. 1979a,b; Wu 1980), DNase I hypersensitive regions are also found in promoters.

The precise structural alterations that render a control region hypersensitive to DNase I have not been defined, but nucleosome remodeling or removal caused by the binding of sequence-specific DNA-binding proteins is likely to be responsible. High-resolution mapping has revealed that some hypersensitive sites immediately flank the binding sites for transcription factors and, therefore, correspond to structural alterations that directly result from factor binding. However, the hypersensitive nucleotides within a control region do not necessarily represent factor-bound sites, because any protein-free nucleotide has the potential to be hypersensitive when compared to the sensitivity of surrounding nucleosome-associated regions.

Before proceeding, it is important to note that, although DNase I hypersensitivity provides evidence that a region of DNA has an important regulatory function, hypersensitive sites have been identified that do not function as transcriptional control regions. The DNase I hypersensitivity of these sites may exist for reasons that have no functional implications. Alternatively, improved assays may be needed to reveal their function.

Advantages and Disadvantages of Using DNase I Hypersensitivity to Identify Control Regions

DNase I hypersensitivity is strongly recommended as a starting point for the identification of distant control regions. A region of 20–100 kbp or more spanning a locus can be scanned rapidly by probing and reprobing genomic Southern blots with various probes derived from phage or cosmid clones. Although it can be quite tedious isolating probes and determining restriction maps throughout a 100-kb region, this step is generally easier and more rapid than scanning a 100-kb region for activity using transient or stable transfection or transgenic assays. If a hypersensitive site is found within the locus, its precise location can be identified by a few additional genomic Southern blots following the cleavage of the DNase-digested DNA with different restriction enzymes. The DNA fragment that contains the hypersensitive region can then be subcloned and studied in functional assays to determine functional relevance.

The most difficult step in the DNase I hypersensitivity procedure is the original restriction mapping of the locus and surrounding sequences, and the isolation of appropriate

Box 6.1

DNase I Hypersensitivity Assay

To perform the DNase I hypersensitivity assay (Fig. 6.2), a source of cells expressing the gene of interest is needed. In general, 1×10^7 cells are desirable for each point of the DNase I titration. The actively growing cells are chilled, washed, and lysed with the nonionic detergent NP-40 to prepare nuclei. Aliquots of the isolated nuclei are then treated with different concentrations of DNase I. Different time points can also be taken. Nuclease digestion is terminated by addition of EDTA and SDS. Proteinase K is then added to digest nuclear and chromosomal proteins. (At

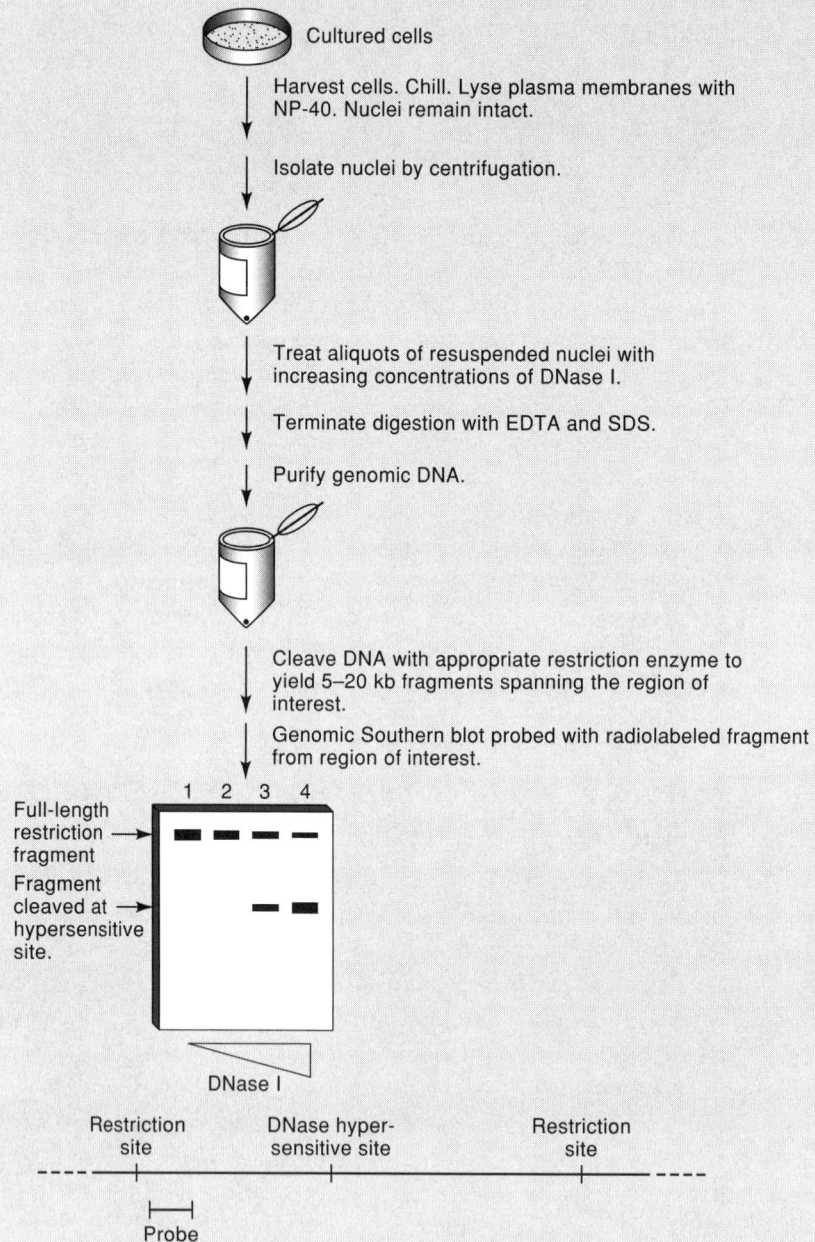

FIGURE 6.2. DNase I hypersensitivity assay.

this point, the approximate degree of DNase I digestion becomes apparent, as the solutions that were treated with low concentrations of DNase or no DNase will be viscous, due to high-molecular-weight DNA molecules. The samples treated with relatively high concentrations of DNase I will be much less viscous.) Following proteinase K treatment, the DNA samples are purified by extraction with phenol and chloroform and by digestion with RNase A. Finally, the DNA is precipitated with ethanol and resuspended in water. One challenge of this procedure is to maintain consistency in preparing DNA from the viscous and less viscous samples, because the viscous samples will be more difficult to extract and to suspend following precipitation.

Aliquots of the purified DNA samples are digested with selected restriction enzymes that yield fragments of 5–20 kb within the region of interest (usually enzymes that recognize 6–8-bp sequences). The fragments are separated by electrophoresis on a low-percentage agarose gel and are transferred to an appropriate membrane for Southern analysis. A single membrane can be probed sequentially using different radiolabeled restriction fragments (or oligonucleotides) spanning as much of the locus as is available. By using different probes, the existence of hypersensitive sites within different restriction fragments spanning the locus can be assessed. Hypersensitive and functional control elements for genes have been identified dozens of kilobases upstream or downstream of the gene, or within the introns of the gene. Detailed protocols can be found in Enver et al. (1985) and the primary research articles cited in the text. Protocol 10.1 in Chapter 10 can also be followed by substituting DNase I for micrococcal nuclease.

The in vivo micrococcal nuclease assay may be equally informative for revealing the locations of critical regulatory regions. This assay, which is discussed in greater depth in Chapter 10 (Protocol 10.1), is performed by the same method as the DNase I hypersensitivity assay. Micrococcal nuclease has different properties from DNase I, in that it preferentially introduces double-stranded DNA breaks at nucleosome linker regions and other regions devoid of nucleosomes. It therefore is a valuable reagent for studying nucleosomal structure and positioning (see Chapter 10). In addition, some important control regions are hypersensitive to cleavage by MNase as well as by DNase I. When performing nuclease hypersensitivity assays, it can be beneficial to try both nucleases because one may yield more compelling results than the other.

probes for hybridization. These objectives are accomplished by basic restriction mapping of phage or cosmid clones spanning the genomic region of interest and by successive subcloning of the genomic DNA to isolate small DNA fragments for labeling and hybridization. Alternatively, portions of the phage or cosmid clones can be sequenced, and then synthetic oligonucleotides or PCR fragments can be obtained for probe preparation.

One important limitation of this approach is that it may fail to identify some or all of the relevant control regions for a gene, because not all control regions yield identifiable hypersensitive sites. For this reason, it is not yet known whether all of the important control regions for any eukaryotic gene have been identified. The best evidence that all of the important control regions for a gene have been identified would be the ability to assemble a synthetic locus that recapitulates wild-type levels of transcription and precise regulation when introduced into the germ line of a transgenic mouse. The synthetic locus would presumably contain distant control regions, the promoter, and the transcription unit for the gene. This objective appears to have been accomplished for a few genes (e.g., the β-globin gene; Grosveld et al. 1987), but the results are inconclusive. To assemble the synthetic construct, it is necessary either to delete large pieces of DNA between the distant control regions and the promoter/transcription unit, or to insert large pieces of unrelated DNA.

Because the deletion or insertion of DNA may alter the activity of the locus, it will be very difficult to determine when all of the control regions for a gene have been found.

Data Interpretation

DNase I hypersensitive sites can be detected as prominent bands or as weak bands that are barely detectable. Moreover, some bands may be sharp, whereas others may be diffuse. Until functional assays are performed with the DNA fragments encompassing the hypersensitive sites, it is impossible to determine their relevance.

To characterize a hypersensitive site further, other cell lines and cell types that express or do not express the gene can be analyzed. If the hypersensitive site is present only in the cell types that express the gene, the probability that it is functionally relevant is likely to be higher. Hypersensitive sites associated with a silencer may be inversely correlated with the expression of the gene. Hypersensitive sites that are ubiquitous may also be critical for regulation, however. The hypersensitive sites in the globin LCR, for example, are detectable in cells from all stages of erythroid development, although the individual genes regulated by the LCR are expressed in only a subset of developmental stages (Crossley and Orkin 1993). Regardless of the cell types in which the hypersensitive site exists, the relevance of the region for regulated expression will need to be determined using functional approaches.

Identification of Matrix Attachment Regions

Basic Principles of the Nuclear Matrix and of MARs and SARs

The concept of MARs or scaffold-associated regions (SARs) originated with evidence that DNA within the nuclei of higher eukaryotes is organized into large chromatin loops (Paulson and Laemmli 1977; Laemmli et al. 1992). The base of each loop was found to be associated with protein components of the nuclear scaffold, or matrix (see Chapter 1). Studying the *Drosophila* histone and heat-shock loci, Mirkovitch et al. (1984) demonstrated that the DNA is attached to the matrix not randomly, but through specific AT-rich sequences in the vicinity of transcribed loci. Gasser and Laemmli (1986) then demonstrated that a subset of the MARs (or SARs) is closely associated with transcriptional control regions, such as enhancers. Numerous MARs from both invertebrates and vertebrates have been identified and characterized, revealing that many, but not all, are in close proximity to transcriptional regulatory sequences (Laemmli et al. 1992).

Unfortunately, the functional significance of matrix association and the mechanism by which MARs contribute to gene regulation have not been clearly established (see Chapter 1). One reason for the uncertainty is that different MARs appear to possess a variety of different functions. As stated above, many MARs are closely associated with transcriptional enhancers or LCRs (see, e.g., Gasser and Laemmli 1986; Jarman and Higgs 1988). However, MARs have been identified that possess no functional activity in the assays which have been employed (see, e.g., Jarman and Higgs 1988).

Advantages and Disadvantages of Using MARs to Identify Distant Control Regions

Localization of MARs as a means of identifying distant control regions is recommended only when DNase I hypersensitivity and functional assays fail. The highly variable and uncertain relationship between MARs and distant control regions is the principal limitation of this approach. If the identification of MARs is used as a method for identifying distant control regions, only a small subset of MARs may actually be in close association with functionally important control regions. In addition, most or all of the distant control regions associated with a gene might not be associated with MARs. Despite this uncertainty, the physical property of matrix association has been used successfully to identify distant control regions for several genes and can be considered if other strategies fail.

Functional Approaches for the Identification of Distant Control Regions

Basic Advantages and Disadvantages of Functional Approaches

DNase I hypersensitivity experiments were strongly recommended above as a starting point for identifying distant control regions. MAR studies, although recommended only as a last resort, may also lead to the identification of a distant control region. After a putative control region is localized, an appropriate functional assay is needed to determine whether it is

---

Box 6.2

Methods for Identifying MARs

The methods used to identify MARs (Fig. 6.3) begin with the isolation of cell nuclei and extraction from the nuclei of most soluble proteins, including histones, using high salt or detergent (Cook and Brazell 1980; Mirkovitch et al. 1984). The genomic DNA is then cleaved with a restriction endonuclease, and the DNA fragments that are not associated with the insoluble matrix are removed by centrifugation and washing of the insoluble pellet. The DNA fragments that remain associated with the insoluble matrix are analyzed by Southern blot, using one or more radiolabeled probes spanning the locus of interest. PCR is also a viable option for determining whether a DNA fragment is preferentially retained in the insoluble matrix. Preferential retention of a DNA fragment in the matrix, as compared to the soluble fraction, suggests that the fragment contains an authentic MAR. Precise mapping of the MAR can be carried out by digesting the matrix fraction with ExoIII after restriction enzyme cleavage. This exonuclease will digest DNA in a 3′ to 5′ direction from the free ends of the restriction fragments, but will be unable to proceed past the nucleotides that are physically associated with matrix proteins. The size of the fragment remaining after ExoIII treatment, as determined by Southern blot, and the radiolabeled probes to which it hybridizes, reveal the precise sequence of the MAR. Most MARs that have been identified using these standard procedures are AT-rich (Laemmli et al. 1992). Therefore, if the sequence identified as a putative MAR is found to be AT-rich following DNA sequencing, the results will be in accord with previous studies.

Perhaps the only key variable in the procedure is the method used to prepare the nuclear matrix fraction. The original methods that were developed included a step involving high-salt washes (2 M NaCl) (Cook and Brazell 1980). This method effectively removed histones and other soluble proteins from the nucleus. However, the high-salt wash also appears, at least in some instances, to alter protein–DNA interactions. To circumvent this potential difficulty, Mirkovitch et al. (1984) developed a matrix preparation method using the detergent lithium diiodosalicylate (LIS) in place of the 2 M NaCl wash. This particular detergent proved to be useful because it can easily be removed from the resulting matrix preparation by washing, thus preventing it from interfering with the subsequent restriction enzyme cleavage.

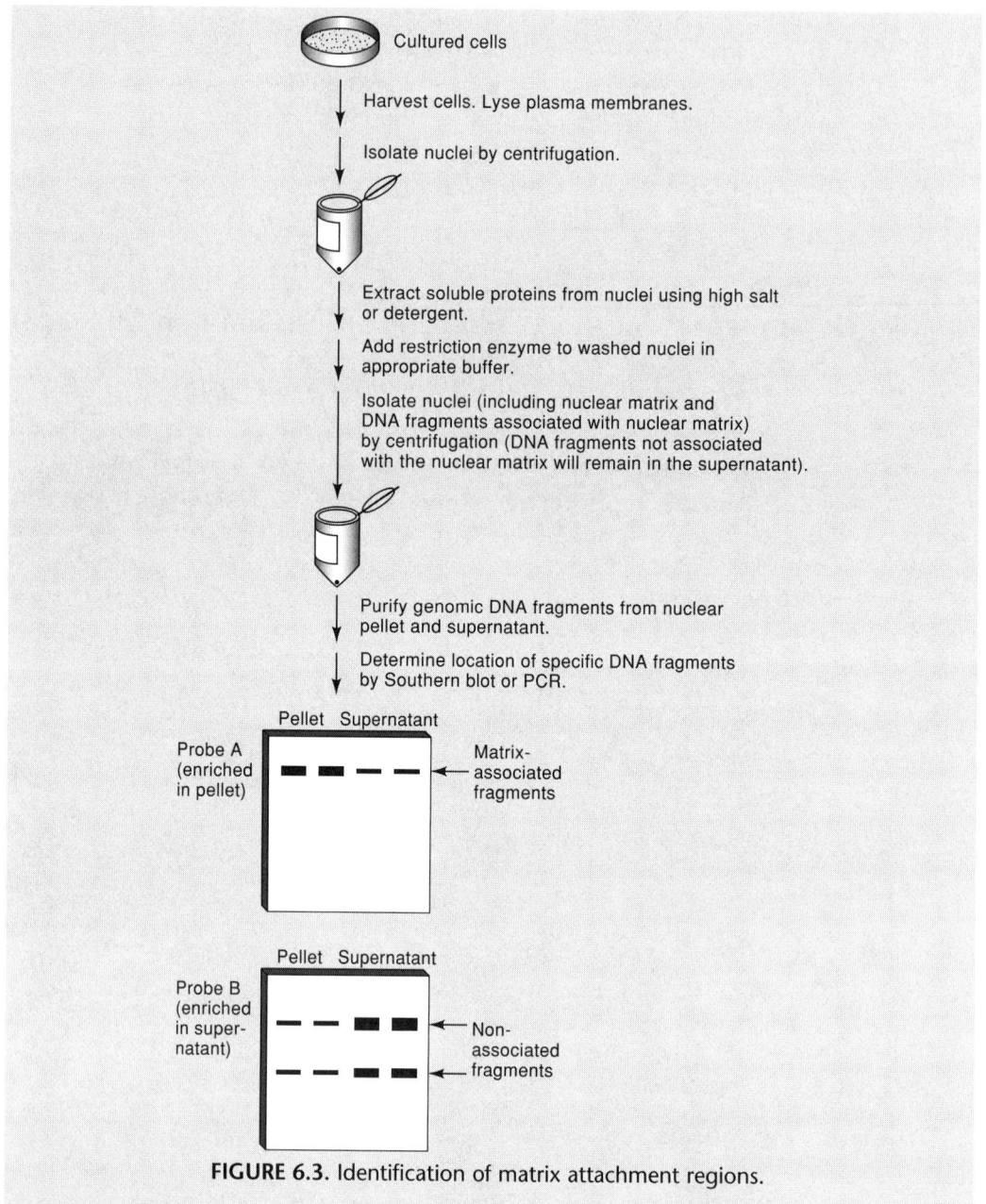

FIGURE 6.3. Identification of matrix attachment regions.

indeed functionally relevant. Before discussing functional assays for the analysis of a control region, we describe two general strategies that can be used in place of DNase hypersensitivity and MAR experiments to identify and localize distant control regions. These functional approaches have the advantage of solving two challenges simultaneously: identifying a potential regulatory region and developing a functional assay for the region. The disadvantage of these strategies is that they may require much more time and effort than would be involved if potential control regions were first localized by DNase I hypersensitivity. In most instances, this disadvantage outweighs the above advantage. Thus, the following functional strategies are primarily recommended if DNase I hypersensitivity experiments fail to identify a functional control region.

Functional Approach Beginning with a Large Genomic DNA Fragment

The two principal approaches that can be employed to identify distant regulatory regions using functional assays are essentially opposed to one another, but can be equally effective. The first approach is to begin by testing a large piece of genomic DNA containing and surrounding the gene of interest (Fig. 6.4). The genomic fragment can be introduced into transgenic mice or into cultured cells by stable transfection. Transcription of the introduced gene can then be monitored to determine whether the DNA fragment contains the control regions that are required for proper regulation. An appropriate starting point for this approach is to obtain a cosmid, P1, or BAC clone containing the gene. After functional activity has been detected, smaller DNA subfragments can be tested to localize the critical regulatory regions, and ultimately, the critical regulatory elements.

The advantage of beginning with a very large DNA fragment is that there is a relatively high likelihood that a positive result will be obtained in the initial experiments. Several challenges must be considered, however, including the need to determine restriction maps for extremely large DNA fragments and the need to prepare subclones of the fragments. Another challenge is that the transcripts derived from the DNA fragment which has been introduced into the cells can be difficult to distinguish from the transcripts derived from the endogenous gene. One solution is to introduce a genomic clone from one species into a closely related species. For example, rat or human genomic clones can be introduced into transgenic mice or into a murine cell line. The transcripts from the two species are likely

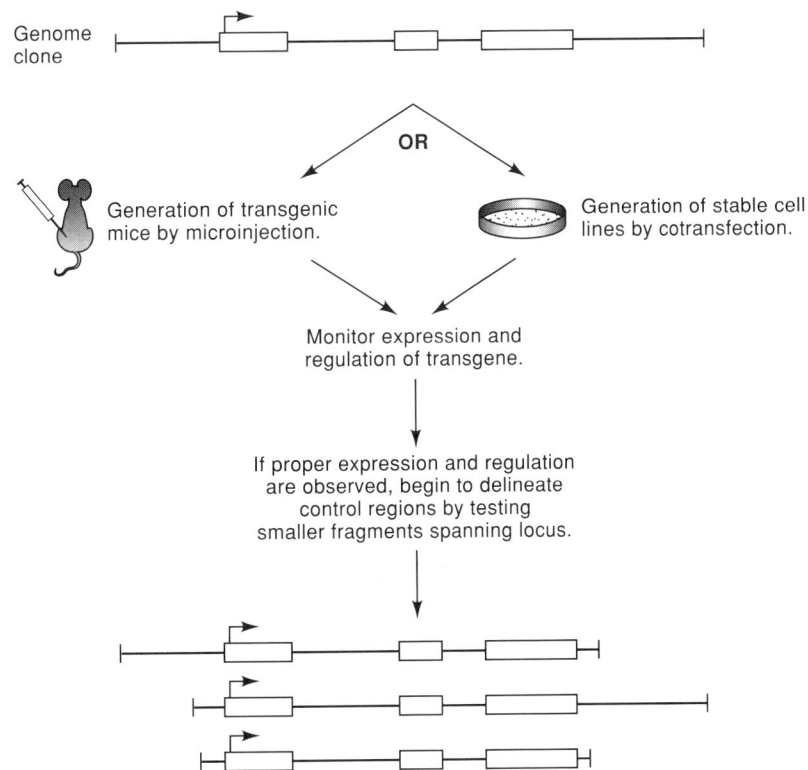

FIGURE 6.4. Functional identification of distant control regions beginning with large DNA fragments.

to contain sufficient sequence differences to allow the rat or human transcripts to be monitored by Northern blot or RT-PCR, using probes that do not cross-react with the endogenous mouse transcripts. In most cases, transcriptional regulatory mechanisms are well-conserved among closely related species. Mouse transcription factors are therefore likely to interact with the rat or human control regions. This strategy has been used routinely to study the regulation of the globin locus, with the human globin genes and control regions introduced into transgenic mice (see, e.g., Blom van Assendelft et al. 1989; Strouboulis et al. 1992).

An alternative to the use of a different species is an allele from the same species containing nucleotide sequence differences that allow transcripts from the endogenous and introduced genes to be distinguished by RT-PCR or Northern blot. Finally, a reporter gene (e.g., GFP or LacZ) or a nucleotide difference (i.e., sequence tag) can be engineered into the exogenous gene, providing a unique sequence for reporter or mRNA analyses. Reporter genes can be introduced into large BACs, for example, by homologous recombination in *E. coli* (Yang et al. 1997).

A final challenge is that the experiments often depend on the use of transgenic mice, where the gene can be introduced by microinjection. Stable transfection assays can also be envisioned, but with very large pieces of genomic DNA, the transfection efficiency may be low and the large DNA fragment may be susceptible to deletion or rearrangement.

Functional Approach Beginning with Smaller Fragments Directing Expression of a Reporter Gene

The alternative to the above approach is a scanning strategy (Fig. 6.5). Numerous smaller fragments of 5–10 kb spanning the locus can be introduced into a reporter vector containing the promoter from the gene of interest or a heterologous promoter, such as an SV40 or HSV-TK promoter. The resulting set of plasmids can then be tested using a transient or stable transfection assay. If one plasmid contains a fragment that enhances reporter gene activity, smaller subfragments can be tested to localize more precisely the relevant control region.

This approach was used to identify and localize the T-cell receptor-α enhancer, which is one of the key paradigms for the study of enhancer function (Ho et al. 1989; Winoto and Baltimore 1989). Initial efforts to identify the enhancer using DNase I hypersensitivity had failed. As an alternative approach, DNA fragments spanning the locus were inserted into a CAT reporter plasmid (Winoto and Baltimore 1989). One plasmid, encompassing a DNA region approximately 5 kb downstream of the gene, yielded greatly enhanced reporter gene activity. Analysis of subfragments from this region precisely localized the enhancer.

There are several advantages of this approach. First, sensitive reporter assays can easily be used to determine when a control region is present that increases or decreases gene transcription. Second, the plasmids generally contain smaller pieces of genomic DNA. If activity is detected, the functional region can rapidly be identified. Third, the activities of some control regions, such as enhancers, some LCRs, and some silencers, can be detected by a simple transient transfection assay, which may be more in line with the resources and capabilities of the laboratory.

In contrast to these advantages, the major disadvantage of this scanning strategy is that considerable effort can be expended constructing and analyzing reporter plasmids containing various DNA fragments spanning the locus of interest. The effort could be minimized by successful DNase I hypersensitivity experiments. A second disadvantage is that the strat-

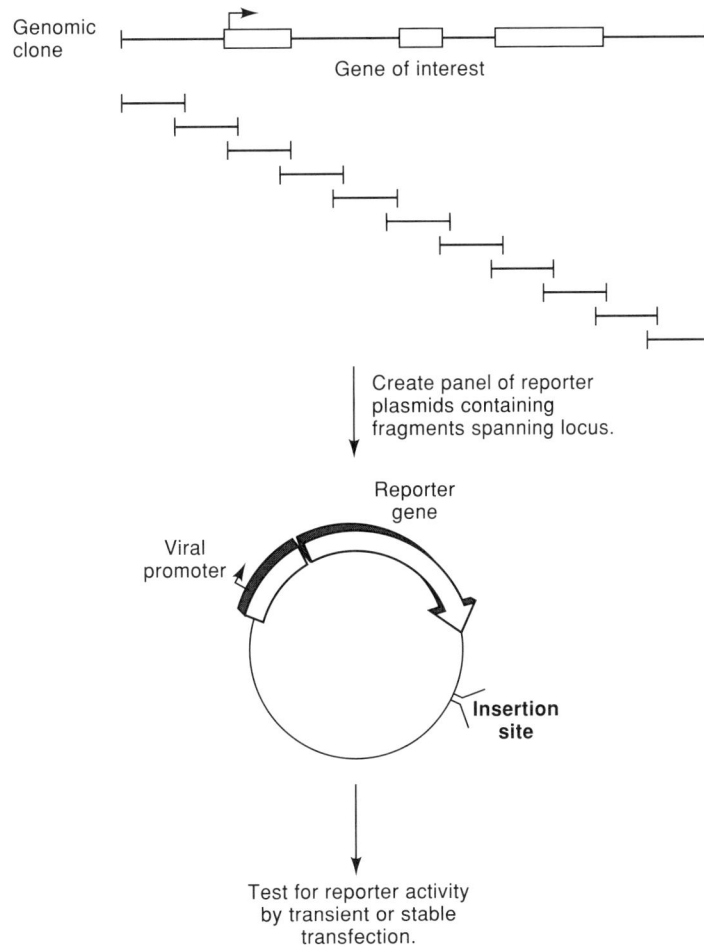

FIGURE 6.5. Functional identification of distant control regions by scanning.

egy is cumbersome when using a transgenic mouse assay because a large number of different plasmids need to be tested. It therefore is most feasible to use this approach in conjunction with transfection assays, which may not reveal the activity of some control regions.

Functional Assays for the Characterization of Distant Control Regions

Transient Transfection Assays

If a putative control region is identified by using DNase I hypersensitivity or MAR experiments, an assay is needed to assess its relevance. An appropriate assay is also required to pursue the functional strategies described above. To develop a transient assay for the analysis of distant control regions, the issues that need to be addressed are discussed in Chapter 5. An appropriate source of cells and appropriate transfection conditions are needed. In addition, an appropriate reporter gene and reporter assay must be chosen. The only new consideration regarding the reporter assay is that it is less important to use an assay that measures the location of a transcription start site when searching for and analyzing distant control regions. For promoter analyses, it is often important to confirm that the control elements within the promoter direct transcription from the correct nucleotide. For analyses of distant control regions, the location of the start site is less relevant, because distant control regions do not directly determine start site positioning. Thus, start site variations are of less consequence for the interpretation of the results.

The reporter plasmids themselves must be designed differently from those used for detection of promoter activity. Usually, the distant DNA fragment of interest is inserted into restriction sites that are conveniently located at the 3′ end of the reporter gene in most commercially available reporter vectors. Vectors are available that contain viral promoters upstream of the reporter gene. These promoters mediate the activity imparted by a DNA fragment of interest inserted downstream of the reporter gene (see Chapter 5).

Because a few examples exist in which a distant control region functions most effectively in concert with its own promoter (Li and Noll 1994; Merli et al. 1996; Ohtsuki et al. 1998), it may be worthwhile considering the use of the promoter from the gene of interest instead of a viral promoter. However, since the vast majority of enhancers, LCRs, and silencers function with any promoter, this alternative is unnecessary except as a last resort if activity is otherwise difficult to detect. It is worth noting that the DNA fragment of interest is usually inserted at the 3′ end of the reporter gene because most distant control elements capable of functioning in a transient assay should have no difficulty functioning from this location. More importantly, if the fragment is inserted immediately upstream of the promoter, fortuitous binding sites for nuclear proteins might cooperate with adjacent control elements within the promoter and influence reporter activity, even if the DNA fragment is not a bona fide enhancer or silencer.

The transient transfection experiments are performed in a straightforward manner, as described for promoter analyses (see Chapter 5). An appropriate internal control is usually included for normalization. Other control plasmids containing viral promoters and enhancers fused to the same reporter gene can be used to normalize the transfection efficiencies obtained with multiple cell lines. Normalization can help to establish whether the control region confers cell-type specific regulation (see Chapter 5).

Stable Transfection Assays

Stable transfection assays can be performed essentially as described in Chapter 5. To provide a brief summary, a plasmid containing a reporter gene regulated by the control region of interest is usually cotransfected into cells along with a second plasmid containing a drug-resistance gene under the control of a viral promoter/enhancer (Fig. 6.6). The two plasmids cointegrate with one another, often with multiple copies of both plasmids, at one or more sites within the genome. Cells that have stably integrated the drug-resistance gene are selected by addition of the drug. The drug-resistant cells usually contain the stably integrated reporter plasmid.

For the analysis of distant control regions like enhancers and LCRs, one new consideration is that the use of a viral enhancer to drive expression of the drug-resistance gene can mask the stimulatory effect of the control region of interest on reporter gene transcription. The reason for this is that the viral enhancer may act on the promoter upstream of the reporter gene. If possible, a drug-resistance gene that is under the control of a relatively weak promoter should be used. This strategy, in which the drug-resistance gene was under the control of an enhancerless HSV-TK promoter, contributed to the success of many globin LCR studies. (see, e.g., Blom van Assendelft et al. 1989). Alternatively, insulator elements can be placed on either side of the reporter cassette to diminish the influence of surrounding control regions on reporter gene transcription (see Chapter 5).

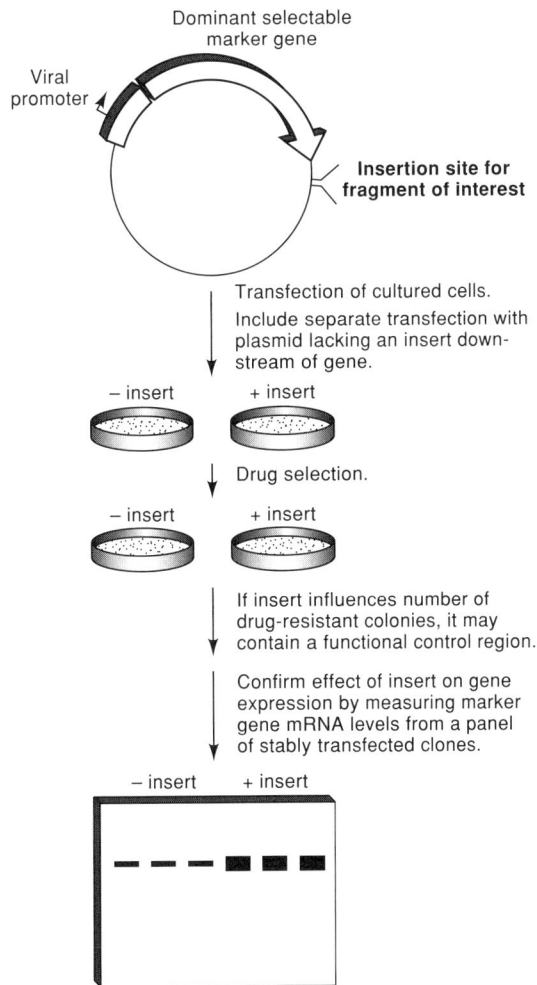

FIGURE 6.6. Functional identification of distant control regions by monitoring regulation of a dominant selectable marker gene.

As another alternative, an assay mentioned only briefly in Chapter 5 can be considered. Instead of cotransfecting a drug-resistance gene regulated by a viral promoter/enhancer with a reporter gene regulated by the control region of interest, the drug-resistance gene (or an oncogene) can be directly regulated by the control region of interest (see, e.g., Fromm and Berg 1982; Zhong and Krangel 1997). A heterologous promoter can be placed upstream of the drug-resistance gene, and the DNA fragment being analyzed for enhancer, silencer, or LCR activity can be inserted downstream. The resulting plasmid can be stably transfected into the appropriate cells, and drug selection can be performed. The number of drug-resistant colonies should be roughly proportional to the strength of the control regions regulating drug-resistance gene transcription. Thus, if the inserted DNA fragment functions as an enhancer or LCR, a larger number of drug-resistant colonies should be obtained than are obtained when the drug-resistance gene is regulated by the promoter alone. For adherent cells, the number of drug-resistant colonies can be determined by plating the cells and selecting on standard tissue-culture dishes. The number of colonies growing within a defined area can be counted. For nonadherent cells, the number of drug-resistant colonies can be determined by plating the transfected cells in 96-well plates and then counting the number of wells that contain colonies after selection.

Demonstration of LCR Activity

Strategies have been presented that are useful for detecting the activities of distant control regions. This broad range of strategies is particularly useful for studying enhancers, which generally function in all of the assays discussed (transient and stable transfection assays, and transgenic assays). Other types of distant control regions, such as LCRs, silencers, and insulators, often function in only a subset of these assays, however. In addition, a particular type of assay may be needed to demonstrate that a distant control region conforms to the definition of an LCR, silencer, or insulator.

To demonstrate conclusively that a control region functions as an LCR, for example, a transgenic mouse assay is needed. The simplest definition of an LCR is a DNA fragment that confers high levels of chromosomal integration-site-independent expression on a linked gene (see Chapter 1 and Fraser and Grosveld 1998). LCRs also stimulate mRNA synthesis at a level that is approximately proportional to the copy number of the integrated transgene. Because these defining properties of an LCR depend on chromosomal integration, transient transfection assays cannot be used, although some LCRs will function as enhancers in a transient assay (Fraser and Grosveld 1998).

Stable transfection results can suggest that a control region functions as an LCR, but the results are usually inconclusive. To use a stable transfection assay for this purpose, several stable clones must be isolated that contain a reporter gene regulated by the region of interest. If a significant number of stably transfected clones do not yield detectable reporter gene activity, yet all possess an integrated reporter gene and control region, the control region is unlikely to be an LCR. In contrast, if consistently high levels of reporter activity are obtained, the results support the hypothesis that the region is an LCR. The results are inconclusive, however, because a drug-resistant colony will not be obtained unless the exogenous DNA integrates into a chromosomal location that is at least partially "accessible." Because the drug-resistance gene and reporter gene almost always cointegrate into the same sites (see Chapter 5), the reporter gene will almost always be integrated into an accessible location in a drug-resistant cell. For this reason, consistently high levels of reporter gene activity might be observed in drug-resistant clones even if the control region is not a true LCR. Thus, although the stable assay has occasionally been used to study LCR activity (Blom van Assendelft et al. 1989), it does not provide compelling evidence that a control region functions as an LCR.

In light of the above, a transgenic assay is the only assay available for demonstrating conclusively that a control region possesses the key property of an LCR (i.e., high levels of integration-site-independent expression). Microinjection of a reporter plasmid containing the control region of interest into a mouse oocyte will result in fairly random chromosomal integration. Because drug selection is not involved, the subsequent results are not biased toward integrants into accessible locations. Several different founder strains containing the germ-line integrated reporter can be analyzed to determine the expression levels obtained in the relevant tissues. If the control region is an LCR, high levels of reporter activity will be detected in virtually all of the founder lines that have integrated the intact plasmid, with expression level roughly proportional to the transgene copy number. In contrast, if the distant control region is an enhancer that lacks the classic properties of an LCR, high levels of reporter gene activity will be detected in a relatively small subset (usually less than 50%) of the founder lines. The remaining lines generally exhibit low levels of reporter activity or no activity.

Demonstration of Silencer Activity

The term "silencer" has been used to define elements that are, in effect, the opposite of both enhancers and LCRs. More specifically, silencers are distant control regions that suppress gene expression in any assay or in only a defined subset of assays.

Ambiguity exists regarding the types of assays in which a silencer functions even for the classic mammalian silencer from the murine CD4 locus. Siu and colleagues have reported that a transgenic mouse assay is needed to detect consistent and compelling CD4 silencer activity (Siu et al. 1994; Duncan et al. 1996). Specifically, in a transgenic mouse, the silencer was capable of suppressing transcription from a reporter gene under the control of the CD4 promoter and enhancer. In contrast, Sawada et al. (1994) reported that the silencer functions in a transient transfection assay when present in multiple tandem copies upstream of a typical promoter and enhancer. Most likely, the transient assay yields relatively modest silencer activity that reflects some of the properties of the silencer. However, to truly understand the mechanism of silencer function, and its possible effects on chromatin structure, the transgenic assay, or perhaps a stable transfection assay, is likely to be necessary.

Because it is not yet clear which types of assays will be needed to study silencer function, it makes most sense, if possible, to begin with a transgenic mouse assay. After the region is identified, stable and transient transfection assays can be attempted to determine whether they are useful for subsequent studies.

Demonstration of Insulator Activity

The defining characteristic of an insulator is its ability to insulate or protect a promoter from the influence of an upstream enhancer. Insulators have been described for only a few genes and have been identified on the basis of nuclease hypersensitivity, matrix attachment, function, and their presence at boundaries between regions of nuclease sensitivity and nuclease resistance (Kellum and Schedl 1991; Chung et al. 1993; Kalos and Fournier 1995; Gdula et al. 1996). Some insulators function as episomal plasmids (Dunaway et al. 1997), but most appear to require chromosomal integration. Because this field is not well-developed, it is best to refer to the primary literature to determine how to identify and characterize a new insulator. To test a control region for insulator function, it usually is inserted into a reporter plasmid between an enhancer and a promoter (Kellum and Schedl 1992). It is important to demonstrate that the region suppresses transcription only when located between the enhancer and promoter, and not when located upstream of both.

Although the basic property of blocking characterizes almost every insulator that has been defined, two insulator properties vary. First, most insulators possess no transcriptional activation ability and merely block the function of distant control regions. In contrast, at least one control region described as an insulator, associated with the apolipoprotein B gene, exhibits properties that overlap with those of an LCR (Kalos and Fournier 1995). This insulator was first identified as a boundary element between the nuclease-sensitive apolipoprotein B locus and the nuclease-resistant flanking region. This property suggests that it insulates the flanking region from chromatin alterations that occur within the locus, and vice versa. Indeed, in transgenic mice, the boundary region insulates a reporter gene from integration site effects. However, this insulator also enhances reporter gene activity, similar to an LCR. The apparent distinction between this region and a true LCR is that LCRs typically enhance nuclease sensitivity in both directions, whereas the apolipo-protein region serves as a boundary between nuclease sensitivity and resistance.

The second variable property among insulators is the types of control regions whose activities can be blocked. The insulators associated with the *Drosophila* 87A7 gene, for example, block the activities of enhancers, but not of LCRs (Kellum and Schedl 1992). In contrast, the insulator found at the 5′ boundary of the chicken β-globin locus blocks the activities of both enhancers and LCRs (Chung et al. 1993). These variable properties must be considered when designing assays to measure the function of a putative insulator region.

In summary, then, regardless of whether a control region is identified by DNase I hypersensitivity, matrix attachment, or a functional approach, an appropriate starting point for the functional studies is to define the range of activities and properties that are associated with the region, by testing it in the various assays. For example:

- Does it function in a transient transfection assay?
- Does it function only in the presence of its own promoter, or is it equally active in the presence of a heterologous promoter?
- Does it function in a stable transfection assay?
- Does it function in an integration site-dependent manner?
- Does it insulate a promoter from the effects of a distant enhancer?
- Does it bind the nuclear matrix?
- Does it correspond to a clear DNase I hypersensitive region, consistent with a nucleosome-free or nucleosome-remodeled region?

By analyzing the range of activities and properties, it is possible to make an educated decision about the type of region being studied. An important outcome will be a decision regarding the appropriate functional assay to use for the subsequent mutant analysis, which will be essential for localizing the important control elements.

An additional goal is to localize the control region as carefully as possible, by using smaller and smaller pieces. Most control regions tend to localize to fragments of a few hundred base pairs. By localizing the control region, the next step, dissection of the critical control elements within the region, can be approached most effectively (see Chapter 7).

REFERENCES

Banerji J., Rusconi S., and Schaffner W. 1981. Expression of a β-globin gene is enhanced by remote SV40 DNA sequences. *Cell* **27:** 299–308.

Blackwood E.M. and Kadonaga J.T. 1998. Going the distance: A current view of enhancer action. *Science* **281:** 61–63.

Blom van Assendelft G., Hanscombe O., Grosveld F., and Greaves D.R. 1989. The β-globin dominant control region activates homologous and heterologous promoters in a tissue-specific manner. *Cell* **56:** 969–977.

Chung J.H., Whiteley M., and Felsenfeld G. 1993. A 5′ element of the chicken β-globin domain serves as an insulator in human erythroid cells and protects against position effect in *Drosophila. Cell* **74:** 505–514.

Cook P.R. and Brazell I.A. 1980. Mapping sequences in loops of nuclear DNA by their progressive detachment from the nuclear cage. *Nucleic Acids Res.* **8:** 2895–2906.

Crossley M. and Orkin S.H. 1993. Regulation of the β globin locus. *Curr. Opin. Genet. Dev.* **3:** 232–237.

Dunaway M., Hwang J.Y., Xiong M., and Yuen H.L. 1997. The activity of the scs and scs′ insulator elements is not dependent on chromosomal context. *Mol. Cell. Biol.* **17:** 182–189.

Duncan D.D., Adlam M., and Siu G. 1996. Asymmetric redundancy in CD4 silencer function. *Immunity* **4:** 301–311.

Elgin S.R. and Weintraub H. 1975. Chromosomal proteins and chromatin structure. *Annu. Rev. Biochem.* **44:** 725–774.

Enver T., Brewer A.C., and Patient R.K. 1985. Simian virus 40-mediated *cis* induction of the *Xenopus* β-globin DNase I hypersensitive site. *Nature* **318:** 680–683.

Fraser P. and Grosveld F. 1998. Locus control regions, chromatin activation and transcription. *Curr. Opin. Cell Biol.* **10:** 361–365.

Fromm M. and Berg P. 1982. Deletion mapping of DNA regions required for SV40 early region promoter function in vivo. *J. Mol. Appl. Genet.* **1:** 457–481.

Gasser S.M. and Laemmli U.K. 1986. Cohabitation of scaffold binding regions with upstream/enhancer elements of three developmentally regulated genes of *D. melanogaster. Cell* **46:** 521–530.

Gdula D.A., Gerasimova T.I., and Corces V.G. 1996. Genetic and molecular analysis of the gypsy chromatin insulator of *Drosophila. Proc. Natl. Acad. Sci.* **93:** 9378–9383.

Grosveld F., van Assendelft G.B., Greaves D.R., and Kollias G. 1987. Position-independent, high-level expression of the human β-globin gene in transgenic mice. *Cell* **51:** 975–985.

Ho I.C., Yang L.H., Morle G., and Leiden J.M. 1989. A T-cell-specific transcriptional enhancer element 3′ of C alpha in the human T-cell receptor alpha locus. *Proc. Natl. Acad. Sci.* **86:** 6714–6718.

Jarman A.P. and Higgs D.R. 1988. Nuclear scaffold attachment sites in the human globin gene complexes. *EMBO J.* **7:** 3337–3344.

Kalos M. and Fournier R.E. 1995. Position-independent transgene expression mediated by boundary elements from the apolipoprotein B chromatin domain. *Mol. Cell. Biol.* **15:** 198–207.

Kellum R. and Schedl P. 1991. A position-effect assay for boundaries of higher order chromosomal domains. *Cell* **64:** 941–950.

———. 1992. A group of scs elements function as domain boundaries in an enhancer-blocking assay. *Mol. Cell. Biol.* **12:** 2424–2431.

Laemmli U.K., Kas E., Poljak L., and Adachi Y. 1992. Scaffold-associated regions: *cis*-acting determinants of chromatin structural loops and functional domains. *Curr. Op. Genet. Dev.* **2:** 275–285.

Levy-Wilson B. and Fortier C. 1989. The limits of the DNase I-sensitive domain of the human apolipoprotein B gene coincide with the locations of chromosomal anchorage loops and define the 5′ and 3′ boundaries of the gene. *J. Biol. Chem.* **264:** 21196–21204.

Li X. and Noll M. 1994. Compatibility between enhancers and promoters determines the transcriptional specificity of gooseberry and gooseberry neuro in the *Drosophila* embryo. *EMBO J.* **13:** 400–406.

McGinnis W., Shermoen A.W., Heemskerk J., and Beckendorf S.K. 1983. DNA sequence changes in an upstream Dnase I-hypersensitive region are correlated with reduced gene expression. *Proc. Natl. Acad. Sci.* **80:** 1063–1067.

Merli C., Bergstrom D.E., Cygan J.A., and Blackman R.K. 1996. Promoter specificity mediates the independent regulation of neighboring genes. *Genes Dev.* **10:** 1260–1270.

Mirkovitch J., Mirault M.E., and Laemmli U.K. 1984. Organization of the highest-order chromatin loop: Specific DNA attachment sites on nuclear scaffold. *Cell* **39:** 223–232.

Ohtsuki S., Levine M., and Cai H.N. 1998. Different core promoters possess distinct regulatory activities in the *Drosophila* embryo. *Genes Dev.* **12:** 547–556.

Paulson J.R. and Laemmli U.K. 1977. The structure of histone-depleted metaphase chromosomes. *Cell* **12:** 817–828.

Picard D. and Schaffner W. 1984. A lymphocyte-specific enhancer in the mouse immunoglobulin κ gene. *Nature* **307:** 80–82.

Sawada S., Scarborough J.D., Killeen N., and Littman D.R. 1994. A lineage-specific transcriptional silencer regulates CD4 gene expression during T lymphocyte development. *Cell* **77:** 917–929.

Siu G., Wurster A.L., Duncan D.D., Soliman T.M., and Hedrick S.M. 1994. A transcriptional silencer controls the developmental expression of the CD4 gene. *EMBO J.* **13:** 3570–3579.

Stalder J., Larsen A., Engel J.D., Dolan M., Groudine M., and Weintraub H. 1980. Tissue-specific DNA cleavages in the globin chromatin domain introduced by DNase I. *Cell* **20:** 451–460.

Strouboulis J., Dillon N., and Grosveld F. 1992. Developmental regulation of a complete 70-kb human β-globin locus in transgenic mice. *Genes Dev.* **6:** 1857–1864.

Talbot D., Collis P., Antoniou M., Vidal M., Grosveld F., and Greaves D.R. 1989. A dominant control region from the human β-globin locus conferring integration site independent gene expression. *Nature* **338:** 352–355.

Udvardy A., Maine E., and Schedl P. 1985. The 87A7 chromomere. Identification of novel chromatin structures flanking the heat shock locus that may define the boundaries of higher order domains. *J. Mol. Biol.* **185:** 341–358.

Weintraub H. and Groudine M. 1976. Chromosomal subunits in active genes have an altered conformation. *Science* **193:** 848–856.

Winoto A. and Baltimore D. 1989. A novel, inducible and T cell-specific enhancer located at the 3′ end of the T cell receptor alpha locus. *EMBO J.* **8:** 729–733.

Wu C. 1980. The 5′ ends of *Drosophila* heat shock genes in chromatin are hypersensitive to DNase I. *Nature* **286:** 854–860.

Wu C., Wong Y.-C., and Elgin S.C.R. 1979a. The chromatin structure of specific genes: II. Disruption of chromatin structure during gene activity. *Cell* **16:** 807–814.

Wu C., Bingham P.M., Livak K.J., Holmgren R., and Elgin S.C. 1979b. The chromatin structure of specific genes: I. Evidence for higher order domains of defined DNA sequence. *Cell* **16:** 797–806.

Yang X.W., Model P., and Heintz N. 1997. Homologous recombination based modification in *Escherichia coli* and germline transmission in transgenic mice of a bacterial artificial chromosome. *Nat. Biotechnol.* **15:** 859–865.

Zhong X.-P. and Krangel M.S. 1997. An enhancer blocking element between α and δ gene segments within the T cell receptor α/δ locus. *Proc. Natl. Acad. Sci.* **94:** 5219–5224.

CHAPTER 7

Identifying *cis*-Acting DNA Elements within a Control Region

Important issues

- *cis*-Acting elements within a control region generally should be identified and characterized before trans-acting regulatory proteins are pursued.

- A comprehensive mutant analysis is the preferred and usually the most effective method for identifying important cis-acting elements.

- *cis*-Acting elements can also be identified by monitoring in vitro or in vivo protein–DNA interactions, or by database analysis, but the information obtained is more limited.

INTRODUCTION

The preceding chapters describe strategies and methods for identifying transcriptional control regions. To begin to elucidate the mechanisms by which a control region regulates transcription, the relevant *cis*-acting DNA sequence elements must next be delineated and the *trans*-acting protein factors that act on those elements defined.

In this chapter, strategies for identifying important DNA elements within an isolated control region are discussed. Strategies for identifying *trans*-acting factors are the topic of Chapter 8. For most studies, it is strongly recommended that a researcher identify DNA elements within a control region before identifying transcription factors that bind to it. Before explaining the reasons for this recommendation, it is worthwhile summarizing our current knowledge of the structure of a typical control region, such as a promoter or an enhancer.

As described in Chapter 1, it is thought that every control region in a metazoan genome contains multiple DNA elements that interact with a defined set of sequence-specific DNA-binding proteins. For a cell-type-specific control region, some elements interact with proteins that are direct mediators of cell-type specificity, whereas others interact with constitutively active proteins. Similarly, for an inducible control region, only a subset of the control elements interacts with proteins that are direct mediators of inducibility. Another relevant concept is that some proteins are likely to bind cooperatively to their recognition sites within a control region, either by directly interacting with one another or by coordinately facilitating chromatin remodeling. After the control region is occupied by the relevant set of DNA-binding proteins, the proteins act in concert to regulate transcription by providing multiple contact points for a single cofactor or general factor or, more likely, by contacting a defined set of cofactors and general factors (see Chapter 1).

On the basis of this general model, one can predict for the following reasons that proteins which interact functionally with a control region may be difficult to identify. First, only a subset of the proteins capable of binding a control region in vitro will be capable of the cooperative interactions (both physical and functional) required for in vivo relevance. Second, some proteins that are relevant in vivo might bind the control region with relatively low affinity and specificity in vitro, and their binding sites might diverge considerably from an experimentally defined high-affinity consensus sequence. Low-affinity protein–DNA interactions can be functionally relevant in vivo because the affinity of binding can be greatly enhanced by cooperative interactions with proteins that recognize adjacent sites. Hence, the highest-affinity and most abundant proteins need not be the functionally relevant proteins.

For these two reasons, the identification of functionally relevant DNA-binding proteins greatly benefits from the prior delineation of the functionally relevant DNA elements within a control region. Knowledge of the relevant DNA elements allows a laboratory to focus on the identification of proteins that bind those elements, and to disregard proteins that bind sequences within the control region that are not functionally relevant. In addition, the importance of a given DNA element relative to other elements within the region provides considerable insight into the relative importance of the factor that binds the element for the regulation of the gene.

To restate the above in more practical terms, a substantial amount of effort is needed to study rigorously the properties and functions of a putative regulatory protein. If the important control elements for the gene of interest are not defined at the beginning of the study, there is a much greater chance that effort will be expended studying a putative regulatory protein with no true relevance for that gene. Most of these types of miscues can be prevented by performing a comprehensive analysis of the *cis*-acting control elements prior to the pursuit of *trans*-acting factors.

In this chapter, common strategies for identifying important sequence elements are discussed. The recommended strategy is to perform a comprehensive mutant analysis of the control region, with each mutant analyzed using one of the functional assays described in Chapters 5 or 6. A second strategy is to identify important control elements based on the ability of proteins to bind specific sites using in vitro or in vivo DNA-binding assays. A third strategy is to search a database of binding sites for known DNA-binding proteins. Although the first strategy is strongly recommended, it can be helped along by the second and third strategies. The notable advantages and disadvantages of each strategy are discussed below. In addition, several methods that can be used to mutate a control region are summarized, but the reader is referred to the literature for detailed protocols.

CONCEPTS AND STRATEGIES

Identification of Control Elements by Comprehensive Mutant Analysis

Rationale for a Comprehensive Analysis

Before describing specific mutagenesis strategies and methods, it is beneficial to discuss in greater detail two key questions raised in the introduction:

- First, why is it important to identify *cis*-acting sequence elements before *trans*-acting factors?

- Second, why is a comprehensive mutant analysis the preferred strategy for defining sequence elements?

These are closely related questions because a comprehensive mutant analysis is the only method that leads to a rigorous functional examination of important sequence elements. The other methods for identifying sequence elements, including database methods and DNA-binding studies, almost always lead to an analysis of transcription factors before the important control elements are functionally defined.

The basic answer to the two questions is the following: If a transcription factor that interacts with a control region is studied before the *cis*-acting sequence elements have been identified, there is a relatively high probability that the factor being studied will not be of central importance for the regulation of the gene. Furthermore, in the absence of a rigorous definition of the control elements, it is extremely difficult to assess the relevance of the transcription factor being studied.

The above assertions can largely be explained by the concepts presented in the introduction to this chapter. A related explanation is derived from the fact that there are large numbers of DNA-binding proteins in eukaryotic cells. Each protein binds DNA with a reasonable amount of sequence specificity, but a range of different sequences can be recognized by each protein with varying affinities. DNA-binding proteins can also be grouped into families, with several members of a family capable of recognizing similar sequences. Therefore, most DNA-binding proteins are certain to bind many irrelevant sites in vitro, as well as relevant sites. The detection of a stable protein–DNA interaction in vitro does not necessarily depend on the functional relevance of the interaction, but rather depends on the affinity of the interaction, the relative abundance of the protein within the nuclear extract, and the precise conditions used for the binding assays, which may facilitate detection of some proteins and prevent detection of others (see also Chapters 8 and 9).

Given these facts, if DNA-binding proteins that interact with a control region are identified by standard in vitro electrophoretic mobility shift analysis (EMSA) and footprinting methods (Chapter 8) in the absence of a comprehensive mutant analysis, only a small fraction of the proteins, and sequence elements to which they bind, are likely to be relevant for the function of the control region. Moreover, some or most of the relevant DNA elements and regulatory proteins are likely to be missed.

If database searches are used to identify control elements, serious problems are also likely to emerge that are based on the large number of DNA-binding proteins within a cell. In particular, given the loose consensus sequences that exist for most DNA-binding proteins, only a subset of putative binding sites indicated in a database search will be functionally relevant. Moreover, the sites that are relevant in vivo may not be the ones that match most closely to the defined consensus sequences; a low-affinity binding site for a protein may be more relevant in vivo than a high-affinity site if binding to the low-affinity site is enhanced by cooperative interactions with proteins that recognize nearby sites. The reliance on database searches for identifying control elements is further complicated by the fact that the databases are composed of experimentally defined recognition sites of variable quality. A final concern about the use of database searches is that they are strongly biased toward known control elements and DNA-binding proteins, or at least known families of proteins. Some of the transcription factors that regulate a gene of interest may be unknown and therefore not represented in the databases.

Given the limitations of in vitro binding and database strategies, it might at first glance appear that in vivo binding assays are more likely to reveal functionally relevant control elements, because these methods indicate which elements are occupied by proteins in an intact cell nucleus. This argument is accurate to some extent, because DNA sequences occupied by proteins in a genomic footprinting assay are likely to play some role in gene regulation. However, the key elements for regulation of the control region might be missed because the genomic footprinting technique rarely shows occupancy of all of the important elements (see below and Chapter 10).

Although in vivo and in vitro binding studies, as well as database searches, have been used successfully in many studies to identify important elements within control regions, relying on these methods in the absence of a comprehensive mutant analysis may lead an investigator to focus on transcription factors that are not the primary regulators of the gene. Unfortunately, this problem continues to be pervasive throughout the field. Comprehensive mutant analyses have been performed for only a small number of control regions. For most of the others, even though investigators have focused on proteins that bind specific elements within a control region, it often has not been established that the proteins being studied are in fact key regulators of the gene of interest.

The Ig μ Gene Example

To reinforce the value of a comprehensive mutant analysis relative to the other methods of identifying control elements, the characterization of the Ig μ enhancer can be considered from an historical perspective (for reviews, see Nelsen and Sen 1992; Calame and Ghosh 1994; Ernst and Smale 1995). A current diagram of the control elements within the enhancer is shown in Figure 7.1. This diagram depicts a series of control elements that are now thought to contribute directly to the B-cell-specific activity of the enhancer by virtue of their restricted expression patterns, activities, or interactions with tissue-specific coactivators. Other control elements are depicted that are needed for optimal enhancer function but that do not appear to interact with cell-type-specific proteins.

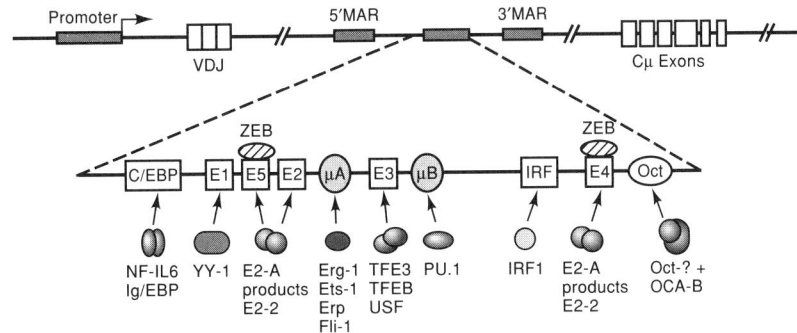

FIGURE 7.1. The murine immunoglobulin μ heavy-chain enhancer. (Adapted, with permission, from Ernst and Smale 1995 [Copyright 1995 Cell Press].)

The first method used to identify potential control elements within the μ enhancer was genomic footprinting (Ephrussi et al. 1985). This method revealed contacts at the Oct element and a subset of the μE elements (E boxes), but not at most of the other functionally important elements shown in Figure 7.1. All of the sites identified by in vivo footprinting are important for enhancer function, as demonstrated by subsequent mutant studies (Lenardo et al. 1987; Kiledjian et al. 1988; Perez-Mutul et al. 1988; Tsao et al. 1988; Jenuwein and Grosschedl 1991; Annweiler et al. 1992). However, if genomic footprinting had been the primary method used to identify Ig μ enhancer elements, some of the most important elements would have been missed.

Before the genomic footprinting experiments were performed on the Ig μ enhancer, the enhancer sequence was examined for binding sites for known transcription factors (see Staudt and Lenardo 1991). This analysis was performed in the mid 1980s, when relatively few transcription factors and consensus binding sites had been identified. It led to the identification of the octamer site and to sequences with homology to functional elements within the SV40 enhancer, known as core binding sites. The octamer site was subsequently shown to be functionally relevant via mutant analysis. In contrast, the core binding sites do not appear to be relevant. Furthermore, many of the sites that are now known to be important (e.g., μA, μB, and IRF) were not identified by these early sequence comparisons.

To evaluate the accuracy of a current database search, we analyzed the Ig μ enhancer sequence using a TESS Combined Search, as accessed through the Baylor College of Medicine Search Launcher (http://kiwi.imgen.bcm.tmc.edu:8088/search-launcher/launcher/html). The results, shown in Figure 7.2, reveal that 7 of the 10 established control elements were identified when using the standard search parameters. This may be an overestimate of the typical rate of success because some of the Ig enhancer elements were present within the database files and therefore were certain to appear. Interestingly, three important control elements were missed in the search, and five recognition sites, which are not thought to be functionally important, were revealed. The elements that were missed can be identified by reducing the stringency of the search, but with reduced stringency, several additional elements appear that are not functionally relevant. Thus, although a database search can provide useful information, it cannot be used as the primary method for identifying important DNA elements.

In vitro footprinting experiments were also performed with the μ enhancer to identify relevant control elements (Peterson et al. 1986; Sen and Baltimore 1986; Singh et al. 1986; Lenardo et al. 1987). These analyses resulted in the identification of proteins binding to the Oct element and a subset of the μE elements. In addition, proteins binding to other

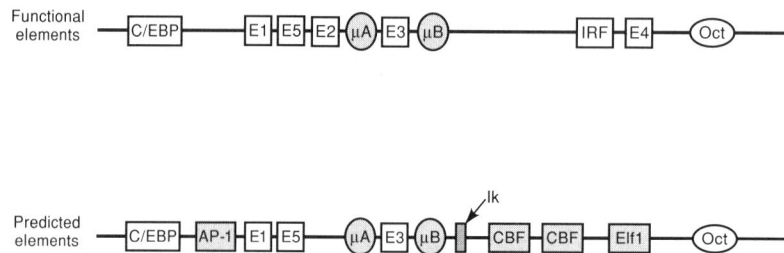

FIGURE 7.2. The murine immunoglobulin μ heavy-chain enhancer.

sequences within the enhancer were identified. The subsequent mutant analyses showed that the Oct and μE elements were functionally important for enhancer function in vivo. However, other protein-binding sites did not appear to be relevant, because mutations in these sites had no effect on enhancer function (the sites that were found to be functionally irrelevant are not shown in Fig. 7.1). More importantly, this approach, which involved both footprinting and EMSA experiments, failed to identify the μB, μA, and IRF elements that are critical for enhancer function. Thus, the in vitro protein–DNA interaction strategy, like the in vivo footprinting and database strategies, identified only a subset of the important elements that were identified by the subsequent mutant analysis.

Another noteworthy experiment was performed to determine whether the basis of the B-cell specificity of the Ig enhancer had been established. The experiment was to prepare a reporter plasmid containing multiple copies of the Oct site upstream of a minimal promoter (Wirth et al. 1987). This plasmid was preferentially active in B cells in transient transfection experiments. A simple hypothesis that was consistent with this result was that the Oct element was the principal determinant of the B-cell specificity of Ig μ enhancer activity.

Our current knowledge of the complicated mechanisms by which B-cell specificity is determined reveals that the multimerization approach is difficult to evaluate in the absence of more extensive knowledge of the control elements within a region of interest. Hypothetically, the Ig μ enhancer analysis might have begun in the mid 1980s with a database search, which at the time would have revealed only the Oct site and μE sites. A mutant in the Oct site would have reduced, but would not have eliminated, enhancer function, and multimerization of the Oct site would have been sufficient for B-cell-specific transcription. On the basis of these results, one might be tempted to conclude that the Oct site is the primary determinant of the B-cell specificity of Ig μ transcription, and might have decided that further characterization of the enhancer was unnecessary. This conclusion would obviously have been premature and unwarranted.

The above discussion is most relevant when the elements missed by the binding and database approaches are considered. These critical elements were not discovered until investigators, in particular Ranjan Sen and colleagues, began to perform a comprehensive mutant analysis of the enhancer. Before describing those experiments, it is important to explain why a comprehensive mutant analysis was not performed at the early stages of the Ig μ enhancer analysis. The primary reason is that the Ig μ enhancer was the first cellular enhancer discovered, and the most effective methods for dissecting enhancers had not been established. In the mid 1980s, it was not known that such a large number of DNA-binding proteins existed and that such a large and diverse number of proteins would be needed for the function of a single enhancer. Thus, when in vivo and in vitro binding experiments revealed contacts at a few sites, database searches revealed a few potential sites, and mutant studies confirmed that many of these sites were functionally relevant, it was considered likely that all of the key elements had been identified. Today, our knowledge of the enormous complexity of transcriptional regulation has revealed the naivete of those conclusions.

A second reason for the reliance on DNA-binding and database methods was that a comprehensive mutant analysis would have been extremely time-consuming and difficult in the mid 1980s. Prior to the common use of PCR methodology for mutagenesis and the availability of kits containing carefully controlled reagents for other types of mutagenesis, only a handful of comprehensive mutant analyses had been performed, all with considerable effort (see, e.g., McKnight and Kingsbury 1982; Eisenberg et al. 1985; Myers et al. 1986). Thus, in retrospect, it is not surprising that a comprehensive mutant analysis was not performed. Furthermore, the absence of such an analysis should in no way diminish the importance of the early studies of the Ig μ enhancer, which was and remains a key paradigm for our current understanding of enhancer function.

In 1990, the comprehensive mutant analysis of the Ig μ enhancer began. As discussed later in this chapter, the strategy required was unusually complicated because considerable redundancy exists within this enhancer. In other words, some elements are redundant with others, such that mutation of only one element at a time has little effect on enhancer activity. Nevertheless, the mutant analysis first revealed the μB element (Nelsen et al. 1990), then the μA element (Nelsen et al. 1993), and, most recently, the IRF site (Dang et al. 1998a). All of these elements interact with proteins that appear to be among the most important regulators of Ig μ enhancer activity. Despite several years of previous work by several laboratories, these elements were not uncovered until a comprehensive mutant analysis was performed. Even today, with our advanced knowledge of the control elements and proteins involved in Ig μ enhancer activity, it is impossible to draw strong conclusions regarding the relative importance of each element and factor for B-cell-specific transcription, largely because redundancy issues have not yet been fully resolved (see below).

Disadvantages of Using Mutagenesis to Identify Control Elements

The advantages of the comprehensive mutant analysis are outlined above. Although these advantages are compelling, they must be balanced against three limitations of this approach. One is that it requires substantially more effort than the other methods. A database search can be performed in a few minutes, and a DNase I footprint can be performed in one day (after the cells are grown and radiolabeled probes prepared). If these techniques identify a small number of potential control elements, those elements can be mutated and the mutants tested in a functional assay within a few weeks. In contrast, a comprehensive mutant analysis minimally requires several months. Fortunately, the time required for a comprehensive analysis is decreasing as new and more rapid methods for mutant production are being developed (see Boxes 7.1–7.6, below).

A second limitation of the comprehensive mutant analysis is equally important: *The control elements identified only include those that are required for the control region to function in the assay being used.* Transient transfection assays may support the activity of a control region, such as a promoter and enhancer, and allow most of the elements required for function to be identified. However, as discussed in Chapters 5 and 6, control elements may exist within a promoter or enhancer that function only in a natural chromatin context, which rarely forms on transiently transfected plasmids. Furthermore, the high plasmid copy number that often exists in a cell following transient transfection may cause some important control elements to be overlooked, or less important elements to predominate. A stable transfection assay may have a better chance of revealing all of the important elements within a control region. However, as described in Chapter 5, this assay is extremely time-consuming. Furthermore, with the stable transfection assay, it is difficult to identify control elements that influence activity by only a few-fold, since activity may vary due to chromosomal integration site differences.

The third limitation of the mutant analysis concerns difficulties caused by redundancy of DNA elements within a control region. If redundancies exist, it can be difficult to determine the boundaries of the control region and to define important DNA elements. This issue is discussed below from the perspective of Ig μ enhancer analysis.

The advantages and disadvantages of the comprehensive mutant analysis are difficult to balance. On one hand, the comprehensive analysis provides more information than the other approaches. On the other hand, the analysis requires more effort and the information obtained may remain incomplete. Therefore, this issue must be evaluated with respect to the overall goals of the analysis, along with the resources and time available for the analysis (see Chapter 2). If an understanding of the regulation of the gene is of central importance to the laboratory and there is a strong desire to truly understand the gene's regulatory mechanisms, a comprehensive mutant analysis is absolutely essential. However, if a commitment to follow through with the analysis cannot be made, due to a lack of resources, time, or interest, the alternative strategies outlined below may lead to the identification of a transcription factor that is important for a biological process of interest. The role of the factor for the biological process can be established by a gene disruption approach, and other experiments can then be performed to further investigate the relevance of the transcription factor for that process. With this latter scenario, one must of course realize that the importance of the factor for the activity of the control region which led to its identification will remain largely unknown (regardless of the phenotype of the knockout mouse) unless a comprehensive mutant analysis is later performed. (See Chapter 9 for further discussion of the relationship between a gene disruption phenotype and a target gene.)

Strategies for a Comprehensive Analysis

Most basic mutagenesis strategies include a combination of deletion and substitution mutants. As an example, we describe a strategy that was employed by one of us to dissect the promoter for the murine interleukin-12 (IL-12) p40 gene (Plevy et al. 1997). Although this promoter is inducible, the initial focus will be on the strategy used to delineate the DNA elements required for activity following induction; this strategy can be applied to any control region, regardless of its mode of regulation. The issue of which elements are directly responsible for induction is discussed later in this chapter. This strategy begins with a deletion analysis, which was performed primarily to define the boundaries of the functional promoter. The individual control elements within those boundaries were then identified by scanning the region with a series of clustered substitution mutations of 6–10 bp each. Small deletions were not used because they have the potential to alter the alignment of control elements, which may be important for proper regulation. The third step was to analyze each of the control elements with more refined substitution mutations of 3 bp, to determine their boundaries and to determine whether each represents a binding site for one protein or a composite site for two or more proteins. The issues considered when designing each step in this strategy are presented, along with the information obtained. Although this example involves the dissection of a promoter, a similar strategy can be envisioned for distant control regions.

The IL-12 p40 gene is expressed in macrophages that have been activated by bacterial products, such as lipopolysaccharide (LPS) and live or heat-killed *Listeria monocytogenes* (HKLM). The initial goal of the analysis was to dissect the mechanism of gene induction by LPS and HKLM. Since most inducible cytokine genes that have been characterized are regulated primarily by promoter sequences, it was considered likely that the IL-12 promoter would, in least in part, be responsible for induction. Prior to our analysis, Trinchieri and colleagues had demonstrated, using nuclear run-ons, that induction is regulated primarily at the level of transcription initiation (Ma et al. 1996). The transcription initiation site for the murine gene had been mapped to a location approximately 25 bp downstream of an AT-rich sequence that was likely to function as a TATA box (Murphy et al. 1995).

A 405-bp promoter fragment, extending from −350 to +55 relative to the start site, was fused to luciferase and chloramphenicol acetyltransferase (CAT) reporter genes in standard reporter vectors (Plevy et al. 1997). The upstream boundary of −350 was chosen because most promoters include key elements within a few hundred base pairs of the start site. The downstream boundary of +55 is near the translation initiation codon and was chosen because some genes contain important promoter elements in the untranslated leader (see Chapter 5). Following transfection of these plasmids into the RAW264.7 macrophage cell line and activation by LPS, the 405-bp fragment was found to be sufficient for strong, inducible promoter activity. Therefore, a mutagenesis strategy was needed to identify the important control elements within this fragment.

Deletion analysis. The first step of the mutant analysis was to generate and analyze a series of 5′ deletion mutants (Plevy et al. 1997). Deletions were prepared by PCR, using a forward primer spanning the desired end point of the deletion and a reverse primer spanning nucleotide +55 (see Box 7.3). Both primers contained sequences that generate restriction sites adjacent to the end points of the promoter fragment, allowing endonuclease cleavage and direct insertion into the reporter vector. Because the mutations were generated by PCR, it was necessary to sequence the final plasmid inserts to ensure that unwanted point mutations were not present.

Deletion mutants were prepared and analyzed before substitution mutants because the deletion analysis defines the minimal sequence that supports full activity. By determining the minimal sequence, the number of substitution mutants that subsequently needs to be prepared is minimized. For example, if the entire 405-bp fragment was essential for activity, 41 10-bp substitution mutations would be needed to scan the region for important elements. In contrast, if 200 bp at the 5′ end of this fragment could be eliminated without a significant effect on activity, only 21 10-bp mutations would be needed to scan the functionally relevant region. Indeed, the results revealed that 100% of the promoter activity was retained with a fragment extending from −200 to +55, and 50% was retained with a fragment extending from −150 to +55 (Fig. 7.3). Furthermore, promoter activity was not significantly enhanced when sequences extending to −800 were included.

A few points regarding the interpretation of these initial deletion results are noteworthy. First, the data in Figure 7.3 show that deletion of sequences between −350 and −150 had small effects on promoter activity. For example, a deletion from −250 to −215 reduced activity twofold, a deletion from −215 to −200 enhanced activity twofold, and a deletion from −200 to −150 again reduced activity twofold. The deleted sequences that led to these twofold effects might contain important positive and negative control elements. However, an alternative possibility is that the twofold differences are irrelevant to promoter activity. Each deletion results in the fusion of a sequence from the IL-12 locus to vector sequences. The vector sequences could have modest effects on promoter activity, either positive or negative, when fused to different nucleotides within the promoter. Thus, it cannot simply be assumed that every twofold difference corresponds to a relevant promoter element. To determine whether a bona fide element exists, for example between −250 and −215, 10-bp substitution mutations scanning this 35-bp region could be introduced into the −350 to +55 promoter fragment. If one or two of these mutations recapitulate the twofold loss in activity, a relevant control element might exist within this region. The putative element could be localized more precisely with smaller substitution mutants and then its mechanism of action could be analyzed in detail. If the twofold effect is not observed with the substitution mutations, the effect observed in the deletion analysis may be irrelevant.

A

B

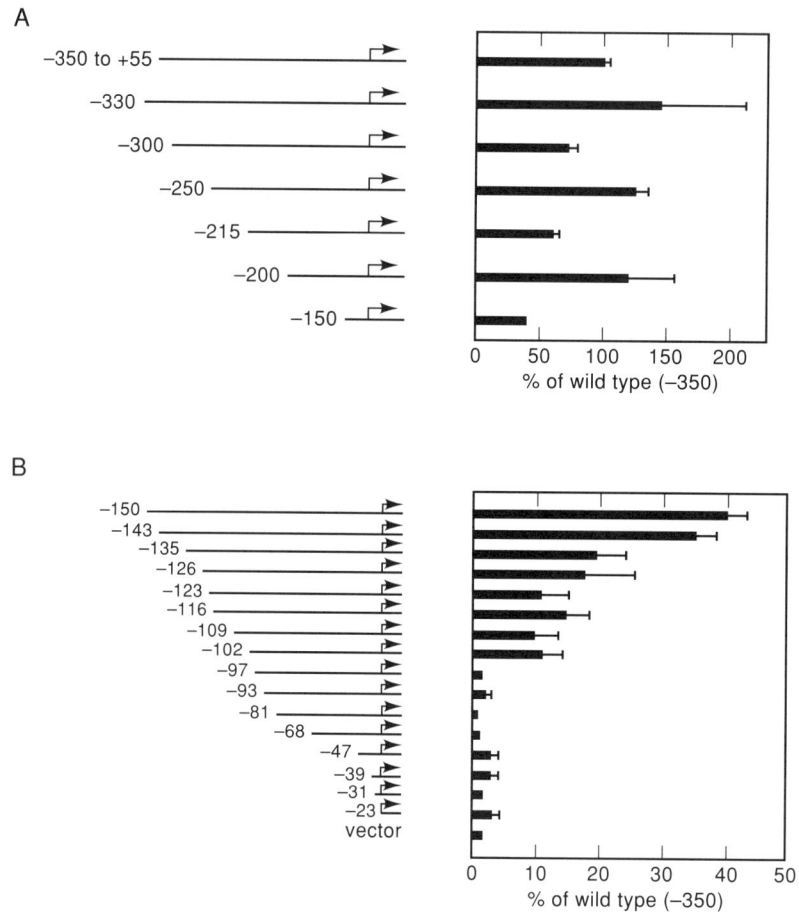

FIGURE 7.3. Basic deletion analysis for the murine IL-12 p40 promoter. (Adapted, with permission, from Plevy et al. 1997 [Copyright American Society for Microbiology].)

A second point regarding the deletion data is that the standard deviations determined for some of the mutants are quite large. The reason for the considerable variability is that data were used from multiple experiments performed on different days with different DNA preparations. The standard deviations would have been much lower if they had been derived from multiple independent transfections performed on the same day or with the same DNA preparation. The reason is that the effect of each mutation varied to some degree from one DNA preparation to another and from day to day. The results obtained with different DNA preparations can vary because the preparations may contain different concentrations of contaminants. Day-to-day variation can result from differences in the growth state, health, and transfection efficiency of the cells. Although larger standard deviations are obtained when these variations are documented, they lead to a more accurate presentation of the data. If the standard deviations had been derived from experiments performed with only one DNA preparation or on only one day, the data could have been less accurate and perhaps misleading.

Finally, the data for each mutant are presented as a percentage of the wild-type promoter activity following induction (i.e., the activity of the induced wild-type promoter is set as 100% with the activity of each mutant following induction determined relative to wild type), not as the fold-activation by LPS. This latter number would rely on the validity of the uninduced signal. Because the uninduced signals were quite close to background, very little confidence can be placed in their validity. Further comments regarding the documentation of inducibility are included below.

The more relevant deletions appeared to be those that extend past −150. As indicated in the graph, deletion from −143 to −135 reduced promoter activity to 20% of wild-type (i.e., the −350 to +55 fragment), deletion from −126 to −123 reduced activity to 10%, and deletion from −102 to −97 reduced activity to near background. These results suggest that important elements are likely to exist between −143 and −97. Furthermore, equally important elements may exist between −97 and +55. Because activity was reduced to background levels following deletion to −97, these data provide no information about the existence of important elements downstream of −97. Substitution mutants or 3′ deletion mutants are needed to determine whether important control elements exist in this downstream region.

For the IL-12 study, deletion mutants were not prepared from the 3′ end of the promoter (i.e., sequentially deleting sequences from +55 toward the transcription start site). The reason for this omission is that the −150 to +55 sequence was determined to be of reasonable size to dissect by substitution mutant analysis. Furthermore, because only 55 bp of untranslated leader was included, only a small number of nucleotides could be deleted from the 3′ end without affecting the core promoter elements, including the TATA box and potential start site sequence. Since deletion of these sequences could influence the ability of the general transcription machinery to form a stable preinitiation complex on the promoter, deletions in this region can be difficult to evaluate. Therefore, it was determined that specific substitution mutations would be more informative. Nevertheless, 3′ deletions would have provided additional information for this study and could be beneficial for other studies.

Substitution mutant analysis. Specific substitution mutations were introduced into the −150 to +55 region to identify important control elements (Figs. 7.4 and 7.5). Although the −150 to +55 fragment retained only 50% of wild-type activity, the activity remained strongly inducible (not shown) and therefore was likely to contain most, if not all, of the key promoter elements involved in inducible transcription. Most of the mutations introduced between −150 and +55 altered 6 bp, although some altered 5 and others 10 (Fig. 7.5).

To construct most of the mutant plasmids, the 5′ deletion constructs described above were used as starting points. To generate a substitution mutant from a deletion mutant, PCR was used to amplify the distal portion of the promoter, which was then fused to the appropriate deletion mutant containing the proximal portion. The plasmid generated from this fusion contained a substitution mutation at the site of the fusion. For example, substitution mutant −99/−94s was generated from deletion mutant −93 (Fig. 7.4). This deletion mutant contains a *Pst*I site immediately upstream of nucleotide −93 of the promoter, with a *Sac*I site immediately upstream of the *Pst*I site. To generate the substitution mutant, the promoter sequence extending from −350 to −100 was amplified by PCR from a full-length promoter template, using an upstream primer containing a *Sac*I restriction site and a downstream primer containing a *Pst*I site. The PCR product was then inserted into the *Sac*I/*Pst*I-cleaved −93 deletion mutant plasmid. The *Pst*I site generated a 6-bp mutation from −99 to −94 in the context of the −350/+55 promoter fragment. Thus, for this analysis, the deletion mutants served as cloning intermediates for many of the substitution mutants. Alternative strategies for generating substitution mutants are presented below (Boxes 7.4–7.6).

Substitution mutations of 5–10 bp were used for this analysis for two reasons. First, it was desirable for the mutations to be sufficiently small so that important control elements could be localized with reasonable precision. Second, it was desirable for them to be sufficiently large so that an unreasonable number of mutants would not be needed. By using 21 5- to 10-bp mutants, almost the entire region from −150 to +55 was scanned for functional elements.

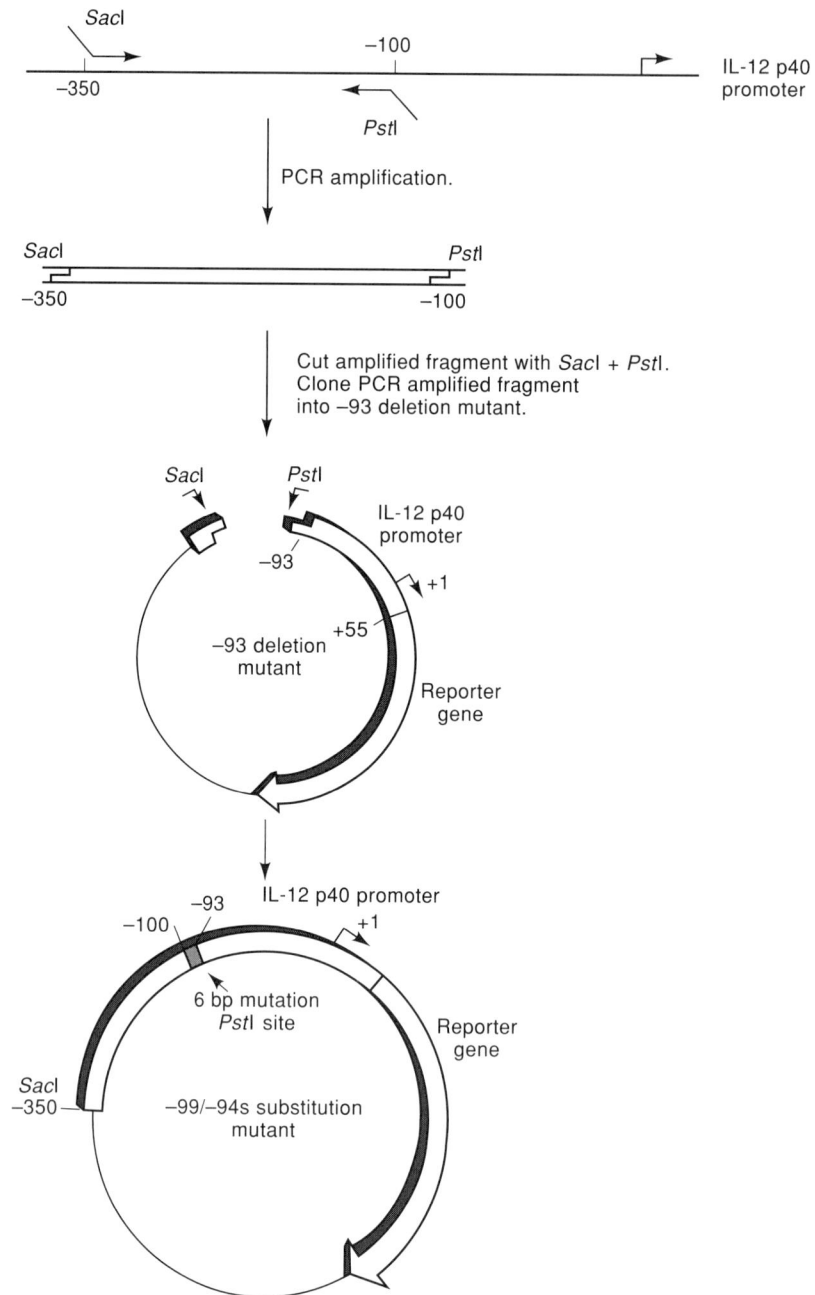

FIGURE 7.4. Strategy for generating substitution mutants from deletion mutant series.

A few gaps in the mutant series shown in Figure 7.5 are apparent. Gaps of 4 bp or less are unlikely to be significant because most sequence-specific DNA-binding proteins recognize sequences of 6 bp or more. Thus, the mutations flanking these small gaps should provide information regarding the existence of a control element. Gaps of 5 bp or more increase the probability that an important control element could be missed. The large gaps remaining after the IL-12 analysis were not intended and have since been analyzed by new mutations (J. Gemberling and S. Plevy, pers. comm.).

By analyzing the substitution mutants in the transient transfection assay, important control elements were identified (Fig. 7.5). The most important element for promoter

A

GGGAGGGAGGAACTTCTTAAAATTCCCCCAGAATGTTTTGACACTAGTTTTCAGTG

-149/-144s   -141/-136s   -132/-127s   -121/-117s   -114/-109s   -107/-102s   -99/-94s

TTGCAATTGAGACTAGTCAGTTTCTACTTTGGGTTTCCATCAGAAAGTTCTGTAGG

-93/-88s   -87/-82s   -80/-75s   -74/69s   -68/-63s   -62/-57s   -56/-51s   -45/40s

AGTAGAG**TATATAA**GCACCAGGAGCAGCCAAGGCAGCAGAAGG**AACAGTGGGTG**

-37/-32s   -29/-24s   -23/-14s   -13/-4s   +8/+17s

B

Construct	% of wild type (−350 to +55)
−350 to +55	~100
−218/−213s	~105
−149/−144s	~55
−141/−136s	~110
−132/−127s	~25
−121/−117s	~85
−114/−109s	~50
−107/−102s	~25
−99/−94s	~10
−93/−88s	~10
−87/−82s	~70
−80/−75s	~45
−74/−69s	~155
−68/−63s	~150
−62/−57s	~80
−56/−51s	~130
−45/−40s	~150
−37/−32s	~95
−29/−24s	~30
−23/−14s	~75
−13/−4s	~95
+8/+17s	~125

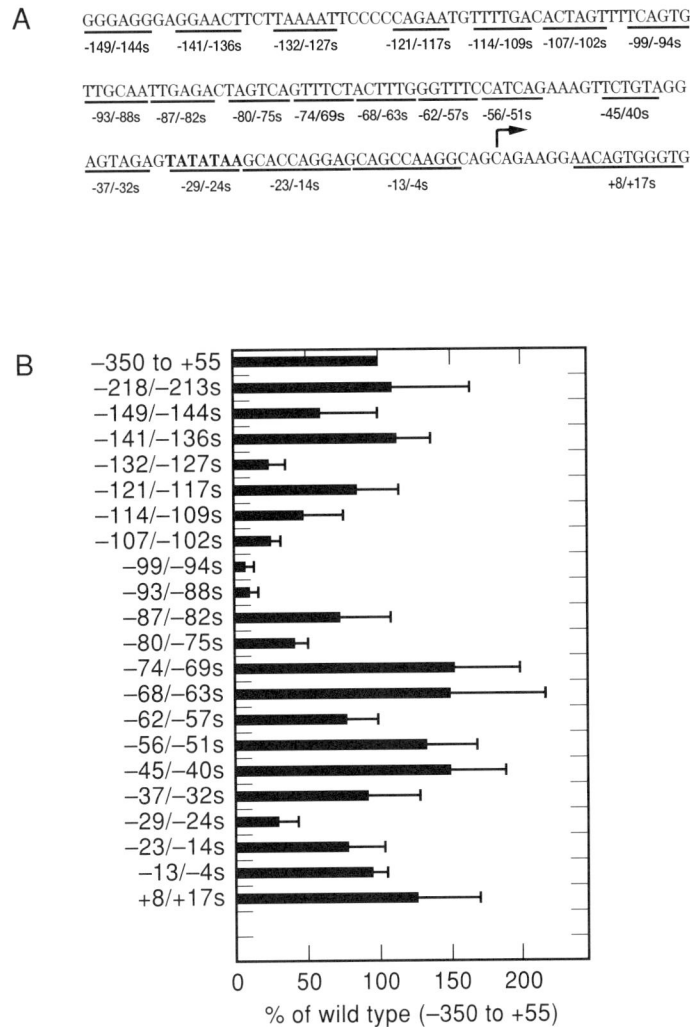

FIGURE 7.5. Substitution mutant analysis of the murine IL-12 p40 promoter. (Adapted, with permission, from Plevy et al. 1997 [Copyright American Society for Microbiology].)

activity in this assay was apparently disrupted by the −99/−94s and −93/−88s mutations. These mutations reduced promoter activity to approximately 10% of wild type. Another mutation, −132/−127s, reduced promoter activity to 25% of wild type, suggesting that it disrupted another important element. Two other mutations, −107/−102s and −29/−24s, reduced activity to approximately 25% of wild type. The latter mutation disrupted the TATA box and the former was immediately adjacent to the two severe mutations, suggesting that it might affect the same element.

Many of the remaining mutants exhibited promoter activities between 50% and 150% of wild type. These small effects suggest the existence of elements that are less important for activity, or elements whose activities are largely redundant (or at least nonsynergistic) with the activities of other elements (see below). Alternatively, as discussed above for the deletion mutants between −350 and −150, these small effects might be due to the introduction of a foreign sequence into a specific site in the promoter. In other words, the sequences mutated might not contain an important element.

Refined substitution mutant analysis. The results in Figure 7.5 suggest that a critical control element exists between approximately −88 and −99, that another may exist between approximately −127 and −132, and, as expected, the TATA box is important for promoter activity. At this point, it could be argued that the mutant analysis has been completed, and experiments should next be performed to identify the proteins that bind the −88/−99 and −127/−132 elements. This argument has some validity. However, before proceeding, it is important to explain why the construction and analysis of additional mutants can provide new and valuable information.

One problem with the results obtained with the 5- to 10-bp substitution mutants is that they provide fairly imprecise information about the boundaries of the important elements. For example, the important nucleotides within the −88/−99 element might extend from −88 to −101, or even further since the −102/−107s mutation reduced promoter activity by fourfold. In addition, the important nucleotides within the −127/−132 element might extend from −122 to −135, since the flanking mutations that had no significant effect on promoter activity were the −121/−117s and −141/−136s mutations.

Before explaining the reason for defining the boundaries of the elements, we describe the strategy and results obtained for the −99/−88 element. The strategy was to generate a series of mutants, each of which alters three adjacent base pairs. These mutations were prepared using the two-step PCR procedure described in Boxes 7.3 and 7.4 (below). The series generated is shown in Figure 7.6.

Analysis of the 3-bp mutant series revealed that three of the mutants exhibited strongly reduced promoter activities (Fig. 7.6). In contrast, the flanking mutations had no significant effect. These findings suggest that the critical nucleotides span a minimum of 5 bp (i.e., −90 to −94) and a maximum of 9 bp (−88 to −96). This is most consistent with the existence of a binding site for one protein. In fact, a binding site database search revealed that the critical 9 bp represent a binding site for CCAAT enhancer-binding protein (C/EBP) family members (Wedel and Ziegler-Heitbrock 1995). Subsequent DNA-binding studies supported the hypothesis that C/EBP proteins functionally interact with the critical element (Plevy et al. 1997; M. Studley and S.T. Smale, unpubl.).

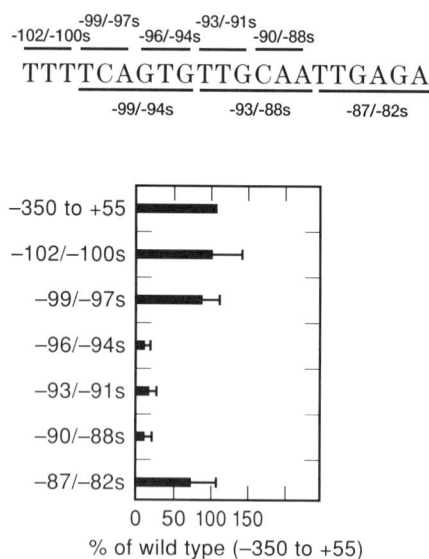

FIGURE 7.6. Refined substitution mutant analysis of the murine IL-12 p40 promoter. (Adapted, with permission, from Plevy et al. 1997 [Copyright American Society for Microbiology].)

We now return to the question, *Why is it worthwhile to localize with precision the boundaries of each element? The answer is that knowledge of the boundaries will help to establish whether each element interacts with one key sequence-specific DNA-binding protein, or represents a composite element containing adjacent sites for two or three proteins.* This is an important issue because composite elements have been found in promoters and distant control regions with considerable frequency (see Chapter 1). Composite elements often contain adjacent binding sites for two or more proteins, which bind cooperatively and/or function in a synergistic manner. If an important element is found to span 5–10 bp, the element most likely interacts with one protein (or is a composite element containing coinciding binding sites for two proteins, such as the NF-κB/HMG-I composite sites) (Thanos and Maniatis 1992). On the other hand, if the element spans 15–20 bp or more, it is more likely to be a composite element. Insight into this important issue can be gained by analysis of a few extra mutations, making the effort worthwhile. Without this information, one would need to rely on the binding studies for insight into the number of proteins that functionally interact with each element. Given the challenge of determining which binding proteins are relevant for a control element (see above and Chapter 9), the sole reliance on binding studies for determining whether a composite element exists is discouraged.

In the case of the –99/–88 analysis, the 3-bp mutant results suggested that the element comprised a maximum of 9 bp. When that 9 bp was found to represent a binding site for C/EBP proteins, the protein studies could be pursued with considerable confidence. If the boundaries of the element had not been determined, uncertainty would have remained regarding the possibility that the element contained binding sites for additional proteins.

It is important to note an added benefit of the 3-bp mutants: When binding activities are subsequently identified, the ability of the protein to bind the different mutants can be assessed. A close correlation between the nucleotides required for protein binding and those required for the function of the element in a transfection experiment provides an important piece of data supporting the functional relevance of the protein (see Chapter 9). In this case, the binding of recombinant C/EBP proteins required precisely the same nucleotides as were required for promoter function, supporting the hypothesis that C/EBP proteins are responsible for the function of the element.

One final issue that should be discussed with regard to the refined mutant analysis is why 3-bp mutants were chosen, rather than single-base-pair mutants. One reason is that three times more mutants would be needed to scan the important region using single-base-pair substitutions. The boundaries of the element would be defined more precisely with single-base-pair mutants, but for many studies, the benefit might not outweigh the additional effort that would be needed. A second reason is that many proteins can tolerate single-base-pair changes at some positions within their binding sites with only a minor loss of activity. Furthermore, if single-base-pair changes were used, the results would have been dependent on the particular nucleotide introduced, because binding proteins often tolerate some substitutions better than others at a given position. The probability that a given 3-bp mutation will be tolerated by a DNA-binding protein is much lower.

Choice of nucleotides for substitution mutants. Unfortunately, there is no foolproof strategy for nucleotide choice when constructing substitution mutants. The possibility will always exist that a mutation will create a fortuitous binding site for another protein that might influence promoter activity. The creation of a new binding site could lead to inaccurate or misleading data. In the 5- to 10-bp substitution mutant series, a restriction enzyme site was inserted in place of the IL-12 promoter sequences. In this case, the restriction site was necessary for the mutagenesis strategy, because the substitution mutants were generated from deletion mutants using a strategy that relied on the presence of a restriction site (see Fig. 7.4). For most mutagenesis strategies, insertion of a restriction site is convenient because it simplifies the process of screening for bacterial colonies containing mutant plasmids; minipreps of the DNAs can simply be analyzed by restriction mapping, rather than DNA sequencing.

For the 3-bp mutants, restriction sites could not be introduced routinely because only 3 bp were altered. Because of the small size of the substitution, it also was important to alter the sequence as severely as possible, to increase the probability that important protein–DNA contacts would be disrupted. To this end, each substitution introduced a transversion, and at the same time, changed the base pair. For example, a C/G base pair was changed to A/T, G/C to T/A, A/T to C/G, and T/A to G/C. These substitutions introduce the most radical structural change possible at a particular site. However, this strategy does not take into account the possibility that the substitution will result in the creation of a binding site for another protein.

To determine whether a binding site for another protein is generated, the mutant sequence can be analyzed by searching a binding-site database. If the mutant sequence is similar to the consensus sequence for a known protein, a different sequence could be introduced. Of course, this strategy will only reveal binding sites for known proteins, based on current knowledge. Since a definitive method is not available for ensuring that a substitution mutation does not introduce another binding site, it is important to keep this possibility in mind during the subsequent steps in the promoter analysis. If a result at any stage of the analysis suggests that a fortuitous binding site was introduced accidentally, the best course of action is to prepare additional substitution mutations at the same location to address this possibility.

Inducibility and cell-type specificity. The mutagenesis strategy outlined above resulted in the identification of two control elements that contribute to IL-12 promoter activity in induced macrophages. The promoter is inducible; thus, both of these elements may bind proteins that directly mediate induction. Alternatively, one of the elements may bind a constitutively active protein that is essential for promoter function, but is not a direct mediator of induction. With this latter scenario, promoter induction would occur when the constitutively active protein carries out an appropriate physical or functional interaction with the induced protein. Because the primary objective of the promoter analysis is to elucidate the mechanism of promoter induction, it eventually will be necessary to distinguish between these two types of elements. Similar issues must be considered when studying cell-type-specific control regions. In those studies, the goal is to distinguish the control elements that bind cell-type-specific proteins from those that bind ubiquitously active proteins.

It can be quite difficult to determine which elements are direct mediators of induction because the activity of a control region following induction will be sensitive to mutations in any control element, including those that do not mediate induction but are merely required for activity. Because the importance of an element following induction provides no significant information regarding inducibility per se, a common strategy for determining which elements are direct mediators of induction is to rely on the importance of each element *prior* to induction. In theory, elements that bind proteins which directly mediate induction will not be involved in basal promoter activity in uninduced cells and will become important only following induction. In contrast, elements that bind constitutively active proteins will be equally important in uninduced and induced cells. Thus, it is thought that the precise role of each element can be determined simply by comparing the "fold-induction" of the wild-type promoter to that of each promoter mutant. In other words, after subtracting background, the induced reporter activity of each construct is divided by the uninduced activity of the same construct, yielding a fold-induction value. If mutation of an element reduces the fold-induction value (i.e., if the element is more important in induced cells than uninduced cells), the element is considered to be a direct mediator of induction. If mutation of an element does not influence the fold-induction value (i.e., if the element is equally important in uninduced and induced cells), the element is unlikely to mediate induction.

The above strategy can be informative for the subset of promoters that yield considerable activity prior to induction, but it is not useful for promoters whose uninduced activities are not *substantially* above background. The activity of the IL-12 promoter, for example, was only slightly greater than background before induction. Although the uninduced activity of the wild-type promoter was statistically significant, mutations in the important control elements reduced the uninduced signal to levels that were not significantly greater than background. This scenario, which is quite common, renders an analysis of the fold-induction values meaningless.

As a specific example of the above concept, the mutation in the Rel site can be considered. This mutation reduced the induced promoter activity to 25% of wild type (see Fig. 7.7). It also reduced the uninduced signal to a value that was only slightly greater than background. Because the uninduced signal for the promoter mutant was nearly zero after subtracting background, the fold-induction remained high. If those values were used as the sole criterion for determining whether an element mediates inducibility, one would conclude that the Rel site was not important for induction, despite considerable evidence that Rel proteins play critical, direct roles during gene induction in macrophages.

The results obtained with mutations in the C/EBP site provide another example. With these mutations, induced promoter activity was reduced to less than 10% of wild type (Fig. 7.7). Uninduced promoter activity was also reduced, but like the Rel mutation, it remained slightly above background. The fold-induction calculations yielded values for the C/EBP mutants that were much lower than for the Rel mutants. One interpretation of these data

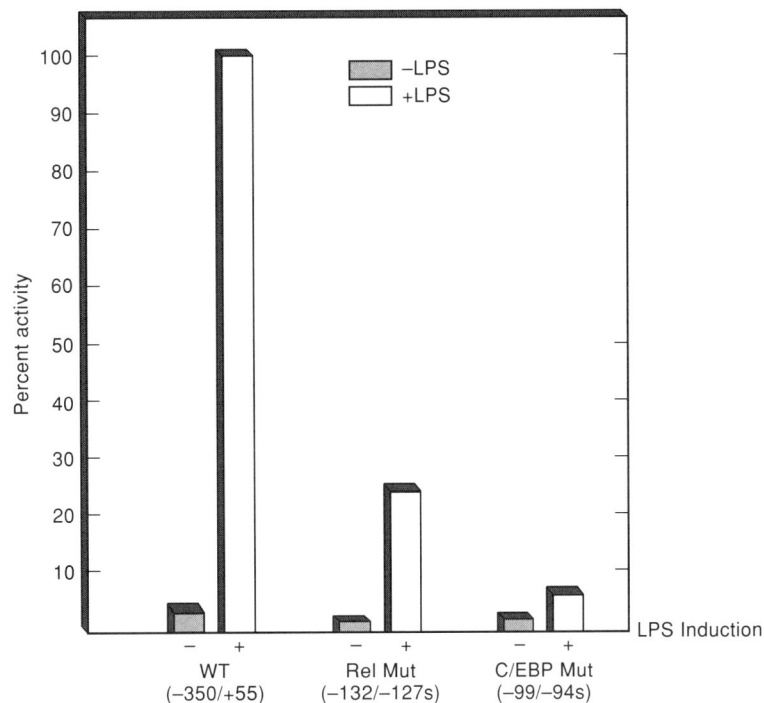

FIGURE 7.7. Effect of IL-12 p40 mutations on promoter induction by lipopolysaccharide.

is that the C/EBP site is the key to induction, with the Rel site much less important. Although this conclusion seems logical, it actually represents a misinterpretation of the data, because the fold-induction values depend on the precision of the uninduced signals. Because the uninduced signals are close to background with both the Rel and C/EBP mutants, as well as with the wild-type promoter, the accuracy of these numbers is difficult to determine, even with statistical analyses. In other words, very small changes in the uninduced signals can have dramatic, but highly questionable, effects on fold-induction values.

Regardless of the results obtained, the activities of promoter mutants rarely provide substantive insight into the issue of which elements are directly responsible for inducibility (or cell-type specificity). To address this issue, the relevant binding proteins must be identified and their properties characterized. If the abundance of the relevant transcription factor increases during cell induction, it may directly contribute to promoter induction. This hypothesis can be tested more rigorously using other approaches, some of which are discussed in Chapter 9. If the transcription factor abundance is not increased during cell induction, the factor may nevertheless play a direct role in induction because it may acquire a posttranslational modification that alters its activity. A careful analysis of each binding site and transcription factor is ultimately needed to determine which ones play direct roles in induction or cell-type specificity.

Transcription start site confirmation. Ideally, the start sites of transcripts should be determined when analyzing a promoter by transient transfection, to confirm that transcription initiates at the correct location (see Chapter 5). The transcription start sites of mutant promoters should also be determined, to confirm that the mutations do not alter the start site or result in the induction of a cryptic start site. Minor start site alterations upon promoter mutagenesis are of little concern, but more severe changes may indicate that a mutation has not simply disrupted a control element. Instead, the mutation may have altered the overall structure and regulation of the promoter. For example, a mutation might lead to the activation of a cryptic TATA-like sequence within the mutant nucleotides or elsewhere in the promoter. If a cryptic TATA-like sequence becomes activated, it may respond to the regulatory elements differently than the authentic core promoter.

Unfortunately, the transient transfection efficiencies of many cell lines are too low to allow start site mapping. This problem was encountered during the IL-12 promoter analysis. In this study, to confirm that the wild-type promoter–reporter plasmids directed transcription from the correct sites, stable transfectants were prepared. Because every cell in the selected lines contained an integrated reporter plasmid, the reporter transcripts were of sufficient abundance for start site mapping by primer extension. This analysis revealed that the major start site was at the expected location. These results provided some confidence that the transcription start sites in the transiently transfected plasmids, at least with the wild-type promoter, were also at the correct location. However, the stable transfection results did not provide conclusive information regarding the transient transfection experiments and did not address the possibility that a mutation altered the start site. Perhaps the only assay that is sufficiently sensitive for the localization of transcription start sites following transient transfection of most cell lines is 5′ RACE, as described in Chapter 4.

Because the locations of transcription start sites can be difficult to determine following transient transfection, this experiment is usually not performed during a typical mutant analysis. For studies that involve the dissection of a core promoter region (i.e., TATA and Inr region), the absence of information regarding the start site location is likely to be problematic. However, for most other studies, the start site analysis is not essential. Nevertheless, one should proceed with considerable caution and remain aware of the fact that the start site has not been confirmed.

Choice of assay. As stated earlier, one key limitation of the mutagenesis strategy is that the only elements identified are those that are important in the functional assay being used. In the IL-12 promoter analysis, which employed a transient transfection assay, the C/EBP site was essential, whereas the Rel site made only a moderate contribution. In contrast, when mutations in these two sites were tested in a stable transfection assay, both were absolutely essential for promoter activity (Plevy et al. 1997). Presumably, the high plasmid copy number in transiently transfected cells, or the episomal nature of the transiently transfected plasmids, diminished the importance of the Rel site (see Chapter 5). It would not be surprising if repetition of the entire mutant analysis using a stable transfection assay resulted in the identification of other essential control elements, which were relatively unimportant in the transient assay. Indeed, several examples of control elements that function in stable, but not transient, transfection assays have been reported (see Chapters 5 and 6).

Some control elements that are important for transcription of the endogenous gene may be missed in both transient and stable transfection assays. This hypothesis is based on the fact that stably transfected plasmids do not necessarily become incorporated into the same chromatin structure as the endogenous gene. Therefore, a subset of the control elements that are important for chromatin remodeling during gene activation or inactivation might be missed.

A recent example of the above is again provided by the Ig μ enhancer analysis (Fig. 7.8). In transgenic mice, the activity of the Ig μ enhancer is strongly stimulated by the adjacent matrix attachment regions (MARs) (Forrester et al. 1994). However, in both transient and stable transfection assays, the MARs have no effect on enhancer activity. Thus, if the Ig μ enhancer had been dissected solely with transfection assays, the importance of the MARs would have been missed. Interestingly, Forrester and Grosschedl have developed a modified stable transfection assay that restores the MAR requirement (Fernandez et al. 1998). For this assay, the reporter plasmid DNA was methylated in vitro prior to transfection and drug selection. Reporter gene activity from the stably integrated, premethylated plasmids requires the presence of the MARs. Apparently, premethylation causes the transfected plasmid to become incorporated into less accessible chromatin, which results in the MAR requirement for transcriptional activation.

On the basis of the above information, a mutant analysis of a new control region would ideally be performed with a transgenic mouse assay, or at least a stable transfection assay. Unfortunately, the time and resources required for a comprehensive mutant analysis by either of these assays makes them impractical for most studies. For this reason, it often is necessary to begin with a transient transfection assay to identify the control elements needed in that assay, and then to proceed to more sophisticated assays when the analysis reaches a more advanced stage.

Redundancy of control elements. A final caveat of the comprehensive mutant analysis outlined above is that it may fail to identify control elements whose activities are redundant (or at least are not strongly synergistic) with the activities of other control elements within the region. Recent data have revealed redundancy within the IL-12 promoter, which caused at least two important control elements to be overlooked in the initial analysis (S. Plevy, pers. comm.). In the analysis described above (Fig. 7.5), mutations at −62/−57 and −80/−75 reduced promoter activity by only about twofold. Because of the small effect of these mutations, it was difficult to determine whether these sites correspond to authentic control elements. Recent experiments have revealed that both of these sites are actually critical for promoter function. Mutation of these sites had modest effects in the original analysis because they are partially redundant with each other and with the Rel site located farther upstream. The apparent redundancy was revealed by analysis of promoters con-

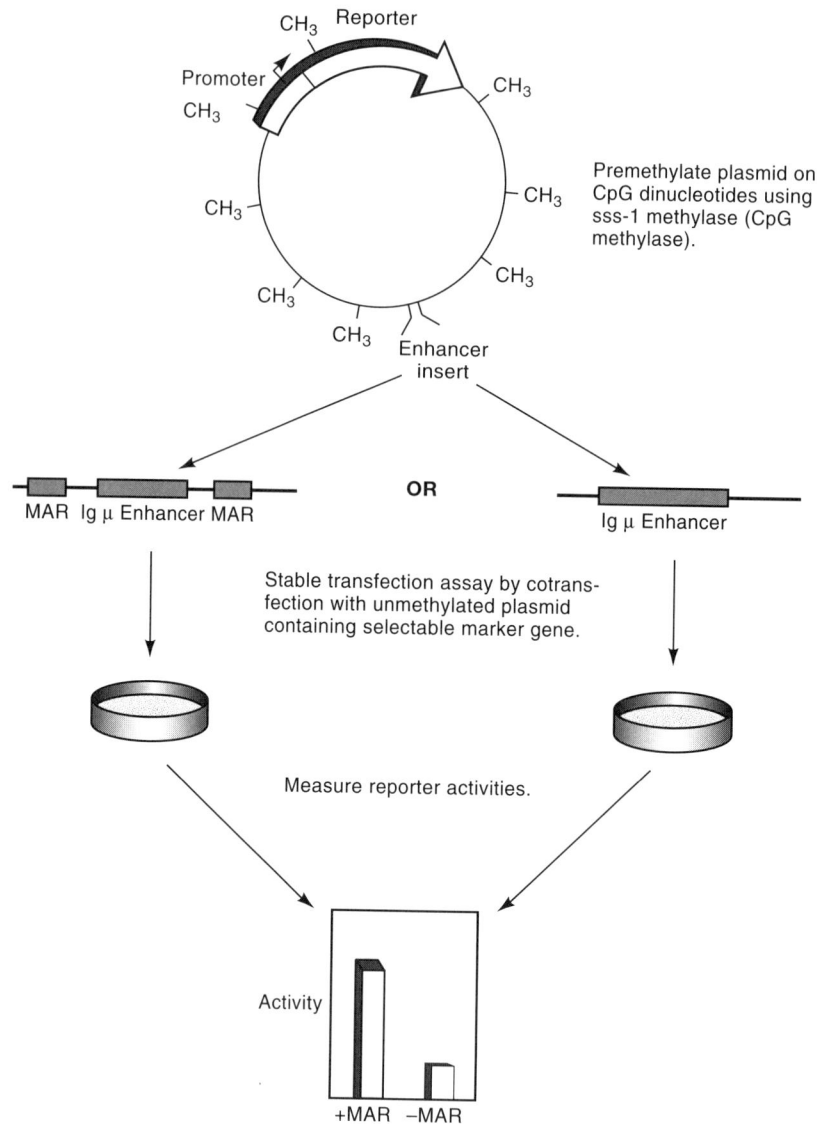

FIGURE 7.8. Detection of MAR function in a stable transfection assay by premethylation of a reporter plasmid.

taining mutations in two elements simultaneously. Simultaneous mutation of the –62/–57 and –80/–75 elements completely inactivated the promoter, revealing that the C/EBP and Rel sites are insufficient for promoter function. Strong promoter activity requires at least one of these other elements, in addition to the C/EBP and Rel elements.

The Ig μ enhancer provides a classic example of redundancy within a control region. Early transfection studies revealed that no substitution mutations reduced activity by more than appoximately twofold (see, e.g., Lenardo et al. 1987). Similar results were obtained in transgenic mouse assays (Jenuwein and Grosschedl 1991; Annweiler et al. 1992), suggesting that the apparent redundancy was not an experimental artifact of the transfection assays. The inability of any mutation to strongly diminish enhancer activity created considerable difficulties for the analysis of the mechanism of Ig μ enhancer function.

Over the past several years, Ranjan Sen and colleagues pursued a strategy for circumventing the redundancy problem and, more recently, for dissecting the molecular basis of

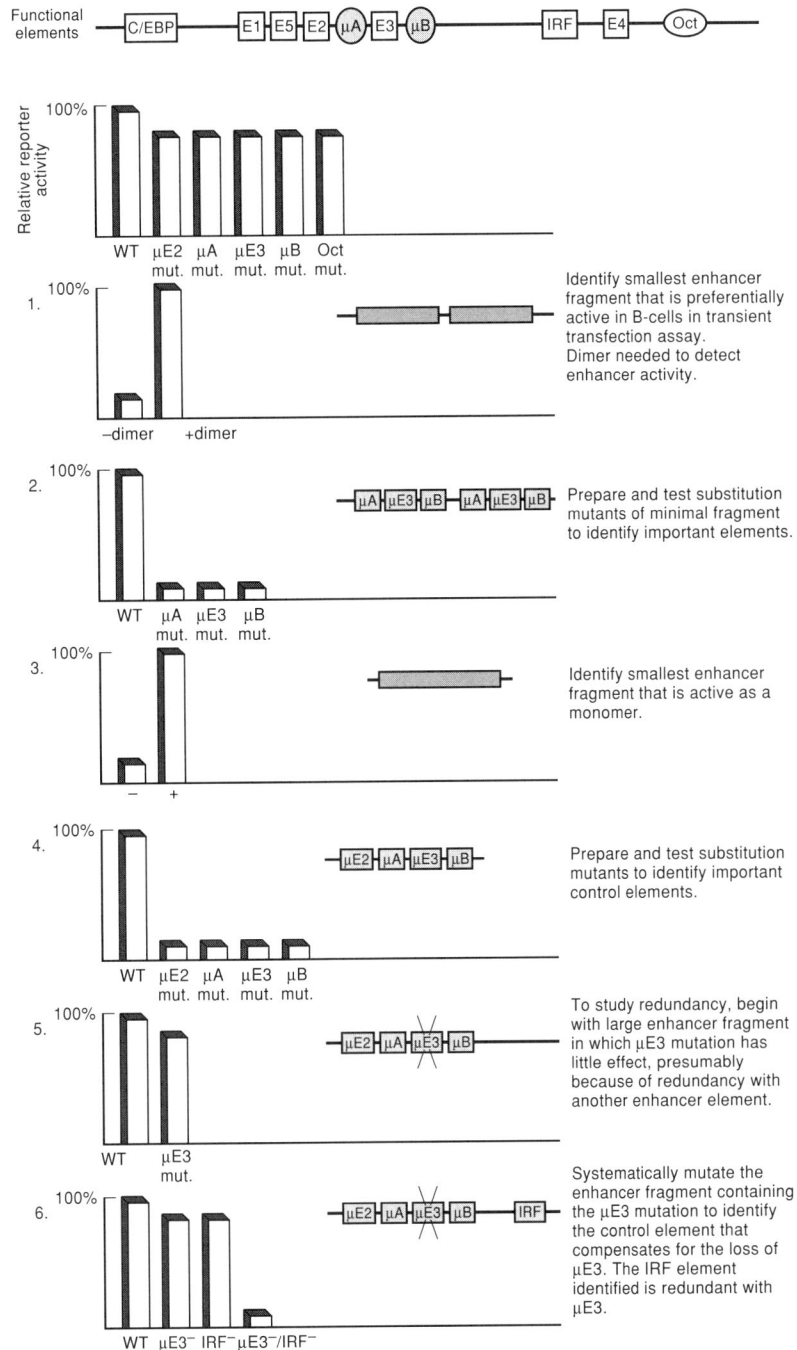

FIGURE 7.9. Strategy for dissecting an enhancer that exhibits considerable redundancy among control elements.

the redundancy (Fig. 7.9). To circumvent the problem, Sen first created deletion mutants to identify the smallest enhancer fragment that supports enhancer function preferentially in B cells (Nelsen et al. 1990, 1993). As expected on the basis of the observed redundancy, several control elements could be deleted with little consequence. Substitution mutations were then introduced into the minimal enhancer fragment, revealing that the remaining control elements were absolutely essential for enhancer function (Nelsen et al. 1990, 1993). Further analysis of the essential control elements led to the identification of proteins that may functionally interact with them (Nelsen et al. 1990, 1993). The molecular mechanism by which these proteins synergize with one another has also been dissected using the minimal enhancer fragment (Erman and Sen 1996; Nikolajczyk et al. 1996, 1997; Rao et al. 1997; Erman et al. 1998).

For the initial studies, Sen used a minimal enhancer fragment that contained only three control elements—μA, μE3, and μB. With this small fragment, it was necessary to fuse dimers to the reporter plasmid to detect activity. After identifying and characterizing the three elements within this fragment, a larger fragment was used, which yielded substantial enhancer activity when present in a single copy (Dang et al. 1998b). Because this larger fragment still lacked the redundant elements, most of the elements remained essential for enhancer activity.

Recently, Sen and colleagues have begun to pursue the molecular basis of redundancy (Fig. 7.9). The strategy employed was to mutate the full-length enhancer systematically to identify the control element that is redundant with μE3. In other words, the goal was to identify the elements critical for function when the enhancer contains a mutant μE3 site. The element that conferred redundancy was a previously undescribed enhancer element, which binds IRF proteins (Dang et al. 1998a). Because the IRF element is largely redundant with the μE3 element, it appears to be just as important for enhancer function, even though it was discovered 13 years after the μE elements were first reported.

Recall that the μE elements were originally identified by in vivo and in vitro protein–DNA interaction studies. In contrast, discovery of the IRF element required a systematic mutant analysis that was sufficiently comprehensive to address the redundancy issue. This systematic analysis can now be extended to determine whether other elements contribute to redundancy within the enhancer. The biological basis for the redundancy remains unknown. One hypothesis is that it allows the enhancer to be activated by distinct combinations of factors at different stages of development, so that the same specific set of factors does not need to be present whenever the enhancer is activated. Alternatively, redundancy may ensure enhancer function in a nuclear milieu of limiting transcription factor concentrations.

The strategy employed by Sen and colleagues is likely to be useful for analyzing redundant (or nonsynergistic) elements in other control regions. Some redundancies may be biologically significant, whereas others may be related to the assay used for the analysis. The Ig μ enhancer appears to be an example of biologically relevant redundancy, since the redundancy was observed in transgenic mouse assays, as well as in transient and stable transfection assays. In contrast, the redundancy observed in the IL-12 promoter may be biologically irrelevant; as stated above, the Rel site is essential for promoter function in a stable transfection assay but is unessential and redundant with the other elements in a transient transfection assay. Thus, in this control region, the redundancy is likely to result from the high copy number and/or episomal nature of the promoter following transient transfection, which allow the promoter to be activated by only a subset of the proteins required for the endogenous promoter.

The type of redundancy observed with the IL-12 promoter may be quite common. It was fortunate that the IL-12 promoter elements were only partially redundant with one another, so that the modest effects of single mutations were sufficient to reveal the existence of the elements. If these elements had been completely redundant with one another, they might still remain unknown, just as the important IRF site in the Ig μ enhancer remained undiscovered for over a decade.

Methodology for Mutating a Control Region

The above discussion argues that a combination of deletion and substitution mutants is most useful for dissecting a control region. Three general methods can be used to prepare deletion mutants. A simple starting point is to identify appropriate restriction sites within the large control region, either from the DNA sequence or by standard restriction enzyme analysis, and then to cleave the DNA with those restriction enzymes and introduce the resulting fragments into an appropriate reporter vector (Box 7.1). A second strategy, which is used relatively infrequently today, is to remove nucleotides from the 5′ or 3′ end of a control region using an exonuclease (Box 7.2). The third and by far the most popular method for generating a series of deletion mutants is PCR (Box 7.3).

A variety of methods are also available for the construction of substitution mutants within a control region. Once again, PCR methods are probably the fastest and most popular (Boxes 7.4 and 7.6). However, oligonucleotide-directed mutagenesis can sometimes be used as effectively and rapidly, particularly if a large number of mutants are being generated (Box 7.5). Each of these methods is summarized at the end of this chapter, along with a discussion of the advantages and disadvantages of each method and appropriate references to the literature and commercial kits.

Identification of Control Elements Using In Vivo or In Vitro Protein–DNA Interaction Methods

Advantages and Disadvantages

As described in the introduction to this chapter, one alternative to the comprehensive mutant analysis for identifying important control elements is to carry out in vivo or in vitro protein–DNA interaction assays and then to introduce mutants into the binding sites identified to test their relevance in a functional assay. One key disadvantage of this approach is that it is likely to reveal only a small subset of the relevant control elements. This disadvantage was apparent from the Ig μ enhancer studies described above. In vivo and in vitro binding studies revealed only a small subset of the important elements that were later identified in the mutant studies. A second disadvantage is that some of the binding sites may not correspond to important elements when mutations in those sites are tested in a functional assay. This disadvantage was also apparent in the Ig μ enhancer analysis, particularly with binding sites identified using in vitro protein–DNA interaction assays (see, e.g., Tsao et al. 1988). A third disadvantage is that it will be difficult to determine the boundaries of the region to analyze for protein–DNA interactions unless minimally a deletion analysis is performed.

A fourth disadvantage of this approach is that it often is difficult to assess the relative importance of the identified site. As an example, consider the IL-12 promoter analysis (Plevy et al. 1997). In vitro DNase I footprinting and, more recently, genomic footprinting (A. Weinmann, unpubl.; see Chapter 10), revealed an inducible protein–DNA interaction at the C/EBP site. If the promoter analysis began with these binding studies, a mutation would have been introduced into the C/EBP site, revealing that the site was important for promoter function in the transient transfection assay. However, in the absence of the comprehensive mutant analysis, it would have been impossible to determine whether the C/EBP site was one of several critical sites within the promoter, or one of the only critical sites, which would suggest a prominent role. Because the protein–DNA interaction studies are likely to miss some of the important interactions, the prominence of the C/EBP interaction would have been difficult to evaluate.

As another example of this concept, consider the Rel site in the IL-12 promoter. Protein–DNA interactions were not observed at this site during in vitro footprinting with crude extracts, or during genomic footprinting. Thus, if the promoter analysis had begun with these assays, the Rel site would have been missed. The reason Rel protein binding was not observed in the footprinting studies remains unknown but is likely due, at least in part, to the fact that the binding site resides within a DNA region that is relatively devoid of DNase cleavage sites.

Rel binding activities can be observed when a small DNA fragment spanning the site is tested in an EMSA (Murphy et al. 1995; Plevy et al. 1997). However, many different EMSA complexes can be observed with probes spanning different regions of the IL-12 promoter (Plevy et al. 1997; S. Plevy, unpubl.). To determine which of the many binding sites are functionally relevant, it would have been necessary to map the nucleotides contacted by each protein, either by methylation interference or by testing probes with mutated nucleotides. Then, it would have been necessary to test mutations in each binding site in a transient transfection assay. The amount of work required to analyze every EMSA complex observed would be comparable to that required for the comprehensive mutant analysis. Furthermore, this approach would be less informative because some important DNA elements may be missed (i.e., those that do not yield a detectable EMSA complex when using a crude nuclear extract).

In the case of the Rel proteins, some evidence that the site was relevant would have been obtained from the observation that the EMSA complex was more abundant in extracts prepared from LPS-induced macrophages. Thus, an investigator could have chosen to characterize only those complexes that were inducible. This criterion for identifying relevant sites is unreliable, however, because many proteins that contribute to gene induction are not induced at the level of their DNA-binding activity.

Despite the disadvantages of using protein–DNA interaction studies as a starting point for identifying important control elements, this approach has two advantages that were implied earlier in this chapter. First, with some luck, it may lead to the identification of an interesting regulatory protein much more rapidly than the comprehensive mutant analysis. If a binding activity is identified that exhibits an interesting expression pattern, one promoter mutant can determine whether the binding site has some relevance for promoter function. If so, an investigator can immediately initiate experiments to identify and characterize the protein. Although the relative importance of the protein for the function of the control region that led to its identification will remain uncertain, the protein may in fact play a critical role in the regulated expression of the gene. Moreover, the protein may be essential for the biological process being studied, which may be of greater importance to the investigator than an understanding of one particular control region. The investigator would be able to rapidly initiate experiments to study the importance of the transcription factor for the biological process using, for example, a gene disruption approach.

The second potential advantage of using protein–DNA interaction studies to dissect a control region is that it may lead to the identification of important regulatory proteins that would be overlooked in the comprehensive mutant analysis, by virtue of the fact that a protein might bind an element whose function is not detected in the artificial assay being used, or whose function is redundant with the function of another protein. Realistically, however, it will be difficult to obtain convincing evidence that the protein identified is important for the regulation of the gene of interest, unless a function for the binding site can be demonstrated using some sort of assay. If the binding site is not important in a transient assay, a mutation in the site can be tested in a stable transfection assay or in a transgenic mouse assay. It must be kept in mind, however, that an overwhelming amount of effort would be needed to use the transgenic assay to test the relevance of every protein–DNA interaction identified.

The strategies and methods for identifying protein–DNA interactions in vitro are discussed in depth in Chapters 8 and 13. In vivo protein–DNA interaction methods are presented in Chapter 10. The focus of those chapters is on strategies used to identify and study protein–DNA interactions at a well-defined control element. Similar strategies can be used to identify and characterize protein–DNA interactions within an entire control region as a starting point for dissecting the control region.

Identification of Control Elements by Database Analysis

Advantages and Disadvantages

A control region can be scanned for recognition sites for known proteins using the database described in Chapter 1. The disadvantages of using a database analysis as a starting point for identifying important control elements within a control region are discussed above. Briefly, only a subset of the binding sites identified in a database analysis are likely to be relevant, and some of the relevant sites are likely to be missed. An additional disadvantage of the database approach is that it is strongly biased toward the identification of binding sites for known proteins. Because the goal of many studies is to identify previously unknown transcriptional regulators, this approach can be counterproductive.

Nevertheless, a database approach as a starting point for dissecting a control region has two distinct advantages. Both advantages are the same as those described above for using protein–DNA interaction studies as a starting point. First, potential control elements can be identified much more rapidly than with a mutant analysis. Second, important elements may be found that are not revealed by the mutant analysis, since the success of the mutant analysis depends on the degree to which the functional assay mimics the regulation of the endogenous gene.

Although in our opinion the disadvantages of the database analysis as a starting point far outweigh its advantages, it can be extremely useful in conjunction with a comprehensive mutant analysis. After the mutant analysis has identified relevant DNA elements, there generally is a need to identify the proteins that interact with them. Protein–DNA interaction studies can be used to identify EMSA complexes and footprints at the relevant sites. However, a database analysis can provide valuable information regarding the identities of the interacting proteins. In the absence of a database analysis (or simply a visual inspection of the site), it would be necessary to clone the gene encoding the interacting protein to identify it. If a database analysis or visual inspection suggests which protein or proteins might interact, antibodies directed against those proteins can be added to the protein–DNA interaction experiments to determine whether they disrupt or alter the interaction (see Chapter 8). Furthermore, recombinant forms of the proteins can be tested for their ability to interact with the site.

A phylogenetic analysis of the DNA element can also be useful at this point because the functionally relevant nucleotides are usually conserved between species. If the two different proteins are predicted to bind an important element, for example, the phylogenetic analysis may reveal that only one of the proteins can bind the corresponding element in related species. Erman et al. (1998) is a study that derived great benefit from a phylogenetic analysis to assess the relevance of factors capable of interacting with an important DNA element.

The reason the database is useful at a relatively late stage of a mutant analysis, but not as a starting point for dissecting a control region, is that the database is searched with a very short DNA sequence at the later stage. This search reveals a relatively small number of potential binding proteins, whose abilities to bind the site can be addressed with a reasonable amount of effort. In contrast, when an entire control region is subjected to a database analysis, it usually results in the identification of a large number of potential binding proteins. Considerable effort would be needed to sort through these proteins to determine which are functionally relevant. Thus, despite the fact that a database analysis provides a relatively simple starting point for the dissection of a promoter, it is unlikely to lead to a broad understanding of a gene's regulatory mechanisms as rapidly and efficiently as a comprehensive mutant analysis supplemented with database information.

Mutagenesis Techniques

Box 7.1

Deletion Mutagenesis Using Restriction Sites

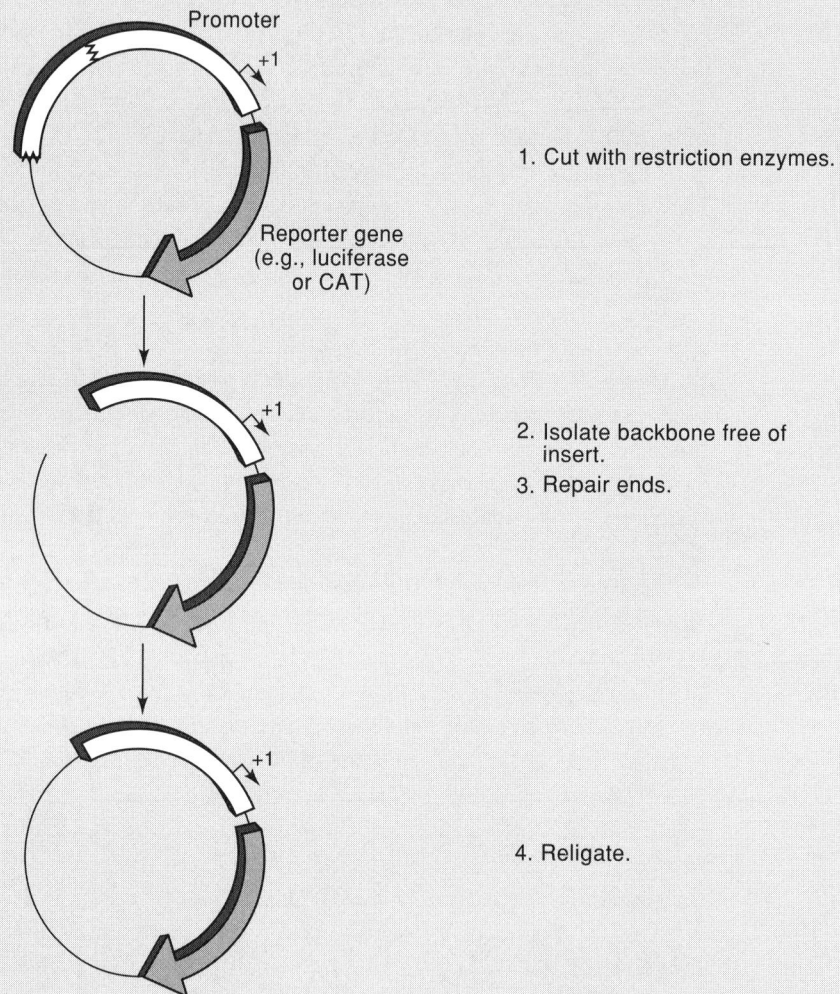

1. Cut with restriction enzymes.

2. Isolate backbone free of insert.
3. Repair ends.

4. Religate.

FIGURE 7.10. Deletion mutagenesis of a promoter by restriction fragment excision.

Often the first method employed by an investigator to study a control region is restriction fragment deletion mutagenesis (Fig. 7.10) (for a review citing the original literature, see Shortle et al. 1981). In the simplest variation of this approach, pairs of endonucleases are employed to excise fragments varying in size from either end of the control region, or from within the control region. In some cases, the deletions are performed in one plasmid to facilitate cloning and site usage and then subcloned into the appropriate reporter vector. The reporter plasmid will contain only a limited number of unique restriction sites for insertion of the fragments, and, as a result, the fragment end points will need to be made compatible with the cleaved reporter. This can be accomplished by a variety of standard molecular biology techniques: (1) A linker containing an appropriate restriction site can be annealed to the end of the promoter fragment; (2) restriction enzymes that create blunt ends can be used, with the fragments inserted into a blunt-ended site; or (3) the reporter vector can be engineered so that it contains all of the required sites. The advantage of this approach is that it is one of the quickest and most efficient methods of producing deletion mutants. A disadvantage is that the locations of the deletion end points are dependent on the availability of convenient restriction sites. In addition, internal deletions can alter the spacing and sometimes the helical relationships between different control elements, which may alter the functions of the elements.

Box 7.2

Deletion Mutagenesis Using Exonucleases

To generate unbiased deletions within a control region, exonucleases such as BAL31 (Gray et al. 1975) and Exonuclease III (*Exo*III) can be employed (Rogers and Weiss 1980). *Exo*III is isolated from *E. coli* and contains a 3′ exonuclease activity that functions on recessed 3′ termini and cleaves one strand of the DNA. BAL31 is an enzyme isolated from *Allermonas espeziani* and contains both a 3′ → 5′ exonuclease activity and a single-stranded endonuclease activity. To produce deletions, the control region is cleaved with a restriction enzyme and the DNA is then incubated with *Exo*III or BAL31 (Fig. 7.11).

By varying the time of cleavage, the investigator can obtain a range of different-sized deletions. In the case of *Exo*III, the DNA must be cleaved subsequently with a single-strand nuclease such as mung bean or S1. BAL31, on the other hand, contains its own single-strand nuclease activity. The DNA is subsequently repaired with T4 DNA polymerase, religated, and transformed into *E. coli*. The relevant DNAs can then be isolated and sequenced to identify the end points of the deletion. There are two forms of BAL31: a slow form and a fast form (Wei et al. 1983). Most commercial suppliers provide the enzyme as a mixture of the two activities. However, short deletions (<100 bp) generally require the slow form.

Deletions can be made unidirectional by *Exo*III by cleaving the DNA at two closely spaced restriction sites; one restriction endonuclease should be chosen to leave a recessed 3′ end, and the other should generate a 3′ overhang (Fig. 7.11C). *Exo*III will only work efficiently on the recessed end, leaving the 3′ overhang largely untouched. BAL31, on the other hand, digests inward in both directions from the cleaved ends (Fig. 7.11A). In such a case deletions can be made unidirectional by cloning in a "stuffer" DNA fragment that recreates unique restriction fragment cleavage sites (1 and 2) (Fig. 7.11B). The fragment is cleaved with enzyme 1 and then digested with BAL31. The BAL31 proceeds bidirectionally but digests the control sequence in one direction and the stuffer in the other. The reaction is terminated, and the DNAs are cleaved with enzyme 2, which removes the undigested portion of the stuffer; thus, the DNA region flanking this undigested portion is left intact. The ends are repaired with Klenow fragment (although this is not always necessary because T4 ligase can ligate the ends) and religated.

One of the major advantages of BAL31 and *Exo*III is the ability to create an unbiased set of deletions. However, these approaches are time-consuming because the deletions must all be sequenced, and the exonuclease digestion requires considerable optimization. BAL31 and *Exo*III are available from a wide variety of commercial sources, as are kits for generating deletions. NEB markets an Exo-Size Deletion kit (NEB cat. #460) and Promega markets the Erase-a-Base system (cat. #E5850); both are based on *Exo*III digestion.

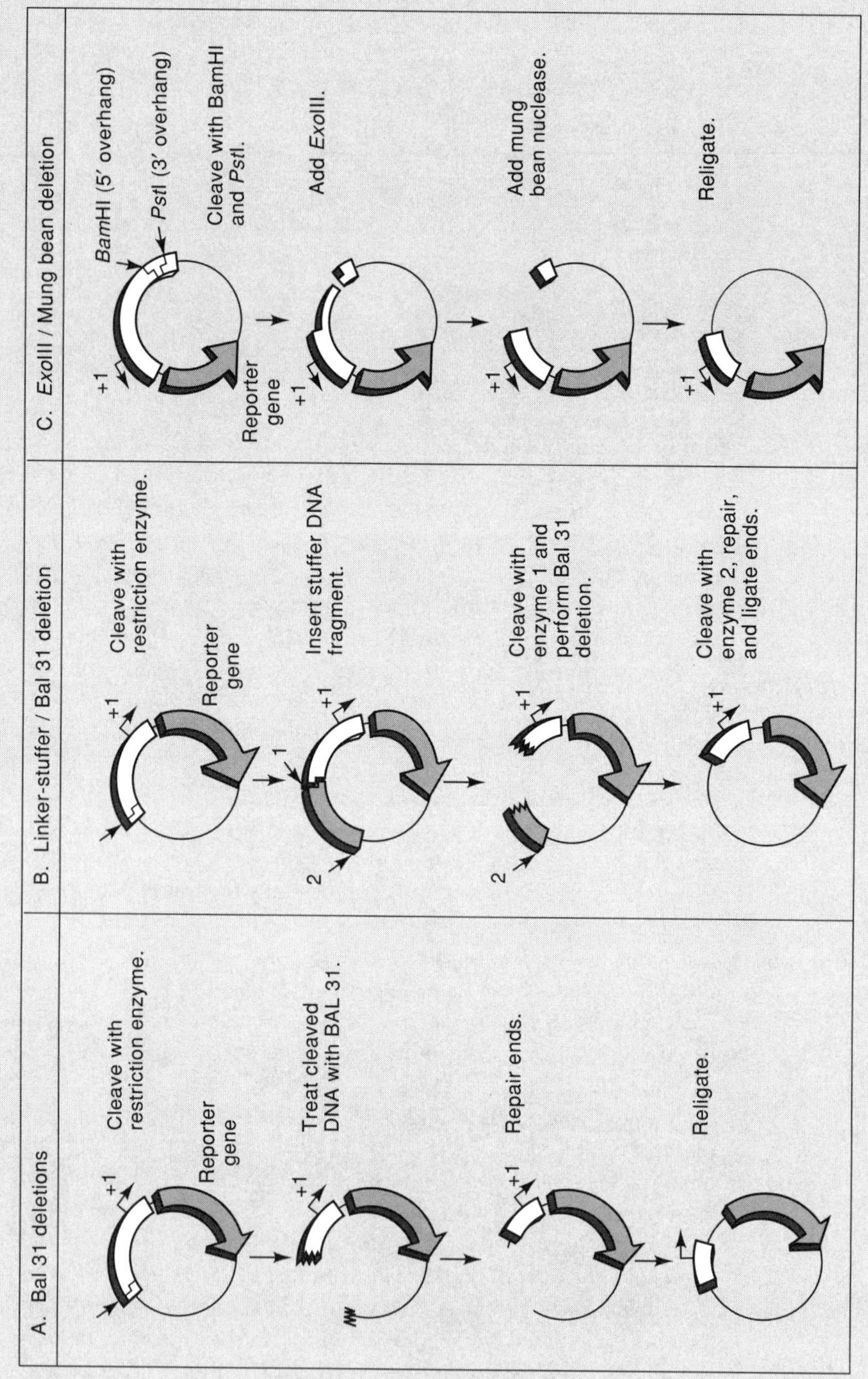

FIGURE 7.11. Deletion mutagenesis using exonucleases.

Box 7.3

Deletion Mutagenesis by PCR

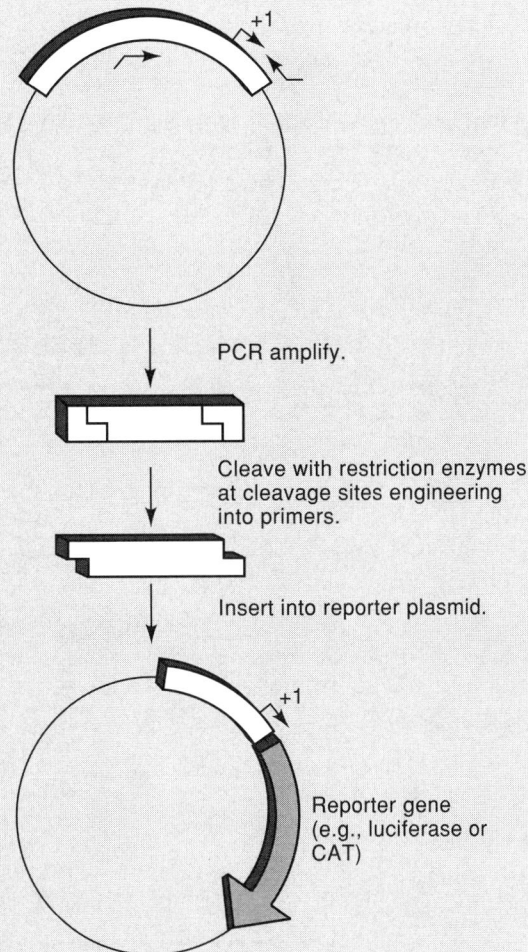

FIGURE 7.12. Generation of promoter deletion mutants by PCR.

The polymerase chain reaction (PCR) is the fastest, easiest, and most popular method for generating deletions (Mullis and Faloona 1987). The disadvantages are the cost of the oligonucleotides and thermostable polymerases (either Taq or Pfu) and the fact that the deletions need to be sequenced to confirm that point mutations were not introduced during the procedure. The PCR method can be used to generate 5′ and 3′ deletions of the control region, or a variation termed PCR-ligation can be employed to generate precise internal deletions (Tomic-Canic et al. 1996).

In simple deletions, two oligonucleotides are generated representing the 5′ and 3′ end points of the deletion (Fig. 7.12). These oligonucleotides need only 18 nucleotides of homology (~ 50% GC). A restriction site is generally added onto the ends of the oligonucleotides to allow subcloning of the resulting fragment into a reporter vector. Because many restriction enzymes cleave very inefficiently when their sites are located at the extreme end of an amplified DNA fragment, the 5′ ends of the two oligonucleotides are designed to contain a 3-nucleotide, GC-rich sequence. This sequence prevents thermal fraying and allows the restriction enzyme to cleave the resultant amplified product.

PCR ligation is an excellent method for generating internal deletions of a control region (and also substitution mutants; see Box 7.4). PCR ligation requires two pairs of primers. Each pair will generate one PCR fragment. The primers are designed so that there is a 15-bp region of overlap between primers 2 and 3 (Fig. 7.13). For example, primer 3 is designed so that it contains 18 nucleotides of homology to the sequence being amplified and 15 nucleotides on its 5′ end, which are complementary to primer 2. The primer pairs (1 and 2, 3 and 4) are then used to separately amplify the intervening fragments of DNA; two fragments will be generated. The downstream end of the first fragment now contains a 15-bp region of overlap with the upstream end of the second fragment. If these fragments are denatured and renatured, the top strand of the first fragment will hybridize with the bottom strand of the second fragment. *Taq* DNA polymerase will "fill in" the ends, thereby linking the two fragments. Primers 1 and 4 can then be used to amplify this hybrid fragment further.

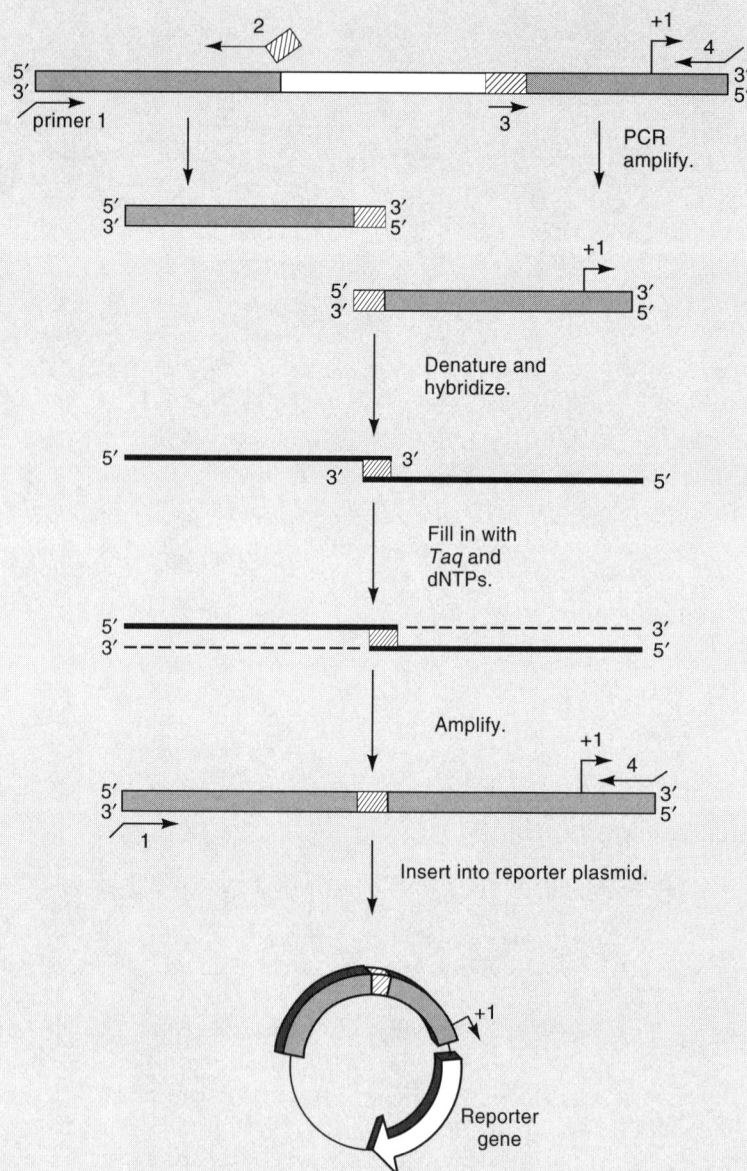

FIGURE 7.13. Generation of internal deletion mutants by PCR.

Although *Taq* DNA polymerase has been used extensively, it is prone to misincorporation at a rate approximating 1 mutation per 10^5 bp. *Pfu* DNA polymerase from *Pyrococcus furiosa* has a 12-fold higher fidelity, minimizing the chance of mutations in long DNA sequences. Nevertheless, sequencing of the constructed mutants to rule out the possibility that unwanted mutations have been introduced is strongly recommended. The availability of a back-up clone from a parallel PCR assay can prevent the unnecessary delay of having to repeat the entire PCR and cloning process. An excellent overview of this and other PCR mutagenesis methods is available in volume 57 of *Methods in Molecular Biology* (Harlow et al. 1996).

Box 7.4

Linker Scanning Mutagenesis by PCR

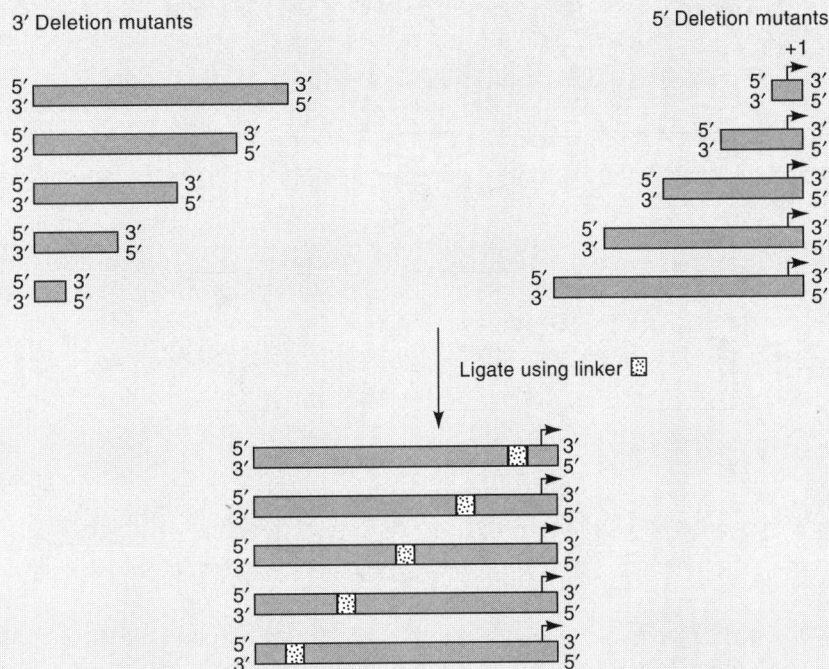

FIGURE 7.14. Classical linker-scanner mutagenesis method.

One essential aspect of control region mapping is to identify individual sequence elements that mediate activity. Although a broad deletion analysis can reveal the boundaries of a regulatory region, it is important to scan the sequences systematically within those boundaries for functional elements using substitution mutations. A typical goal is to generate a series of mutants, in which each mutation alters a small cluster of 5–10 bp (see Fig. 7.5). The linker scanning approach, developed in 1982 by Steve McKnight, was the first developed to generate a series of clustered substitution mutants. In the original method, 5′ and 3′ nested sets of deletions were combined via an oligonucleotide linker in such a way as to replace the 9-bp wild-type sequence with the linker (Fig. 7.14) (McKnight and Kingsbury 1982). Thus, two members from a set are chosen that, when joined, give rise to the complete wild-type control region minus 9 bp. The two deletions are then joined by a linker, which generates the clustered 9-bp mutation. By systematically placing such linkers throughout the control region, one can scan the entire region

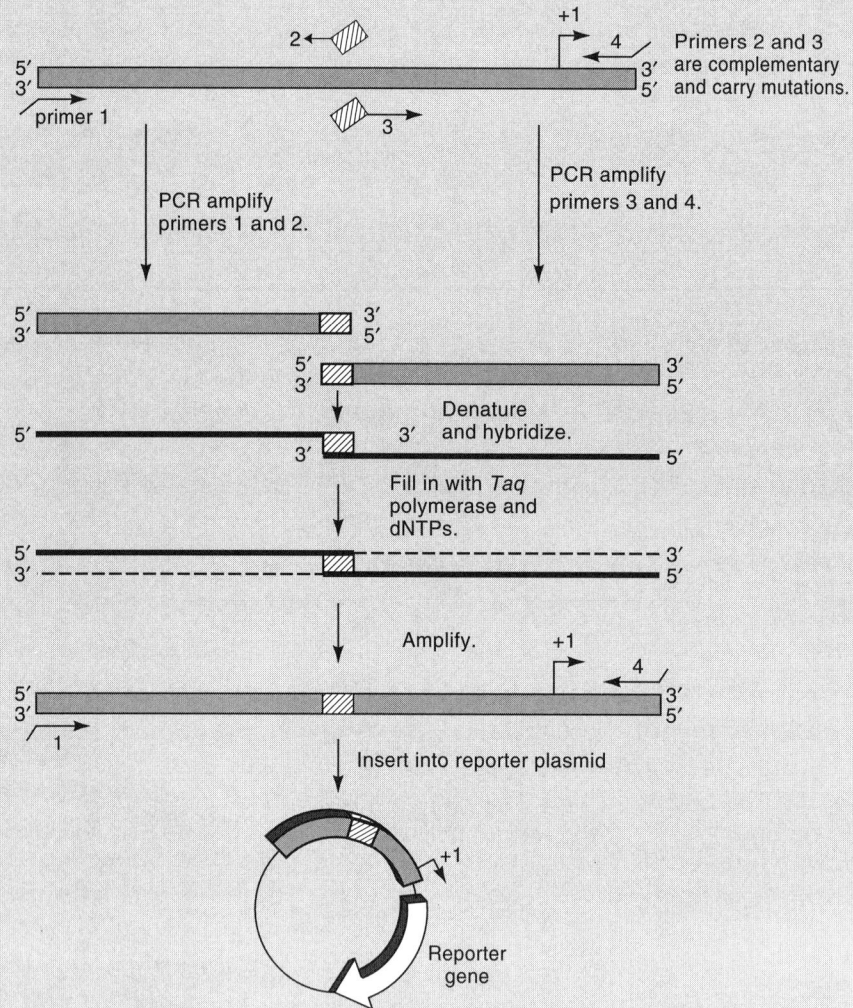

FIGURE 7.15. Generation of substitution mutants by PCR.

for functional sequence elements without altering the relative positioning or helical phasing relationships of the remaining regulatory sites (Fig. 7.14). A variation of this method was used to generate the substitution mutants of the IL-12 p40 promoter (see Fig. 7.4).

Today, a recommended method for generating substitution mutants employs PCR ligation or overlap extension to generate the mismatch (Fig. 7.15) (Ho et al. 1989; Harlow et al. 1996). The method is essentially as described above for generating internal deletions (Box 7.3). The ends of the internal primers are designed to hybridize 10 bp apart but to include 10-bp nucleotide complementary tails containing the desired mutant sequence (if the goal is to introduce a 10-bp mutation). The mutant is synthesized in two steps. In step 1, the internal primers and the two flanking primers are used to generate two PCR fragments, which are purified, mixed, and PCR-ligated as described above (Box 7.3). Although 10-bp mutations were used as the example, an advantage of this approach is that mutations of any size, from one to dozens of base pairs, can be generated with appropriate primers. The cost of primers can be excessive for an extensive mutagenesis, but the time commitment is often considerably less than in the traditional approach.

Box 7.5

Oligonucleotide-directed Mutagenesis

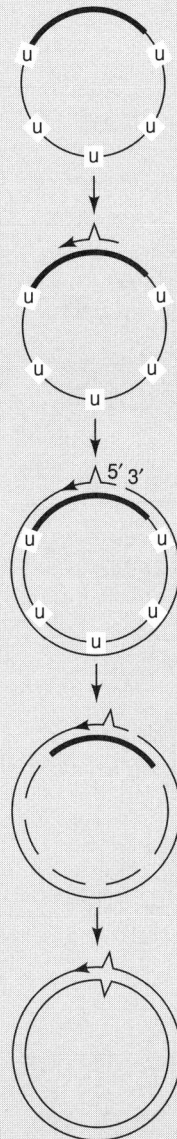

1. Isolate single-stranded phagemid DNA from dut⁻/ung⁻ strain of *E. coli*.

2. Anneal primer with mismatch.

3. Extend primer with DNA polymerase and dNTPs.

4. Transform wild-type *E. coli* strain with plasmid. Uracil-N-glycosylase (Ung) degrades uracil-containing strand.

5. Replicate mutagenized DNA. Screen for positive colonies by digestion with restriction enzyme.

FIGURE 7.16. Oligonucleotide-directed mutagenesis.

The methodology for oligonucleotide-directed mutagenesis was originally developed by Michael Smith (1985). In the simplest variation, an oligonucleotide incorporating a site-directed mutation is hybridized to a wild-type control region, generally in the form of a single-stranded phagemid DNA. The oligonucleotide is then elongated to generate a double-stranded circle containing a mismatch. The double-stranded plasmid is then treated with DNA ligase and introduced into *E. coli*. Individual colonies are screened to identify those that contain the mutation. Typically, there is considerable background because the polymerase often reads through the 3′ end of the mutagenic oligonucleotide, displacing it to create a wild-type second strand. Furthermore, both the mutant and wild-type strands can replicate in *E. coli*, and the unmethylated mutant strand is often preferentially repaired by the mismatch repair system. Several methods have been developed that minimize background, including the use of dut/ung selection and the use of a second primer, which restores the coding sequence of an antibiotic marker in the mutagenized DNA.

In the dut/ung approach, a single-stranded phagemid containing the control region is grown up in a dut⁻/ung⁻ strain of *E. coli*. dut encodes dUTPase, which degrades intracellular pools of dUTP (Fig. 7.16) (Kunkel et al. 1987). In its absence, cellular pools of dUTP become so high that U residues become incorporated into the DNA. ung is uracil *N*-glycosylase, which degrades DNA containing U residues. In its absence, the U is not removed from the phagemid DNA. The single-stranded DNA produced in a dut/ung strain contains many U residues incorporated in place of Ts. The single-stranded DNA is then hybridized to a primer bearing a mismatch at the position to be mutagenized. The mismatch is usually placed in the center of the primer. The hybridized primer is extended, in vitro, with a DNA polymerase and dNTPs. DNA polymerase replicates through the U's, incorporating A at the position with high fidelity. The double-stranded DNA is then introduced into a wild-type strain of *E. coli* bearing dUTPase and uracil *N*-glycosylase. ung preferentially degrades the original U-rich strand bearing the wild-type control region. The other strand bearing the mutation is then preferentially replicated. DNA is isolated and sequenced to confirm the mutation. Although one primer is generally used, multiple primers can be employed simultaneously. In such a case, it has been found that molecules which have successfully incorporated the sequence of one primer often incorporate the sequence of another primer on the same DNA molecule. Thus, different mutations can be introduced simultaneously, or the phagemid can be designed so that one primer incorporates a mutation while another rescues an antibiotic-resistance phenotype (Lewis and Thompson 1990) on the phagemid so it can be selected in a subsequent transformation.

Recently, site-directed mutagenesis strategies similar to those described above have been developed that employ double-stranded DNA templates. Stratagene markets two kits that make use of double-stranded templates: the Chameleon Double-Stranded, Site-Ditected Mutagenesis Kit (#200509) and the ExSite PCR-based Site-Directed Mutagenesis Kit (#200502). Promega markets the Gene Editor In Vitro Site-Directed Mutagenesis System (#Q9280) and the Altered Sites II In Vitro Mutagenesis System (#Q6080).

Box 7.6

Site-Directed Mutagenesis by Linear Amplification

Several different PCR methods can be used to introduce substitution mutations into DNA. (Weiner et al. 1994). In one recommended approach, PCR ligation or overlap extension is used to generate mutants differing in the sequence at the junction of two amplified fragments (Boxes 7.3 and 7.4). A second, newer approach is also highly recommended. This approach, developed by scientists at Stratagene, uses full plasmid linear amplification (QuikChange Site-Directed Mutagenesis Kit, cat. #200518) to generate substitution mutants. A detailed description and protocol are provided in Chapter 12 because this method has proven to be particularly useful for introducing small substitution mutations into the cDNAs for transcription factors for the purpose of detailed structure–function studies.

REFERENCES

Annweiler A., Muller U., and Wirth T. 1992. Functional analysis of defined mutations in the immunoglobulin heavy-chain enhancer in transgenic mice. *Nucleic Acids Res.* **20:** 1503–1509.

Calame K. and Ghosh S. 1995. Regulation of immunoglobulin gene transcription. In *Immunoglobulin genes* (ed. Honjo T. and Alt F.W.). Academic Press, San Diego, California; London, United Kingdom; New York, New York.

Costa G.L., Bauer J.C., McGowan B., Angert M., and Weiner M.P. 1996. Site-directed mutagenesis using a rapid PCR-based method. *Methods Mol. Biol.* **57:** 239–248.

Dang W., Nikolajczyk B.S., and Sen R. 1998a. Exploring functional redundancy in the immunoglobulin μ heavy-chain gene enhancer. *Mol. Cell. Biol.* **18:** 6870–6878.

Dang W., Sun X.H., and Sen R. 1998b. ETS-mediated cooperation between basic helix-loop-helix motifs of the immunoglobulin μ heavy-chain gene enhancer. *Mol. Cell. Biol.* **18:** 1477–1488.

Eisenberg S.P., Coen D.M., and McKnight S.L. 1985. Promoter domains required for expression of plasmid-borne copies of the herpes simplex virus thymidine kinase gene in virus-infected mouse fibroblasts and microinjected frog oocytes. *Mol. Cell. Biol.* **5:** 1940–1947.

Ephrussi A., Church G.M., Tonegawa S., and Gilbert W. 1985. B lineage-specific interactions of an immunoglobulin enhancer with cellular factors in vivo. *Science* **227:** 134–140.

Erman B. and Sen R. 1996. Context dependent transactivation domains activate the immunoglobulin mu heavy chain gene enhancer. *EMBO J.* **15:** 4565–4575.

Erman B., Cortes M., Nikolajczyk B.S., Speck N.A., and Sen R. 1998. ETS-core binding factor: A common composite motif in antigen receptor gene enhancers. *Mol. Cell. Biol.* **18:** 1322–1330.

Ernst P. and Small S.T. 1995. Combinatorial regulation of transcription II: The immunoglobulin μ heavy chain gene. *Immunity* **2:** 427–438.

Fernandez L.A., Winkler M., Forrester W., Jenuwein T., and Grosschedl R. 1998. Nuclear matrix attachments regions confer long-range function upon the immunoglobulin μ enhancer. *Cold Spring Laboratory Symp. Quant. Biol.* **63:** 151–524.

Forrester W.C., van Genderen C., Jenuwein T., and Grosschedl R. 1994. Dependence of enhancer-mediated transcription of the immunoglobulin μ gene on nuclear matrix attachment regions. *Science* **265:** 1221–1225.

Gray H.B. Jr., Ostrander D.A., Hodnett J.L., Legerski R.J., and Robberson D.L. 1975. Extracellular nucleases of *Pseudomonas* BAL 31. I. Characterization of single strand-specific deoxyriboendonuclease and double-strand deoxyriboexonuclease activities. *Nucleic Acids Res.* **2:** 1459–1492.

Harlow P.P., Hobson G.M., and Benfield P.A. 1996. Construction of linker-scanning mutations using PCR. *Methods Mol. Biol.* **57:** 287–295.

Ho S.N., Hunt H.D., Horton R.M., Pullen J.K., and Pease L.R. 1989. Site-directed mutagenesis by overlap extension using the polymerase chain reaction [see comments]. *Gene* **77:** 51–59.

Jenuwein T. and Grosschedl R. 1991. Complex pattern of immunoglobulin μ gene expression in normal and transgenic mice: Nonoverlapping regulatory sequences govern distinct tissue specificities. *Genes Dev.* **5:** 932–943.

Kiledjian M., Su L.-K., and Kadesch T. 1988. Identification and characterization of two functional domains within the murine heavy-chain enhancer. *Mol. Cell. Biol.* **8:** 145–152.

Kunkel T.A., Roberts J.D., and Zakour R.A. 1987. Rapid and efficient site-specific mutagenesis without phenotypic selection. *Methods Enzymol.* **154:** 367–382.

Lenardo M., Pierce J.W., and Baltimore D. 1987. Protein-binding sites in Ig gene enhancers determine transcriptional activity and inducibility. *Science* **236:** 1573–1577.

Lewis M.K. and Thompson D.V. 1990. Efficient site directed in vitro mutagenesis using ampicillin selection. *Nucleic Acids Res.* **18:** 3439–3443.

Ma X., Chow J.M., Gri G., Carra G., Gerosa F., Wolf S.F., Dzialo R., and Trinchieri G. 1996. The interleukin 12 p40 gene promoter is primed by interferon gamma in monocytic cells. *J. Exp. Med.* **183:** 147–157.

McKnight S.L. and Kingsbury R. 1982. Transcriptional control signals of a eukaryotic protein-coding gene. *Science* **217:** 316–324.

Mullis K.B. and Faloona F.A. 1987. Specific synthesis of DNA in vitro via a polymerase-catalyzed chain reaction. *Methods Enzymol.* **155:** 335–350.

Murphy T.L., Cleveland M.G., Kulesza P., Magram J., and Murphy K.M. 1995. Regulation of interleukin 12 p40 expression through an NF-κB half-site. *Mol. Cell. Biol.* **15:** 5258–5267.

Myers R.M., Tilly K., and Maniatis T. 1986. Fine structure genetic analysis of a β-globin promoter. *Science* **232:** 613–618.

Nelsen B. and Sen R. 1992. Regulation of immunoglobulin gene transcription. *Int. Rev. Cytol.* **133:** 121–149.

Nelsen B., Kadesch T., and Sen R. 1990. Complex regulation of the immunoglobulin μ heavy-chain gene enhancer: μB, a new determinant of enhancer function. *Mol. Cell. Biol.* **10:** 3145–3154.

Nelsen B., Tian G., Erman B., Gregoire J., Maki R., Graves B., and Sen R. 1993. Regulation of lymphoid-specific immunoglobulin μ heavy chain gene enhancer by ETS-domain proteins. *Science* **261:** 82–86.

Nikolajczyk B.S., Nelsen B., and Sen R. 1996. Precise alignment of sites required for μ enhancer activation in B cells. *Mol. Cell. Biol.* **16:** 4544–4554.

Nikolajczyk B.S., Cortes M., Feinman R., and Sen R. 1997. Combinatorial determinants of tissue-specific transcription in B cells and macrophages. *Mol. Cell. Biol.* **17:** 3527–3535.

Perez-Mutul J., Macchi M., and Wasylyk B. 1988. Mutational analysis of the contribution of sequence motifs within the IgH enhancer to tissue specific transcriptional activation. *Nucleic Acids Res.* **16:** 6085–6096.

Peterson C.L., Orth K., and Calame K.L. 1996. Binding in vitro of multiple cellular proteins to immunoglobulin heavy-chain enhancer DNA. *Mol. Cell. Biol.* **6:** 4168–4178.

Plevy S.E., Gemberling J.H., Hsu S., Dorner A.J., and Smale S.T. 1997. Multiple control elements mediate activation of the murine and human interleukin 12 p40 promoters: Evidence of functional synergy between C/EBP and Rel proteins. *Mol. Cell. Biol.* **17:** 4572–4588.

Rao E., Dang W., Tian G., and Sen R. 1997. A three-protein-DNA complex on a B cell-specific domain of the immunoglobulin μ heavy chain gene enhancer. *J. Biol. Chem.* **272:** 6722–6732.

Rogers S.G. and Weiss B. 1980. Exonuclease III of *Escherichia coli* K-12, an AP endonuclease. *Methods Enzymol.* **65:** 201–211.

Sen R. and Baltimore D. 1986. Multiple nuclear factors interact with the immunoglobulin enhancer sequences. *Cell* **46:** 705–716.

Shortle D., DiMaio D., and Nathans D. 1981. Directed mutagenesis. *Annu. Rev. Genet.* **15:** 265–294.

Singh H., Sen R., Baltimore D., and Sharp P.A. 1986. A nuclear factor that binds to a conserved sequence motif in transcriptional control elements of immunoglobulin genes. *Nature* **319:** 154–158.

Smith M. 1985. In vitro mutagenesis. *Annu. Rev. Genet.* **19:** 423–462.

Staudt L.M. and Lenardo M.J. 1991. Immunoglobulin gene transcription. *Annu. Rev. Immunol.* **9:** 373–398.

Thanos D. and Maniatis T. 1992. The high mobility group protein HMG I(Y) is required for NF-κB-dependent virus induction of the human IFN-β gene. *Cell* **71:** 777–789.

Tomic-Canic M., Bernerd F., and Blumenberg M. 1996. A simple method to introduce internal deletions or mutations into any position of a target DNA sequence. *Methods Mol. Biol.* **57:** 249–257.

Tsao B., Wang X.-F., Peterson C., and Calame K. 1988. In vivo functional analysis of in vitro protein binding sites in the immunoglobulin heavy chain enhancer. *Nucleic Acids Res.* **16:** 3239–3253.

Wedel A. and Ziegler-Heitbrock H.W. 1995. The C/EBP family of transcription factors. *Immunobiology* **193:** 171–185.

Wei C.F., Alianell G.A., Bencen G.H., and Gray H.B. Jr. 1983. Isolation and comparison of two molecular species of the BAL 31 nuclease from *Alteromonas espejiana* with distinct kinetic properties. *J. Biol. Chem.* **258:** 13506–13512.

Weiner M.P., Costa G.L., Schoettlin W., Cline J., Mathur E., and Bauer J.C. 1994. Site-directed mutagenesis of double-stranded DNA by the polymerase chain reaction. *Gene* **151:** 119–123.

Wirth T., Staudt L., and Baltimore D. 1987. An octamer oligonucleotide upstream of a TATA motif is sufficient for lymphoid-specific promoter activity. *Nature* **329:** 174–178.

CHAPTER 8

Identification of DNA-binding Proteins and Isolation of Their Genes

Important issues

- *DNA-binding proteins are most commonly identified by electrophoretic mobility shift assay or DNase I footprinting; each method has distinct advantages and limitations.*

- *Database methods can help to identify proteins that bind a DNA element.*

- *To isolate the gene encoding a protein detected by EMSA or DNase footprinting, the protein often needs to be purified.*

- *Other methods, including the one-hybrid screen, in vitro expression library screen, mammalian expression cloning methods, degenerate PCR, and database methods, are attractive alternatives for cloning genes encoding DNA-binding proteins.*

INTRODUCTION

The previous chapter described methods for delineating important *cis*-acting DNA sequence elements within a control region of interest. After this step has been completed, DNA-binding proteins that interact with the DNA elements can be identified and, ultimately, cloned and characterized.

There are three recommended strategies for identifying DNA-binding proteins. To enhance the probability that the functionally relevant protein will be identified, all three strategies can be pursued simultaneously. The first begins with a database search to predict which known proteins and protein families might be capable of binding a particular DNA sequence (Fig. 8.1). If candidate proteins are identified, a protein–DNA interaction assay, such as an electrophoretic mobility shift assay (EMSA) or DNase I footprinting assay, can be performed with the predicted recombinant or in-vitro-translated proteins to determine whether they are indeed capable of binding the element.

The second strategy begins with an analysis of extracts from appropriate cells by EMSA and DNase I footprinting (Fig. 8.1, strategy 2). Binding assays can be performed with extracts from different tissues and with mutant DNA probes to determine whether the activities have the potential to be functionally relevant in vivo (i.e., responsible for the in vivo function of the endogenous control element). If viable DNA–protein interactions are identified, antibodies can be used to determine whether they require known proteins. The antibodies used are generally directed against proteins predicted to bind the site on the basis of a database search. If the protein that binds the element in an EMSA or DNase I footprinting assay cannot be established as a known protein using antibodies, the gene encoding the protein can be cloned. To clone the gene, the protein can be purified. Following purification, a partial amino acid sequence must be determined and used to predict the coding sequence, allowing a gene fragment to be isolated by PCR. The gene fragment can be used to isolate a full-length cDNA, which in turn can be used to prepare recombinant protein. Alternative strategies for cloning genes encoding DNA-binding proteins include the yeast one-hybrid screen, in vitro expression library screen, mammalian expression cloning methods, degenerate PCR, and genome database methods.

The third strategy for identifying proteins that bind a defined control element is to clone genes encoding specific DNA-binding proteins directly in the absence of preliminary EMSA or DNase I footprinting experiments (Fig. 8.1, strategy 3). The yeast one-hybrid screen is often the preferred strategy because of its speed and simplicity. However, any of the other strategies described above, with the exception of protein purification, can be used for direct cloning. Thus, it may be possible to identify and clone candidate binding proteins in one step.

These three general strategies are the topic of this chapter. The advantages and limitations of each strategy are described. Detailed information is also provided about the development of a basic protein–DNA interaction assay using crude cell extracts. Furthermore, we discuss several methods that can be used to isolate a gene encoding a DNA-binding protein that has been detected, or for the direct isolation of a gene encoding a protein that interacts with a DNA element of interest. Protocols for each of these cloning methods are not included because of space limitations, but appropriate sources are cited.

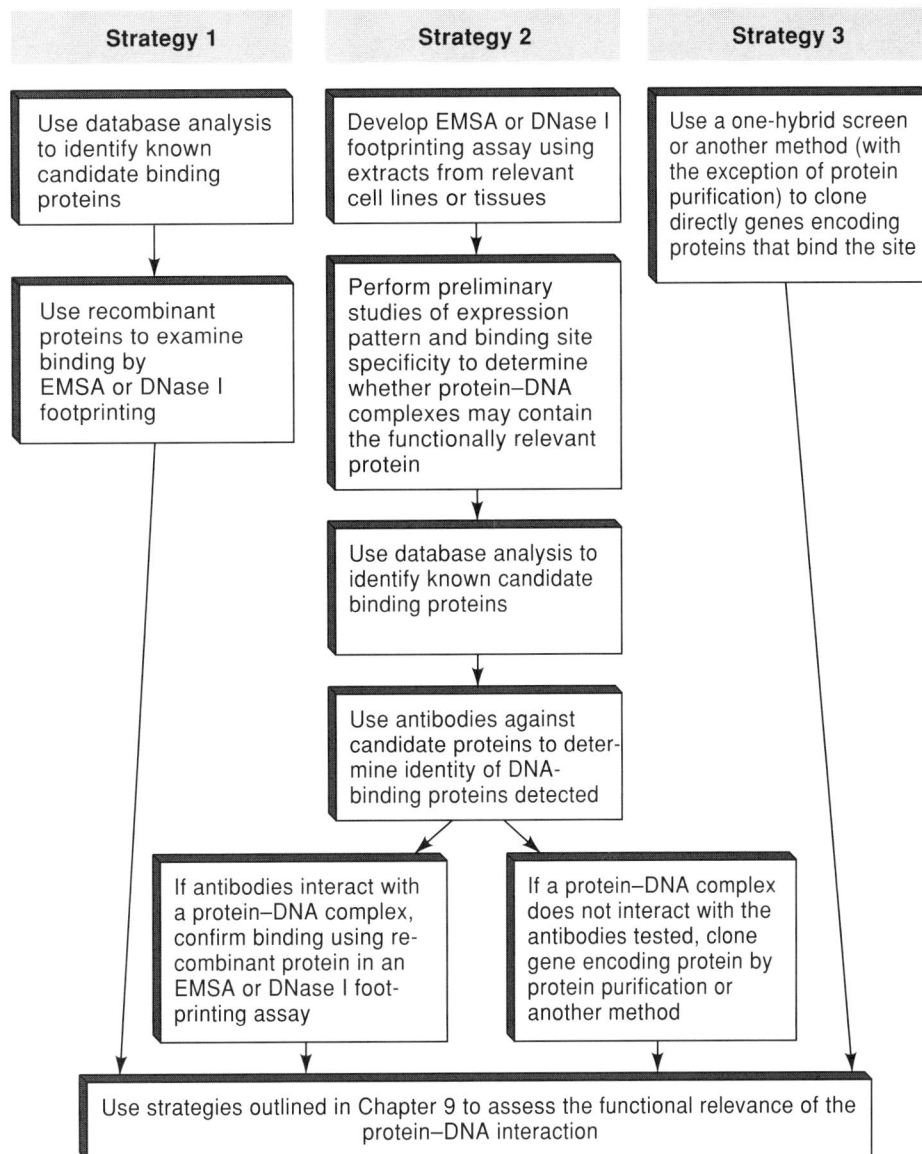

FIGURE 8.1. Recommended strategy for characterizing proteins that bind to a defined control element.

It is important to stress that each of the three strategies discussed in this chapter may lead to the identification of one or more proteins that bind a DNA element of interest. However, none of the strategies will necessarily lead to the protein that is the functionally relevant regulator of the control element in the context of the endogenous locus. Regardless of how a DNA-binding protein is isolated and cloned, it merely becomes a candidate regulator of the gene of interest. The relevance of each candidate will need to be evaluated using the strategies presented in Chapter 9.

CONCEPTS AND STRATEGIES FOR THE IDENTIFICATION OF DNA-BINDING PROTEINS

Database Methods

In Chapters 1 and 7, we describe database methods as a means of identifying important DNA sequence elements within a control region of interest. As discussed, an appropriate starting point is the TRANSFAC database (http://transfac.gbf.de/index.html). These methods can predict which known proteins interact with the control region, but they do not substitute for a comprehensive mutant analysis because of the strong probability that critical DNA elements will be missed and that some of the control elements identified will be functionally irrelevant. Although database methods have limited utility when a control region is first characterized, they can be extremely valuable after important DNA elements within a control region have been defined. If a database search provides information about the specific protein or family of proteins that binds an element, considerable time and effort can be saved. Experiments can immediately be initiated to determine whether the protein revealed by the database search is a relevant regulator (see Chapter 9). If the protein is not relevant, another member of the same protein family may be relevant, since other family members may bind the same sequence. To address this latter possibility, experiments could be initiated to identify or clone the other family members (see degenerate PCR and genome database methods below). The key disadvantage of the database strategy is, of course, the substantial probability that it will fail to identify the relevant proteins, either because the protein is novel and not represented in the database, or because the recognition sequence information in the database is incomplete or inaccurate. When the complete human and murine genome sequences become available, the situation will be quite different. Database information will yield all of the members of a protein family of interest. However, recognition sequence information will remain incomplete for several years. Furthermore, the strategies described in Chapter 9 will remain essential to determine which protein family and which member of that family is functionally relevant.

To perform a database search with an individual control element, the precise DNA sequence that is known to be functionally important should be analyzed (see Chapters 1 and 7). If the mutant analysis was sufficiently detailed, the sequence used will be 10–15 bp long. With a sequence of this length, the search can be performed at a low stringency to identify as many known proteins as possible that may be capable of binding the element. The reduced stringency can be critical for a successful search, because only a small subset of relevant binding sites for a protein will precisely match its reported consensus recognition sequence. It is important to note that the consensus sequences used in the database programs are often derived from a simple comparison of several known binding sites for a protein, or from a binding site selection analysis. The former approach is usually inaccurate because only a small subset of binding sites are compared and because relative affinities are rarely taken into account. The binding site selection approach is also incomplete because it is biased toward the highest-affinity sites and often overlooks recognition sequences of slightly lower affinity that may be functionally relevant.

If candidate DNA-binding proteins are revealed by the database search, the ability of the predicted proteins to bind the DNA element in vitro can usually be determined with little difficulty. In most cases, a cDNA or expression plasmid for the protein can be obtained from an appropriate commercial or academic source. If a cDNA is not available, it can be isolated by reverse transcription-polymerase chain reaction (RT-PCR) or by screening a cDNA library. The recombinant protein can then be expressed and purified as described in Chapter 11, and tested for its ability to bind the DNA element by EMSA or

DNase I footprinting (see Chapter 13). Alternatively, the protein can be produced by in vitro transcription–translation (Chapter 11; see also Ausubel et al. 1994, Unit 10.17). If the protein is indeed capable of binding the DNA element in vitro, the strategies described in Chapter 9 can be considered to assess the in vivo relevance of the protein.

Development of a Protein–DNA Interaction Assay for Crude Cell Lysates

To complement the database strategy, an in vitro assay capable of detecting DNA-binding activities in crude extracts can be developed as a starting point toward the second strategy outlined in Figure 8.1. A primary advantage of this strategy is that it has the potential to identify either known or novel DNA-binding proteins. A second advantage is that preliminary experiments to test the functional relevance of a protein–DNA interaction that is detected can be performed (see Fig. 8.1 and Chapter 9.) If the results of those experiments demonstrate that the protein is not a viable candidate, further analysis of the protein can be avoided. The primary disadvantage of this strategy is that the functionally relevant DNA-binding protein may be undetectable in crude cell lysates analyzed by EMSA and DNase I footprinting (see below).

Standard Methods for Detecting Protein–DNA Interactions

Several methods have been used for the in vitro detection and characterization of protein–DNA interactions, including EMSA, DNase I footprinting, exonuclease III (*Exo*III) footprinting, Southwestern blotting, various chemical protection and interference assays, and UV crosslinking. However, EMSA and DNase I footprinting are by far the most frequently used assays, largely because they are the most straightforward and have proven to be the most successful for detecting and characterizing specific protein–DNA interactions at an early stage of an analysis. Therefore, we largely limit our discussion in this chapter to these two techniques. Most of the other techniques that can be used to identify novel protein–DNA interactions are described in Chapter 13.

In a sense, the EMSA and DNase I footprinting assays are complementary to one another, and thus, both assays should be used during the early stages of an analysis. Because of their different characteristics, they may lead to the identification of different proteins that interact with the same DNA element. All proteins identified must initially be considered as candidates for the functional regulator.

Basic EMSA and DNase I footprinting strategies. The EMSA and DNase I footprinting assays are described in detail in Chapter 13 from the perspective of an investigator interested in characterizing a purified recombinant protein. Briefly, the EMSA is based on the principle that a protein–DNA complex migrates through a native gel more slowly than the free DNA, with the mobility of the protein–DNA complex determined primarily by the size, shape, charge, and multimeric state of the protein (see Chapter 13, Protocol 13.5). Thus, proteins within a crude cell extract that specifically recognize a given control element can be identified by first incubating a small radiolabeled DNA fragment (i.e., probe) with the extract to allow the formation of protein–DNA complexes; the mixture is then applied to a native polyacrylamide gel, which upon electrophoresis will separate the free radiolabeled probe molecules from the molecules bound by proteins. The free and bound DNA molecules are detected by autoradiography or phosphorimager analysis.

For the DNase I footprinting assay, a radiolabeled DNA probe is first incubated with a cell extract to allow the formation of protein–DNA complexes (see Box 13.3, Protocol 13.1). An important difference between this binding reaction and that used for the EMSA is that the double-stranded, radiolabeled DNA probe used for footprinting must be labeled on only one strand and on only one end. (In contrast, for the EMSA, the locations and number of radiolabeled nucleotides are unimportant.) The solution containing probe and extract is then treated with a limiting concentration of DNase I for a short, defined period of time. The DNase I treatment parameters are chosen so that each probe molecule is cleaved, on average, only once. Because cleavage is largely random, it will generate a nested set of DNA molecules varying in length from mono- and di-nucleotides to full-length molecules that have not been cleaved. The DNA fragments are analyzed by denaturing polyacrylamide gel electrophoresis (PAGE) followed by autoradiography or phosphorimager analysis. If appropriate DNase I digestion parameters were employed, the sample lacking the cell extract will yield a ladder of bands extending from the bottom to the top of the image. If a protein in the extract is capable of binding a specific DNA sequence found on the probe, it will protect nucleotides within that region from digestion. On the resulting image, the protection will appear as a region devoid of bands or a region containing very weak bands relative to those observed in the absence of extract. The protected region is referred to as a DNase I footprint. DNA sequencing markers can be electrophoresed adjacent to the footprinting reactions, allowing a determination of the nucleotides protected by the protein.

Differential detection of proteins by EMSA and DNase I footprinting. Although many proteins can be detected with both assays, some DNA-binding proteins can be detected with only one. For most of the examples in which a protein is detected by only one of the two assays, the precise reason has not been determined. However, a summary of the key differences between the two assays will provide insight into the possible explanations (Table 8.1).

First and foremost among the differences is that the EMSA is more sensitive and can reveal a specific protein–DNA complex even when the protein is at a low concentration within the extract. The reason for this sensitivity is that a protein bound to only one or a few percent of the probe molecules will result in a complex that migrates more slowly than the free probe; if the specific activity of the probe is sufficiently high, this complex can easily be detected by autoradiography or phosphorimager analysis. In contrast, to detect a protein–DNA complex by DNase I footprinting, a majority of the probe molecules must be bound by the protein. Efficient binding is needed because the basic footprinting procedure does not involve a step that separates the bound and free probe molecules. On the gel image, bands must be diminished to a considerable extent to yield a convincing footprint. If only 50% of the probe molecules are bound by protein, for example, the intensities of the protected bands will be diminished by only twofold relative to the bands observed in the absence of extract. This small difference is usually difficult to evaluate. Because of this difference in sensitivity, some proteins, primarily those present at low concentrations within an extract, may be detectable by EMSA, but not by DNase I footprinting.

A second key difference in the two assays is that DNase I footprinting is performed entirely in solution, whereas the EMSA requires maintenance of the protein–DNA interac-

TABLE 8.1. *Characteristic differences between the EMSA and DNase I footprinting assays*

EMSA	DNase I footprinting
Effective when protein is at low concentrations	Effective primarily when protein is at high concentrations
Protein–DNA interaction must survive gel electrophoresis	Binding must be stable only in solution
Large amount of nonspecific competitor can disrupt specific interactions	$MgCl_2$ and $CaCl_2$, which are required for DNase I activity, may disrupt specific interactions

tion during gel electrophoresis. Few proteins bind DNA with dissociation rates that are compatible with the typical 2- to 4-hour electrophoresis time period. For most proteins, the success of the EMSA relies on a process referred to as "caging," in which the gel matrix keeps the protein and DNA in close proximity to each other when they dissociate during electrophoresis, allowing them to reassociate rapidly and migrate through the gel as a complex. Although the caging process facilitates the analysis of many protein–DNA complexes, some protein–DNA complexes appear to be maintained poorly. Evidence that a complex does not tolerate gel electrophoresis is usually provided by the observation that a protein yields strong protection in a DNase I footprinting assay, with the same amount of protein and probe yielding a weak or undetectable EMSA complex.

A third difference between the EMSA and DNase I footprinting assays is noteworthy: the amount of nonspecific competitor DNA that is typically needed for the assay when analyzing a crude cell extract. In both assays, a competitor DNA, such as poly(dI:dC) or poly(dA:dT), is usually included in excess to prevent nonspecific and low-specificity DNA-binding proteins within the extract from binding the probe and obscuring the specific protein–DNA complexes. For DNase I footprinting reactions, 1 μg of poly(dI:dC) competitor can usually be used with 10–100 μg of crude cell extract. In contrast, 2–8 μg of poly(dI:dC) is often needed for EMSA reactions with 5–10 μg of crude extract. Relatively large concentrations of competitor are needed for EMSA reactions because modest amounts of nonspecific nucleic-acid-binding proteins can lead to a smear of radioactive probe on the gel image that prevents detection of the desired complexes. With the DNase I footprinting assay, the nonspecific binding proteins are distributed throughout the probe and are therefore less likely to affect significantly the detection of specific protein–DNA complexes. This difference can influence the degree of success of each assay if the DNA-binding protein of interest possesses a moderate affinity for the competitor DNA; the high concentrations of competitor used for EMSA reactions may compete with the probe for binding of the protein of interest, diminishing or abolishing the specific protein–DNA complex. A fourth difference is that the DNase I footprinting assay requires the addition of $MgCl_2$ and $CaCl_2$ for DNase I activity. Although most protein–DNA interactions are unaffected or even enhanced by these ions, some interactions may be weakened.

In light of the differences between the two assays described above, it should not be surprising that the activities of some DNA-binding proteins are more easily detected with one of the two assays. If a protein is detected only by EMSA, it usually is not sufficiently concentrated to bind the high percentage of probe molecules required for a DNase I footprint. NF-κB is one example of a protein that is much easier to detect by EMSA than by DNase I footprinting because of its relatively low concentration in nuclear extracts. Sp1 and Ikaros are examples of proteins that are more easily detected by DNase I footprinting. Both proteins can be detected by EMSA when using cell extracts, but they yield much stronger footprints. In fact, both the Sp1 and Ikaros binding activities were originally detected by DNase I footprinting and were difficult to detect by EMSA until the binding and electrophoresis conditions were later optimized (Dynan et al. 1983; Lo et al. 1991; Hahm et al. 1994). To reiterate, the use of both assays is strongly recommended during initial attempts to detect a DNA-binding protein. The use of both assays greatly enhances the probability that the relevant protein will be detected.

Information provided by EMSA and DNase I footprinting. When developing an assay for the initial analysis of DNA-binding activities, one should be aware of the unique information that it provides (Fig. 8.2). The EMSA, for example, provides little information about the location of a protein's recognition sequence on the probe, whereas DNase I footprinting reveals the approximate location of the recognition site. In addition, if multiple

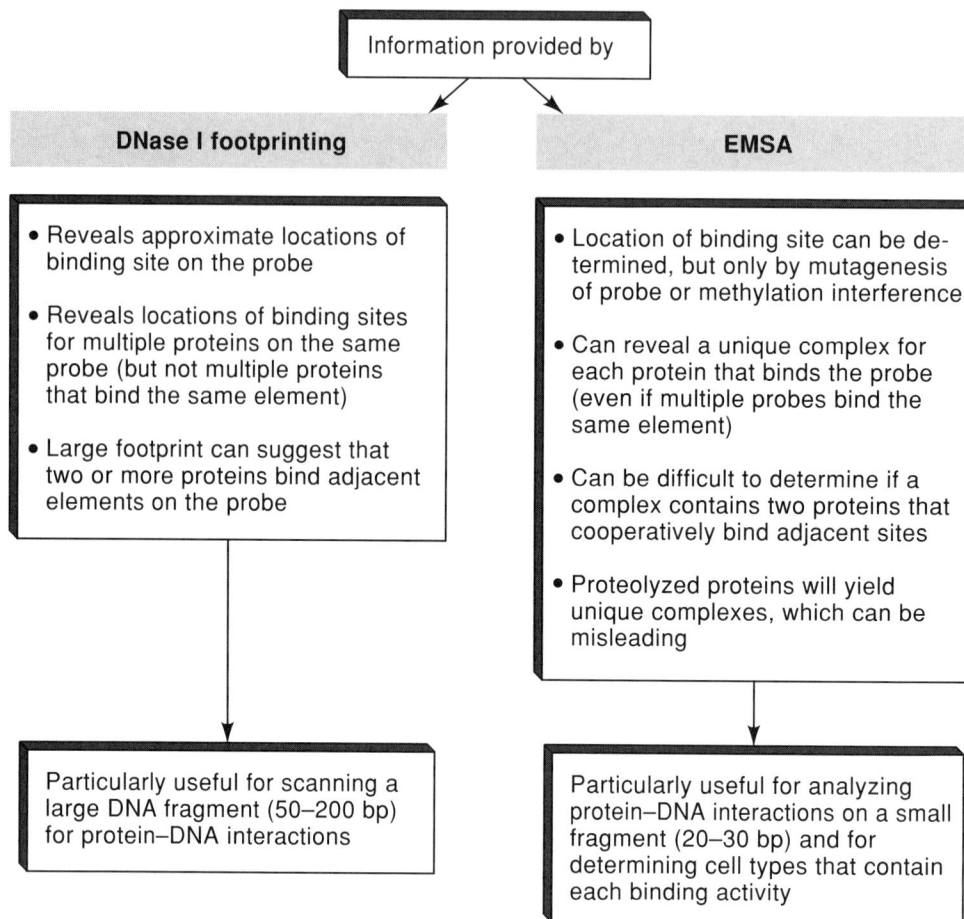

FIGURE 8.2. Information provided by DNase I footprinting and EMSA.

proteins bind the probe, the DNase I footprinting assay, but not the EMSA, will reveal the approximate locations of each binding site. A related limitation of the EMSA is that, in the initial experiments, it usually cannot distinguish between a complex containing a protein bound to one DNA element and a complex containing proteins bound cooperatively to two elements on the probe. To address these issues, a mutant analysis of the probe or a methylation interference assay (see below and Chapter 13) must be performed in conjunction with the EMSA. A final limitation of the EMSA is that partially proteolyzed proteins can yield protein–DNA complexes with unique migrations. If two or more cell types are being compared to determine the relative abundance of each complex that is observed, differential proteolysis in the various extracts can yield misleading results.

The practical outcome of these limitations of the EMSA is that the DNase I footprinting assay is often more informative when using one or two probes to scan a relatively large (e.g., 200–300 bp) control region for protein–DNA interactions. If protein–DNA interactions are observed in the footprinting assay, one can immediately determine the location of each interaction and will therefore know whether it occurs in the vicinity of important DNA elements (which presumably were defined by mutagenesis as described in Chapter 7). Experiments can then be initiated to characterize the proteins that bind the important elements.

Despite the limitations of the EMSA, it possesses unique features that make it useful for initial studies of protein–DNA interactions, particularly when short probes containing well-defined DNA elements are employed. Two unique features are its sensitivity, as described above, and its ability to distinguish proteins of different size and shape within an extract that bind the same DNA element. Because of this unique feature, the EMSA is particularly useful for determining whether one, but not all, of the proteins that bind an element is cell-type-specific or subject to induction by a relevant agent. The DNase I footprinting assay also can provide information about cell specificity and inducibility. However, because most DNA-binding proteins are members of multiprotein families, with several members of a family capable of recognizing a similar sequence, the DNase I footprinting assay might fail to reveal cell-type-specificity or inducibility of a particular family member (if, for example, another member of the same family binds the element in a different cell type).

Electrophoretic Mobility Shift Assay

As described above, the EMSA is a common and relatively straightforward method for studying protein–DNA interactions. This method is described in detail in Chapter 13 (Protocol 13.5) from the perspective of an analysis of a pure recombinant protein. Below, we focus on specific issues that must be considered when using EMSA experiments for the initial identification and characterization of proteins within crude extracts that bind a defined DNA element.

Radiolabeled probe design. The design of the radiolabeled EMSA probe can influence the quality of the results obtained. For many proteins detected in crude extracts, short, double-stranded oligonucleotides of approximately 20–25 bp yield consistent, high-quality results. For example, EMSA experiments to detect NF-κB complexes are often performed with 19-bp oligonucleotides (Pierce et al. 1988). An oligonucleotide probe should minimally contain 5–10 bp on each side of the functionally defined control element, because the adjacent sequences, although not specifically recognized by the protein, may be required for a stable protein–DNA interaction.

To prepare the double-stranded fragment, two single-stranded oligonucleotides can be synthesized and annealed (see Ausubel et al. 1994, Unit 6.4). Alternatively, one long oligonucleotide containing both complements of the recognition sequence can be prepared, allowing self-annealing with an intervening loop. Crude oligonucleotides are usually purified by gel electrophoresis or high-performance liquid chromatography (HPLC) to remove contaminants that can inhibit end-labeling or protein binding. The purified oligonucleotides are then annealed and radiolabeled either by 5′ end-labeling with [γ-^{32}P]ATP and T4 polynucleotide kinase (see Protocol 4.1), or by polymerization at a 5′ overhang using [α-^{32}P]nucleoside triphosphates and *E. coli* DNA polymerase Klenow fragment (Ausubel et al. 1994, Unit 3.5). The latter method has the advantage of labeling only those oligonucleotides that have annealed into a double-stranded molecule, as Klenow requires a template for nucleotide incorporation.

For some DNA-binding proteins, short oligonucleotides of 20–25 bp do not support stable protein–DNA interactions, making it necessary to use longer probes. The failure of short probes could result from the protein's requirement for a longer stretch of adjacent DNA for stable binding. Alternatively, short probes containing a particular recognition sequence might anneal improperly, dissociate during the binding reaction, or possess secondary structures that prevent binding. Several methods can be used to generate longer probes, one of which is simply to use longer synthetic oligonucleotides. Another method is to excise and radiolabel a restriction fragment from a plasmid containing the DNA element (Protocol 13.6).

PCR with radiolabeled primers provides an increasingly attractive method for preparing EMSA probes (Ausubel et al. 1994, Unit 15.2). One advantage of PCR is that wild-type and mutant probes can be prepared with similar specific activities if the same radiolabeled primer is used on the wild-type and mutant templates. If a series of mutations in a control element have been generated for a mutant analysis in a transfection assay, the same two PCR primers can be used to prepare probes containing the wild-type sequence and each mutant sequence for EMSA (see, e.g., Emami et al. 1997).

Extract preparation. For most studies of in vitro protein–DNA interactions, nuclear extracts prepared by the method of Dignam et al. (1983) are used. This method is described in detail in Chapter 14 (Protocol 14.1). (Variations of the Dignam method that are better suited for small quantities of cells have also been developed [see Protocol 14.1].) Briefly, cultured cells are swelled and then lysed in a hypotonic buffer by physical disruption with a Dounce homogenizer. An alternative method involves lysis with a low concentration of the nonionic detergent NP-40. With either physical or detergent lysis, the nuclei from most mammalian cell types remain reasonably intact and can be separated from the cytoplasmic proteins by low-speed centrifugation. To extract most DNA-binding proteins from the nuclei, the nuclear pellet is resuspended and stirred in a buffer containing a high concentration of NaCl or KCl. Dignam and Roeder determined empirically that a NaCl concentration of 0.42 M was optimal for the preparation of nuclear extracts that support in vitro transcription from the adenovirus major late promoter. This same salt concentration is used by most investigators to prepare nuclear extracts for the analysis of DNA-binding proteins. However, it may be beneficial to vary the salt concentration used for extraction of the nuclei when pursuing new DNA-binding activities. Lower salt concentrations may prevent inhibitors and nonspecific binding proteins from being extracted, increasing the success of the assay. Higher salt concentrations might be required for efficient extraction of a protein of interest.

If attempts to detect DNA-binding activities by EMSA fail, the nuclear extract can be further concentrated or fractionated to enhance the probability of success. The protein concentration of the extract can be increased by ammonium sulfate precipitation (see Chapter 14, Protocol 14.1) followed by extensive dialysis. A DNA-binding activity of interest may be detected more readily when a more concentrated extract is added to the binding reactions. Dignam and Roeder found that extract precipitation in the presence of 53% ammonium sulfate was optimal for enhancing the activity of in vitro transcription extracts, but a different percentage may be optimal for a particular DNA-binding protein.

A second strategy for improving the detection of DNA-binding activities is to fractionate the extract (see Chapter 14; Ausubel et al. 1994, Unit 10.10). The negatively charged affinity resins heparin agarose and heparin Sepharose are useful for this purpose because, if extracts are applied to these columns in a buffer containing 0.1 M KCl or NaCl, most DNA-binding proteins will be eluted with a 0.4 M salt step. A 1 M salt step elution should also be performed in case the protein binds the resin with an unusually high affinity. The fractionation procedure will separate the DNA-binding proteins from other proteins that might inhibit a binding reaction. Furthermore, the specific proteins within the 0.4 M and 1 M salt fractions may be more concentrated than in the crude extracts. Therefore, DNA-binding activities may be detected with these fractions that were not detected with the crude extract.

If all attempts to detect candidate binding activities fail using nuclear extracts, whole-cell extracts, cytoplasmic extracts, or other nuclear extract preparation methods can be tested (see Chapter 14 and Gorski et al. 1986; Shapiro et al.1988; Soeller et al. 1988). Some nuclear proteins leak into the cytoplasmic fraction during the preparation of a nuclear extract, leading to the possibility that DNA-binding activity may be more abundant in a cytoplasmic or whole-cell extract.

Competitor DNA and other components of binding reactions. The nonspecific competitor DNA added to the binding reaction can be critical for the success of an EMSA experiment, especially when a crude nuclear extract is used. As stated above, the competitor DNA prevents nonspecific and low-specificity nucleic-acid-binding proteins from stably interacting with the radiolabeled probe. In the absence of an appropriate nonspecific competitor, numerous proteins within the extract, including abundant RNA-binding proteins, will bind the probe, leading to a radioactive smear on the gel image. Increasing concentrations of the competitor sequester increasing numbers of protein molecules from the radiolabeled probe, leaving primarily those proteins that have the highest specific affinity for the probe relative to the unlabeled competitor. Although high competitor concentrations often enhance the quality and clarity of the results, they also increase the probability that the relevant DNA-binding protein will be sequestered from the labeled probe.

Poly(dI.dC) is the nonspecific competitor that is used most frequently. However, other competitors, including poly(dG.dC), poly(dA.dT), or sheared genomic DNA from *E. coli* or salmon sperm, may yield better, or at least different, results in some instances. If the protein of interest binds the competitor with a substantial affinity, it may not yield a detectable complex on the EMSA gel. Instead, complexes containing proteins that are not relevant to the function of the control element may be observed, if those proteins bind the competitor with a lower affinity. For this reason, it can be beneficial to test different competitors when attempting to identify candidates for the relevant regulator of a DNA element. If different protein–DNA complexes are observed, the strategies described in Chapter 9 can be followed for each complex to determine which one is relevant.

In addition to the competitor DNA, other components of the binding reactions can be varied to enhance detection of protein–DNA complexes. For example, the concentration of monovalent or divalent cations can be varied. Some protein–DNA complexes may benefit from inclusion of a low concentration (e.g., 0.01%) of the nonionic detergent NP-40 in the binding reaction. Other complexes may benefit from inclusion of polyvinyl alcohol (2%), which can increase the effective concentration of proteins in the extract. Additional recommendations for optimizing binding reactions for EMSA experiments can be found in Protocol 13.5 and Ausubel et al. (1994; Unit 12.2). Furthermore, an EMSA kit is available from Promega (cat. # E3050) and may be beneficial for an investigator performing this method for the first time.

Gel electrophoresis conditions. When developing an EMSA for a DNA element of interest, pilot experiments with different gel electrophoresis buffers are strongly recommended, as detection of some protein–DNA complexes depends on the specific buffer used. The three most common EMSA gel buffers are Tris-borate-EDTA (TBE), Tris-acetate-EDTA (TAE), and Tris-glycine. Inclusion of glycerol (5%) can also enhance the detection of some binding activities. Furthermore, electrophoresis can be performed at room temperature or at 4°C. More information about different gel electrophoresis buffers can be found in Chapter 14, Protocol 14.5, and in Sambrook et al. (1989; Unit 6.6-6.7) and Ausubel et al. (1994; Unit 12.2).

Design of initial experiments and interpretation of results. The initial EMSA experiments performed with a radiolabeled probe containing a control element of interest may result in the detection of one or more protein–DNA complexes on the gel image. The initial experiment is usually performed with a few different extract amounts (between 2 and

20 μg) and, for each extract amount, different amounts of nonspecific competitor (1–10 μg). If complexes are not detected, the binding reaction and gel conditions can be modified following the above recommendations. For each complex that is detected, two questions are generally addressed during the initial stages of the analysis: (1) Is the protein–DNA complex specific for the probe being tested? and (2) Is there a significant probability that the protein–DNA complex is functionally relevant (i.e., responsible for the in vivo function of the DNA element in the context of the endogenous control region)?

These two questions are generally addressed by additional EMSA experiments performed in the presence of specific competitor DNAs or with mutant probes. The most straightforward information is usually provided by a comparison of the complexes obtained with the radiolabeled wild-type probe to those obtained with a radiolabeled mutant probe. If a protein–DNA complex is not observed with a probe containing nucleotide substitutions in the control element of interest, the protein–DNA interaction almost certainly involves those nucleotides. Therefore, this result answers the two questions posed above, at least in part, by demonstrating that the interaction is sequence-specific and involves functionally relevant nucleotides. To obtain further evidence that the complex might be responsible for the function of the control element, probes can be prepared that contain other substitution mutations. If nucleotide substitutions that selectively disrupt the function of the control element (e.g., in a transfection experiment) correspond to those that disrupt the EMSA complex, the probability that the complex is functionally relevant would be enhanced (see Chapter 9).

As with all standard EMSA experiments, the mutant analysis should be performed with low concentrations of probe (1–10 fmoles per reaction) and protein concentrations that yield complexes within a linear range (i.e., 1–50% of the probe associated with protein). Unfortunately, the relative affinities of a protein for wild-type and mutant sequences cannot be determined simply by comparing the abundance of a complex formed with wild-type and mutant probes; a slight decrease in the affinity of a protein for a mutant probe might lead to a dramatic decrease in the amount of complex detected (relative to wild-type) if the complex formed on the mutant probe is not sufficiently stable to survive gel electrophoresis. Thus, if a mutation diminishes the abundance of a complex, one can conclude only that the protein has a lower affinity for the mutant probe than for the wild-type probe; the magnitude of the difference cannot be discerned. On the other hand, if a mutation has no effect on complex formation, the protein most likely binds the wild-type and mutant probes with comparable affinities. Overall, despite the lack of quantitative data, a comparison of wild-type and mutant probes can provide valuable information about (1) the specificity of an interaction, (2) the nucleotides required for the interaction, and (3) the relationship between the required nucleotides and those required for function of the DNA element.

One final issue regarding a comparison of wild-type and mutant probes is that the comparison should be performed with radiolabeled probes possessing similar specific activities. If the mutant probe has a lower specific activity than the wild-type probe, the protein–DNA complex will be less abundant, even if the mutation has no effect on the affinity of the interaction. Probe preparation by PCR using radiolabeled primers provides the best chance of obtaining comparable specific activities (see above).

A second method for answering the two questions posed above regarding the specificity and functional relevance of a protein–DNA complex is to perform competition experiments with wild-type and mutant DNA fragments that are not radiolabeled (see Box 8.1). Unlabeled DNA fragments containing wild-type or mutant sequences can be annealed oligonucleotides, PCR products, or restriction fragments. For the results of competition experiments to be meaningful, the wild-type and mutant competitors must be carefully quantified to ensure that their concentrations are comparable.

A competition experiment is performed by including a radiolabeled wild-type probe in a series of binding reactions containing increasing concentrations of unlabeled wild-type or mutant competitor DNAs. Radiolabeled complex abundance decreases in the presence of unlabeled competitor because the DNA-binding protein distributes itself randomly among the probe and competitor molecules. Because the wild-type competitor has a higher affinity for the protein than does the mutant competitor, lower concentrations are required to reduce the abundance of the complex.

Valuable information about the sequence-specificity and functional relevance of a protein–DNA interaction can be obtained by comparing a wild-type competitor to a series of mutant competitors (Box 8.1). If an EMSA complex is diminished to a greater extent by the wild-type competitor than a mutant, the protein must be binding DNA in a sequence-specific manner. If the mutations that diminish the extent of competition correspond to the mutations that reduce the function of the DNA element (e.g., in a transfection assay), the results suggest that the protein–DNA interaction may be functionally relevant.

The principal advantage of assessing specificity by using mutant competitor DNAs instead of radiolabeled mutant probes is that the former provides more accurate information about the relative affinities of a protein for the wild-type and mutant sequences. As explained above, a comparison of radiolabeled wild-type and mutant probes provides limited information about relative binding affinities. Competition experiments are not subject to the same caveat because the key events involved in the comparison of wild-type and mutant sequences occur in solution during the initial binding reaction; gel electrophoresis is performed AFTER the competition has been completed. If the affinity of the protein for the mutant probe is reduced by 2-fold, an approximately 2-fold higher concentration of the mutant competitor relative to the wild-type competitor should reduce complex abundance by a given amount. Thus, the competition strategy has greater potential to provide meaningful quantitative information about relative affinities.

Although the above considerations appear straightforward, the results of competition experiments can be complicated. For example, a 2-fold reduction often requires a large excess of wild-type competitor. The reason for this is that most EMSA experiments contain excess protein, despite the presence of free probe at the bottom of the gel. Thus, competition occurs only when the concentration of competitor is high enough to exceed the K_d of the interaction. Further discussion of K_d measurements can be found in Chapter 13 and Box 8.1.

Although this issue makes it difficult to predict the effect of a given concentration of wild-type competitor, it does not completely invalidate the comparison of wild-type and mutant competitors. Even if a 50-fold excess of the wild-type competitor is needed to achieve a 2-fold reduction in the abundance of a protein–DNA complex, the effect of the mutant competitor can provide insight into the relative affinities of the protein for the wild-type and mutant sequences. If a 50-fold excess of the mutant competitor results in a similar 2-fold reduction in the protein–DNA complex, one can conclude that the mutation has no significant effect on the affinity of the protein for the DNA. If a 100-fold excess of the mutant competitor is needed for a 2-fold reduction in the protein–DNA complex, the affinity of the protein for the mutant sequence is reduced by approximately 2-fold relative to the wild-type sequence.

Because the two strategies for assessing specificity and functional relevance of a protein–DNA interaction possess unique advantages and limitations, the use of both strategies to characterize a protein–DNA complex of interest is strongly recommended.

Box 8.1

Hypothetical EMSA Results

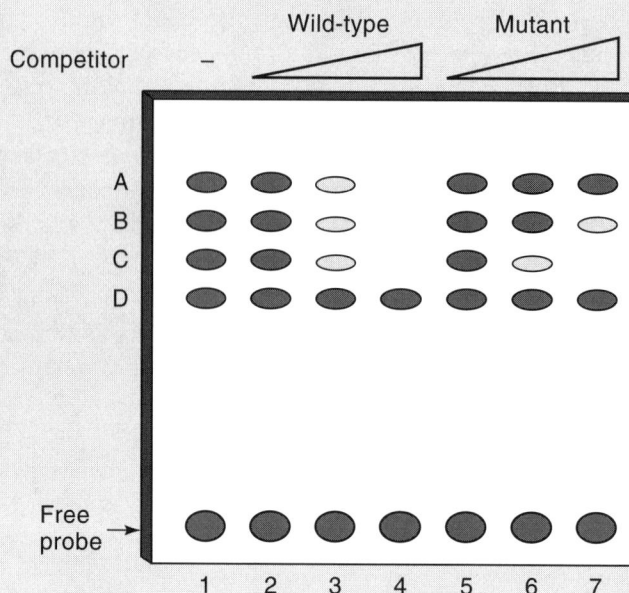

FIGURE 8.3. Hypothetical EMSA competition experiment.

The basic concepts underlying EMSA competition experiments are described in the text. To illustrate these concepts, the hypothetical competition experiment in Figure 8.3 is presented. In the absence of a specific competitor, four complexes, A through D, were obtained with a crude extract and the hypothetical wild-type probe (lane 1). Presumably, this binding reaction included a considerable quantity of a nonspecific competitor, such as poly(dI.dC), to diminish complexes containing nonspecific nucleic acid-binding proteins.

In the presence of increasing concentrations of wild-type competitor, complex A is diminished (lanes 2–4). In contrast, the abundance of complex A is unaffected by the mutant competitor (lanes 5–7). This result suggests that the protein responsible for complex A is a sequence-specific DNA-binding protein, which binds the mutant competitor with a substantially lower affinity than the wild-type competitor. If the mutation alters nucleotides that were shown to be important for the function of the control element (e.g., in a transfection assay), the results provide evidence that the protein binds at least a subset of the functionally important nucleotides within the probe.

Complex B is diminished by both the wild-type (lanes 2–4) and mutant competitors (lanes 5–7), although higher concentrations of the mutant competitor are required to diminish the abundance of the complex by a given amount (compare lanes 2 and 7). This result is more typical than that observed with complex A, as the protein responsible for complex B, like most sequence-specific DNA-binding proteins, binds to a wide array of DNA sequences with a low affinity. Therefore, it is not uncommon for a relatively high concentration of a mutant competitor to diminish a specific protein–DNA complex. Nevertheless, the behavior of complex B suggests that, like complex A, it contains a sequence-specific DNA-binding protein that contacts functionally important nucleotides.

Complex C is reduced to a similar extent by the wild-type and mutant competitors. This result suggests that protein binding does not require the nucleotides that were altered in the mutant probe. Nevertheless, the protein is likely to be a sequence-specific DNA-binding protein that recognizes another sequence on the two probe molecules. An alternative, but less likely, explanation for this result is that the protein binds DNA nonspecifically. A nonspecific DNA-binding protein would be expected to bind any radiolabeled probe. This possibility can be examined by testing a number of labeled probes possessing unrelated DNA sequences. Alternatively, additional mutants of the probe of interest can be tested.

Results like those obtained with complex D are frequently misinterpreted; this complex is not reduced by either the wild-type or mutant competitor. Numerous published manuscripts have suggested that this behavior demonstrates that the complex contains a nonspecific DNA-binding protein. Instead, the result merely shows that the protein is unusually abundant within the extract and is likely to be binding the probe with a relatively low affinity: No significant information is provided about the sequence-specificity of the protein. Regardless of whether a protein is a sequence-specific DNA-binding protein or a nonspecific nucleic acid-binding protein, it should be possible to diminish the complex by addition of sufficient quantities of an unlabeled wild-type competitor. If the complex is not diminished, the result merely demonstrates that a higher concentration of specific competitor is needed (or a lower concentration of extract). As mentioned above, the sequence-specificity of the protein can also be addressed by analyzing its ability to bind a number of radiolabeled probes containing mutant or unrelated sequences. The only reason the behavior of complex D raises the possibility of a nonspecific nucleic acid-binding protein is that many nonspecific proteins are abundant. Most typical transcriptional activators and repressors are present at relatively low concentrations in cells and cell extracts. Thus, when a complex is not diminished by a considerable excess of a wild-type competitor, it probably is not the transcriptional activator or repressor that one hopes to isolate. Nevertheless, as researchers begin to pursue studies of chromatin structure and nuclear organization, their level of interest in highly abundant, sequence-specific DNA-binding proteins may increase.

Determining the K_d of a protein using EMSA and oligonucleotide competition

As described above, oligonucleotide competition can be used to assess the specificity of a protein–DNA complex and to determine the relative affinity of a protein for a wild-type versus a mutant DNA site. However, the competition method can also be employed to determine the K_d of a protein and its site under a given set of reaction conditions. In this book, we describe two methods for determining K_d using EMSA and oligonucleotide competition. One method requires a highly purified and concentrated protein sample and is described along with the general theory of protein–DNA interactions in Chapter 13. Another method, which can be performed with both crude extracts and dilute protein samples, is described below. In crude extracts, this method generates a useful approximation but is less accurate than using pure proteins. The measured K_d will be for the competitor oligonucleotide so that different DNA competitors with different affinities can be measured (i.e., wild-type vs. mutant). Below, we will describe systematically how the equation was derived and will present the final derivation in Equation 11. The final equation can be used to determine K_d based on the concentration of added oligonucleotide competitor and the ratios of bound and free probe, before and after oligonucleotide competition.

The interaction between a protein and its site is described by the equilibrium:

$$P+S \leftrightarrow PS \tag{1}$$

Where P is the concentration of free protein, S is the concentration of free DNA site, and PS is the concentration of the complex.

The total protein concentration, P_T, in the reaction can then be described by the following equation:

$$P_T = [P] + [PS] \tag{2}$$

Similarly, the total DNA concentration, S_T, is described by the equation

$$S_T = [S] + [PS] \tag{3}$$

The affinity or K_d for the binding reaction is described by the equation

$$K_d = [P][S]/[PS] \tag{4}$$

The ratio of [S] to [PS] (or [PS] to [S]) is an experimental variable that is determined by quantitating the EMSA gel—[S] is the free probe and [PS] is the bound probe. Derivation of the K_d will depend on scanning the gel and determining the starting and ending ratio of [S] over [PS]. We will denote the ratio of [S]/[PS] in the absence and presence of oligonucleotide competitor as 1/X and 1/Y, respectively. By substituting these values into Equation 4 the following relationships emerge:

$$\text{Before oligonucleotide competition, the } K_d = [P]/X \text{ or } [P] = XK_d \tag{5}$$

$$\text{After oligonucleotide competition, the } K_d = [P]/Y \text{ or } [P] = YK_d \tag{6}$$

In a EMSA before adding competitor oligonucleotide, the initial ^{32}P-labeled probe or S_T is generally much less than the K_d. If the investigator is using 1 fmole of ^{32}P-probe, for example, in a 10-μl reaction, the concentration of probe will be 10^{-10} M, which is below the typical K_d for a DNA-binding protein ($\sim 10^{-9}$ M). Under these conditions the DNA is considered to be limiting. Because the DNA concentration is limiting, very little of the total protein is present in the form of PS. Therefore, the total protein is approximately equal to the free protein, or $P_T = P$. By making this assumption and substituting P_T for P in Equation 5, the amount of DNA-binding protein added to the reaction can be described in terms of K_d units multiplied by X:

$$P_T = K_d X \tag{7}$$

Therefore, even though the precise concentration of DNA-binding protein in the crude extract is not known, the amount added to the binding reaction can be represented in a quantitative fashion using K_d units and the ratio of [S]/[PS] from scanning the gel. This information will be used later to derive the K_d after oligonucleotide competition.

To compete the protein effectively from its site, the unlabeled probe concentration must be raised above the K_d. As the protein becomes distributed among the unlabeled and labeled probe, the ratio of [S]/[PS] increases from 1/X to 1/Y. At this stage, the added oligonucleotide must be in vast excess over the ^{32}P-labeled probe. Under these conditions S_T is approximately equal to the added oligonucleotide concentration. Because $S_T = [S] + [PS]$ (Eq. 3), if [S] and [PS] are expressed in terms of X, Y, and K_d, *the K_d can be determined because X and Y are known values derived from scanning the gel and S_T is equal to the competitor oligonucleotide concentration.*

To solve for [PS] after oligonucleotide competition, one returns to Equation 2. Because $P_T = [P] + [PS]$, substituting Equations 6 and 7 for [P] and P_T, respectively ($P_T = XK_d$ and $[P]=YK_d$), generates the following representation of [PS] in terms of X and Y:

$$[PS] = XK_d - YK_d \text{ or } (X - Y)\, K_d \tag{8}$$

By substituting this value of PS into Equation 3

$$S_T = [S] + (X - Y)\, K_d \tag{9}$$

Now to complete the derivation, we must solve for [S] in terms of K_d. According to Equation 4, [S] can be expressed in terms of K_d, [P], and [PS]. $[S] = K_d [PS]/[P]$. By substituting in Equations 6 and 8 for [P] and [PS], respectively, we obtain

$$[S] = K_d(X-Y)K_d/K_dY = (X-Y)\ K_d/Y \tag{10}$$

By substituting Equation 10 for [S] in Equation 9 we derive the final equation, which allows one to solve for K_d based on the known values of X, Y, and oligonucleotide competitor.

S_T = competitor oligo concentration = $[(X-Y)\ K_d/Y] + (X-Y)\ K_d = [(X/Y) + X - Y - 1]\ K_d$

Therefore,

$$K_d = \text{oligo concentration}/[(X/Y) + X - Y - 1] \tag{11}$$

For example, imagine that addition of a certain amount of crude extract forms a complex that represents 90% bound probe and 10% unbound probe (i.e., 90% would be shifted). In this case, [S]/[PS] = 1/X = 1/9 or X = 9. After addition of 50 nM oligonucleotide, the amount of bound probe decreases so that the ratio of [S]/[PS] = 1/Y = 1/1 (i.e., 50% probe would be bound and 50% unbound) or Y = 1. By substituting the values into Equation 11, one obtains the following result:

50 nM = $[(9/1) + 9 - 1 - 1]\ K_d$

which allows one to solve for K_d: 50 nM = $16\ K_d$

K_d = 50 nM/16 = 3 nM

Several additional considerations bear on the utility of the approach described above. First, the DNA must be limiting before beginning a competition experiment. This condition is generally met if addition of a few-fold excess of unlabeled oligonucleotide probe has no effect on the ratio of [S]/[PS]. Second, we have not taken into account the affinity of the protein for nonspecific DNA. This is a straightforward manipulation but requires additional steps, which may not be necessary when the method is used simply to compare relative affinities of a protein for different sites under a given set of binding conditions. Third, the method does not take into account the multimerization status of a protein. If a protein must dimerize before binding DNA, the method may only detect protein that is already in the dimeric form. If the DNA actively promotes dimerization, the kinetics will be altered by adding competitor. Finally, in many cases a single complex of protein and DNA is not observed, particularly in crude extracts. For example, in Figure 8.3 several complexes are apparent. In such cases the amounts of nonspecific complexes are added onto the concentration of free probe to determine the value of [S]. If the specific complexes are due to different proteins, each complex must be considered separately. In mixtures with multiple complexes the method is most accurate when, before competition, the amount of each complex is small relative to the amount of free probe.

Analysis of EMSA complexes by methylation interference. The experiments discussed above provide information about the specificity and nucleotide requirements of a protein–DNA interaction detected by EMSA. Another technique, methylation interference, can be coupled to the EMSA to provide information about the specific nucleotide contacts involved in the interaction. The methylation interference method is described in detail in Chapter 13 and Protocol 13.4. Briefly, this assay begins with an EMSA probe labeled on only one end, which, after labeling and purification, is modified on guanine bases with dimethylsulfate (DMS). (For this assay to succeed, it is often necessary to prepare probes from restriction fragments or by PCR, rather than using synthetic oligonucleotides.) A concentration of DMS and time of incubation are chosen that result in an average of one methylguanine per probe molecule. The modified probe is incubated with extract in a standard EMSA binding reaction. DNA-binding proteins bind randomly to the modified and unmodified probe molecules. However, a protein may be incapable of binding probe molecules in which a guanine that must be in close contact with the protein is methylated. The EMSA gel is used to separate the free probe molecules from the protein–DNA complexes. Following autoradiography, the film and gel are aligned, allowing excision of polyacrylamide gel slices containing the free probe and the protein–DNA complexes. The DNAs are eluted from the two gel slices and are then incubated with piperidine, which cleaves each probe molecule at the nucleotide(s) containing methylguanine. The resulting DNA fragments are analyzed by polyacrylamide gel electrophoresis, followed by autoradiography or phosphorimager analysis.

If methylation of a particular guanine prevents DNA binding, DNA fragments cleaved at that guanine will be absent in the sample derived from the protein–DNA complex, but will be abundant in the sample derived from the free DNA probe. Because methyl groups are quite small, the inhibition of protein binding by a particular methylguanine residue suggests that the protein is in close proximity with that guanine.

Methylation interference results can provide information that, in many respects, is similar to that provided by the competition experiments and mutants studies. The methylation results can demonstrate that binding is sequence-specific and can reveal the location of the binding site. One limitation of methylation interference results, however, is that they provide only partial information about the nucleotides required for the protein–DNA interaction; a careful mutant analysis can provide much more detailed information, which can be correlated more effectively with the results of a functional analysis to address the relevance of the protein–DNA interaction for the function of the control element (see Chapter 9). An additional limitation of the methylation interference technique is that it is technically challenging if the protein–DNA complex is not abundant. For weak complexes, it may not be possible to obtain compelling results, making the competition and mutant studies essential. For strong complexes, the use of all three approaches (methylation interference, competition studies, and analysis of radiolabeled mutant probes) is recommended to provide a complete picture of a protein–DNA interaction. Chapter 13 describes additional techniques that can be used to model a protein–DNA interaction more rigorously; this detailed analysis is usually performed with recombinant proteins after the gene encoding the protein has been cloned.

Analysis of previously described proteins in an EMSA complex. After the sequence-specificity and nucleotide requirements for a protein–DNA interaction have been determined, it often is important to address the possibility that the protein responsible for the interaction has been described previously. The first step of this analysis is to search a binding-site database, as described earlier in this chapter (see Fig. 8.1), to determine whether the DNA element is similar to the recognition site for one or more known proteins.

If candidate proteins are identified by the database search, the possibility that they are responsible for the EMSA complex can be evaluated using antibodies. Antibodies directed against many known DNA-binding proteins are available from a variety of commercial and academic sources. Otherwise, antibodies can be prepared against a synthetic peptide or a bacterially expressed fusion protein (see Harlow and Lane 1999).

To determine whether an EMSA complex contains a known protein, an antibody can be added directly to the binding reaction or preincubated with the extract prior to addition of radiolabeled probe. Preincubation may permit more efficient formation of the antibody–antigen complex if the DNA-binding domain is recognized by the antibody. In such a case, the antibody may prevent DNA binding, leading to a reduction in the amount of protein–DNA complex observed on the EMSA gel image (Fig. 8.4A). If the antibody recognizes a domain that is distinct from the DNA-binding domain, it is more likely to

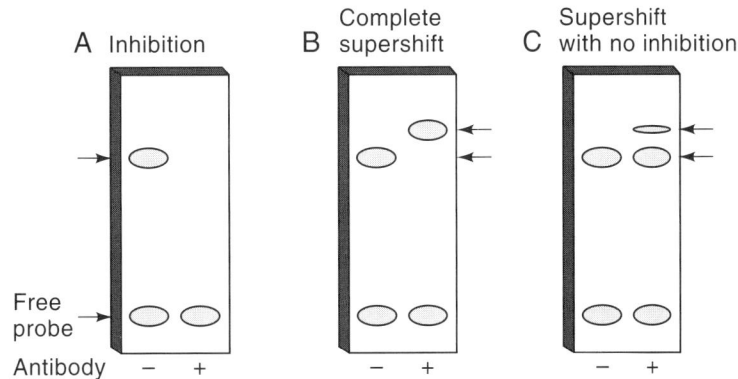

FIGURE 8.4. Effects of antibodies on EMSA complexes.

"supershift" the complex, meaning the complex will migrate more slowly through the EMSA gel because its molecular weight will be higher when bound to the antibody (Fig. 8.4B). An antibody also has the potential to stabilize a protein–DNA interaction by stabilizing the protein in a conformation that is competent for DNA binding. This results in a supershifted band that is more abundant than the original band observed in the absence of antibody.

Antibodies are valuable reagents for determining the identity of a protein within an EMSA complex. However, they must be used with caution, primarily because most polyclonal and monoclonal antibody preparations are impure. The contaminants within the preparations may inhibit complex formation as effectively as a specific antibody–antigen interaction. For this reason, controls must be included and carefully designed, particularly if the antibody inhibits, rather than supershifts, a complex. Appropriate controls for a polyclonal antibody are preimmune serum or a polyclonal preparation directed against an unrelated protein. It also is helpful to test the effect of an antibody preparation on an EMSA complex obtained with an unrelated probe bound by a protein that should not be recognized by the antibody. The specificity of the antibody effects can sometimes be greatly improved by purifying the antibodies by protein A or protein G Sepharose chromatography, or by antigen affinity chromatography (see Harlow and Lane 1999). If purified polyclonal antibodies are used in conjunction with the EMSA experiments, control antibodies should be subjected to the same purification protocol and analyzed along with the test antibodies. If monoclonal antibodies are used, a control monoclonal prepared by the same method and directed against an unrelated protein should be included.

If the antibodies supershift rather than inhibit the complex, the controls described above are somewhat less important. However, supershift results must be interpreted with equal caution. In some instances, the abundance of a protein–DNA complex of interest will not be diminished by the antibody, even though a supershifted complex of weak or moderate intensity appears on the gel image (Fig. 8.4C). This result does not establish that the protein recognized by the antibody is a component of the original complex. Because some antibodies stabilize protein–DNA interactions (see above), the supershifted complex might correspond to a stabilized form of a complex that was not detected in the absence of antibody. This complex might be unrelated to the original protein–DNA complex. Because of this possibility, evidence that a complex contains a particular protein is provided only when the abundance of the complex decreases as the supershifted complex appears. Another caveat of antibody supershifts is that an antibody has the potential to promote cooperative binding of a protein to two adjacent sites.

If the antibody tested does not disrupt or supershift the complex of interest, the protein recognized by the antibody may not be present within the complex. Alternatively, the antibody may be too dilute or may bind with a low affinity to the target protein. To confirm that the antibody is sufficiently concentrated and binds with a sufficient affinity, a positive control is needed, in which the antibody is shown to disrupt or supershift an EMSA complex known to contain the protein.

If the protein–DNA complex is convincingly disrupted or supershifted by an antibody against a known protein, the complex is likely to contain that protein. To test this hypothesis further, a cDNA for the protein can be obtained and analyzed. The full-length protein can often be produced by in vitro transcription/translation (Ausubel et al. 1994; Unit 10.17) and tested by EMSA side by side with the crude cell extract. If the complex detected with the crude extract contains only one protein, it should comigrate with the EMSA complex containing the in vitro translated protein. If not, it may be necessary to reevaluate the antibody results. Alternatively, the protein may bind as a heterodimer with another protein or may bind cooperatively with a protein that recognizes an adjacent site on the probe. Knowledge of the general properties of the protein family, and detailed knowledge of the nucleotides within the probe required for complex formation, should allow these possibilities to be evaluated.

If the results of these types of studies suggest that a known protein binds the control element of interest, the relevance of the protein for the function of the element can be evaluated as described in Chapter 9. In contrast, if the results suggest that the protein has not been previously described, the gene encoding the protein can be pursued by the methods described below. A final possibility is that the complex contains a heterodimer of a known protein and a novel partner, or cooperatively bound known and novel proteins. To identify the unknown components, the proteins that form the complex may need to be purified and their genes cloned.

DNase I Footprinting

The DNase I footprinting assay is described briefly above and in great detail in Chapter 13. In this section, specific issues are discussed that must be considered when using a DNase I footprinting assay with crude extracts to identify protein–DNA interactions.

Radiolabeled probe design. Two important issues are the length of the radiolabeled footprinting probe and the location of the relevant DNA element(s) within the probe. As described above and in Chapter 13, the probe must be radiolabeled on only one end and on only one strand. Such probes can be prepared by direct labeling of restriction fragments (Protocol 13.6) or by PCR with one radiolabeled primer and one unlabeled primer (Ausubel et al. 1994, Unit 15.2.6). Synthetic oligonucleotide probes are used less frequently for DNase I footprinting than for EMSAs because longer probes are generally preferred.

For DNase I footprinting with crude extracts, probes generally range from approximately 100 to several hundred base pairs in length. It is important that the DNA element(s) of interest not be within 40 bp of either end of the probe. One reason for this restriction is that some extracts contain abundant DNA end-binding proteins, which can obscure detection of a specific protein–DNA interaction at either end of a probe. Furthermore, an accurate evaluation of a DNase I footprinting result depends on the detection of cleavage products on both sides of the protected region that are unaffected by protein binding; if the protected region is too close to the end of the probe, it may be difficult to detect unaffected cleavage products at one end.

To evaluate DNase I footprinting data accurately, it also is beneficial for the DNA elements of interest to be within approximately 200 bp of the labeled end. Probes of 500 bp or more can be used successfully in the footprinting assay, but protein–DNA interactions that occur more than 200 bp from the radiolabel are often difficult to detect because of compression of bands at the top of the gel. Considering these restrictions, the DNase I footprinting procedure can usually provide information about protein–DNA interactions that occur within a region of approximately 160 bp, corresponding to the region that is 40–200 bp from the labeled end of the probe. Thus, if the control region being analyzed is approximately 300 bp in length, it may be possible to scan the entire region for specific protein–DNA interactions with only two probes.

To increase further the probability that the protein–DNA interaction of interest will be detected, an independent set of probes labeled on the opposite strand can be tested. Some protein–DNA interactions are detected more readily with a probe labeled on one strand than the other. Furthermore, analysis of both strands reduces the probability that an important protein–DNA interaction will occur in a region that is resistant to DNase I cleavage. The nuclease activity of DNase I nicks one strand of a double-stranded DNA fragment. Strand nicking occurs at a large number of sites on a naked DNA probe, but is not completely random. Thus, regions of 20 bp or more that are resistant to DNase I nicking may be observed. If a relevant protein–DNA interaction occurs within such a region, it will not be detected. By testing two probes labeled on opposite strands, the probability that an interaction will be missed because a long stretch of DNA on one strand or the other is resistant to DNase I cleavage will be reduced.

It is important to note that the location of a protein–DNA interaction detected by DNase I footprinting is determined by comparing the protected region on the gel image to DNA size markers that are run in an adjacent lane (see Chapter 13). Radiolabeled restriction fragments can provide some indication of the distance from the labeled end of the probe to the protected region, but DNA sequencing markers provide more accurate information.

Extract preparation. Extract preparation methods and strategies for optimizing and concentrating extracts for DNase I footprinting experiments are similar to those described above for EMSA experiments. However, for DNase I footprinting, it is more frequently necessary to concentrate extracts by ammonium sulfate precipitation or heparin-Sepharose chromatography (see Chapter 14). Extract concentration is more important for DNase I footprinting than for the EMSA because a much larger percentage of the probe molecules must be bound by a specific protein to detect a footprint (see above).

Binding reactions. In general, the binding reactions for DNase I footprinting with crude extracts are similar to those used for EMSA, but the concentrations of some of the components differ. The concentration of nonspecific competitor DNA, such as poly (dI.dC), is generally lower, and the concentration of extract is generally higher in a DNase I footprinting assay. A standard concentration of poly(dI.dC) for DNase I footprinting with a crude extract is 1 μg/50 μl binding reaction. Optimal crude extract concentrations range from 50 to 250 μg/50 μl reaction. It is particularly important to test different concentrations of extract. A concentration that is too high can result in probe degradation or dephosphorylation by proteins within the extract, or complete resistance of the probe to DNase I cleavage.

DNase I cleavage reactions are described in Chapter 13 from the perspective of an analysis of purified recombinant DNA-binding proteins. The only notable difference when analyzing DNA-binding activities in crude extracts is that a higher concentration of DNase I may be required. Therefore, the optimal concentration of DNase I must be determined separately for control reactions lacking protein and for the reactions containing each concentration of extract. When the extract concentration in the reaction is increased, it may be necessary to increase the concentration of DNase I further. For each set of reaction conditions, the goal is to identify a DNase I concentration that yields bands of similar intensity toward the top and bottom of the gel image.

Two final tips for improving DNase I footprinting results when using crude extracts are as follows: First, if protection is not observed at the DNA element of interest, the sensitivity of the assay can often be enhanced by adding 2% polyvinyl alcohol to the binding reactions (from a 10% stock solution). This polymer increases the effective concentration of the proteins within the extract, which may enhance the efficiency of a relevant protein–DNA interaction. Second, the results of DNase I footprinting experiments are sometimes more consistent if the binding reactions and DNase I digestions are performed on ice rather than at room temperature (unless the protein only binds DNA at room temperature or a higher temperature). For a given extract, both temperatures should be tested to determine whether one yields significantly better results than the other. A higher concentration of DNase I is required when the digestions are performed on ice because the rate of cleavage is reduced. For an investigator performing a DNase I footprinting experiment for the first time, it may be beneficial to use the kit from Promega (cat. # E3730), which includes standard reagents and positive controls.

Interpretation of initial results. DNase I footprinting results obtained with crude extracts are often visually pleasing, in part because the detection of one or more strong footprints provides immediate and compelling evidence that sequence-specific DNA-binding proteins within the extract can bind specific regions of the probe. A reaction performed in the absence of DNA-binding proteins is run on the gel adjacent to the reactions performed in the presence of extract (see Protocol 13.1). To simplify interpretation of the data, the bands in the control lane and experimental lanes should be of similar intensity throughout the gel, except at specific regions of protection. If the general band intensities in the two lanes are significantly different, it may be necessary to alter the concentration of DNase I used in one of the reactions.

When analyzing crude extracts, one additional control that should be performed is to incubate the extract and probe in a typical binding reaction, and then to perform a "mock" nuclease digestion by adding $MgCl_2$ and $CaCl_2$, but no DNase I. This control reaction will provide information about cleavage products resulting from nucleases within the extract, which may cleave at different sites than DNase I, thereby complicating the interpretation of the results. Nucleases can be even more problematic if the standard binding buffer also contains $MgCl_2$. Because most DNA-binding proteins do not require Mg^{++} for binding, the initial experiments should be attempted with binding buffers both containing and lacking $MgCl_2$. If substantial nuclease activity is detected in the extract, it may be necessary to fractionate the extract by column chromatography (e.g., heparin-Sepharose chromatography) prior to footprinting analysis in an attempt to separate the nucleases from the majority of the DNA-binding proteins.

In addition to the cleavage sites that are protected from DNase I digestion by bound proteins, other cleavage sites can be hypersensitive to DNase I cleavage. Hypersensitivity is frequently observed near the edges of protected regions and results from conformational changes in the DNA upon protein binding. At least one protein family, the Ets family, induces a characteristic hypersensitive site at the center of the protected region (see, e.g., Ernst et al. 1993). Hypersensitive sites can be highly informative because they can appear when only a small fraction of the probe molecules are bound by protein, whereas compelling protection appears only when the vast majority of the probe molecules are occupied. Thus, the presence of a hypersensitive site in a region that does not exhibit protection suggests that a protein within the extract binds nearby, but the protein may not be sufficiently concentrated for protection. Two alternative explanations for the presence of isolated hypersensitive sites are as follows: (1) A nuclease within the extract can lead to the appearance of hypersensitive cleavage sites because those sites may not be cleaved by DNase I in the control reaction that lacks extract and (2) structural alterations that enhance DNase I cleavage may occur upon binding of proteins to distant sites. For example, two proteins bound at sites that are distant from one another may interact, inducing structural alterations and enhanced DNase I cleavage in the intervening sequence.

Some of the issues discussed above for interpreting initial EMSA experiments are relevant to DNase I footprinting experiments, but others are less relevant. In particular, the specificity of a protein–DNA interaction is much easier to establish with the DNase I footprinting assay. If a discrete footprint is observed, the protein–DNA interaction is specific for a sequence within the protected region; competition experiments and mutant studies generally are not needed to establish the specificity of the interaction. Experiments must be performed, however, to determine whether the protein–DNA interaction involves nucleotides that are relevant for the function of the control element. As with the EMSAs, the nucleotide requirements for the interaction can be addressed by analyzing radiolabeled probes containing mutations of specific nucleotides, or by competition of the footprint using wild-type or mutant oligonucleotides. The former strategy is generally preferred because it is much less susceptible to misinterpretation. If radiolabeled wild-type and mutant probes are prepared by PCR using the same radiolabeled primer, the probes will have comparable specific activities. If a protein–DNA interaction is disrupted by a mutation, the results provide strong evidence that the nucleotides mutated are required.

Competition experiments are advantageous for the EMSA because they provide more accurate information about relative affinities of the protein–DNA interaction on wild-type and mutant probes (see above). Because DNase I footprinting reactions are performed entirely in solution (i.e., the protein–DNA interaction does not need to survive a gel electrophoresis step), relative affinities determined by comparing radiolabeled wild-type and mutant probes can be as accurate, if not more so, than relative affinities determined by competition experiments. Because there is no significant advantage of competition experiments when performing DNase I footprinting, the direct analysis of radiolabeled wild-type and mutant probes is recommended.

As with the EMSA approach, a precise correlation between the nucleotides required for the protein–DNA interaction and those required for the function of the control element can provide evidence that the protein detected is functionally relevant. Further evidence of functional relevance can be obtained using the strategies outlined in Chapter 9.

Antibodies were described above as powerful reagents for determining whether an EMSA complex contains a previously described protein. Antibodies can also be used to determine whether the protein responsible for a DNase I footprint has been described previously. However, the antibodies are useful only if they disrupt the DNA-binding activity of the protein, usually by binding the DNA-binding domain. Antibodies that do not disrupt DNA binding usually have no noticeable effect in a DNase I footprinting experiment, in contrast to the supershift observed in an EMSA.

Finally, it is important to emphasize that different types of information are provided by the DNase I footprinting assay and methylation interference assay, which can be coupled to the EMSA (see above). The protected region observed during DNase I footprinting corresponds to the nucleotides that are inaccessible to DNase I cleavage when the protein is bound. Because DNase I is a protein with a substantial mass, and because the DNA-binding protein usually covers nucleotides adjacent to those that are directly contacted, the protected region almost always includes at least a few nucleotides on both sides of the critical nucleotides. The critical nucleotides are usually toward the center of the protected region, but, in the absence of mutant or competition studies, they are difficult to predict. In contrast, the methylation interference results provide more precise information about the protein–DNA contacts because the DNA is modified by a small methyl group prior to protein binding. Methyl groups interfere with protein binding only if they modify a guanine that directly contacts the protein or is in very close proximity to an inflexible region of the protein (see also Chapter 13).

CONCEPTS AND STRATEGIES FOR CLONING GENES ENCODING DNA-BINDING PROTEINS

The strategies that are used most frequently for identifying proteins that bind a DNA element of interest are outlined above. If a binding activity is detected by EMSA or DNase I footprinting, experiments can be performed with antibodies to determine whether the protein has been described previously. If the identity of the protein cannot be determined, the protein–DNA interaction assay can be used to perform a basic characterization. For example, extracts from a number of different cell types or developmental stages can be analyzed to determine whether the binding activity is ubiquitous or developmentally regulated. In addition, binding site mutants can be analyzed to identify the nucleotides that are required for the protein–DNA interaction. A more extensive characterization of the protein will depend on the cloning of its gene, however. With the gene in hand, the relevance of the protein for a biological process can be assessed by gene disruption, the domains of the protein responsible for its activities can be determined by mutagenesis, and its mechanism of action and mode of regulation can be analyzed.

For some studies, the preferred method for cloning the gene encoding a protein responsible for a particular EMSA or DNase I footprinting activity is to purify the protein and obtain a partial amino acid sequence (Box 8.2). From the amino acid sequence, degenerate primers for PCR can be designed and used to amplify a gene fragment. The gene fragment can then be used to isolate a full-length cDNA from a cDNA library or by 5′ and 3′ rapid amplification of cDNA ends (RACE). The purification strategy is sometimes preferred because, following each purification step, the column fractions containing the binding activity can be monitored using the EMSA or DNase I footprinting assay. One's ability to follow the binding activity through the purification procedure ensures that the protein originally detected by EMSA or footprinting is being purified and, ultimately, cloned (Fig. 8.5). The other cloning strategies described in this section have a lower probability of yielding the gene encoding a binding activity observed in an EMSA or DNase I footprinting experiment. A second advantage of the protein purification strategy is that binding activities which depend on two or more proteins or subunits can be purified and cloned. In contrast, most of the other methods described in this chapter will succeed only if the binding activity is a monomer, homodimer, or homomultimer.

The primary disadvantage of the protein purification strategy is that it can be challenging for laboratories with protein purification experience and overwhelming for labo-

Methods for cloning genes encoding DNA-binding proteins

1. Protein purification

Advantages
- Most likely to result in the cloning of the gene encoding a DNA-binding protein detected by EMSA or DNase I footprinting
- Will allow the cloning of genes encoding heterodimeric proteins

Disadvantages
- Very difficult for laboratories with little or no protein purification experience
- Time consuming

2. Yeast one-hybrid screen

3. In vitro expression library screen

4. Mammalian expression cloning

5. Genome database and degenerate PCR methods

Advantages
- Relatively rapid
- Generally easier than protein purification
- Can be performed without preliminary EMSA or DNase I footprinting experiments

Disadvantages
- Less likely to result in the cloning of the gene encoding a DNA-binding protein detected by EMSA or DNase I footprinting
- Not useful for proteins that must bind as heterodimers

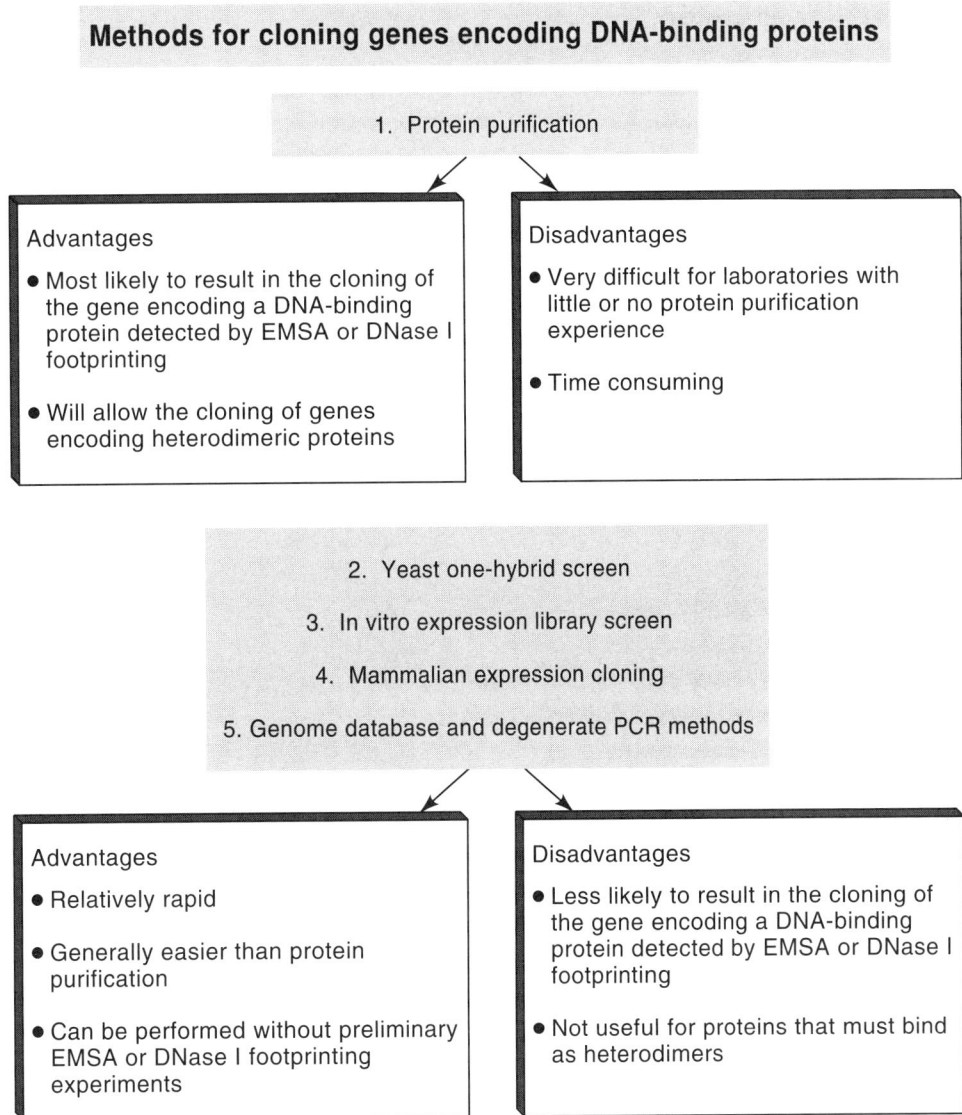

FIGURE 8.5. Methods for cloning genes encoding DNA-binding proteins.

ratories with no experience (Fig. 8.5). The other strategies are relatively easy. A second limitation of the protein purification strategy is that an abundant source of cells is usually required. The other strategies require only a cDNA library.

Alternative strategies for cloning genes encoding DNA-binding proteins include one-hybrid screens, in vitro expression library screens, mammalian expression cloning methods, database strategies, and degenerate PCR (Fig. 8.5). Each of these strategies is outlined below. A common feature is that they do not rely on the initial development of a protein–DNA interaction assay. In other words, if an important DNA element has been identified in a promoter or distant control region, these strategies can be used to clone genes encoding proteins that specifically bind the element, without the need for preliminary EMSA or DNase I footprinting experiments with crude extracts.

Box 8.2

Cloning by Protein Purification

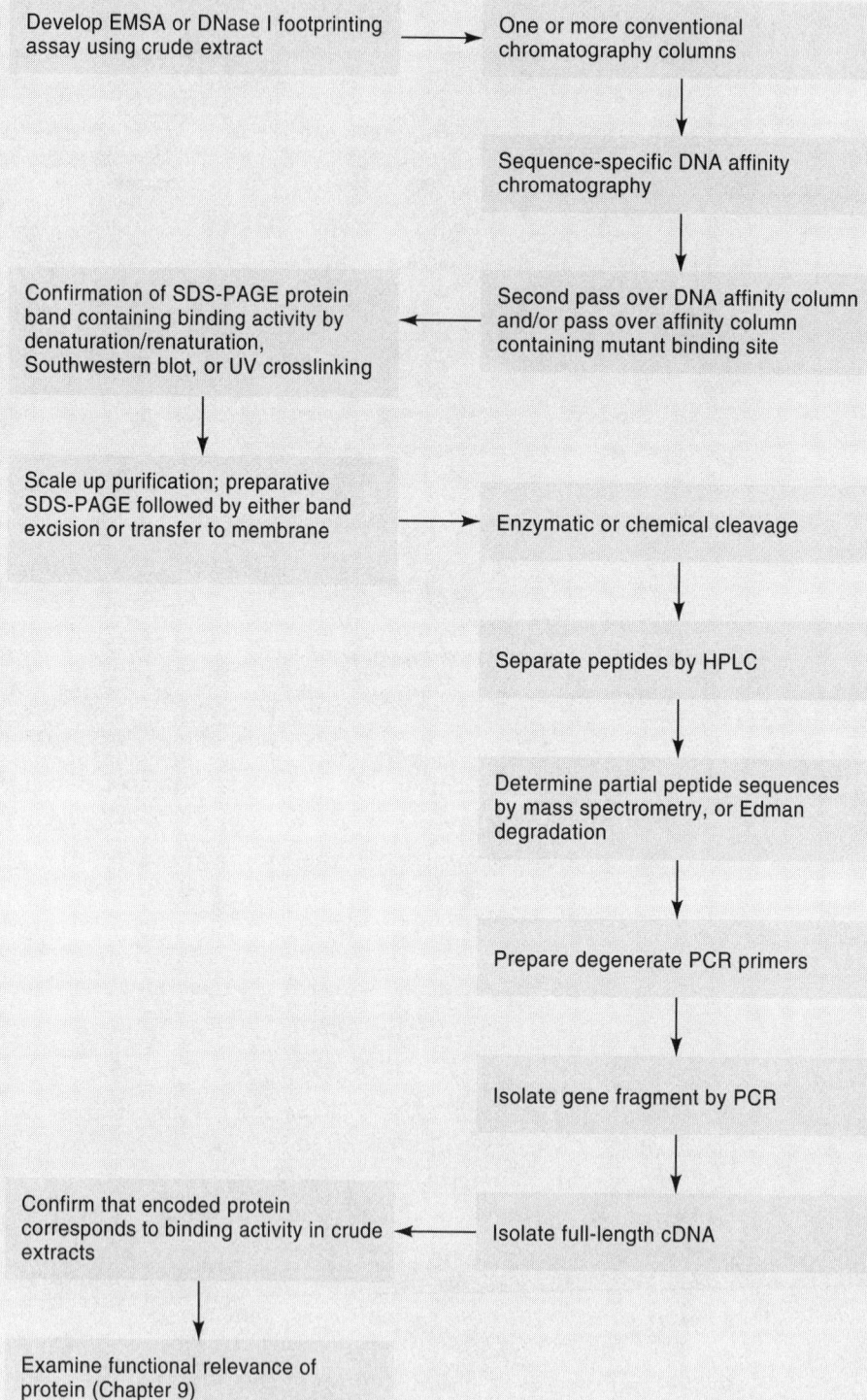

Develop EMSA or DNase I footprinting assay using crude extract ⟶ One or more conventional chromatography columns

↓

Sequence-specific DNA affinity chromatography

↓

Confirmation of SDS-PAGE protein band containing binding activity by denaturation/renaturation, Southwestern blot, or UV crosslinking ⟵ Second pass over DNA affinity column and/or pass over affinity column containing mutant binding site

↓

Scale up purification; preparative SDS-PAGE followed by either band excision or transfer to membrane ⟶ Enzymatic or chemical cleavage

↓

Separate peptides by HPLC

↓

Determine partial peptide sequences by mass spectrometry, or Edman degradation

↓

Prepare degenerate PCR primers

↓

Isolate gene fragment by PCR

↓

Confirm that encoded protein corresponds to binding activity in crude extracts ⟵ Isolate full-length cDNA

↓

Examine functional relevance of protein (Chapter 9)

FIGURE 8.6. Cloning by protein purification.

To purify and clone a DNA-binding protein by sequence-specific DNA affinity chromatography (Fig. 8.6), an abundant source of cells for extract preparation is needed. The extract is usually concentrated by ammonium sulfate precipitation, which also serves as a crude fractionation step. The concentrated extract is partially purified by one or more conventional chromatography columns to help separate the protein of interest from nucleases, proteases, and other DNA-binding proteins that tend to interfere with purification. The binding activity is monitored in the fractions that elute from each column by EMSA or DNase I footprinting. The fractions containing the majority of the binding activity are pooled, mixed with a nonspecific competitor DNA, and applied to the DNA affinity column. The DNA affinity resin contains covalently linked oligonucleotide multimers of the binding site of interest. The nonspecific competitor DNA prevents unwanted nucleic acid-binding proteins from stably binding the resin. The DNA-binding protein of interest should bind to its high-affinity sites when passed through the resin. The column is then washed and the bound proteins are eluted with buffers containing gradually increasing salt concentrations. If necessary, further purification can be achieved by a second pass over the affinity column or by chromatography on an affinity column containing a mutant oligonucleotide sequence. The proteins within the purified fractions are detected by SDS-PAGE, followed by silver-staining. The protein band responsible for the binding activity of interest is determined by denaturation/renaturation of proteins from gel slices, by Southwestern blot, or by UV crosslinking (see text).

To clone the gene encoding the protein, amino acid sequences of proteolytic peptides are determined by Edman degradation or mass spectrometry. (An alternative is to prepare antibodies against the pure protein, which can be used to screen an expression library.) The amino acid sequences are used to design degenerate oligonucleotides, which are used to isolate a fragment of the gene by RT-PCR. The gene fragment is then used to isolate a full-length cDNA by screening a cDNA library or by 5′ and 3′ RACE.

This feature can be both an advantage and a disadvantage. It is an advantage because considerable time and effort can be saved if a protein–DNA interaction assay does not need to be developed and optimized. With a defined DNA element, cloning experiments can be initiated immediately. A second advantage is that these strategies can be used even if preliminary EMSA and DNase I footprinting experiments fail to detect a binding activity.

The principal disadvantage is that a gene and encoded protein that are isolated possess a somewhat lower probability of being relevant for the function of the control element of interest (although there is a good chance that the relevant protein will indeed be found). To explain this concept, consider the following example of a DNA sequence element that has been carefully analyzed by mutagenesis in a functional assay, as described in Chapter 7. If an EMSA complex is detected using crude extracts, mutations in the control element can be analyzed to determine whether complex formation requires the functionally important nucleotides. A preliminary analysis of the expression pattern of the protein can also be performed to determine whether it corresponds to the anticipated pattern. If these preliminary studies provide compelling evidence that the protein is not relevant, the EMSA conditions can be varied until a more attractive candidate for the relevant DNA-binding protein is identified. One's ability to perform a preliminary characterization can enhance the probability that the protein that is purified and cloned will correspond to the functionally relevant protein. In contrast, if a one-hybrid screen is used to clone the gene encoding a DNA-binding protein, little evidence will be obtained that the protein recognizes functionally relevant nucleotides and exhibits the anticipated expression pattern until the cDNA and recombinant protein have been characterized. Of course, for most laboratories, several cDNAs isolated in a one-hybrid screen can be characterized more rapidly than a protein can be purified.

On the basis of the above considerations, the protein purification strategy is recommended if a particularly attractive candidate for a relevant DNA-binding protein has been identified by EMSA or DNase I footprinting, *and* if the laboratory has experience with protein purification procedures. In addition, the protein purification strategy may be essential if the protein–DNA interaction requires a heterodimeric or heteromultimeric protein. On the other hand, if an attractive candidate has not been identified by EMSA or footprinting, or if the laboratory has little experience with protein purification, the other approaches, in particular the one-hybrid screen, are recommended as a starting point. The functional relevance of candidate genes identified by these methods can be assessed using the strategies described in Chapter 9. If the candidates turn out to be irrelevant, the more difficult protein purification strategy can then be initiated.

In the following sections, the principal strategies that are used to clone genes encoding DNA-binding proteins are described (see Fig. 8.5). Protocols are not included because of space limitations. However, appropriate sources of detailed protocols are provided.

Cloning by Protein Purification and Peptide Sequence Analysis

The strategy that has proven to be the most successful for purifying mammalian DNA-binding proteins makes use of column chromatography resins with covalently linked oligonucleotide multimers containing the DNA sequence of interest. This technique, known as sequence-specific DNA affinity chromatography, was developed by Kadonaga and Tjian for the purification of Sp1 (Kadonaga and Tjian 1986; Kadonaga 1991; Ausubel et al. 1994, Unit 12.10; Marshak et al. 1996) (see Box 8.2).

The sequence-specific DNA affinity chromatography strategy has been used to purify and obtain peptide sequences for over 100 proteins (Marshak et al. 1996). A few mammalian DNA-binding proteins have been purified in the absence of a DNA affinity chromatography step (see, e.g., Landschulz et al. 1988). However, because of the generally low abundance of mammalian transcription factors, purification by conventional methods is more difficult and rarely successful. Detailed descriptions of the DNA affinity chromatography method can be found in Kadonaga and Tjian (1986), Kadonaga (1991), Ausubel et al. (1994; Unit 12.10), and Marshak et al. (1996). Central issues and supplementary suggestions are discussed here, but these references should be consulted for specific protocols and additional advice.

Amount of Starting Material

One important issue is the amount of starting material needed to purify and obtain peptide sequences for a mammalian DNA-binding protein. Until recently, 50–100 pmoles of a purified protein were needed to obtain sufficient peptide sequence information for gene cloning. Theoretically, if 1000 copies of a protein are present in each cell, 30 liters of cultured cells grown to a density of 1×10^6 cells/ml are needed to obtain 50 pmoles of the protein. In practice, however, the yields obtained during purification by DNA affinity chromatography are quite low, as described below. In addition, DNA affinity columns are generally much more effective when large amounts of protein are applied. Thus, it is not unusual for the purification and peptide sequencing of a protein to require one hundred or more liters of cultured cells.

Animal tissues provide an alternative source of starting material that can be quite beneficial when large quantities are required. The purification and cloning of NF-κB, for example, greatly benefited from the use of rabbit lung as a source of starting material (Ghosh et al. 1990). A disadvantage of animal tissues is that many contain high concentrations of proteases. Because of these considerations, the availability of an appropriate source of cells and the resources for obtaining the cells should be evaluated before pursuing the purification of a DNA-binding protein.

Another consideration regarding the amount of starting material is that the methodology available for sequencing peptides has greatly improved. The sensitivity of current Edman degradation technology and the emergence of mass spectroscopy methods for peptide sequencing have made it possible to sometimes obtain sequences from less than 1 pmole of a purified protein. If a researcher has access to a modern sequencing facility, the amount of starting material often can be greatly diminished.

Conventional Chromatography Steps

As indicated in Box 8.2, a conventional chromatography step is usually included before the DNA affinity column. Gel filtration may be the most effective conventional step (Kadonaga and Tjian 1986; Kadonaga 1991; Marshak et al. 1996). Gel filtration is useful because it separates proteins on the basis of a property (i.e., size) that is unlinked to the DNA-binding activity of the protein. In other words, the protein of interest will be separated from most other nucleic-acid-binding proteins during gel filtration chromatography. Ion-exchange resins, such as heparin and DEAE resins, are easier to use and have been used successfully for the initial fractionation (see, e.g., Hahm et al. 1994). However, many DNA-binding proteins bind these resins with comparable affinities. Therefore, they usually do not separate the protein of interest from the majority of other nucleic-acid-binding proteins in the extract. Because nucleic-acid-binding proteins are the most troublesome contaminants of DNA affinity column eluates, techniques like gel filtration that separate these proteins from the protein of interest are preferred. The disadvantage of the gel filtration step, however, is that a large column is generally needed. Because of the challenge of pouring and running a large gel filtration column, ion-exchange columns are more commonly used as a first step.

The number of columns needed prior to an affinity column is variable. Some proteins, including Sp1, AP-1, and Ikaros, required only one gel filtration or ion-exchange column prior to affinity chromatography (Kadonaga and Tjian 1986; Lee et al. 1987; Hahm et al. 1994). However, other proteins, including NF-κB, required several (Ghosh et al. 1990). In the case of NF-κB, its unusually low abundance may explain why additional columns were needed to achieve sufficient purity.

DNA Affinity Chromatography

The methodology for preparing and running DNA affinity columns is described in detail in Kadonaga and Tjian (1986), Kadonaga (1991), Ausubel et al. (1994; Unit 12.10), and Marshak et al. (1996). One relatively new advance that may improve the success of the procedure is the generation of oligonucleotide multimers by PCR (Hemat and McEntee 1994). The traditional method for generating multimers is to anneal, phosphorylate, and ligate oligonucleotide monomers until long multimers are observed on an ethidium bromide-stained agarose gel (Kadonaga and Tjian 1986). This method has been used with considerable success, but long multimers can be difficult to obtain for reasons that are not well understood. To prepare multimers by PCR, complementary oligonucleotides are synthesized that contain dimers of the binding site. During each PCR cycle, a fraction of the oligonucleotides will anneal in a staggered manner, allowing the gradual generation of long multimers.

DNA affinity chromatography procedures are similar to conventional chromatography procedures. One difference, however, is that the success of a DNA affinity step can usually be enhanced by overloading the column with protein. Whereas 10–40 mg of protein/ml is generally applied to a conventional chromatography column, 100 mg of protein/ml is sometimes applied to a DNA affinity column. Application of such a large amount of protein can result in overloading, in that a significant fraction of the protein of interest will be unable to bind the column. However, overloading can help to saturate the column with the protein of interest, which increases the concentration of the protein in the high-salt eluates. The relatively higher concentration can contribute to the detection and stability of the protein. Overloading greatly reduces the overall yield of pure protein. However, overloading may be necessary to achieve the final objective of purifying the protein to a sufficient extent for peptide sequencing. In other words, yield may need to be sacrificed for purity.

A second difference between conventional chromatography columns and DNA affinity columns is that small-scale (i.e., pilot) DNA affinity columns are often unsuccessful. When developing a conventional chromatography procedure, it is quite common to establish conditions by loading relatively small amounts of protein onto a small column, in some cases less than the recommended amount per milliliter of resin. With DNA affinity columns, this strategy is often unsuccessful because the protein within the eluates may not be sufficiently concentrated or stable for detection. If binding activity is not obtained during small-scale pilot experiments, the solution may simply be to repeat the experiment at a larger scale.

The addition of nonspecific competitor DNA is another feature of the DNA affinity chromatography procedure that leads to a sacrifice in overall yield to achieve purity. Competitor DNA is usually preincubated with the protein sample prior to its application to the column. The procedure for determining the amount of competitor to add is described in Kadonaga (1991) and Marshak et al. (1996). The competitor is bound by DNA-binding proteins that do not specifically bind the oligonucleotides linked to the column, thereby preventing them from efficiently binding the column. In practice, the competitor also prevents the protein of interest from binding the column to some extent. The yield of the protein of interest can sometimes be enhanced by reducing the amount of competitor. However, a high concentration of competitor may be required for optimal purity.

Identification of the Relevant Band following SDS-PAGE

After the protein has been purified to a point in which one or a few abundant bands are observed on a silver-stained gel, it often is useful to determine which of the bands (if any) is responsible for the binding activity. If only one band is observed, it is useful to determine whether that band is indeed responsible for the binding activity or whether it is a contaminant. It is tempting to assume that the single band observed corresponds to the correct protein. However, that assumption can lead to considerable frustration if it is later found to be the wrong protein.

It is not uncommon for researchers to focus on a contaminant rather than the correct band, in part because the amount of protein needed for strong EMSA activity can be quite small. Thus, a few microliters of an affinity-purified protein can yield a strong EMSA complex, but 40 μl of the same protein may be difficult to detect on a silver-stained gel. The approximate amount of protein in an EMSA complex can be determined by estimating the number of moles of probe in the reaction and the number of moles of probe shifted into complex. The amount of protein in a silver-stained band can be determined by comparing the intensity of the band to the intensities of a serially diluted protein standard, such as BSA. If the amount of protein detected by silver staining appears to be substantially greater than the amount that gives rise to the EMSA complex, the silver-stained protein may not be the protein of interest.

The three techniques that are most commonly used for confirming which band on a silver-stained gel corresponds to the DNA-binding protein of interest are denaturation/renaturation (Box 8.3), Southwestern blot analysis (see, e.g., Gunther et al. 1990), and crosslinking (see Ausubel et al. 1994, Unit 12.5). The Southwestern blot and crosslinking techniques provide information about the sizes of proteins that specifically bind a radiolabeled probe. These techniques can be quite powerful for confirming that a particular protein observed on a silver-stained gel specifically binds a radiolabeled oligonucleotide. However, the denaturation/renaturation technique is recommended because it provides more information.

Denaturation/renaturation (see Box 8.3 and Fig. 8.7) allows the specific DNA-binding activity to be monitored using the EMSA or DNase I footprinting assay that has been used throughout the purification procedure. Thus, if the proteins eluted and renatured from a particular gel slice yield an EMSA complex that comigrates with the EMSA complex of interest, that gel slice almost certainly contains the protein of interest. Similarly, if the proteins renaturing from a particular gel slice yield a DNase I footprint with a pattern that matches the pattern generated by the protein of interest, that gel slice contains the relevant protein. In contrast, with Southwestern and crosslinking experiments, the evidence of specific binding by a protein is provided by a radioactive band on an SDS-PAGE blot or gel. If a contaminant in the protein preparation binds with reasonable affinity and specificity to the probe, a band will appear even if it does not represent the protein of interest. If the contaminant is more abundant than the protein of interest or binds with higher affinity, the only band observed may correspond to the contaminant. The limitations of the Southwestern and crosslinking approaches can be overcome to some extent by performing careful controls with radiolabeled mutant probes or unlabeled competitors. However, it will be difficult to achieve the same level of confidence as found with the denaturation/renaturation procedure.

An additional limitation of all of these procedures is that it is much more difficult to obtain meaningful results if the DNA-binding activity of interest requires multiple distinct subunits of different sizes. If two or more proteins are required, the Southwestern procedure has virtually no chance of success. The denaturation/renaturation procedure may succeed if the samples eluted from the gel slices are mixed in the various combinations either before or after renaturation. The crosslinking procedure may have the best chance of success because crosslinking is performed before the protein complex is denatured and separated by gel electrophoresis. However, it is possible that only one subunit of the binding complex will efficiently crosslink to the probe, causing the other protein(s) to be overlooked.

Amino Acid Sequence Analysis and Gene Cloning

When reasonable evidence has been obtained that a particular band is responsible for the binding activity, a partial peptide sequence of the protein can be obtained. An alternative is to use the pure protein to generate antibodies, which can then be used to screen an expression library. Either method can work, although the peptide sequence strategy is the most commonly used. An example of the antibody strategy is the cloning of the TAFs within the TFIID complex (Hoey et al. 1993).

Box 8.3

Denaturation/Renaturation

Subject purified or partially
purified protein to SDS page

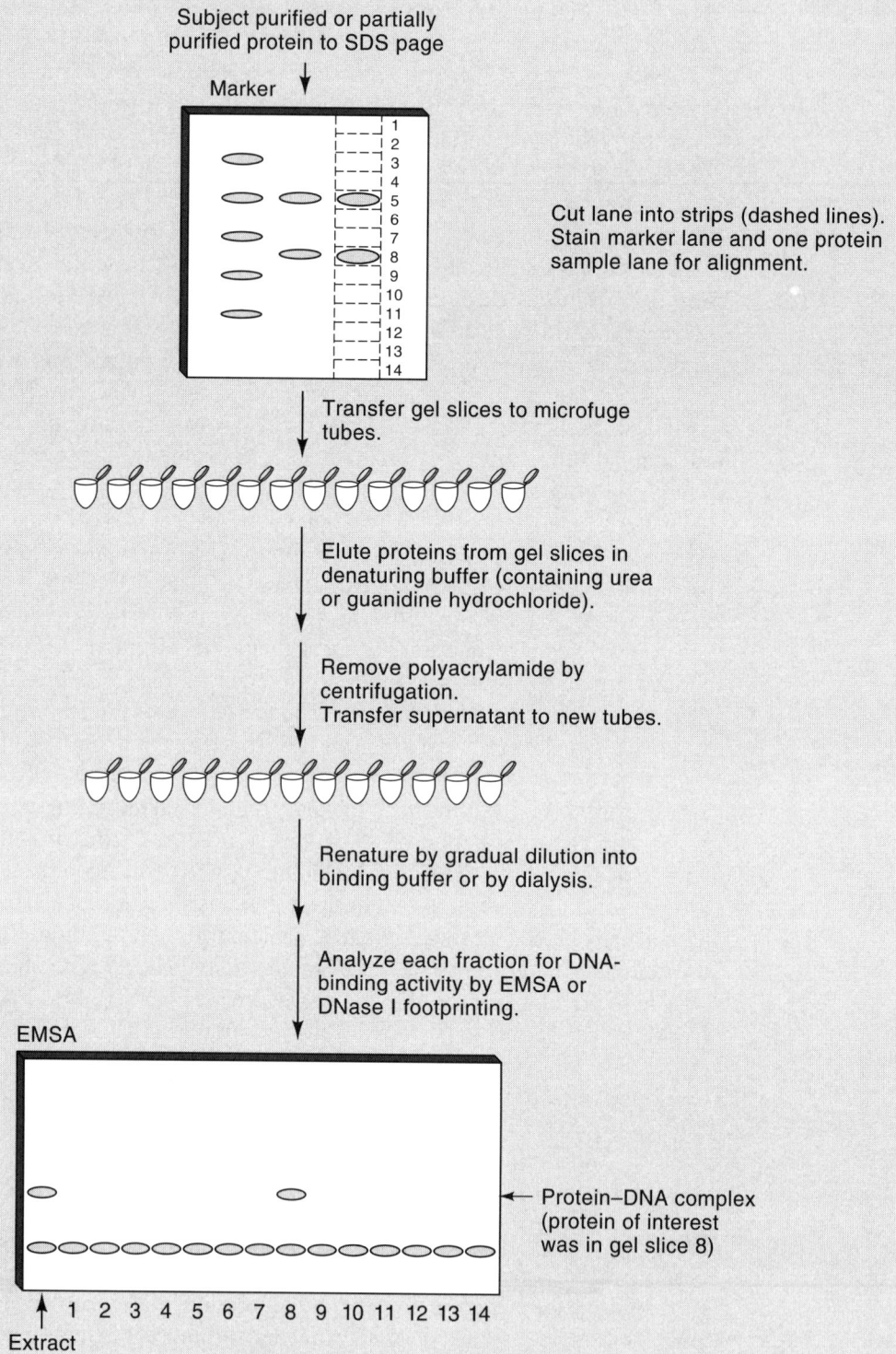

Cut lane into strips (dashed lines).
Stain marker lane and one protein
sample lane for alignment.

Transfer gel slices to microfuge
tubes.

Elute proteins from gel slices in
denaturing buffer (containing urea
or guanidine hydrochloride).

Remove polyacrylamide by
centrifugation.
Transfer supernatant to new tubes.

Renature by gradual dilution into
binding buffer or by dialysis.

Analyze each fraction for DNA-
binding activity by EMSA or
DNase I footprinting.

← Protein–DNA complex
(protein of interest
was in gel slice 8)

FIGURE 8.7. Denaturation/renaturation procedure.

The first step in the denaturation/renaturation procedure (Fig. 8.7) is to fractionate the proteins in the DNA affinity-purified sample by SDS-PAGE. It can be beneficial to apply as much protein as possible to the gel because only a small fraction of the original DNA-binding activity may be recovered. After electrophoresis, the relevant lane of the gel is separated into several gel slices using a razor blade. Molecular-weight markers and another aliquot of the purified protein can be run on adjacent lanes and stained (prestained markers can also be used) to determine the size range of each gel slice that was excised. The proteins within each gel slice are denatured and eluted from the polyacrylamide by soaking in an appropriate buffer. After elution, the proteins can be renatured by gradually diluting the denaturing solution into a non-denaturing buffer. Alternatively, the proteins can be dialyzed into a non-denaturing buffer. The presence of the specific DNA-binding activity can be determined by analyzing each sample in an EMSA or DNase I footprinting assay. The efficiency of renaturation varies considerably from protein to protein, with smaller proteins generally renaturing more efficiently than large proteins. Several different protocols for denaturation/renaturation have been published (see, e.g., Baeuerle and Baltimore 1988). To optimize renaturation for a protein of interest, it may be necessary to test multiple protocols, because different proteins may renature more effectively with one particular procedure.

Partial peptide sequences are usually obtained by collaboration with an established facility that uses automated Edman degradation or mass spectrometry. For Edman degradation analysis, it usually is necessary to cleave the protein chemically or enzymatically into small peptides. Small peptides must usually be sequenced because the amino terminus of most mammalian proteins is blocked by posttranslational modifications and therefore is resistant to sequence analysis. After proteolysis, the peptides are usually separated by reverse-phase high-performance liquid chromatography (HPLC) prior to amino acid sequence analysis of individual peptides. A useful discussion of sample preparation for peptide sequence analysis can be found in Ausubel et al. (1994; Unit 10.19). The Edman degradation and mass spectrometry technologies are not discussed further, as they are beyond the scope of this manual. In addition, the technology is improving rapidly, and specific procedures vary from facility to facility. For example, some facilities prefer to perform the proteolysis step on proteins within an excised polyacrylamide gel slice, whereas others prefer to have the protein transferred from the gel to a membrane for proteolysis. The facility to be used should be consulted for specific procedures.

After the peptide sequences are obtained, they can first be analyzed against a database to determine whether they correspond to a known protein or an expressed sequence tag (EST) (an appropriate starting point for searching EST databases is the National Center for Biotechnology Information website http://www.ncbi.nlm.nih.gov/). If so, the gene or gene fragment can be obtained from an appropriate source or isolated by RT-PCR. If the peptides are not represented in the databases, it will be necessary to clone the gene. The preferred strategy for cloning on the basis of peptide sequences is to perform PCR using primers representing the degeneracy of the DNA sequence encoding the peptide sequences (see Ausubel et al. 1994, Unit 7.7, and for codon usage tables, see http://www.dna.affrc.go.jp./%7Enakamura/CUTG.html). It is important to note that the probability of creating degenerate primers capable of amplifying a gene fragment will increase with increasing amounts of amino acid sequence information. Amino acid sequence information from several peptides can be beneficial because the degree of degeneracy varies from amino acid to amino acid. With considerable sequence information, one will be more likely to identify sequences of 4–7 amino acids that possess relatively low degeneracy.

Several different PCR strategies can be used to isolate gene fragments using degenerate primers. Perhaps the most common strategy is to perform PCR with primers oriented in opposite directions from two independent peptide sequences. Because the relative locations of the two peptides are usually unknown, consideration must be given to the possibility that either peptide is amino-terminal of the other within the protein. Thus, degenerate primers complementary to both strands should be prepared and tested. If possible, the primers should be designed so that the amplified DNA adjacent to the primer encodes a few amino acids of known sequence from the same peptide. In other words, a degenerate primer directed toward the 3′ end of a gene should not include sequences encoding the carboxy-terminal amino acids of the sequenced peptide. This primer design strategy has an important advantage, in that it will be easy to determine whether an amplified PCR product is truly relevant; if the product is relevant, it should contain the nucleotides encoding the known amino acids that were not represented in the primer. An alternative strategy for PCR is to use a single degenerate primer for 5′ or 3′ RACE. This strategy eliminates the need for "good" primers from two different peptides. Another alternative that was more common before PCR was developed is to use a radiolabeled degenerate primer to screen a cDNA library.

One common source of DNA for degenerate PCR is to simply perform a reverse transcriptase reaction with mRNA from the cell line of interest using an oligo(dT) primer or random priming. A second common source of DNA is a phage cDNA library stock. The PCR-amplified products can be detected on an agarose gel stained with ethidium bromide. In most cases, the expected size of the PCR product will be unknown. The PCR products can be analyzed by direct sequencing or by cloning into a standard vector and then sequencing. As mentioned above, the best initial evidence that a particular PCR product corresponds to a fragment of the gene of interest can be obtained from the finding that the PCR product encodes known amino acids from the peptides that were not encoded by the degenerate primers. If a relevant gene fragment is successfully amplified, it can be used to isolate a full-length cDNA by screening a cDNA library or by 5′ and 3′ RACE.

Confirmation That the Gene Isolated Encodes the DNA-binding Activity of Interest

Conclusive evidence that the gene isolated encodes the protein that was purified can be obtained if additional peptide sequences are available that were not used during the cloning steps. If these additional sequences are encoded by the gene that was isolated, one can safely conclude that the gene encodes the protein that was purified and sequenced.

Additional experiments are needed, however, to confirm that the protein encoded by the gene is responsible for the DNA-binding activity that was originally detected and characterized in crude extracts. In most instances, the only evidence linking the pure protein to the original binding activity at this point in the analysis will be a denaturation/renaturation, Southwestern, or crosslinking experiment. If two proteins comigrated on the gel, one the protein of interest and the other a contaminant, it is possible that the peptide sequence information was obtained for the contaminant, leading to cloning of the wrong gene.

To confirm that the cloned gene encodes the protein of interest, two experimental strategies are needed. First, the recombinant protein can be produced (by in vitro translation or expression in *E. coli*, baculovirus, etc.) and its DNA-binding properties compared to those of the protein originally detected in crude extracts. The recombinant protein should bind the same DNA sequence with the same nucleotide requirements as the protein that was purified. Furthermore, if the recombinant protein is full length, it may generate an EMSA complex that comigrates with the complex detected in crude extract, and it may generate a DNase I footprint or methylation interference pattern that is indistinguishable from that generated by the crude or purified protein. Strategies for preparing and analyzing recombinant proteins are described in Chapters 11 and 13.

The second strategy is to prepare antibodies against the recombinant protein, or against a synthetic peptide, and show that they react with the crude and purified protein. The antibodies should be capable of supershifting or disrupting the EMSA complex that was observed with the crude extract. If these experiments confirm that the gene encodes the protein of interest, the strategies described in Chapter 9 can be considered to assess the relevance of the encoded protein for the function of the control element to which it binds.

Cloning by Methods That Do Not Require a Protein–DNA Interaction Assay

One-hybrid Screen

The one-hybrid screen, typically performed in yeast, is rapidly emerging as the method of choice for identifying genes encoding proteins that bind a DNA element of interest (Fields and Song 1989; Li and Herskowitz 1993; Wang and Reed 1993; Inouye et al. 1994). The procedure begins with the construction of a reporter plasmid containing multiple copies of the binding site of interest upstream of a TATA box and a reporter gene, which is often the HIS3 or *lacZ* genes (Fig. 8.8). This reporter plasmid is stably integrated into the yeast genome. Next, a library containing cDNAs from an appropriate cell source is prepared. The library is designed so that the encoded proteins are synthesized as fusions with a strong transcriptional activation domain. The yeast cells harboring the reporter gene are transformed with the library. Yeast cells expressing a fusion protein capable of binding the multimerized site express the reporter gene. Reporter gene expression is usually monitored by a selection strategy. For example, expression of a HIS3 reporter is monitored by the growth of cell colonies on minimal medium lacking histidine. Finally, the cDNA can be isolated from the selected cells and further characterized to confirm that the encoded protein binds the DNA sequence of interest.

The one-hybrid approach has been used to clone several important DNA-binding proteins and it possesses several advantages. First, it is relatively straightforward and rapid. Second, the proteins are screened for binding under relatively native in vivo conditions. In contrast, protein purification and the expression library screening method described below depend on binding in vitro. Although mammalian proteins expressed in yeast are unlikely to acquire all of the posttranslational modifications found in mammalian cells, they may acquire one or more essential modifications. A final advantage is that the method is often extremely sensitive and can identify proteins that bind with only modest affinity.

These advantages need to be balanced against two notable disadvantages that were discussed above. Namely, the method will generally succeed only if binding requires a protein monomer, homodimer, or homomultimer, and the procedure will not necessarily lead to the identification of the gene encoding a protein that was detected in an EMSA or DNase I footprinting assay using a crude mammalian extract. An additional caveat of the one-hybrid approach is that the binding site used must not interact with an endogenous yeast activator protein. If a yeast activator binds the site, reporter gene expression will be observed in the absence of the relevant cDNA expression plasmid. To overcome this prob-

**Multiple copies
of DNA
element
of
interest**

TATA

Reporter gene
(e.g., HIS3)

cDNA library

cDNA A | Act-domain

cDNA B | Act-domain

Stably integrate into
yeast genome.

Transform yeast cells with cDNA
library.

Select for reporter gene
expression (e.g., by growth on
minimal medium. Lacking histidine
for the HIS3 reporter gene).

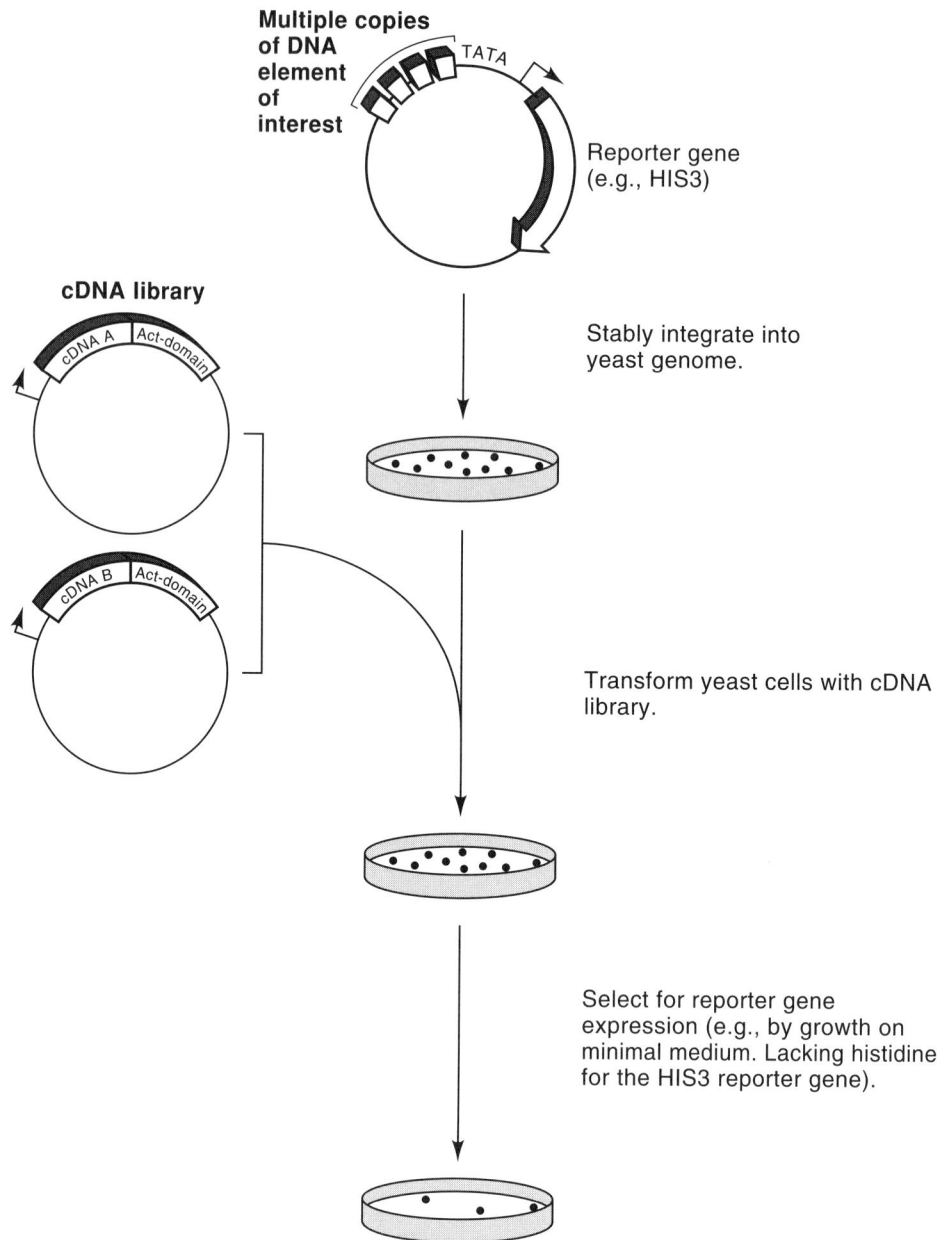

FIGURE 8.8. One-hybrid screening strategy.

lem, it may be possible to use variants of the binding site that retain activity in their natural context, but do not bind yeast activators.

Despite the above disadvantages, the one-hybrid procedure is strongly recommended as a starting point for identifying a gene if there is a reasonable chance the site can be recognized by a monomer, homodimer, or homomultimer. Detailed procedures for performing a one-hybrid screen can be found in the articles cited above or can be obtained from CLONTECH, which markets a kit for performing one-hybrid screens (MATCHMAKER One-Hybrid System, Cat #K1603-1) and a large number of premade cDNA libraries. It also offers custom libraries.

In Vitro Expression Library Screening with DNA or Antibody Probes

Before the one-hybrid screening method was developed, a common method for cloning genes encoding novel DNA-binding proteins involved an expression library screen using radiolabeled DNA probes. This method was developed in 1988 by Harinder Singh, Phillip Sharp, and colleagues (Singh et al. 1988) and was an important advance in gene cloning technology.

To perform this technique, a cDNA library prepared with mRNA from the cell type of interest is needed. The library must be constructed in a lambda phage vector that allows inducible expression of the protein encoded by the inserted cDNA after infection of *E. coli*. The recommended vector is λgt11, which expresses proteins as β-galactosidase fusion proteins. Successful results have been obtained with libraries prepared by priming with either oligo(dT) or random oligonucleotides. The phage library is plated on bacteria under lytic growth conditions (Fig. 8.9). When plaques appear, transcription of the cDNA is induced by treatment with the inducing agent isopropyl-β-D-thiogalactoside (IPTG). This is accomplished by placing a nitrocellulose filter soaked in an IPTG solution on the plates. As the phage lyse, the induced proteins adhere to the nitrocellulose filters. After removal from the plates, the filters are incubated in a blocking solution to prevent nonspecific interactions between the probe and nitrocellulose filter. Then, the filters are probed with a radiolabeled DNA fragment containing multiple copies of the binding site of interest. The probe should bind specifically to plaques containing sequence-specific DNA-binding proteins. A nonspecific competitor DNA is included in this binding reaction to suppress nonspecific interactions by the radiolabeled probe. In some procedures (e.g., Vinson et al. 1988), the expressed proteins on the filter are first subjected to a denaturation/renaturation procedure, with the hope that the number of properly folded protein molecules capable of binding the probe will be increased. After the "positive" plaques are isolated and rescreened for the purpose of plaque purification, lysogen extracts can be prepared to allow the recombinant fusion protein to be characterized by EMSA.

Like the one-hybrid approach, this approach will not succeed if the protein must bind DNA as a heteromer and may not lead to the identification of the gene encoding the protein identified in an EMSA or DNase I footprinting assay. One additional challenge of the expression library screening method has been the generation of probe molecules with sufficiently high specific activity and a sufficient number of binding-site copies. In the original protocol, probes were prepared by excising from an appropriate plasmid a restriction fragment containing multiple copies of the binding site, followed by phosphorylation (Singh et al. 1988). New PCR methodology for preparation of long multimeric probes, as described above for the preparation of DNA affinity columns, may greatly enhance the success achieved with this approach (Hemat and McEntee 1994; Schmitt et al. 1996).

For most studies, the one-hybrid screen is probably more attractive than the in vitro expression library screen for three reasons. First, the one-hybrid screen does not require the time-consuming phage titration steps that are generally required before an in vitro expression library screen can be performed. Second, the one-hybrid screen does not require large amounts of radioactivity. Third, the one-hybrid screen is less susceptible to the protein folding and stability problems that can be encountered during the in vitro screen. For these reasons, the in vitro expression library screen is recommended primarily if the one-hybrid screen fails to identify the relevant activator or if the one-hybrid screen is not feasible because an endogenous yeast activator binds the site of interest. A detailed description of the expression library screening method can be found in Ausubel et al. (1994; Unit 12.7).

Infect *E. coli* with cDNA expression library and plate on lawn of *E. coli* under lytic growth conditions.

Induce transcription of cDNA (under the control of the lac promoter/operator) by adding a nitrocellulose filter soaked in IPTG.

Remove filters.

Incubate in blocking solution (to block nonspecific probe–nitrocellulose interaction).

^{32}P radiolabeled probe

Incubate filters with radiolabled DNA probe containing multimers of DNA element of interest.

Wash filters.

Analyze by autoradiography.

Excise positive plaques from plate.

Perform secondary screen to purify plaques.

Prepare lysogen extract to analyze protein–DNA interaction.

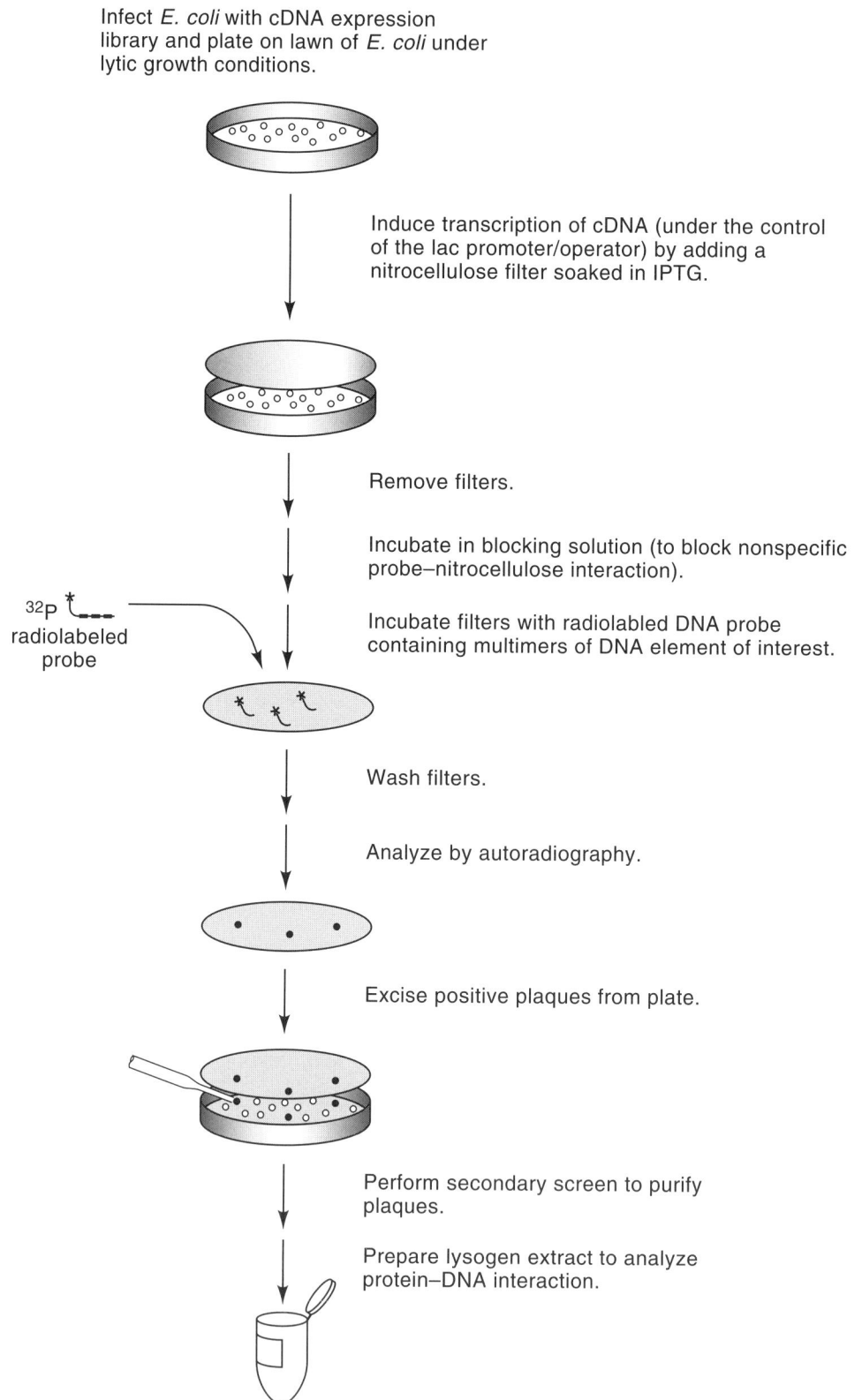

FIGURE 8.9. In vitro expression cloning.

Mammalian Expression Cloning Methods

Mammalian expression cloning methods have been used to identify new genes for over two decades. The basic principle is that a gene of interest is transferred to a recipient mammalian cell line, usually by transfection with sheared genomic DNA or a cDNA expression library. The cells that take up the gene of interest are identified by either selection or screening.

An early example of the use of a selection strategy for expression cloning was the identification of proto-oncogenes on the basis of their ability to confer a transformed phenotype on recipient cells (see, e.g., Shih et al. 1979). Screening strategies have been used to isolate dozens of new genes, including several genes encoding cell-surface proteins. For one common screening procedure, antibodies directed against the protein of interest and a cDNA expression library in a plasmid vector are needed (see Seed 1987; Ausubel et al. 1994, Unit 6.11). The expression library contains the cDNAs downstream of a strong promoter and enhancer. The plasmids also contain an SV40 replication origin, which allows replication to high copy number in SV40 T-antigen-expressing COS-7 monkey kidney cells. The library is divided into several pools, each containing a diverse mixture of clones. The pools are then introduced into the COS-7 cells, resulting in plasmid replication and cDNA expression. To isolate the gene encoding a cell-surface protein, each transfected cell population is analyzed for protein expression by panning. In panning, a specific antibody is immobilized on a plastic dish, facilitating the adherence of cells expressing the cell-surface protein and the removal of nonexpressing cells. Plasmid DNAs from the adherent cells are isolated and amplified in *E. coli*. The resulting plasmids are again divided into pools and the entire procedure (i.e., transfection, panning, plasmid recovery, and amplification) is repeated several times, eventually leading to a single plasmid species containing the cDNA of interest.

Because of the success of the mammalian expression cloning strategies described above, several laboratories attempted to use similar procedures to clone transcription factors that bind defined control elements or control regions. For some experiments, synthetic promoters containing multiple copies of the site of interest were used to drive expression of a dominant selectable marker gene. For others, the gene encoding a cell-surface protein was placed downstream of a synthetic promoter, with the hope that the panning procedure described above could be used to isolate a cDNA encoding the relevant transcription factor. Native tissue-specific promoters were also tested, with the expectation that the recipient cells would contain all of the ubiquitous factors needed for transcription, allowing the promoter to function when a cDNA encoding an essential tissue-specific factor was provided. Unfortunately, these strategies, to our knowledge, were never successful. Their failure was presumably due to a combination of factors, the most serious of which may have been signal-to-noise difficulties resulting from the inability of a single transcription factor to induce transcription to a sufficient extent over background.

Although the above strategies have not been used successfully to clone transcription factors that bind specific control elements, they occasionally have been used to clone genes encoding other proteins that are important for gene regulation. One example is the cloning of the gene encoding SCAP (SREBP cleavage-activating protein; Hua et al. 1996), which regulates the function of SREBP-1, the sterol response element binding protein (Yokoyama et al. 1993). To clone SCAP, a plasmid cDNA library was transfected into human 293 cells in pools of about 1000 clones, along with a plasmid containing a luciferase reporter gene under the control of multiple SREBP-1 binding sites. The library was prepared from mRNA from sterol-resistant cells, and the goal of the study was to isolate the gene responsible for sterol resistance. In response to sterol overload, normal cells down-regulate the expression of SREBP-1-responsive genes. Thus, cells that express the protein that confers sterol resistance should exhibit higher luciferase activity than normal 293 cells in the presence of high concentrations of sterol. Indeed, a plasmid pool was identified that yielded high luciferase activity. Sequential analysis of plasmid subpools led to the cloning of the SCAP gene. Functional analysis of SCAP revealed that it cleaves membrane-bound SREBP-1, allowing it to translocate to the nucleus and activate transcription. The isolation of the SCAP gene by expression cloning raises the possibility that improved technology for preparing and analyzing plasmid libraries will allow the cloning of other transcriptional regulators by similar methods.

Genome Database Methods and Degenerate PCR

A chapter on transcription factor cloning methods is not complete without a brief discussion of the use of genome databases and degenerate PCR for the identification of new genes that may functionally interact with a DNA element of interest. These approaches may become useful after a gene encoding a protein that binds the DNA element has been identified by any of the strategies described above. One scenario is that a careful analysis of the gene, using the strategies described in Chapter 9, will reveal that it does not carry out a functional interaction with the DNA element. However, the results may be consistent with the possibility that another member of the same protein family is the functionally relevant regulator. Therefore, the new goal may be to isolate novel members of the same protein family. Genome database methods and degenerate PCR are usually the easiest and fastest route toward achieving this goal.

EST databases and other genome databases are likely to reveal genes encoding novel proteins with homology to the DNA-binding domain of the protein that was originally isolated. The National Center for Biotechnology Information web site (http://www.ncbi.nlm.nih.gov/) provides an appropriate starting point for a database analysis, as mentioned earlier in this chapter. A second approach is to isolate new genes by PCR, using primers representing the degeneracy of the DNA encoding a specific amino acid sequence within the DNA-binding domain (see Ausubel et al. 1994, Unit 7.7, and the following Web sites: http://blocks. fhcrc.org/blocks/codehop.html and http://bibiserv.techfak.uni-bielefeld.de/genefisher/main.html). Because the DNA-binding domains of several transcription factor families are highly conserved among members of the family or subfamily, degenerate PCR often allows the isolation of new genes encoding family members capable of binding similar DNA sequences. One potential advantage of degenerate PCR relative to the database search is that the PCR assays can be performed with cDNAs prepared from any cell type. In contrast, EST databases currently contain information from only a handful of cell types. In the near future, however, the entire human and murine genome sequences will be available, making degenerate PCR obsolete and greatly enhancing the power of database approaches.

A preliminary analysis of the expression patterns of the new genes identified by degenerate PCR or a database search may allow some or most to be ruled out as viable candidates because they may not be expressed in the appropriate cell types. The candidates that remain can be expressed in *E. coli* or by in vitro transcription/translation and tested for their ability to bind the control element of interest. If binding is observed, the functional relevance of each candidate gene can be assessed using the strategies described in the next chapter.

REFERENCES

Ausubel F.M., Brent R.E., Kingston E., Moore D.D., Seidman J.G., Smith J.A., and Struhl K. 1994. *Current protocols in molecular biology.* John Wiley and Sons, New York.

Baeuerle P.A. and Baltimore D. 1988. Activation of DNA-binding activity in an apparently cytoplasmic precursor of the NF-κB transcription factor. *Cell* 53: 211–217.

Dignam J.D., Lebovitz R.M., and Roeder R.G. 1983. Accurate transcription initiation by RNA polymerase II in a soluble extract from isolated mammalian nuclei. *Nucleic Acids Res.* 11: 1475–1489.

Dynan W.S. and Tjian R. 1983. The promoter-specific transcription factor Sp1 binds to upstream sequences in the SV40 early promoter. *Cell* 35: 79–87.

Emami K.H., Jain A., and Smale S.T. 1997. Mechanism of synergy between TATA and initiator: Synergistic binding of TFIID following a putative TFIIA-induced isomerization. *Genes Dev.* 11: 3007–3019.

Ernst P., Hahm K., and Smale S.T. 1993. Both LyF-1 and an Ets protein interact with a critical promoter element in the murine terminal transferase gene. *Mol. Cell. Biol.* 13: 2982–2992.

Fields S. and Song O. 1989. A novel genetic system to detect protein-protein interactions. *Nature* 340: 245–246.

Ghosh S., Gifford A.M., Riviere L.R., Tempst P., Nolan G.P., and Baltimore D. 1990. Cloning of the p50 DNA binding subunit of NF-κB: Homology to rel and dorsal. *Cell* 62: 1019–1029.

Gorski K., Carneiro M., and Schibler U. 1986. Tissue-specific in vitro transcription from the mouse albumin promoter. *Cell* 47: 767–776.

Gunther C.V., Nye J.A., Bryner R.S., and Graves B.J. 1990. Sequence-specific DNA binding of the proto-oncoprotein ets-1 defines a trancriptional activator sequence within the long terminal repeat of the Moloney murine sarcoma virus. *Genes Dev.* 4: 667–678.

Hahm K., Ernst P., Lo K., Kim G.S., Turck C., and Smale S.T. 1994. The lymphoid transcription factor LyF-1 is encoded by specific, alternatively spliced mRNAs derived from the Ikaros gene. *Mol. Cell. Biol.* 14: 7111–7123.

Harlow E. and Lane D. 1999. *Using antibodies: A laboratory manual.* Cold Spring Harbor Laboratory Press, Cold Spring Harbor, New York.

Hemat F. and McEntee K. 1994. A rapid and efficient PCR-based method for synthesizing high-molecular-weight multimers of oligonucleotides. *Biochem. Biophys. Res. Comm.* 205: 475–481.

Hoey T., Weinzierl R.O., Gill G., Chen J.L., Dynlacht B.D., and Tjian R. 1993. Molecular cloning and functional analysis of *Drosophila* TAF110 reveal properties expected of coactivators. *Cell* 72: 247–260.

Hua X., Nohturfft A., Goldstein J.L., and Brown M.S. 1996. Sterol resistance in CHO cells traced to point mutation in SREBP cleavage-activating protein. *Cell* 87: 415–426.

Inouye C., Remondelli P., Karin M., and Elledge S. 1994. Isolation of a cDNA encoding a metal response element binding protein using a novel expression cloning procedure: The one hybrid system. *DNA Cell Biol.* 13: 731–742.

Kadonaga J.T. 1991. Purification of sequence-specific binding proteins by DNA affinity chromatography. *Methods Enzymol.* 208: 10–23.

Kadonaga J.T. and Tjian R. 1986. Affinity purification of sequence-specific DNA binding proteins. *Proc. Natl. Acad. Sci.* 83: 5889–5893.

Landschulz W.H., Johnson P.F., Adashi E.Y., Graves B.J., and McKnight S.L. 1988. Isolation of a recombinant copy of the gene encoding C/EBP. *Genes Dev.* 2: 786–800.

Lee W., Mitchell P., and Tjian R. 1987. Purified transcription factor AP-1 interacts with TPA-inducible enhancer elements. *Cell* 49: 741–752.

Li J.J. and Herskowitz I. 1993. Isolation of ORC6, a component of the yeast origin recognition complex by a one-hybrid system. *Science* 262: 1870–1874.

Lo K., Landau N.R., and Smale S.T. 1991. LyF-1, a transcriptional regulator that interacts with a novel class of promoters for lymphocyte-specific genes. *Mol. Cell. Biol.* **11:** 5229–5243.

Marshak D.R., Kadonaga J.T., Burgess R.R, Knuth M.W., Brennan W.A. Jr., and Lin S.-H. 1996. *Strategies for protein purification and characterization: A laboratory manual.* Cold Spring Harbor Laboratory Press, Cold Spring Harbor, New York.

Pierce J.W., Lenardo M., and Baltimore D. 1988. Oligonucleotide that binds nuclear factor NF-κB acts as a lymphoid-specific and inducible enhancer element. *Proc. Natl. Acad. Sci.* **85:** 1482–1486.

Sambrook J., Fritsch E.F., and Maniatis T. 1989. *Molecular cloning: A laboratory manual.* Cold Spring Harbor Laboratory Press, Cold Spring Harbor, New York.

Schmitt A.P. and McEntee K.1996. Msn2p, a zinc finger DNA-binding protein, is the transcriptional activator of the multistress response in *Saccharomyces cerevisiae. Proc. Natl. Acad. Sci.* **93:** 5777–5782.

Seed B. 1987. An LFA-3 cDNA encodes a phospholipid-linked membrane protein homologous to its receptor CD2. *Nature* **329:** 840–842.

Shapiro D.J., Sharp P.A., Wahli W.W., and Keller M.J. 1988. A high-efficiency HeLa cell nuclear transcription extract. *DNA* **7:** 47–55.

Shih C., Shilo B.Z., Goldfarb M.P., Dannenberg A., and Weinberg R.A. 1979. Passage of phenotypes of chemically transformed cells via transfection of DNA and chromatin. *Proc. Natl. Acad. Sci.* **76:** 5714–5718.

Singh H., LeBowitz J.H., Baldwin A.S. Jr., and Sharp P.A. 1988. Molecular cloning of an enhancer binding protein: Isolation by screening of an expression library with a recognition site DNA. *Cell* **52:** 415–423.

Soeller W.C., Poole S.J., and Kornberg T. 1988. In vitro transcription of the *Drosophila* engrailed gene. *Genes Dev.* **2:** 68–81.

Vinson C.R., LaMarco K.L., Johnson P.F., Landschulz W.H., and McKnight S.L. 1988. In situ detection of sequence-specific DNA binding activity specified by a recombinant bacteriophage. *Genes Dev.* **2:** 801–806.

Wang M.M. and Reed R.R. 1993. Molecular cloning of the olfactory neuronal transcription factor Olf-1 by genetic selection in yeast. *Nature* **364:** 121–126.

Yokoyama C., Wang X., Briggs M.R., Admon A., Wu J., Hua X., Goldstein J.L., and Brown M.S. 1993. SREBP-1, a basic-helix-loop-helix-leucine zipper protein that controls transcription of the low density lipoprotein receptor gene. *Cell* **75:**187–197.

CHAPTER 9

Confirming the Functional Importance of a Protein–DNA Interaction

Important issues

- *The functional relevance of a protein–DNA interaction is difficult to establish.*

- *The hypothesis that an interaction is relevant can be tested by several different experiments, although none by itself is conclusive.*

- *The combined results of several experimental approaches are needed to rigorously examine the relevance of a protein–DNA interaction.*

INTRODUCTION

Discovering a gene encoding a novel DNA-binding protein has become a relatively straightforward task in the modern era of molecular biology, but definitively establishing that the protein directly regulates a target gene by binding to a defined control element is among the most difficult of tasks.

The preceding chapters described experimental strategies that lead to the identification of important *cis*-acting sequence elements, as well as proteins that bind those elements. In most instances, a protein identified has been implicated as a potential regulator of the gene of interest because it interacts with an important control element in an electrophoretic mobility shift assay (EMSA) or DNase I footprinting experiment. The identification of a specific DNA-binding protein provides a significant advance because it allows one to hypothesize that the protein is responsible for the function of the control element in vivo. However, by itself, the identification of a protein–DNA interaction does not demonstrate its relevance.

The detection of a protein–DNA interaction in a crude cell extract reflects many factors: (1) the abundance of the protein in the cells from which the extract was prepared, (2) the efficiency with which the protein was extracted from the cells, (3) the stability of the active protein within the extract, (4) the maintenance of essential posttranslational modifications during extract preparation, (5) the conditions used for the in vitro DNA-binding assay, and (6) the affinity of the protein for the isolated control element (Table 9.1).

The above criteria for detecting a protein–DNA interaction in vitro are very different from the criteria that determine which protein functionally interacts with the control element in vivo (i.e., which protein regulates the endogenous gene by binding to the control element of interest) (Table 9.1). These criteria include (1) the abundance and stability of the protein in the cell nucleus, (2) the affinity of the protein for the site, (3) the ability of the protein to carry out appropriate interactions with other proteins bound to adjacent sites and with non-DNA-binding cofactors, (4) appropriate posttranslational modifications that allow the protein to carry out the necessary protein–DNA and protein–protein interactions, and (5) the appropriate subnuclear localization of the protein.

When considering the above points, it is readily apparent that the detection of a protein–DNA interaction in vitro provides only weak evidence that it is relevant in vivo, even when the DNA sequence element to which the protein binds is known to be important (i.e., by mutagenesis). As discussed in Chapter 7, most DNA-binding proteins are capable of recognizing a wide range of DNA sequences with a wide range of affinities.

TABLE 9.1. *Factors that influence the in vitro detection and in vivo relevance of protein–DNA interactions*

Detection in vitro	Functional relevance in vivo
Abundance of protein in cell	Abundance and stability of protein in cell nucleus
Efficiency of protein extraction	Affinity of protein for the site
Stability of protein in extract	Ability of protein to carry out appropriate interactions with other proteins
Maintenance of essential post-translational modifications during extract preparation	Posttranslational modifications that allow the protein to carry out necessary interactions
Conditions used for in vitro DNA-binding assay	Appropriate subcellular and subnuclear localization of the protein
Affinity of the protein for the isolated control element	

Furthermore, most DNA-binding proteins are members of multiprotein families, with each cell type containing several family members that recognize similar DNA sequences. On the basis of these considerations, there is a high probability that multiple proteins will be capable of binding a defined control element in vitro, including several members of a particular protein family, and perhaps members of another family that recognize a similar or overlapping sequence. The difficult challenge is to determine which of these proteins is capable of carrying out the protein–protein and protein–DNA interactions that allow it to regulate the endogenous gene. If by chance only one predominant DNA-binding protein is detected in vitro, it may or may not be the one that is responsible for the function of the control element in vivo; other proteins within the cells will almost certainly be capable of binding the same element, even if they were not detected in the initial EMSA or DNase I footprinting studies.

A definitive approach for confirming the functional importance of a protein–DNA interaction in mammalian cells has not yet been developed. In fact, there are few examples of protein–DNA interactions that have been shown conclusively to be relevant. However, the situation is not as bleak as the above discussion might make it appear. Athough the functional relevance of most protein–DNA interactions that have been reported in higher eukaryotes remains tenuous (including some of those studied in the laboratories of the authors), several have been shown to be relevant beyond reasonable doubt. In addition, it should be noted that relevant binding sites for some classes of transcription factors, such as ligand-inducible nuclear hormone receptors, are relatively easy to identify.

In the absence of a definitive experiment (i.e., an experiment that allows one to visualize directly a protein binding to a control element within its natural chromosomal location and regulating transcription of the linked gene), the only viable approach is to hypothesize that a protein–DNA interaction is relevant, and then to subject that hypothesis to as many rigorous tests as possible. This approach, of course, is not new, because it is a central tenet of the scientific method. If a number of independent experimental tests support the hypothesis, confidence that it might be correct is enhanced. The point at which a hypothesis of this sort has been confirmed beyond a reasonable doubt depends on a subjective evaluation and, therefore, is best left to the judgment of the scientific community.

In this chapter, twelve different approaches which can be used to test a hypothesis that a specific protein–DNA interaction is functionally important are presented. We describe the information gained from each approach and explain why each yields inconclusive results. The approaches vary widely with respect to the amount of effort required and the quality of information obtained. If a major goal is to establish the functional importance of an interaction, as many of these approaches as possible should be pursued.

Before proceeding, it is important to emphasize that a rigorous examination of the relevance of a protein–DNA interaction might not be an important objective of the laboratory. The primary interest may instead be to determine whether the DNA-binding protein that was discovered is important for a biological process. In this instance, the next step is more obvious: The gene encoding the protein can be disrupted in a cell or animal. If an interesting phenotype is observed, the laboratory would likely choose to begin a broadly based analysis of the DNA-binding protein, regardless of whether it carries out a functional interaction with the control element that originally was used for its identification. This is a common and valid course of action. However, if the long-term goal is to link a DNA-binding protein to its relevant target genes, or to carefully dissect the mechanism by which a target gene is induced by a constellation of DNA-binding proteins, cofactors, and general transcription factors, the issues discussed in this chapter will ultimately need to be considered.

CONCEPTS AND STRATEGIES

Abundance of a Protein–DNA Complex In Vitro

As stated above, the abundance of a protein–DNA complex in an in vitro binding assay, such as an EMSA or DNase I footprinting assay, is often used to develop an initial hypothesis that a given protein is responsible for the function of a control element. A predominant protein–DNA complex might be observed, for example, in an EMSA using a radiolabeled probe containing an important control element and a nuclear extract prepared from a relevant cell line or tissue. Although one can hypothesize that the interaction is relevant, the hypothesis is only weakly supported by this result for the reasons discussed in the introduction to this chapter.

The interleukin-12 (IL-12) promoter analysis discussed in Chapter 7 provides an example of this concept. To determine which protein is responsible for the function of the IL-12 promoter element between −88 and −99, an EMSA experiment was employed using extracts from uninduced and induced macrophages (Plevy et al. 1997). Several distinct protein–DNA complexes were detected. By adding specific antibodies to the binding reactions, the most abundant complexes were found to contain the C/EBPβ protein. Less abundant complexes contained C/EBPδ and C/EBPα. These results led to the hypothesis that C/EBPβ is responsible for the function of the −88/−99 element. However, the data do not rule out the possibility that, instead, C/EBPδ, C/EBPα, or another binding protein that does not yield a detectable EMSA complex is functionally relevant. C/EBPβ may yield the most abundant complex because it (1) is the most abundant protein in the cell extracts, (2) binds to the isolated element with highest affinity, (3) is extracted from the cells more efficiently than other proteins, (4) is more stable in the extract than other proteins, (5) maintains posttranslational modifications that enhance DNA binding more effectively than other proteins, or (6) is more compatible with the particular ESMA conditions used than the other proteins. However, a different protein may be functionally relevant in vivo for any of a number of reasons, including the following: (1) A different protein may carry out optimal physical and functional interactions with other proteins that bind the promoter, (2) a protein that was less abundant than C/EBPβ in the cell extract, and therefore undetectable, may bind the element with higher affinity in vivo, or (3) a protein that cannot be detected with the in vitro EMSA conditions used may preferentially bind the control element in vivo.

For the IL-12 promoter analysis, it will be particularly difficult to determine which C/EBP family member is functionally relevant. Although the most abundant EMSA complex contains C/EBPβ, C/EBPδ binds the same sequence, and like C/EBPβ, its abundance increases upon macrophage activation. The challenge of distinguishing between members of a multiprotein family is discussed frequently in this chapter.

Although multiprotein families contribute considerably to the challenge of establishing the functional relevance of a specific protein–DNA interaction, it can also be difficult to identify the relevant protein from among different families. The analysis of the terminal transferase (TdT) gene, which is expressed in immature lymphocytes, provides an example of this issue. DNase I footprinting studies led to the identification of a protein in nuclear extracts from immature lymphocytes that bound a critical control element in the TdT promoter, called the D′ element (Lo et al. 1991). Upon purification and cloning, this protein was found to be Ikaros, suggesting that Ikaros or an Ikaros family member was the relevant activator of TdT transcription through the D′ element (Georgopoulos et al. 1992; Hahm et al. 1994). However, subsequent experiments revealed that members of the Ets family of transcription factors are capable of binding the D′ element with considerable affinity, even though these proteins were not easily detected in EMSA or footprinting experiments with crude nuclear extracts (Ernst et al. 1993). By subjecting the Ikaros-D′ and Ets-D′ interactions to many of the tests described below, the relevance of the Ets-D′ interaction for TdT activation was strongly supported, and the relevance of the Ikaros-D′ interaction was not (Ernst et al. 1996). It remains unknown why Ikaros is the predominant D′-binding protein in nuclear extracts from TdT-expressing cells, despite several lines of evidence suggesting that it is not the relevant activator.

Relative Expression Patterns of the DNA-binding Protein and Target Gene

A second straightforward test of the hypothesis that a protein–DNA interaction is functionally relevant is a comparison of the expression patterns of the DNA-binding protein and putative target gene. This test can be carried out with transformed cell lines and primary cells by first quantifying steady-state mRNA levels or nascent transcripts (see Chapter 3) from the putative target gene. Target gene transcription can then be compared to the expression pattern of the proposed regulatory protein, which can be monitored by measuring steady-state protein levels (using immunoblot, immunoprecipitation, immunofluorescence, flow cytometry, or a biochemical assay).

The methods that are available for monitoring each DNA-binding protein and target gene vary. However, the preferred comparison would involve assays that are most meaningful for demonstrating a relationship between the DNA-binding protein and target gene; namely, a biochemical assay to quantify the DNA-binding activity (or capacity for transcriptional activation) of the protein, and a nuclear run-on assay to monitor nascent transcription of the target gene. If these assays are impractical because of the large number of cells needed to monitor DNA binding and the difficulty of the nuclear run-on, an alternative is to quantify the abundance of the DNA-binding protein and the steady-state mRNA abundance for the target gene.

The information gained by carefully comparing the expression patterns of the DNA-binding protein and target gene can either support or help to rule out a hypothesis. If the DNA-binding protein of interest is present in all of the cell types that express the target gene, the results would support the hypothesis that the protein is a relevant activator of the target. If, on the other hand, cells are identified that express the target gene, but not the putative activator, the hypothesis would be weakened. The hypothesis would not be negated, however, because some genes are activated by different sets of factors in different cell types (see, e.g., Lauring and Schlissel 1999).

It is important to note that a target gene is rarely, if ever, expressed in all of the cell types that express a relevant transcriptional activator. More likely, it is expressed in only a subset of the cell types expressing an activator. According to basic combinatorial principles of gene regulation (see Chapter 1), a given DNA-binding protein contributes to the transcription of many genes with varying expression patterns by acting in conjunction with several other transcription factors, each possessing its own unique expression pattern. In addition, numerous transcription factors are regulated by posttranscriptional mechanisms, allowing them to be present in an inactive state (or in an inappropriate subcellular compartment) in cells that do not express relevant target genes (e.g., NF-κB, nuclear factor of activated T cells [NFAT], nuclear hormone receptors). For these reasons, a perfect correlation between activator and target gene is almost never observed; the most one can expect is the presence of the activator in all cells that express the target.

One additional limitation of the correlation between activator and target gene is that it does not provide a direct functional link between the two. The correlation may be fortuitous or the activator may indirectly regulate the target gene by regulating one of the gene's direct regulators.

For genes that are inducible in a cell line or tissue, the basic expression pattern analysis can be enhanced by a comparison of the kinetics of induction of the transcription factor and putative target gene. The cells can be treated with an appropriate inducing agent and the time course of induction of the DNA-binding activity in nuclear extracts can be compared to the time course of induction of the target gene (or a chromosomally integrated reporter gene). If the induction kinetics are similar or if the DNA-binding activity is induced slightly earlier than the target gene, the results would be consistent with the hypothesis that the protein regulates the gene. For some classes of transcription factors, such as nuclear hormone receptors, kinetic experiments can provide strong evidence that a protein–DNA interaction is relevant.

One caveat to the interpretation of kinetic experiments is that the precise concentration of a DNA-binding protein that needs to be present for a target gene to be activated is usually unknown. Perhaps the target gene can be activated when the DNA-binding protein reaches a concentration that is only 10% of its maximum concentration. If a target gene is transcribed when its putative activator is present at only 10% of its maximal concentration, the results at first glance would suggest that target gene induction is independent of the activator. In some instances, it has been found that a high threshold concentration of a critical DNA-binding protein must be present for target gene activation (Fiering et al. 1990). If the threshold concentration of a protein is indeed high, a kinetic analysis may be informative. However, in the absence of information about the threshold concentration, it is difficult to evaluate experiments that attempt to correlate the kinetics of induction of a transcription factor and target gene.

Correlation between Nucleotides Required for Protein Binding and Those Required for Activity of the Control Element

A third fundamental test of the hypothesis that a protein–DNA interaction is functionally relevant involves a detailed comparison of the nucleotides required for the function of the control element and the nucleotides required for binding of the putative transcription factor. This is an extremely powerful and underutilized method that can provide relatively strong support for the hypothesis that a particular DNA-binding protein, or at least a member of a particular family of DNA-binding proteins, is responsible for the activity of the control element.

The TdT promoter analysis provides an example of this strategy. As mentioned previously, the important D′ element can interact with both Ikaros and Ets proteins. Ikaros was the predominant D′-binding protein observed in extracts from TdT-expressing cells, leading to the hypothesis that it might be the functional activator. The expression patterns of Ikaros and Ets proteins did not help to determine which protein was the functional activator, because proteins of both families were expressed in all TdT-expressing cells examined. A detailed mutant analysis of the D′ element appears to have been a useful strategy for identifying the relevant protein family (Table 9.2) (Ernst et al. 1996). A series of single- and double-base substitutions in the TdT D′ element were constructed and tested for their effect on promoter function in transient and stable transfection experiments, and for their effect on the binding of Ikaros and Ets proteins. As summarized in Table 9.2, the results revealed that the nucleotides required for Ikaros binding were significantly different from the nucleotides required for promoter activity in both transfection assays. Two different

TABLE 9.2. *Effect of substitution mutations on promoter function and protein binding to the TdT D´ element*

D´ Sequence	Promoter function	Ets binding	Ikaros binding
WT G C A G G A A G T T G	+	+	+
m83 G A A G G A A G T T G	+/–	+/–	+
m84 G C C G G A A G T T G	+++	++	–
m85 G C A T T A A G T T G	–	–	–
m86 G C A G G C C G T T G	–	–	–
m87 G C A G G A A T T T G	–	–	+
m88 A C C G G A A G T A G	+++	++	–

mutations enhanced promoter activity while abolishing Ikaros binding, and another mutation abolished promoter activity without affecting Ikaros binding. In contrast, the nucleotides required for promoter activity matched precisely the nucleotides required for the binding of various proteins of the Ets family.

The above results support the hypothesis that an Ets family protein is a functional activator of TdT transcription. However, this strategy has significant limitations. In particular, although the strategy can be useful for addressing the relevance of different families of proteins, it usually will provide little distinction between members of the same family. A second limitation is that the assays used to monitor the effect of mutations, such as the transfection assays, are artificial (see Chapter 7). In the TdT promoter analysis, the nucleotide requirements for activity of the D´ element in the transfection assays are likely to reflect the requirements for D´ activity in the endogenous gene. However, the high plasmid copy number during the transfection assay and the removal of the control region from its natural context have the potential to alter the outcome of the analysis. A final limitation is that careful mutant studies of this type can be much more difficult to evaluate if the function of the element depends on the simultaneous binding of two or more proteins. In this instance, it may be very difficult to demonstrate a correlation between the nucleotides required for activity and the nucleotides required for the binding of an individual protein. Despite these limitations, this strategy often provides a useful test of a hypothesis.

trans-Activation of a Reporter Gene or Endogenous Gene by Overexpression of a DNA-binding Protein

The ability of an overexpressed or ectopically expressed protein to *trans*-activate a reporter gene regulated by the control region of interest, or to *trans*-activate an endogenous gene, can provide support for the functional relevance of a protein–DNA interaction. However, these experiments are difficult to interpret when the DNA-binding protein is expressed at a higher concentration than is found in a normal cell. The presence of multiple copies of a reporter plasmid in a transfected cell can lead to similar interpretation problems.

An experiment that is commonly performed begins with the insertion of a cDNA encoding the DNA-binding protein into a vector that drives expression following introduction into cultured cells (Fig. 9.1). For mammalian cells, common expression vectors contain a strong viral promoter/enhancer, such as that derived from cytomegalovirus. Cultured cells are then cotransfected with this expression plasmid and a reporter plasmid regulated by the control region of interest; the reporter assay is used to monitor the effect

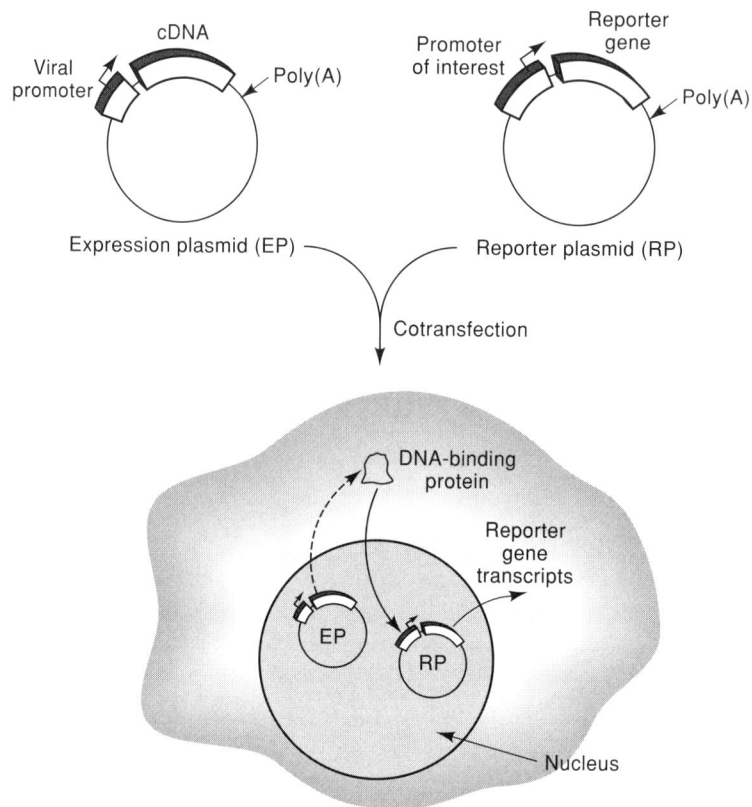

FIGURE 9.1. *trans*-Activation of reporter gene expression by ectopic overexpression of a transcription factor. (From Molecular Cell Biology by Lodish et al. Copyright 1986, 1990, 1996, by Scientific American Books, Inc. Used, with permission, by W.H. Freeman and Company.)

of the overexpressed protein on the activity of the control region. If overexpression results in activation of the control region, the requirement for the protein's binding site can be assessed by repeating the experiment with a reporter plasmid containing a binding site mutant. In this experiment, *trans*-activation should not be observed.

A positive result with this type of experiment strongly suggests that the DNA-binding protein can activate the control region *when both the DNA-binding protein and control region are overexpressed*. However, the result provides little evidence that the protein, when expressed at physiological concentrations, can regulate the two copies of the endogenous target gene present in diploid cells. Current models suggest that genes are regulated by multiple protein–DNA and protein–protein interactions (see Chapter 1). By substantially increasing the concentration of a protein that is not normally involved in regulating a gene, aberrant protein–protein and/or protein–DNA interactions might take place that are sufficient for the gene to be activated or repressed. Overexpression of the reporter plasmid containing the control region of interest can also enhance protein–DNA interactions that do not normally occur.

One example of this experimental approach is provided by the IL-12 promoter analysis (Plevy et al. 1997). To test the hypothesis that C/EBPβ is a relevant activator through the –88/–99 element, uninduced macrophages were cotransfected with a C/EBPβ expression plasmid and an IL-12 promoter-reporter plasmid. Overexpressed C/EBPβ enhanced promoter activity in uninduced cells to a level comparable to that observed in induced cells in the absence of overexpression. Mutation of the C/EBPβ-binding site abolished the *trans*-

activation. At first glance, these results appear to suggest that C/EBPβ is a relevant activator of the IL-12 promoter. However, they merely show that the promoter can be activated when C/EBPβ and the IL-12 promoter-reporter plasmid are present at unusually high concentrations. The results provide no significant evidence that C/EBPβ, when present at physiological concentrations, is a relevant activator of the endogenous IL-12 p40 gene.

These same issues must be considered when an overexpressed transcription factor is found to *trans*-activate an endogenous gene. However, in this instance, a positive result may provide a modest level of support for the hypothesis being tested. First, if the overexpressed DNA-binding protein is capable of *trans*-activating the endogenous gene in a cell type that does not normally express the gene, it must be capable of carrying out protein–DNA and protein–protein interactions that are of sufficient specificity and affinity to overcome chromatin barriers, etc. Second, because only two copies of the endogenous gene exist in a diploid cell, strong activation by a protein that is not physiologically relevant seems less likely; to achieve significant levels of transcription from an endogenous gene, highly specific interactions with other proteins may be required.

One potential solution to the overexpression problem is to express the protein ectopically at a concentration comparable to that found in normal cells. This might be possible by stably expressing the protein. Individual cell clones can then be isolated and tested for protein expression level. Clones that express the protein at a level comparable to that found in a cell line that naturally expresses it can be used for further analysis of target gene transcription. Although this strategy may lead to informative results, it may fail; ectopic expression of a single DNA-binding protein at a normal concentration is unlikely to be sufficient to activate a target gene, unless it is truly the only regulator of that gene missing from the cells being used. If other tissue-specific proteins are needed, the gene will not be efficiently activated.

A second set of experiments that can lessen the concern about protein overexpression is to compare the activities of several members of a transcription factor family when each is overexpressed to a similar extent. If only one family member can *trans*-activate a control region, it must bind the control element with an unusually high affinity, or it must be capable of carrying out specific interactions with other proteins needed for the control region to function. In the IL-12 promoter analysis, for example, Murphy et al. (1995) determined the extent of *trans*-activation following overexpression of various combinations of Rel proteins. The simultaneous overexpression of p50 and c-Rel resulted in much stronger *trans*-activation than overexpression of other Rel proteins, either alone or in other combinations. An important control for this experiment is to show that all of the proteins are expressed at similar levels and are similarly active with a reporter plasmid that does not exhibit a preference for p50/c-Rel binding. Although these experiments used overexpressed proteins, the specificity observed suggests that a p50/c-Rel heterodimer may indeed be a functional activator of IL-12 transcription.

Cooperative Binding and Synergistic Function of Proteins Bound to Adjacent Control Elements

Support for the hypothesis that a protein–DNA interaction is functionally relevant can sometimes be provided by the selective ability of the protein to bind cooperatively with other proteins that interact with the control region. Support can also be provided by the selective ability of the protein to synergize functionally with other proteins (functional synergy does not necessarily involve cooperative binding; see Chapter 1).

The specific interaction between PU.1 and Pip at the Ig κ 3′ enhancer provides an example of how a hypothesis can be supported by cooperative binding (Pongubala et al. 1992; Eisenbeis et al. 1995). Two of the critical elements within this enhancer are immediately adjacent to one another and appear to act in synergy. EMSA experiments performed with nuclear extracts and a radiolabeled probe spanning the two elements yielded a complex containing proteins bound to both sites (Fig. 9.2, lane 1) (Pongubala et al. 1992). DNA binding by the two proteins was highly cooperative as mutation of either site strongly reduced protein binding to the probe (Fig. 9.2, lanes 4–8). One protein within the complex was found to be an Ets protein named PU.1, and the other was found to be a member of the IRF family, named Pip or NF-EM5 (Eisenbeis et al. 1995).

The detection of a stable complex containing proteins bound cooperatively to two functionally important sites provides strong evidence a priori that these proteins (and not other members of the Ets and IRF families) are the relevant regulators of the Ig-κ enhancer. As discussed earlier, the detection of an EMSA complex with an isolated control element provides relatively weak evidence that the protein is functionally relevant because the criteria for the in vitro detection of a protein–DNA interaction are very different from the criteria for a relevant in vivo interaction. Nevertheless, if an EMSA complex observed with a crude extract contains proteins bound cooperatively to two sites that are functionally synergistic in vivo, the probability that the proteins within the complex are functionally relevant is substantially increased.

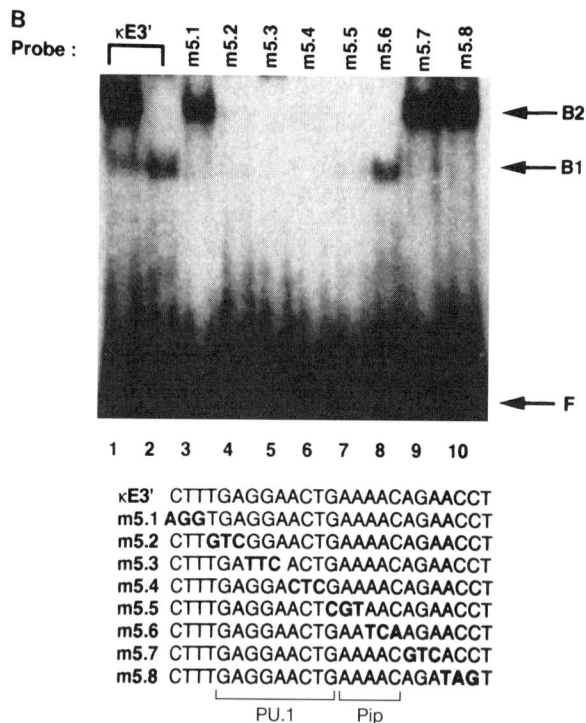

FIGURE 9.2. Highly cooperative binding of PU.1 and Pip to the Ig κ 3′ enhancer. EMSA experiments were performed with radiolabeled oligonucleotides containing the wild-type or mutant enhancer sequences shown at the bottom. Binding reactions in lanes 1 and 3–10 were perfomed with nuclear extracts from the S194 cell line. The binding reaction in lane 2 contains in vitro translated PU.1. Band B2 contains PU.1 and Pip cooperatively bound to the probe. Band B1 contains PU.1 alone. Band F corresponds to free probe. The locations of the PU.1 and Pip-binding sites are depicted at the bottom. (Adapted, with permission, from Pongubala et al. 1992 [Copyright American Society for Microbiology].)

A key limitation of this assay is that strong, cooperative binding and functional synergy are observed with only a small subset of protein–DNA interactions. If cooperative binding to the control region is not observed in vitro, this strategy will not be useful. The
PU.1/Pip example is particularly powerful because the cooperatively bound proteins are
easily detected in EMSA experiments using crude nuclear extracts. If cooperative binding
was observed only when the two recombinant proteins were added to an EMSA experiment, the results would provide much weaker support for the hypothesis. In this case, a
demonstration of the selectivity of the cooperative binding would be needed, by comparing various Ets and IRF family members.

A second example of the cooperative binding and synergistic activation strategy, which
highlights its limitations, recently emerged from studies of the interferon-β (IFN–β)
enhancer by the Maniatis laboratory (Wathelet et al. 1998). Previous studies had found that
an IRF family member, IRF-1, interacts with a functionally important enhancer element
and binds cooperatively to the enhancer with other transcription factors, including ATF-
2/c-jun and NF-κB (Thanos and Maniatis 1995; Kim and Maniatis 1997). IRF-1 also can
trans-activate the enhancer in synergy with those factors. The ability of IRF-1 to carry out
cooperative and synergistic interactions at the enhancer suggested that it was a relevant
activator of IFN-β transcription. The recent study by Wathelet et al. (1998), however,
appears to have disproven that hypothesis. In this study, a protein complex was found to
interact with the IRF recognition element following induction of IFN-β transcription. The
inducible protein complex did not contain IRF-1, but instead contained two other IRF
family members, IRF-3 and IRF-7. Several additional experiments confirmed that IRF-3
and IRF-7, but not IRF-1, contribute to the activation of IFN-β transcription in vivo. In
particular, a single-base-pair substitution in the enhancer was identified that abolishes
IRF-1 binding, but has no effect on IRF-3/IRF-7 binding or on enhancer activity. Thus,
despite the initial evidence that IRF-1 can bind cooperatively to the enhancer with other
relevant factors, and synergistically activate the enhancer in concert with those factors, it
appears to be irrelevant for enhancer activity or IFN-β transcription.

The IRF-1 studies were misleading primarily because recombinant proteins were initially used to demonstrate cooperative binding and synergistic activation. This contrasts with
the PU.1/Pip studies where the cooperative interactions were first observed in crude nuclear
extracts from cells that express the Ig κ gene; recombinant proteins and overexpressed proteins were not needed. When cooperative interactions were finally observed with the IFN-β
enhancer in crude nuclear extracts from induced cells (in the absence of protein overexpression), the complex was found to contain IRF-3 and IRF-7 instead of IRF-1.

Comparison of Genomic and In Vitro Footprinting Patterns

A straightforward test of the hypothesis that a specific protein–DNA interaction is relevant
is to compare the footprinting pattern observed in vitro with a known DNA-binding protein to the pattern observed in genomic footprinting experiments with cells that actively
express the putative target gene. This comparison can be carried out using any of the
genomic footprinting methods described in Chapter 10, including DNase I footprinting or
dimethyl sulfate (DMS) protection. If the footprinting pattern observed in vitro matches
the pattern observed in vivo, the protein analyzed in vitro might actually be the protein
bound to the endogenous gene when it is active, strongly suggesting that it is the relevant
activator. If DNase I footprinting is used to perform this comparison, it usually will be necessary to use ligation-mediated PCR (LM-PCR) technology (see Chapter 10) for both the
in vitro and genomic footprinting procedures, because it is difficult to compare footprint-

ing results obtained by LM-PCR with those obtained using radiolabeled probes. In contrast, if DMS footprinting is used, the results obtained by LM-PCR in vivo can be compared directly to the results obtained in vitro using a radiolabeled probe.

This strategy was used during the analysis of the lymphocyte-specific RAG-2 promoter (Lauring and Schlissel 1999). In vitro EMSA studies demonstrated that the B-cell activator, BSAP, can interact with an important control element. These results suggested that BSAP is an important activator of RAG-2 transcription in B cells. An in vivo interaction at the element was observed by DMS genomic footprinting, but in vivo footprinting results do not reveal the identity of an interacting protein. To provide evidence that the protein bound to the site in vivo was BSAP, a DMS protection analysis was performed in vitro using recombinant BSAP. The in vivo and in vitro interaction patterns were identical, suggesting that BSAP is indeed the protein that occupies the endogenous element in B cells.

The primary limitation of this strategy is the following: Even if the in vitro and genomic footprinting patterns are identical, the results do not usually distinguish between the various members of a protein family, because most family members lead to a similar footprint. Furthermore, some DNA-binding proteins do not yield footprinting patterns that are sufficiently unique, making it difficult to determine whether the protein bound to the endogenous locus in vivo is related to the known protein being tested in vitro.

Nevertheless, if clear genomic and in vitro footprints with characteristic properties are obtained, this comparison can provide compelling support for the relevance of a particular family of proteins for the function of a defined control element.

Relative Affinity of a Protein–DNA Interaction

One property of a protein–DNA interaction that may support its relevance is its affinity, relative to the affinities of other protein–DNA interactions that can occur at the same site. To our knowledge, the relationship between affinity and functional relevance of a transcription factor has not been examined carefully and systematically (although, in model systems, the affinity of a transcription factor for its target site is proportional to activation; see, e.g., Mauxion et al. 1991; Lehman et al. 1998). In some cases, such as the PU.1/Pip example described above, low-affinity interactions by the individual DNA-binding proteins are preferred so that the activity of the composite element remains dependent on cooperative binding by two or more proteins. Nevertheless, it seems possible that in many instances, the protein that functionally interacts with a control element will bind with an unusually high affinity, relative to the affinities of irrelevant proteins capable of binding the same site.

Affinity may be a particularly useful criterion for determining which member of a multiprotein family is most likely to carry out a functional interaction with an element. This assertion is based on the notion that the various members of a multiprotein family, although recognizing very similar sequences, have subtle sequence preferences. These sequence preferences may play a major role in determining which factor acts on a given control element. If this hypothesis is correct, a careful comparison of binding affinities of the members of a multiprotein family for a site might provide insight into the functionally important family member.

To compare relative affinities of a variety of proteins for a given site, several methods can be considered. One method is to obtain the pure recombinant proteins to be compared, and then to carry out careful K_d measurements as described in Chapter 13. This method will provide valuable information, but it has two disadvantages. First, it is difficult to prepare and accurately quantify recombinant forms of all the proteins that should be tested, in particular when working with a large multiprotein family. More importantly, the affinity of a recombinant protein for a DNA sequence may be significantly different from that of the native protein. One reason for this potential difference is that protein–DNA interaction affinities can be altered by posttranslational modifications or by interactions with other proteins. In particular, a growing number of proteins are being found to possess autoinhibitory domains that reduce their affinities for DNA unless appropriately modified (see, e.g., Jonsen et al. 1996).

Thus, a more informative experiment would be to compare relative affinities in a cell extract, where the proteins are more likely to exist in their native state. If most of the proteins of interest within an extract are sufficiently concentrated to permit their detection in a basic EMSA, a rough comparison of their affinities for a site can be obtained. One approach is to determine the susceptibility of each protein–DNA complex to nonspecific competitor DNA, salt, or detergent by titrating each into the binding reactions. Higher-affinity interactions are likely to be more resistant to these reagents, particularly if one is comparing members of a family. A limitation of this approach, however, is that it will result in a comparison only of those proteins that are sufficiently concentrated to detect in a basic EMSA experiment. Proteins that are of relatively low abundance, or whose binding properties are incompatible with the EMSA conditions being used, cannot be tested.

An alternative approach was carried out in one of our laboratories with the TdT D′ element discussed above (Fig. 9.3) (Ernst et al. 1996). The goal was to determine which Ets protein within an extract from a TdT-expressing cell line binds to the D′ element with highest affinity, to identify a candidate for the functional activator of TdT transcription. Several known, and perhaps other novel, Ets proteins are present in TdT-expressing cells, but only a few EMSA complexes containing Ets proteins can be detected using nuclear extracts. To compare the relative affinities of proteins within the extract for the D′ element, in the absence of a bias toward known proteins or proteins that could be detected by EMSA using crude extracts, sequence-specific DNA affinity chromatography was employed (see Chapter 8). The expectation was that the highest-affinity proteins would elute from an affinity column with the highest salt concentration. Indeed, following affinity chromatography, a single abundant EMSA complex was observed in the high-salt eluates, with all other EMSA complexes more abundant in the lower-salt eluates. A silver-stained protein gel led to the identification of the protein responsible for the complex (complex Y), and peptide sequencing revealed that the protein was an Ets protein called Elf-1. Immunoblot experiment of column fractions confirmed that Elf-1 eluted at higher salt concentrations than several other Ets family members. Subsequent experiments, like those described in this chapter, provided additional support for the hypothesis that Elf-1 is a functional activator of TdT transcription through the D′ element. However, the hypothesis remains unproven. Thus, it is not yet known whether affinity provides a valid criterion for assessing functional relevance.

The affinity chromatography strategy possesses several advantages relative to the other two strategies mentioned above. First, the native proteins in the crude extract may retain posttranslation modifications that were present in the intact cells. Second, this method is not biased toward the most abundant proteins or previously described proteins. In fact, this strategy should result in the purification of any protein of reasonable abundance that binds the site with high affinity. Finally, if the high-affinity binding protein is novel, it can be identified or cloned by microsequencing (see Chapter 8).

A notable limitation of this approach is that the posttranslational modifications present in the intact cell might not be retained in the extract. Furthermore, the use of affinity chromatography as a measure of relative affinities might be somewhat inaccurate if the different proteins that bind a site rely to different extents on electrostatic interactions between the DNA-binding domain and DNA. If the electrostatic interactions are substantially different, salt elution from an affinity column will not provide an accurate assessment of relative affinities.

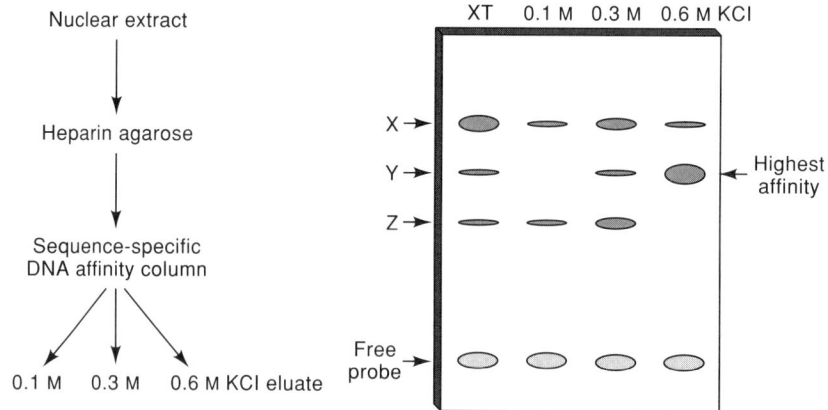

FIGURE 9.3. Comparison of protein–DNA interaction affinities in crude nuclear extracts by sequence-specific DNA affinity chromatography. Nuclear extracts from the murine T-cell line RLm11 were initially fractionated by heparin agarose to remove nucleases. Sequence-specific DNA affinity chromatography was then performed using a resin with covalently linked oligonucleotides containing the D′ element from the TdT promoter. The sample was applied to the column in a buffer containing 0.1 M KCl. Proteins were then eluted with 0.3 and 0.6 M KCl. At the right, an idealized EMSA result is depicted; three protein–DNA complexes were detected when the extract was used (lane *1*), but only one (complex Y) was preferentially retained on the column until the 0.6 M KCl elution (lane *4*).

Gene Disruption or Antisense Experiments

A DNA-binding protein can be linked to a putative target gene by abolishing expression of the protein in a cell line or animal, using homologous recombination or antisense RNA technology (Capecchi 1989; Branch 1998; Stein 1998). The absence of the DNA-binding protein might result in a reduction of target gene expression, implicating the protein as a regulator of the gene.

Although gene disruption via homologous recombination and antisense has proven to be invaluable for assessing the importance of DNA-binding proteins for specific biological processes, the information provided about primary target genes is much less compelling. The phenotype of a mouse or cell line that lacks expression of a DNA-binding protein undoubtedly results from the altered expression of one or more genes. However, in most instances, it is difficult to determine which genes are directly responsible for the phenotype and which are direct targets of the DNA-binding protein. If the expression of a particular gene is diminished in a mutant animal, the gene could be a direct or indirect target of the DNA-binding protein. In other words, the transcription factor may directly regulate the gene, or it may regulate the expression of other genes that influence the expression of the gene of interest. If the absence of the DNA-binding protein does not disrupt the expression of a gene of interest, the gene may still be a direct target of the protein. Functional redundancy between two DNA-binding proteins might obscure their functions. Alternatively, a related DNA-binding protein might be up-regulated, allowing it to compensate for the disrupted protein.

To provide an example of these concepts, we can return to the TdT promoter and its potential regulation by Ikaros. As described above, Ikaros binds with high affinity to the critical D′ element within the TdT promoter. On the basis of this result, Ikaros was originally considered a likely activator of TdT transcription. Support for this hypothesis was provided by an Ikaros gene disruption experiment (Georgopoulos et al. 1994). The mutant mice lack all cells of the B- and T-lymphocyte lineages, including the earliest progenitors of both lineages. Because TdT is one of the first genes activated when a hematopoietic stem cell becomes committed to the B- and T-cell lineages, the phenotype supported the hypothesis that TdT is a functionally relevant target of Ikaros: (1) Ikaros binds to a critical control element in the TdT promoter, (2) the TdT gene is activated in progenitor B and T cells, and (3) the Ikaros mutant mice specifically lack the progenitor B and T cells that express the TdT gene.

Despite the fact that this model was attractive and supported by the available data, the experiments described earlier in this chapter suggest that Ikaros is not an activator of TdT transcription. In fact, recent data suggest that Ikaros is actually a repressor of TdT transcription at a later stage of lymphocyte development. Thus, the absence of TdT expression in Ikaros mutant mice is likely to be an indirect effect.

An example of the effect of transcription factor redundancy on the interpretation of gene disruption results was provided by studies of the LEF-1 and TCF-1 proteins implicated in regulation of the T-cell receptor α (TCRα) enhancer (Clevers and Grosschedl 1996; Okamura et al. 1998). LEF-1 and TCF-1 are highly homologous members of the HMG-1 family of proteins. Both proteins can bind a critical element within the TCRα enhancer and can *trans*-activate the enhancer in concert with DNA-binding proteins that bind adjacent elements, with LEF-1 *trans*-activation approximately 10 times stronger than TCF-1 *trans*-activation. Disruption of either the LEF-1 or TCF-1 genes revealed little effect on TCRα gene transcription, however (van Genderen et al. 1994; Verbeek et al. 1995). If only one of these proteins were known to exist, it would have been tempting to speculate, on the basis of the gene disruption result, that the protein was not a relevant activator of the TCRα enhancer. Because both proteins had been discovered, however, the possibility of redundancy was considered and examined by generating mice in which both genes were disrupted. The LEF-1$^{-/-}$TCF-1$^{-/-}$ mice exhibited a severe defect in thymocyte development and were deficient for TCRα gene transcription, strongly suggesting that the two proteins are at least partially redundant.

These results highlight the fact that gene disruption and antisense phenotypes must be interpreted cautiously, and that other approaches are needed to evaluate rigorously the relevance of a DNA-binding protein for the regulation of a putative target gene.

Dominant-negative Mutants

By definition, a dominant-negative mutant of a protein is a protein variant that, when expressed in a cell containing the wild-type protein, disrupts the functions of that protein. Disruption usually occurs because the mutant retains some, but not all, of the wild-type protein's activities, allowing it to compete, albeit nonproductively, for an important target or substrate of the wild-type protein. Dominant-negative mutants of a DNA-binding protein can sometimes provide information about the functional importance of a protein–DNA interaction. However, the information provided by these experiments is limited, and the type of dominant negative used has a notable impact on the interpretation of the data.

A common method for performing dominant-negative studies is to cotransfect cells with a reporter plasmid containing a standard reporter gene regulated by the control region of interest and an expression plasmid for a dominant-negative version of the DNA-binding protein of interest (Fig. 9.4). The dominant-negative protein that is expressed is often designed to retain the capacity to bind DNA but not to carry out transcriptional activation functions. In other words, the protein contains its DNA-binding domain and lacks its transcriptional activation domains (see Chapter 12). The expectation is that this mutant protein will compete with the wild-type endogenous protein, thereby preventing the wild-type protein from activating reporter gene transcription (Fig. 9.4). Inhibition of transcription with this type of dominant-negative protein is often used to support the hypothesis that the wild-type protein is a functional activator of the control region fused to the reporter.

As a recent example, this approach was used by Plevy et al. (1997) to support the hypothesis that C/EBPβ is a relevant activator of the IL-12 p40 promoter in lipopolysaccharide (LPS)-activated macrophages. Increasing concentrations of an expression plasmid for a dominant-negative form of C/EBPβ, called LIP (Descombes and Schibler 1991), were cotransfected into the macrophages along with IL-12/CAT reporter plasmid. LIP contains the bZIP DNA-binding and dimerization domains of C/EBPβ, but lacks its transcriptional activation domains. The cells were then activated with LPS, and the effect of LIP on IL-12 promoter activity was monitored using the CAT assay. The results showed that promoter activity was suppressed by LIP. Important control experiments showed that LIP expression had no effect on reporter plasmids containing promoters lacking C/EBP-binding sites (e.g., a CMV-CAT reporter).

Dominant-negative experiments of this type can support a hypothesis regarding the functional relevance of a DNA-binding protein, such as C/EBPβ. However, a careful con-

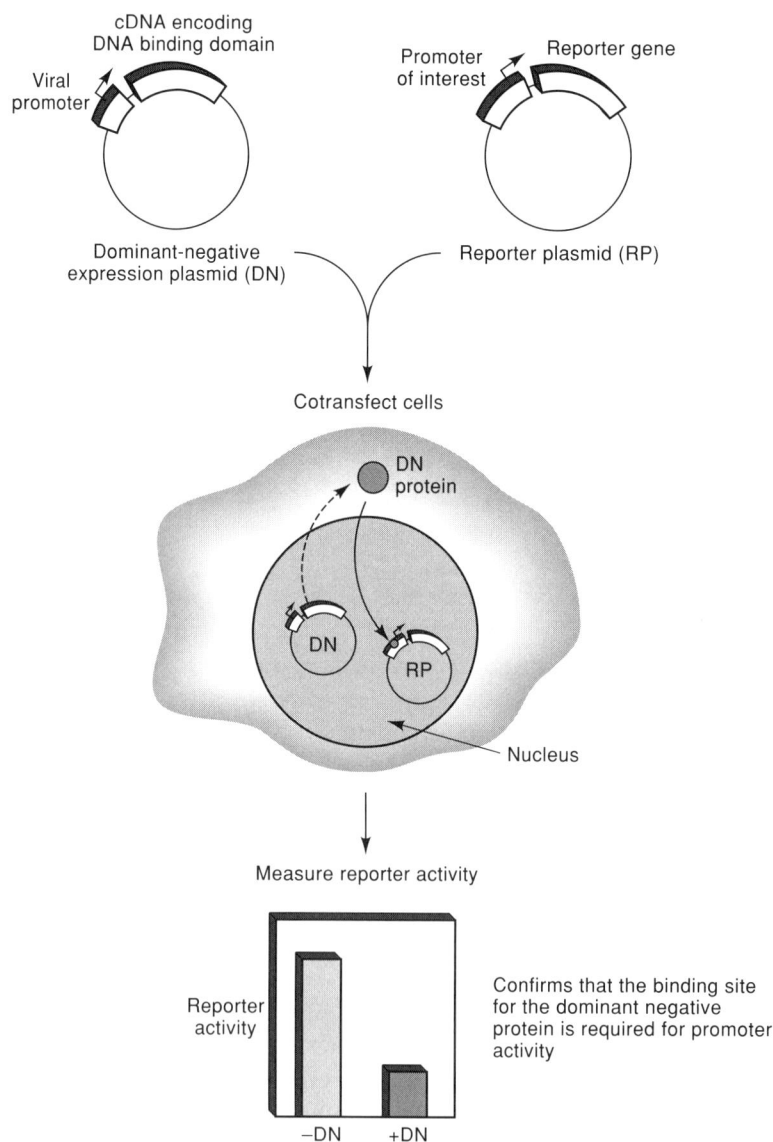

FIGURE 9.4. Inhibition of reporter activity by a dominant-negative version of a transcription factor that retains its DNA-binding domain.

sideration of the experiment reveals that the result merely confirms that the DNA-binding domain, when overexpressed, is capable of binding to the important control element, where it blocks activation by the functional activator. Regardless of the identity of the functional activator, the overexpressed dominant-negative mutant may block the important protein–DNA interaction, simply by occupying the binding site. Thus, the result provides two relatively modest pieces of information: (1) It demonstrates that the control element to which the dominant-negative protein binds is important for function of the control region, a result already established by the promoter mutant analysis; and (2) it demonstrates that the DNA-binding domain within the dominant-negative protein is capable of binding to that control element in vivo when overexpressed.

An alternative experiment is to test the effect of the same dominant-negative mutant on expression of the endogenous gene. This experiment provides an opportunity to confirm that the control element found to be important in an artificial transfection assay is important for regulation of the endogenous gene. However, this experiment suffers from most of the above limitations and from two additional caveats. First, it is difficult to rule out the possibility that the dominant-negative protein indirectly inhibits transcription from the endogenous target gene by altering expression of other genes within the cell. For example, the dominant-negative protein may inhibit expression of cellular genes that are needed for survival, leading to toxicity that might indirectly inhibit transcription from the target gene of interest. (The possibility of indirect effects must also be considered when analyzing the effect of a dominant negative on a transfected reporter plasmid, but this possibility can be addressed, at least in part, by testing a mutant reporter.) The second limitation of analyzing an endogenous target gene is that it is technically more difficult than analysis of a transiently cotransfected dominant-negative mutant and reporter plasmid. In the cotransfection assay, virtually every cell that takes up the reporter plasmid takes up the dominant-negative expression plasmid. Thus, the cell population can be analyzed for an effect of the dominant-negative protein on reporter gene transcription. In contrast, to monitor the effect of the dominant-negative protein on endogenous gene transcription, one must keep in mind that only a fraction of the cells are likely to be transfected and express the dominant-negative protein. Because the majority of cells will not be transfected, it may be difficult to observe the effect of the dominant negative on target gene expression.

Despite this added difficulty, there are several fairly straightforward solutions. First, the dominant-negative protein can be expressed from a high-titer retroviral vector that can infect all of the cells (see Chapter 11), perhaps leading to a measurable effect on transcription of the putative target gene. Second, the dominant-negative protein can be expressed in the cells by stable transfection. Preferably, the dominant-negative protein should be under the control of an inducible promoter, so that any toxicity of the protein is not manifested during the selection process (see Chapter 11). Third, the cells that are transiently transfected with the dominant-negative expression plasmid could be distinguished from the untransfected cells by flow cytometry or a related immunologic method (see Chapter 5, Box 5.5).

Although the dominant-negative proteins described above have limited value, other types of dominant-negative proteins may be more useful for specific types of DNA-binding protein families. In particular, dominant-negative mutants can be used to monitor the importance of proteins that bind DNA as dimers, such as b-ZIP and bHLH proteins. For these classes of proteins, dominant-negative versions of the protein can be expressed that retain the dimerization domain but lack the DNA-binding domain. If these dominant-negative proteins inhibit transcription from the putative target gene, the result would suggest that a protein capable of dimerizing with the dominant-negative protein is the functional activator through the control element of interest. Depending on the nature of the dimerization domain, this result may implicate an entire family of DNA-binding proteins, or it may suggest that a protein among a specific subset of family members is likely to play a role. To confirm the relevance of this result, it may be necessary to test the dimerization domain's capacity to interact with various members of the protein family.

A classic example of a dominant-negative protein that acts by dimerizing with specific transcription factors is the HLH protein, Id1 (Benezra et al. 1990). Id1 and other Id family members contain helix–loop–helix domains that allow them to form heterodimers with bHLH activators, but they lack the basic region needed for DNA binding. The Id proteins therefore inhibit transcriptional activation by bHLH proteins by preventing them from binding to DNA. Thus, Id proteins can be used to assess the relevance of the bHLH family of proteins for the function of a control element. However, because each Id protein can dimerize with multiple bHLH proteins (Langlands et al. 1997), an inhibition of transcriptional activation provides limited insight into the identity of the functionally relevant family member.

In Vitro Transcription Strategies

In vitro transcription experiments can be used to support the relevance of a protein–DNA interaction. The basic approach begins with the development of an in vitro transcription assay for the control region, which can often be quite challenging (see Chapter 5 for cited examples of regulated transcription in nuclear extracts, and Chapter 14 for in vitro transcription methodology). Nucleosome reconstitution may aid in the assay's development (see Chapters 5 and 14). To determine whether the DNA element of interest contributes to the in vitro activity, mutations in the element can be tested. If the assay is found to be dependent on the DNA element, it may be possible to use the assay to assess the relevance of a particular DNA-binding protein for the element's function. As a starting point, monoclonal or polyclonal antibodies directed against the candidate protein can be added to the in vitro reactions to determine whether they block the function of the control element (Fig. 9.5A). Concentrated, affinity-purified antibodies may be needed because it may be difficult to add sufficient antibody to neutralize the protein. To determine whether sufficient antibody has been added, its effect on an EMSA complex can be monitored. Control experiments should also be performed to determine whether the antibody affects transcription from an unrelated promoter that is independent of the DNA-binding protein. Additional controls are also needed, including reactions monitoring the effect of an unrelated antibody prepared by a similar method.

If antibody addition does not inhibit transcriptional activity, the candidate binding protein may not be essential for the activity of the control element. Alternatively, the antibody may not bind to an essential epitope of the protein or may not be sufficiently concentrated to neutralize all of the protein molecules within the extract.

An alternative strategy, which may be more successful, is to deplete the protein from the extract by immunoprecipitation or immunoaffinity chromatography, or by sequence-specific DNA affinity chromatography (Fig. 9.5B) (see also Chapter 8). The immunoprecipitation and immunoaffinity methods should allow the efficient depletion of a protein from an extract, unless the critical epitope of the protein is blocked by its tight association with other proteins. DNA affinity chromatography results in the depletion of all proteins that bind the element of interest, providing little insight into the identity of the protein that functionally interacts with the element. However, that insight can be provided by the subsequent addition of a recombinant or pure form of the putative regulator. If efficient transcription is restored, the results support the hypothesis that the protein is a relevant regulator of the gene.

A. **Transcription Factor Neutralization**

Mix nuclear extract and antibody against specific transcription factor.

↓

Add DNA template, NTPs, and buffer.

B. **Transcription Factor Depletion**

Mix nuclear extract and antibody against specific transcription factor.

↓

Add Protein A-Sepharose.

↓

Remove antibody–protein complexes by centrifugation.

↓

Add supernatant to in vitro reaction containing DNA template, NTPs, and buffer.

Analyze transcripts.

−Ab   +Ab   −Ab   +Ab

Promoter of interest   Control promoter

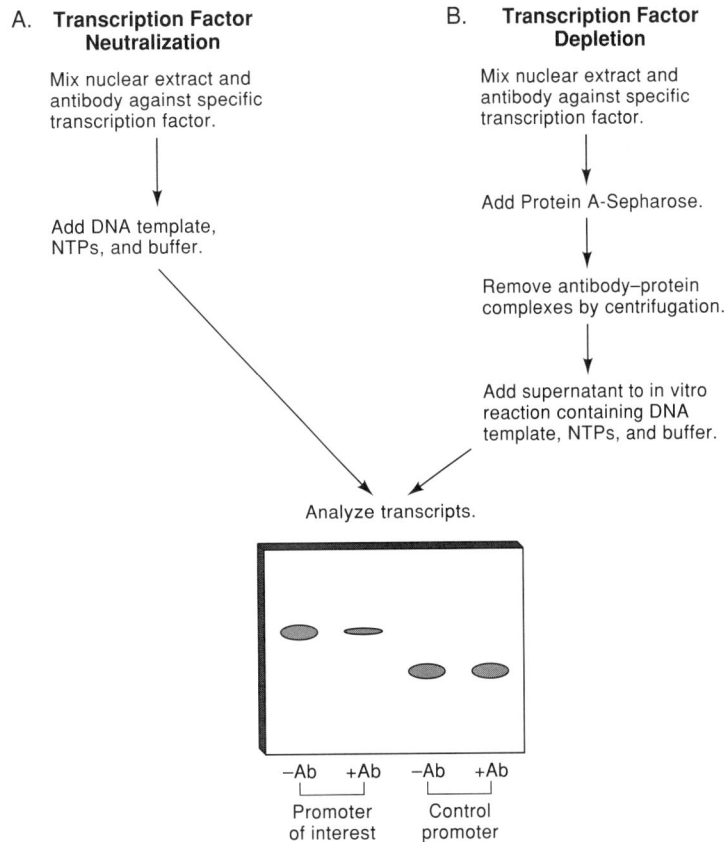

FIGURE 9.5. Assessing the relevance of a protein–DNA interaction by antibody inhibition of an in vitro transcription reaction.

In practice, depletion experiments to monitor the requirement for a protein in an in vitro transcription assay have proven to be tedious and difficult. Depletion often inhibits the in vitro transcription assay nonspecifically because of extract dilution or inactivation of a general transcription factor. The protein bound by the antibody or DNA resin also has the potential to bind or form aggregates with other proteins that are essential for in vitro transcription, resulting in the co-depletion of those proteins. These difficulties are not unusual, because the extracts must be maintained in a concentrated state during depletion to function in the subsequent in vitro reactions.

Because of the potential problems with these types of experiments, careful controls are needed, including immunodepletion controls with antibodies that should have no effect on transcription from the promoter of interest, as well as control templates that should not be affected by the depletion. Furthermore, as mentioned above, to demonstrate that the specific inhibition observed is due solely to depletion of the protein of interest, it should be possible to restore transcription by complementing the reaction with a recombinant or pure form of the protein. The results obtained using these in vitro approaches can support the functional importance of a protein for transcription of a target gene. However, the relevance of the protein would require additional support because a protein that functionally interacts with a control element in vitro is not necessarily responsible for its function in vivo.

An in vitro transcription analysis of the IFN-β enhancer provides an example of transcription factor depletion by immunoaffinity and DNA affinity chromatography (Kim and Maniatis 1997). The activity of this enhancer in vivo is thought to require ATF2/c-jun, an

IRF family member, NF-κB, and HMGI(Y) (see Chapter 1). To study the specific protein requirements for enhancer function in vitro, an in vitro transcription assay dependent on the binding site for each factor was developed. Then, the first three factors were depleted by sequence-specific DNA affinity chromatography. Excess HMGI(Y) was depleted by immunoaffinity chromatography. To assess the function of each protein, recombinant forms of the proteins were added back to the depleted extracts, resulting in a restoration of enhancer activity. These results provided evidence that the specific proteins used are capable of activating the enhancer. However, it should be noted that recombinant IRF-1 was used for these studies, whereas more recent findings strongly suggest that other IRF family members, in particular IRF-3 and IRF-7, are actually the relevant activators of IFN-β transcription (see above and Wathelet et al. 1998). Thus, as mentioned above, the in vitro transcription method by itself is unreliable for demonstrating the relevance of a protein–DNA interaction.

Although this section has focused on strategies that make use of protein neutralization and depletion in crude extracts, in vitro transcription can also be used as an assay for the unbiased purification of a protein that activates a promoter (see Chapter 14). A classic example of this strategy led to the identification, purification, and cloning of the Sp1 protein that stimulates the SV40 early promoter (Dynan and Tjian 1983a,b; Kadonaga et al. 1987). An in vitro transcription assay was first developed that supported accurate transcription from the promoter. The cell extract was then fractionated to identify an essential promoter-specific transcription factor. The factor identified, Sp1, was found to bind specific DNA-sequence elements within the promoter. Sp1 was then purified, a partial peptide sequence was obtained, and its gene was cloned. The isolation of Sp1 as a protein that selectively activates the SV40 promoter provides reasonably strong support for the hypothesis that it is a relevant activator of the promoter in vivo.

In Vivo Protein–DNA Crosslinking

One of the strategies discussed above for confirming the importance of a protein–DNA interaction is to compare a genomic footprinting pattern to a footprinting pattern observed in vitro with a purified protein. This strategy can provide some evidence that a particular protein is bound to control elements within the context of the endogenous locus. One limitation of this approach, however, is that it usually cannot distinguish between the various members of a multiprotein family, which often yield indistinguishable footprinting patterns.

An alternative strategy for identifying the specific protein that associates with a control element in the context of an endogenous allele is to use in vivo protein–DNA crosslinking (Fig. 9.6), which was first discussed in Chapter 3. This method, originally developed as a means of determining whether RNA polymerase is paused on the transcribed leader of a gene prior to induction (see Chapter 3; Gilmour and Lis 1984, 1985, 1986), has recently been used to monitor the association of specific transcription factors with DNA (Walter et al. 1994; Boyd and Farnham 1997; Boyd et al. 1998; Wathelet et al. 1998).

In brief, growing cells are treated with ultraviolet light or formaldehyde to crosslink DNA-binding proteins to their target sites (Orlando et al. 1997; Walter and Biggin 1997). The cells are then lysed, and the DNA is cleaved into fragments by digestion with a restriction enzyme or by shearing. Protein–DNA complexes are then purified by immunoprecipitation with antibodies directed against the DNA-binding protein of interest. To determine whether the protein was crosslinked to a putative target element, the immunoprecipitate is analyzed by Southern blot or PCR for the presence of a DNA fragment encompassing the

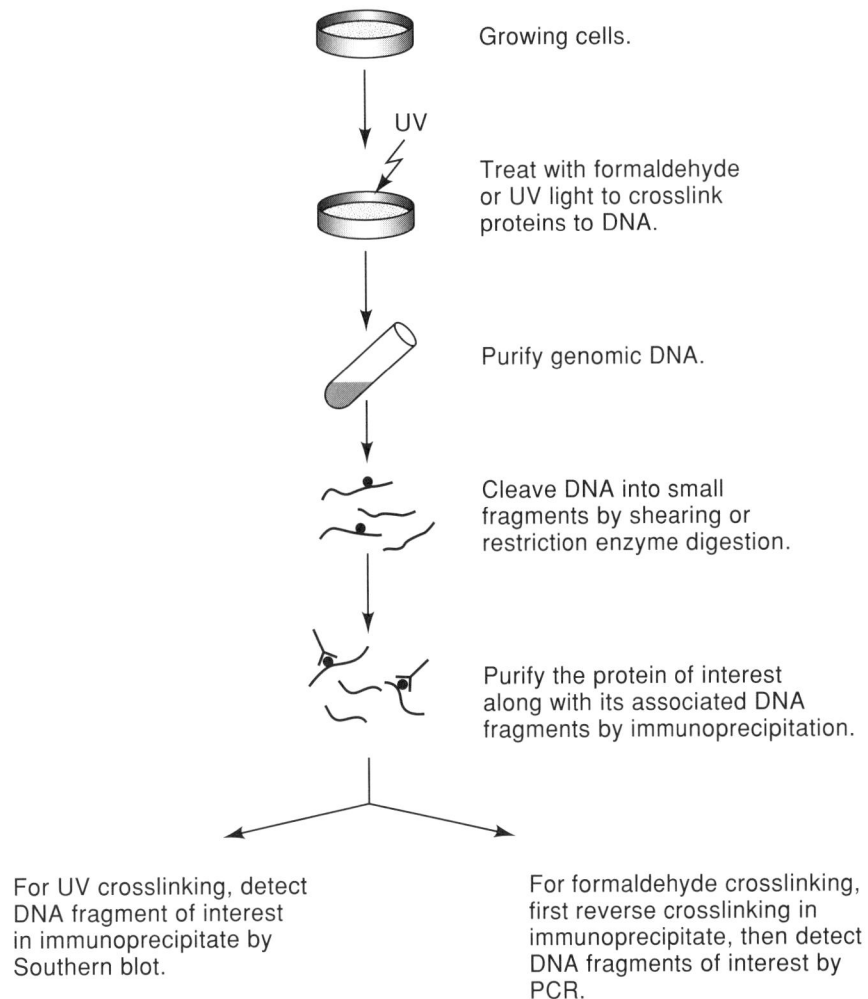

FIGURE 9.6. In vivo crosslinking of proteins to DNA-binding sites.

element. PCR is most useful as a detection method when coupled to formaldehyde crosslinking, which can be reversed following immunoprecipitation, preventing the linked protein from interfering with the PCR.

The IFN-β enhancer analysis by the Maniatis laboratory provides an example of the crosslinking strategy (Fig. 9.7) (Wathelet et al. 1998). Other experiments carried out by the laboratory had implicated two IRF family members, IRF-3 and IRF-7, as relevant activators of the IFN-β enhancer, which is induced by virus infection. To strengthen the hypothesis that these two proteins are indeed relevant activators, the authors used the in vivo cross-linking strategy in mock-infected and virus-infected cells. Following formaldehyde crosslinking, extract preparation, and DNA fragmentation, antibodies against various IRF family members were used for immunoprecipitation. PCR analysis of the DNA within the immunoprecipitates revealed that the IFN-β enhancer fragment was present following immunoprecipitation with IRF-3 antibodies and IRF-7 antibodies, but not following immunoprecipitation with IRF-1 antibodies (Fig. 9.7). Importantly, the IFN-β enhancer fragment was only present in the immunoprecipitates from virus-infected cells and not

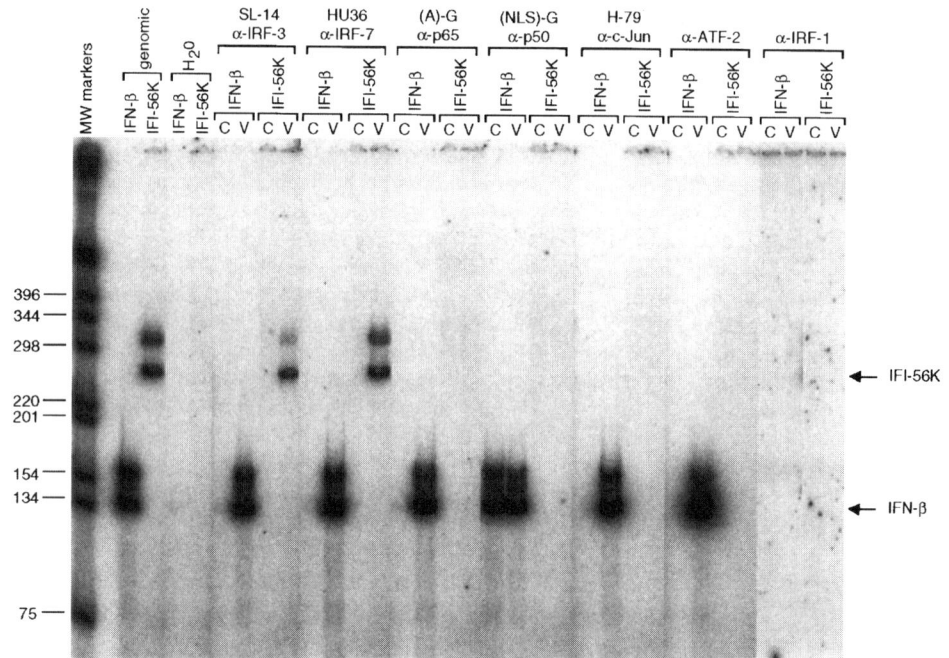

FIGURE 9.7. Association of specific transcriptional activator proteins with virus-inducible promoters in vivo. HeLa cells were mock (C)- or SV (V)-infected for 6 hours, and treated with formaldehyde to crosslink proteins bound to DNA. Crosslinks were purified and immunoprecipitated with the indicated antibodies as described previously (Orlando et al. 1997). After reversal of the crosslinking, the DNA was amplified using primers for the IFI-56K (–237/–219 and +12/+31) and IFN-β (–129/–117 and –1/–19) promoters (PCRs, 20 cycles of 1 min at 94°C, 55°C, and 72°C each). Molecular-weight marker and control amplification using genomic DNA and water are shown on the left. (Reprinted, with permission, from Wathelet et al. 1998 [Copyright 1998 Cell Press].)

from mock-infected cells. Immunoprecipitation with antibodies against other DNA-binding proteins (p65, p50, c-Jun, and ATF-2) confirmed that they also interact with the IFN-β promoter, but not with a control promoter (IFI-56K, Fig. 9.7). These results provide strong evidence that IRF-3 and IRF-7 are indeed relevant activators of IFN-β enhancer activity.

The principal strength of the in vivo crosslinking strategy is that it is the only method currently available for directly "visualizing" an in vivo interaction between a specific protein and control element. If compelling data are obtained, the method can provide strong and direct evidence that a site is occupied by a specific protein in growing cells.

A limitation of the approach is that it is difficult to obtain compelling data. In fact, the procedure had been attempted in mammalian cells for several years, but manuscripts making use of it have only recently begun to appear in the literature (see, e.g., Boyd and Farnham 1997; Boyd et al. 1998; Wathelet et al. 1998). The principal controls needed for convincing results are (1) experiments performed with multiple unrelated antibodies, which should not precipitate the DNA fragment of interest, and (2) experiments with PCR primer sets or Southern blot probes directed against several other chromosomal regions that should not precipitate with the specific antibodies. An additional limitation of the crosslinking strategy is that association of a protein with a specific DNA fragment does not provide conclusive evidence that the association is functionally relevant. In fact, Walter et al. (1994) have reported that a subset of *Drosophila* homeodomain proteins crosslink to a large number of genomic sites, whereas other homeodomain proteins crosslink to a small number of genes that are thought to be direct targets. The reason for the relatively ubiquitous crosslinking of some proteins remains unknown. However, these results emphasize the need for careful controls and for the cautious interpretation of crosslinking data.

Altered Specificity Experiments

The final strategy that is discussed for testing the relevance of a protein–DNA interaction is an altered specificity strategy (Fig. 9.8). Although an ideal altered-specificity experiment is difficult to design and perform, it has the potential to provide more compelling evidence that a protein–DNA interaction is relevant than any of the other strategies described above. To perform this type of experiment, the DNA-binding domain of the protein of interest is first mutated so that it recognizes a different DNA sequence. The new sequence recognized by the altered protein is then inserted into the control region of interest in place of the sequence element recognized by the wild-type protein. It is important for the new recognition sequence to be fairly unique, so that it is not recognized with a significant affinity by other DNA-binding proteins within the cell. The altered-specificity DNA-binding protein is then expressed in cells containing an endogenous gene or reporter gene regulated by the altered control region, and its capacity to regulate transcription is monitored.

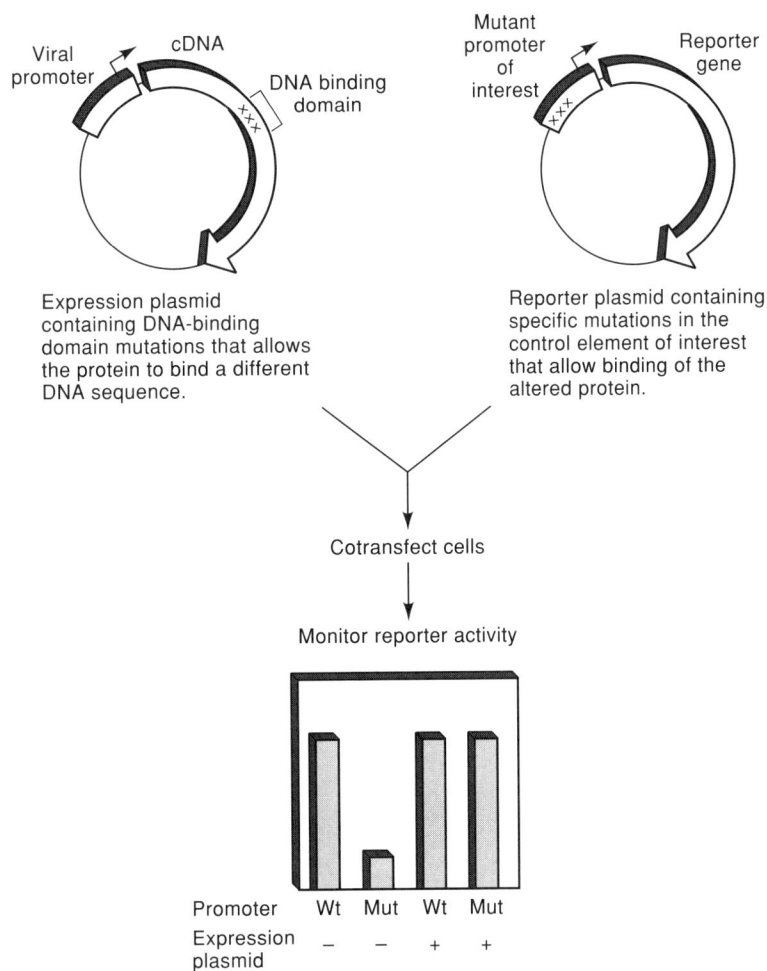

FIGURE 9.8. Altered specificity strategy.

A particularly elegant example of an altered-specificity experiment, which reveals its strengths and weaknesses, was reported by Shah et al. (1997). (Another important example was recently reported by Gillemans et al. 1998.) The goal of this study was to determine which of two POU-domain proteins, OCT-1 or OCT-2, is a functional activator of the Ig heavy- and light-chain gene promoters. Both proteins can bind with similar affinity to the octamer elements within the Ig promoters, both proteins are expressed in B lymphocytes, and gene disruption experiments have been uninformative. Therefore, it has been difficult to identify the relevant activator of the Ig promoters.

To create an altered-specificity OCT protein, Shah et al. (1997) examined the known crystal structure of the OCT–DNA complex. They focused their attention on a particular amino acid sequence within the POU domain that contacted one of the nucleotides within the octamer DNA sequence. They first mutated that nucleotide, which disrupted binding by the wild-type OCT proteins. Then, they set out to isolate an altered OCT POU domain capable of binding the altered DNA sequence with high affinity. To achieve this goal, they generated a phage expression library for the POU domain, in which each phage expressed a POU domain with a randomly generated amino acid sequence in the region that was predicted to be in close proximity to the altered base pair. To isolate phage that express a mutant POU domain capable of binding the altered DNA sequence, the expression library was probed with a radiolabeled oligonucleotide containing the altered sequence. The amino acids within the selected POU domain, which allowed high-affinity binding, were then determined and introduced into mammalian expression plasmids for both OCT-1 and OCT-2.

The altered OCT-1 and OCT-2 proteins were tested in B cells for their ability to *trans*-activate reporter plasmids under the control of Ig promoters containing the altered octamer DNA sequence. By using the altered octamer sequence, endogenous OCT-1 and OCT-2 proteins within the B cells were rendered nonfunctional on the Ig promoters; only the altered-specificity OCT-1 or OCT-2 introduced into the cells could bind. Because the experiments could be performed in B cells, it was anticipated that the altered proteins would be capable of functionally interacting with the other proteins needed for B-cell-specific Ig promoter and enhancer activity.

The results revealed that the altered OCT-1 and OCT-2 proteins were equally capable of stimulating Ig promoter activity. Surprisingly, however, when the reporter plasmids included an Ig enhancer in addition to the Ig promoter, OCT-1 was found to be a much more potent activator. These results are consistent with a hypothesis in which OCT-1 is the functional activator of endogenous Ig genes, because only OCT-1 can carry out the protein–protein interactions needed for promoter- and enhancer-dependent transcription.

The altered specificity strategy can provide compelling evidence that a DNA-binding protein acts at a particular target site. In effect, this strategy can provide much of the same information as would be provided by a gene disruption experiment (see above), but the key limitations of a gene disruption experiment are eliminated. One important limitation of gene disruption is that loss of the DNA-binding protein might result in a loss of cell viability. A second limitation is that it is extremely difficult to distinguish between direct and indirect effects of the binding protein on the candidate target gene. A third limitation is that the DNA-binding protein may be redundant with a related protein, such that its disruption has no effect on target gene transcription. None of these limitations is relevant in the altered-specificity experiment: The cells remain viable because the endogenous DNA-binding protein is still expressed; indirect effects are less likely because the altered recognition site for the altered-specificity protein has been introduced only into the control region of interest, and redundancy is not observed because the altered DNA-binding domain is present in only one protein.

Despite these considerable advantages, the altered-specificity approach has three limitations. One limitation is that it typically involves overexpression of the altered DNA-binding protein by either transient or stable transfection. As described above, results obtained with overexpressed proteins are inconclusive and difficult to interpret. To enhance the degree to which altered-specificity results can be interpreted, the altered protein should be expressed at a concentration similar to that of the endogenous, wild-type protein. To achieve this goal, different stable cell lines expressing the altered protein can be examined, or a variety of expression vectors containing different promoters or enhancers can be tested.

The second limitation arises when the altered control region is analyzed in the context of a transiently transfected reporter plasmid. The high copy number of the reporter plasmid and its removal from its natural chromosomal environment could influence its ability to respond to the altered-specificity protein. In an ideal experiment, the substitution mutation creating the altered control element would be introduced into the endogenous gene by homologous recombination. This rigorous approach would provide the strongest evidence that a particular DNA-binding protein acts on a target gene. Of course, this experiment would require much more effort than the basic transfection experiment. The use of a stably transfected reporter plasmid would be preferred to the use of a transiently transfected reporter, but the results would remain inconclusive.

The third limitation, which cannot be overcome, is that the transcriptional activation function of the wild-type DNA-binding protein might depend on the specific amino acids that are altered. These amino acids could be directly involved in transcriptional activation, or the protein–DNA interaction could lead to a conformational change in another surface of the protein involved in transcriptional activation. In most altered-specificity DNA-binding proteins, the amino acids responsible for DNA binding are unimportant for transcriptional activation, but examples of a conformational link between DNA binding and activation have been described (Lefstin and Yamamoto 1998).

Nevertheless, despite this limitation and the others mentioned above, the altered-specificity strategy can provide valuable information regarding the relevance of a protein–DNA interaction and holds an important position among the arsenal of strategies described in this chapter.

To summarize, we have described twelve general strategies that can be used to test the hypothesis that a DNA-binding protein is a relevant regulator of a gene by binding a defined control element. Some of these strategies may not be feasible for analysis of some DNA-binding proteins, and other strategies that were not described can be envisioned (e.g., in vivo antibody microinjection experiments). As stated in the introduction, no single strategy can conclusively establish the functional relevance of a protein–DNA interaction. However, a hypothesis can be greatly strengthened by subjecting it to as many rigorous tests as possible.

REFERENCES

Benezra R., Davis R.L., Lockshon D., Turner D.L., and Weintraub H. 1990. The protein Id: A negative regulator of helix-loop-helix binding proteins. *Cell* **61:** 49–59.

Boyd K.E. and Farnham P.J. 1997. Myc versus USF: Discrimination at the cad gene is determined by core promoter elements. *Mol. Cell. Biol.* **17:** 2529–2537.

Boyd K.E., Wells J., Gutman J., Bartley S.M., and Farnham P.J. 1998. c-Myc target gene specificity is determined by a post-DNA binding mechanism. *Proc. Natl. Acad. Sci.* **95:** 13887–13892.

Branch A.D. 1998. A good antisense molecule is hard to find. *Trends Biochem. Sci.* **23:** 45–50.

Capecchi M.R. 1989. Altering the genome by homologous recombination. *Science* **244:** 1288–1292.

Clevers H.C. and Grosschedl R. 1996. Transcriptional control of lymphoid development: Lessons from gene targeting. *Immunol. Today* **17:** 336–343.

Descombes P. and Schibler U. 1991. A liver-enriched transcriptional activator protein, LAP, and a transcriptional inhibitory protein, LIP, are translated from the same mRNA. *Cell* **67:** 569–579.

Dynan W.S. and Tjian R. 1983a. Isolation of transcription factors that discriminate between different promoters recognized by RNA polymerase II. *Cell* **32:** 669–680.

———.1983b. The promoter-specific transcription factor Sp1 binds to upstream sequences in the SV40 early promoter. *Cell* **35:** 79–87.

Eisenbeis C.F., Singh H., and Storb U. 1995. Pip, a novel IRF family member, is a lymphoid-specific, PU.1-dependent transcriptional activator. *Genes Dev.* **9:** 1377–1387.

Ernst P., Hahm K., and Smale S.T. 1993. Both LyF-1 and an Ets protein interact with a critical promoter element in the murine terminal transferase gene. *Mol. Cell. Biol.* **13:** 2982–2992.

Ernst P., Hahm K., Trinh L., Davis J.N., Roussel M.F., Turck C.W., and Smale S.T. 1996. A potential role for Elf-1 in terminal transferase gene regulation. *Mol. Cell. Biol.* **16:** 6121–6131.

Fiering S., Northrop J.P., Nolan G.P., Mattila P.S., Crabtree G.R., and Herzenberg L.A. 1990. Single cell assay of a transcription factor reveals a threshold in transcription activated by signals emanating from the T-cell antigen receptor. *Genes Dev.* **4:** 1823–1834.

Georgopoulos K., Moore D.D., and Derfler B. 1992. Ikaros, an early lymphoid-specific transcription factor and a putative mediator for T cell commitment. *Science* **258:** 808–812.

Georgopoulos K., Bigby M., Wang J.H., Molnar A., Wu P., Winandy S., and Sharpe A. 1994. The Ikaros gene is required for the development of all lymphoid lineages. *Cell* **79:** 143–156.

Gillemans N., Tewari R., Lindeboom F., Rottier R., de Wit T., Wijgerde M., Grosveld F., and Philipsen S. 1998. Altered DNA-binding specificity mutants of EKLF and Sp1 show that EKLF is an activator of the β-globin locus control region in vivo. *Genes Dev.* **12:** 2863–2873.

Gilmour D.S. and Lis J.T. 1984. Detecting protein-DNA interactions in vivo: Distribution of RNA polymerase on specific bacterial genes. *Proc. Natl. Acad. Sci.* **81:** 4275–4279.

———.1985. In vivo interactions of RNA polymerase II with genes of *Drosophila melanogaster*. *Mol. Cell. Biol.* **5:** 2009–2018.

———.1986. RNA polymerase II interacts with the promoter region of the noninduced hsp70 gene in *Drosophila melanogaster* cells. *Mol. Cell. Biol.* **6:** 3984–3989.

Hahm K., Ernst P., Lo K., Kim G.S., Turck C., and Smale S.T. 1994. The lymphoid transcription factor LyF-1 is encoded by specific, alternatively-spliced mRNAs derived from the *Ikaros* gene. *Mol. Cell. Biol.* **14:** 7111–7123.

Jonsen M.D., Petersen J.M., Xu Q.P., and Graves B.J. 1996. Characterization of the cooperative function of inhibitory sequences in Ets-1. *Mol. Cell. Biol.* **16:** 2065–2073.

Kadonaga J.T., Carner K.R., Masiarz F.R., and Tjian R. 1987. Isolation of cDNA encoding transcription factor Sp1 and functional analysis of the DNA binding domain. *Cell* **51:** 1079–1090.

Kim T.K. and Maniatis M. 1997. The mechanism of transcriptional synergy of an in vitro assembled interferon-β enhanceosome. *Mol. Cell* **1:** 119–129.

Langlands, K., Yin X., Anand G., and Prochownik E.V. 1997. Differential interactions of Id proteins with basic helix-loop-helix transcription factors. *J. Biol. Chem.* **272:** 19785–19793.

Lauring J. and Schlissel M.S. 1999. Distinct factors regulate the murine RAG-2 promoter in B and T cell lines. *Mol. Cell. Biol.* **19:** 2601–2612.

Lefstin J.A. and Yamamoto K.R. 1998. Allosteric effects of DNA on transcriptional regulators. *Nature* **392:** 885–888.

Lehman A.M., Ellwood K.B., Middleton B.E., and Carey M. 1998. Compensatory energetic relationships between upstream activators and the RNA polymerase II general transcription machinery. *J. Biol. Chem.* **273:** 932–939.

Lo K., Landau N.R., and Smale S.T. 1991. LyF-1, a transcriptional regulator that interacts with a novel class of promoters for lymphocyte-specific genes. *Mol. Cell. Biol.* **11:** 5229-5243.

Lodish H., Baltimore D., Berk A., Zipursky S.L., and Darnell J. 1996. *Molecular cell biology*, 3rd edition. W.H. Freeman, New York.

Mauxion F., Jamieson C., Yoshida M., Arai K., and Sen R. 1991. Comparison of constitutive and inducible transcriptional enhancement mediated by κ B-related sequences: Modulation of activity in B cells by human T-cell leukemia virus type I tax gene. *Proc. Natl. Acad. Sci.* **88:** 2141–2145.

Murphy T.L., Cleveland M.G., Kulesza P., Magram J., and Murphy K.M. 1995. Regulation of interleukin 12 p40 expression through an NF-κ B half-site. *Mol. Cell. Biol.* **15:** 5258–5267.

Okamura R.M., Sigvardsson M., Galceran J., Verbeek S., Clevers H., and Grosschedl R. 1998. Redundant regulation of T cell differentiation and TCRα gene expression by the transcription factors LEF-1 and TCF-1. *Immunity* **8:** 11–20.

Orlando V., Strutt H., and Paro R. 1997. Analysis of chromatin structure by in vivo formaldehyde cross-linking. *Methods* **11:** 205–214.

Plevy S.E., Gemberling J., Hsu S., Dorner A.J., and Smale S.T. 1997. Multiple control elements mediate activation of the murine and human IL-12 p40 promoters: Evidence of functional synergy between C/EBP and Rel proteins. *Mol. Cell. Biol.* **17:** 4572-4588.

Pongubala J.M., Nagulapalli S., Klemsz M.J., McKercher S.R., Maki R.A., and Atchison M.L. 1992. PU.1 recruits a second nuclear factor to a site important for immunoglobulin κ 3′ enhancer activity. *Mol. Cell. Biol.* **12:** 368–378.

Shah P.C., Bertolino E., and Singh H. 1997. Using altered specificity Oct-1 and Oct-2 mutants to analyze the regulation of immunoglobulin gene transcription. *EMBO J.* **16:** 7105–7117.

Stein C.A. 1998. How to design an antisense oligodeoxynucleotide experiment: A consensus approach. *Antisense Nucleic Acid Drug Dev.* **8:** 129–132.

Thanos D. and Maniatis T. 1995. NF-κB: A lesson in family values. *Cell* **80:** 529–532.

van Genderen C., Okamura R.M., Farinas I., Quo R.G., Parslow T.G., Bruhn L., and Grosschedl R. 1994. Development of several organs that require inductive epithelial-mesenchymal interactions is impaired in LEF-1-deficient mice. *Genes Dev.* **8:** 2691–2703.

Verbeek S., Izon D., Hofhuis F., Robanus-Maandag E., te Riele H., van de Wetering M., Oosterwegel M., Wilson A., MacDonald H.R., and Clevers H. 1995. An HMG-box-containing T-cell factor required for thymocyte differentiation. *Nature* **374:** 70–74.

Walter J. and Biggin M.D. 1997. Measurement of in vivo DNA binding by sequence-specific transcription factors using UV cross-linking. *Methods* **11:** 215–224.

Walter J., Dever C.A., and Biggin M.D. 1994. Two homeo domain proteins bind with similar specificity to a wide range of DNA sites in *Drosophila* embryos. *Genes Dev.* **8:** 1678–1692.

Wathelet M.G., Lin C.H., Parekh B.S., Ronco L.V., Howley P.M., and Maniatis T. 1998. Virus infection induces the assembly of coordinately activated transcription factors on the IFN-β enhancer in vivo. *Mol. Cell* **1:** 507–518.

CHAPTER 10

In Vivo Analysis of an Endogenous Control Region

Important Issues

- *Genomic footprinting and in vivo protein–DNA crosslinking can help to establish the relevance of specific protein–DNA interactions at an endogenous control region.*

- *Low-resolution or high-resolution micrococcal nuclease assays can be used to identify positioned nucleosomes that may be important for the function of an endogenous control region.*

- *Restriction enzyme accessibility and a chromatin immunoprecipitation assay using antibodies specific for acetylated histones can provide valuable information about nucleosome remodeling and histone acetylation.*

- *Further insight into the regulatory strategies for a gene can be provided by a determination of its methylation status and subnuclear localization.*

INTRODUCTION

The preceding chapters described strategies for identifying important control regions for a gene of interest and for characterizing relevant protein–DNA interactions within a control region. These strategies are essential for elucidating transcriptional regulatory mechanisms, but they usually rely on the use of artificial functional assays and in vitro protein–DNA interaction assays. Artificial functional assays are essential because it is too difficult and time-consuming to scan a control region for important DNA elements by introducing specific mutations into an endogenous locus. Protein–DNA interaction assays must be performed in vitro because rigorous biochemical analyses are not possible in vivo. Because the artificial aspects of these approaches limit the information obtained, the dissection of a control region can greatly benefit from an analysis of the endogenous locus using the methods described in this chapter. Such an analysis may reveal, for example, specific protein–DNA interactions that were not detected during a mutant analysis using an artificial transfection assay. An examination of the endogenous locus can also help to confirm the relevance of protein–DNA interactions detected using in vitro binding assays, and can provide insight into regulatory contributions from nucleosomes, DNA methylation, and subnuclear localization.

The preceding chapters presented several methods that can provide information about an endogenous locus, including:

- nuclear run-on experiments to assess the regulation of an endogenous gene at the level of transcription initiation and elongation (Chapter 3)

- potassium permanganate experiments to test the possibility that an inducible gene contains a paused polymerase molecule immediately downstream of the transcription start site prior to gene induction (Chapter 3)

- DNase I sensitivity studies to assess the global chromatin structure of a locus (Chapter 6)

- DNase I hypersensitivity and matrix attachment region (MAR) studies to localize distant control regions (Chapter 6)

- genomic footprinting to identify important protein–DNA interactions (Chapter 7)

- genomic footprinting and in vivo crosslinking/immunoprecipitation studies to help assess the relevance of a specific protein–DNA interaction (Chapter 9)

In this chapter, we discuss some of these assays and others from the perspective of a comprehensive analysis of an endogenous gene or control region.

The chapter begins with a discussion of methods for detecting interactions between sequence-specific DNA-binding proteins and an endogenous control region (Fig. 10.1). Then, basic strategies for analyzing the chromatin status of a locus are discussed, including methods for determining the presence, positioning, and remodeling of nucleosomes. Approaches that can be used to determine the methylation status and subnuclear localization of a locus are briefly discussed. Finally, specific protocols for several relevant procedures are presented.

Sequence-specific protein–DNA interactions

- DNase I genomic footprinting
- DMS genomic footprinting
- Crosslinking/immunoprecipitation

DNA melting

- Potassium permanganate genomic footprinting

Nucleosome positioning

- MNase-Southern blot
- MNase-LM-PCR
- DNase I genomic footprinting (rotational phasing)

Nucleosome remodeling

- DNase I sensitivity (Chapter 6)
- DNase I hypersensitivity (Chapter 6)
- Restriction enzyme accessibility
- Crosslinking/immunoprecipitation with histone antibodies

DNA methylation

- Methylation-sensitive restriction enzyme analysis
- Bisulfite-induced modification assay

Subnuclear localization

- MAR association
- ImmunoFISH analysis

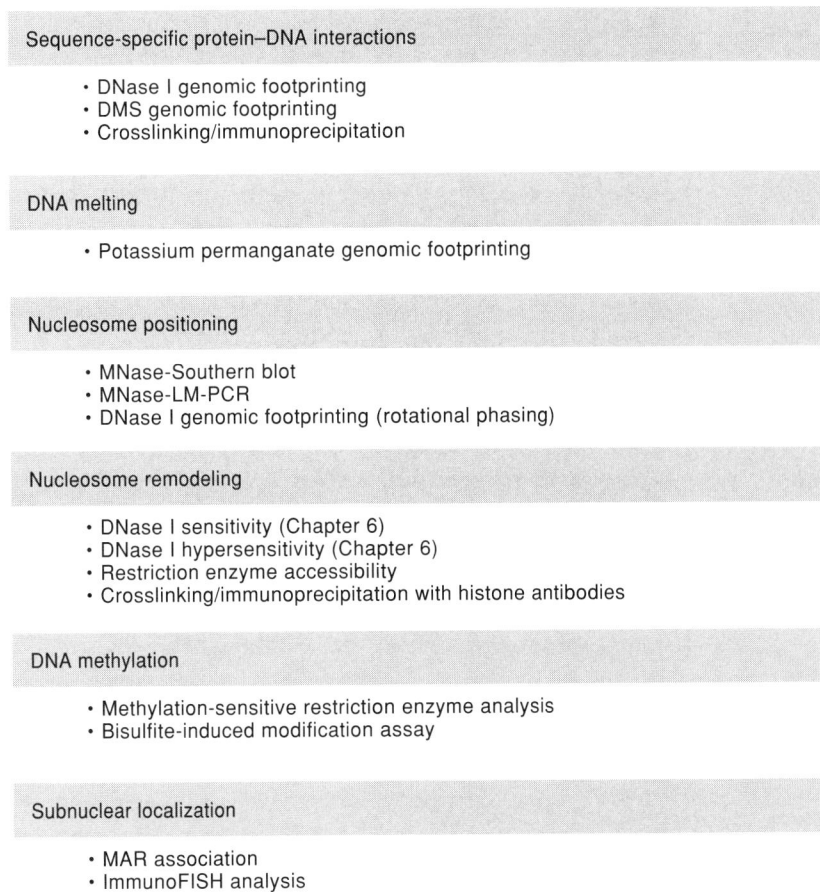

FIGURE 10.1. Methods for analyzing an endogenous control region.

CONCEPTS AND STRATEGIES

In Vivo Analysis of Sequence-specific Protein–DNA Interactions

DNase I and DMS Genomic Footprinting

DNase I and DMS genomic footprinting are preferred methods for detecting protein–DNA interactions at an endogenous control region (Mueller et al. 1988; Mueller and Wold 1989; Garrity and Wold 1992; Garrity et al. 1995). As stated in Chapters 7 and 9 and in the introduction to this chapter, these methods can reveal important protein–DNA interactions that were not detected using other methods and can help confirm the importance of an interaction. These procedures and others described in this chapter rely on a technique known as ligation-mediated PCR (LM-PCR; see Box 10.1). LM-PCR can be employed in concert with a variety of agents that cleave or modify genomic DNA within an intact cell or nucleus, with each agent providing unique insight into the regulation of a gene.

To perform a genomic footprinting experiment, genomic DNA can be cleaved with DNase I in isolated nuclei (Fig. 10.2A and Protocol 10.2). The cleaved genomic DNA is then purified and analyzed by LM-PCR. Alternatively, genomic DNA can be modified at guanosine and adenosine residues in intact cells by dimethylsulfate (DMS; Fig. 10.2B and Protocol 10.2; see also Chapter 13). The modified DNA is then purified, treated with piperidine to catalyze hydrolysis adjacent to the modified bases, and analyzed by LM-PCR. For both the DNase I and DMS methods, control reactions are needed, in which purified genomic DNA or a genomic clone containing the locus is analyzed by the same procedure. This provides the cleavage pattern for unbound DNA for comparison with the pattern observed following DMS modification in cells or DNase I cleavage in nuclei. Protein–DNA interactions appear as regions of protection and/or hypersensitivity (see Chapters 8 and 13 for general discussion of footprinting theory).

As with in vitro DNase I footprinting, genomic footprinting using DNase I provides information about the DNA sequences that are inaccessible to digestion by the nuclease because of occupancy by a DNA-binding protein (see Chapters 8 and 13). The DMS protection technique, like the comparable in vitro technique (see Chapter 13), provides information about nucleotides that are inaccessible to modification with DMS because of their close association with bound proteins. Because DMS is a small chemical, the DNA region protected from modification by a bound protein is much shorter than the region protected from DNase I digestion and is likely to correspond more closely to the nucleotides required for protein binding (see Chapter 13).

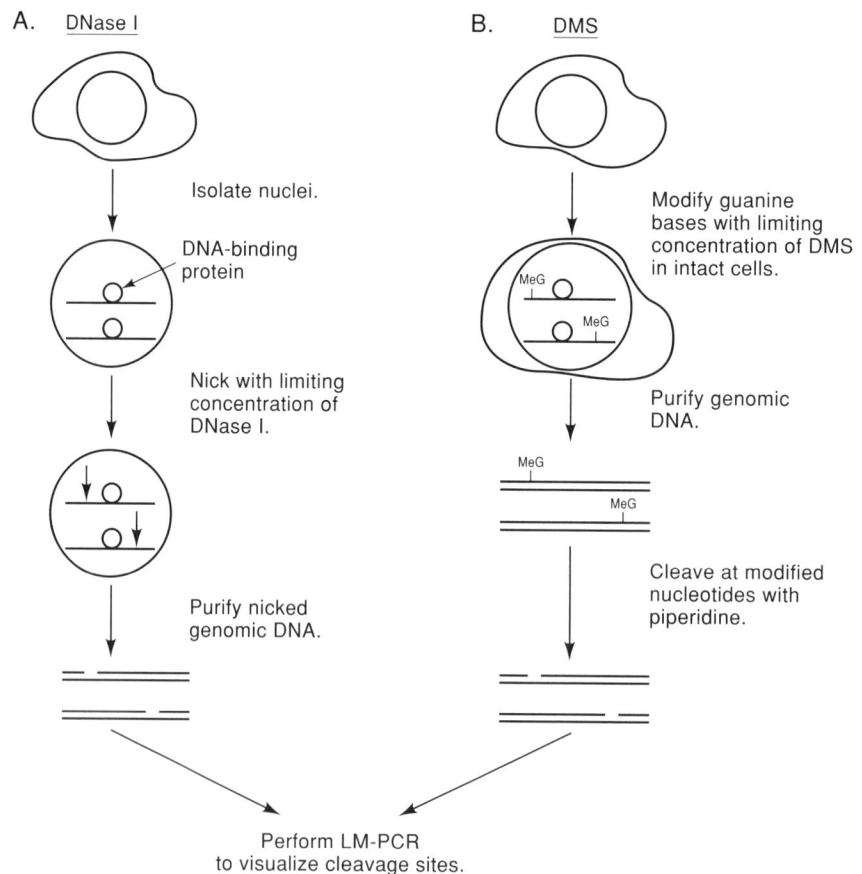

FIGURE 10.2. Genomic footprinting.

Box 10.1

Ligation-mediated PCR

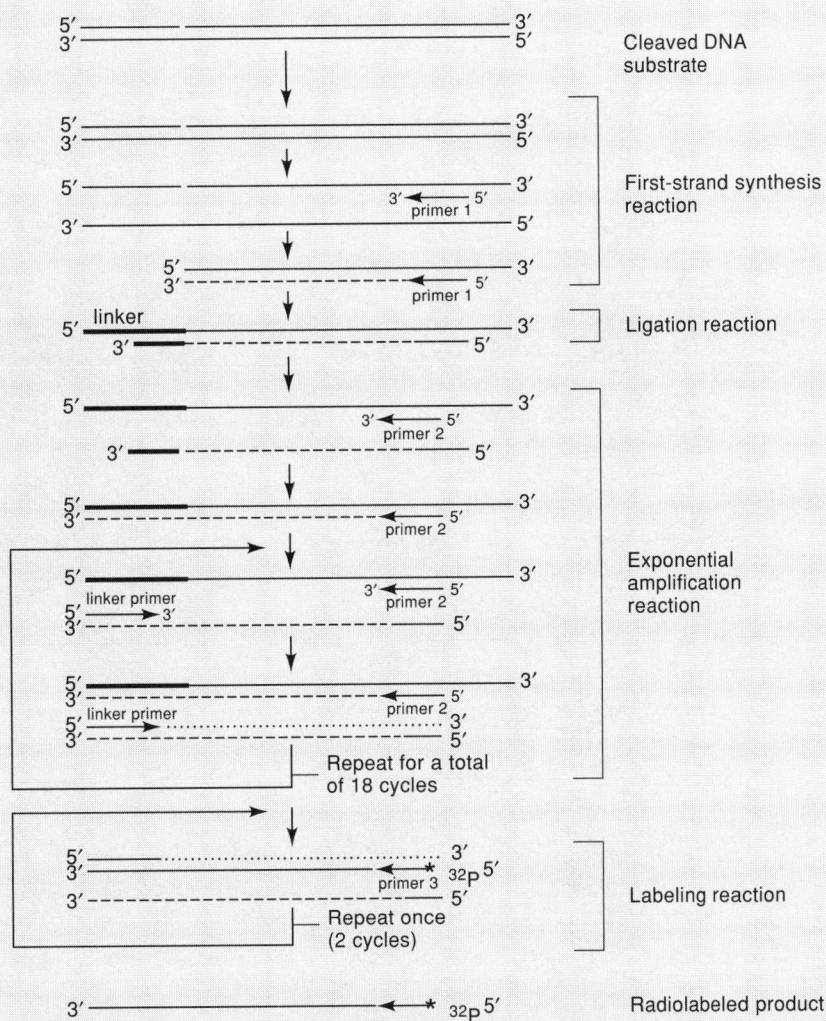

FIGURE 10.3. Standard LM-PCR. (Adapted, with permission, from Garrity and Wold 1992 [Copyright National Academy of Sciences].)

LM-PCR is a powerful technique for detecting DNA strand breaks within a complex sample. The technique is well suited for this purpose because of its extreme sensitivity and specificity. LM-PCR (Fig. 10.3) was first described by Barbara Wold and colleagues (Mueller et al. 1988; Mueller and Wold 1989), with more recent technical improvements by the same laboratory (Garrity and Wold 1992; Garrity et al. 1995; see also Protocol 10.2). The technique is typically performed on purified genomic or plasmid DNA that has previously been cleaved with limiting concentrations of a nuclease (e.g., DNase I, micrococcal nuclease, or a restriction endonuclease; Fig. 10.3). Alternatively, LM-PCR can be performed on DNA that had been modified with DMS or potassium permanganate, followed by strand cleavage at the modified nucleotides by piperidine.

The first step in most LM-PCR procedures is to anneal a synthetic oligonucleotide primer (primer #1) to a specific sequence of the nuclease- or piperidine-cleaved and denatured DNA. Primer #1 usually anneals at least 100 bp from the region of interest (see Protocol 10.2). The primer molecules are extended to the end (i.e., cleavage site) with a thermostable DNA polymerase, yielding a blunt end. An oligonucleotide linker is then ligated to the blunt-ended DNA fragments. The linker employed for this step contains one blunt end and one staggered end so that it always ligates in the same orientation. (Note that the first steps in the LM-PCR procedure following MNase cleavage are somewhat different; see Box 10.3 and Protocol 10.2). Because the sequence of the linker is known, a primer, termed the linker primer, can be designed to anneal to that sequence for use in PCR.

The first PCR step is performed with the linker primer and primer #2. Primer #2 usually hybridizes to a sequence that overlaps the primer #1 sequence, but is slightly "internal" to that sequence, ensuring that only those DNA fragments derived from the gene of interest will be amplified. For the genomic footprinting procedures, this amplification step results in a nested set of double-stranded DNA fragments. Following the first amplification step, a second PCR step is performed, using the same linker primer and yet another primer (#3) that is slightly internal to primer #2. Before use, primer #3 is 5′ end-labeled with T4 polynucleotide kinase and [γ-^{32}P]ATP. This final PCR results in the amplification of radiolabeled DNA fragments, whose sizes correspond to the distance from the 5′ end of primer #3 to the sites of nuclease or chemical cleavage, plus the length of the linker primer. Because the starting point was a nested set of DNA fragments, the final radiolabeled products should span a wide range of sizes. The fragments can be analyzed on a denaturing polyacrylamide gel, followed by autoradiography or phosphorimager analysis.

Uses and limitations of genomic footprinting. As discussed in Chapter 7, genomic footprinting with DNase I or DMS can be used as a starting point for the dissection of a control region because it can provide insight into the locations of functionally important DNA elements. However, it generally cannot substitute for a comprehensive mutant analysis because only a subset of the important control elements are likely to yield genomic footprints. In addition, genomic footprinting provides only limited information about the identity of the relevant binding protein and about its importance for gene regulation relative to other proteins that regulate the same control region (see Chapter 7). As discussed in Chapter 9, genomic footprinting can provide support for the functional relevance of a candidate regulatory protein. If the genomic footprinting pattern observed at an endogenous control element matches the footprinting pattern observed in vitro using a purified recombinant protein (when the two reactions are performed side by side), that protein may indeed be responsible for the function of the endogenous element. One caveat of this experiment, however, is that the basic cleavage pattern observed in vivo can be quite different from that observed in vitro if the in vivo DNA is assembled into nucleosomes; nucleosomal DNA will exhibit a very different pattern of cleavage products, making a comparison of in vivo and in vitro footprints difficult. A third general use for genomic footprinting is to determine whether the occupancy of a DNA element within an endogenous control region is inducible or subject to cell-type-specific regulation.

One issue to discuss in greater detail is why some important protein–DNA interactions do not yield detectable genomic footprints. This issue is poorly understood, but several possible reasons can be considered. For DNase I genomic footprinting, one possibility is that an important protein might interact with a DNA sequence that is intrinsically resistant to DNase I cleavage in the absence of protein binding (see Chapters 8 and 13). If DNase I does not efficiently cleave a region of the protein-free control template, it will be difficult to detect protection in the presence of protein. Another possibility is that a protein may dissociate from the DNA during the preparation of nuclei. For DMS footprinting, the DMS may directly modify the DNA-binding protein, causing it to dissociate from the DNA. Another possibility is that the protein may not closely contact major groove guanines or minor groove adenines, which are the targets of DMS. Yet another possibility is that DNase I or DMS protection may be obscured by the large number of PCR cycles, which may preferentially amplify background cleavage products or weak cleavage products within the protected regions.

In addition to these purely technical considerations, a protein–DNA interaction may be undetectable by genomic footprinting for other reasons. For example, the interaction may occur in only a fraction of the cells within the population used for the experiment. This is a common problem for inducible genes, which often are induced in only a fraction of cells, even when a clonal cell line is used for the experiment. The interleukin-2 (IL-2) and IL-12 p40 genes are two examples of inducible genes that are induced in only a small percentage of cells within a clonal population (Garrity et al. 1993; A. Weinmann and S.T. Smale, unpublished data). To address this possibility, the expression efficiency can first be monitored by flow cytometry or immunofluorescence. If these experiments confirm that the gene is expressed in only a small percentage of cells, methods to separate expressing cells from nonexpressing cells (e.g., fluorescence-activated cell sorting [FACS] or magnetic beads) can be considered. It is important to keep in mind that a region of protection is difficult to detect unless at least 75–80% of the sites within the cell population are occupied by a protein (see Chapters 8 and 13). If efficient expression is not observed and if a method is unavailable for isolating expressing cells, compelling genomic footprinting results may be difficult to obtain.

Yet another reason a protein–DNA interaction may not be detectable by genomic footprinting is that the site may be occupied only during a specific stage of the cell cycle. If cell cycle regulation is responsible for expression in only a fraction of the cells, cell synchronization experiments may be required to demonstrate the cell cycle regulation and to isolate a homogeneous population. Finally, some DNA-binding proteins may dissociate from the DNA after each round of transcription initiation and may then reassociate prior to the next round. This possibility is purely speculative, however, as little is known about the dynamics of protein–DNA interactions at a eukaryotic locus.

Despite the inability to detect some protein–DNA interactions, DNase I and DMS genomic footprinting can provide valuable information about the interactions that occur at an endogenous control region. When combined with functional assays and studies of specific DNA-binding proteins, this information can lead to a more complete understanding of the mechanism of gene regulation. Protocols for DNase I and DMS genomic footprinting are provided at the end of this chapter (Protocol 10.2).

Potassium permanganate genomic footprinting. Potassium permanganate is a third reagent that can be coupled to LM-PCR to analyze an endogenous locus. Genomic footprinting with potassium permanganate was first mentioned in Chapter 3 as a method for analyzing RNA polymerase pausing as a mode of transcriptional regulation. The theoretical basis of permanganate footprinting is discussed in detail in Chapter 15, in a discussion of in vitro assays for dissecting transcriptional activation mechanisms. In brief, potassium permanganate can enter intact cells and modify thymines, tagging them for subsequent cleavage by piperidine. Permanganate is unique, however, in that it modifies thymines within single-stranded DNA much more efficiently than within double-stranded DNA. Regions of a genomic locus that are melted are therefore hypersensitive to cleavage with this reagent. Because of this unique property, permanganate footprinting is primarily used to identify melted DNA. Nevertheless, regions of protection can also be observed, providing information about protein–DNA interactions similar to that provided by the DNase I and DMS genomic footprinting methods. Permanganate footprinting for this latter purpose has been particularly useful for detecting TBP–TATA interactions due to the preference of permanganate for Ts.

With regard to melted DNA, potassium permanganate genomic footprinting has primarily been used to establish that inducible genes in eukaryotic cells possess a paused RNA polymerase molecule downstream from their transcription start sites prior to transcriptional induction. In Chapter 3, the nuclear run-on assay was described as the principal method for demonstrating that polymerase pausing contributes to the regulation of particular genes (e.g., the HIV-1, c-*myc*, and *Drosophila* hsp70 genes). Permanganate genomic footprinting is a second method that can provide evidence of paused polymerase molecules. When the polymerase pauses a few dozen base pairs downstream of the transcription start site, the DNA template in the vicinity of the polymerase molecule remains melted. Thus, hypersensitivity to permanganate modification in uninduced cells can support the hypothesis that a paused polymerase molecule is present. As stated in Chapter 3, neither the nuclear run-on assay nor permanganate genomic footprinting provides proof that an unactivated gene contains a paused polymerase molecule, because both assays are technically difficult. To provide strong support for the hypothesis, data from both assays are needed. Additional support can be provided by an in vivo protein–DNA crosslinking/immunoprecipitation assay using antibodies directed against the RNA polymerase (see Chapter 3 and below).

In Vivo Protein–DNA Crosslinking/Immunoprecipitation

As described above, genomic footprinting can reveal the locations of protein–DNA interactions at an endogenous control element. However, an important limitation of genomic footprinting is that it cannot provide strong evidence regarding the identity of a bound protein. A second limitation is that a genomic footprint depends on efficient occupancy of the site. The in vivo protein–DNA crosslinking/immunoprecipitation assay has great potential to overcome both of these limitations, although it is relatively new to researchers studying gene regulation in mammalian cells. This assay was already discussed in considerable detail in Chapter 9 (see Figs. 9.6 and 9.7).

It is important to emphasize that the assay complements, but does not substitute for, genomic footprinting because the two assays provide different types of information. Genomic footprinting provides detailed information about the DNA sequences within an endogenous control region that are occupied by proteins, but limited information about the identities of the proteins. In contrast, crosslinking/immunoprecipitation can provide strong evidence that a particular protein contacts a control region, but limited evidence about the precise location of the binding site. A final distinction is that the former assay can provide unbiased evidence of protein–DNA interactions, whereas the latter assay provides evidence of binding only by those proteins for which high-quality antibodies are available.

Nucleosome Positioning and Remodeling

Model Systems

The important role played by chromatin in the regulation of gene expression has become increasingly apparent. However, as described in Chapter 1, most of the recent advances have involved in vitro or in vivo studies of artificial model genes. These approaches are very important, and essential for dissecting the fundamental mechanisms of nucleosome remodeling, but a complete understanding of the regulation of a gene of interest requires studies of nucleosome positioning and remodeling at the endogenous locus. Nucleosomes have been studied to some extent at several endogenous control regions, but detailed studies have been limited to a surprisingly small number of model genes. The long-term goal of all of these studies is to determine the precise order of events that lead to transcriptional activation. In other words, which protein–DNA interactions stimulate the initial chromatin alterations at a locus? Which remodeling complexes or histone acetylases are responsible for the alterations? Which nucleosomes are remodeled? What is the consequence of remodeling with respect to the capacity of other factors to bind the control region and stimulate (1) additional remodeling events or (2) the recuitment of coactivators and the general transcription machinery?

Three of the well-characterized models are the *Saccharomyces cerevisiae* Pho5, human immunodeficiency virus (HIV), and murine mammary tumor virus (MMTV) promoters (Fig. 10.4). The Pho5 promoter is contained within positioned nucleosomes, with an extended linker region between two of the nucleosomes containing a binding site for the PHO4 transcriptional activator (Almer and Horz 1986; Lohr 1997). Upon induction by low phosphate, PHO4 is dephosphorylated, allowing for translocation to the nucleus, where it binds DNA and facilitates the remodeling of four nucleosomes containing and surrounding the promoter (Almer et al. 1986; Fascher et al. 1993; Kaffman et al. 1998; McAndrew et al. 1998). The importance of nucleosomes for Pho5 regulation was established by genetic studies in which nucleosome loss by disruption of histone H4 expression was sufficient for a substantial induction of Pho5 transcription (Han and Grunstein 1988). This result suggests that, in the absence of nucleosomes, general transcription factors can access the promoter to some extent in a Pho4-independent manner.

Most of the important DNA elements within the HIV-1 promoter are in a region that is devoid of nucleosomes, but nucleosomes are positioned upstream of the promoter and

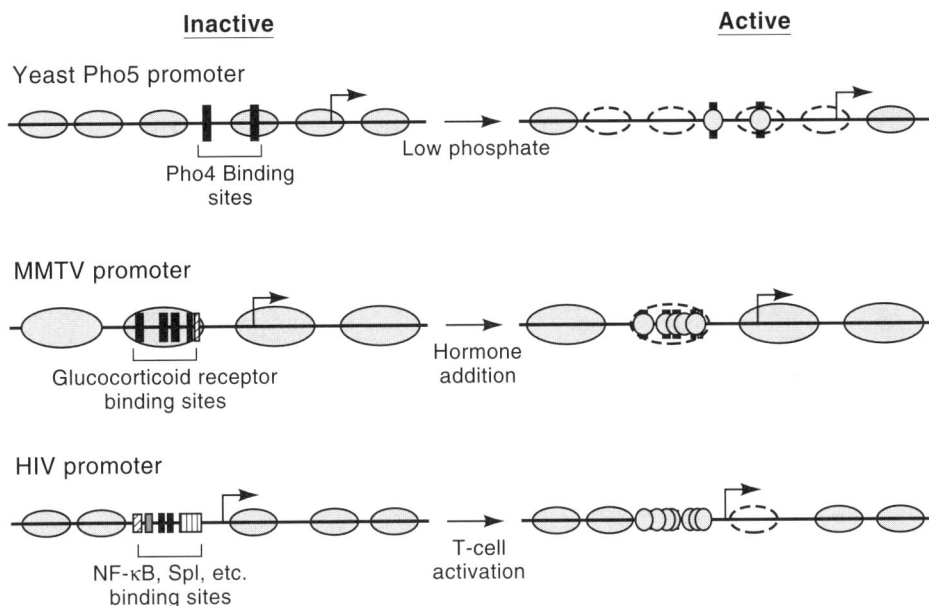

FIGURE 10.4. Model genes for analysis of nucleosome positioning and remodeling.

immediately downstream of the transcription start site (Verdin et al. 1993). Upon T-cell activation, transcription factors bind the promoter and appear to stimulate the selective remodeling of the downstream nucleosome. Like the Pho5 promoter, nucleosome disruption (in this case by treatment with a deacetylase inhibitor) is sufficient for strong transcription in uninduced cells (Van Lint et al. 1996).

The MMTV promoter, which is induced by the glucocortocoid receptor, is contained within positioned nucleosomes, with one nucleosome spanning the receptor binding sites (Richard-Foy and Hager 1987; Fragoso et al. 1995). Upon ligand activation of the receptor, this nucleosome is selectively remodeled, facilitating the binding of additional factors that are necessary for transcription. An interaction between the glucocortocoid receptor and a nucleosome remodeling protein, BRG1, is essential for remodeling, with different receptor domains needed for the subsequent transcriptional activation (Fryer and Archer 1998).

The papers cited above, and other papers from the laboratories involved in the characterization of these and other promoters, should be consulted before initiating an in vivo analysis of nucleosome positioning and remodeling at a control region. The specific strategies used for each promoter are far from uniform, but a review of the literature should provide insight into the strengths and weaknesses of each approach. A few basic methods are discussed below that are likely to be necessary for the early stages of an analysis of nucleosomes at an endogenous locus. Protocols for three of these methods, micrococcal nuclease (MNase)-Southern blot, MNase-LM-PCR, and restriction enzyme accessibility-LM-PCR, are provided at the end of the chapter (Protocols 10.1 and 10.2).

Low-resolution Analysis of Nucleosome Positioning by the MNase-Southern Blot Method

When beginning a chromatin analysis, a reasonable starting point is to determine whether the endogenous control region of interest is assembled into nucleosomes or is devoid of nucleosomes. A second goal is to determine whether the nucleosomes are positioned at consistent locations within a cell population. In other words, are the nucleosomes similarly positioned at the locus in every cell within a population? Consistent nucleosome positioning is thought to be critical for appropriate regulation of some genes (see Chapter 1). The determinants of nucleosome positioning are not understood in detail, but specific examples of DNA sequences and proteins capable of directing positioning have been reported (Travers and Drew 1997; Vermaak and Wolffe 1998).

The classic strategy for addressing the two issues mentioned above is to take advantage of the unique properties of MNase. This nuclease preferentially introduces double-strand breaks into nucleosomal linker regions or DNA regions that are devoid of nucleosomes. If limiting concentrations of MNase are added to cell nuclei, the majority of the genomic DNA will be digested into fragments that correspond in size to multiples of the nucleosome core plus linker region (approximately 200 bp). After purifying the DNA, this nucleosome ladder can be detected by agarose gel electrophoresis followed by staining with ethidium bromide (see Fig. 10.5). If higher concentrations of MNase are added to the cell nuclei, a larger fraction of the DNA will be digested into fragments corresponding in size to a core mononucleosome (147 bp).

To determine whether a DNA fragment of interest is assembled into nucleosomes, MNase-digested DNA can be analyzed by Southern blot, using a radiolabeled probe containing the DNA fragment of interest. If the probe hybridizes to a ladder of bands that correspond to the nucleosome ladder, the fragment of interest may be assembled into nucleosomes. If a radioactive smear is observed instead of a ladder, or if no hybridization is apparent, the region may be devoid of nucleosomes.

Despite the relative ease of this procedure, there are two important caveats. First, because most of the DNA on the gel is in the form of MNase-resistant nucleosomes, one must be careful to demonstrate that the probe is hybridizing specifically to the genomic fragment of interest, rather than nonspecifically to the bulk DNA. One method for establishing the specificity of hybridization is to include a sample of MNase-treated DNA derived from cells of a different species, which should exhibit the same nucleosome ladder upon ethidium bromide staining, but should not hybridize to the radiolabeled probe (assuming the probe is derived from a DNA fragment that is sufficiently diverged between the species). Another method is to follow the procedure described below for determining whether the nucleosomes are positioned. This method, which is only slightly different from that described above, is much less susceptible to artifacts caused by nonspecific hybridization.

The second caveat is that MNase resistance does not provide definitive evidence that a fragment is nucleosomal. MNase-resistant fragments of approximately the correct size have been observed that are thought to result from a tightly packed array of sequence-specific DNA-binding proteins (see, e.g., Verdin et al. 1993). Therefore, additional experiments are needed to support the hypothesis that an MNase-resistant fragment is indeed nucleosomal. The experiments described below for analyzing nucleosome positioning and remodeling may be beneficial for this purpose. For example, if the putative nucleosome is precisely positioned and is flanked by other nucleosomes at intervals that are consistent with a typical beads-on-a-string array, the MNase-resistant fragment may indeed represent a nucleosome. Evidence of a nucleosome remodeling event upon gene activation can also support the hypothesis that a fragment is nucleosomal because the assays used to monitor remodeling (e.g., the restriction enzyme accessibility assay, see below) will usually yield negative results if the region is devoid of nucleosomes.

To determine whether the nucleosomes associated with a particular DNA fragment are consistently positioned, the MNase-Southern blot method can be modified. The modification is to digest the MNase-digested genomic DNA further with a restriction enzyme prior to Southern blot analysis. This method is described in greater detail in Box 10.2. The greatest limitation of this assay is that the boundaries of the nucleosomes can only be determined within approximately 40 bp because of the resolution of an agarose gel and the variability of cleavage sites within the linker. The low resolution of this technique therefore can limit the accuracy of determining the locations of positioned nucleosomes relative to important control elements.

High-resolution Analysis of Nucleosome Positioning by an MNase-LM-PCR Method and DNase I Genomic Footprinting

A method that provides information about nucleosome positioning at higher resolution than the MNase-Southern blot method described above involves a combination of MNase digestion and LM-PCR (Box 10.3). A typical result from this type of analysis performed with the IL-12 p40 promoter is shown in Figure 10.6. Two tightly packed clusters of hypersensitive cleavage sites are apparent, which presumably correspond to the internucleosomal regions where MNase has introduced double-stranded breaks. The 135-bp protected region between these sites is approximately the size of a nucleosome core. By comparing the sizes of the hypersensitive fragments to the DNA sequence, and with knowledge of the location of the radiolabeled primer, the locations of these putative internucleosomal cleavages can be determined with considerable accuracy (to nucleotides −350 and −485 relative to the transcription start site in this case). It is important to note, however, that absolute precision is not possible. First, the locations of the MNase hypersensitive sites are deter-

Box 10.2

Analysis of Nucleosomal Positioning by MNase-Southern Blot Assay

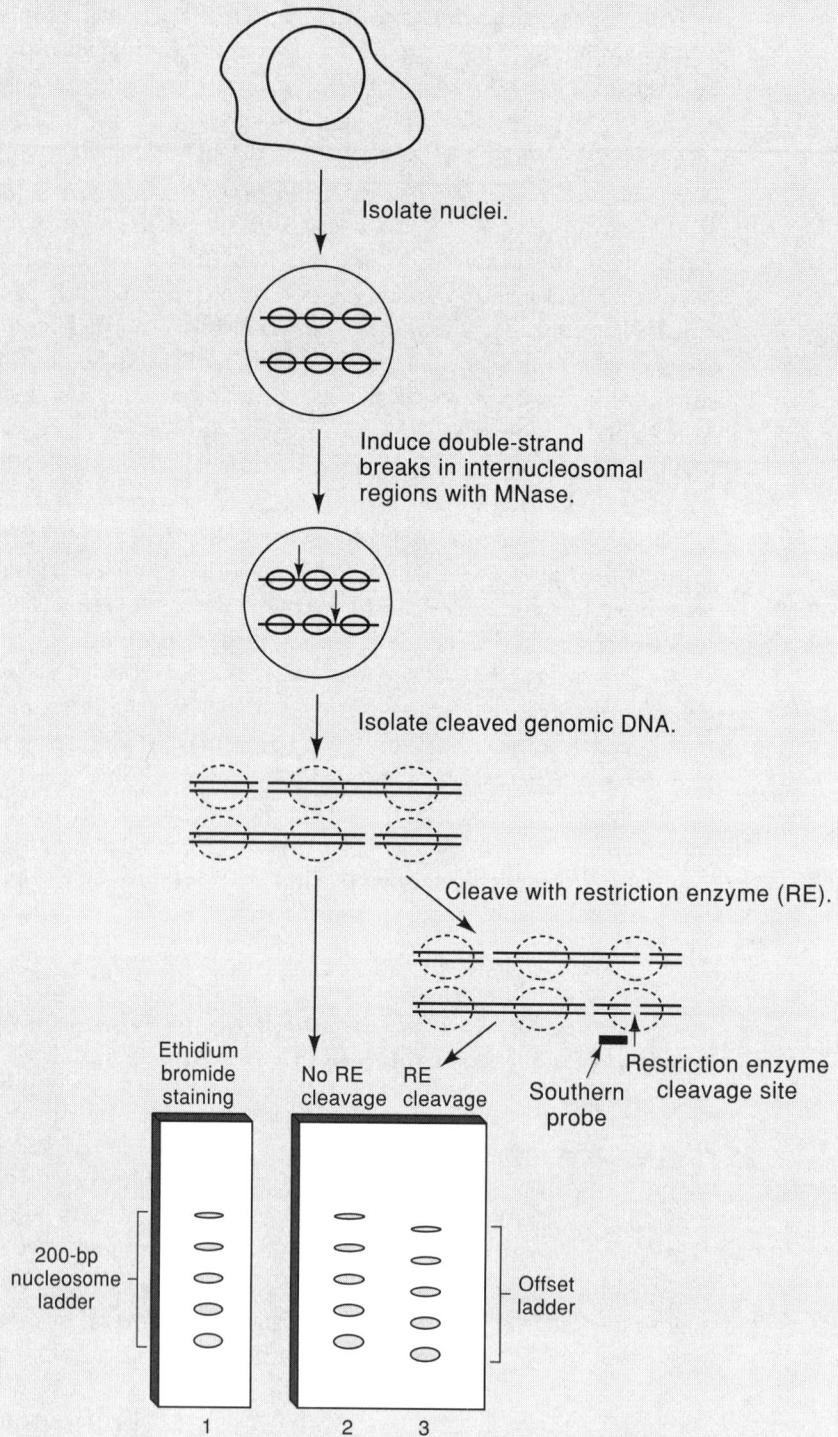

FIGURE 10.5. MNase-Southern blot assay.

The MNase-Southern blot technique is recommended as a first step toward an analysis of nucleosome positioning and remodeling at a control region or gene of interest. This technique can help to establish whether a DNA region contains a typical array of positioned nucleosomes. To determine whether a DNA region is assembled into positioned nucleosomes, nuclei are first prepared from the appropriate cells by lysing the plasma membranes with a nonionic detergent like NP-40 or Triton X-100. After the nuclei are isolated by centrifugation, aliquots are then treated with different concentrations of MNase. MNase digestion is terminated with a solution containing SDS, EDTA, and proteinase K. The genomic DNA is then purified and cleaved with an appropriate restriction enzyme, followed by Southern blot analysis. To choose appropriate restriction enzymes, the restriction map of the locus must be known. The desired enzymes are those that cleave near the DNA region of interest. A radiolabeled probe is then needed that hybridizes to DNA sequences only on one side of the restriction site and as close to the restriction site as possible. Hybridization of a probe immediately adjacent to a site of cleavage is referred to as "indirect end-labeling" because the results are similar to those obtained if the cleaved DNA molecules were directly labeled with ^{32}P.

In the absence of restriction enzyme cleavage, the MNase-digested DNA reveals a typical nucleosome ladder at 200-bp intervals upon probe hybridization if the region is assembled into nucleosomes. If the nucleosomes are consistently positioned, the samples cleaved with a restriction enzyme yield a new ladder that is offset from the standard ladder by an amount that depends on the location of the restriction site relative to the nucleosome. For example, if the restriction site is at the center of the nucleosome, the new products should be 100, 300, 500 bp, etc., which will migrate half-way between the standard nucleosome products of 200, 400, and 600 bp. If the restriction site is within the internucleosomal region, the restriction enzyme-cleaved products may be difficult to distinguish from the standard nucleosome ladder, and it will be necessary to repeat the experiment with different restriction enzymes that cleave within a nucleosome. The results become more complicated if the region does not contain a standard beads-on-a-string nucleosomal array. If the nucleosomes are not consistently positioned, restriction enzyme cleavage should result in increased smearing of the Southern blot image. If multiple restriction enzymes fail to yield bands suggestive of positioning, the region may be devoid of nucleosomes or may contain nucleosomes that are inconsistently positioned.

Box 10.3

Analysis of Nucleosome Positioning by MNase-LM-PCR and DNase I Genomic Footprinting

The high-resolution method for determining nucleosome positioning involves a combination of the MNase technique described above (Box 10.2) and the LM-PCR technique (Box 10.1). This method provides information about the locations of specific internucleosomal regions which may be located within a few hundred base pairs of the LM-PCR primer.

When used in conjunction with MNase cleavage, the LM-PCR technique is modified slightly relative to that used in conjunction with DNase I cleavage. First, the purified genomic DNA must be phosphorylated at the 5′ end using T4 polynucleotide kinase prior to the ligation of the linker. This step is necessary because MNase digestion of DNA results in 5′ ends containing hydroxyl groups rather than phosphates; T4 DNA ligase requires the presence of 5′ phosphates. Second, the first primer extension step is omitted when MNase is employed. With the DNase I method, this step is included to create a blunt-ended fragment at each point where

DNase I has created a nick. MNase creates blunt-ended double-strand breaks when it cleaves within linker regions, making the primer extension step unnecessary. More importantly, the procedure will not work if the first step is included because MNase efficiently nicks DNA within nucleosomes. In other words, MNase actually cleaves DNA throughout the nucleosomal DNA region, but it only cleaves in a double-stranded manner in the linker regions. A detailed procedure for this technique is provided in Protocol 10.2.

mined in part by the DNA sequence. MNase does not cleave randomly within internucleosomal regions, but possesses intrinsic sequence preferences. Thus, one can conclude with some confidence that the hypersensitive MNase cleavage sites are within an internucleosomal region, but the precise boundaries of the flanking nucleosomes cannot be determined. Second, nucleosome positioning appears to be flexible and therefore may not be tightly restricted to a precise location (Fragoso et al. 1995).

In addition to the MNase-LM-PCR method, DNase I genomic footprinting can contribute to an understanding of nucleosome positioning. DNase I preferentially nicks helical DNA that is facing outward from the histone octamer. Thus, as the helical DNA wraps around the nucleosome, a slightly hypersensitive cleavage is observed at approximately 10-bp intervals. These cleavages, if observed, can provide information about the rotational phasing of the nucleosome.

In Vivo Methods for Analyzing Nucleosome Remodeling

As described in Chapter 1, nucleosome remodeling plays an important role in the regulation of most, if not all, genes. To monitor nucleosome remodeling at an endogenous locus, a few different methods can be employed. Two common methods are the DNase I sensitivity and hypersensitivity assays described in Chapter 6. The DNase I sensitivity method was originally developed to monitor changes in the overall chromatin structure of a locus. DNase I hypersensitivity is a useful method for identifying transcriptional control regions, which are sometimes nucleosome-free or associated with remodeled nucleosomes. Therefore, changes in DNase I sensitivity or hypersensitivity can be used to monitor changes in chromatin structure upon gene activation.

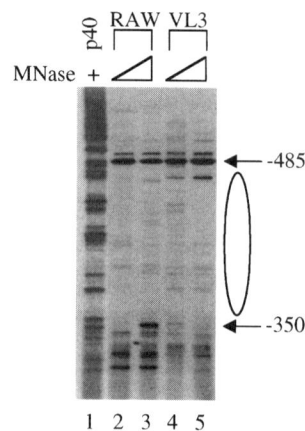

FIGURE 10.6. Typical result from MNase-LM-PCR analysis. This experiment was performed with the inactive IL-12 p40 promoter. (Reprinted, with permission, from Weinmann et al. 2000.)

Although the DNase I methods can be useful, they are not generally recommended for a detailed analysis of nucleosome organization and remodeling because they provide relatively limited information about the remodeling event. DNA sensitivity is useful primarily for analyzing an entire locus and is not generally used to monitor changes in nucleosome structure at individual control regions. DNase I hypersensitive sites often correspond to individual control regions. However, the appearance of a hypersensitive site can sometimes provide ambiguous information because hypersensitivity can result from either an altered chromatin structure or the binding of a sequence-specific DNA-binding protein.

The MNase-Southern blot or MNase-LM-PCR assays can provide more compelling evidence of nucleosome remodeling upon gene activation. Cells that express the gene can be compared to cells that lack expression. Evidence of nucleosome remodeling can sometimes be provided by enhanced sensitivity to MNase cleavage in the expressing cells. In other instances, enhanced MNase sensitivity may not be observed, however. For example, if a gene is inefficiently induced in a cell line, which is quite common, the enhanced cleavage in a small fraction of the cells may be obscured by the absence of an effect in the majority of the cells. One example of the use of the MNase-LM-PCR assay to study remodeling of the IL-12 p40 promoter is shown in Figure 10.7. In this example, MNase cleavage within the nucleosome is only slightly enhanced upon gene induction, yielding results that are suggestive, but inconclusive. A likely reason for the modest effect is that the IL-12 gene is expressed in only about 25% of the cells within this clonal cell line.

An assay that overcomes the limitations of the DNase I and MNase methods for monitoring nucleosome remodeling is the "restriction enzyme accessibility" assay (Box 10.4).

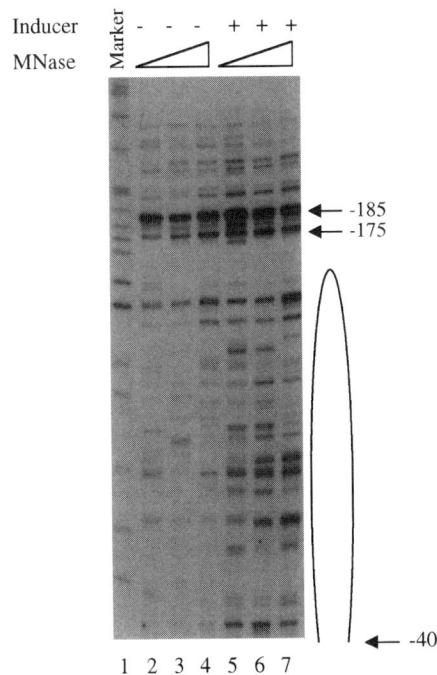

FIGURE 10.7. This figure depicts the MNase cleavage pattern over the functional IL-12 p40 promoter region before and after transcription induction by LPS. Before induction, a putative internucleosomal region is revealed by the hypersensitive cleavage sites at nucleotides –175 and –185 (lanes 2–4). After induction, the region downstream of these sites, between –40 and –175, becomes more sensitive to MNase cleavage, consistent with remodeling of the nucleosome (lanes 5–7). (A. Weinmann and S.T. Smale, unpublished data.)

This assay monitors the degree to which a given enzyme can access and cleave its recognition site in isolated nuclei. In a typical nucleosome prior to remodeling, most sites cannot be recognized efficiently and cleaved. However, upon remodeling, structural alterations occur that enhance accessibility to endonuclease cleavage.

The restriction enzyme accessibility assay can be coupled to the LM-PCR technique, like all of the other nuclease and chemical cleavage assays described above. Because restriction enyzme digestion usually results in a relatively small number of cleavage products, the restriction enzyme accessibility assay is easier and more sensitive than the genomic footprinting and MNase-LM-PCR assays described above. An example of the restriction enzyme accessibility assay, performed with the IL-12 p40 promoter, is shown in Figure

Box 10.4

Restriction Enzyme Accessibility Assay

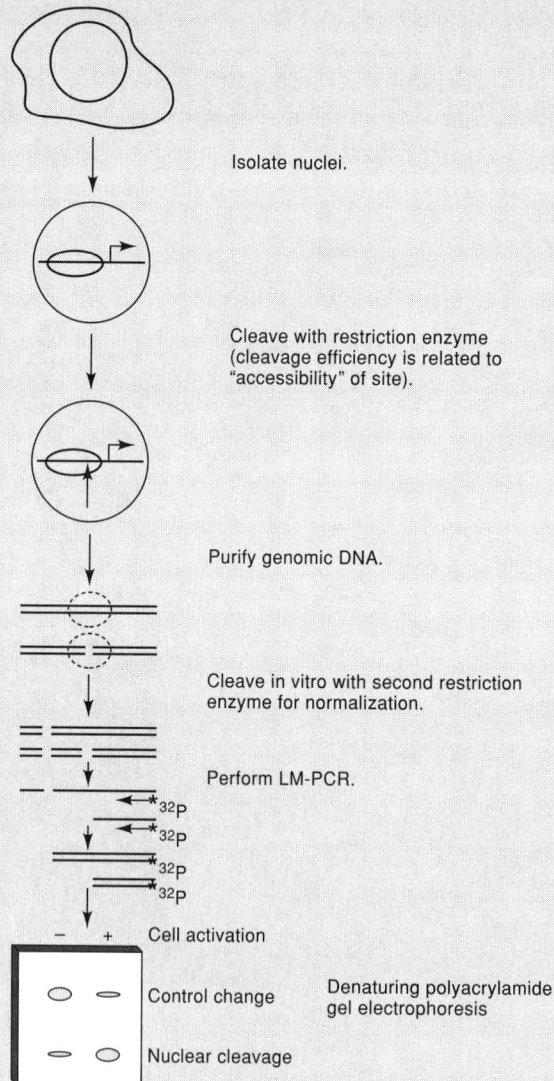

FIGURE 10.8. Restriction enzyme accessibility assay.

The restriction enzyme accessibility assay (Fig. 10.8) is strongly recommended as a method for monitoring nucleosome remodeling events at specific nucleosomes. This assay involves the preparation of intact nuclei by detergent lysis, as described above (Box 10.2). The nuclei are then incubated with a limiting concentration of a restriction enzyme in a typical restriction enzyme digestion buffer. It is important to perform the digestion for a relatively short time period to minimize the possibility that the integrity of the nucleosome will be disrupted. After purifying the cleaved genomic DNA, it is cleaved to completion with another restriction enzyme that serves as an internal standard. This enzyme should be one that cleaves relatively close to the first enzyme, so that the two in vitro cleavage products can be monitored in the same LM-PCR assays. Following the LM-PCR procedure (Box 10.1), accessibility is often presented as a fraction, by dividing the radioactivity in the nuclear cleavage product by the total radioactivity (i.e., in the nuclear cleavage product plus the in vitro cleavage product). This ratio may not provide an accurate reflection of the fraction that was cleaved in the nuclei (because the LM-PCR technique is not linear), but a comparison of the results obtained in cells expressing the gene and lacking expression of the gene should be informative.

10.9. The sensitivity and utility of this assay, relative to the MNase assay, should be apparent from a comparison of Figures 10.7 and 10.9.

A final assay for monitoring nucleosome remodeling at an endogenous control region is an in vivo crosslinking/immunoprecipitation (i.e., chromatin immunoprecipitation) assay (see Chapter 9, Fig. 9.6) to determine the extent of histone acetylation. Antibodies that specifically bind acetylated histones are commercially available from Upstate Biotechnology. In addition, a kit for this procedure is being marketed by Upstate Biotechnology. As of this writing, only a few studies employing this technique for the analysis of mammalian genes have been reported (e.g., Parekh and Maniatis 1999). However, it is likely to become a common method for obtaining evidence of nucleosome remodeling and, more specifically, demonstrating that histone acetylation accompanies remodeling.

If the accumulated evidence suggests that nucleosome remodeling contributes to gene regulation, several more advanced studies can be considered. One important goal is to determine which transcription factors contribute to remodeling. This goal can be achieved by developing a functional assay for testing mutants in the control region. Stable transfection assays performed with episomally maintained plasmids have been useful for the chromatin studies of the MMTV promoter because these plasmids, unlike transiently transfected plasmids, appear to be assembled into chromatin that resembles the endogenous chromatin. The transcription factors required for remodeling at an endogenous locus can alternatively be pursued by blocking expression of specific transcription factors and monitoring the effect on remodeling. After the important transcription factors have been determined, their relevant interactions with remodeling complexes and histone acetylases and deacetylases can be pursued. Although in vitro studies have dominated the chromatin field in recent years (see Chapter 1), these types of in vivo strategies remain at the forefront of the chromatin field and can be followed in recent literature.

DNA Methylation

The techniques described in the above sections provide information about the properties of an endogenous locus, including specific protein–DNA interactions, nucleosome posi-

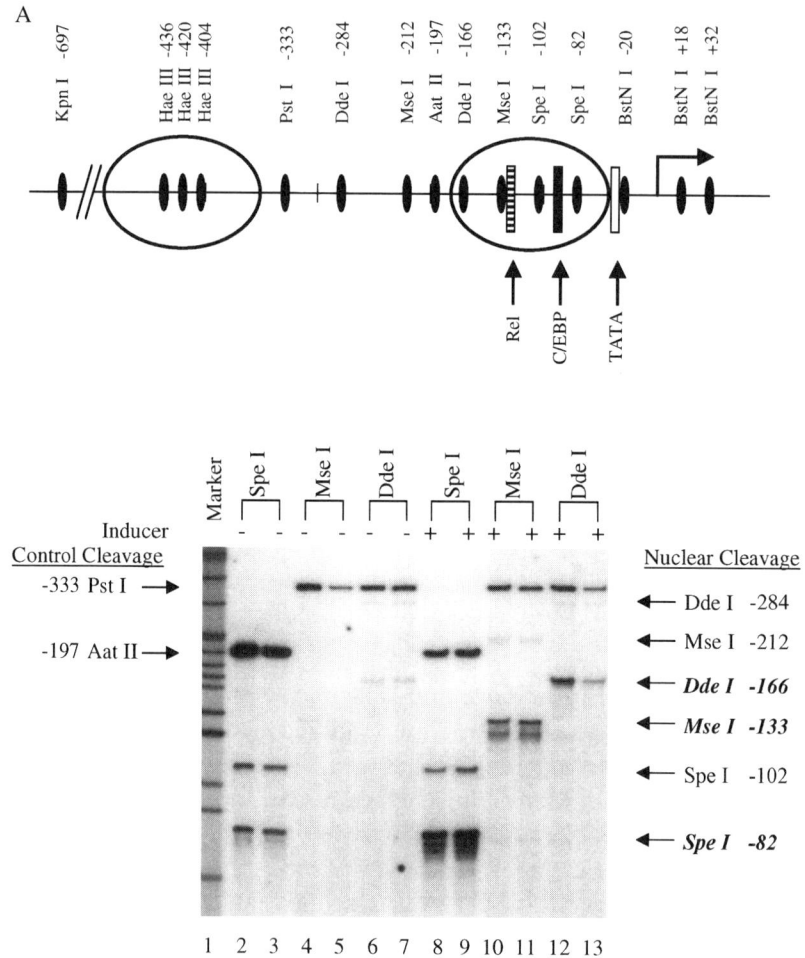

FIGURE 10.9. An example of the restriction enzyme accessibility assay, performed with the IL-12 p40 promoter. The results reveal that restriction enzyme cleavage in nuclei from activated cells (with the enzymes *Spe*I, *Mse*I, and *Dde*I; lanes *8–13*) yields LM-PCR products that are approximately 10 times more abundant than observed following cleavage with the same enzymes in nuclei from unactivated cells (lanes *2–7*), following normalization to the in vitro controls (*Pst*I-333 and *Aat*II-197). (Reprinted, with permission, from Weinmann et al. 2000.)

tioning, and nucleosome remodeling. For a complete analysis of a gene, it may be useful to obtain insight into other properties and changes that may occur when the gene is activated or repressed. One of these properties is the gene's methylation status (see Chapter 1). DNA methylation within mammalian genomes is restricted to cytosines within a subset of CpG dinucleotides. The traditional method for monitoring cytosine methylation has been to cleave genomic DNA with methylation-sensitive restriction enzymes, following by Southern blot analysis using a probe that will detect the cleavage products of interest. Restriction enzyme cleavage products can also be analyzed by PCR (see, e.g., Mostoslavsky et al. 1998).

One limitation of the use of methylation-sensitive restriction enzymes is that they can only be used to study the status of CpG dinucleotides that fall within restriction enzyme recognition sequences. To obtain comprehensive information about the methylation status of a gene or control region, a different PCR-based technique must be employed (Frommer et al. 1992). This technique begins with the treatment of genomic DNA with bisulfite. Under appropriate reaction conditions, bisulfite will convert cytosine to uracil, but 5-methylcytosine remains nonreactive. After bisulfite treatment, the DNA region of interest is amplified by PCR and the PCR products are sequenced (either directly or after cloning into a plasmid vector). By comparison with the known DNA sequence, the cytosines that were methylated within the genomic DNA can be identified because they will be the only cytosines that were not converted to uracil. If the PCR products are sequenced directly, or if multiple independent clones are sequenced, insight can also be obtained into the approximate percentage of genomic DNA molecules within a population that contains 5-methyl-cytosine at a given site.

Subnuclear Localization of a Gene

Models have been proposed based on genetic, biochemical, and cytological studies that genes may be organized at discrete subnuclear locations when they are active or, conversely, repressed (for review, see Lamond and Earnshaw 1998). Two strategies that can be considered for monitoring the subcellular localization of an endogenous gene are association with the nuclear matrix and immunofish analysis. The methods for determining whether a control region associates with the nuclear matrix are described in Chapter 6. In that chapter, the goal was to identify regions surrounding a locus that might associate with the matrix and that therefore might correspond to functional regulatory regions. If a control region has been identified by a different approach, its ability to associate physically with the matrix can be determined. This might help establish models for the function of the control region. Knowledge of nuclear matrix association has limited utility, however, as the general significance of this property has not been clearly established (see Chapters 1 and 6).

A second technique may prove to be generally useful for monitoring the subnuclear localization of particular endogenous loci (Brown et al. 1997, 1999). This technique combines confocal immunofluorescence with fluorescent in situ hybridization (FISH) to determine the relative location of a gene and a particular protein or nuclear structure. Brown et al. (1997, 1999) used this technique to demonstrate that, in lymphocytes, specific inactive genes monitored by FISH colocalize with centromeric heterochromatin, as monitored by immunofluorescence with antibodies against a protein that is predominantly localized to centromeric foci. These results suggest that gene inactivation may sometimes require recruitment to centromeric foci, which may assemble the locus into heterochromatin.

In conclusion, the arsenal of experimental strategies discussed in this chapter can lead to broadly based models for a gene's regulatory mechanisms, especially when these studies of the endogenous locus are performed in concert with the studies discussed in preceding chapters. The models that result from these studies will aid in the design of advanced biochemical studies as described in Chapters 11 through 15.

TECHNIQUES

PROTOCOL 10.1

MNase-Southern Blot Assay

SUMMARY

MNase is unique among nucleases in its relative ability to induce double-strand breaks within nucleosomal linker regions, but only single-stranded nicks within the nucleosome itself. Because of this property, MNase can be used to determine whether a DNA fragment of interest is nucleosomal. In addition, MNase can be used to determine the approximate positions of nucleosomes at a region of DNA, if the nucleosomes are consistently positioned (see Box 10.2). In brief, the cells are lysed and nuclei are isolated by centrifugation. Limiting concentrations of MNase are then added to the nuclei, resulting in cleavage at nucleosome linker regions (preferably cleavage at two sites per DNA molecule). The cleavage reactions are stopped and the genomic DNA is purified. Agarose gel electrophoresis and ethidium bromide staining of the purified DNA should result in a ladder of bands corresponding in size to multiples of the nucleosome core plus linker (approximately 200 bp). To determine whether a DNA fragment of interest is nucleosomal, the genomic DNA can be subjected to Southern blot analysis. If a probe derived from the DNA fragment hybridizes to the ladder of nucleosomal bands, the fragment may indeed be assembled into nucleosomes. To determine nucleosome positioning, the purified genomic DNA must be cleaved with a restriction enzyme prior to gel electrophoresis and Southern blot analysis (see Box 10.2). The following protocol was adapted from the protocols of Richard-Foy and Hager (1987) and Enver et al. (1985).

TIME LINE AND ORGANIZATION

Before beginning this protocol, the only special reagents that are needed, in addition to those listed below, are cells and a DNA fragment for Southern blot probe preparation. With these reagents in hand, the complete protocol will require several days to complete. On the first day, buffers are prepared, cells are harvested, and the nuclei are prepared. The nuclei are then digested with MNase, followed by an overnight incubation with proteinase K to degrade the MNase and nuclear proteins. The next day, the genomic DNA is purified and cleaved with appropriate restriction enzymes, which proceeds for several hours or overnight. The Southern blot is then performed, including gel electrophoresis, transfer to a membrane, and hybridization to a radiolabeled probe. After the blot is washed, it is exposed to film or a phosphorimager screen.

OUTLINE

Low-resolution in vivo MNase analysis

Step 1: Prepare buffers (1 hour)
Step 2: Harvest cells and prepare nuclei (1 hour)
Step 3: Digest with MNase (5 minutes followed by overnight incubation)

Step 4: Purify genomic DNA (1 day)

Step 5: Digest purified MNase-treated DNA with restriction enzyme in vitro (overnight)

Step 6: Agarose gel electrophoresis and Southern blot transfer (1 day)

Step 7: Prepare radiolabeled Southern probe (1 hour)

Step 8: Southern blot prehybridization and hybridization (1 day)

Step 9: Wash membrane and expose to film or phosphorimager screen (2-4 hours to overnight)

Step 10: (optional): Strip membrane (1 hour)

PROCEDURE

CAUTIONS: *CaCl$_2$, Chloroform, Ethanol, Ethidium bromide, Formamide, HCl, KCl, MgCl$_2$, NaOH, Phenol, Polyvinylpyrrolidine, Radioactive substances, SDS, UV radiation. See Appendix I.*

Step 1: Prepare buffers

NP-40 lysis buffer (store at 4°C):
> 10 mM Tris-HCl (pH 7.4)
> 10 mM NaCl
> 3 mM MgCl$_2$
> 0.5% NP-40 (Nonidet P-40)
> 0.15 mM spermine
> 0.5 mM spermidine

MNase digestion buffer (store at 4°C):
> 10 mM Tris-HCl (pH 7.4)
> 15 mM NaCl
> 60 mM KCl
> 0.15 mM spermine
> 0.5 mM spermidine

MNase stop buffer (store at room temperature):
> 100 mM EDTA
> 10 mM EGTA
> Adjust to pH 7.5

Denaturing solution (prepare fresh):
> 0.5 M NaOH
> 1.5 M NaCl

Neutralizing solution (prepare fresh):
> 1 M Tris-HCl (pH 8)
> 1.5 M NaCl

20x SSC (store at room temperature):

175.3 g of NaCl
88.2 g of sodium citrate
dH$_2$O to 1 liter
Adjust pH to 7.2

20x Denhardt's (store at –20°C):

2 g of ficoll
2 g of polyvinylpyrrolidine
2 g of BSA (Pentax fraction V)
dH$_2$O to 500 ml
Filter through 0.45-μm membrane

Prehybridization buffer (make fresh before use):

4x SSC
1x Denhardt's
1% SDS
100 μg/ml sonicated, denatured salmon sperm DNA

Hybridization solution (per membrane, make fresh before use):

0.4 ml of 3 mg/ml sonicated denatured salmon sperm DNA
4.6 ml of H$_2$O
1.2 ml of 10% SDS
3 ml of 40% sodium dextran sulfate
0.6 ml of 20x Denhardt's
2.4 ml of 20x SSC

Step 2: Harvest cells and prepare nuclei

Note: Begin with approximately 100 million cells (for up to nine different MNase digestion conditions).

1. Pellet cells at 1500 rpm at 4°C for 10 minutes in disposable 50-ml conical tubes.

2. Discard the supernatant. Wash the cell pellet with ice-cold 1x PBS (10 ml). Pellet cells at 1500 rpm at 4°C for 10 minutes.

3. Resuspend the cell pellet in 5 ml of ice-cold NP-40 lysis buffer. Incubate on ice for 5 minutes.

   Note: The nuclei should be kept cold throughout the procedure. Manipulations and transport to the centrifuge should be on ice.

4. Pellet the nuclei at 1000 rpm for 10 minutes at 4°C. Discard the supernatant.

   Note: The nuclei will form a loose pellet. Care should be taken not to disturb the pellet when removing supernatant. Decant the supernatant or remove with a pasteur pipet. Do not vacuum aspirate from this point forward.

Step 3: Digest with MNase

1. Wash the nuclei with 2.5 ml of MNase digestion buffer.

2. Pellet the nuclei at 1000 rpm for 10 minutes at 4°C. Carefully discard the supernatant. Resuspend the nuclei in 1 ml of MNase digestion buffer containing 1 mM CaCl$_2$.

   Note: Add 1 μl of 1 M CaCl$_2$ per 1 ml of MNase digestion buffer.

3. Transfer 100 μl of nuclei to a series of microfuge tubes containing diluted MNase (Pharmacia & Upjohn Diagnostics).

   Note: A titration of MNase should be performed. As a starting range, 0.25–75 units can be used. The MNase concentration will need to be optimized according to results.

4. Incubate samples at room temperature for 5 minutes.

5. Add 80 μl of MNase digestion buffer and 20 μl of MNase stop buffer to each sample.

6. Add 3 μl of proteinase K (25 mg/ml) and 10 μl of 20% SDS.

   Note: Steps 5 and 6 stop the MNase digestion reaction and should be performed rapidly.

7. Incubate overnight at 37°C.

Step 4: Purify genomic DNA

1. Extract samples with 200 μl of phenol/chloroform, pH 8.

   Note: To avoid extensive shearing of the genomic DNA samples, do not vortex samples. Instead, gently mix by flicking the tubes or rocking.

2. Spin samples at high speed in microfuge for 5 minutes. Carefully transfer aqueous layer to new microfuge tube.

   Note: The genomic DNA may appear as insoluble "strings." Slowly remove the aqueous layer with a pipet, including the "strings," while minimizing disturbance of the phenol/chloroform interface.

3. Extract samples with 200 μl of chloroform.

   Note: Use same precautions as noted in Step 1.

4. Spin samples at high speed in microfuge for 5 minutes. Carefully transfer aqueous layer to new microfuge tube.

   Note: Use same precautions as noted in Step 2.

5. Add 2 μl of heat-treated RNase A (10 mg/ml) to each sample. Incubate at 37°C for 2 hours.

   Note: This step will remove RNA from the genomic DNA preparations.

6. Extract samples with 200 μl of phenol/chloroform, pH 8.

   Note: Use same precautions as noted in Step 1.

7. Spin samples at high speed in microfuge for 5 minutes. Carefully transfer the aqueous layer to a new microfuge tube.

   Note: Use same precautions as noted in Step 2. At this stage, the interface is usually cloudy. Avoid taking the interface, but if a small amount is retained, this can be removed in the chloroform extraction.

8. Extract samples with 200 µl of chloroform.

   Note: Use same precautions as noted in Step 1.

9. Spin samples at high speed in microfuge for 5 minutes. Carefully transfer the aqueous layer to a new microfuge tube.

   Note: Do not retain any of the cloudy interface at this stage; it is preferable to leave behind some of the aqueous layer.

10. Precipitate the digested, genomic DNA by adding 1/10 volume 3 M sodium acetate, pH 5.2, and 2.5 volumes ice-cold ethanol. Incubate samples at –20°C for about 1 hour.

11. Pellet DNA at high speed in microfuge for 10 minutes at 4°C.

12. Discard the supernatants. Wash the DNA pellets with ice-cold 70% ethanol. Spin the samples at high speed for 1–2 minutes at 4°C.

13. Discard the supernatants. Dry the DNA on the bench top or use a SpeedVac with no heat.

    Note: Be careful not to over-dry the DNA, which makes solubilization difficult.

14. Resuspend the pellet in 100 µl of H$_2$O or TE (10 mM Tris-HCl, pH 7.5, 1 mM EDTA).

    Note: Resuspend the DNA gently. Do not vortex or pipet vigorously. It may be necessary to allow the DNA to dissolve overnight at 4°C.

Step 5: Digest purified MNase-treated DNA with restriction enzyme in vitro

Note: Multiple restriction enzyme sites should be tested surrounding the region of interest. Adjacent restriction enzyme sites within the distance of 1 nucleosome (147 bp) are recommended.

1. Determine the DNA concentration using a spectrophotometer. Use 5–20 µg of DNA for in vitro restriction enzyme digestion.

   Note: If the RNase A digestion was not complete, the calculated DNA concentration may be misleading.

2. Digest DNA in 35 µl total volume at 37°C overnight. Use an excess of restriction enzyme in its optimal buffer to ensure complete digestion.

Step 6: Agarose gel electrophoresis and Southern blot transfer

1. Prepare 1% agarose gel approximately 25 cm long containing ethidium bromide (see Sambrook et al. 1989, Chapter 6).

   Note: A higher percentage of agarose (e.g., 1.4%) can enhance resolution, but can reduce the efficiency of DNA transfer.

2. Add 7 µl of 6X gel-loading buffer (type II, Sambrook et al. 1989; Chapter 6) to the overnight restriction enzyme digests for a final concentration of 1X.

3. Load entire sample (42 µl) into one well of the agarose gel. Load DNA size standards in an adjacent well.

4. Proceed with electrophoresis at <1 V/cm.

   Note: The agarose gel is run slowly for tighter migration of DNA fragments.

5. Photograph the agarose gel on a UV light box with a ruler alongside. Remove the wells

with a razor blade, making sure to retain a reference for later estimation of DNA fragment sizes.

Note: The size standards can strongly and nonspecifically hybridize to the radiolabeled probe, obscuring the adjacent sample lanes. Therefore, it can be helpful to remove the size standard lane with a razor blade.

6. Irradiate the gel with short-wave UV for 2–5 minutes.

7. Soak the gel in denaturing solution for 45 minutes to 1 hour on a shaker.

   Note: Prepare the denaturing solution fresh each time.

8. Remove the denaturing solution and briefly rinse the gel with deionized water. Soak the gel in neutralizing solution for 45 minutes to 1 hour on a shaker.

9. Prepare the nylon membrane for transfer. First, cut the membrane to the size of the gel. For orientation purposes, cut a notch in one corner of the membrane with a corresponding notch in the gel. Soak the membrane in deionized H_2O for 5 minutes. Next, soak the membrane in transfer buffer (10x or 20x SSC) for at least 5 minutes.

   Note: Nylon membranes, particularly positively charged nylon membranes, are recommended. Nitrocellulose is not recommended because of several disadvantages (Sambrook et al. 1989, pp. 9.34–9.36; Ausubel et al. 1994, pp. 2.9.11–2.9.12).

10. Prepare capillary transfer apparatus as described in Sambrook et al. (1989, pp. 9.34–9.46) or Ausubel et al. (1994, pp. 2.9.7–2.9.8). Proceed with overnight capillary transfer of DNA to the nylon membrane in neutral conditions as described previously (Sambrook et al. 1989, pp. 9.34–9.46; Ausubel et al. 1994, pp. 2.9.7–2.9.8).

    Note: The efficiency of transfer can be examined indirectly by restaining the agarose gel with ethidium bromide and determining the amount of DNA still present.

11. UV crosslink (254 nm) the DNA to the nylon membrane as described in Ausubel et al. (1994, pp. 2.9.5–2.9.6).

    Note: Membranes can be stored after the UV crosslinking step. For short-term storage, the dry membrane should be kept between pieces of Whatman paper at room temperature. For long-term storage, the membranes should be kept between pieces of Whatman paper in a desiccator at 4°C.

Step 7: Prepare radiolabeled Southern probe

Note: When designing the DNA probe, several issues should be considered. First, the probe should ideally be designed to end at the restriction enzyme site used for the in vitro digestion. If multiple restriction enzyme digests are analyzed on the same blot, the probe should hybridize only on one side of each restriction enzyme site. Second, to ensure that the probe will detect only MNase-digested fragments associated with an in vitro restriction enzyme cleavage, the probe should be limited to the approximate size of one nucleosome and linker (about 200 bp).

1. Prepare radiolabeled probe by either nick translation (see Sambrook et al. 1989, pp. 10.6–10.12; Ausubel et al. 1994, pp. 3.5.4–3.5.6) or random priming (Sambrook et al. 1989, pp. 10.13–10.17; Ausubel et al. 1994, pp. 3.5.9–3.5.10).

2. Purify radiolabeled primer from free nucleotides.

   Note: This can be accomplished by using a Stratagene push column (# 400701, NucTrap columns; see Protocol 4.1) or a similar method.

Step 8: Southern blot prehybridization and hybridization

1. Soak the membrane in 6x SSC for 2 minutes. Allow the membrane to float on the surface of the 6x SSC first before submerging.

2. Place the membrane in a sealable bag (if hybridization is to be carried out in a water bath) or in tubes (if using a hybridization oven). Completely cover the membrane with prehybridization solution. If using sealable bags, remove bubbles before sealing. Bubbles will interfere with prehybridization and hybridization.

   Note: An alternative prehybridization solution prepared with formamide (see Sambrook et al. 1989, p. 9.52) can be used.

3. Carry out prehybridization for 2 hours at 65°C. If using formamide prehybridization solution, prehybridize for 2 hours at 42°C.

4. Prepare hybridization solution (see Step 1). Alternate hybridization buffer compositions (with and without formamide) can be found in Sambrook et al. (1989, p. 9.52) and Ausubel et al. (1994, p. 2.10.7).

5. If using double-stranded DNA probe, boil probe for 5 minutes.

   Note: Use $40–100 \times 10^6$ cpm of radiolabeled probe per 10 ml of hybridization solution.

6. Centrifuge the probe briefly to collect condensation and immediately place on ice.

7. Remove the prehybridization solution and add 10 ml of hybridization solution. Add boiled, radiolabeled probe to the hybridization solution and carefully seal. Mix contents to ensure even distribution of probe.

   Note: If using sealable bags, be careful not to incorporate bubbles upon sealing.

8. Proceed with hybridization at 65°C overnight.

   Note: Hybridization temperature may vary according to the specific probe. If using hybridization buffer containing formamide, carry out hybridization at 42°C.

Step 9: Wash membrane and expose to film or phosphorimager screen

1. Remove hybridization solution with caution and wash filter with wash buffer 1 (2x SSC, 0.1% SDS) at room temperature for 5 minutes.

   Note: It is convenient to perform washes in containers with lids. Place the container on a shaker for gentle agitation. Appropriate care should be used when handling radioactive liquid/solids.

2. Remove wash buffer 1 and replace with fresh wash buffer 1. Incubate at room temperature for 15 minutes on shaker.

   Note: Monitor the wash buffers for radioactivity. The first washes are likely to contain significant amounts of radioactivity.

3. Remove wash buffer 1 and replace with wash buffer 2 (1x SSC, 0.1% SDS). Incubate at room temperature for 15 minutes on shaker.

   Note: Washing stringency increases with decreasing SSC concentration. This will remove probe that hybridized nonspecifically.

4. Remove wash buffer 2 and replace with wash buffer 3 (0.5x SSC, 0.1% SDS). Incubate at room temperature for 15 minutes on shaker.

5. Remove wash buffer 3 and monitor the membrane for radioactivity.

   Note: It is helpful to monitor the membrane as washing stringency increases to determine the rate at which radioactive counts are being lost.

6. Add wash buffer 4 (0.2x SSC, 0.1% SDS) if a significant amount of radioactivity is still retained on the membrane. Incubate at room temperature for 15 minutes on shaker.

7. Remove wash buffer and monitor for radioactivity. If significant counts remain, add wash buffer 2 and incubate at 42°C for 15 minutes.

   Note: Washing stringency increases with increasing temperature. Be careful to monitor membrane during increased temperature washes.

8. Repeat steps 4–6 with 42°C incubation if needed.

9. If significant radioactivity remains, repeat Steps 3–6 with incubation at 65°C.

10. Remove excess liquid from membrane by blotting with Whatman paper. Wrap membrane in plastic wrap and expose to film or phosphorimager screen.

    Note: Do not let membrane completely dry.

Step 10 (optional): Strip membrane

1. Place blot in stripping solution (0.2 N NaOH, 0.1x SSC, and 1% SDS). Incubate for 30 minutes at 65°C.

   Note: This step will remove the hybridized radioactive probe from the membrane. A different probe can then be hybridized.

2. Wash membrane with distilled H_2O.

3. Wash membrane two times with a solution containing 4x SSC and 1% SDS.

4. Membrane can be rehybridized or stored for later use (see above).

ADDITIONAL CONSIDERATIONS

1. It is helpful to run a sample of MNase-digested DNA that has not been digested in vitro with a restriction enzyme along with the restriction enzyme-digested samples. If the DNA sequence of interest is contained within nucleosomes, this sample will appear as a ladder corresponding to multiples of the nucleosome core plus linker. This sample will also show the location of the standard nucleosomal ladder for comparison to the ladders obtained with restriction-enzyme-digested DNA. The retention of a ladder upon in vitro restriction enzyme digestion suggests that the nucleosomes within that DNA region are specifically positioned. The shifting of the ladder with in vitro restriction enzyme digestion as compared to the uncut ladder suggests that the restriction enzyme site is located within a positioned nucleosome (see Box 10.2).

2. It often is more convenient to use probes that span more than a single nucleosome. If this is the case or the probe does not directly abut the restriction enzyme site, the data need to be interpreted more cautiously. For instance, the hybridization may result in two distinct nucleosome ladders: one corresponding to the uncut MNase-digested DNA ladder and one shifted in comparison. This occurs because the restriction enzyme cleavage site is at one end of only a subset of the DNA fragments that hybridize to the probe.

3. DNase I sensitivity and DNase I hypersensitivity studies can be performed using this same procedure, simply by substituting the DNase I cleavage steps described in Protocol 10.2 for the MNase cleavage steps described here.

TROUBLESHOOTING

Weak or no hybridization

Possible cause: Conditions for hybridization or membrane washing are too stringent.
Solution: It often is necessary to determine empirically the conditions used for hybridization. Lower the hybridization temperature to reduce stringency. It is also possible that specific hybridization was lost during the washing steps. Monitor the membrane during washes and stop washing steps when radioactivity retained on the membrane decreases.

Possible cause: Inefficient transfer of DNA to membrane.
Solution: One method for monitoring transfer is to determine whether the DNA is present in the agarose gel after transfer. If a high proportion of the DNA is retained, the transfer was inefficient. Even if the DNA is no longer detected in the agarose gel, this does not necessarily indicate the transfer was efficient. Methylene blue can be used to stain the membrane (Ausubel et al. 1994, p. 2.10.14). Different blotting techniques can also be considered to alleviate transfer problems (see Sambrook et al. 1989, pp. 9.34–9.51; Ausubel et al. 1994, Unit 2.9).

Possible cause: Inefficient probe labeling.
Solution: A high-specific-activity probe is usually needed for a genomic Southern blot. The quality of the probe is critical in attaining results. The probe should be "body labeled" to enhance the specific activity. End-labeled probes are not recommended.

High background

Possible cause: Hybridization conditions not sufficiently stringent.
Solution: The probe could be hybridizing to the DNA nonspecifically, in which case increasing the temperature of hybridization may help. Alternatively, hybridization stringency can be increased by using a 50% formamide buffer for hybridization (see Sambrook et al. 1989, pp. 9.52; Ausubel et al. 1994, p. 2.10.7 for buffer compositions).

Possible cause: Insufficient stringency of washes.
Solution: Washing stringency increases with increasing temperature, decreasing SSC concentration, and increasing SDS concentration. Perform washes with more stringent conditions.

Possible cause: Prehybridization/hybridization blocking was inefficient.
Solution: The blocking reagents used (Denhardt's solution, salmon sperm DNA, or others) are critical and must be of high quality. Therefore, new reagents can be tested. The time of prehybridization can also be increased. Alternative blocking reagents can be considered (see Sambrook et al. 1989; p. 9.49; Ausubel et al. 1994, p. 2.10.16).

Additional troubleshooting

There are many areas where problems can arise during the Southern blot technique. For a more extensive list of possible problems and solutions, see Ausubel et al. 1994 (Unit 2.10).

PROTOCOL 10.2

LM-PCR Methods

DNase I genomic footprinting

MNase mapping of nucleosome positioning

Restriction enzyme accessibility to monitor nucleosome remodeling

DMS genomic footprinting

SUMMARY

LM-PCR methodology was developed by Wold and colleagues in 1988 (Mueller et al. 1988) as a sensitive technique for detecting DNA strand breaks within a complex sample (see Box 10.1). This general method has proven to be invaluable for a number of purposes, in particular an analysis of the properties of an endogenous DNA locus. As described in the text and the protocols below, LM-PCR can be coupled to a number of DNA modification and cleavage reagents. Coupling to DNase I digestion or DMS modification results in assays that are analogous to in vitro DNase I footprinting or methylation protection (see Chapter 13). Coupling to MNase digestion allows for the high-resolution analysis of nucleosome positioning. Coupling to potassium permanganate modification reveals regions of DNA that are melted. Coupling to restriction enyzme digestion in isolated nuclei provides evidence of nucleosome remodeling. These assays, with the exception of permanganate modification, are described below.

In brief, for DNase I, MNase, and restriction enzyme analyses, nuclei must first be isolated by lysis of the cells with a nonionic detergent, followed by centrifugation. The isolated nuclei are then treated with the appropriate nuclease, followed by purification of the genomic DNA. In contrast to the above, DMS is added directly to intact cells. After quenching the modification reaction, genomic DNA is purified, and cleavage adjacent to modified bases is induced by piperidine treatment. LM-PCR is then performed on the genomic DNA cleaved by any of the above methods. A general outline of the LM-PCR strategy can be found in Figure 10.3.

The protocols described below are derived from a number of sources. The LM-PCR method is a modified version of that developed by Wold and coworkers (Mueller et al. 1988; Garrity and Wold 1992; Ausubel et al. 1994, Chapter 15.5; Garrity et al. 1995). The DNase I method was derived from Enver et al. (1985), the MNase method from Richard-Foy and Hager (1987) and McPherson et al. (1993), the restriction enzyme method from Reik et al. (1991), and the DMS method from Mueller et al. (1988) and Ausubel et al. (1994). For a more detailed description of the DMS protocol, Ausubel et al. (1994, Unit 15.5) is strongly recommended.

TIME LINE AND ORGANIZATION

Ligation-mediated PCR (LM-PCR) usually requires the use of three gene-specific primers and two complementary oligonucleotides that are annealed to form the double-stranded linker; one of these linker oligonucleotides is also used as the linker primer for the PCR steps. The linker provides a constant 5′ sequence despite the variable nature of the sequence created by the in vivo digest. The preparation of the linker requires an overnight incubation. Thus, it should be prepared in advance of the LM-PCR portion of the protocol. The instructions for linker preparation are included in Step 6 (LM-PCR day #1), but should be completed before this time.

The design and integrity of the gene-specific primers are critical for the success of the LM-PCR procedure. The following guidelines are useful to consider:

- It is convenient for gene-specific primers 1 and 2 to be 25-mers and primer 3 to be a 30-mer.

- The G/C content of the gene-specific primers should increase with each successive primer. The increasing G/C content will result in increasing T_ms. G/C contents of 48% (primer 1), 56% (primer 2), and 53–60% (primer 3) have worked well.

- Primer 3 should hybridize 50–200 bp from the region of interest (50–350 bp for restriction enzyme accessibility).

- Both the linker and gene-specific primers should be gel purified. This will ensure that the primers are full length and pure.

- It is ideal to have primer 3 overlap primer 2 by at least half of its sequence. The exclusion of primer 2 during the labeling cycles is thought to be more effective with the overlapping sequences.

- When calculating the size of the final LM-PCR products, remember that 25 bp must be added to account for the ligated linker sequence.

OUTLINE

In vivo DNase I, MNase, and restriction enzyme footprinting

Step 1: Prepare buffers (2 hours)
Step 2: Harvest cells and prepare nuclei (1 hour)
Step 3: Digest with nuclease (15 minutes and overnight)
Step 4: Purify DNA (1 day)
Step 5: Prepare DNA for LM-PCR (2–4 hours)
Step 6: LM-PCR day #1 (1 day)
Step 7: LM-PCR day #2 (1 day)

In vivo DMS footprinting

Step 1: Prepare buffers (2 hours)
Step 2: Treat cells with DMS and harvest cells (5 hours)
Step 3: Purify DNA (1 day)
Step 4: Treat with piperidine (4 hours)
Step 5: Prepare DNA for LM-PCR (2–4 hours)
Steps 6 and 7: As above

PROCEDURE FOR DNASE I, MNASE, AND RESTRICTION ENZYME FOOTPRINTING

CAUTIONS: β-*Mercaptoethanol, Bromophenol blue, Chloroform, Ethanol, Formamide, HCl, KCl, MgCl₂, MgSO₄, Phenol, SDS, Sodium acetate. See Appendix I.*

Step 1: Prepare buffers

Cell lysis and enzyme digestion reagents

NP-40 lysis buffer (store at 4°C):

    10 mM Tris-HCl (pH 7.4)
    10 mM NaCl
    3 mM $MgCl_2$
    0.5% NP-40 (Nonidet P-40)
    0.15 mM spermine
    0.5 mM spermidine

Buffer A (DNase I digestion buffer) (store at 4°C):

    100 mM NaCl
    50 mM Tris-HCl (pH 8.0)
    3 mM $MgCl_2$
    0.15 mM spermine
    0.5 mM spermidine

Micrococcal nuclease (MNase) digestion buffer (store at 4°C):

    10 mM Tris-HCl (pH 7.4)
    15 mM NaCl
    60 mM KCl
    0.15 mM spermine
    0.5 mM spermidine

MNase stop buffer (store at room temperature):

    100 mM EDTA
    10 mM EGTA
    Adjust to pH 7.5

Restriction enzyme digestion buffer (store at 4°C):

    10 mM Tris-HCl (pH 7.4)
    50 mM NaCl
    10 mM $MgCl_2$
    0.2 mM EDTA
    0.2 mM EGTA
    0.15 mM spermine
    0.5 mM spermidine
    1 mM β-mercaptoethanol

2× Proteinase K buffer (store at room temperature):

> 100 mM Tris-HCl (pH 7.5)
> 200 mM NaCl
> 2 mM EDTA
> 1% SDS

LM-PCR reagents

Linker oligonucleotide (LM-PCR 1):

> (5′-3′): GCGGTGACCCGGGAGATCTGAATTC

Linker oligonucleotide (LM-PCR 2):

> (5′-3′): GAATTCAGATC

5× First strand buffer (store at –20°C):

> 200 mM NaCl
> 50 mM Tris-HCl (pH 8.9)
> 25 mM $MgSO_4$
> 0.05% gelatin

25 mM dNTP mix (store at –20°C):

> 25 mM dATP
> 25 mM dCTP
> 25 mM dGTP
> 25 mM dTTP

5× Amplification buffer (store at –20°C):

> 200 mM NaCl
> 100 mM Tris-HCl (pH 8.9)
> 25 mM $MgSO_4$
> 0.05% gelatin
> 0.5% Triton X-100

Stop solution (make fresh):

> 10 mM Tris-HCl (pH 7.5)
> 4 mM EDTA
> 260 mM sodium acetate (pH 7.0)
> 67 μg/ml yeast tRNA

Step 2: Harvest cells and prepare nuclei

Begin with approximately 3–4 million cells per digestion reaction.

Note: The nucleus digestion procedure can be scaled according to the number of samples needed. The procedure illustrated here is for approximately 15–20 million cells (5 digestion reactions).

1. Pellet cells at 1500 rpm at 4°C for 10 minutes in disposable conical tubes.

2. Discard the supernatant. Wash the cell pellet with ice-cold 1× PBS. Pellet the cells at 1500 rpm for 10 minutes at 4°C.

3. Resuspend the cell pellet in 2.5 ml of NP-40 lysis buffer (cold) and incubate on ice for 5 minutes.

   Note: The polyvalent cations, spermine and spermidine, are included in the buffers to prevent nuclei clumping. It is important to keep the samples cold throughout the rest of the procedure; keep on ice during manipulations and transport to centrifuge. All subsequent buffers should be ice-cold.

4. Pellet the nuclei at 1000 rpm for 10 minutes at 4°C. Discard the supernatant.

   Note: The nuclei will form a loose pellet. Carefully remove the supernatant; do not vacuum aspirate from this point forward.

Step 3: Digest with nuclease

Note: After the nuclei are prepared, the procedure diverges according to specific digestion type. Each type of digestion is outlined (A = DNase I, B = MNase, and C = restriction enzyme). The DMS procedure, which involves DMS treatment of intact cells, is included at the end of this protocol.

DNase I digestion:

1A. Wash the nuclei pellet with 1.3 ml of Buffer A.

2A. Pellet the nuclei at 1000 rpm for 10 minutes at 4°C. Carefully discard the supernatant. Resuspend the nuclei in 500 µl of Buffer A containing 1 mM $CaCl_2$.

   Note: Add 1 µl of 1 M $CaCl_2$ per 1 ml of Buffer A. Calcium ions are required for DNase I activity.

3A. Transfer 100 µl of nuclei to microfuge tubes containing diluted DNase I.

   Note: DNase I should be stored at –80°C in aliquots at a concentration of 2.5 mg/ml. Prior to use, DNase I should be diluted to a final concentration of 0.15 µg/µl. As a starting point, a titration range from 0.15 µg/reaction to 1.5 µg/reaction can be attempted. The amount of DNase I will need to be optimized according to LM-PCR results.

4A. Incubate samples at 37°C for 2 minutes.

   Note: Time and temperature of DNase I digestion may need to be optimized.

5A. Add 2 µl of 0.5 M EDTA and 100 µl of Buffer A to each sample.

   Note: EDTA chelates the calcium and magnesium ions.

6A. Add 3 µl of proteinase K (25 mg/ml) and 10 µl of 20% SDS to each sample.

   Note: Steps 5A and 6A stop the DNase I digestion reaction and should be performed rapidly.

7A. Incubate the samples overnight at 37°C.

MNase digestion:

1B. Wash the nuclei with 1.3 ml of MNase digestion buffer.

2B. Pellet the nuclei at 1000 rpm for 10 minutes at 4°C. Carefully discard the supernatant. Resuspend the nuclei in 500 µl of MNase digestion buffer containing 1 mM $CaCl_2$.

   Note: Add 1 µl of 1 M $CaCl_2$ per 1 ml of MNase digestion buffer.

3B. Transfer 100 µl of nuclei to microfuge tubes containing MNase (Pharmacia & Upjohn Diagnosis).

Note: MNase should be stored in aliquots according to units of activity at –20°C. As a convenient starting point, a titration ranging from 0.5 unit to 50 unit can be used. The amount of MNase will need to be optimized according to LM-PCR results.

4B. Incubate samples at room temperature for 5 minutes.

5B. Add 80 μl of MNase digestion buffer and 20 μl of MNase stop buffer to each sample.

6B. Add 3 μl of proteinase K (25 mg/ml) and 10 μl of 20% SDS.

Note: Steps 5B and 6B stop the MNase digestion and should be done rapidly.

7B. Incubate overnight at 37°C.

Restriction enzyme digestion:

1C. Wash nuclei with 1.3 ml of restriction enzyme digestion buffer.

2C. Pellet nuclei at 1000 rpm for 10 minutes at 4°C. Carefully discard the supernatant. Resuspend nuclei in 250 μl of the optimal restriction enzyme digestion buffer (1x) supplied with the enzyme.

Note: Dilute the 10x buffer supplied or recommended by the company from which the restriction enzyme was purchased for use in this step. If multiple restriction enzymes are to be analyzed, use a compatible buffer or separate the samples at Step 1C into multiple aliquots.

3C. Transfer 50 μl of nuclei to microfuge tubes containing the restriction enzyme.

Note: It is convenient to first try 2 μl of concentrated restriction enzyme (10 units/μl) for digestion. The goal is to add an amount of enzyme that is sufficient for detection of a product following the LM-PCR assay. However, in cells in which the control region is inactive, the LM-PCR band resulting from this nuclear cleavage should be considerably less intense than the band resulting from the subsequent in vitro control cleavage (see step 5 below). If nuclear cleavage is too efficient, the amount of enzyme used should be reduced.

4C. Incubate the samples at 37°C for 10 minutes.

Note: The time of digestion and the amount of enzyme may need to be optimized to ensure a limiting restriction enzyme digestion.

5C. Add 50 μl of 2x proteinase K buffer. Incubate at 55°C for 1 hour.

Note: This step stops the restriction enzyme digestion.

6C. Add 50 μl of 2x proteinase K buffer, 50 μl of restriction enzyme digestion buffer, and 3 μl of proteinase K (25 mg/ml).

7C. Incubate overnight at 37°C.

Step 4: Purify DNA

1. Extract samples with 200 μl of phenol/chloroform, pH 8.

Note: To avoid shearing the genomic DNA, do not vortex samples. Instead, mix by flicking the tubes or rocking to mix the aqueous and phenol/chloroform layers thoroughly.

2. Spin samples at high speed in a microfuge for 5 minutes. Carefully transfer the aqueous layer to a new microfuge tube.

Note: The genomic DNA may appear as "strings." Slowly pipet up the aqueous layer, keeping the "strings," but trying to minimize disturbance of the phenol/chloroform interface.

3. Extract samples with 200 μl of chloroform.

   Note: Take the same precautions as noted in Step 1.

4. Spin samples at high speed in a microfuge for 5 minutes. Carefully transfer the aqueous layer to a new microfuge tube.

   Note: Take the same precautions as noted in Step 2.

5. Add 2 μl of heat-treated RNase A (10 mg/ml) to each sample. Incubate at 37°C for 2 hours.

6. Extract samples with 200 μl of phenol/chloroform, pH 8.

   Note: Take the same precautions as noted in Step 1.

7. Spin samples at high speed in a microfuge for 5 minutes. Carefully transfer the aqueous layer to a new microfuge tube.

   Note: Take the same precautions as noted in Step 2. At this stage, the interface is usually cloudy. Avoid retaining the interface. If a small amount is retained, this can be removed in the chloroform extraction.

8. Extract samples with 200 μl of chloroform.

   Note: Take the same precautions as noted in Step 1.

9. Spin samples at high speed in a microfuge for 5 minutes. Carefully transfer the aqueous layer to a new microfuge tube.

   Note: Do not retain any of the cloudy interface at this stage; it is better to leave behind some of the aqueous layer.

10. Precipitate the digested, genomic DNA by adding 1:10 volume 3 M sodium acetate, pH 5.2, and 2.5 volumes of ice-cold ethanol. Incubate the samples at –20°C for about 1 hour.

11. Pellet DNA at high speed in a microfuge for 10 minutes at 4°C.

12. Discard the supernatant. Wash the DNA pellet with ice-cold 70% ethanol. Centrifuge the samples at high speed for 1–2 minutes at 4°C.

13. Discard the supernatant. Dry the DNA on the bench top or use a SpeedVac with no heat.

    Note: Be careful not to over-dry the DNA, which makes redissolving difficult.

14. Resuspend the DNA in 50 μl of H_2O or TE (10 mM Tris-HCl, pH 7.5, 1 mM EDTA).

    Note: Resuspend the DNA gently; do not vortex or pipet vigorously. It may be necessary to allow the DNA to dissolve overnight at 4°C. If the solution becomes highly viscous, it can be diluted with 25 μl of H_2O.

15. For MNase-treated DNA, repeat Steps 5–14.

    Note: It is critical to remove the RNA completely before the kinasing reaction. It may be necessary to check the DNA on an agarose gel to determine whether RNA is still present.

Step 5: Prepare DNA for LM-PCR

Note: Preparation of DNA for LM-PCR varies according to the in vivo digestion (A = DNase I, B = MNase, and C = restriction enzyme).

DNase I-digested DNA

1A. DNA is ready for LM-PCR day #1. Determine the DNA concentration using a spectrophotometer (OD 260/280). Use 1 μg of DNA for LM-PCR.

MNase-digested DNA

1B. Determine the DNA concentration using a spectrophotometer. Use 1 μg of DNA in a phosphorylation reaction.

Note: MNase digestion creates blunt-ended cleavages within nucleosome-free DNA and nucleosome linkers. The 5′ phosphate group is removed in the cleavage reaction. To perform the subsequent unidirectional linker ligation, the phosphate must be restored.

2B. Perform the phosphorylation reaction in 50 μl total volume with 1 mM ATP.

Note: The phosphorylation reaction can be performed in the company buffer supplied for ligation if it is compatible with the T4 polynucleotide kinase. These buffers usually already contain ATP, whereas kinase buffers supplied by companies lack ATP. Therefore, the phosphorylation reaction consists of 5 μl of 10x ligase buffer, 1 μg of MNase-digested DNA, H_2O, and 1.5 μl of T4 polynucleotide kinase (New England Biolabs, 10 U/μl) in 50 μl total volume.

3B. Incubate the reaction at 37°C for 1 hour.

4B. Add 150 μl of H_2O to the mixtures and extract with 200 μl of phenol/chloroform, pH 8.

Note: Do not vortex the DNA samples; mix gently but thoroughly.

5B. Spin samples at high speed in a microfuge for 5 minutes. Transfer the aqueous phase to a new microfuge tube.

6B. Extract the samples with 200 μl of chloroform.

Note: Do not vortex the DNA samples; mix gently but thoroughly.

7B. Spin samples at high speed in a microfuge for 5 minutes. Transfer the aqueous phase to a new microfuge tube.

8B. Precipitate the DNA with 1:10 volume sodium acetate, pH 5.2, and 2.5 volumes ice-cold ethanol. Place samples on dry ice for 10 minutes.

9B. Pellet DNA at high speed in a microfuge for 10 minutes at 4°C. Discard the supernatant. Wash the DNA pellet with 70% ethanol.

10B. Centrifuge samples at high speed in a microfuge for 1–2 minutes. Carefully remove the supernatant. Dry the DNA on the bench top or use a SpeedVac with no heat.

Note: Be careful not to over-dry the DNA.

11B. Resuspend the DNA in 5 μl of H_2O or TE (10 mM Tris-HCl, pH 7.5, 1 mM EDTA). Proceed directly to the ligation reaction.

Note: To detect MNase digestion of protein-free regions, first-strand synthesis is **not** performed. MNase can nick DNA within nucleosomes. First-strand synthesis will detect these nicks, creating confusing results. As stated earlier, MNase will create blunt ends within protein-free regions. Thus, ligation of the unidirectional linker at this stage will result in detection of these cleavages.

Restriction enzyme-digested DNA

1C. Determine the DNA concentration using a spectrophotometer (OD 260/280). Use 1 μg of DNA for in vitro restriction enzyme digestion.

Note: The in vitro restriction enzyme digestion is for normalization of input DNA following PCR amplification. The restriction enzyme cleavage site chosen for the in vitro digestion should lie upstream of the in vivo restriction enzyme site relative to the three primers used for LM-PCR.

2C. Perform restriction enzyme digestion in a total volume of 20 μl. Incubate digestion reaction at 37°C for approximately 2 hours.

Note: The in vitro restriction enzyme digestion should be performed to completion.

3C. Heat-inactivate the restriction enzyme at 65°C for 20 minutes.

4C. Use 5 μl of the digestion reaction (0.25 μg of DNA) for LM-PCR first-strand synthesis.

Step 6: LM-PCR day #1

Prepare 20 μM unidirectional linker solution

1. Gel-purify or HPLC purify the LM-PCR 1 and LM-PCR 2 oligonucleotides.

   Note: It is convenient to gel-purify all the oligonucleotides for LM-PCR at the same time. Primers 1–3 should be gel-purified as well. The LM-PCR1 and LM-PCR2 oligonucleotides are different lengths so that only one end is blunt; this allows for unidirectional ligation to the blunt-ended products of the first-strand synthesis step.

2. Prepare 20 μM unidirectional linker mix: 20 μM LM-PCR 1, 20 μM LM-PCR 2, and 250 mM Tris-HCl, pH 7.7. A large volume of this mix can be prepared and stored in aliquots at –20°C for future use.

3. Denature the linker mix at 95°C for 5 minutes.

4. Centrifuge briefly to collect condensation. Place the 20 μM unidirectional linker mix in a 70°C heat block, remove block from heat source, and allow to cool slowly to room temperature (will take approximately 1 hour).

5. Incubate linker mix in block at 4°C overnight.

6. Prepare aliquots for storage at –20°C.

   Note: Always thaw linker mix on ice.

End-label primer 3

1. Phosphorylate primer 3 with [γ-^{32}P]ATP (6000 Ci / mmole). The kinasing reaction contains the following components: 4.0 μl of 10x polynucleotide kinase buffer, 7.0 μl (70 pmoles) of primer 3, 4.0 μl of [γ-^{32}P]ATP, 23.5 μl of H$_2$O, and 1.5 μl of T4 polynucleotide kinase (New England Biolabs, 10 U/μl).

2. Incubate reaction mixture at 37°C for 30 minutes to 1 hour.

3. Purify the radiolabeled primer from free [γ-^{32}P]ATP.

   Note: This can be accomplished using a Stratagene push column (see Protocol 4.1) or a similar method.

First-strand synthesis reaction

Note: **DO NOT** perform the first-strand synthesis reaction with MNase-treated DNA. (Instead combine mock first-strand mix, dilution mix, and ligation mix for immediate ligation.) The first-strand synthesis will allow for the detection of single-stranded cuts within the

DNA sequence of interest. The single round of PCR will transform the single-stranded cuts within the sequence surrounding the gene-specific primer into blunt-ended double-stranded DNA fragments. These DNA fragments can subsequently participate in the ligation reaction with the unidirectional linker.

1. Add 1–2 µg of in vivo-digested DNA to a PCR tube in 5 µl total volume. Place the samples on ice.

   Note: The amount of DNA added to the first-strand synthesis reaction from the restriction enzyme digestion will be 0.25 µg (5 µl).

2. Prepare the first-strand synthesis reaction mix on ice: (per reaction) 6.0 µl of 5x first-strand buffer, 0.3 µl (0.3 pmole) of primer 1, 0.25 µl of 25 mM dNTP mix, and 18.2 µl of H₂O. Immediately before use, add 0.25 µl of pfu polymerase. Mix well and centrifuge briefly.

   Note: The polymerase used for LM-PCR is critical. A comparison of various DNA polymerases was performed by Garrity and Wold (1992). The DNA polymerase pfu was not examined in this study, but we have found that it works well for LM-PCR. This protocol will employ pfu as the DNA polymerase.

3. Add 25 µl of ice-cold first-strand synthesis reaction mix to each sample.

   Note: The samples should be kept on ice to prevent spurious activity of the DNA polymerase.

4. Place the samples in an automated thermal cycler, and perform first-strand synthesis under the following conditions: 5 minutes at 95°C (denaturing), 30 minutes at 60°C (annealing), and 10 minutes at 76°C (extension).

   Note: For MNase samples, prepare mock first-strand buffer (25 µl per reaction). Add (per reaction): 19.0 µl of H₂O and 6.0 µl of 5x first-strand buffer. Add 25 µl to each sample on ice and proceed as described below.

5. Prepare dilution solution (20 µl per reaction) on ice. Add (per reaction): 2.2 µl of 1 M Tris-HCl, pH 7.5, 0.35 µl of 1 M MgCl₂, 2.0 µl of 0.5 M DTT, 0.25 µl of DNase-free bovine serum albumin (BSA) (10 mg/ml), and 15.2 µl of H₂O.

6. Prepare ligation solution (25 µl per reaction) on ice. Add (per reaction): 0.25 µl of 1 M MgCl₂, 1.0 µl of 0.5 M DTT, 0.75 µl of 100 mM rATP, 0.13 µl of BSA (10 mg/ml), and 16.87 µl of H₂O. Immediately before use, add (per reaction): 5.0 µl of 20 µM unidirectional linker and 1.0 µl (3 Units) of T4 DNA ligase.

7. Add 20 µl of ice-cold dilution solution to each sample.

8. Add 25 µl of ice-cold ligation solution to each sample.

9. Incubate the samples at 17°C overnight (12+ hours).

Step 7: LM-PCR day #2

1. Prepare precipitation mix on ice: Add (per reaction) 8.4 µl of 3 M sodium acetate, pH 7, and 1.0 µl of yeast tRNA (10 mg/ml).

2. Transfer ligated samples to new microfuge tubes. Add 9.4 µl of precipitation mix and 220 µl of ice-cold ethanol to each sample. Incubate on dry ice for 10 minutes or at –20°C for 2 hours.

3. Pellet DNA at high speed in microfuge for 10 minutes at 4°C.

4. Discard supernatant. Wash the DNA pellet with ice-cold 70% ethanol. Centrifuge the samples at high speed for 1–2 minutes at 4°C.

5. Discard the supernatant and dry the DNA pellet on the bench top or use a SpeedVac with no heat.

   Note: Be careful not to over-dry the DNA.

6. Resuspend the DNA in 70 µl of H_2O. Leave samples at room temperature until the DNA dissolves (about 15–30 minutes).

7. Prepare amplification mix on ice (30 µl per reaction). Add (per reaction) 20 µl of 5x amplification buffer, 1.0 µl (10 pmoles) of LM-PCR 1, 1.0 µl (10 pmoles) of primer 2, 0.8 µl of 25 mM dNTP mix, and 7.2 µl of H_2O.

8. Combine 30 µl of amplification mix and 70 µl of sample in new PCR tubes on ice.

9. Prepare pfu dilution mix on ice. Add (per reaction) 0.2 µl of 5x amplification buffer, 0.63 µl of H_2O, and 0.17 µl of pfu. Add 1.0 µl of pfu dilution mix to each sample on ice. Mix the samples and centrifuge briefly.

   Note: The amount of pfu polymerase may need to be titrated because either too much or too little can be problematic.

10. Proceed with PCR as follows: DNase I (21 cycles), MNase (22–23 cycles), and restriction enzyme (18 cycles). Use the following conditions: first cycle, denature for 4 minutes at 95°C, with the remaining cycles denatured at 95°C for 1 minute. The annealing step should be performed for 2 minutes at the optimal hybridization temperature for primer 2. Perform extension at 76°C for 5 minutes for the first cycle with an additional 15 seconds in each subsequent cycle. After the final extension, an additional 5-minute extension at 76°C is performed.

    Note: Ideally, the annealing temperature used for primer 2 should be 2–5°C above the calculated T_m of primer 2 (or above the T_m for the linker primer, whichever is lower), but the optimal annealing temperature needs to be determined empirically. We have found an annealing temperature of 63–64°C for a primer (25-mer) with a G/C content of 56% has worked well, but this step needs to be optimized for each primer.

11. Prepare labeling mix (4.5 µl per reaction) on ice: Add (per reaction) 1.0 µl of 5x amplification buffer, 2.0 µl (2.0 pmoles) of end-labeled primer 3, 0.4 µl of 25 mM dNTP mix, and 0.8 µl of H_2O. Immediately before use add (per reaction) 0.3 µl of pfu.

12. On ice add 4.5 µl of labeling mix to each sample. Proceed with three cycles of labeling PCR: For the first cycle, denature for 4 minutes at 95°C; for cycles 2 and 3, denature for 1 minute at 95°C; anneal for 2 minutes at the optimal annealing temperature for primer 3; extend for 10 minutes at 76°C.

    Note: If the background or amount of radiolabeled amplification product is too high, labeling PCR can be done in two cycles. Also, the annealing temperature for primer 3 needs to be higher than for primer 2. This will exclude primer 2 from annealing to the template during the labeling cycles.

13. Transfer LM-PCR samples to new microfuge tubes and add 300 µl of stop solution to each sample.

14. Extract samples with 400 µl of phenol/chloroform, pH 8. Spin in microfuge for 5 minutes at room temperature. Transfer the aqueous phase to a new microfuge tube.

Note: This step can be omitted, but if the general "background" radioactivity in the gel lanes is high, it should be performed.

15. In a fresh microfuge tube, add 94 μl of sample and 235 μl of ice-cold ethanol. Precipitate the DNA on dry ice for 10 minutes. Save the remainder of each sample at –20°C with proper shielding.

16. Spin the samples in a microfuge at high speed for 10 minutes at 4°C. Discard the supernatant (radioactive) and wash the DNA pellet with ice-cold 70% ethanol.

17. Centrifuge the samples at high speed for 1–2 minutes at 4°C. Discard the supernatant (radioactive) and dry the DNA on the bench top or use a SpeedVac with no heat.

18. Resuspend the DNA pellet in 6.0 μl of formamide loading buffer.

19. Denature the samples by boiling for 2 minutes. (Using Lid Loks will prevent the caps from opening during boiling.) Centrifuge briefly to collect condensation.

20. Load all 6.0 μl of each sample on a DNA sequencing gel (8% denaturing polyacrylamide) with 6-mm wells. Electrophorese at 60 W for approximately 1.5 hours.

    Note: The radioactive primer should be at the bottom of the gel after 1.5 hours; thus, the bottom portion of the gel (slightly above the bromophenol blue dye) can be cut off to prevent this radioactivity from obscuring the desired signals. Discard in radioactive waste.

21. Dry the gel and expose to film or phosphorimager screen.

PROCEDURE FOR IN VIVO DMS FOOTPRINTING

CAUTIONS: β-*Mercaptoethanol, Chloroform, DMS, Ethanol, Phenol, Piperidine, SDS, Sodium acetate. See Appendix I.*

Step 1: Prepare buffers

DMS stop buffer:

      1.5 M sodium acetate, pH 7.0
      1 M β-mercaptoethanol
      100 μg/ml yeast tRNA

Lysis buffer (prepare immediately before use):

      1 mM Tris-HCl, pH 7.5
      400 mM NaCl
      2 mM EDTA
      0.2% SDS
      0.2 mg/ml proteinase K

LM-PCR reagents:

As in Step 1 above (Procedure for DNase I, MNase, and restriction enzyme footprinting)

Step 2: Treat cells with DMS

For adherent cells (15 cm plate)

1. Prewarm medium and PBS at 37°C.

   Note: The medium for the DMS treatment is the same as the cell culture medium.

2. Remove the medium from the cells.

3. In a fume hood, add 1 µl of DMS per 1 ml of medium for a final concentration of 0.1% DMS. Mix and immediately add to the cells. Use a sufficient amount of the solution to easily cover the cells. Incubate for 2 minutes at room temperature.

   Note: Prepare the DMS solution immediately before use. A titration of the DMS concentration may need to be performed to ensure optimal reactivity.

4. Remove the DMS-containing medium with a disposable pipet and wash the cells with PBS.

   Note: The time of DMS treatment can be varied when optimizing the DMS footprint. Be sure to dispose of the liquid and solid DMS waste appropriately.

5. Remove the PBS with a disposable pipet or vacuum aspirator (in the fume hood) and repeat the PBS wash three times.

   Note: Gently rock the cells for 30 seconds during each wash. The cells can be taken outside of the fume hood following the final wash.

6. After the final PBS wash, lyse the cells with 1.5 ml of lysis buffer. Carefully rock the cells to spread the solution over the cells. Incubate at room temperature for an additional 5 minutes.

7. Remove the cell lysate from the plate by scraping and transfer to a 15-ml disposable, polypropylene tube.

8. Incubate the cell lysate for 3–5 hours at 37°C with periodic mixing.

For suspension cells

1. Aliquot 49 ml of ice-cold PBS into two 50-ml disposable conical tubes for each sample. Place a 37°C water bath or heat block in a fume hood for DMS treatment. Prepare 2.7 ml of lysis solution for each sample.

2. Place suspension cells (approximately 1×10^8 cells) in a 50-ml conical tube and pellet by centrifugation (5 minutes at 1500 rpm).

3. Resuspend cell pellet in 1 ml of medium (prewarmed to 37°C) and transfer to a 1.5-ml microfuge tube. Place in a 37°C water bath or heat block.

4. In a fume hood, make a 10% DMS solution by dissolving DMS in 100% ethanol. Mix solution by vortexing followed by a brief spin. This solution should be prepared immediately before use.

5. Add 10 µl of the 10% DMS solution to each sample. Mix by inverting the closed tube and incubate at 37°C for 1 minute.

   Note: The amount of DMS added and time of treatment can be varied to optimize the reactivity and DMS footprint.

6. Transfer the DMS-treated samples to the 49-ml ice-cold PBS aliquot and mix. Pellet the cells at 1500 rpm for 5 minutes. Remove the supernatant and dispose of the waste appropriately.

7. Resuspend the cell pellet in 1 ml of ice-cold PBS and repeat Step 6, using the second 49-ml PBS aliquot.

8. Resuspend the cell pellet in 300 μl of ice-cold PBS. Place each sample in a 15-ml polypropylene tube and add 2.7 ml of lysis buffer. Mix the samples gently but thoroughly.

9. Incubate the cell lysate for 3–5 hours at 37°C with periodic mixing by inversion.

Step 3: Purify DNA

1. Extract the samples with an equal volume phenol/chloroform. Mix the samples thoroughly by gently inverting the tubes repeatedly. Spin the samples at 3000 rpm in a tabletop centrifuge for 10 minutes.

   Note: To avoid shearing the genomic DNA, do not vortex.

2. Remove phenol/chloroform (bottom) layer and discard in phenol waste. Repeat the phenol/chloroform extraction of the aqueous layer.

   Note: The organic layer can be removed by placing a pasteur pipet through the aqueous layer into the phenol/chloroform layer and carefully pipetting. This may be easier than removing the aqueous layer to a new tube because the genomic DNA will be viscous. This method will also minimize shearing.

3. Remove the phenol/chloroform layer and discard in phenol waste. Extract the samples twice with equal volumes of chloroform.

4. Precipitate the DNA with equal volume of isopropanol.

   Note: DNA will be visible as a white, string-like precipitate.

5. Collect the DNA by spooling, using a sealed pasteur pipet.

   Note: For a detailed description of DNA spooling, refer to Ausubel et al. (1994, Unit 15.5).

6. Gently rock the spooled DNA in 3 ml of TE (10 mM Tris-HCl, pH 7.5, 1 mM EDTA) for resuspension.

   Note: It may take a few hours to overnight for the DNA to go back into solution. Do not vortex.

7. Precipitate DNA with 1:10 volume of 3 M sodium acetate, pH 5.2, and 2 volumes of ethanol. Collect DNA by spooling.

8. Resuspend spooled DNA in 200–500 μl of TE, pH 7.5, at 4°C.

   Note: It may take a few hours to overnight for the DNA to go back into solution. Do not vortex.

9. Determine the DNA concentration using a spectrophotometer and adjust to a final concentration of 1–1.5 mg/ml.

   Note: A large fraction of this preparation is RNA, which will be eliminated during the piperidine cleavage steps. Store the samples at 4°C.

Step 4: Treat DNA with piperidine

1. Precipitate 200 μl of DNA with 50 μl of DMS stop solution and 750 μl of ethanol. Place samples on dry ice for 30 minutes. Spin samples in microfuge for 10 minutes at 4°C.

2. Wash the DNA pellet with 75% ethanol. Spin samples for 1–2 minutes at 4°C and remove the supernatant.

   Note: Do not allow the DNA pellet to dry completely at this stage.

3. Prepare a diluted stock (1 M) of piperidine by adding 1:10 volume of piperidine to water. Add 200 μl of diluted piperidine to each sample and resuspend by vortexing.

   Note: It may take 15–30 minutes for the DNA to dissolve in the piperidine. Vortex periodically during this time.

4. Briefly spin samples in a microfuge and place Lid Loks on tubes to prevent caps from opening during heating. Place tubes in a 90°C heat block in a fume hood for 30 minutes.

   Note: This step will cleave the DNA at methylated guanines.

5. Briefly spin samples in a microfuge to collect condensation. Place samples on dry ice for 10 minutes.

6. Remove piperidine by evaporation in a SpeedVac concentrator for 1–2 hours at room temperature.

7. Resuspend the DNA in 360 μl of TE, pH 7.5. Precipitate with 1:10 volume 3 M sodium acetate, pH 7.0, and 2.5 volumes of ethanol. Place samples on dry ice for 10 minutes.

8. Spin samples at high speed in a microfuge for 15 minutes at 4°C.

9. Discard the supernatant. Resuspend the pellet in 500 μl of TE, pH 7.5. Precipitate DNA with 170 μl of 8 M ammonium acetate and 670 μl of isopropanol. Incubate at –20°C for 2 hours.

10. Spin samples at high speed in a microfuge for 15 minutes at 4°C.

11. Discard the supernatant. Wash the DNA pellet with ice-cold 75% ethanol. Spin the samples at high speed in a microfuge for 1–2 minutes.

12. Discard the supernatant. Resuspend in 50 μl of dH_2O.

13. Remove the dH_2O by evaporation in a SpeedVac concentrator for 1 hour with heat.

14. Resuspend the DNA in TE, pH 7.5, to a final concentration of approximately 1 μg/ml.

15. Spin samples at high speed in a microfuge for 10 minutes at room temperature. Remove the supernatant to a new tube, leaving behind a clear pellet.

Step 5: Prepare DNA for LM-PCR

1. Determine the DNA concentration using a spectrophotometer. Use 1 μg of DNA for LM-PCR.

Steps 6 and 7: LM-PCR day #1 and LM-PCR day #2

Proceed as described in Steps 6 and 7 above (Procedure for DNase I, MNase, and restriction enzyme footprinting).

ADDITIONAL CONSIDERATIONS

1. A control that must be performed for the interpretation of in vivo DNase I, MNase, and DMS data is to subject purified DNA to the same procedure. Purified genomic DNA or a plasmid genomic clone can be used. MNase and DNase I possess significant sequence preferences; therefore, the purified DNA control is critical for the identification of protected and hypersensitive nucleotides. The purified DNA control for DNase I footprinting can be somewhat difficult to interpret because histones, in addition to sequence-specific DNA-binding proteins, can alter the DNase I digestion pattern. If an analysis of transcription factor binding is the goal of an in vivo footprinting study, a comparison between the transcriptionally silent and active states of the gene may be more useful than a comparison with purified DNA.

2. An important issue to consider when interpreting in vivo footprinting data is the nature of the cell population being studied. In particular, the percentage of cells within a population that express the gene can dramatically influence the quality and validity of the results. If only 25% of the cells are transcriptionally active, it will be very difficult to detect the occupancy of control elements by DNA-binding proteins.

3. It is important to keep in mind that in vivo footprinting techniques, unlike standard in vitro footprinting techniques, require PCR amplification. The PCR amplification can enhance or diminish the extent of protection or hypersensitivity that is observed. If interactions are not observed, changing the number of amplification cycles may prove to be beneficial.

4. The requirement for PCR amplification greatly increases the need for an objective evaluation of in vivo footprinting data, and for many repetitions of each experiment to demonstrate reproducibility. In most isolated experiments with a given primer set, several nucleotides that appear to be protected or hypersensitive will be observed. However, only a subset of those changes will be reproducible from experiment to experiment to an extent sufficient to reach strong conclusions.

5. It is important to remember that the LM-PCR products possess an additional 25 bp that is contributed by the ligated and amplified linker sequence. This 25 bp needs to be taken into account when comparing protected and hypersensitive sites to DNA sequencing markers during the calculations involved in determining the positions of those sites.

TROUBLESHOOTING

High background

Possible cause: Annealing temperature of primers may not be sufficiently stringent.
Solution: Different annealing temperatures for primers 2 and 3 should be tested. It is important to use the highest possible temperatures to minimize nonspecific annealing and to enhance the exclusion of primer 2 during the labeling reaction.
Possible cause: PCR overamplification.
Solution: If the lanes become overexposed, it may be necessary to decrease the number of PCR cycles. If the purified DNA control appears overamplified in comparison to the in-vivo-digested DNA, decrease the amount of template in the PCR.

Blank lanes

Possible cause: For the MNase procedure, the kinasing of the MNase-treated DNA may have been inefficient.

Solution: Kinasing efficiency is decreased in the presence of high amounts of RNA. Run the MNase-treated samples on an agarose gel to determine if RNA is present. If so, repeat the RNase A digestion. Also, kinasing large quantities of DNA will decrease the overall efficiency of the reaction.

Possible cause: The level of in vivo digestion was too high or low.

Solution: A wider titration of the enzyme will increase the probability of finding the appropriate range. The digested DNA can be analyzed on an agarose gel to determine the level of digestion.

Possible cause: The reagents were of poor quality.

Solution: It is extremely important to prepare the reagents precisely and to use high-quality components.

Possible cause: The annealing temperature was too high or the primer combination was incompatible.

Solution: First try adjusting the annealing temperature for primers 2 and 3. If this does not help, the primers may be incompatible; design new primers.

REFERENCES

Almer A. and Horz W. 1986. Nuclease hypersensitive regions with adjacent positioned nucleosomes mark the gene boundaries of the PHO5/PHO3 locus in yeast. *EMBO J.* **5:** 2681–2687.

Almer A., Rudolph H., Hinnen A., and Horz W. 1986. Removal of positioned nucleosomes from the yeast PHO5 promoter upon PHO5 induction releases additional upstream activation DNA elements. *EMBO J.* **5:** 2689–2696.

Ausubel F.M., Brent R.E., Kingston E., Moore D.D., Seidman J.G., Smith J.A., and Struhl K. 1994. *Current protocols in molecular biology.* John Wiley and Sons, New York.

Brown K.E., Guest S.S., Smale S.T., Hahm K., Merkenschlager M., and Fisher A.G. 1997. Association of transcriptionally silent genes with Ikaros complexes at centromeric heterochromatin. *Cell 91: 845–854.*

Brown K.E., Baxter J., Graf D., Merkenschlager M., and Fisher A.G. 1999. Dynamic repositioning of genes in the nucleus of lymphocytes preparing for cell division. *Mol. Cell 3:* 207–217.

Enver T., Brewer A.C., and Patient R.K. 1985. Simian virus 40-mediated cis induction of the Xenopus beta-globin DNase I hypersensitive site. *Nature* **318:** 680–683.

Fascher K.D., Schmitz J., and Horz W. 1993. Structural and functional requirements for the chromatin transition at the PHO5 promoter in Saccharomyces cerevisiae upon PHO5 activation. *J. Mol. Biol.* **231:** 658–667.

Fragoso G., John S., Roberts M.S., and Hager G.L. 1995. Nucleosome positioning on the MMTV LTR results from the frequency-biased occupancy of multiple frames. *Genes Dev.* **9:** 1933–1947.

Frommer M., McDonald L.E., Millar D.S., Collis C.M. Watt F., Grigg G.W., Molloy P.L., and Paul C.L. 1992. A genomic sequencing protocol that yields a positive display of 5-methylcytosine residues in individual DNA strands. *Proc. Natl. Acad. Sci.* **89:** 1827–1831.

Fryer C.J. and Archer T.K. 1998. Chromatin remodelling by the glucocorticoid receptor requires the BRG1 complex. *Nature* **393:** 88–91.

Garrity P.A. and Wold B.J. 1992. Effects of different DNA polymerases in ligation mediated PCR: Enhanced genomic sequencing and in vivo footprinting. *Proc. Natl. Acad. Sci.* **89:** 1021–1025.

Garrity P.A., Chen D., Rothenberg E.V., and Wold B.J. 1994. Interleukin-2 transcription is regulated in vivo at the level of coordinated binding of both constitutive and regulated factors. *Mol. Cell. Biol.* **14:** 2159–2169.

Garrity P.A., Wold B., and Mueller P.R. 1995. Ligation-mediated PCR. In *PCR 2: A practical approach* (ed. McPherson M.J., Hames B.D., and Taylor G.R.), p. 309. IRL Press at Oxford University Press, New York.

Han M. and Grunstein M. 1988. Nucleosome loss activates yeast downstream promoters in vivo. *Cell* **55:** 1137–1145.

Kaffman A., Rank N.M., and O'Shea E.K. 1998. Phosphorylation regulates association of the transcription factor Pho4 with its import receptor Pse1/Kap121. *Genes Dev.* **12:** 2673–2683.

Lamond A.I. and Earnshaw W.C. 1998. Structure and function in the nucleus. *Science* **280:** 547–553.

Lohr D. 1997. Nucleosome transactions on the promoters of the yeast GAL and PHO genes. *J. Biol. Chem.* **272:** 26795–26798.

McAndrew P.C., Svaren J., Martin S.R., Horz W., and Goding C.R. 1998. Requirements for chromatin modulation and transcription activation by the Pho4 acidic activation domain. *Mol. Cell. Biol.* **18:** 5818–5827.

McPherson C.E., Shim E.Y., Friedman D.S., and Zaret K.S. 1993. An active tissue-specific enhancer and bound transcription factors existing in a precisely positioned nucleosomal array. *Cell* **75:** 387–398.

Mostoslavsky R., Singh N., Kirillov A., Pelanda R., Cedar H., Chess A., and Bergn Y. 1998. Kappa chain monoallelic demethylation and the establishment of allelic exclusion. *Genes Dev.* **12:** 1801–1811.

Mueller P.R. and Wold B. 1989. In vivo footprinting of a muscle specific enhancer by ligation mediated PCR. *Science* **246:** 780–786.

Mueller P.R., Salser S.J., and Wold B. 1988. Constitutive and metal-inducible protein:DNA interactions at the mouse metallothionein I promoter examined by in vivo and in vitro footprinting. *Genes Dev.* **2:** 412–427.

Parekh B.S. and Maniatis T. 1999. Virus infection leads to localized hyperacetylation of histones H3 and H4 at the IFN-β promoter. *Mol. Cell* **3:** 125–129.

Pfeifer G.P., Steigerwald S.D., Mueller P.R., Wold B, and Riggs A.D. 1989. Genomic sequencing and methylation analysis by ligation mediated PCR. *Science* **10:** 246: 810–813.

Reik A., Schütz G., and Stewart A.F. 1991. Glucocorticoids are required for establishment and maintenance of an alteration in chromatin structure: Induction leads to a reversible disruption of nucleosomes over an enhancer. *EMBO J.* **10:** 2569–2576.

Richard-Foy H. and Hager G.L. 1987. Sequence-specific positioning of nucleosomes over the steroid-inducible MMTV promoter. *EMBO J.* **6:** 2321–2328.

Sambrook J., Fritsch E.F., and Maniatis T. 1989. *Molecular cloning: A laboratory manual.* Cold Spring Harbor Laboratory Press, Cold Spring Harbor, New York.

Travers A. and Drew H. 1997. DNA recognition and nucleosome organization. *Biopolymers* **44:** 423–433.

Van Lint C., Emiliani S., Ott M., and Verdin E. 1996. Transcriptional activation and chromatin remodeling of the HIV-1 promoter in response to histone acetylation. *EMBO J.* **15:** 1112–1120.

Verdin E., Paras P., and Van Lint C. 1993. Chromatin disruption in the promoter of human immunodeficiency virus type 1 during transcriptional activation. *EMBO J.* **12:** 3249–3259.

Vermaak D. and Wolfe A.P. 1998. Chromatin and chromosomal controls in development. *Devel. Genet.* **22:** 1–6.

Weinmann A., Plevy S.E., and Smale S.T. 2000. Rapid and selective remodeling of a positioned nucleosome during the induction of IL-12 p40 transcription. *Immunity* (in press.)

CHAPTER 11

Approaches for the Synthesis of Recombinant Transcription Factors

Important issues

- *There are many methods that can be employed to overproduce a transcription factor for a particular application.*

- *The advantages and disadvantages of the different expression systems depend on the application for the recombinant protein.*

- *The addition of sequence tags to a protein aids in its purification or visualization.*

- *Several expression systems can be employed to overproduce and purify macromolecular complexes.*

INTRODUCTION

The synthesis of recombinant proteins is essential for any biochemical study on gene control. Although in vivo experiments have powerful applications, most of the key mechanistic issues must ultimately be resolved biochemically to confirm their direct nature and to study the molecular details of a protein's action. Such studies, however, are generally preceded by transfection experiments, which are used to define the activity and functional domains of a regulatory protein and to provide an in vivo system for comparison with the biochemistry. The intact regulatory factor or important regulatory domains are then synthesized in recombinant form and studied in vitro.

The choice of a system for overexpressing a protein depends on the particular application, but will include consideration of how much protein is needed for the analysis, whether the protein has to be purified to homogeneity, and whether the protein's activity is dependent on posttranslational modifications. For example, crystallography studies typically require hundred-milligram quantities of highly pure protein, in vitro transcription and transcription complex assembly experiments require microgram–milligram quantities, and analyses of mutants for DNA binding or protein–protein interaction studies require only tens of nanograms of material. Therefore, the key issues before initiating a study are identifying the experimental goals and then determining the amount of effort they will entail and the feasibility of obtaining sufficient quantities of usable protein to reach those goals. We discuss the methods as they apply to a typical academic laboratory, where the resources and facilities for large-scale bacterial fermentation or cell culture are not readily available.

The methods available for synthesis of recombinant proteins can be subdivided into three categories (see also Table 11.1):

1. *Eukaryotic expression systems* include the widely popular insect baculovirus system, the mammalian vaccinia virus system, the yeast systems (e.g., *Saccharomyces cerevisiae* or *Pichia pastoris*), tissue-culture plasmid transfections, and retrovirus infections. These systems have the advantage of being able to synthesize large recombinant regulatory proteins (>75 kD) in modest quantities (i.e., 10 μg–10 mg), although the yield varies considerably depending on the system and protein.

2. *Prokaryotic systems* include, but are not restricted to, the phage T7 and Tac expression systems in *Escherichia coli*. These systems are ideal for synthesis of large (i.e., 10–100 mg) quantities of small- to moderate-sized (<75 kD) recombinant proteins not requiring posttranslational modifications.

3. *In vitro transcription/translation reactions* are performed in a test tube and employ translation extracts supplemented with exogenous amino acids to translate synthetic mRNAs, which are generated by in vitro transcription using bacteriophage-derived RNA polymerases (T7, SP6, and T3). The in vitro systems can support synthesis of large recombinant factors, but in very small (i.e., 1–100 ng) amounts.

This chapter focuses on approaches used widely and successfully in the eukaryotic transcription field. We discuss strategies employed for synthesis of recombinant regulatory proteins or their domains, general transcription factors, coactivators, and intact protein complexes. The development of the technology for synthesis of recombinant factors has in the past decade shifted from academia to industry. Thus, although the intent is to provide an overview of the strategies and available systems, improvements are occurring rapidly and it is important to consult the literature, recent catalogs, or web sites of the suppliers to become familiar with the latest technologies. A partial listing of these commercial

TABLE 11.1. *Comparing efficiency of different systems*

	Difficulty	Typical yields for overexpression (% recombinant protein)
Eukaryotic		
Baculovirus	+++	>10–20%
Vaccinia virus	++++	~5%
Yeast	+++	1–5%
Plasmids	++	<1%
Prokaryotic		
Tac promoter	++	~ 10%
T7 polymerase/promoter system	++	10–30%
In vitro transcription/ translation		
T7, T3, and SP6 RNA polymerases/ wheat germ or rabbit reticulocyte translation extracts	+	<1%

resources is included at the end of the chapter. The BioSupplyNet Sourcebook can be consulted for additional suppliers. The journal *Current Opinion in Biotechnology* reviews advances in expression systems annually in the October issue; recent articles on many of the systems described in this chapter can be found in that journal. Detailed protocols are available from many sources including the manuals accompanying commercial kits, the primary literature, *Current Protocols in Molecular Biology* (Ausubel et al. 1994), and *Current Protocols in Protein Science* (Coligan et al. 1996).

CONCEPTS AND STRATEGIES

Prokaryotic Expression Systems

The *E. coli* expression systems are ideally suited for synthesis of intact, folded proteins under 75 kD and for synthesizing functional domains of regulatory factors. Most sequence-specific regulatory proteins, for example, have small modular DNA-binding and activation/repression domains. These domains reside within 50–150 amino acid stretches of protein. When fragments are of this size, *E. coli* presents the most efficient, as well as most commonly used and inexpensive, expression system for obtaining milligram quantities of protein. For example, recent crystal structures of complexes between DNA-binding domains of eukaryotic regulatory proteins and their sites, including GAL4 (Marmorstein et al. 1992), the Fos/Jun heterodimer (Glover and Harrison 1995), glucocorticoid receptor (Luisi et al., 1991), NF-κB (Ghosh et al. 1995; Chen et al. 1998), p53 (Cho et al. 1994), and many others, were obtained with proteins purified from *E. coli* expression systems.

This is not to say that *E. coli* is used only for expression of sequence-specific regulatory proteins. Many general factors and coactivators have also been synthesized in *E. coli*. These include TFIIA (DeJong and Roeder 1993; Ozer et al. 1994; Sun et al. 1994), TFIIB (Ha et al. 1991), TATA-binding protein (TBP) (Hahn et al. 1989), TFIIE (Peterson et al. 1990), TFIIF (Sopta et al. 1989; Finkelstein et al. 1992), some TBP-associated factors (TAFs) (Klemm et al. 1995), and certain coactivators like PC4 (Ge and Roeder 1994). All of these proteins are near or below 75 kD in size and can be synthesized intact under the appropriate conditions. The *E. coli* system can also be used to generate large numbers of transcription factor derivatives for biochemical studies. Using the T7 system coupled with a method for tagging and purifying the proteins (see Box 11.3), Abate and colleagues were able to purify over a dozen variants of Fos and Jun for EMSA, footprinting, and in vitro transcriptional analyses (Abate et al. 1991).

Box 11.1

Two Common *E. coli* Expression Systems

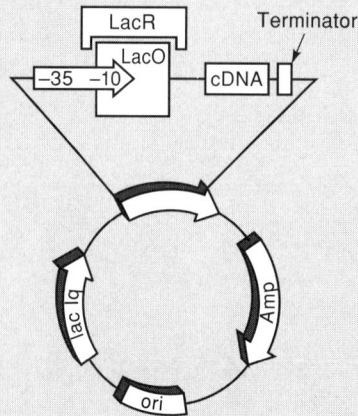

1. Clone gene of interest into tac expression vector.

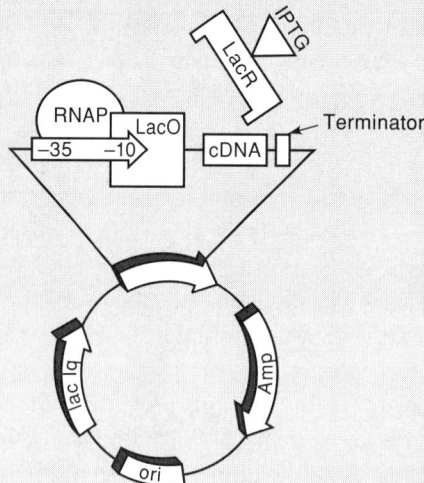

2. Induce expression by adding IPTG. IPTG binds to the Lac repressor, causing dissociation from the operator and transcription.

FIGURE 11.1. An expression vector containing the Tac promoter. (LacO) Lac operator, or binding site; (LacR) Lac repressor.

Tac ("Trp-Lac") promoters are a fusion of the –35 consensus region of the Trp promoter (with sequence TTGACA) and the –10 consensus region of the Lac UV5 promoter (also called the Pribnow box; its sequence is TATAAT (Fig. 11.1) (Amann et al. 1983; de Boer et al. 1983). This particular combination of sequences has a high affinity for *E. coli* RNA polymerase and, once induced from a multicopy plasmid, synthesizes the bulk of cellular mRNA. In a normal logarithmically growing cell, the promoter is repressed by the Lac repressor, the *lacI* gene product. The LacI protein binds tightly to its operator within the promoter, but it can be dissociated by lactose, allolactose, and other metabolites found in wild-type *E. coli*, or by the galactose analog isopropyl-β-D-thiogalactoside (IPTG). A cDNA can be cloned into a plasmid downstream of the Tac promoter (step 1, Fig. 11.1), and the expression of the cDNA can be induced by IPTG in transformed *E. coli* grown to mid-late log phase (step 2).

Uninduced or leaky expression of the downstream cDNA can often be fatal to the cell. Although the phenomenon is not well understood, leaky expression of toxic proteins can lead to very low expression levels of the encoded protein. It is believed in such cases that the plasmid incurs mutations that render the promoter or the downstream cDNA inactive, allowing the host cell to survive. To maintain the ability to induce expression from Tac vectors, the cloning of the cDNA must be carried out in strains containing a *lacIq* allele, which is an overexpressing allele of the *lacI* gene (Brent and Ptashne 1981). Overexpression of the LacI protein using this allele prevents leaky expression of the Tac promoter. In some cases, such as in the Xa-90 strain, the *lacIq* allele is located on an episome in the cell, whereas in other cases, *lacIq* is conveniently cloned into the same plasmid as the Tac promoter–cDNA fusion gene. Addition of 0.2 mM IPTG to the culture medium is sufficient to dissociate the repressor, permitting binding of RNA polymerase and expression of the downstream gene.

In a typical protocol, the transformed cells are grown to an A_{600} of 0.5–0.7 and IPTG is added for 2–10 hours, depending on the stability of the protein. Most proteins reach a plateau of expression after 2 hours, but some proteins induce more slowly because of their toxic effects on cell growth.

Strain choice is important when employing the Tac system. If the expression plasmid bears its own *lacIq* allele, it can be introduced into virtually any strain. This allows one to take advantage of the many protease-deficient strains of *E. coli* that are better suited for expression of proteolytically sensitive proteins.

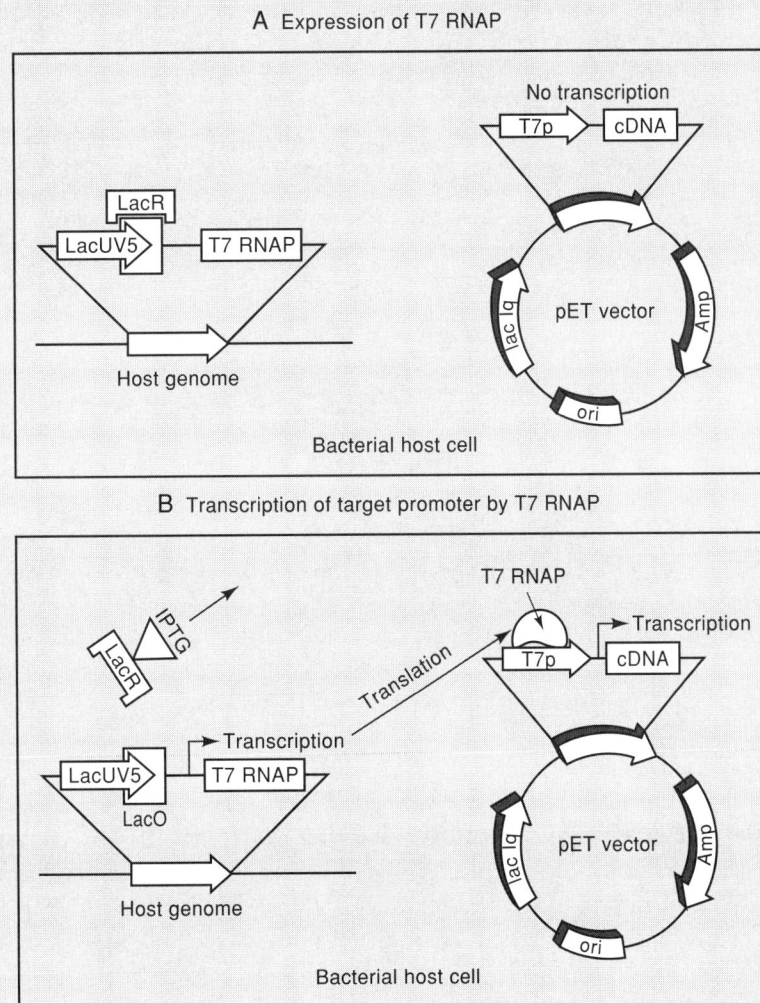

FIGURE 11.2. The T7 promoter system. (LacO) Lac operator, or binding site; (LacR) Lac repressor.

Amersham Pharmacia Biotech provides the pKK223-3 Tac-containing expression vector (Amersham Pharmacia Biotech # 27-4935-01), although many variations of the Tac system are available from academic sources.

The T7 RNA polymerase systems developed by Studier and colleagues (Studier and Moffatt 1986; Rosenberg et al. 1987; Studier et al. 1990) are described as being significantly more efficient than Tac or other systems for protein production in *E. coli*. The T7 system comprises two components (Fig. 11.2). The first component is a special *E. coli* strain bearing the gene for T7 RNA polymerase under control of the Lac promoter, often integrated into the genome through a bacteriophage lysogen (Fig. 11.2A). Recombinant bacteriophage lysogens, such as the one carrying the T7 RNA polymerase gene, can be produced easily using phage molecular genetic techniques. The second component is a plasmid containing a T7 promoter and terminator.

In a typical case, the cDNA is cloned downstream of the T7 promoter into a multiple cloning site. To avoid the potential for leaky expression, discussed above for the Tac system, cloning is performed in a strain lacking T7 polymerase. The plasmid is then introduced into the appropriate strain (e.g., BL21[DE3]). Transcription of the cDNA is induced by adding IPTG, which results in synthesis of T7 RNA polymerase. T7 RNA polymerase, in turn, binds to its promoter in the expression vector and transcribes the cDNA (Fig. 11.2B). T7 is an extremely potent RNA polymerase and synthesizes large quantities of mRNA, leading to a very high level of over-expression of the protein. However, as in the Tac system, in uninduced cells T7 RNA polymerase, even when expressed at very low levels, can synthesize enough mRNA to disrupt cell growth and survival. There are three potential solutions to this problem.

1. The first solution is to overexpress the Lac repressor using *lacIq*. Overexpression of Lac repressor down-regulates the IPTG-inducible T7 RNA polymerase promoter in the host strain (Dubendorff and Studier 1991). Many newer T7 expression plasmids now encode the *lacIq* allele (Fig. 11.2A). In addition, many newer vectors also contain a binding site for Lac repressor immediately downstream of the T7 promoter. In these vectors, *lacIq* can efficiently repress both T7 RNA polymerase expression and T7 polymerase transcription from the T7 promoter. This dual system of repression is probably the simplest and most efficient method for preventing leaky expression.

2. A second approach is to express moderate levels of phage T7 lysozyme, which apparently binds stoichiometrically to T7 RNA polymerase and inhibits its activity (Studier 1991). Overexpression of T7 RNA polymerase in the presence of IPTG would be sufficient to overcome the stoichiometric inhibition and express the cDNA.

3. A third, less frequently used, strategy is to provide the T7 RNA polymerase by phage infection. In this case, cells bearing the expression plasmid but lacking T7 RNA polymerase are grown to mid-late log phase and then superinfected with a special strain of T7 phage over-expressing the RNA polymerase. Because the cells initially lack T7 RNA polymerase, there should be no leaky expression of the cDNA until the superinfection phase.

The T7 expression system is marketed by Novagen, which provides an excellent selection of strains, cloning vectors (the pET series), and reagents.

The most significant advantage of *E. coli* is the short time required to subclone, measure expression, and purify the protein. Together these steps can ideally be carried out in 2 weeks or less. In a typical procedure, a PCR fragment encoding the polypeptide of interest is subcloned into a vector, and then the cells bearing the expression vector are grown to mid-log phase at 37°C and induced (see Box 11.1). Ideally at this point, one should have an assay available to measure activity of the protein to ensure it is being folded properly and is functional after purification.

Prior to large-scale synthesis, a time course is performed on 1–2 ml of culture to monitor the induction and determine the optimum expression time. Most overexpressed proteins can be directly visualized by resuspending the pellet from 100 µl of cells in an SDS-cell lysis dye mix, fractionating the extract on a SDS-polyacrylamide gel, and staining the gel with Coomassie Blue. It is rarely necessary to resort to immunoblotting unless expression problems are encountered.

After the optimum induction time is determined, the cells are grown in larger scale using either a fermentor, if large quantities of protein are required, or 100-ml–2-liter flasks if smaller amounts are needed. A good degree of aeration is necessary for efficient growth of the *E. coli* and for high levels of induction. The cells are harvested by centrifugation and extracts are typically prepared by either detergent lysis, sonication, or French press (see Marshak et al. 1996, Coligan et al. 1996, or Scopes 1994, for a summary of protein purification methodology). The extracts are very crude, and it is advisable to remove as much nucleic acid and contaminating protein as possible before continuing on to conventional chromatography.

Contaminating cellular and plasmid DNA can be removed by polyethyleneimine (PEI) (0.10–0.5%) precipitation. Alternatively, a major RNA contaminant, rRNA, can be removed by precipitation with streptomycin sulfate (2%). The highly basic PEI also precipitates some acidic proteins. Therefore, if the protein of interest is acidic and precipitates with PEI, this step can result in substantial purification. In such instances, the protein can be selectively extracted from the precipitate by resuspending the pellet in buffer containing high salt concentrations followed by a short mixing period and recentrifugation. Similarly, if desired, raising the salt concentration to modest levels (0.2–0.5 M salt, for example) prior to PEI addition may allow the protein to be retained in the soluble extract during precipitation. The protein can then be purified from the extract.

After the nucleic acid is removed, ammonium sulfate precipitations can be performed on the resulting extract to remove much of the contaminating protein before chromatography. Ammonium sulfate precipitates protein by sequestering water and promoting hydrophobic interactions that lead to aggregation and insolubility. The amount of ammonium sulfate needed to precipitate a protein is a unique function of both the hydrophobicity of the protein and its concentration in the extract. Therefore, different proteins precipitate at different ammonium sulfate concentrations. It is our experience that many abundantly expressed recombinant proteins in *E. coli* extracts precipitate at low ammonium sulfate concentrations (30–40%), resulting in significant purification from contaminants in the extract. Scopes (1994) provides a chart and an excellent description of the methods for performing ammonium sulfate cuts.

The ammonium sulfate precipitates are collected by centrifugation. The pellets are resuspended in buffer and the excess ammonium sulfate is removed by dialysis or gel filtration. Once desalted, the extract can be subjected to ion-exchange or affinity chromatography. For DNA-binding proteins, the first step usually involves chromatography over phosphocellulose, CM- or SP-Sepharose, heparin-Sepharose, and, occasionally, DNA cellulose or specific DNA affinity resins.

It is possible to bypass the preliminary steps by attaching a purification tag to the protein. This strategy is particularly useful if the protein is expressed poorly or is insoluble (see below), or if multiple proteins are being isolated in parallel and there is a need to purify them rapidly. The common purification tags fused to eukaryotic transcription factors expressed in *E. coli* systems are His_6, glutathione-*S*-transferase (GST), and maltose-binding protein (MBP), although thioredoxin and others have been successfully employed. In such cases, a PCR fragment encoding the polypeptide is cloned into a vector encoding an in-frame purification tag, and the fusion gene is expressed (Box 11.2). If it is necessary to remove the tags to facilitate biochemical analysis of the protein, many vectors encode cleavage sites for sequence-specific proteases. These sites are between the purification tag and the fused protein (Box 11.2).

Box 11.2

Common Purification and Detection Tags

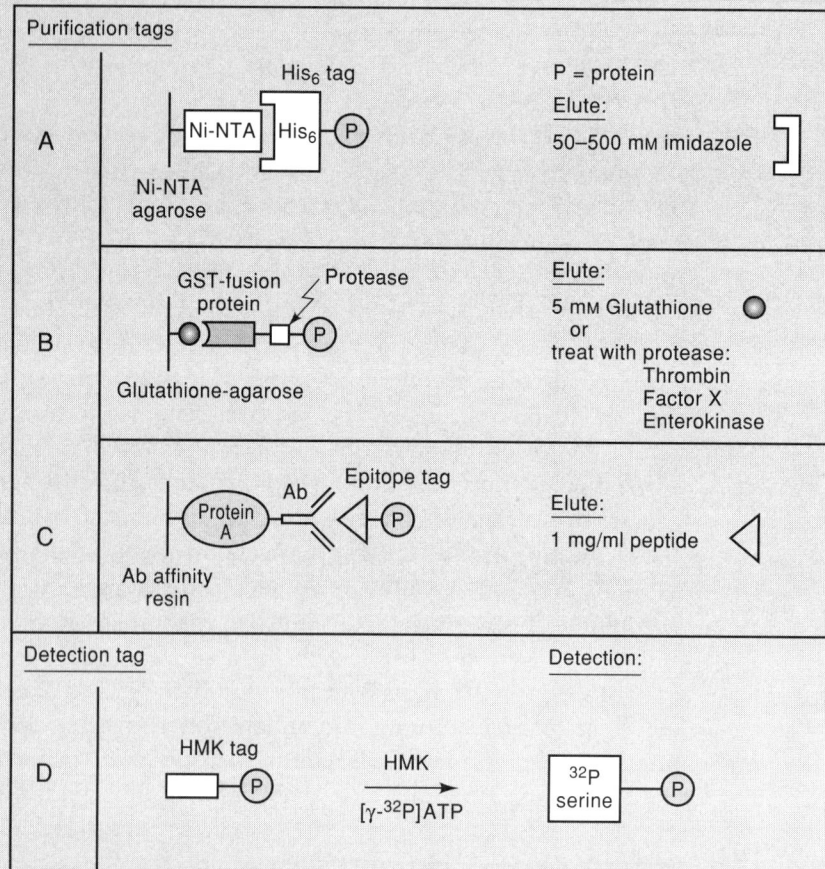

FIGURE 11.3. Common purification and detection tags.

Histidine tags. To facilitate purification of proteins overproduced in bacterial or eukaryotic systems, many expression vectors encode six tandem histidines at either the 5′ or 3′ ends of the polylinker into which a cDNA is cloned. Bacterial lysates containing the fusion protein can then be fractionated over a Sepharose or silica column complexed with nickel-nitrilotriacetic acid (Ni-NTA). QIAGEN and Invitrogen, among others, market nickel-purification column kits. The Talon metal resin (CLONTECH # 8901-1) is a newer matrix, which the manufacturer claims yields better purification results. The histidines on the expressed protein complex with the metal on the resin, allowing the protein to adhere to the column matrix (Fig. 11.3A). Most *E. coli* proteins do not bind the nickel-Sepharose and pass through the column. The His_6 protein that has adhered can then be eluted from the column using an imidazole buffer (or, depending on the protocol, a guanidine hydrochloride buffer at pH 5). The charged imidazole competes with the histidines for the metal and disrupts the interaction, thereby eluting the protein from the resin. His_6 tags are also a component of many eukaryotic vectors. The increased protein complexity of the eukaryotic extracts means that some proteins have natural stretches of histidines that adhere to the resin. In such cases, the proteins eluting from the nickel column are generally cruder and must be purified further (either before or after nickel-Sepharose chromatography). The large subunit of TFIIA is an example of such a contaminant. It contains a contiguous stretch of histidines and often contaminates His_6-tagged protein preparations from eukaryotic nuclear extracts. Indeed, purification of native TFIIA from HeLa extracts involves a nickel chromatography step, and nickel chromatography can be used to deplete a HeLa extract of endogenous TFIIA quantitatively for complementation studies (see Ozer et al. 1994).

A major advantage of the His_6 tags is that the chromatography can be carried out under denaturing conditions, facilitating purification of proteins that are insoluble in *E. coli* or other expression systems.

GST fusions. GST is a 26-kD protein. In the most commonly used vectors, those from the Amersham Pharmacia Biotech pGEX collection, the *Schistosoma japonicum* GST is fused to the cDNA in a *lacIq*-controlled Tac promoter. The induced cell extracts bearing the chimeric protein are passed over a glutathione-Sepharose or agarose column (or the binding can be done in batch). The GST portion of the fusion protein adheres to the matrix and can be eluted by addition of 5 mM free glutathione (Fig. 11.3B). Glutathione is acidic and the elution requires a slightly basic or neutral pH, so care must be taken to ensure that the elution buffer has been properly buffered.

GST has advantages in addition to its purification applications. For example, GST apparently has a stabilizing effect on some proteins. We have found that several protein fragments that were not synthesized to detectable levels in *E. coli* by themselves were synthesized when fused to GST. Additionally, certain protein fragments that were insoluble in isolation became solubilized as a consequence of their fusion to GST. GST from various sources has been reported to be a dimer (Singh et al. 1987; Smith and Johnson 1988; Tiu et al. 1988; Lai et al. 1989), and any studies on dimerization of recombinant factors should take this observation into account. Amersham Pharmacia Biotech provides a complete kit and reference manual for GST purification, including the widely used pGEX series of vectors containing the Tac promoter, the *lac Iq* gene, and prescission, thrombin or Factor Xa-protease cleavage sites (cat. # 27-4570-01).

MBP fusions. The *E. coli* MBP, encoded by *malE*, can be fused to a cDNA and expressed under the Tac promoter. The MBP fusion is subsequently bound to amylose-affinity resins (capacity ~ 3 mg/ml) and eluted with excess soluble maltose. A complete MBP kit is available from New England Biolab (cat # 800).

Protease cleavage sequences. The cleavage sites for thrombin (Leu-Val-Pro-Arg-Gly-Ser), enterokinase (Asp-Asp-Asp-Asp-Lys), and factor Xa (Ile-Glu-Gly-Arg), among others, are often placed between a fusion partner, such as the His_6 or GST tags, and the recombinant protein. This permits facile purification of the transcription factor away from the fusion partner. The protease is incubated with the fusion protein while it is attached to the column matrix. The eluted protein ends up in the supernatant and is isolated simply by separating the supernatant from the column matrix by centrifugation or filtration.

Immunotags. Immunotags have become a convenient and highly specific means of purifying a fusion protein or identifying it in crude extracts from expressing cells. The immunotags are particularly useful for purifying proteins from eukaryotic systems where the protein complexity is much greater than in *E. coli*. Due to the high affinity and specificity of a monoclonal antibody for its epitope, there is little binding of contaminating protein to the immunoaffinity resins at moderate salt concentrations. The amount of resin necessary for immunodepletion must be determined empirically because crude extracts have a moderate inhibitory effect on binding of the tagged protein. Immunotags are rarely used in *E. coli* expression systems because the antibody resins are expensive and offer little advantage for purification when other tags like

GST or His$_6$ are widely available. In the absence of an antibody against the native protein, however, immunotags can be useful for detection during purification or in experimental procedures using *E. coli*-synthesized proteins (Fig. 11.3C).

The influenza virus hemagglutinin (HA) tag (Tyr-Pro-Tyr-Asp-Val-Pro-Asp-Tyr-Ala) and the FLAG tag (Asp-Tyr-Lys-Asp-Asp-Asp-Asp-Lys) are short peptide epitopes fused to either the amino or carboxyl terminus of a recombinant protein (others include the vesicular stomatitis virus [VSV] and Myc tags). The tags bind specifically and tightly to monoclonal antibody affinity resins and can be eluted efficiently with the corresponding peptide. Mammalian and bacterial expression vectors encoding the HA, VSV, and Myc tags, their corresponding immunoaffinity resins, and peptides for elution are available through Roche Molecular Biochemicals. The FLAG tag resins and peptides are available through Kodak and are marketed by Sigma (or Babco for HA). Typically, the 12CA5 (HA) or M1/M2 (FLAG) antibodies are covalently cross-linked to either protein A–Sepharose (or agarose) or activated affinity resins such as CNBr-activated Sepharose beads. The cellular extracts bearing the fusion protein are then passed over the affinity resin via column or batch chromatography. The resin is washed extensively with high-salt buffers to remove nonspecifically bound protein. The bound protein is eluted with ~1 mg/ml peptide corresponding to the epitope and moderate to high salt concentrations. The FLAG M1 antibody requires Ca^{++} for binding, and its removal by EGTA causes antibody dissociation from the FLAG epitope, providing a more convenient method of elution.

Heart muscle kinase (HMK) tags. The HMK tags offer no benefits for purification, although they are useful for detection and manipulation of the recombinant proteins. HMK is heart muscle kinase, the catalytic subunit of protein kinase A. It recognizes and phosphorylates a serine located within a highly specific 5-amino-acid sequence sometimes called the HMK tag (Arg-Arg-Ala-Ser-Val) (Fig. 11.3D). This sequence can be fused onto the 5′ or 3′ end of the gene encoding the cDNA so that the resulting protein will contain the sequence. This tag enables one to label the carboxyl or amino terminus of the pure protein using HMK and [γ-^{32}P]ATP. The labeled protein has many applications, including as a probe for screening expression libraries. It can also be used in numerous protein–protein interaction studies (Blanar and Rutter 1992; Hori and Carey 1997). Vectors from many commercial sources contain the HMK tag.

Strategies for Overcoming Expression Problems in *E. coli*

Overexpression of foreign proteins in *E. coli* can sometimes be problematic. The protein of interest may be expressed at low levels, become proteolyzed, or be insoluble in the cytoplasm such that it fractionates with the membranous material during extract preparation. Some of these issues are covered in recent reviews whereas others require searching of the literature to examine purification strategies used by others studying similar proteins (for review, see Hockney 1994; Makrides 1996). We summarize the more common problems and potential solutions below.

Low expression. The first indication that a protein is expressed at low levels is that it cannot be visualized by staining on SDS-polyacrylamide gel of the extracts after induction. The possible causes for such behavior are that the promoter being employed is too weak, the expression cassette is improperly designed, poor *E. coli* codon usage, or the protein is becoming proteolyzed.

Typically, an expression cassette contains a highly inducible promoter with a low basal level of expression, a translational enhancer such as a Shine-Dalgarno sequence or ribosome binding site, and a transcription termination site. If any one of these features is missing, mRNA levels will be extremely low. Northern blotting may be useful in ascertaining the mRNA levels, although *E. coli* messages typically have short half-lives. In the event that mRNA levels cannot be measured directly, fusion of a marker gene such as β-galactosidase to the promoter can be useful in determining whether the vector is active. A large number of different expression systems with numerous built-in features are available commercially and from academic sources (Makrides 1996). Box 11.1 discusses two highly active expression systems based on the *E. coli* Tac and phage T7 promoters.

Often a gene is transcribed at high levels, but the protein is not detectably expressed. This scenario is evident when an mRNA or expression measurement method (see Chapter 5) reveals that the promoter is active, but immunoblotting or staining of SDS gels shows that the protein is not being produced. In this case, the low expression may be due to (1) unusual secondary structures in the mRNA, (2) poor *E. coli* codon usage, or (3) proteolysis.

1. Unusual secondary structures are typically very difficult to identify and to deal with in *E. coli*. Ideally when cloning the cDNA into the vector, it is essential to minimize the amount of sequence flanking the actual coding region to enhance expression levels.

2. Codon usage problems, on the other hand, are identified by scanning the coding sequence to determine if the mammalian gene is employing codons that are rare in *E. coli* (Makrides 1996 provides a chart of problem codons). Two strategies are pursued in such cases. In the first, the eukaryotic codon is altered by site-directed mutagenesis to one more frequently used in *E. coli*. In the second, an expression plasmid encoding the rare tRNA is co-transformed with the expression vector encoding the cDNA of interest (for review, see Makrides 1996).

3. Low levels of induced protein are occasionally due to proteolysis. Typically, endogenous *E. coli* proteases cleave the protein at discrete sites, generating fragments that can be observed directly on Coomassie Blue-stained SDS gels or by immunoblotting. The proteolysis frequently occurs in vitro during extract preparation. There are several approaches to solving this problem. First, the protease sites on the protein can be identified by protein sequencing of the fragments. The cleavage site can be altered by site-directed mutagenesis if the change does not interfere with the biochemical activity of the protein. Alternatively, selected protease-deficient strains (i.e., BL21) are available and can be tested to determine if they increase the yield of intact protein (see Makrides 1996). Frequently, however, proteolysis occurs during extract preparation even if the cells are lysed immediately in SDS-loading buffer and analyzed on a gel. Therefore, it is essential to include protease inhibitors in the lysis buffers used during extract preparation or in the lysis buffer used to fractionate cells by SDS-polyacrylamide gel electrophoresis (PAGE). The most common inhibitors used in preparing *E. coli* extracts are benzamidine and phenylmethylsulfonyl fluoride (PMSF). Roche Molecular Biochemicals markets a wide range of protease inhibitors, and their catalog contains a chart explaining their applications.

Insolubility. In some cases, the protein is synthesized at high levels as measured by SDS-PAGE, but due to improper folding, it becomes embedded in insoluble vesicle-like structures called inclusion bodies. Typically, insolubility in inclusion bodies is manifested when, after cell lysis and centrifugation to remove the membranous material, the protein pellets with the membranes. The inability of a protein to fold properly in *E. coli* has many causes. For intact polypeptides, *E. coli* chaperones may be overwhelmed by the high levels of expression or may not be designed to fold eukaryotic proteins with certain structural features. Alternately, the eukaryotic protein may expose surfaces that render the protein insoluble because these surfaces normally are not exposed inside eukaryotic cells. For example, the protein may normally be part of a macromolecular complex and may be missing a crucial partner that maintains its structure in eukaryotic cells. Alternately, in the case of protein fragments, an essential domain may be missing that either is normally packed up against the fragment or plays a kinetic role in mediating folding. Because the determinants of protein folding are not well understood in *E. coli*, or any other system for that matter, there are no strict rules for solving insolubility problems. Therefore, solubilizing the protein for purification and biochemical studies may require a good deal of trial and error.

In the event that the protein or protein fragment is insoluble, there are several initial options that can be pursued by the investigator, including (1) resolubilizing the protein directly from the inclusion bodies; (2) denaturing the protein, purifying it, and then renaturing; (3) adding the appropriate protein ligand during induction; (4) altering *E. coli* growth conditions to generate soluble protein; and (5) modifying the protein to enhance solubility.

1. First, very little contaminating protein, with the exception of membrane proteins, fractionates with the inclusion bodies. If buffers containing the chaotrophic agents urea or guanidinium HCl can solubilize the material and it can be properly refolded by dialysis against decreasing chaotroph gradient (e.g., see Ranish et al. 1992), the insolubility actually offers a quick and efficient method of purification. Several reviews describe different methods for purifying and solubilizing proteins from inclusion bodies (see, e.g., Lin and Cheng 1991, and Shi et al. 1997; Burgess 1996; Lilie et al. 1998).

2. In some cases the protein can be purified first in a denatured state and then renatured. Ni-chelate chromatography of His$_6$-tagged recombinant proteins can be performed under urea/guanidinium denaturing conditions (Ozer et al. 1994). The insoluble pellet is generally resolubilized in urea or guanidinium HCl-containing buffers and bound to the Ni-affinity resins. A detailed protocol for the solubilization can often be found in the literature accompanying the Ni-chelate resins. The bound protein is eluted from the resin under denaturing conditions with a decreasing pH gradient or increasing imidazole concentrations. The eluted protein is subsequently renatured by step dialysis with buffers containing decreasing urea or guanidinium HCl concentrations. Such procedures, although they do not generally result in yields comparable to those obtained with soluble proteins, nevertheless generate enough protein for extensive biochemical analysis. Proteins containing cysteines are particularly sensitive to such conditions, and addition of reducing agents such as DTT or 2-mercaptoethanol is essential. However, high concentrations of thiols can inhibit binding of His$_6$ proteins to the Ni-columns. Consult the manufacturers' literature for guidelines on the proper amounts to add.

3. It is important to remember that many eukaryotic proteins fold improperly in *E. coli* because they lack the proper chaperones, or simply the high level of overexpression overwhelms the *E. coli* protein-folding machinery. There is one class of protein, however, where the insolubility problem is easier to tackle. Many eukaryotic transcription factors encode a metal- or ligand-binding domain such as a zinc finger (e.g., nuclear receptors, Sp1, GAL4, EGR-1). In such cases, the addition of ligand (e.g., ZnCl$_2$) can enhance solubility during induction. In the case of small GAL4 fragments, the addition of ZnCl$_2$ or CdCl$_2$ dramatically enhanced solubility of the resulting protein in *E. coli* (Marmorstein et al. 1992).

4. The next simplest option for dealing with insolubility is to alter the *E. coli* growth conditions to favor solubility. Makrides (1996) outlines a number of ways this has been achieved. The most common method has been to induce expression at reduced temperatures ranging from 15°C to 30°C. Although the overall amount of induced protein in these cases is lower, many investigators have found that the amount of soluble protein (versus 37°C) is greater.

5. If the growth alteration approach fails, it has been noted that in many cases purification/detection tags can enhance protein solubility and may be added to the protein if they are not already present (Makrides 1996). Alternatively, varying the amount of flanking sequence can result in enhanced solubility although there are no strict rules. Typically a number of derivatives are synthesized and compared side by side for their solubility behavior.

Synthesizing Large Regulatory Proteins

Because the size of eukaryotic regulatory proteins can vary greatly, the choice of system for overexpression must be determined empirically. In any event, it is useful to attempt expression in *E. coli* first because the effort, cost, and required resources are minimal. If *E. coli* fails, however, because the protein is either too large, insoluble, proteolyzed, or otherwise, most investigators employ the eukaryotic systems, which include yeast, baculovirus, vaccinia virus, and retroviruses.

There are several advantages to the eukaryotic systems. For one, eukaryotic transcription factors, particularly large ones, are more likely to be folded properly and synthesized intact in their native systems. Additionally, with the insect baculovirus and the mammalian retrovirus and vaccinia virus systems, the host cells generally support exogenous modifications that may be necessary for a protein's activity (e.g., glycosylation, phosphorylation, acetylation, acylation, and carboxylation). The baculovirus and mammalian systems display differences in the types and extents of modifications they are capable of producing. Therefore, it is necessary to consult the literature to determine whether a particular modification is supported by the system being employed. (Invitrogen provides a table in their baculovirus manual listing posttranslational modification events.)

Box 11.3

Yeast Expression Systems

Saccharomyces cerevisiae. The best characterized yeast system is *S. cerevisiae*. Two highly inducible, strong yeast promoters are ADH1 and GAL1. Both of these promoters have been used for induction and purification of eukaryotic proteins. Expression vectors are typical *E. coli* shuttle vectors that contain a promoter upstream of a multiple cloning site (i.e., pYES2, Invitrogen # V825-20). The vectors also contain the plasmid 2μm origin of replication, which allows the expression vector to replicate to high copy number in yeast, and a selectable marker (e.g., HIS2, URA3, ADE2) for transformation into a yeast auxotroph.

For the GAL1 promoter, the transformed cells are grown in glycerol to maintain uninducing conditions. In uninduced cells, the GAL80 repressor binds directly to the GAL4 *trans*-activator, four dimers of which are bound to the Gal upstream activating sequence (UAS$_G$) 189 bp upstream of the GAL1 TATA box. Expression is induced by adding galactose to the growth medium which dissociates GAL80 and allows GAL4 function (Foreman and Davis 1994). The ADH1 promoter has been used in many applications and can be induced by alcohol or other non-glucose carbon sources, although it is almost always constitutively active at some level (for an overview of these and other yeast systems, see Sudbery 1996).

Pichia pastoris. This is a methylotrophic yeast that utilizes methanol in the absence of its standard carbon source, glucose. The first stage in the metabolism of methanol involves an enzyme called alcohol oxidase (AOX1) (for review, see Hollenberg and Gellissen 1997).

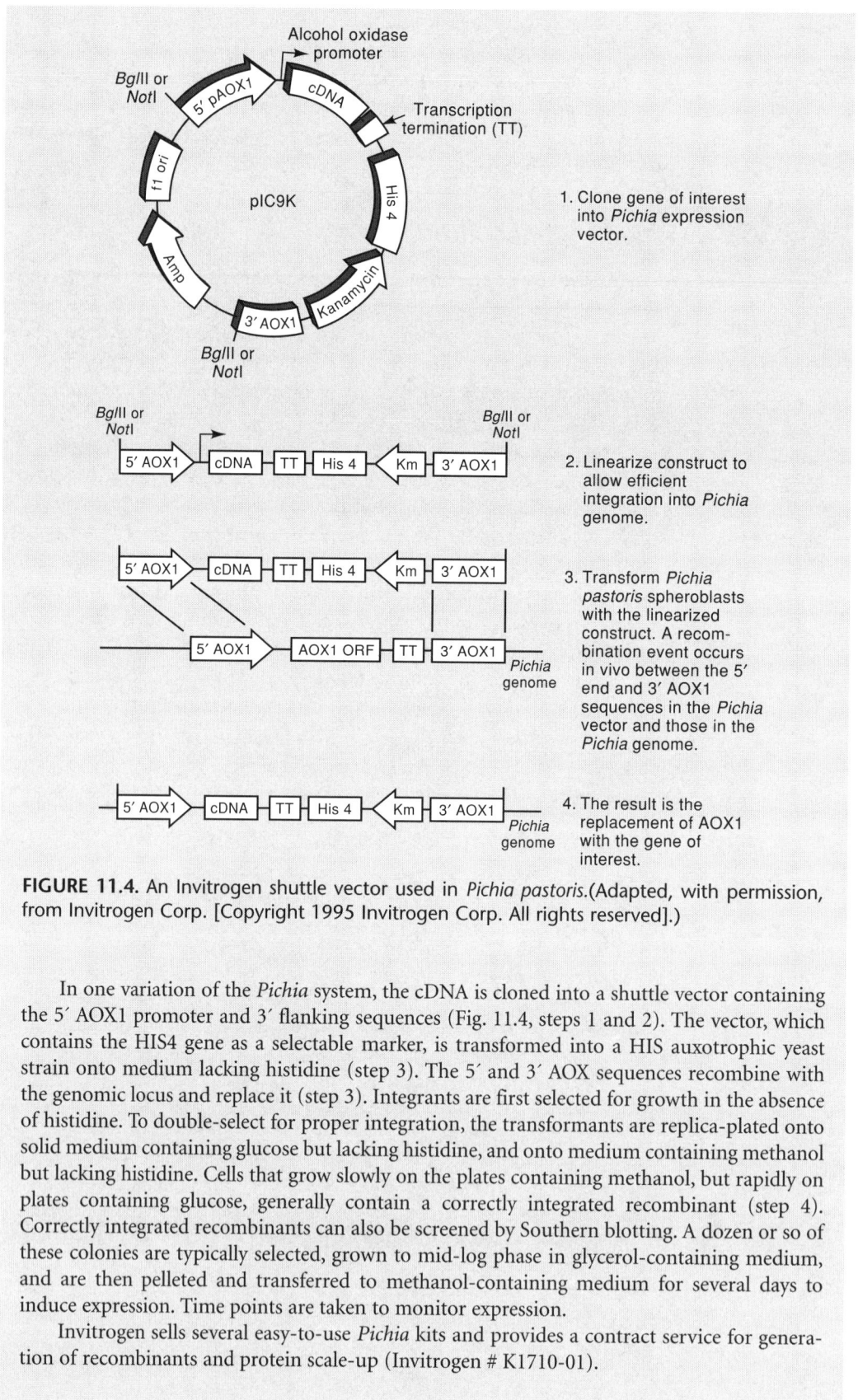

FIGURE 11.4. An Invitrogen shuttle vector used in *Pichia pastoris.*(Adapted, with permission, from Invitrogen Corp. [Copyright 1995 Invitrogen Corp. All rights reserved].)

In one variation of the *Pichia* system, the cDNA is cloned into a shuttle vector containing the 5′ AOX1 promoter and 3′ flanking sequences (Fig. 11.4, steps 1 and 2). The vector, which contains the HIS4 gene as a selectable marker, is transformed into a HIS auxotrophic yeast strain onto medium lacking histidine (step 3). The 5′ and 3′ AOX sequences recombine with the genomic locus and replace it (step 3). Integrants are first selected for growth in the absence of histidine. To double-select for proper integration, the transformants are replica-plated onto solid medium containing glucose but lacking histidine, and onto medium containing methanol but lacking histidine. Cells that grow slowly on the plates containing methanol, but rapidly on plates containing glucose, generally contain a correctly integrated recombinant (step 4). Correctly integrated recombinants can also be screened by Southern blotting. A dozen or so of these colonies are typically selected, grown to mid-log phase in glycerol-containing medium, and are then pelleted and transferred to methanol-containing medium for several days to induce expression. Time points are taken to monitor expression.

Invitrogen sells several easy-to-use *Pichia* kits and provides a contract service for generation of recombinants and protein scale-up (Invitrogen # K1710-01).

Yeast. The main advantages of the yeast systems are that they are simple to set up, inexpensive, and require no special equipment (see Box 11.3) (Sudbery 1996). The two common yeast systems are based on *S. cerevisiae* and *P. pastoris.* Typically, the cDNA is cloned in front of a strong constitutive or inducible promoter and the vector is introduced into yeast. The vector can be recombined into the genome or exist as a stable replicating single or multicopy plasmid. Yeast cells can be grown up in 1–2-liter flasks at 30°C. The protein then can be purified from cell extracts. Yeast systems have not been widely used for expression of mammalian transcription factors.

Box 11.4

The Baculovirus Expression System

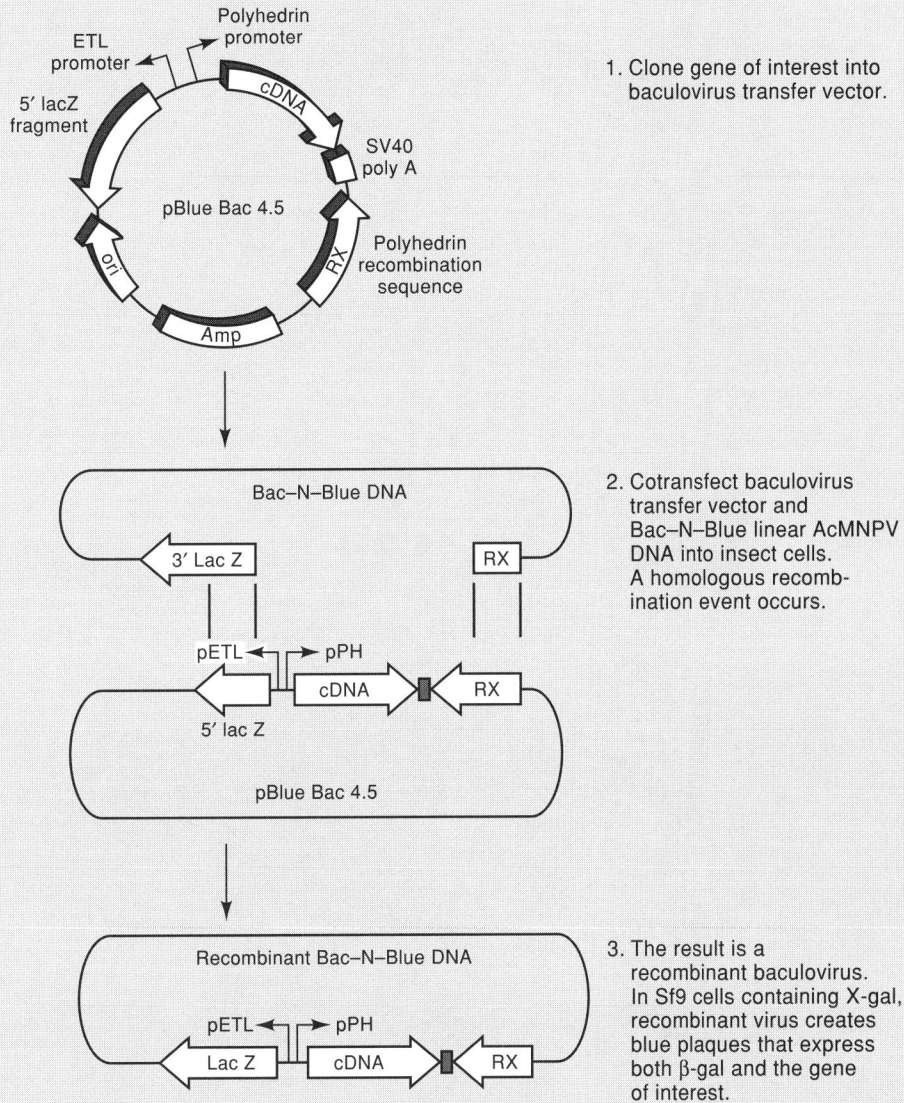

FIGURE 11.5. An Invitrogen shuttle vector for baculovirus: Recombination procedure. (β-gal) β-Galactosidase; (Amp) ampicillin. (Redrawn, with modifications, with permission from Invitrogen Corp. [Copyright 1993, 1996 Invitrogen Corp. All rights reserved.].)

The commonly used baculovirus, *Autographa californica* polyhedrosis virus (AcPNV), is an insect virus containing a double-stranded 134-kb DNA genome that can infect and lyse insect cells in culture. During infection, prior to lysis and cell death, the virus particles initially bud from the cell, but then the virus accumulates and begins to form a large polyhedral-shaped aggregate, an occlusion body, in the nucleus. The major viral protein polyhedrin is responsible for this aggregation; however, it is not required for the actual infection of insect cells in tissue culture. Hence, a cDNA encoding a mammalian regulatory factor can be inserted downstream of the polyhedrin promoter, replacing and disrupting the normal polyhedrin gene and generating a recombinant baculovirus (Smith et al. 1983). The cDNA is then expressed at high levels during a viral infection. Several other viral promoters can be manipulated in a similar fashion.

It is difficult, but not impossible, to introduce foreign genes directly into the baculovirus genome (Vialard et al. 1995), although newer vectors do permit direct ligation (for review, see Possee 1997). More commonly, the synthesis of a recombinant virus is accomplished using a baculovirus shuttle vector which, for cloning purposes, is designed to be maintained as a plasmid in *E. coli* (Fig. 11.5, step 1). The cloned gene within the vector is then inserted into the virus by homologous recombination in vivo in the insect cells (step 2). The process entails co-transfection of the shuttle vector with the virus or with viral DNA into *Sf9* cells (a caterpillar cell line) plated onto standard 6- or 10-cm polarized dishes. Recombination occurs at a low frequency, often only at about 1:1000, and the recombinant viruses do not form occlusion bodies. Therefore, when screening plates for recombinant plaques, dull, nonrefractile plaques are selected because refractile plaques are a product of nonrecombinant virus that still has its polyhedrin gene intact. The baculovirus genome begins to replicate at around 6 hours post-infection (hpi), and infected cells begin to release extracellular virus particles as early as 10 hpi, although occlusion bodies from wild-type virus do not begin to form until about 2 days post infection. Plaques can be screened beginning at about 5 days post-infection (step 3). Figure 11.5 shows a standard shuttle vector. The newest vectors (e.g., the Invitrogen pBlueBac-4 series of vectors, cat. # V1995-20) have incorporated the β-galactosidase gene into the shuttle vector under control of another promoter called ETL (early-to-late promoter) so that occlusion-negative plaques appear blue when overlaid with agarose containing X-gal. This forms another method of selection. Another method marketed by PharMingen involves disruption of the β-gal gene cloned into the virus, where clear plaques are screened (PharMingen cat. # 21201P).

Despite the apparent advantage of a color-based screening assay, the frequency of recombination is still low, and only a fraction of the virus generated in a transfection is recombinant, necessitating several tedious rounds of plaque purification (three rounds of plaque purification can take a month to complete). The most efficient systems employ a viral DNA bearing a lethal deletion that prevents viral replication. The ability to replicate is then dependent on recombination with the shuttle vector, which restores the missing sequences (for review, see Possee 1997). In such cases, more than 99% of the virus recovered from the procedure is recombinant, although one round of plaque purification is recommended (PharMingen, Inc. and CLONTECH market such systems). Other newer improvements include vectors with multiple promoters and cloning sites so that several proteins can be synthesized simultaneously. Several companies market such systems and vectors, including Invitrogen, PharMingen, and CLONTECH. Additionally, Life Technologies Inc. offers a baculovirus system (Bac-to-Bac, cat. # 10359-016) where the viral genome is cloned into a Bac vector and the recombination event is performed and selected for in *E. coli*. The resulting viral recombinants are directly transfected into insect cells.

After the recombinant virus is isolated, it is necessary to titer and scale up the virus prior to expression of the recombinant transcription factor. The scale-up of virus must be done very carefully because multiplicities of infection (moi) greater than 1 generate viral recombinants that express the encoded protein at much lower levels. Large-scale production of protein is preceded by pilot studies optimizing the moi and the timing (often 36–48 hours). During such studies, the protein is generally monitored by immunoblotting. Infection can be onto standard *Sf9* cell monolayers in tissue culture flasks or directly into caterpillars. *Sf9* cells are easily adapted to spinner cultures, which grow at or near room temperature, and where larger-scale production can be carried out. Cells are harvested, and standard nuclear and cytoplasmic extracts (see Chapter 14) are used as a starting source for the purification of recombinant proteins. Typically 1–10% of the protein in the infected extract is recombinant, although this varies widely. For those labs lacking the necessary facilities, many companies provide contract services (i.e., Invitrogen).

Baculovirus. The baculovirus system has been used for synthesizing numerous recombinant eukaryotic transcription factors of varying size and has an excellent track record (see Box 11.4) (see Richardson 1995). The disadvantages of the baculovirus systems are the tissue-culture expense, the time it takes to generate viral recombinants, and the occasional poor yields or insoluble protein. Nevertheless, the baculovirus system represents the first choice of many investigators when *E. coli* fails.

The basic baculovirus procedure involves cloning a cDNA into a shuttle vector containing a strong baculoviral promoter, usually the polyhedrin promoter, and flanking sequences. The resulting plasmid is then transfected into insect cells with intact virus. The flanking sequences in the shuttle vector recombine with the virus at the polyhedrin gene locus to generate a recombinant virus. Polyhedrin is not an essential baculovirus gene, so the recombination event does not have a deleterious effect on viral growth in culture. Recombination occurs at low frequency, and the recombinant viruses are isolated using a variety of clever selection methods (Box 11.4). The viruses are scaled up and used to infect insect cells. Insect cells have been adapted for growth in spinner culture, allowing large-scale infections that under ideal conditions can generate 1–10-mg quantities of proteins per liter of culture. Prior to scale-up, a time course must be performed to determine the optimum induction kinetics. Standard nuclear and cytoplasmic extracts (see Chapter 14) are prepared, and the protein is purified from these extracts.

Recent advances in the design of baculovirus shuttle vectors have resulted in facile methods for producing recombinant virus. For one method, the cDNA is cloned downstream of the viral polyhedrin promoter flanked by viral sequences. The shuttle vector is then co-transfected with viral DNA containing a deletion in a critical gene. The viral DNA will not grow on its own, whereas the shuttle vector restores the missing sequence when it recombines with the virus (see Box 11.4). The newest vectors available commercially contain all of the amenities found in the *E. coli* expression systems, including purification and detection tags.

Mammalian expression systems. Mammalian expression systems, including vaccinia virus, retrovirus, and plasmid transfections, may be necessary if a transcription factor possesses sophisticated structural features or requires posttranslational modifications supported only by mammalian cells. For example, the androgen receptor (AR), a member of the steroid subfamily of nuclear receptors, contains unusual structural domains, undergoes ligand-induced conformational changes, and interacts with numerous partners ranging from heat shock proteins to coactivators (nuclear receptors are reviewed in Beato et al. 1995). AR synthesized using the baculovirus system was found to be insoluble and had to be resolublized in the presence of Zn or $CdCl_2$ (Chang et al. 1992; Xie et al. 1992). However, His_6-tagged AR synthesized in HeLa cells using the vaccinia system (see Box 11.5; reviewed in Carroll and Moss 1997) was found to be highly soluble and active for DNA binding and *trans*-activation in vitro without any manipulations (De Vos et al. 1994). To take another example, the transcription factor Sp1 is found only in mammalian cells. The protein is phosphorylated and glycosylated on specific residues and requires those modifications for full activity. Sp1 synthesized using a vaccinia virus expression system was shown to be properly modified and active for DNA binding and transcriptional activation in vitro (Jackson et al. 1990).

1. Vaccinia virus: The vaccinia virus system represents only one of the mammalian systems that can generate biochemically amenable quantities (e.g., microgram to milligram amounts) of recombinant protein. In the vaccinia system, a shuttle vector under control of a viral or heterologous promoter is recombined with intact vaccinia virus to generate a recombinant virus (see Box 11.5). The recombinant virus is scaled up and used to infect mammalian tissue-culture cells transiently. The expressed protein is then purified from the cellular extracts.

Box 11.5

The Vaccinia Virus Expression Systems

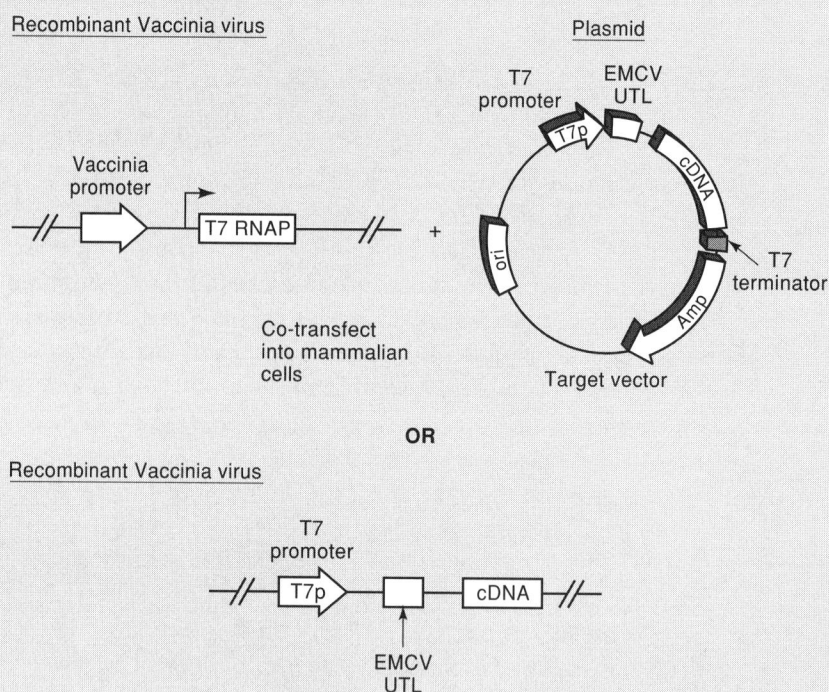

FIGURE 11.6. Schematic of vaccinia system. (Amp) Ampicillin.

Vaccinia virus is a double-stranded DNA pox virus that infects a variety of mammalian cell types in culture. The virus replicates in the cytoplasm and encodes its own DNA and RNA polymerases. There are several variants of this system, originally developed by Moss and colleagues (Elroy-Stein et al. 1989; Wyatt et al. 1995; Ramsey-Ewing and Moss 1996).

The older systems involve generating recombinant viruses by using shuttle vectors that bear vaccinia genomic sequences flanking a vaccinia promoter that drives expression of the downstream cDNA. The most common method of selection is inactivation of the viral thymidine kinase (TK) gene by recombination followed by bromodeoxyuridine (BrdU) selection. The vaccinia virus RNA polymerase and 5′ capping enzymes synthesize the synthetic mRNAs in the cytoplasm, which are then translated. The most advanced system employs a recombinant vaccinia virus, which synthesizes the prokaryotic phage T7 RNA polymerase under control of a vaccinia virus promoter (Fig. 11.6). T7 RNA polymerase can recognize its promoter with high affinity and specificity even in eukaryotic cells (for a recent review, see Carroll and Moss 1997). It is more efficient than most eukaryotic RNA polymerases and can localize and function in either the cytoplasm or nucleus of a eukaryotic cell. Co-transfection of the T7 RNA polymerase virus with either a plasmid, or another recombinant virus containing the transcription factor cDNA under control of the T7 promoter, results in the production of a synthetic mRNA (Fuerst et al. 1986). The cDNA under control of the T7 promoter is not capped, and normally would not be translated efficiently in a eukaryotic cell. To overcome this problem, the encephalomyocarditis virus (EMCV) untranslated leader (UTL) is fused to the 5′ end of the cDNA. EMCV is a picornavirus whose mRNAs are uncapped but remain competent for translation by containing a 5′ untranslated leader sequence with a strong ribosome-binding site. These cap-independent translational enhancers are sometimes called CITEs.

The vaccinia systems are ideal for production of recombinant human proteins, although the construction of recombinant vaccinia viruses is a lengthier task than is the construction of recombinant baculoviruses. Some newer vectors allow direct ligation of cDNA into the genome (see Carroll and Moss 1997). Detailed methods for generation of recombinant viruses and for infection and protein scale-up can be found in the chapter by Moss in *Current Protocols in Molecular Biology* (Ausubel et al. 1994, Unit 16.15).

The advantages of the vaccinia virus system are that the genome can bear very large insertions up to 25 kb in size, expression is performed transiently, and the system is capable of synthesizing milligram quantities of protein when facilities for large-scale cell growth are available. The disadvantages are that it is difficult to generate recombinant viruses, and that vaccinia virus infects humans. It is often advisable for the researcher to consult the local biosafety committee for rules and regulations for vaccinia usage. Safer strains that require minimal precautions have recently become available (see Carroll and Moss 1997).

2. *Retroviruses*: Retroviral vectors represent yet another system for expressing mammalian transcription factors (see Box 11.6, p. 384). They are most frequently employed to express and assess the biological importance of transcription factors in a given tissue or cell line. They have not been used for analysis of *cis*-acting elements in promoters because the long terminal repeat (LTR) of the retrovirus is a strong enhancer that influences downstream reporter gene expression. However, with new vectors in which the enhancer has been disabled, it may be possible to perform reporter gene studies (Hofmann et al. 1996).

In some systems, retroviral infection is the preferred method for introducing genes into cells or tissues, because infection efficiency can be much higher (up to 100%) than with simple plasmid transfections. Retroviral vectors can be replication-competent (containing *gag, pol*, and *env* genes), but today almost all are replication-defective (missing one or, in most cases, all of these genes). After transfection into a producer cell line that expresses gag, pol, and env proteins, or co-transfection with a plasmid expressing these proteins, virus particles whose RNA genome encodes the cDNA of choice bud off into the culture supernatant. The supernatant is then collected and stored in workable aliquots.

Box 11.6

Retroviral Expression Systems

Retroviruses have become a popular method for establishing stable lines expressing mammalian proteins because of their ability to integrate in efficiently and stably single copies into infected mammalian cells. Retroviruses require several elements to confer infectivity: LTR regions provide the promoter, poly A sites, and sequences for integration into the host genome; *gag, pol*, and *env* genes for replication of the viral genome and production of the viral coat proteins; and the Psi element, which is required for packaging the genome into the viral envelope. Replication-defective virions cannot infect neighboring cells once the virus has entered one host cell. The virus must first be prepared in a packaging cell line or co-transfected with a vector expressing packaging proteins. Excellent reviews of the methodology by Warren Pear and Constance Cepko can be found in *Current Protocols in Molecular Biology* (Ausubel et al. 1994, Unit 9.9.1–9.14.6). Most packaging cell lines contain a plasmid encoding the GAG, POL, and ENV proteins under control of the CMV promoter. The plasmid lacks the Psi element, thus this genome cannot be packaged into a mature virus particle. The packaging-deficient genome can either be stably expressed in the production cell line, or it can be transiently transfected into a cell, along with the viral genome expressing the cDNA of choice (Pear et al. 1993). Other packaging manipulations can be made. For instance, to increase viral titer and expand the viral host range, the env protein of Moloney murine leukemia virus (MMLV) (one of the simplest and most commonly used retroviruses) can be replaced with the P-glycoprotein of VSV, which also permits concentration of the virus by centrifugation (Burns et al. 1993).

As with transfection procedures, the retroviral vector must contain a marker to screen for infected cells. The marker can either be a reporter gene (β-galactosidase, alkaline phosphatase, and green fluorescent protein are common reporters) or a selectable drug-resistance gene (like neomycin, puromycin, or hygromycin). Some retroviral vectors contain both types of markers, and it is up to the researchers to determine which markers are most convenient for their needs. Remember that selection by drug resistance may take up to several weeks, but this is the standard way to select positive colonies in tissue culture. In contrast, the use of reporter genes permits rapid screening of infected cells 1–2 days post-infection for determination of viral titer.

Retroviruses can be pantropic (infecting many hosts), ecotropic (infecting only rodents), amphotropic (infecting rodents and primates, including humans), or xenotropic (infecting hosts other than rodents). Furthermore, a virus may only be able to infect a certain cell type within an organism, and most retroviruses (lentiviruses excepted; lentivirus systems are reviewed in Naldini 1998) can only infect dividing cells. As mentioned before, some viruses can be produced in higher titers than others. All of these characteristics must be taken into consideration when designing an infection protocol, especially if lab safety is an issue. If a researcher decides to use an amphotropic virus, for instance, he must ensure that the virus is properly handled, contained, and disposed of so as to pose no undue risk of infection to members of his lab. New vector systems have resulted in safe and efficient regulated expression of exogenous proteins (Naviaux et al. 1996; Lindemann et al. 1997).

In the past, researchers requested retroviral vectors from collaborators in academic laboratories. CLONTECH Laboratories, Inc. (Palo Alto, CA) now offers a series of retroviral vectors (Retro-X System, CLONTECH # K1060-1). For proteins that are toxic when constitutively expressed, CLONTECH also markets vectors incorporating components of the Tet-inducible system (Retro-Off and Retro-On, CLONTECH # K1626-1, #K1627-1). However, the researcher should compare these commercially available vectors with vectors from a laboratory that studies similar biological phenomena in a similar system.

For both replication-competent and -defective vectors, virus titer should be determined before proceeding with the overexpression. This is accomplished by incubating cells in culture with small volumes of serially diluted culture supernatant. Normally, infection is performed in the presence of Polybrene (a cationic amine), and the supernatant is left on the cells for several hours before it is replaced with fresh medium. Infected cells are screened for expression of a marker gene (i.e., GFP) several days post-infection. Once the titer is determined, the virus can be used to infect cells transiently in culture or to generate stable cell lines where the viral cDNA integrates into the genome. Stable cell lines typically express lower levels of proteins than transient infections. Typically the cDNA encodes a purification tag (Fig. 11.3) and the expressed protein can be purified quite easily from extracts. For labs with the capacity to grow liter quantities of cells in spinner culture, one can obtain minimally 3–5 µg of protein or protein complex from 1 liter of cells, depending on the toxicity of the protein.

The main disadvantage of retroviral vectors is that the viruses cannot tolerate cDNAs larger than a few kilobases in size, and the yield of virus decreases significantly as the size of the insert increases. Furthermore, with large inserts it is difficult to obtain high virus titers for transient studies, which may be necessary if the protein is toxic and stable cell lines cannot be generated. Even when stable cell lines can be generated, facilities for large-scale tissue culture must be in place to obtain large amounts of protein. Finally, certain retroviruses are designed to infect human cells in culture; they can also infect humans, and the research must be carried out in biosafety cabinets following institutional guidelines.

One potential disadvantage of many retrovirus-based systems is that constitutive expression of a protein may be toxic to the cell. In such a case, there are several systems that support inducible expression. Three of the newest systems show the greatest potential and flexibility: the tetracycline-inducible system, the ecdysone receptor-based system, and the rapamycin-inducible system (see Box 11.7, p. 386). All have been adapted for use both in tissue culture and in animals (Hofmann et al. 1996; Amara et al. 1997; Magari et al. 1997; Saez et al. 1997).

Synthesizing Small Quantities of Crude Protein

Expression systems are needed to generate not only large quantities of pure protein but also small amounts of a recombinant protein for DNA binding or for measuring protein–protein interactions. In such cases, in vitro transcription and translation systems (Box 11.6) or small-scale cell culture transfection can be employed. The main advantage of in vitro transcription/translation systems is their ease of use and the fact that several companies produce kits that simplify the process so that no special equipment or cell or bacterial culture facilities are needed. The systems are capable of generating both large and small recombinant factors. The main limitation is the small amount of protein synthesized in these systems and the large amounts of contaminating protein in the translation extracts. The advantages of small-scale transfection are that some cell lines such as 293T are highly transfectable using calcium phosphate, and the protein produced should be modified correctly and in its native form.

In vitro transcription/translation. The in vitro transcription/translation approach was first applied in the eukaryotic transcription field to study DNA binding by the yeast GCN4 protein (Hope and Struhl 1985). In the in vitro systems the cDNA is typically cloned into

Box 11.7

Specialized Inducible Expression Systems

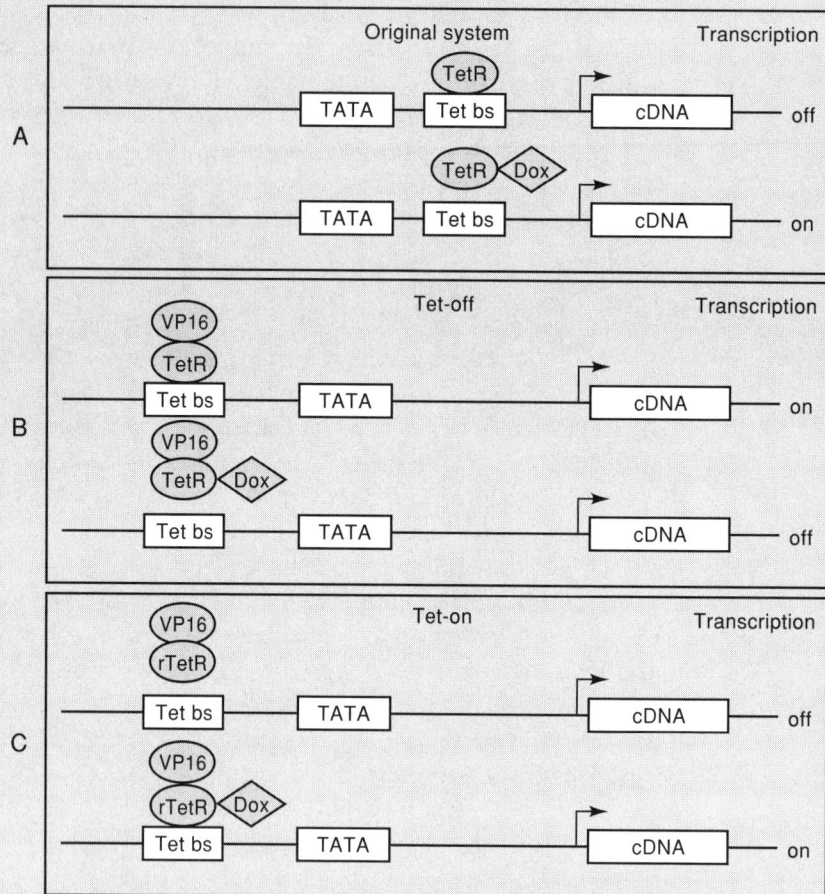

FIGURE 11.7. Three typical tetracyline-inducible systems. (Tc) Tetracycline; (DOX) doxycycline. (Part B adapted, with permission, from Invitrogen Corp. [Copyright Invitrogen Corp. All rights reserved.].)

It is often important to regulate expression of cloned cDNAs introduced into mammalian systems either because constitutive expression is toxic to the cell or to examine the effect of the protein under different experimental conditions. Numerous systems have been devised for this, including systems based on the nuclear receptors, the bacterial Tet repressor, and rapamycin-induced dimerization (for review, see Saez et al. 1997).

The Tet system. The tetracycline-inducible (Tet) system is based on the prokaryotic Tet repressor (TetR), which binds to a specific 19-bp inverted repeat upstream of the Tet-responsive promoter in the *E. coli* transposon Tn*10* and represses it in the absence of tetracycline or its analogs (Fig. 11.9). Tet directly binds TetR and dissociates it from DNA, allowing the downstream gene to be turned on. This system has now been adapted for mammalian cells and there are several variations in use: In TetR-repressible pol II promoters, TetR-binding sites are placed near the transcription control elements (TATA and start site) and TetR represses the gene. Addition of tetracycline to the medium causes dissociation of TetR, resulting in activation of the promoter. However, mutants in Tet have been isolated where Tet causes binding rather than dissociation. Another version of this approach employs a fusion of TetR to an activation domain such as that of the herpes simplex virus (HSV) VP16 *trans*-activator. The TetR-VP16 chimera is placed under either a tissue-specific promoter or a constitutive one according to the experiment. Tet-binding sites are then situated upstream of a core promoter, driving expression of the cDNA. The continued presence of tetracycline inhibits expression by preventing binding of the TetR-VP16 chimera to the expression vector. However, removal of tetracycline from the medium permits TetR-VP16 binding and induces synthesis of the downstream cDNA (Gossen et al. 1995). A third and potentially more powerful version from a technical standpoint is the development of a TetR-KRAB domain fusion. The KRAB domain is a potent eukaryotic trans-repressor, which when placed upstream of a strong enhancer, such as that from CMV, can inhibit its expression. Thus, in the absence of tetracycline, the downstream cDNA is repressed. Addition of tetracycline causes dissociation of the TetR-KRAB fusion, permitting expression of the downstream cDNA (Deuschle et al. 1995). In this latter case, the tetracycline is acting as a typical induction drug and is conceptually similar to the prokaryotic systems described below. The Tet system is marketed by CLONTECH, Life Technologies Inc. and others. Consult Rossi and Blau (1998) for a recent review.

Rapamycin-inducible system. The rapamycin-inducible system is one of the newest and most effective systems for controlling the induction of a gene (reviewed in Rossi and Blau 1998). In this system, rapamycin is used as a dimerization agent that links a DNA-binding domain and an activation domain. A DNA-binding domain such as GAL4 is fused to the immunophilin FKBP12, while VP16 is fused to the FKBP-rapamycin-binding domain of FRP/RAFT1. Addition of rapamycin links the GAL4 to the VP16 and activates the GAL4-responsive promoter driving the cDNA (Ho et al. 1996). The reagents are available from ARIAD Pharmaceuticals.

Ecdysone-inducible systems. These systems are based on the highly inducible nuclear receptors (for review, see Beato et al. 1995). The ecdysone receptor is an insect nuclear receptor that binds the hormone ecdysone. It can form a heterodimer with the mammalian RXR receptor and in the presence of ecdysone or the analog muristerone A it will bind to its sites and activate transcription of a downstream cDNA. The ecdysone receptor functions in mammalian cells even though there is no mammalian equivalent, and it is highly inducible upon addition of hormone. In addition, the hormone is inert in mammalian cells in the absence of the receptor, minimizing any potentially toxic side effects. Typically, an expression vector bearing a selectable marker and synthesizing both ecdysone and RXR under a constitutive promoter is transfected into a mammalian cell line and stable integrants are selected. The vector expressing the cDNA under control of an ecdysone-responsive promoter is then transfected either stably using a second selectable marker or transiently. Double integrant cell lines are constructed and various clones are tested for expression of the cDNA in the presence of ecdysone. Invitrogen markets an ecdysone-inducible mammalian expression system (Invitrogen # K1000-01).

a vector bearing either the phage T7, T3, or SP6 promoter. Many companies market suitable vectors. For example, Promega sells the pGEM series bearing convergent T7 and SP6 promoters separated by a multiple cloning site. Phage RNA polymerases are then used to generate synthetic mRNAs, which are added to translation extracts. In certain variations, the transcription and translation are coupled (see Box 11.8). These systems often generate 10–100-ng quantities of protein sufficient for small-scale EMSA analysis and occasionally DNase I footprinting (Johnson and Herskowitz 1985), although the extracts appear to contain nucleases, which can hinder the footprinting in some cases. Improvements in translation efficiency can be accomplished by use of capped mRNA or addition of CITEs, such as the EMCV UTL sequence. These modifications have been used to generate quantities of

Box 11.8

In Vitro Transcription/Translation Systems

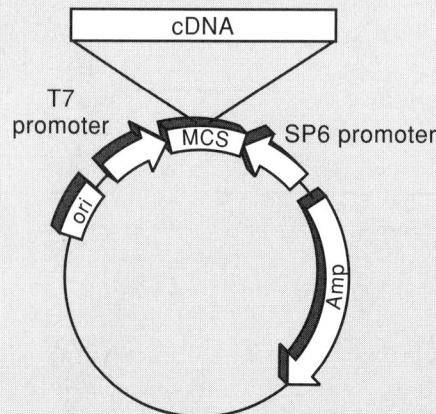

FIGURE 11.8. In vitro transcription/translation vector. (Amp) Ampicillin; (MCS) multiple cloning site.

The in vitro transcription/translation protocols employ DNA templates in which the cDNA has been cloned downstream of promoters for the phage SP6, T7, or T3 RNA polymerases (Fig. 11.8). Purified SP6, T7, or T3 RNA polymerase can then be added with nucleoside triphosphates (NTPs) to generate synthetic mRNAs. The synthetic mRNAs are then added to an in vitro translation system, such as a wheat germ lysate or a rabbit reticulocyte extract, to synthesize small amounts of the protein. If so desired, the protein can be radiolabeled internally by adding [^{35}S]methionine and cysteine to the translation reaction. The transcription and translation reactions can be coupled, and several companies, including Promega, Novagen, and Roche Molecular Biochemicals sell vectors and kits for this process. Promega's TnT system features a one-step process in which transcription and translation reagents are added together and incubated at 30°C for 60–90 minutes (Promega # L5020). Novagen's Single Tube Protein System 3 (Novagen # 70192-3) is a two-step process, where transcription is performed for 15 minutes at 30°C, and then translation mix is added to the tube and incubated for an additional 60–90 minutes. The translation extracts are very crude (the reticulocyte system contains 50 mg/ml of protein, mainly hemoglobin) and the resulting protein is relatively impure. Nevertheless, the technique is simple and can be used to synthesize a number of mutant derivatives simultaneously when desired. These mutant derivatives can be employed in some analytical biochemical experiments such as EMSA reactions, protein–protein interaction assays, and, in some cases, in vitro transcription. For larger quantities of protein or when a high degree of purity is desired, one must use the other expression systems described above.

Generally, the amount of protein synthesized in these systems is less than 100 ng. Furthermore, with mRNAs encoding large proteins, translation sometimes begins at internal methionines, generating a ladder of smaller products. The efficiency of translation can be increased by adding a 7-Me-guanosine cap during the transcription reaction or by cloning the cDNA downstream of the picornavirus UTL or other eukaryotic transcriptional enhancers. The vectors can be obtained from numerous commercial and academic sources (e.g., Novagen's pCITE-4 and -5 [cat. # 69913-3, 69914-3] and pT7Blue [cat. # 70174-3] vectors contain sequences that improve translation efficiency).

protein suitable for extensive DNA binding and dimer analysis by the *Xenopus* UBF Pol I transcription factor (McStay et al. 1991). One minor problem with the in vitro translation systems is that if large proteins are being generated, there is a tendency for the in vitro system to produce truncated products resulting from translation initiation at internal methionines. Thus, when the resulting protein is labeled with radioactive amino acids such as [^{35}S]methionine and cysteine, the internal translation products often appear as a ladder of lower-molecular-weight products. Again, capping and the use of CITEs minimize this problem. Another potential problem is that eukaryotic translation extracts may be contaminated with cofactors that enhance or inhibit a protein's activity. One example of this problem is TFIIA, which is thought to contaminate the reticulocyte extracts and influence binding by recombinant TBP (A. Berk, pers. comm.).

Small-scale transfections. If small amounts of a native protein are required for EMSA or affinity chromatography studies, and there is no concern over the potential for contaminants, cell-culture transfections can be used to generate extracts bearing the protein of interest. There are two primary cell lines that researchers have employed for such purposes: COS and 293T cells.

COS-1 and -7 cells have an integrated copy of the SV40 T antigen (Gluzman 1981). T antigen is required for SV40 DNA replication, and plasmid vectors encoding the SV40 origin are amplified considerably in COS cells (see Geisse et al. 1996). This amplification, in combination with strong viral enhancers such as cytomegalovirus (CMV), SV40, Rous sarcoma virus (RSV), or others, can lead to high levels of overexpression. Extracts from the transfected cells can be used in EMSA studies or can be used in affinity studies to analyze protein–protein interactions. The human 293T cell line, which also contains large T antigen, has become widely used recently because of its high transfection efficiency. Some studies have employed this line in place of COS cells to generate small amounts of protein for EMSA (Box 11.9).

Box 11.9

Mammalian Expression Vectors

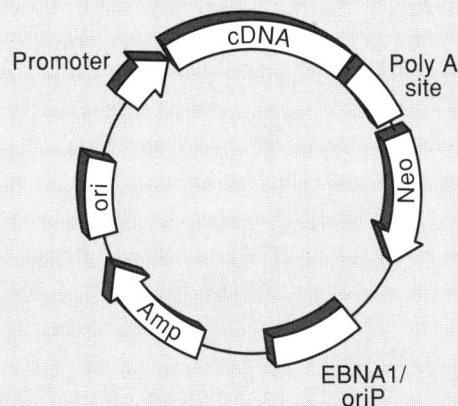

FIGURE 11.9. A typical mammalian expression vector bearing a neomycin (G418)-selectable marker. The ori contains the binding site for Ad1. The vector also contains a strong mammalian promoter/enhancer, a multiple cloning site for the cDNA, and a poly(A) termination site.

In addition to in vitro translation, small amounts of protein can be generated by transient or stable transfection (see Chapter 6) into mammalian cells. There are numerous expression vectors marketed by various companies for such applications. A typical mammalian expression vector is shown in Figure 11.9. Features of mammalian expression systems are given in Table 11.1 and Web site listings of some commercially available expression vectors in Table 11.2.

TABLE 11.2. *Web sources for commercial expression systems and vectors*

Vectors
http://www.clontech.com/clontech/Vectors/Vectors.html
http://www.promega.com/vectors/
http://www.invitrogen.com/vectors.html
http://www.neb.com/neb/frame_cat.html
http://www.apbiotech.com
http://www.stratagene.com/vectors/index.htm
http://www.pharmingen.com/
http://www.novagen.com/vectfram.html

Expression systems
http://www.clontech.com/clontech/Catalog/GeneExpressiontoc.html
http://www.promega.com/expression/invivo.htm
http://www.invitrogen.com/expressions/index.html
http://www.neb.com/neb/frame_cat.html
http://www.apbiotech.com/
http://www.stratagene.com/vectors/index2.htm
http://www.pharmingen.com/
http://www.novagen.com/sercfram.html

Synthesis and Purification of Macromolecular Complexes

Recent studies in the mammalian transcription field have shown that many polypeptides constituting the transcription machinery assemble into stable macromolecular complexes. TFIID, for example, is composed of TBP and numerous TAFs that generally copurify as a complex. It is often desirable to isolate the complexes in pure form and study their biochemical properties. The synthesis of large multisubunit protein complexes like TFIID often involves sophisticated expression systems. There are multiple approaches to isolating such complexes, and again the method chosen depends on the goals of the study. One approach is to synthesize the recombinant subunits individually and then assemble them into a complex in vitro. This approach permits various subcomplexes to be assembled and their functions examined biochemically. It also permits different mutant subunits to be incorporated into the complex and studied. The alternative approach is to devise methods for assembly of the complexes or subcomplexes in vivo coupled with methods that allow purification of the intact complexes.

The TAFs of TFIID have all been expressed as baculovirus-synthesized proteins, purified, and then assembled into recombinant TFIID in vitro. This approach was employed by Tjian and colleagues to examine and compare the ability of various TFIID subcomplexes to respond to different activators in vitro (Chen et al. 1994).

The complexes can also be assembled in vivo. Different subunits of a multicomponent complex can be cloned independently into baculovirus and co-infected into insect cells to generate large multisubunit proteins (see, e.g., Crute et al. 1991). Certain baculovirus shuttle vectors even contain divergent promoters that allow at least two proteins to be synthesized simultaneously from the same baculovirus vector. In the case of the yeast origin recognition complex (ORC), multiple baculoviruses were co-infected into insect cells and the complicated 6-subunit ORC could be purified intact from the infected cell extracts.

This permitted both the synthesis of large amounts of recombinant ORC and the ability to incorporate mutant subunits and examine their properties (Klemm et al. 1997).

Another approach for isolation of intact recombinant multisubunit complexes is to create an immunotagged (see Box 11.2, p. 372) version of a single subunit in the complex and introduce this gene into a retrovirus or a mammalian expression vector containing a selectable marker. After stable integrants are isolated by selection against the marker, colonies of cells are scaled up, extracts are assayed by immunoblotting against the tag to identify clonal populations expressing the recombinant protein at the desired levels. The protein and its associated factors can then be isolated by immunoaffinity chromatography. The advantage of immunoaffinity chromatography with the HA and FLAG tags is the ability to elute the proteins with peptide from the resin for subsequent biochemical analysis.

Examples of such a procedure include isolation of TFIID and the thyroid receptor co-activator complex. Berk and colleagues created a fusion of the HA epitope to the TBP subunit of TFIID. A retrovirus encoding the fusion was used to generate HeLa cell lines expressing immunotagged TBP. These cells were then grown in large scale, extracts were prepared, and intact TFIID bearing the tagged TBP subunit was purified by immunoaffinity chromatography against the HA tag (Zhou et al. 1992). Not only did this approach generate a facile method for TFIID purification, but it also allowed Berk and colleagues to test the ability of a mutant form of TBP, lacking its amino terminus, to assemble into the TAF complex and function in gene activation (Zhou and Berk 1995).

Roeder and colleagues fused the FLAG tag onto the thyroid receptor (TR) and introduced this chimeric gene into mammalian cells using a retroviral vector. After immunoaffinity chromatography of infected cell extracts over a FLAG monoclonal affinity column, the TR was assayed by SDS-PAGE and functional assays. TR was shown to associate with a series of TR-associated proteins (TRAPs) that were found to be transcriptional coactivators (Fondell et al. 1996).

Monoclonal antibodies to an exposed epitope on any protein can be employed for immunopurification of complexes. An excellent example was the use of monoclonal antibodies to co-precipitate the adenoviral E1A protein with its cellular targets. The retinoblastoma and p300 proteins, both important transcriptional regulators, were originally identified in this experiment (Harlow et al. 1986). In such cases, however, elution conditions that maintain the integrity of the complex must be employed if the complex is to be subjected to further biochemical analysis. For example, a monoclonal antibody to the carboxy-terminal domain (CTD) of Pol II has been used to purify active Pol II from crude extracts (Thompson and Burgess 1996). In this case, the antibody–CTD interaction was sensitive to polyols and Pol II could be eluted with moderate concentration of polyethylene glycol under native conditions. In situations where peptides against natural epitopes or native buffers cannot be employed, well-characterized immunotags provide the simplest approach for purifying a multiprotein complex from crude cell extracts. Immunotags must be positioned such that the epitope is exposed in the native complex and that it does not interfere with the protein's activity.

Choosing an Appropriate System

There are few guidelines for choosing an expression system because the requirements for particular research projects vary greatly. However, it is instructive to describe why and how the choices of expression systems were arrived at in the well-characterized GAL4 system. GAL4 is an 881-amino-acid protein from *S. cerevisiae* that activates transcription in almost all eukaryotic organisms tested to date. At its natural expression levels, GAL4 is produced in low amounts in yeast and is difficult to purify in bulk. To study its DNA-binding and *trans*-activation properties in vitro, the gene was initially cloned into an *E. coli* Tac expression vector. However, despite numerous attempts, the protein could not be synthesized intact in *E. coli*. An immunoblot of induced cell extracts revealed that GAL4 was being induced to high levels but that the protein was extensively degraded, likely intracellularly. At around the same time, yeast transformation experiments on GAL4 deletion mutants had demonstrated that GAL4, like most activators, could be subdivided into distinct DNA-binding and activation domains. Furthermore, the activation domain of heterologous activators could be fused to the minimal DNA-binding domain of GAL4 to create chimeric activators with new biological properties.

The first GAL4 derivative containing the intact DNA-binding domain was synthesized by taking advantage of a restriction site near amino acid 147. Thus, the gene encoding GAL4 amino acids 1–147 was cloned into a Tac expression vector and synthesized in *E. coli* to exceptionally high levels in a strain called Xa-90, which bears an episomal *lacIq* allele. The protein fragment was soluble and could be purified to homogeneity (Carey et al. 1989). Deletion analysis of GAL4(1–147) showed that the actual DNA-binding domain was located between amino acids 1 and 94 and that this region could be further subdivided into a DNA recognition domain containing amino acids 1–65 and a dimer domain located between amino acids 65 and 94. The DNA recognition region, which was a monomer in solution, although it retained DNA-dependent dimer capabilities, was crystallized and its structure was solved in 1992 (Marmorstein et al. 1992). One point of interest is that the GAL4 DNA-binding domain requires zinc for proper folding. Six cysteines between amino acids 11 and 38 chelate two zinc ions to fold into a structure containing two perpendicularly disposed alpha-helices, one of which binds DNA whereas the other serves as a protein scaffold. Although bacterial medium contains some zinc, supplementing it with 10–100 μM $ZnCl_2$ or zinc acetate during induction with IPTG substantially increased yields of active protein and, in some instances, improved solubility dramatically. Certain other metals such as cadmium chloride could substitute for zinc.

The availability of small GAL4-derived activators such as GAL4-VP16, containing amino acids 1–147 of GAL4 and 411–490 of VP16, made the synthesis of these chimeric proteins in *E. coli* an easy task (Carey et al. 1989). Even natural GAL4 derivatives containing amino acids 1–147 fused to the activation domain 768–881 were shown to be stably expressed and soluble in *E. coli* (Reece et al. 1993). Many of the initial in vitro transcription studies employed these chimeric derivatives because they were easier to synthesize and manipulate.

Later biochemical studies required intact GAL4. The intact gene was cloned independently by two groups into baculovirus and yeast expression vectors (Parthun and Jaehning 1990; 1992; Kang et al. 1993). Both systems were successful in generating large enough quantities of materials for biochemical analysis. Unfortunately, GAL4 was somewhat insoluble in the baculovirus system, even when detergents were used to aid solubility. However, the high degree of overexpression made the amount of soluble protein that was produced satisfactory for the analyses being performed. By synthesizing the intact protein, it was found that intact GAL4, but not the DNA-binding domain alone, was able to bind cooperatively to the four sites constituting the UAS_G (Kang et al. 1993).

What was learned from this case study? As discussed above, the set of expression systems chosen for the purification of GAL4 and its derivatives was based on the most pressing needs of the research project at hand. The *E. coli* Tac system was used first because it is inexpensive, straightforward, and not too time-consuming. GAL4 derivatives were made in this system when the original full-length GAL4 could not be otherwise synthesized. Researchers moved on to completely different expression systems, the baculovirus and yeast systems, only when it was realized that full-length GAL4 was needed to further our understanding of its interactions with DNA and with itself. Therefore, choosing a system is not so much a trial-and-error process, but rather a careful analysis of the costs, benefits, and results of each available system.

REFERENCES

Abate C., Luk D., and Curran T. 1991. Transcriptional regulation by Fos and Jun in vitro: Interaction among multiple activator and regulatory domains. *Mol. Cell. Biol.* **11:** 3624–3632.

Amann E., Brosius J., and Ptashne M. 1983. Vectors bearing a hybrid trp-lac promoter useful for regulated expression of cloned genes in *Escherichia coli. Gene* **25:** 167–178.

Amara J.F., Clackson T., Rivera V.M., Guo T., Keenan T., Natesan S., Pollock R., Yang W., Courage N.L., Holt D.A., et al. 1997. A versatile synthetic dimerizer for the regulation of protein-protein interactions. *Proc. Natl. Acad. Sci.* **20:** 10618–10623.

Ausubel F.M., Brent R.E., Kingston E., Moore D.D., Seidman J.G., Smith J.A., and Struhl K., eds. 1994. *Current protocols in molecular biology.* John Wiley and Sons, New York.

Beato M., Herrlich P., and Schutz G. 1995. Steroid hormone receptors: Many actors in search of a plot. *Cell* **83:** 851–857.

Blanar M.A. and Rutter W.J. 1992. Interaction cloning: Identification of a helix-loop-helix zipper protein that interacts with c-Fos. *Science* **256:** 1014–1018.

Brent R. and Ptashne M. 1981. Mechanism of action of the lexA gene product. *Proc. Natl. Acad. Sci.* **78:** 4204–4208.

Burgess RR. 1996. Purification of overproduced *Escherichia coli* RNA polymerase sigma factors by solubilizing inclusion bodies and refolding from Sarkosyl. *Methods Enzymol.* **273:** 145–149.

Burns J.C., Friedmann T., Driever W., Burrascano M., and Yee J.K. 1993. Vesicular stomatitis virus G glycoprotein pseudotyped retroviral vectors: Concentration to very high titer and efficient gene transfer into mammalian and nonmammalian cells. *Proc. Natl. Acad. Sci.* **90:** 8033–8037.

Carey M., Kakidani H., Leatherwood J., Mostashari F., and Ptashne M. 1989. An amino-terminal fragment of GAL4 binds DNA as a dimer. *J. Mol. Biol.* **209:** 423–432.

Carroll M.W. and Moss B. 1997. Poxviruses as expression vectors. *Curr. Opin. Biotechnol.* **8:** 573–577.

Chang C., Wang C., DeLuca H.F., Ross T.K., and Shih C.C. 1992. Characterization of human androgen receptor overexpressed in the baculovirus system. *Proc. Natl. Acad. Sci.* **89:** 5946–5950.

Chen F.E., Huang D.B., Chen Y.Q., and Ghosh G. 1998. Crystal structure of p50/p65 heterodimer of transcription factor NF-κB bound to DNA. *Nature* **391:** 410–413.

Chen J.L., Attardi L.D., Verrijzer C.P., Yokomori K., and Tjian R. 1994. Assembly of recombinant TFIID reveals differential coactivator requirements for distinct transcriptional activators. *Cell* **79:** 93–105.

Cho Y., Gorina S., Jeffrey P.D., and Pavletich N.P. 1994. Crystal structure of a p53 tumor suppressor-DNA complex: Understanding tumorigenic mutations [see comments]. *Science* **265:** 346–355.

Coligan J.E., Dunn B.M., Ploeqh H.L., Speicher D.W., and Wingfield P.T., eds. 1996. *Current protocols in protein science* (series ed. Benson Chanda V.) John Wiley and Sons, New York.

Crute J.J., Bruckner R.C., Dodson M.S., and Lehman I.R. Herpes simplex-1 helicase-primase. Identification of two nucleoside triphosphatase sites that promote DNA helicase action. *J. Biol. Chem.* **266:** 21252–21256.

Current Opinion in Biotechnology (ISSN0958-1669). Published bimonthly by Current Biology Ltd., 34-42 Cleveland St., London W1P#GLB United Kingdom.

de Boer H.A., Comstock L.J., and Vasser M. 1983. The tac promoter: A functional hybrid derived from the trp and lac promoters. *Proc. Natl. Acad. Sci.* **80:** 21.

DeJong J. and Roeder R.G. 1993. A single cDNA, hTFIIA/alpha, encodes both the p35 and p19 subunits of human TFIIA. *Genes Dev.* **7:** 2220–2234.

Deuschle U., Meyer W.K., and Thiesen H.J. 1995. Tetracycline-reversible silencing of eukaryotic promoters. *Mol. Cell. Biol.* **15:** 1907–1914.

De Vos P., Schmitt J., Verhoeven G., and Stunnenberg H.G. 1994. Human androgen receptor expressed in HeLa cells activates transcription in vitro. *Nucleic Acids Res.* **22:** 1161–1166.

Dubendorff J.W. and Studier F.W. 1991. Controlling basal expression in an inducible T7 expression system by blocking the target T7 promoter with lac repressor. *J. Mol. Biol.* **219:** 45–59.

Elroy-Stein O., Fuerst T.R., and Moss B. 1989. Cap-independent translation of mRNA conferred by encephalomyocarditis virus 5′ sequence improves the performance of the vaccinia virus/bacteriophage T7 hybrid expression system. *Proc. Natl. Acad. Sci.* **86:** 6126–6130.

Finkelstein A., Kostrub C.F., Li J., Chavez D.P., Wang B.Q., Fang S.M., Greenblatt J., and Burton Z.F. 1992. A cDNA encoding RAP74, a general initiation factor for transcription by RNA polymerase II. *Nature* **355:** 464–467.

Fondell J.D., Ge H., and Roeder RG. 1996. Ligand induction of a transcriptionally active thyroid hormone receptor coactivator complex. *Proc. Natl. Acad. Sci.* **93:** 8329–8333.

Foreman P.K. and Davis R.W. 1994. Cloning vectors for the synthesis of epitope-tagged, truncated and chimeric proteins in *Saccharomyces cerevisiae. Gene* **144:** 63–68.

Fuerst T.R., Niles E.G., Studier F.W., and Moss B. 1986. Eukaryote transient-expression system based on recombinant vaccinia virus that synthesizes bacteriophage T7 TNA. *Proc. Natl. Acad. Sci.* **83:** 8122–8126.

Ge H. and Roeder R.G. 1994. Purification, cloning, and characterization of a human coactivator, PC4, that mediates transcriptional activation of class II genes. *Cell* **78:** 513–523.

Geisse S., Gram H., Kleuser B., and Kocher H.P. 1996. Eukaryotic expression systems: A comparison. *Protein Expr. Purif.* **8:** 271–282.

Ghosh G., van Duyne G., Ghosh S., and Sigler P.B. 1995. Structure of NF-kappa B p50 homodimer bound to a kappa B site [see comments]. *Nature* **373:** 303–310.

Glover J.N. and Harrison S.C. 1995. Crystal structure of the heterodimeric bZIP transcription factor c-Fos-c-Jun bound to DNA. *Nature* **373:** 257–261.

Gluzman Y. 1981. SV40-transformed simian cells support the replication of early SV40 mutants. *Cell* **23:** 175–182.

Gossen M., Freundlieb S., Bender G., Muller G., Hillen W., and Bujard H. 1995. Transcriptional activation by tetracyclines in mammalian cells. *Science* **268:** 1766–1769.

Ha I., Lane W.S., and Reinberg D. 1991. Cloning of a human gene encoding the general transcription initiation factor IIB. *Nature* **352:** 689–695.

Hahn S., Buratowski S., Sharp P.A., and Guarente L. 1989. Isolation of the gene encoding the yeast TATA binding protein TFIID: A gene identical to the SPT15 suppressor of Ty element insertions. *Cell* **58:** 1173–1181.

Harlow E., Whyte P., Franza B.R. Jr., and Schley C. 1986. Association of adenovirus early-region 1A proteins with cellular polypeptides. *Mol. Cell. Biol.* **6:** 1579–1589.

Ho S.N., Biggar S.R., Spencer D.M., Schreiber S.L., and Crabtree G.R. 1996. Dimeric ligands define a role for transcriptional activation domains in reinitiation. *Nature* **382:** 822–826.

Hockney R.C. 1994. Recent developments in heterologous protein production in *Escherichia coli. Trends Biotechnol.* **12:** 456–463.

Hofmann A., Nolan G.P., and Blau H.M. 1996. Rapid retroviral delivery of tetracycline-inducible genes in a single autoregulatory cassette [see comments]. *Proc. Natl. Acad. Sci.* **93:** 5185–5190.

Hollenberg C.P. and Gellissen G. 1997. Production of recombinant proteins by methylotrophic yeasts. *Curr. Opin. Biotechnol.* **8:** 554–560.

Hope I.A. and Struhl K. 1985. GCN4 protein, synthesized in vitro, binds HIS3 regulatory sequences: Implications for general control of amino acid biosynthetic genes in yeast. *Cell* **43:** 177–188.

Hori R. and Carey M. 1997. Protease footprinting analysis of ternary complex formation by human TFIIA. *J. Biol. Chem.* **272:** 1180–1187.

Jackson S.P.; MacDonald J.J.; Lees-Miller S., and Tjian R. 1990. GC box binding induces phosphorylation of Sp1 by a DNA-dependent protein kinase. *Cell* **63:** 155–165.

Johnson A.D. and Herskowitz I. A repressor (MAT alpha 2 Product) and its operator control expression of a set of cell type specific genes in yeast. *Cell* 1985 **42:** 237–247.

Kang T., Martins T., and Sadowski, I. (1993). Wild type GAL4 binds cooperatively to the GAL1-10 UASG in vitro. *J. Biol. Chem.* **268:** 9629–9635.

Klemm R.D., Austin R.J., and Bell S.P. 1997. Coordinate binding of ATP and origin DNA regulates the ATPase activity of the origin recognition complex. *Cell* **88:** 493–502.

Klemm R.D., Goodrich J.A., Zhou S., and Tjian R. 1995. Molecular cloning and expression of the 32-kDa subunit of human TFIID reveals interactions with VP16 and TFIIB that mediate transcriptional activation. *Proc. Natl. Acad. Sci.* **2:** 5788–5792.

Lai H.C., Qian B., and Tu C.P. 1989. Characterization of a variant rat glutathione S-transferase by cDNA expression in *Escherichia coli. Arch. Biochem. Biophys.* **273:** 423–432.

Lilie H., Schwarz E., and Rudolf R. 1998. Advances in refolding of proteins produced in *E. coli. Curr. Opin. Biotechnol.* **9:** 497–501.

Lin K.H. and Cheng S.Y. 1991. An efficient method to purify active eukaryotic proteins from the inclusion bodies in *Escherichia coli. Biotechniques* **11:** 748, 750, 752–753.

Lindemann D., Patriquin E., Feng S., and Mulligan R.C. 1997. Versatile retrovirus vector systems for regulated gene expression in vitro and in vivo. *Mol. Med.* **3:** 466–476.

Luisi B.F., Xu W.X., Otwinowski Z., Freedman L.P., Yamamoto K.R., and Sigler P.B. 1991. Crystallographic analysis of the interaction of the glucocorticoid receptor with DNA [see comments]. *Nature* **352:** 497–505.

Magari S.R., Rivera V.M., Iuliucci J.D., Gilman M., and Cerasoli F., Jr. 1997. Pharmacologic control of a humanized gene therapy system implanted into nude mice. *J. Clin. Investig.* **100:** 2865–2872.

Makrides S.C. 1996. Strategies for achieving high-level expression of genes in *Escherichia coli. Microbiol. Rev.* **60:** 512–538.

Marmorstein R., Carey M., Ptashne M., and Harrison S.C. 1992. DNA recognition by GAL4: Structure of a protein-DNA complex [see comments]. *Nature.* **356:** 408–414.

Marshak D.R., Kadonaga J.T., Burgess R.R., Knuth M.W., Brennan W.A., Jr., and Lin S.-H. 1996. *Strategies for protein purification and characterization: A laboratory manual.* Cold Spring Harbor Laboratory Press, Cold Spring Harbor, New York.

McStay B., Hu C.H., Pikaard C.S., Reeder R.H. 1991. xUBF and Rib 1 are both required for formation of a stable polymerase I promoter complex in *X. laevis. EMBO J.* **10:** 2297–2303.

Naldini L. 1998. Lentiviruses as gene transfer agents for delivery to non-dividing cells. *Curr. Opin. Biotechnol.* **9:** 457–463.

Naviaux R.K., Costanzi E., Haas M., and Verma I.M. 1996. The pCL vector system: Rapid production of helper-free, high-titer, recombinant retroviruses. *J. Virol.* **70:** 5701–5705.

Ozer J., Moore P.A., Bolden A.H., Lee A., Rosen C.A., and Lieberman P.M. 1994. Molecular cloning of the small (γ) subunit of human TFIIA reveals functions critical for activated transcription. *Genes Dev.* **8:** 2324–2335.

Parthun M.R. and Jaehning J.A. 1990. Purification and characterization of the yeast transcriptional activator GAL4. *J. Biol. Chem.* **265:** 209–213.

———1992. A transcriptionally active form of GAL4 is phosphorylated and associated with GAL80. *Mol. Cell. Biol.* **12:** 4981–4987.

Pear W.S., Nolan G.P., Scott M.L., and Baltimore D. 1993. Production of high-titer helper-free retroviruses by transient transfection. *Proc. Natl. Acad. Sci.* **90:** 8392–8396.

Peterson M.G., Tanese N., Pugh B.F., and Tjian R. 1990. Functional domains and upstream activation properties of cloned human TATA binding protein [published erratum appears in *Science* 1990 Aug 24; **249:** 844]. *Science* **248:** 1625–16230.

Pfeifer T.A. 1998. Expression of heterologous proteins in stable insect culture. *Curr. Opin. Biotechnol.* **9:** 518–521.

Possee R.D. 1997. Baculoviruses as expression vectors. *Curr. Opin. Biotechnol.* **8:** 569–572.

Ramsey-Ewing A. and Moss B. 1996. Recombinant protein synthesis in Chinese hamster ovary cells using a vaccinia virus/bacteriophage T7 hybrid expression system. *J. Biol. Chem.* **271:** 16962–16966.

Ranish J.A., Lane W.S., and Hahn S. 1992. Isolation of two genes that encode subunits of the yeast transcription factor IIA. *Science* **255:** 1127–1129.

Reece R. J., Rickles R. J., and Ptashne M. 1993. Overproduction and single-step purification of GAL4 fusion proteins from *Escherichia coli. Gene* **126:** 105–107.

Richardson C.D. 1995. Baculovirus expression protocols. In *Methods in molecular biology,* J.M. Walker, ed. Totowa, New Jersey: Humana Press, p. 418.

Rosenberg A.H., Lade B.N., Chui D.S., Lin S.W., Dunn J.J., and Studier F.W. 1987. Vectors for selective expression of cloned DNAs by T7 RNA polymerase. *Gene* **56:** 125–135.

Rossi F.M. and Blau H.M. 1998. Recent advances in inducible gene expression systems. *Curr. Opin. Biotechnol.* **9:** 451–456.

Saez E., No D., West A., and Evans R.M. 1997. Inducible gene expression in mammalian cells and transgenic mice. *Curr. Opin. Biotechnol.* **8:** 608–616.

Scopes Robert K. 1994. *Protein purification: Principles and practice,* 3rd ed. New York: Springer-Verlag.

Shi P.Y., Maizels N., and Weiner A.M. 1997. Recovery of soluble, active recombinant protein from inclusion bodies. *Biotechniques* **23:** 1036–1038.

Singh S.V., Leal T., Ansari G.A., and Awasthi Y.C. 1987. Purification and characterization of glutathione S-transferases of human kidney. *Biochem. J.* **246:** 179–186.

Smith D.B. and Johnson K.S. 1988. Single-step purification of polypeptides expressed in *Escherichia coli* as fusions with glutathione S-transferase. *Gene* **67:** 31–40.

Smith G.E., Summers M.D., and Fraser M.J. 1983. Production of human β interferon in insect cells infected with a baculovirus expression vector. *Mol. Cell. Biol.* **3:** 2156–2165.

Sopta M., Burton Z.E., and Greenblatt J. 1989. Structure and associated DNA-helicase activity of a general transcription initiation factor that binds to RNA polymerase II. *Nature* **341:** 410–414

Studier F.W., and Moffatt B.A. 1986. Use of bacteriophage T7 RNA polymerase to direct selective high-level expression of cloned genes. *J. Mol. Biol.* **189:** 113–130.

Studier F. W., Rosenberg A. H., Dunn J. J., and Dubendorff J. W. 1990. Use of T7 RNA polymerase to direct expression of cloned genes. *Methods Enzymol.* **185:** 60–89.

Studier F.W. 1991. Use of bacteriophage T7 lysozyme to improve an inducible T7 expression system. *J. Mol. Biol.* **219:** 37–44.

Sudbery P.E. 1996. The expression of recombinant proteins in yeasts. *Curr. Opin. Biotechnol.* **7:** 517–524.

Sun X., Ma D., Sheldon M., Yeung K., and Reinberg D. 1994. Reconstitution of human TFIIA activity from recombinant polypeptides: A role in TFIID-mediated transcription. *Genes Dev.* **8:** 2336–2348.

Tiu W.U., Davern K.M., Wright M.D., Board P.G., and Mitchell G.F. 1988. Molecular and serological characteristics of the glutathione S-transferases of *Schistosoma japonicum* and *Schistosoma mansoni. Parasite Immunol.* **10:** 693–706.

Thompson N.E. and Burgess R.R. 1996. Immunoaffinity purification of RNA polymerase II and transcription factors using polyol-responsive monoclonal antibodies. *Methods Enzymol.* **274:** 513–526.

Vialard J.E., Arif B.M., and Richardson C.D. 1995. Introduction to the molecular biology of baculoviruses. *Methods Mol. Biol.* **39:** 1–24.

Wyatt L.S., Moss B., and Rozenblatt S. 1995. Replication-deficient vaccinia virus encoding bacteriophage T7 RNA polymerase for transient gene expression in mammalian cells. *Virology* **210:** 202–205.

Xie Y.B., Sui Y.P., Shan L.X., Palvimo J.J., Phillips D.M. and Janne O.A. 1992. Expression of androgen receptor in insect cells. Purification of the receptor and renaturation of its steroid- and DNA-binding functions. *J. Biol. Chem.* **267:** 4939–4948.

Zhou Q. and Berk A.J. 1995. The yeast TATA-binding protein (TBP) core domain assembles with human TBP-associated factors into a functional TFIID complex. *Mol. Cell. Biol.* **15:** 534–539.

Zhou Q., Lieberman P.M., Boyer T.G., and Berk A.J. 1992. Holo-TFIID supports transcriptional stimulation by diverse activators and from a TATA-less promoter. *Genes Dev.* **6:** 1964–1974.

LISTED CATALOGS

Amersham Pharmacia Biotech, Inc. BioDirectory 1999. Call 1(800) 526-3593. Address: 800 Centennial Ave., P.O. Box 1327, Piscataway, NJ 08855-1327.

ARIAD Pharmaceuticals, Inc. Call (617) 494-0400. Address: 26 Landsdowne Street, Cambridge, MA 02139.

Bio-Rad Laboratories. Life Science Catalog 1998/99. Call 1(800) 424-6723. Address: 2000 Alfred Nobel Drive, Hercules, CA 94547.

CLONTECH Laboratories Inc. Call 1(800) 662-2566. Address: 1020 E. Meadow Cir., Palo Alto, CA 94303-42300.

Eastman Kodak Co. Scientific Imaging Systems. 4 Science Park, New Haven, CT 06511. Call 1(877) SIS-HELP (toll free in U.S. and Canada) or see Web site http://www.kodak.com.

Invitrogen, Inc. 1997 Product Catalog. Call 1(800) 955-6288. Address: 1600 Faraday Ave., Carlsbad, CA 92008.

Life Technologies Inc. (GIBCO BRL) 1996 Catalog. Call 1(800) 828-6686. Address: Grand Island, NY.

New England Biolabs, Inc. 1996-97 Product Catalog. Call 1(800) 632-5227. Address: 32 Tozer Road, Beverly, MA 01915-5599,

Novagen, Inc. 1998-99 Catalog. Call 1(800) 526-7319. Address: 601 Science Drive, Madison, WI 53711.

PharMingen, Inc. 1999 Research Products Catalog. Call 1(800) 848-6227. Address: 10975 Torreyana Road, San Diego, CA 92121. Baculovirus information provided in PharMingen publication by Gruenwald S. and Heitz J. (1993). Baculovirus Expression Vector System: Procedures and Methods Manual. 2nd Edition

Promega, Inc. 1999 Catalog. In the U.S., call 1(800) 356-9526. Address: 2800 Woods Hollow Road, Madison, WI 53711-5399.

QIAGEN, Inc. Product Guide 1996. In the U.S., call 1(800) 426-8157. Address: 9600 De Soto Avenue, Chatsworth, CA 91311.

Roche Molecular Biochemicals Catalog 1999. Call 1(800) 262-1640. Address: P.O. Box 50414, Indianapolis, IN 46250-0414.

Stratagene, Inc. 1999 Product Catalog. Call 1(800) 424-5444. Address: 11011 N. Torrey Pines Rd., La Jolla, CA 92037.

CHAPTER 12

Identifying and Characterizing Transcription Factor Domains

Important issues

- *Identification of a regulatory factor's functional domains is important for understanding its mechanism.*

- *Domain swaps are a useful approach for determining the function of a domain.*

- *Site-directed mutagenesis is useful in particular cases when structures are available to make specific experimental predictions.*

- *Context-dependent interactions must be considered when analyzing the domains of a regulatory factor.*

- *The physiological relevance of transcription factor interaction can be validated in a mammalian system without the use of "classical" genetics.*

INTRODUCTION

Understanding the molecular details of the RNA polymerase II (Pol II) transcription complex assembly is one of the most challenging problems in the field of mammalian gene regulation. In large part, this challenge is due to the technical limitations in deciphering interactions among transcription factors and DNA occurring within multicomponent, macromolecular assemblies. However, to obtain a detailed picture of these interactions, mutagenesis must be combined with functional studies. First, the players in a particular regulatory context, such as the DNA elements regulating expression, the activators and repressors controlling a gene, and the cofactors necessary for communicating these effects to the general machinery and chromatin, must be identified. This issue has been discussed at length in earlier chapters. Second, an attempt must be made to understand the domains of the factors that carry out a particular function such as DNA binding and activation, a point we cover in this chapter. After completing such an analysis, one would begin to understand the precise molecular mechanism of the interactions through detailed mutagenesis and functional studies. What are the targets of the activation domains, what amino acids are involved in the interactions, and how does the interaction affect gene expression? Obtaining the complete picture allows a description of the process of regulation in molecular detail as well as employment of that information to manipulate gene expression for experimental or therapeutic purposes.

The typical strategy for domain analysis in mammalian cells has been to characterize activators or general factors by coupling mutagenesis with cell culture transfection and DNA–protein or protein–protein interaction studies. Mutants deficient in certain interactions in vitro are correlated with altered transcription rates in vitro and in vivo. However, the inability to characterize further the interactions using classical genetics leads to uncertain conclusions. New advances in the development of mammalian quasi-genetic systems are lending credence to the biochemical studies and paving the way to the study of complicated protein–protein interactions in macromolecular assemblies.

This chapter presents an overview of general approaches for identifying functional domains of mammalian regulatory proteins. A special emphasis is placed on the concepts and methodology employed in structure–function analysis, particularly recent studies involving mutagenesis of the general transcription machinery. In such cases, because the structure is known, very specific mechanistic inferences can be made based on the results of the mutagenesis. Experimental details relating to the mutagenesis are covered in Chapter 7. Chapter 13 examines the details and theory for studying DNA binding. Chapters 14 and 15 concern the establishment of sophisticated systems and assays for studying individual activators and their effects on transcription complex assembly; in the ensuing discussion, we refer to some of the concepts in these chapters.

CONCEPTS AND STRATEGIES: DEFINING DOMAINS

Basic Mutagenesis Principles

As described in Chapter 7, there are several facile methods available for mutagenesis of a DNA sequence, be it a promoter or a coding region. For proteins, deletion analysis using restriction endonucleases, PCR, Exonuclease III (*Exo*III), or BAL31 can be employed as a starting point to provide information on the global positioning of domains within a regulatory protein. Once deletion analysis has identified a potentially interesting domain, that region is subjected to domain swap analysis and/or site-directed mutagenesis.

In domain swap analysis, the domain of one regulatory protein is swapped for that of another to create a chimeric regulator with new biological properties. The ability of a domain to function in a heterologous context demonstrates that the domain contains a sufficient amount of sequence information to execute a particular function. Domain swap analysis is feasible because of the modular design of eukaryotic regulatory proteins as discussed in Chapter 1. It has been most successful when applied to activators and repressors.

Site-directed mutagenesis is a variation of point mutation analysis where only defined amino acids, either individually or in clusters, are altered within a protein. Site-directed mutational analysis allows one to identify the precise amino acids within the domain that are necessary for function. Site-directed mutagenesis can be used to delineate further the regions identified by deletions or can be used in combination with a crystal structure to test models for function.

In general, there are two widely used variations of site-directed mutagenesis in proteins: alanine and radical substitution. In alanine substitution, the codon for alanine, a small hydrophobic amino acid with a methyl group side chain, is introduced in place of the pre-existing amino acid (Fig. 12.1, cf. A and B) (Cunningham and Wells 1989). The alanine lacks carbons beyond the C_β of the side chain and is considered to represent the equivalent of a side-chain truncation or loss-of-contact mutant. When placed on the solvent-exposed surface of a protein, the alanine is assumed to have little overall (i.e., energetically neutral) effect on the structure. In some cases, the energetic consequences of a loss-of-contact mutant may be too small to produce a phenotype. In these instances, radical substitution is employed. In radical substitution, a chemically diverse amino acid with a protruding side chain is inserted in place of the wild-type residue (e.g., arginine to aspartic acid) (Fig. 12.1, cf. A and C). The concept is to insert an amino acid that both removes the original contact

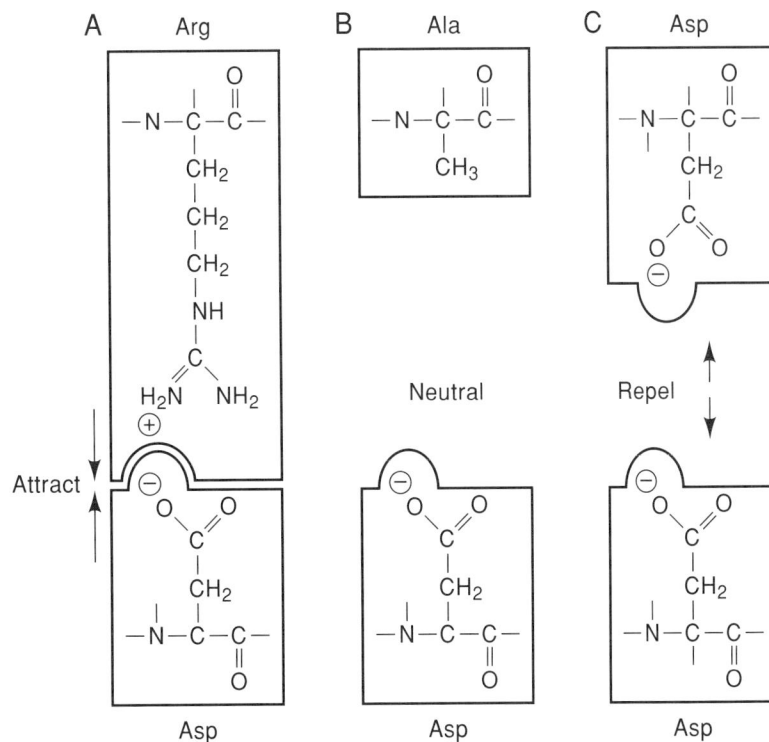

FIGURE 12.1. Alanine scanning versus radical substitution.

and sterically or chemically disrupts (i.e., a charge-repelling interaction) binding of any other protein to that surface.

Site-directed mutagenesis is most powerful when combined with crystal structures or used to mutagenize conserved domains of related factors. For example, if the crystal structure of a DNA-binding domain is known, then site-directed mutagenesis can be employed to remove critical amino acid–nucleotide interactions to validate their importance in solution assays or in vivo. Alternatively, if the structure of a related transcription factor is known, the crystal structure of the first factor can be used to guide a mutagenesis study. By mutagenizing residues occupying similar positions in the related proteins, one can determine if the modes of binding are similar. In such a case, it is often necessary to build models of the protein using the α-carbon backbone trace of the crystal structure using structural computer programs such as Insight II (Molecular Simulations, Inc.). In the absence of a structure, the sequence conservation of a protein can be used to identify important amino acids. For example, two related regulatory proteins might share sequences that perform critical functions. If mutants in one protein are known to inactivate a domain, similar mutants can be employed to study function in the other family member. We have included a simple and straightforward protocol for site-directed mutagenesis at the end of this chapter because of the importance this method plays in understanding structure–function relationships.

One question to pose at the outset is whether a structure–function analysis will yield valuable information regarding the system being studied or reveal new insights into the mechanism of regulation. Numerous studies on activators and repressors have been performed that have never progressed past the initial stages of an analysis. The reason is that subsequent stages are, from a technical standpoint, much more difficult to execute, and even in the transcription field, these steps have not been well defined. However, given the large body of basic information on activators, additional studies will add to our knowledge of mechanism when they either reveal a novel mode of action or are part of a concerted effort to decipher the biology of promoter function in a regulated system.

Domains of a Gene Activator

A typical activator can be divided into several functional domains, each one of which can be further subdivided. (The reader should be advised that the concept of domain employed here refers to a region with a particular function. The classical definition of domain, as defined through analysis of the crystal structure or proteolysis—a region of polypeptide that can fold independently—has not been rigorously established in many cases.) In general, one of these domains targets the protein to a DNA site within the appropriate promoter context, whereas another mediates activation of transcription, and yet others play roles dependent on the cell context (for review, see Ptashne 1992). The functional domains may be part of the same polypeptide or they may be present in different subunits of an activator complex. In some cases, two functions are closely intermingled and can only be delineated by detailed site-directed mutagenesis. Such is the case with certain activation functions in MyoD and the glucocorticoid receptor (Schena et al. 1989; Davis et al. 1990).

The DNA-binding domain is often further subdivided into two or more distinct subdomains (Fig. 12.2). One subdomain recognizes and tethers the protein to a specific DNA site, another mediates oligomerization (i.e., dimer, tetramer, etc.), and others mediate cooperative binding with nearby activators. As discussed in Chapters 1 and 13, it is the sum of these interactions that allows an activator to bind to the correct sequence within the appropriate

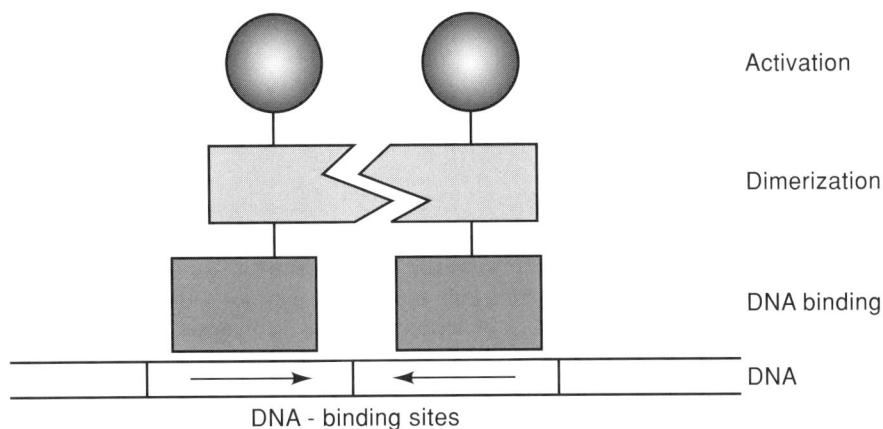

FIGURE 12.2. Domains of a typical activator.

promoter context. We know much about the structure of DNA-binding domains because many of these domains fall into distinct families whose structures have been solved by X-ray crystallography or nuclear magnetic resonance (NMR) (see Chapters 1 and 13).

The activation domain, on the other hand, is responsible for stimulating transcription. Activation domains have been described in some detail in Chapter 1 and mediate their action by interacting directly with the transcriptional machinery, or with coactivators that act as intermediaries in interactions with the general machinery or chromatin. Although a sequence-specific activator bears an individual activation domain, it most often acts synergistically in conjunction with other activators in the form of an enhanceosome (Carey 1998). Domains involved in enhanceosome assembly (e.g., cooperative DNA binding) could be incorrectly interpreted or scored as activation domains in certain mutational analyses (e.g., when using natural promoters as reporter templates) because their removal would abolish enhanceosome assembly and, hence, transcription. Detailed biochemical experiments and transfection assays designed to compare mutants on natural and model promoters are necessary to distinguish among these possibilities. Unlike DNA-binding domains, little is known of the structure of activation domains, making it difficult to classify them into discrete families.

Separating DNA-binding and Activation Domains of an Activator

General Considerations

The initial strategy for dissecting an activator in a mammalian system is to develop *trans*-activation and DNA-binding assays for the activator being analyzed. In the *trans*-activation assay, a natural promoter, known to be responsive to the activator, or an artificial promoter, bearing tandem binding sites for the activator adjacent to a minimal TATA box, is used to drive expression of luciferase, chloramphenicol acetyltransferase (CAT), or some other easily measurable reporter gene (see Chapter 5 for details). In cases where a precise measure of mRNA synthesis is necessary, primer extension, RNase protection, S1 nuclease, or other suitable methods can be employed (see Chapter 4). The activator and its mutants are then placed under control of a strong promoter/enhancer and co-transfected with the reporter into a suitable cell line. Because mutagenesis can dramatically alter the stability of a protein, it is crucial in the transfection studies to establish an assay where the levels of the protein can be measured (i.e., immunoblotting, immunoprecipitation, or EMSA). In the event that a phenotype is observed, one must be able to confirm whether or not the wild-type protein and its mutants were synthesized at comparable levels. If an antibody against the protein being studied is not available, then immunotagging (e.g., with the FLAG or hemagglutinin [HA] epitopes) as described in Chapter 11 can provide a means of measuring the protein levels by immunoprecipitation or blotting.

It is important to point out that the "loss of function" associated with mutation of a protein in a *trans*-activation or DNA-binding assay does not always imply that the region mutated mediates the effect. The rationale is that although the mutation could directly affect DNA binding or activation by removing critical amino acids, it might also result in a structural perturbation that indirectly alters activity. To confirm that the mutation has directly affected an important function, additional experiments must be performed. For example, a domain swap could be used to show that the domain alone possesses DNA-binding or *trans*-activation capabilities. Alternatively, as knowledge of the domain structure increases, studies coupling the mutagenesis with biochemical experiments will reveal additional insights into the function.

DNA Binding

A first approximation of the DNA-binding domain is the region of a protein that when deleted abolishes DNA binding in a biochemical assay. The binding assay, usually a DNase I footprint or electrophoretic mobility shift, could be performed with either in-vitro-translated proteins, protein synthesized in *Escherichia coli*, or proteins from transfected cell extracts. The minimal DNA-binding domain is defined as the smallest polypeptide fragment that binds DNA in vitro (see Chapter 11).

After the ability of the domain to bind specifically has been established, a key issue is whether the biochemical situation accurately reflects what occurs in the cell. Loss of a dimerization domain, for example, may still allow DNA binding in vitro if the determinants of sequence specificity are still present. Lowering the binding stringency (i.e., salt concentration) or raising the protein concentration can, in some cases, compensate for the loss of the dimerization domain and permit binding in vitro, whereas the same protein would be incapable of binding in vivo.

There are three approaches for establishing that a minimal DNA-binding domain is capable of binding in vivo. The first method is in vivo dimethyl sulfate (DMS) footprinting (Chapter 10), which involves a considerable amount of special technology and expertise. However, this technique is more appropriate for use in the advanced stage of an analysis and is rarely used simply to map the DNA-binding domain. A second approach is to show that the DNA-binding domain alone can act as a "transdominant negative," or inhibitor, of the intact activator in a co-transfection/*trans*-activation assay. High concentrations of the DNA-binding domain should compete with the intact activator for promoter sites and reduce its activation (or repression) capabilities. A third, more elegant, approach is through the use of domain swaps (see Brent and Ptashne 1985).

In a typical DNA-binding domain swap (Fig. 12.3), the minimal DNA-binding domain is fused to a heterologous activation domain like that of VP16. The three components of VP16 activator complex on herpes simplex virus (HSV) early genes include Oct-1, which tethers the complex to the DNA site, host cell factor (HCF), and the HSV-VP16 protein, which contains the activation domain (see Box 12.1; Chapter 1). The activation domain of VP16 can be fused to the transcriptionally inactive GAL4 DNA-binding domain (amino acids 1–94) to generate GAL4-VP16. Because it is unlikely that the heterologous activation

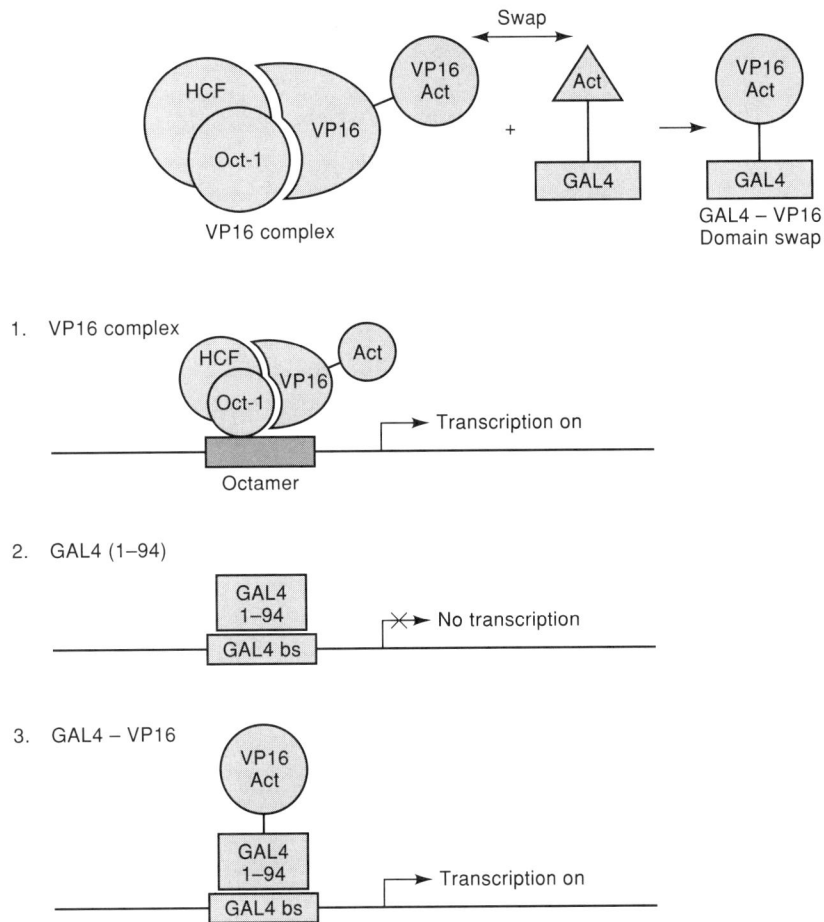

FIGURE 12.3. Domain swaps.

domain could have contributed protein elements important for DNA binding, *trans*-activation by the chimera of a reporter bearing binding sites for the activator in question would imply that the DNA-binding domain was independently capable of recognizing its sites in vivo.

Note that in some cases the ability of DNA-binding domains to recognize their sites is controlled by other regions of the protein or, indeed, by other proteins altogether. In several well-studied cases, another domain directly or indirectly masks the ability of the DNA-binding domain to bind DNA unless an appropriate ligand is present or unless the domain is interacting with an appropriate site or the correct protein partners in an enhanceosome. In some cases, the removal of the inhibitory domain might enhance DNA-binding affinity in vitro (see, e.g., Petersen et al. 1995) or activation potential in vivo. The key point is that such domains may not be apparent in domain swap experiments and that other approaches must be used to reveal its nature. Examples of these inhibitory interactions include intra- or intermolecular domains that regulate the nuclear localization of the activator, such as the ligand-binding domain of the steroid receptors (Beato et al. 1995), or IκB in the case of NF-κB (Stancovski and Baltimore 1997). Alternatively, the DNA-binding domain may be coupled to an autoinhibitory domain, as in the case of the Ets-1 (Petersen et al. 1995).

Activation

Analysis of the activation domain is more problematic. Two issues must be considered. First, a general feature of activators is that they contain several regions that contribute to activation (see Chapter 1). Deletion of any one of these may be insufficient to decrease activation significantly. Conversely, decreased stimulatory potential in transcription assays could be due to causes unrelated to the mechanism of activation, including decreased protein stability or aberrant folding. Hence, unlike the case of the DNA-binding domain, the definition of an activation domain in many cases rests on the ability of that domain to function in the context of a domain swap experiment (this general domain swap approach also works for identifying repression domains; see Saha et al. 1993; Tzamarias and Struhl 1994).

In cases where the activation domain is modular and separable from the other domains of the protein, an activation domain swap involves constructing a chimeric protein containing a transcriptionally inactive DNA-binding domain (i.e., LexA or the DNA-binding domain of GAL4) and the putative activation domain of the protein being studied (Fig. 12.3). The chimera is co-transfected into a mammalian tissue-culture cell with a reporter template bearing tandem, oligomerized LexA or GAL4 sites positioned upstream of a minimal promoter and reporter. The activity of the DNA-binding domain alone is then compared with the chimera, which should stimulate transcription if the domain is active. Once the domain is identified, the important residues can be identified by point mutational analysis.

Limitations of the Domain Swap

Despite the power of the domain swap, recent studies have underscored the fact that activation and DNA-binding domains are quite complex, and cell-, promoter-, and site-dependent contextual interactions often regulate the ability of the activation domain to function (see Chapter 1). Furthermore, the domain swap has the capacity to identify portions of a protein that play no physiological role, which, when taken out of context, can act as nonspecific activation domains. We discuss below some of the limitations and considerations that must be applied when employing domain swap analysis.

The first and foremost consideration is the observation that some protein domains within an activator interact with either coactivators or other nearby activators. These interactions contribute both to the activator's binding specificity and to its ability to function in a promoter-specific context. One example is the LEF-1 protein. In the T-cell receptor α (TCR-α) enhancer, LEF-1 both bends the DNA to permit other activators to interact within the context of the enhanceosome and contains an activation domain whose function, although not well understood, is dependent on the TCR-α enhancer. Thus, a simple reporter bearing multimerized LEF-1 sites fails to function on a model reporter template bearing multimerized LEF-1 sites (Giese and Grosschedl 1993).

Another possibility is that the activation domain communicates with the DNA-binding domain in a manner dependent on the particular binding sites. NF-κB, for example, binds several different sequences and appears to adopt unique conformations that influence its activity (Fujita et al. 1992). Glucocorticoid receptor also binds to several different sites in responsive promoters. Biochemical and site-directed mutagenesis experiments suggest that the DNA-binding domain communicates with the activation domain either to control its activity in a positive fashion or to convert it to a repressor (Lefstin et al. 1994). Such activation domains may not be easily scored in domain swaps, and their identification will require more sophisticated methodologies (Lefstin and Yamamoto 1998).

In addition to complex activation domains where the activity may not be measurable in a domain swap, there exists the opposite problem where a nonphysiological domain activates transcription. In experiments performed in yeast, 5% of random DNA sequences from the *E. coli* genome were capable of generating activation domains when fused to the GAL4 DNA-binding domain (Ma and Ptashne 1987b). Since it is unlikely that 5% of any genome codes for activation sequences, this observation suggests that there is great potential for identifying nonphysiological sequences as activation domains in swap experiments. For example, domain analysis of GATA-1, a transcription factor important for erythroid differentiation, revealed an activation domain essential for GATA-1 *trans*-activation in nonerythroid cells on model reporter templates. The domain, however, was found not to be essential in a more physiological assay measuring the ability of GATA-1 to promote differentiation of immature erythroblasts into terminal erythroid cells (Weiss et al. 1997). Instead, another cell-specific domain was required. This result suggested that the orginal activation domain, although functional in a swap assay, was not necessary for activation in a natural context.

The growing number of activation domains that are displaying site, cell, and promoter specificity suggests that, in addition to simple domain swaps and deletion analyses performed on model reporter templates, new approaches must be taken to identify context-dependent domains properly. These approaches include, but are not limited to, performing the initial deletion analysis of a regulatory protein on its naturally responsive promoters, measuring phenotypic effects of the deletion mutants on a global activity of the protein (i.e., cell differentiation in the case of GATA-1), and, finally, performing the experiments either in cells where the activator is normally active or in a closely related cell type. In cases where the domain analysis is performed on a model promoter, it should be confirmed that the same domains are active on a natural promoter.

Some investigators have even gone so far as to perform domain swaps in the context of a natural promoter. In such studies, the activator's natural binding site within a promoter is substituted for a GAL4 or LexA site. Chimeric regulators bearing LexA and GAL4 fused to the domains of interest are used either to identify or to confirm the relevance of context-specific activation domains (see, e.g., Rellahan et al. 1998). Alternatively, one could ask

---

Box 12.1

VP16: A Case Study

Early studies on activators in yeast showed that the activation domain was largely refractory to point mutational analysis (Hope and Struhl 1986). Unlike the dramatic effects that certain point mutants in a DNA-binding domain generally have, point mutants in activation domains showed subtle effects and could only be identified when small segments of the activation domain were studied in isolation (see Cress and Triezenberg 1991 or Gill and Ptashne 1987). Nevertheless, it is instructive to consider how VP16, one of the prototypic activation domains, was identified and characterized, a project that is still in progress (Triezenberg 1995).

VP16 is a *trans*-activator that activates the immediate-early genes of herpes simplex virus type 1 (HSV-1). The protein is a structural component of the viral capsid, and its action requires no viral protein synthesis during the initial infection. After infection and viral uncoating, VP16 binds to early viral promoters in a complex with the cellular proteins Oct-1 and HCF. The binding specificity comes largely from interaction of Oct-1 with the octamer-binding site, called the TATGARAT (R is a purine), although studies by the Herr and Sharp laboratories have shown that VP16 does interact, albeit weakly, with the DNA in a sequence-specific fashion (see, e.g., Kristie and Sharp 1990). This marginal specificity adds enough to the binding energy of Oct-1 to direct the complex to specific octamer sites in the viral rather than in cellular promoters. Oct-1 is also known to associate with other coactivators such as the B-cell-specific protein OcaB/Bob1, which also binds DNA weakly, to direct it to cellular promoters (Luo and Roeder 1995; Gstaiger et al. 1996). The concept of a coactivator programming the ultimate specificity of a multiprotein complex is an emerging theme in the gene expression field. This theme, like that of the enhanceosome, provides another explanation for how common sets of activators can be employed with different combinatorial partners to effect new patterns of gene expression.

Deletion analysis by Triezenberg, McKnight, and colleagues originally localized the VP16 *trans*-activation domain to the carboxy-terminal 78 amino acids of the 490-amino-acid protein (Triezenberg et al. 1988). Subsequent fusion of that region to the GAL4 DNA-binding domain created a chimera, called GAL4-VP16. GAL4-VP16 was found to activate transcription to high levels from a GAL4 site-responsive reporter template (Sadowski et al. 1988). This established the idea that the carboxy-terminal domain was sufficient for activation.

Later studies showed that VP16 in the context of the chimera could be further subdivided, into two subdomains called VP16 (N), amino acids 413–456, and VP16 (C), amino acids 457–490. Both domains functioned in the context of the virus and could support viral replication, although more weakly than the intact domain. It was initially thought that the two domains were simply reiterated segments of the same motif (Emami and Carey 1992; Walker et al. 1993; Tanaka et al. 1994). However, detailed point mutagenesis by Triezenberg and colleagues revealed that similarly positioned amino acids in the two subdomains responded differently to mutagenesis (Cress and Triezenberg 1991; Regier et al. 1993). This and other mutagenesis studies led to the proposal that the two subdomains functioned by different mechanisms. Indeed, the amino and carboxyl domains were shown to interact preferentially with different targets in affinity chromatography experiments (see, e.g., Goodrich et al. 1993). Although this alternate target hypothesis is somewhat attractive, the physiological meaning of the binding studies is still unclear because reiterating either the amino or the carboxyl domain, or segments thereof, creates extremely potent activators that can function independently of the other domain. Thus, even if the domains contact different targets, either one of the interactions is sufficient for activation (Emami and Carey 1992).

Although the amino and carboxyl subdomains are functional in the context of an artificial promoter and in the form of a chimeric GAL4-VP16 protein, they may play unique promoter-specific roles, which would only be manifested by assaying the chimeras in a variety of promoter and cellular contexts. Indeed, in natural HSV promoters, VP16 only activates from a promoter-proximal position. Attempts to obtain distal activation, like that obtained in artificial systems (Sadowski et al. 1988; Carey et al. 1990), have failed (Hagmann et al. 1997), demonstrating that the model systems do not accurately recapitulate regulation of activation domain function.

Recent spectroscopic studies have begun to reveal some insight into the structural transitions of the VP16 activation domains in the presence of its targets. A study using fluorescence anisotropy concluded that VP16 undergoes a dramatic structural change when it interacts with TATA-binding protein (TBP), a possible target (Shen et al. 1996a,b). In such studies, point mutants (Cress and Triezenberg 1991) were essential controls for determining the specificity of the structural changes. Another study, using nuclear magnetic resonance (NMR) to investigate interactions of VP16 complexed with one potential target $TAF_{II}32$, provided support for the idea that the conformational change is from a random coil to an α helix (Uesugi et al. 1997).

The idea that activation domains might be amphipathic helices was originally explored by Giniger and Ptashne (1987) using short peptide activation domains. Indeed, the first crystal structure of an activation domain, that of the τ domain of the nuclear receptors (for review, see Mangelsdorf and Evans 1995), revealed it to be an amphipathic helix that formed part of the ligand-binding domain. Future studies, possibly combining structural methods, targets, and mutants, will be required to establish the bona fide physiological interaction surfaces of VP16.

the reciprocal question, whether the natural DNA-binding domain could be fused to a heterologous activation sequence and retain function (see, e.g., Driever et al. 1989).

Even with the proper promoter, however, some domains may require the action of other cell-specific activators or coactivators to function. It may also be necessary for the co-transfection analysis to employ cells already expressing the activator in question. This latter scenario raises the potential problem that the endogenous activator may increase background transcription so high as to obscure an effect of the transfected activator and its mutants. However, new transfection methods have raised the transfection efficiency to the point where it is often possible to overexpress a transfected activator well enough above endogenous levels to observe an effect. Additionally, closely related cell types not expressing the activator might also be employed. Such experiments will more accurately reflect the activity of the activator in its natural context, with the caveat that the investigator must be aware that overexpression may also influence the outcome of the assay.

Finally, despite the success of the overall deletion and domain swap approaches, only in rare cases have activators with activation domain swaps been shown to substitute stably for the natural gene product to drive the life cycle of an organism (Driever et al. 1989; Baumann et al. 1993, 1995). Thus, although the approach described above yields results with analytical value, it is important not to overinterpret the physiological implications of such experiments.

Subdividing DNA Recognition and Oligomerization Subdomains

Identification of dimer domains is an important step in the analysis of a protein, as disrupting the dimerization interface is one method for inactivating a factor for regulatory or therapeutic purposes. The Id protein, for example, is a helix-loop-helix (HLH) domain that lacks a basic region (Benezra et al. 1990). Id and related family members heterodimerize with basic HLH proteins and block them from binding DNA and activating transcription during myogenesis and hematopoeisis.

There are many different classes of dimer motifs, of which the bZIP coiled-coil motif found in Jun and Fos (Landschultz et al. 1988; O'Shea et al. 1989; Ellenberger et al. 1994) and the HLH domain found in E26 (Murre et al. 1989; Ellenberger et al. 1994) represent only a small subset. Indeed, some proteins, such as MyoD, contain both HLHs and leucine zippers (Ma et al. 1994). Identifying these motifs and separating them from the DNA recognition subdomain requires extensive biochemical analysis and analytical methods that extend beyond simple DNA-binding assays.

Amino-terminal fragments of GAL4 and λ repressor, for example, still retain minimal, low-affinity DNA-binding ability, even after removal of the dimer subdomain (Pabo et al. 1979; Carey et al. 1989). Although each polypeptide fragment remains a monomer in solution — unlike the intact dimeric proteins — the truncated derivatives still bind to DNA as dimers by employing residual dimer interfaces. These interfaces, identified by crystallography, may or may not be physiologically relevant, yet they are enough to drive binding and obscure the domain analysis.

Dimerization can be measured directly by several methods, including heterodimer analysis (see Chapter 13), gel filtration (Li et al. 1997), and glutaraldehyde cross-linking (Smith et al. 1990; Wagner and Green 1993). However, in the initial analysis of a DNA-binding domain using an assay like an electrophoretic mobility shift assay (EMSA), as the dimerization subdomain is gradually disrupted (e.g., by progressive deletions), it is common to still observe specific binding (i.e., when comparing a mutant and wild-type recognition site) even while observing gradual decline in the affinity of the protein for its site.

This lowered affinity is often the first piece of evidence that the dimer domain has been compromised.

If it is suspected that the dimer domain has been compromised, a dimer domain swap experiment can be performed (Fig. 12.4). In such an experiment, a heterologous dimer domain (i.e., a coiled coil) is fused to a polypeptide displaying minimal DNA affinity. The affinity of that chimera is then compared with that of the unfused protein in a DNA-binding assay. If the fusion enhances that protein's affinity in an EMSA, one might suspect that reduced affinity of the unfused protein was because the natural dimer domain has been compromised or removed. To establish that the mutated region indeed contained a dimerization domain, a reciprocal swap can be performed in which the putative dimer domain is fused to a minimal DNA-binding domain. This reciprocal swap is then assayed to determine whether it enhances binding. Hu and Sauer, for example, showed that the bZIP coiled coil enhances binding of the amino-terminal DNA-binding module of λ repressor in vitro and in vivo (Hu et al. 1990).

CONCEPTS AND STRATEGIES: PROTEIN–PROTEIN INTERACTIONS

Interaction of Activation Domains with Coactivators and General Factors

Defining the interaction partner for an activation domain is a common and important goal for understanding how the domain regulates transcription. Unlike the strong interactions among subunits in Pol II or TFIID, activator–target interactions are relatively weak. Thus, defining their interactions is a challenging task. There are many approaches and considerations that can be highlighted by discussing interactions between activators and general factors. The goal in a structure–function analysis of an activation region is to use the mutants as tools to identify the protein targets of an activator. There are several biochemical and genetic methods employed to identify and characterize individual protein–protein interactions outside the context of a macromolecular complex. These methods include the use of specific ligand-affinity columns (Greenblatt and Ingles 1996), co-immunoprecipitation (Hisatake et al. 1995), chemical crosslinking (Coleman et al. 1995), and the yeast two-hybrid assay (Fields and Song 1989). Additional assays such as protease footprinting (Hori et al. 1995), fluorescent anisotropy (Heyduk et al. 1996), and surface plasmon resonance (Bushnell et al. 1996; Wu et al. 1996) are highly specialized and can be employed to refine the analysis.

There are two philosophical approaches: (1) model-oriented approaches, in which the activation domain is selected for its interaction with likely targets, and (2) random approaches, in which genetic or biochemical techniques are used to identify a putative target from a collection of potential targets. The model-oriented approach assumes that a step is important and influenced by the activator; e.g., that an activator affects transcription by influencing TBP binding, because that is the earliest step in complex assembly. This notion is then tested by affinity chromatography and validated using activation domain mutants. Despite its relative simplicity, the model-oriented approach can be a dangerous starting point because the researcher may be guessing that a protein–protein interaction is important. Such studies can result in a model that will not withstand subsequent tests of its validity. The random approach is more cautious and often more prudent. In such an approach, once the importance of a domain is established, broad attempts are made to identify its tar-

DNA binding

Jun:

Coiled-coil
dimer

On

Basic DNA-binding
domain

DNA

λ repressor:

Carboxy-terminal
dimerization domain

On

Amino-terminal DNA binding
Helix-turn-helix motif

DNA

DNA-binding domain:

λ repressor:

Off

DNA

Dimer swaps:

On

DNA

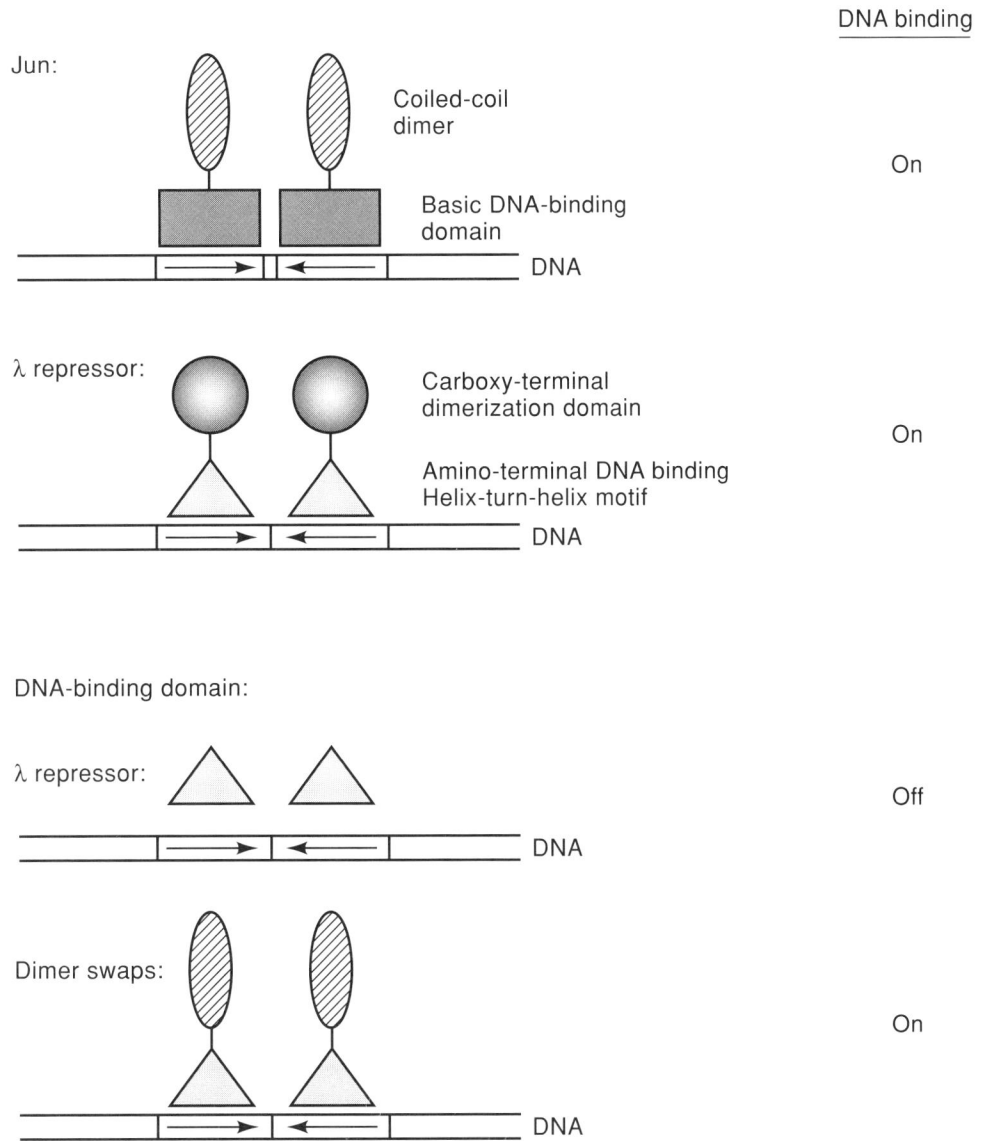

FIGURE 12.4. Dimer swaps.

Box 12.2

GAL4: A Case Study

To illustrate the DNA-binding and dimerization issues, one can turn to GAL4, which served as
a prototype for many of the strategies described above. As described in Chapter 13, GAL4 binds

upstream of the galactose metabolism genes in yeast and activates them in the presence of galactose (for review, see Johnston 1987). Initially, it was not clear whether GAL4 bound DNA at all because it had been isolated in a broad genetic screen of genes important for galactose catabolism. Thus, in the early 1980s, the only hint that GAL4 bound DNA was that it contained a DNA sequence that might encode a relative of the zinc finger recently discovered by Klug and colleagues (Miller et al. 1985). In vivo footprinting by Ptashne and colleagues with wild-type and mutant *gal4* yeast showed that GAL4 was responsible for generating a DMS footprint over the four DNA sites constituting the galactose upstream activating sequence (UAS$_G$) (Giniger et al. 1985). This observation prompted Ptashne and colleagues to analyze various deletions of GAL4 generated by using convenient restriction endonuclease cleavage sites within the coding region. The deletions were synthesized in *E. coli* as fusions to β-galactosidase (Keegan et al. 1986). The resulting proteins were then analyzed by DNase I footprinting and EMSA approaches.

Fragments bearing the amino-terminal 74 or 147 amino acids of GAL4 bound the UAS in vitro by footprinting analyses. In vivo assays on a suitable reporter demonstrated that derivatives containing the amino terminus but lacking the carboxyl terminus failed to activate. It was assumed that these derivatives were indeed binding in vivo but simply lacked the activation domain. This assumption was proven by showing that fusion of a wide array of heterologous activation domains, including VP16 to the DNA-binding domain identified in vitro, restored the ability of the DNA-binding region to activate transcription in vivo (Ma and Ptashne 1987a,b; Sadowski et al. 1988). Conversely, the amino acids encompassing the deleted region on the carboxyl terminus, when fused onto the transcriptionally inactive LexA DNA-binding domain, conferred the ability to activate transcription from a LexA site in vivo (Brent and Ptashne 1985). These domain swaps were the first examples of the modularity of eukaryotic activators and, with experiments being done concurrently on GCN4 (Hope and Struhl 1986), provided the paradigm for analyzing eukaryotic activators.

The situation was considerably more complex when analyzing the dimerization interface. Three experiments suggested that GAL4 bound DNA as a dimer: the dyad characteristics of the 17-bp site, the symmetric contacts with the site revealed by chemical and nuclease footprinting techniques, and finally the heterodimer analysis (see Chapter 13; Carey et al. 1989). The earliest analyses of GAL4 employed mutants generated by restriction enzyme deletions of coding fragments. The DNA-binding domain contained 6 cysteines between amino acids 11 and 38 that resembled at the time the zinc-finger DNA-binding domain, but any attempt to delineate further the domain by deleting from the amino terminus abolished the DNA-binding capabilities of the protein. Therefore, to identify the minimal DNA-binding domain, mutants containing progressive carboxy-terminal deletions from amino acids 147 toward the GAL4 amino terminus were created. It had been noticed earlier that GAL4 amino acids 1–74 bound DNA less tightly than 1–147, so it was assumed that position 74 lay within the hypothetical dimer region. To compensate potentially for loss of the dimer interface during deletion, the resulting mutants were fused to the dimer domain of λ repressor. Using this approach, the minimal fragment that would specifically and efficiently bind DNA was found to be amino acids 1–65. Comparison of the affinity of 1–65 alone versus 1–65 fused to the λ repressor revealed a 100-fold or more difference in affinity. Indeed, deletion of 1–147 on its own showed little effect until amino acid 94, whereafter the affinity dropped gradually until it reached a minimum at 1–65. Note that although the region between 65 and 94 was implicated in dimerization by the above argument, it was never validated by the sufficiency test; i.e., shown to function in the context of a chimeric protein and to confer dimer capabilities onto a monomer (i.e., analogous to the approach of Hu et al. 1990). It is important to point out that this type of analysis is of considerable value when attempting to generate proteins for structural study and was the original rationale for choosing GAL4(1–65) for crystallographic analysis (Marmorstein et al. 1992).

get using two-hybrid analyses, genetic tests, or affinity chromatography of crude extracts or large collections of pure factors.

We discuss the use of simple affinity resins and the drawbacks of such approaches because most of the caveats hold for more sophisticated methodologies. It is important to emphasize that the sophistication of an approach is not necessarily a measure of its validity. Surface plasmon resonance can just as easily give an aberrant result as affinity chromatography, although the result will be more quantitative.

One of the key issues to consider is that interactions of an activator with "target" proteins (i.e., coactivators, or the general machinery) may occur on surfaces displayed only when the activator, general factors, and coactivators are assembled into a transcription complex. These interactions, in the context of the complex, may be part of a highly cooperative, stereospecific network of contacts. However, individually, in isolation, the interactions might be very weak. If the interaction is indeed weak it may make it difficult both to measure and to validate its physiological relevance. For this reason, it is imperative eventually to perform functional biochemical studies correlating an interaction observed in a simple two-component binding assay with an effect on transcription and transcription complex assembly in vitro. We elaborate on this point below and argue why a deleterious effect on binding in a two-component interaction assay and activation in a transfection assay is not by itself a significant validation of a biochemical interaction.

Affinity Chromatography

Principles

Historically, affinity chromatography has proven to be a powerful tool for identifying interacting protein partners. It has been used extensively to study protein–protein interactions among activation domains and general transcription factors (e.g., see Greenblatt and Ingles 1996). One of the classical approaches for identifying transcription factor partners was developed by Greenblatt and colleagues, who covalently linked RNA polymerase II to a resin and used the resulting affinity matrix to isolate RAP30 and RAP74 from crude extracts. Remarkably, RAP30 and RAP74 turned out later to be the small and large subunits of TFIIF (Sopta et al. 1985). The approach is very flexible and was adopted by Green and colleagues to study interaction of the VP16 activation domain with the general machinery (Lin et al. 1991), and more recently, by Rosenfeld and colleagues to understand the interactions between different activators and coactivators like p300 (Korzus et al. 1998). In the experiments by Lin et al. (Fig. 12.5), extracts bearing all of the general factors were passed over GST–VP16 affinity matrices. After washing, most of the free factors were removed from the matrix and only TFIIB remained bound. TFIIB could then be eluted with high salt concentrations. These data suggest that TFIIB might be a target of VP16.

In cases where sufficient quantities of a recombinant protein are available, that protein can be easily crosslinked to commercially available resins to generate affinity columns. Examples include CNBr-activated Sepharose available from Amersham Pharmacia Biotech (cat. #17-0820-01) or the Affi-Gel resins available from Bio-Rad (cat. #153-6101). The Affi-Gels contain an N-hydroxysuccinimide-activated matrix that crosslinks to the primary amines present either in lysines or at the amino terminus of the protein. The buffers must be devoid of primary amines to prevent a decrease in crosslinking efficiency. (HEPES, for example, is a common buffer used in crosslinking applications whereas Tris, hydroxylamine, and ethanolamine are not appropriate because they react with the chemical crosslinking group.) Typically, the purified protein to be crosslinked is incubated with the

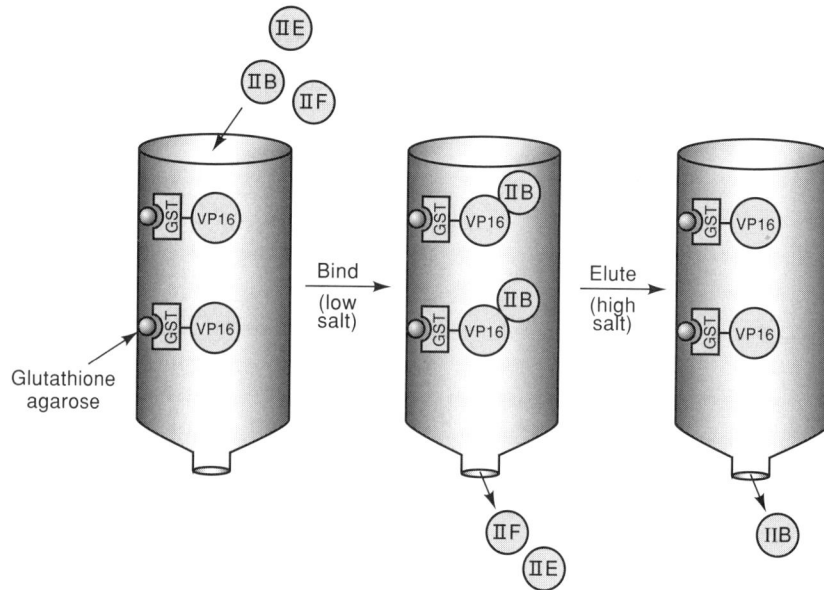

FIGURE 12.5. Affinity chromatography.

activated resin at densities from 1 mg to 20 mg of protein per milliliter of added resin. The mixtures are rocked gently for 1 hour to overnight to allow crosslinking, and the unbound protein is washed away with buffer. The affinity matrix is then neutralized by addition of ethanolamine and prepared for affinity binding assays by pre-equilibrating in the binding buffer. A preservative such as EDTA or azide is added for long-term storage in the refrigerator. Some column matrices cannot withstand freeze-thaw cycles, which collapse the beads; thus, freezing the matrix should be avoided.

It is important to vary the density of the ligand on the matrix and to examine its consequences on binding of the target. The ligand density has been shown to have a significant effect on the amount and type of protein bound (see Greenblatt and Ingles 1996). In some cases, lower-density matrices are unable to bind ligand, whereas the higher-density matrices may begin to exhibit significant nonspecific interactions. Because it is well established that many prokaryotic and eukaryotic proteins bind nonspecifically to agarose and cellulose matrices, it is necessary in these studies to generate a control column containing the resin alone or the resin linked to a mutant ligand. These controls should be in hand prior to initiating the analysis.

Rather than covalently linking the protein to a resin, it can instead be fused to a purification tag (see Chapter 11) like glutathione-S-transferase (GST), maltose-binding protein (MBP), or His$_6$ (see Chapter 11). The resulting fusion proteins can then be attached to affinity matrices such as glutathione-agarose, amylose, or nickel-NTA/Sepharose, respectively.

Once a suitable matrix is available, it is incubated with the putative target protein or a crude protein mix (i.e., a nuclear extract) under various binding conditions. The matrix is washed with buffer to remove unbound protein, and the bound protein is then eluted with either high salt concentrations, a ligand, or specialized buffers. The bound and unbound fractions can then be examined by sodium dodecyl sulfate polyacrylamide gel electrophoresis (SDS-PAGE) and immunoblotting. The incubation with the matrix can be performed by allowing the mixture of target proteins to slowly pass through the column. Alternately, the binding reaction can be performed in batch in a test tube, using low-speed centrifugation to pellet the beads after incubation and/or washing steps.

Caveats of the Affinity Approach

Although the affinity approach is simple and appealing, it is fraught with difficulty. The correct control for specificity in the affinity assays is to compare binding of both wild-type and mutant activators (or general factors) to a putative target protein or mixture of proteins. In the case of eukaryotic activation domains, however, we do not fully understand the nature of activation domain mutants and cannot be sure that the mutation is specifically affecting amino acids important for a protein–protein interaction or whether the mutation is having a global effect on structure.

Contrast this case with that of the classic positive control (pc) mutants found in prokaryotic activators such as λ repressor (see Ptashne 1992). The pc mutants contain a domain comprising both the activation and DNA-binding functions of repressor. The ability of the pc mutant to bind DNA demonstrates that the domain is largely intact from a structural standpoint. It can be inferred, therefore, that the deleterious effect of the pc mutation on activation is due to the loss of a critical contact.

Disappointingly few comparable examples are found among eukaryotic activators due to the modularity of the activator. Many typical eukaryotic activation domains lack obvious structure (see Chapter 1) until they interact with their targets, making it difficult to ascertain whether the mutation is influencing the overall folding of the domain or simply weakening a critical protein interface. Thus, the absence of binding of a target protein to a mutant activator may not be interpretable as it is in the prokaryotic case, because it cannot be determined whether the mutation affected the overall structure of the activation domain or simply removed a critical contact as in the λ repressor pc mutants. Even in the case where a protein–protein interaction is of low affinity and nonspecific (e.g., purely a nonspecific ionic interaction), these interactions may also be influenced by a global change in the structure of a protein bound to an affinity matrix. Therefore, affinity binding assays represent a minimal starting point. The results of these and other protein interaction methods are not conclusive without the use of additional biochemical or genetic assays to demonstrate the relevance of the interaction.

One example of the danger of overinterpreting the affinity assay is the abundance of studies focusing largely on TBP–activator interactions from the early 1990s. Because TBP was cloned and widely available, many activators were studied for their interaction with TBP without considering the dozens of other polypeptides in the transcription complex. Wild-type activators were shown to interact with TBP, whereas deletion or point mutants were shown not to interact in affinity binding assays. In literally dozens of publications, these data alone were taken as the sole evidence for the validity of the interaction. However, subsequent studies suggested that some of the activation domains were interacting with the basic underside of TBP. Such an interaction should have inhibited rather than promoted binding of TBP to DNA. However, because the effect of the activator on TBP binding was not measured, the biochemical consequences of the interaction were never established. In other cases it was clear from subsequent studies that the activation domain "stuck" to many different transcriptional components, making the original observations less certain. In most studies the interactions have still not been sorted out because they were never pursued past the initial stages of the analysis. In addition to ignoring the biochemical consequences of an interaction on transcription complex assembly, few of the studies ever addressed whether there existed reciprocal mutants in TBP that abolished both interaction with the activator and response to the activator in a transcription assay (see, e.g., Tansey and Herr 1995). Such assays would have revealed the important surfaces on the target and aided in determining whether the interactions made sense from a physiological point of view.

Taken together, reciprocal mutational approaches and biochemistry should always be employed to support the original affinity observation. Such fundamental issues are rarely addressed but are critical for confirming the relevance of an interaction, given the intrinsic uncertainty surrounding protein–protein interactions in mammalian transcription.

Although there is some controversy over the physiological target of VP16 (Gupta et al. 1996), the VP16–TFIIB interaction studies by Green and colleagues provide an excellent paradigm for the overall philosophy toward analyzing activator-target interactions. VP16 was initially shown to affect a step in transcription complex assembly involving TFIIB (Lin and Green 1991). TFIIB was then shown to bind to wild-type but not to a point mutant of VP16 (Lin et al. 1991). Further studies suggested that there were mutants in TFIIB that abolished the interaction with VP16 but were still able to mediate basal transcription (Roberts et al. 1993). Finally, VP16 appears to induce a change in the conformation of TFIIB, again supporting the importance of the interaction (Roberts and Green 1994). Taken together, the data constituted an argument for the validity of the VP16-TFIIB contact. This argument can be further refined by higher-resolution mutant analyses as described below.

In summary, the main problem encountered in studying protein–protein interactions in mammalian systems has been that it is difficult, if not impossible, to study activator interactions with the coactivators, chromatin, or the general transcriptional machinery in vivo in a physiological milieu due to interference from endogenous proteins. Because there is no tractable genetic system, sophisticated biochemical assays with multiple general factors must be performed to analyze the validity of an interaction. Although the biochemical assays are of great use, it is possible to manipulate the assays to emphasize one reaction step over another. Therefore, although the approach has value, it nevertheless suffers from the inability to validate biochemical models genetically. The close relationship between the genetically tractable yeast system and the human general machinery has allowed a limited comparison of the mechanisms of activation. In general, many of the mechanisms studied in yeast support the current biochemical view of the mammalian transcription process. The challenge over the next decade, however, will be to establish new methodologies that permit genetic analyses to be performed in mammalian systems.

Altered Specificity Genetic Systems

The altered specificity approach, discussed for DNA-binding proteins in Chapter 9, is an excellent case in which ingenious experimental design has been used to circumvent the lack of genetics in the mammalian system. The approach allows one to dissect the relevance of protein–protein interactions in cells that, by necessity, express endogenous versions of the protein. Were it not for the altered specificity, the endogenous proteins would interfere with the analysis. The basic premise of altered specificity is analogous to the concept of allele-specific suppression in genetics. In allele-specific suppression, a mutation in one component is suppressed by a specific compensatory mutation in the other component. The compensatory mutation, in principle, is specific to the original mutation and functions to restore direct contacts among two interacting and complementary amino acids, or an amino acid and a base pair. In practice, perfect allele-specific suppressors are rarely observed, but the concept remains valid as certain suppressors will be somewhat allele-specific and can nevertheless change the specificity of the interaction. The allele-specific suppressor approach has been used to construct DNA-binding proteins with new sequence specificities and to understand protein–protein interactions in prokaryotic transcription systems (Li et al. 1994). To date, the approach has been applied in eukaryotic transcription

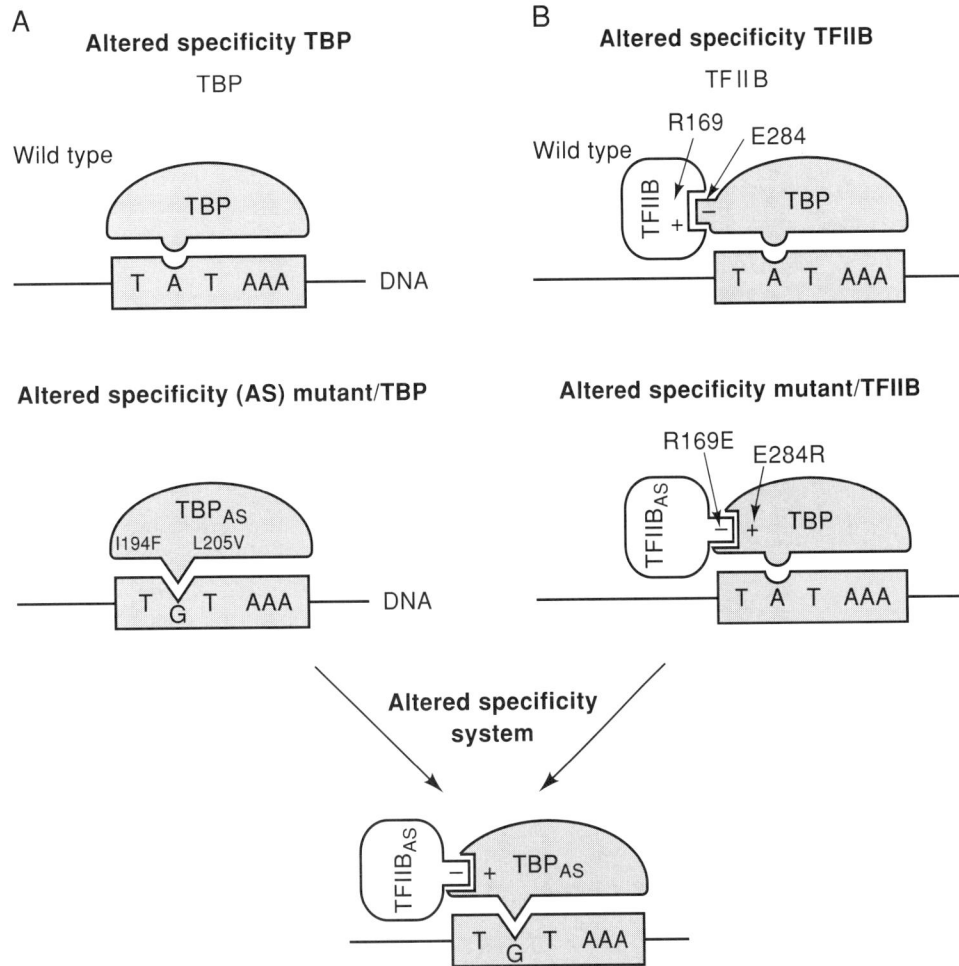

FIGURE 12.6. Altered specificity mutants.

systems to several DNA-binding proteins and to two components of the general transcription machinery, TBP and TFIIB, although in theory it will be applicable to other general factors.

In the case of TBP, certain amino acid changes in the DNA-binding surface will alter the specificity of TBP (TBP$_{AS}$) so that it recognizes TGTAAA with high affinity (Strubin and Struhl 1992). This interaction is proscribed for wild-type TBP due to a steric clash between the exocyclic minor groove imido group on guanine and the DNA-binding surface of TBP. The mutant amino acids apparently create a pocket that can accommodate the imido group (S. Burley, pers. comm.). Although this mutant also binds to a wild-type TATAAA box, the ability to bind to TGTAAA means that an artificial promoter can be created that will respond only to TBP$_{AS}$ in vivo. In such a system, a reporter template bearing activator binding sites upstream of the TGTAAA can be co-transfected with TBP$_{AS}$ and the activator in question. Because only TBP$_{AS}$ binds the TGTAAA, interactions between the activator and TBP$_{AS}$ can be studied (Fig. 12.6A). This system, developed originally in yeast, also functions well in mammalian cells. The system has been used, for example, to analyze the surfaces of TBP that interact with the general machinery in vivo, a topic we discuss below (Bryant et al. 1996).

An altered specificity version of TFIIB has also been constructed that can be coupled with the TBP$_{AS}$ to study activator–TFIIB interactions (Fig. 12.6) (Tansey and Herr 1997). In the crystal structure of the TATA:TBP:TFIIB complex, the glutamic acid at position 284 in TBP interacts with the arginine at position 169 in TFIIB. Mutation of the glutamic acid at amino acid 284 in TBP to arginine abolishes the interaction. However, a suppressor mutation in TFIIB of arginine to glutamic acid restores the interaction (Fig. 12.6B). This system was used to study how different activators interfaced with TFIIB during transcrip-

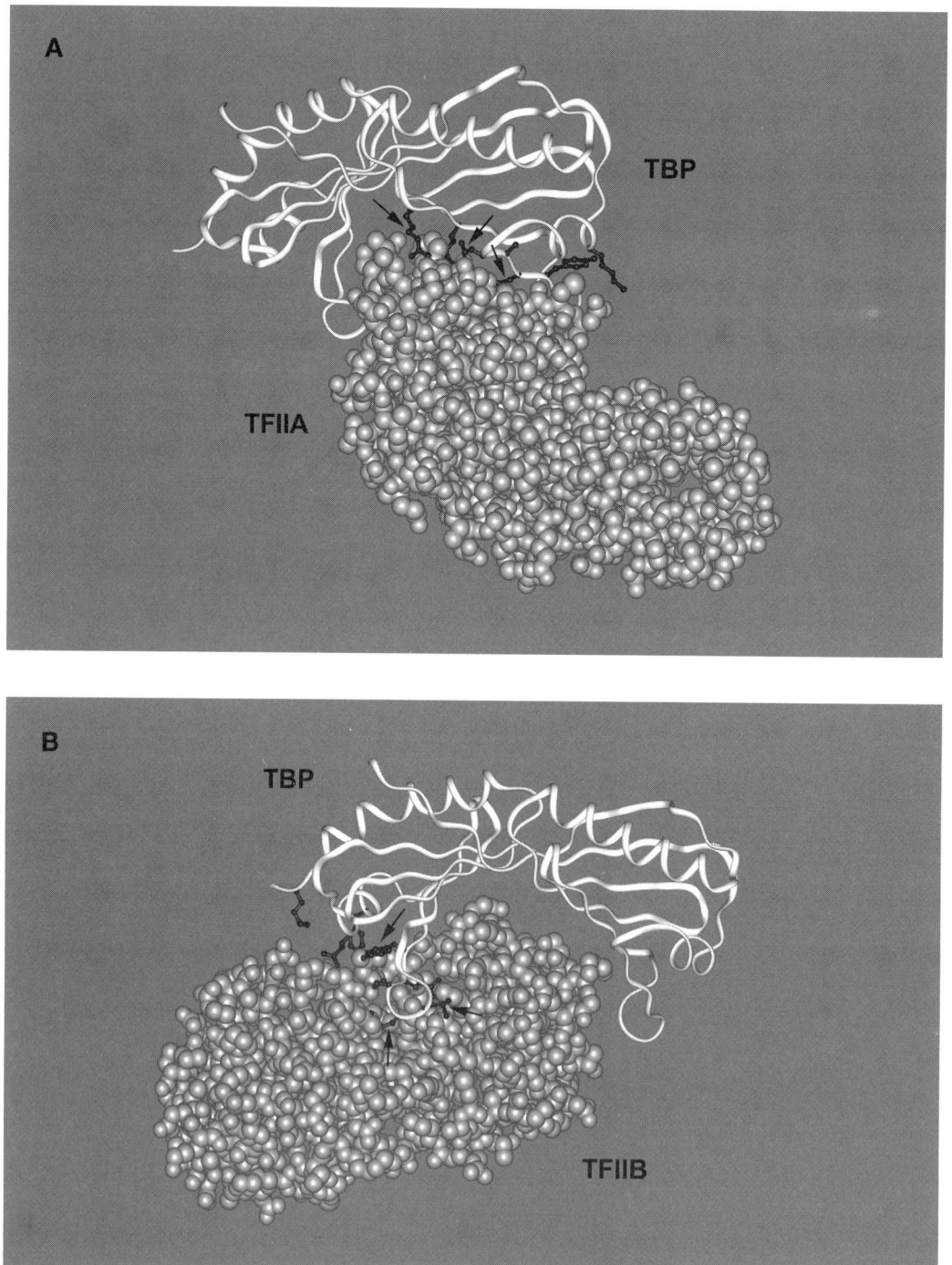

FIGURE 12.7. (*See facing page for legend*).

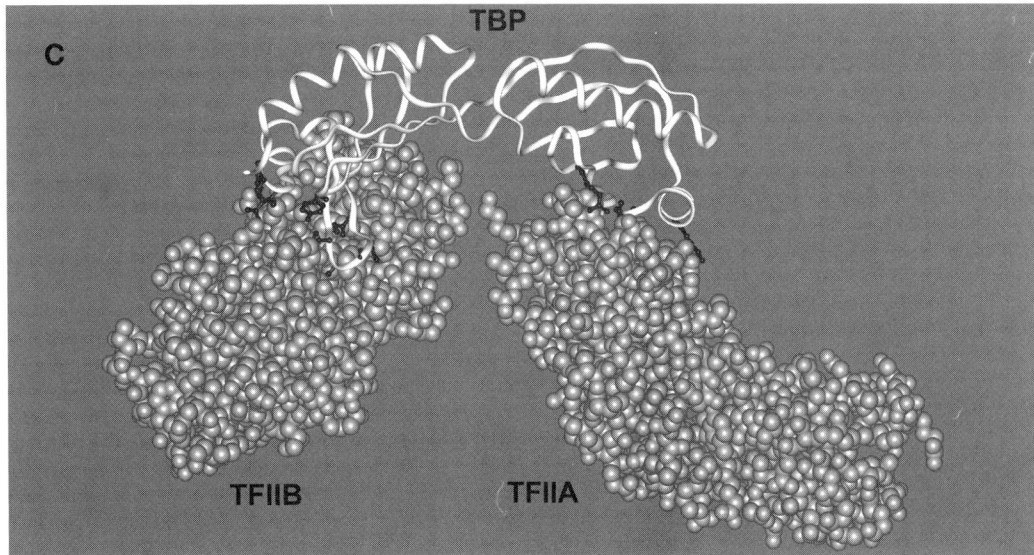

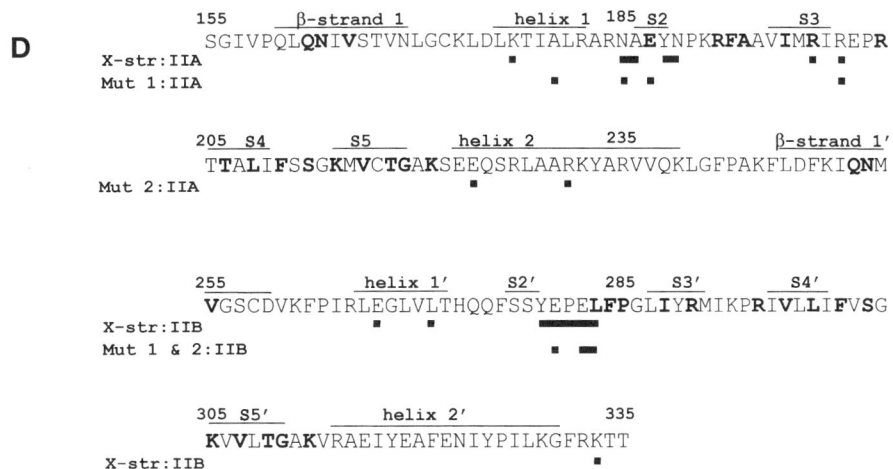

FIGURE 12.7. Structure–function study of TBP. *A*, *B*, and *C* show the structure of TBP as a ribbon and the structures of TFIIA and TFIIB as van der Waals renderings. DNA was omitted for clarity. The residues of TBP shown to interact in the crystal structures with TFIIA (*A*, *C*) and TFIIB (*B*, *C*) are shown in black ball-and-stick representations. Arrows point to residues that affect binding of TFIIA and TFIIB in mutagenesis studies described in the text. The sequence of TBP in *D* employs the numbering scheme of Nikolov et al. 1996. The structures are reviewed in Nikolov and Burley 1997. The positions of helices 1 and 2, and 1' and 2', and sheets 1–5 and sheets 1'–5' in the crystal structure of TBP are shown as lines above the sequence. Residues in bold contact DNA in the crystal. X-structure IIA refers to the residues of TBP underlined in bold that interact with TFIIA in the yeast co-complex of TBP, TFIIA, and DNA (Tan et al. 1996). The contacts were superimposed on the homologous positions in the human TBP structure. X-structure TFIIB refers to the residues of TBP that interact with TFIIB in the crystal structure of the TBP, TFIIB, and TATA (Nikolov et al. 1996). Mutagenesis 1 is the study by Berk and colleagues and illustrates the residues implicated in TFIIA and TFIIB binding by radical substitution (Bryant et al. 1996). Mutagenesis 2 is the study by Ebright and colleagues (Tang et al. 1996) and illustrates residues implicated in TFIIA and TFIIB binding by alanine substitution. The mutagenesis studies by and large identified residues in the TFIIA and TFIIB interfaces identified by crystallography, with the exception of mutagenesis 1, which implicated residues in helix 2 that purportedly interacted with TFIIA but were not indicated in the crystal structure. Differences between the human and yeast TBPs, as well as the possibility that residues important for helix 2 interaction were missing from the yeast structure (not all of TFIIA was crystallized), may account for the discrepancy. (Crystal structures rendered by Michael Haykinson [UCLA] using the Molecular Graphics structure modeling compputer program Insight II.)

tional activation. By continually building on this system, it should eventually be feasible to analyze genetically a wide array of potential contacts in vivo in mammalian cells.

Structure–Function Analysis of the General Transcriptional Machinery

Mutational analysis of the mammalian general machinery has, until recently, been performed in biochemical systems using in vitro transcription and affinity for other general factors as a readout. The reason, as described above, was that because the general transcription machinery is functional in all cells, the endogenous general factors are likely to interfere with any attempt to perform mutational analysis in vivo. Deletion analysis was initially used to identify domains of different general factors required for function. Although the approach proved tractable for some proteins like TFIIB (Ha et al. 1993), the results were difficult to interpret for others, and attention focused on the use of point mutants.

The most exciting and interpretable use of site-directed mutagenesis, however, has been when the structure of the protein is known and the mutagenesis can be used to ask very specific questions about the mechanism. The best-studied examples to date are mutagenesis studies of TBP, designed to understand which surfaces of TBP participate in interactions with the other general factors (Fig. 12.7) (Bryant et al. 1996; Tang et al. 1996). In the future, the approaches will likely be generalized to activator–coactivator interactions.

Using the crystal structure of TBP as a guide (for review, see Burley and Roeder 1996), Ebright and colleagues coupled alanine scanning with EMSA analysis to identify human TBP mutants that failed to interact with general transcription factors in vitro (Tang et al. 1996). The overall strategy was to use alanine scanning only for the residues in the DNA–protein complex that were accessible to solvent. DNA-binding residues and residues involved in packing of the structure were not mutated. The mutants eventually included 81 of the non-proline surface residues on TBP. The authors screened for phenotypes that resulted in a greater than fivefold reduction in affinity of a general factor for the TBP:TATA complex. They calculated that this was equivalent to 1 kcal of energy, which was claimed to be an average side-chain interaction. The goal was to identify mutations that had little effect on TBP binding to the TATA box but instead disrupted the binding of other general factors in EMSA experiments.

Whereas different mutants had significant effects on binding of TFIIA, TFIIB, TFIIF, and Pol II, they had only negligible effects on binding of TFIIE and TFIIH. By superimposing the positions of the mutations onto the surface of TBP, the study was able to formulate a model for how the factors assembled onto TBP to form a basal transcription complex, an issue we discuss in more detail in Chapter 15. Although the study was performed before the crystal structures of the TBP-TFIIA and TBP-TFIIB ternary complexes with DNA were solved, the TBP-TFIIB interface inferred from the mutagenesis closely agreed with the TFIIB-binding site on TBP identified by crystallography (for review, see Burley and Roeder 1996). In contrast, the TFIIA-binding site on TBP inferred from the study did not agree with the subsequent crystal structure of the TATA:TBP:TFIIA ternary complex (Tan et al. 1996).

In another excellent example of a comprehensive structure–function study of human TBP by Berk and colleagues (Bryant et al. 1996), radical substitutions were used to assay 89 surface-accessible residues of TBP. The mutagenesis was coupled not only with in vitro binding studies to TFIIA and TFIIB, but also with cell culture transfection assays, in which activated transcription was measured using TBP$_{AS}$-responsive promoters. Radical mutations in the previously identified TFIIB interface on TBP were found to lower both basal and activated transcription in agreement with both the co-crystal structure and the independent study by Ebright and colleagues (Tang et al. 1996). In addition, however, the Berk study also identified residues involved in binding TFIIA, which agreed well with the published crystal structure. Two other clusters of mutants were identified on the TBP surface that represent potential binding sites for proteins important for activated transcription — possibly either TBP-associated factors (TAFs) or other unidentified co-factors. It is not clear why one study correctly identified the TFIIA-binding site (Bryant et al. 1996) while the other failed to do so (Tang et al. 1996). Two important differences are that (1) one study used radical substitution mutants, which actively disrupt the interface, whereas the other study used alanine scanning, which is considered a loss of contact; and (2) one study coupled the in vitro assays with functional in vivo transcription, which may have helped to resolve ambiguities in the binding experiments.

In summary, there are a wide range of mutagenesis approaches and methods for analyzing the function of a *trans*-activator and its targets. The validation of all site-directed mutagenesis studies does not rely solely on loss of function, but on demonstrating that the mutants affect very specific biochemical processes while leaving the overall structural integrity of the proteins intact.

TECHNIQUES

PROTOCOL 12.1

PCR-mediated Site-directed Mutagenesis

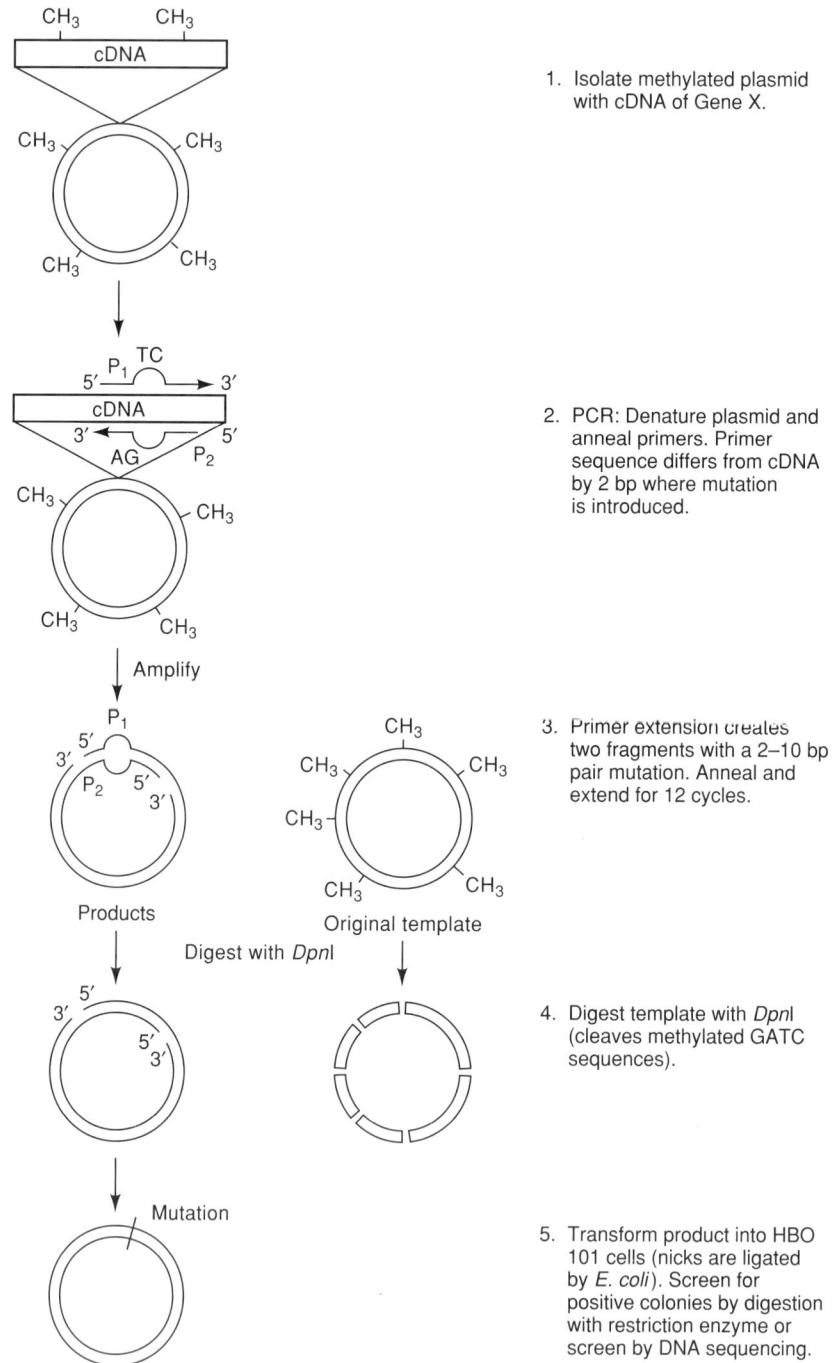

1. Isolate methylated plasmid with cDNA of Gene X.

2. PCR: Denature plasmid and anneal primers. Primer sequence differs from cDNA by 2 bp where mutation is introduced.

3. Primer extension creates two fragments with a 2–10 bp pair mutation. Anneal and extend for 12 cycles.

4. Digest template with *Dpn*I (cleaves methylated GATC sequences).

5. Transform product into HBO 101 cells (nicks are ligated by *E. coli*). Screen for positive colonies by digestion with restriction enzyme or screen by DNA sequencing.

FIGURE 12.8. PCR-mediated site-directed mutagenesis.

SUMMARY

Unlike traditional site-directed mutagenesis, the PCR-mediated mutagenesis method only requires a single PCR step (Fig. 12.8) (Weiner et al. 1994), employing full plasmid amplification to generate point mutants. The method can be employed to introduce small mutations into promoter sites and is even better suited for introduction of single or double mutations into proteins. The mutagenesis method is elegant in its simplicity and, because of this advantage, it can be applied quite easily by any lab using standard protein expression vectors.

A pair of primers encoding the mutant of interest and flanked by 15 bp of homologous coding sequence on each side of the mutation (which is usually restricted to 3 bp or less) is synthesized. It is essential that the melting temperature of the final oligonucleotide be 78°C or greater to ensure tight binding during the procedure. It is convenient if the mutation can be used to generate or remove a restriction site to simplify analysis of the resulting clones. The primers are added to the plasmid, which is produced in a methylation-competent strain of E. coli (DNA from most common strains is methylated) and amplified using Pfu DNA polymerase. Pfu polymerase is used in place of Taq because of its proofreading functions and the decreased likelihood of a random mutation.

The Pfu polymerase travels the length of the plasmid, generating a series of complementary, full-length, single-stranded linear DNA molecules initiating at the primer and terminating at the nucleotide immediately preceding the 5′ end of the primer. The two complementary linear molecules hybridize to form an intact plasmid circle bearing the mutation. The hybrids also contain two nicks, preceding the 5′ ends of each primer. The 30 bp or so of complementarity of the primers is enough to allow the plasmid circles to remain circularized and to transform E. coli. The recombinant hybrid is unmethylated, whereas the parental DNA obtained from a dam$^+$ strain of E. coli is heavily methylated at certain restriction sites. For 12 cycles, a maximum of 12 times the original amount of plasmid is produced. (Note that the amplification is not logarithmic.) Thus, analysis and detection of the products on a gel may not reveal significant amplification.

The two molecules are then subjected to cleavage with *Dpn*I, which cleaves only dam-methylated DNA at the sequence 5′G^{m6}ATC3′. This effectively removes the methylated parental DNA from the mixture. The uncleaved sibling mutant molecules are transformed into E. coli and screened. It is crucial to perform a mock reaction without the primers to ensure that the *Dpn*I completely removes the parental molecules. It is also necessary to ensure that the 5′ end of the primer is stably hybridized to the plasmid during the elongation phase, or the Pfu polymerase will displace it, generating background parental molecules that can transform E. coli. Usually 15 bp is sufficient for the primer, but if more can be introduced this will increase the efficiency of the mutant output. By introducing or removing a restriction site by mutagenesis, the products obtained from miniprep molecules can be easily screened to identify the correct ones. When a convenient site cannot be introduced, sequencing analysis must be performed to identify recombinant molecules. Typically at least 20%, and often 90–100%, of the products are mutagenized.

TIME LINE AND ORGANIZATION

The mutagenesis method is simple and does not require any special manipulations. Typically, after obtaining the primers, the amplification, digestion of parental molecules, and subsequent transformation can be completed in a single day. The colonies can be grown, and recombinant plasmids can be prepared the next day. Ideally, a restriction site is added or removed by mutation to facilitate screening. Otherwise, sequencing will have to be performed. As with all PCR mutagenesis methods, it is prudent to perform the reaction in duplicate to have a backup clone available in the event that the PCR process introduces additional mutations into the plasmid. This is becoming increasingly rare as thermostable DNA polymerases like Pfu with proofreading functions are identified.

This method is suited for large-scale point mutagenesis and is limited only by the cost of primers and the time it takes to screen. In the dut/ung method (Kunkel 1985) marketed by Bio-Rad (Muta-Gene M13 In Vitro Mutagenesis Kit, #1703580), multiple site-directed mutations can be generated in the same plasmid by simultaneously using different primers. To our knowledge, this is not feasible with PCR-mediated site-directed mutagenesis, and thus a plasmid with one mutation must be generated before the next can be added.

Preparation: Primer selection and plasmid preparation

PCR-mediated site-directed mutagenesis

Day 1: Mutagenesis
Day 2: Plasmid preparation and restriction analysis
Day 3: Sequence analysis

OUTLINE

Preparation

Step 1: Primer selection (1 day in addition to time required to obtain primers)
Step 2: Plasmid preparation (1 day by QIAGEN method #27104)

Mutagenesis (time commitment: 4–5 hours)

Step 1: Generate mutant plasmid DNA using PCR (1 hour 30 minutes)
Step 2: Cleavage of parental DNA with *Dpn*I(1 hour 30 minutes)
Step 3: Transformation of suitable *E. coli* host (1–2 hours and grow overnight)

Product analysis (time commitment: 2 days)

Step 1: Plasmid preparation (mini-prep) and restriction analysis (time commitment: 1 day and overnight)
Step 2: Screening by DNA sequencing (5–6 hours)

PROCEDURE

CAUTIONS: *Ethidium bromide, UV light. See Appendix I.*

PREPARATION

Step 1: Primer selection

Primer 1 and Primer 2 are usually about 30–35 nucleotides in length and must be purified by FPLC or by gel electrophoresis before use. The engineered mismatch should not be more than 3–4 nucleotides long (optimally, mismatches of 2 nucleotides are most easily incorporated; primers containing more than 2-nucleotide mismatches are less efficient). The mismatch should be located in the middle of the primer. Furthermore, the T_m for each primer must be equal to or greater than 78°C for this method to work well. This generally means that the primers must contain at least 50% G-C content. There are some commercially available programs that can determine the optimal T_m for a primer designed for a certain region. Alternately, the T_m of a given primer can be calculated roughly using this equation:

$$T_m = 81.5 + 0.41\ (\%GC) - 675/N - \%\ \text{mismatch}$$
$$N \text{ is the primer length in base pairs.}$$

Stratagene markets a kit (QuikChange Site-Directed Mutagenesis Kit, cat. #200518) similar to our protocol shown below, although the method can easily be performed with commercially available reagents.

Select suitable primers by computer analysis and availability of restriction sites to facilitate screening. Order or synthesize primers.

Step 2: Plasmid preparation

A plasmid bearing the region of interest is grown in a dam-methylase-positive *E. coli* strain (e.g., HB101). If unsure about the strain used to amplify a preexisting plasmid, simply check cleavage of plasmid by *Dpn*I prior to performing the mutagenesis. The plasmid is isolated by either the CsCl density method or QIAGEN columns (#27104). DNA prepared by either of these methods is suitable for the PCR mutagenesis procedure.

MUTAGENESIS

Step 1: Generate mutant plasmid DNA using PCR

1. Prepare the reactions in 0.6-ml siliconized eppendorf tubes or special PCR tubes. The reaction mixtures should be overlaid with mineral oil except when using thermocyclers bearing insulated chambers. The mixture below is for a single reaction. Perform the reactions in duplicate and perform one mock reaction lacking primers. This will be essential in later steps to determine the efficiency of cleavage of parental DNA by *Dpn*I. The reactions should be assembled on ice to prevent background synthesis or primer cleavage by Pfu DNA polymerase (Stratagene #600135).

   Mix:

10x Pfu reaction buffer	5.00 µl
Primer 1 (100 pmole/µl)	1.25 µl
Primer 2 (100 pmole/µl)	1.25 µl
dH$_2$O	39.50 µl
plasmid containing cDNA (100 ng/µl)	1.00 µl
Pfu DNA polymerase (2.5 units/µl)	1.00 µl
dNTP mix (25 mM)	1.00 µl
Total	50.00 µl

2. Perform the linear amplification in a thermocycler. It is important that the temperature of the elongation step is well below the T_m of the primers.

| Step 1: | Denature: | 96°C, 30 seconds |

Subsequent steps (12 cycles)

Step 1:	Denature:	96°C, 30 seconds
Step 2:	Anneal:	50°C, 1 minute
Step 3:	Elongate:	68°C, 12 minutes for full circle

Step 2: Cleavage of PCR amplified product with *Dpn*I

At this point the mixtures can be transferred to another tube if necessary, and 1–5 units of *Dpn*I can be added. The mixtures are incubated 1–2 hours at 37°C in a water bath. Heat digestion reactions at 65°C for 10 minutes.

Step 3: Transformation of a suitable *E. coli* host

There is no need for phenol extraction and precipitation prior to transformation of *E. coli*. Transform 45 μl of the restriction endonuclease-cleaved mixture into 100 μl of HB101-competent cells. The transformation efficiencies may be very low in some cases due to the nature of the mutagenesis. Before transformation, check the literature to be sure that the restriction enzyme you will be using to screen for positive colonies is not methylation-sensitive. If it is, transform the plasmid into a dam-minus strain of *E. coli*. Always include three controls for transformation. Control #1 contains supercoiled plasmid to check for the competency of cells. Control #2 consists of a mock reaction (e.g., lacking primers or Pfu polymerase) to check for the efficiency of *Dpn*I digestion. Control #3 contains no plasmid to make sure that the cells alone do not grow in the presence of the selective antibiotic used. Be careful not to carry mineral oil over into the transformation, as this inhibits the efficiency.

PRODUCT ANALYSIS

Analyze the signal-to-noise ratio. If there are a sufficiently large number of colonies on the plates bearing cells transformed with the mock reaction products (i.e., lacking primers), it may not be wise to proceed any further because this generally is indicative of inefficient cleavage by *Dpn*I. If there is a two- to threefold increase in colonies on the plates transformed with the mutagenesis mixtures, continue to the screening stage.

Step 1: Plasmid preparation and restriction analysis

1. Perform the mini-prep analysis (QIAGEN).

2. Digest the DNA with an appropriate restriction enzyme to screen for loss or gain of a restriction site due to the mutation.

3. Analyze the restriction products on a 1% agarose gel containing ethidium bromide and visualize the products by UV light.

Step 2: Screen the mutation by DNA sequencing

ADDITIONAL CONSIDERATIONS

Screening recombinants

Clearly, there may not always be a restriction site compatible with the mutation being introduced, particularly if one is performing a comprehensive alanine scanning analysis, for example (see this chapter for details). We have found that within a short time, and by following the primer T_m rules, almost all of the clones arising from a given mutagenesis are positive for the mutation. This permits the large-scale use of the technique for generating site-directed mutants (see Chapter 7), which, like the dut/ung method, must eventually be confirmed by DNA sequencing.

Dam methylation-sensitive restriction sites

Some enzymes are unable to cleave methylated DNA sites from plasmids grown in dam⁺ strains. For example, the *Bcl*I site will not be recognized by the *Bcl*I restriction enzyme if the DNA is methylated in the N^6 position of adenine by dam methylase. Methylated adenines are not recognized by a number of common restriction enzymes (check supplier's catalog for a complete list). Transforming the mutagenized plasmid into a dam-deficient strain of *E. coli* will allow subsequent screening of plasmid DNA by dam-methylation-sensitive enzymes. It is crucial, however, to grow the plasmid DNA to be mutagenized in a dam⁺ strain.

Controls

The most essential control for every mutagenesis is comparing the mutants with the mock reactions.

TROUBLESHOOTING

Recover only parental plasmids

A good signal-to-noise ratio is observed when screening colonies from mutagenesis versus mock transformation, but parental plasmid is recovered when screening.

Possible cause: This is apparently due to a low primer T_m. We do not fully understand the problem but speculate that at too low a T_m the primer can be displaced by Pfu after it has traveled full circle.

Solution: Resynthesize a new primer with correct T_m.

No transformants

Possible cause: The transformation efficiency of the cells is low.
Solution: Check the transformation efficiency of the cells using the parental plasmid.

Possible cause: If transformation efficiency is not the problem, the result may be due to inefficient amplification of the parental DNA, of which there are several potential causes.

One is having too few amplification cycles. Another is inefficient amplification of large plasmids. Finally, as the size of the mutation goes beyond two or three nucleotides, the efficiency of the procedure decreases dramatically, as it does with many mutagenesis methods. *Possible solutions:* Increase the number of cycles and repeat, increase the time of extension, and, if all else fails, introduce the mutation by PCR ligation and subclone into the appropriate region of the cDNA.

REFERENCES

Baumann R., Warren G., and Askovic S. 1995. Restoration of the Epstein-Barr virus ZEBRA protein's capacity to disrupt latency by the addition of heterologous activation regions. *Virology* **211:** 64–72.

Baumann R., Grogan E., Ptashne M., and Miller G. l993. Changing Epstein-Barr viral ZEBRA protein into a more powerful activator enhances its capacity to disrupt latency. *Proc. Natl. Acad. Sci.* **90:** 4436–4440.

Beato M., Herrlich P., and Schutz G. 1995. Steroid hormone receptors: Many actors in search of a plot. *Cell* **83:** 851–857.

Benezra R., Davis R.L., Lockshon D., Turner D.L., and Weintraub H. 1990. The protein Id: A negative regulator of helix-loop-helix DNA binding proteins. *Cell* **61:** 49–59.

Brent R. and Ptashne M. 1985. A eukaryotic transcriptional activator bearing the DNA specificity of a prokaryotic repressor. *Cell* **43:** 729–736.

Bryant G.O., Martel L.S., Burley S.K., and Berk A.J. 1996. Radical mutations reveal TATA-box binding protein surfaces required for activated transcription in vivo. *Genes Dev.* **10:** 2491–2504.

Burley S.K. and Roeder R.G. 1996. Biochemistry and structural biology of transcription factor IID (TFIID). *Annu. Rev. Biochem.* **65:** 769–799.

Bushnell D.A., Bamdad C., and Kornberg R.D. 1996. A minimal set of RNA polymerase II transcription protein interactions. *J. Biol. Chem.* **271:** 20170–20174.

Carey M. 1998. The enhanceosome and transcriptional synergy. *Cell* **92:** 5–8.

Carey M., Kakidani H., Leatherwood J., Mostashari F., and Ptashne M. 1989. An amino-terminal fragment of GAL4 binds DNA as a dimer. *J. Mol. Biol.* **209:** 423–432.

Carey M., Leatherwood J., and Ptashne M. 1990. A potent GAL4 derivative activates transcription at a distance in vitro. *Science* **247:** 710–712.

Coleman R.A., Taggart A.K., Benjamin L.R., and Pugh B.F. 1995. Dimerization of the TATA binding protein. *J. Biol. Chem.* **270:** 13842–13849.

Cress W.D. and Triezenberg S.J. 1991. Critical structural elements of the VP16 transcriptional activation domain. *Science* **251:** 87–90.

Cunningham B.C. and Wells J.A. 1989. High-resolution epitope mapping of hGH-receptor interactions by alanine-scanning mutagenesis. *Science* **244:** 1081–1085.

Davis R.L., Cheng P.F., Lassar A.B., and Weintraub H. 1990. The MyoD DNA binding domain contains a recognition code for muscle-specific gene activation. *Cell* **60:** 733–746.

Driever W., Ma J., Nusslein-Volhard C., and Ptashne M. 1989. Rescue of bicoid mutant Drosophila embryos by bicoid fusion proteins containing heterologous activating sequences. *Nature* **342:** 149–154.

Ellenberger T., Fass D. Arnaud M., and Harrison S.C. 1994. Crystal structure of transcription factor E47: E-box recognition by a basic region helix-loop-helix dimer. *Genes Dev.* **8:** 970–980.

Emami K.H. and Carey M. 1992. A synergistic increase in potency of a multimerized VP16 transcriptional activation domain. *EMBO J.* **11:** 5005–5012.

Fields S. and Song O. 1989. A novel genetic system to detect protein-protein interactions. *Nature* **340:** 245–246.

Fujita T., Nolan G.P., Ghosh S., and Baltimore D. 1992. Independent modes of transcriptional activation by the p50 and p65 subunits of NF-κB. *Genes Dev.* **6:** 775–787.

Giese K. and Grosschedl R. 1993. LEF-1 contains an activation domain that stimulates transcription only in a specific context of factor-binding sites. *EMBO J.* **12:** 4667–4676.

Gill G. and Ptashne M. 1987. Mutants of GAL4 protein altered in an activation function. *Cell* **51:** 121–126.

Giniger E. and Ptashne M. 1987. Transcription in yeast activated by a putative amphipathic alpha helix linked to a DNA binding unit. *Nature* **330:** 670–672.

Giniger E., Varnum S.M., and Ptashne M. 1985. Specific DNA binding of GAL4, a positive regulatory protein of yeast. *Cell* **40:** 767–774.

Goodrich J.A., Hoey T., Thut C.J., Admon A., and Tjian R. 1993. Drosophila TAFII40 interacts with both a VP16 activation domain and the basal transcription factor TFIIB. *Cell* **75:** 519–530.

Greenblatt J. and Ingles C.J. 1996. Interaction between acidic transcriptional activation domains of herpes simplex virus activator protein VP16 and transcriptional initiation factor IID. *Methods Enzymol.* **274:** 120–133.

Gstaiger M., Georgiev O., van Leeuwen H., van der Vliet P., and Schaffner W. 1996. The B cell coactivator Bob1 shows DNA sequence-dependent complex formation with Oct-1/Oct-2 factors, leading to differential promoter activation. *EMBO J.* **15:** 2781–2790.

Gupta R., Emili A., Pan G., Xiao H., Shales M., Greenblatt J., and Ingles C.J. 1996. Characterization of the interaction between the acidic activation domain of VP16 and the RNA polymerase II initiation factor TFIIB. *Nucleic Acids Res.* **24:** 2324–2330.

Ha I., Roberts S., Maldonado E., Sun X., Kim L.U., Green M., and Reinberg D. 1993. Multiple functional domains of human transcription factor IIB: Distinct interactions with two general transcription factors and RNA polymerase II. *Genes Dev.* **7:** 1021–1032.

Hagmann M., Georgiev O., and Schaffner W. 1997. The VP16 paradox: Herpes simplex virus VP16 contains a long-range activation domain but within the natural multiprotein complex activates only from promoter-proximal positions. *J. Virol.* **71:** 5952–5962.

Heyduk T., Ma Y., Tang H., and Ebright R.H. 1996. Fluorescence anisotropy: Rapid, quantitative assay for protein-DNA and protein-protein interaction. *Methods Enzymol.* **274:** 492–503.

Hisatake K., Ohta T., Takada R., Guermah M., Horikoshi M., Nakatani Y., and Roeder R.G. 1995. Evolutionary conservation of human TATA-binding-polypeptide-associated factors TAFII31 and TAFII80 and interactions of TAFII80 with other TAFs and with general transcription factors. *Proc. Natl. Acad. Sci.* **92:** 8195–8199.

Hope I.A. and Struhl K. 1996. Functional dissection of a eukaryotic transcriptional activator protein, GCN4 of yeast. *Cell* **46:** 885–894.

Hori R., Pyo S., and Carey M. 1995. Protease footprinting reveals a surface on transcription factor TFIIB that serves as an interface for activators and coactivators. *Proc. Natl. Acad. Sci.* **92:** 6047–6051.

Hu J.C., O'Shea E.K., Kim P.S., and Sauer R.T. 1990. Sequence requirements for coiled-coils: Analysis with lambda repressor-GCN4 leucine zipper fusions. *Science* **250:** 1400–1403.

Johnston M. 1987. A model fungal gene regulatory mechanism: The GAL genes of *Saccharomyces cerevisiae*. *Microbiol. Rev.* **51:** 458–476.

Keegan L., Gill G. and Ptashne M. 1986. Separation of DNA binding from the transcription-activating function of a eukaryotic regulatory protein. *Science* **231:** 699–704.

Korzus E., Torchia J., Rose D.W., Xu L., Kurokawa R., McInerney E.M., Mullen T.M., Glass C.K., and Rosenfeld M.G. 1998. Transcription factor-specific requirements for coactivators and their acetyltransferase functions. *Science* **279:** 703–707.

Kristie T.M. and Sharp P.A. 1990. Interactions of the Oct-1 POU subdomains with specific DNA sequences and with the HSV alpha-trans-activator protein. *Genes Dev.* **4:** 2383–2396.

Kunkel T.A. 1985. Rapid and efficient site-specific mutagenesis without phenotypic selection. *Proc. Natl. Acad. Sci.* **82:** 488–492.

Landschulz W.H., Johnson P.F., and McKnight S.L. 1988. The leucine zipper: A hypothetical structure common to a new class of DNA binding proteins. *Science* **240:** 1759–1764.

Lefstin J.A., Thomas J.R., and Yamamoto K.R. 1994. Influence of a steroid receptor DNA-binding domain on transcriptional regulatory functions. *Genes Dev.* **8:** 2842–2856.

Lefstin J.A. and Yamamoto K.R. 1998. Allosteric effects of DNA on transcriptional regulators. *Nature* **392:** 885–888.

Li X., Lopez-Guisa J.M., Ninan N., Weiner E.J., Rauscher F.J. 3rd, and Marmorstein R. 1997. Over-expression, purification, characterization, and crystallization of the BTB/POZ domain from the PLZF oncoprotein. *J. Biol. Chem.* **272:** 27324–27329.

Li M., Moyle H., and Susskind M.M. 1994. Target of the transcriptional activation function of phage lambda cI protein. *Science* **263:** 75–77.

Lin Y.S., Ha I. Maldonado E., Reinberg D., and Green M.R. 1991. Binding of general transcription factor TFIIB to an acidic activating region. *Nature* **353:** 569–571.

Lin Y.S. and Green M.R. 1991. Mechanism of action of an acidic transcriptional activator in vitro. *Cell* **64:** 971–981.

Luo Y. and Roeder R.G. 1995. Cloning, functional characterization, and mechanism of action of the B-cell-specific transcriptional coactivator OCA-B. *Mol. Cell. Biol.* **15:** 4115–4124.

Ma P.C., Rould M.A., Weintraub H., and Pabo C.O. 1994. Crystal structure of MyoD bHLH domain-DNA complex: Perspectives on DNA recognition and implications for transcriptional activation. *Cell* **77:** 451–459.

Ma J. and Ptashne M. 1987a. Deletion analysis of GAL4 defines two transcriptional activating segments. *Cell* **48:** 847–853.

———. 1987b. A new class of yeast transcriptional activators. *Cell* **51:** 113–119.

Mangelsdorf D.J. and Evans R.M. 1995. The RXR heterodimers and orphan receptors. *Cell* **83:** 841–850.

Marmorstein R., Carey M., Ptashne M., and Harrison S.C. 1992. DNA recognition by GAL4: Structure of a protein-DNA complex [see comments]. *Nature* **356:** 408–414.

Miller J., McLachlan A.D., and Klug A. 1985. Repetitive zinc-binding domains in the protein transcription factor IIIA from *Xenopus* oocytes. *EMBO J.* **4:** 1609–1614.

Murre C., McCaw P.S., and Baltimore D. 1989. A new DNA binding and dimerization motif in immunoglobulin enhancer binding, daughterless, MyoD, and myc proteins. *Cell* **56:** 777–783.

Nikolov D.B. and Burley S.K. 1997. RNA polymerase II transcription initiation: A structural view. *Proc. Natl. Acad. Sci.* **94:** 15–22.

Nikolov D.B., Chen H., Halay E.D., Hoffman A., Roeder R.G., and Burley S.K. 1996. Crystal structure of a human TATA box-binding protein/TATA element complex. *Proc. Natl. Acad. Sci.* **93:** 4862–4867.

O'Shea E.K., Rutkowski R., and Kim P.S. 1989. Evidence that the leucine zipper is a coiled coil. *Science* **243:** 538–542.

Pabo C.O., Sauer R.T., Sturtevant J.M., and Ptashne M. 1979. The lambda repressor contains two domains. *Proc. Natl. Acad. Sci.* **76:** 1608–1612.

Petersen J.M., Skalicky J.J., Donaldson L.W., McIntosh I.P., Alber T., and Graves B.J. 1995. Modulation of transcription factor Ets-1 DNA binding: DNA-induced unfolding of an alpha helix. *Science* **269:** 1866–1869.

Ptashne M. 1992. *A Genetic Switch*, 2nd ed. Cell Press and Blackwell Science, Malden, Massachusetts.

Regier J.L., Shen F., and Triezenberg S.J. 1993. Pattern of aromatic and hydrophobic amino acids critical for one of two subdomains of the VP16 transcriptional activator. Proc. *Natl. Acad. Sci.* **90:** 883–887.

Rellahan B.L., Jensen J.P., Howcroft T.K., Singer D.S., Bonvini E., and Weissman A.M. 1998. Elf-1 regulates basal expression from the T cell antigen receptor zeta-chain gene promoter. *J. Immunol.* **160:** 2794–2801.

Roberts S.G. and Green M.R. 1994. Activator-induced conformational change in general transcription factor TFIIB. *Nature* **371:** 717–720.

Roberts S.G., Ha I., Maldonado E., Reinberg D., and Green M.R. 1993. Interaction between an acidic activator and transcription factor TFIIB is required for transcriptional activation. *Nature* **363:** 741–744.

Sadowski I., Ma J., Triezenberg S., and Ptashne M. 1988. GAL4-VP16 is an unusually potent transcriptional activator. *Nature* **335:** 563–564.

Saha S., Brickman J.M., Lehming N., and Ptashne M. 1993. New eukaryotic transcriptional repressors. *Nature* **363:** 648–652.

Schena M., Freedman L.P., and Yamamoto K.R. 1989. Mutations in the glucocorticoid receptor zinc finger region that distinguish interdigitated DNA binding and transcriptional enhancement activities. *Genes Dev.* **3:** 1590–1601.

Shen E., Triezenberg S.J., Hensley P., Porter D., and Knutson J.R. 1996a. Critical amino acids in the transcriptional activation domain of the herpesvirus protein VP16 are solvent-exposed in highly mobile protein segments. An intrinsic fluorescence study. *J. Biol. Chem.* **271:** 4819–4826.

———. 1996b. Transcriptional activation domain of the herpesvirus protein VP16 becomes conformationally constrained upon interaction with basal transcription factors. *J. Biol. Chem.* **271:** 4827–4837.

Smith M.J., Charron-Prochownik D.C., and Prochownik E.V. 1990. The leucine zipper of c-Myc is required for full inhibition of erythroleukemia differentiation. *Mol. Cell. Biol.* **10:** 5333–5339.

Sopta M., Carthew R.W., and Greenblatt J. 1985. Isolation of three proteins that bind to mammalian RNA polymerase II. *J. Biol. Chem.* **260:** 10353–10360.

Stancovski I. and Baltimore D. 1997. NF-κB activation: The IκB kinase revealed? *Cell* **91:** 299–302.

Strubin M. and Struhl K. 1992. Yeast and human TFIID with altered DNA-binding specificity for TATA elements. *Cell* **68:** 721–730.

Tan S., Hunziker Y., Sargent D.F., and Richmond T.J. 1996. Crystal structure of a yeast TFIIA/TBP/DNA complex [see comments]. *Nature* **381:** 127–151.

Tanaka M., Clouston W.M., and Herr W. 1994. The Oct-2 glutamine-rich and proline-rich activation domains can synergize with each other or duplicates of themselves to activate transcription. *Mol. Cell. Biol.* **14:** 6046–6055.

Tang H., Sun X., Reinberg D., and Ebright R.H. 1996. Protein-protein interactions in eukaryotic transcription initiation: Structure of the preinitiation complex. *Proc. Natl. Acad. Sci.* **93:** 1119–1124.

Tansey W.P. and Herr W. 1995. The ability to associate with activation domains in vitro is not required for the TATA box-binding protein to support activated transcription in vivo. *Proc. Natl. Acad. Sci.* **92:** 10550–10554.

———. 1997. Selective use of TBP and TFIIB revealed by a TATA-TBP-TFIIB array with altered specificity. *Science* **275:** 829–831.

Triezenberg S.J. 1995. Structure and function of transcriptional activation domains. *Curr. Opin. Genet. Dev.* **5:** 190–196.

Triezenberg S.J., Kingsbury R.C., and McKnight S.L. 1988. Functional dissection of VP16, the transactivator of herpes simplex virus immediate early gene expression. *Genes Dev.* **2:** 718–729.

Tzamarias D. and Struhl K. 1994. Functional dissection of the yeast Cyc8-Tup1 transcriptional corepressor complex. *Nature* **369:** 758–761.

Uesugi M., Nyanguile O., Lu H., Levine A.J., and Verdine G.L. 1997. Induced alpha helix in the VP16 activation domain upon binding to a human TAF. *Science* **277:** 1310–1313.

Walker S., Greaves R., and O'Hare P. 1993. Transcriptional activation by the acidic domain of Vmw65 requires the integrity of the domain and involves additional determinants distinct from those necessary for TFIIB binding. *Mol. Cell. Biol.* **13:** 5233–5244.

Wagner S. and Green M.R. HTLV-I Tax protein stimulation of DNA binding of bZIP proteins by enhancing dimerization. *Science* **262:** 395–399.

Weiner M.P., Costa G.L., Schoettlin W., Cline J., Mathur E., and Bauer J.C. 1994. Site-directed mutagenesis of double-stranded DNA by the polymerase chain reaction. *Gene* **151:** 119–23.

Weiss M.J., Yu C., and Orkin S.H. 1997. Erythroid-cell-specific properties of transcription factor GATA-1 revealed by phenotypic rescue of a gene-targeted cell line. *Mol. Cell. Biol.* **17:** 1642–1651.

Wu Y., Reece R.J., and Ptashne M. 1996. Quantitation of putative activator-target affinities predicts transcriptional activating potentials. EMBO *J.* **15:** 3951–3963.

CHAPTER 13

Theory, Characterization, and Modeling of DNA Binding by Regulatory Transcription Factors

Important issues

- *Transcription factors recognize their DNA sites using a variety of different mechanisms.*
- *Parameters besides the DNA recognition sequence regulate binding in vivo.*
- *DNA binding is measured and quantitated using several simple assays.*
- *Modeling DNA–protein interactions is necessary for understanding the mechanism of DNA binding.*

INTRODUCTION

A mechanistic analysis of a promoter generally involves experiments to determine how sequence-specific transcriptional regulatory proteins recognize and bind to their DNA sites, both alone and in combinations. Chapter 9 discussed the criteria for determining the physiological relevance of a DNA–protein interaction. Here we discuss the theory of DNA recognition, how to identify a high-affinity recognition site for a DNA-binding protein, and finally, how to study and model a DNA–protein interaction using chemical and nuclease probes. We then elaborate on some simple principles and methods for studying the formation of multi-activator complexes or enhanceosomes.

The current model is that a eukaryotic DNA-binding protein binds to its physiological sites by continually colliding with nuclear DNA until it encounters a functional site within a promoter. A functional site is one that mediates the physiological action of a transcription factor in the context of a regulated promoter. Nonspecific sites, on the other hand, comprise random sequence or, as we discuss below, specific recognition sequences in an incorrect context. The cell has devised three strategies to expedite and confer specificity to the search. First, much of the untranscribed DNA in a cell is packaged into chromatin, which is largely inaccessible to the regulatory molecule. Second, the concentration of regulatory molecules in the nucleus is raised sufficiently high to overcome any significant competition from nonspecific DNA (i.e., most activators and repressors appear to be expressed at levels of 1000–50,000 molecules/nucleus, likely in excess of their specific sites; this issue is discussed in Ptashne 1992). Finally, a substantial amount of the binding energy is derived from protein–protein interactions that occur only in the proper promoter context.

Context-dependent DNA–protein and protein–protein interactions are central to locating a site. There are two issues that must be considered. First, the actual number of sites which an activator can recognize in naked genomic DNA far exceeds the number of physiological sites. Imagine, for example, a factor that recognizes and contacts a 6-bp site. Statistically, this 6-bp site is present 732,422 times in the human genome (i.e., 3 billion bp divided by 4^6, the number of combinatorial possibilities for a 6-bp site). Because there are only 50,000–100,000 protein-coding genes in the cell, and it is unlikely that any regulatory factor, with the exception of the general machinery, binds to all of them, the actual number of recognition sites for any given factor very likely exceeds the number of physiologically relevant or functional sites. Second, transcription factors often fall into families that recognize related or identical sites in vitro (see Luisi 1995). Because the carefully orchestrated action of transcription factor family members on distinct promoters is critical to the proper functioning and development of eukaryotic cells, these issues raise the question of how physiological specificity is imparted on a DNA–protein interaction.

The enhanceosome theory has been invoked to explain how a protein is able to achieve the proper specificity (see Chapter 1; Echols 1986; Grosschedl 1995; Carey 1998). Figure 13.1 shows the prototypic IFN-β enhancer complex and schematically illustrates its docking with the general machinery. The concept is that the arrangements of sites within a promoter/enhancer and the specific repertoire of regulatory proteins that bind these sites generate a unique network of protein–DNA and protein–protein interactions. The energy or stability of the final structure is dependent on the accurate placement of binding sites and binding of the correct regulatory factors to these sites. The ultimate goal is to assemble a stable complex with the lowest free energy, much like the assembly of a puzzle from its component pieces. In the case shown, the c-Jun/ATF heterodimer binds cooperatively with IRF-3, IRF-7, and NF-κB to generate an enhanceosome complex.

This view would clarify how an activator distinguishes its physiological sites from nonphysiological sites. First, by cooperating with other proteins in a complex as in Figure 13.1A, an activator has a higher affinity for its physiological sites. Presumably, under physiological conditions, only the combination of factors shown in Figure 13.1A could bind and assemble the enhanceosome due to the correct balance of activator concentration and protein–protein interactions. Second, it solves the paradox of related sequence preferences. Although several regulatory proteins may recognize an identical sequence, the subsequent stereospecific protein–protein interactions and the final free energy of the complex would "select" for the correct factor. Another level of selectivity is that the enhanceosome itself would generate a surface complementary to a surface on the Pol II general machinery; only when the correct interface was formed would the enhanceosome loop out the DNA and recruit Pol II, coactivators, and the general factors to the promoter (Fig. 13.1B). Indeed, as discussed in Chapter 1, under such a mechanism the general machinery would assist in assembling the enhanceosome via reciprocal cooperative interactions.

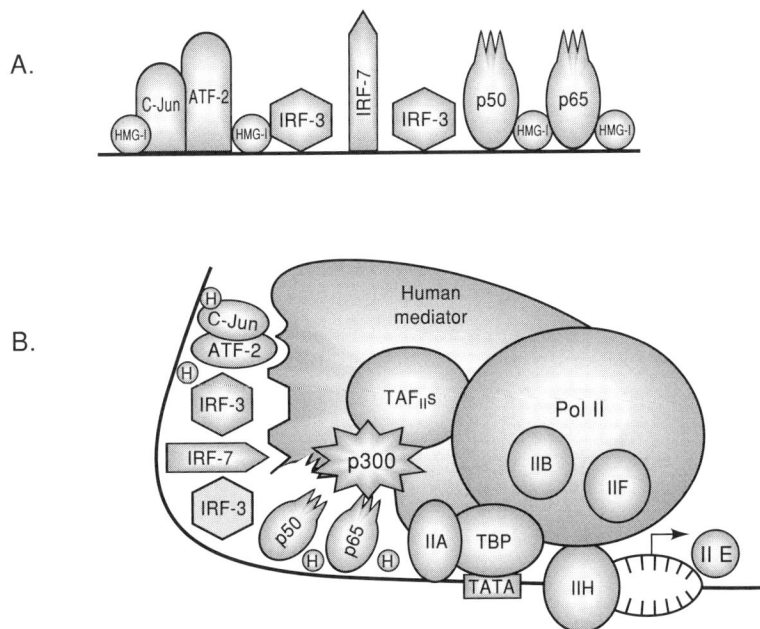

FIGURE 13.1. (A) Schematic of the IFN-β enhanceosome. (Adapted, with permission, from Carey 1998 [Copyright 1998, Cell Press].) (B) Docking with general machinery. (Adapted, with permission, from Ptashne et al. 1998 [Copyright 1998, Elsevier Science].)

According to the hypothesis above, cooperative, promoter context-dependent interactions are the driving force behind distinguishing a functional site from a nonphysiological site. Proteins that deform or bend the DNA, such as HMG-I in Figure 13.1B, may be a necessary component of such complexes. These proteins could permit certain combinatorial protein–protein interactions otherwise restricted by the limited flexibility of the intervening DNA within its persistence length and the size and flexibility of the bound proteins.

This chapter examines biochemical methods and strategies for understanding basic aspects of promoter recognition. We initially focus on how a single cloned regulatory protein recognizes a site, followed by a summary of methods for studying DNA bending and cooperative binding, two phenomena necessary to generate more sophisticated and specific enhanceosome complexes.

CONCEPTS AND STRATEGIES

General Theory and Examples of DNA–Protein Interactions
Theory of DNA Recognition

DNA site recognition by a regulatory protein is generally influenced by both specific interactions with the bases and nonspecific interactions with the phosphate/sugar backbone. Although there is some mild controversy surrounding the use of this "simplified" terminology to describe DNA binding, it nevertheless provides a framework that can be refined on the basis of the regulatory context.

In a typical interaction the regulatory protein (P) and DNA site (S) are in reversible equilibrium with the protein-site (PS) complex. The equilibrium is represented as PS↔P+S with an equilibrium dissociation constant of $K_d = k_1/k_2 = [P][S]/[PS]$. This equilibrium constant, also defined as the ratio of the forward (k_1) and reverse (k_2) rate constants, takes into account all of the enthalpic and entropic energies contributing to binding, including the cost of locating the site ($\Delta G = \Delta H - T\Delta S$). This K_d can also be defined in terms of free energy, using the Gibbs free-energy equation $\Delta G = -RT\ln K_d$ (for a discussion of how this equation bears on biological reactions, see Dill 1997).

The energy and specificity of a protein–DNA interaction are generated by a unique stereospecific array of amino acid side chains that are chemically and spatially complementary to an array of chemical groups displayed by the bases in the major or minor groove of the DNA. Each chemical interaction provides a quantum of free energy; each deviation from the ideal site generates a dramatic reduction in site affinity due to the logarithmic relationship between free energy (ΔG) and K_d as described above by the Gibbs equation. This logarithmic relationship is one mechanism for enhancing specificity. The form of complementarity described above is called direct readout.

Another form of DNA recognition is called indirect readout, and it concerns the ability of a protein to bind a specific sequence based on the DNA secondary structure, or conformation. If the recognition site deviates from the optimum, the inherent deformability of the DNA at the site may be affected. The resulting change in K_d may be much greater than would be predicted by the loss of energy from simple chemical interactions. The *Eco*RI GAATTC recognition site, for example, depends on a specific sequence array to accommodate a deformation. Substitution of a single base within this site alters the ΔG of binding and subsequently raises the K_d (Lesser et al. 1990).

In addition to the specific interactions of amino acid side chains with the exposed groups of base pairs, nonspecific interactions with the relatively uniform phosphate backbone of the B-DNA helix also contribute to the binding energy or K_d. Crystallography studies suggest that these interactions often provide the bulk of the free energy within the recognition complex, although they do not generate specificity (Pabo and Sauer 1992).

In addition to the noncovalent enthalpic effects described above, entropic contributions derived from the release of ordered water and salt ions from a site upon binding are believed to be significant driving forces in protein–DNA interactions (Ha et al. 1989).

Studies on the partitioning of energy in site-specific DNA recognition have not yet yielded a satisfactory general understanding of the problem. However, significant advances and a renewed interest in the chemistry and physics of site recognition suggest a solution in the future.

Chemical Basis of the Interactions

We now examine more closely the primary enthalpic contributions to DNA binding—the interactions between the amino acid side chains, and occasionally, backbone amide and carbonyl groups, with both the phosphate/sugar backbone and the exposed chemical groups of base pairs displayed in the major or minor grooves. The four classes of specific interactions in the major and minor groove are:

1. The C5-methyl group of thymines and the C5-hydrogen group of cytosine participate in van der Waals contacts with the aliphatic amino acids (Fig. 13.2A).

2. Specific H-bonds with the exposed edges of base pairs and the phosphates along the DNA helix (Fig. 13.2B, C). Amino acid–DNA H-bonding interactions are supported by a variety of amino acid side-chains (Fig. 13B) and both the amide and carbonyl groups of the peptide backbone (Fig. 13C). All specific protein–DNA complexes employ this strategy.

3. Water molecules can serve as a hydrogen-bonding bridge between an amino acid and a base pair or phosphate (e.g., the *Eco*RI and Trp-R co-complexes: Otwinowski et al. 1988; Narayana et al. 1991; Shakked et al. 1994) (Fig. 13.2D). This concept has arisen as a result of higher-resolution protein–DNA structures.

4. Occasionally an amino acid side chain will intercalate between two bases. Observed mainly in DNA-bending proteins, this interaction is exemplified by the TBP, which inserts two phenylalanines between the first and last base pairs of the TATA box. The intercalation generates an 80° bend, unwinds the helix, and widens the minor groove. HMG proteins employ a similar strategy (Werner and Burley 1997). We cover this topic in more detail later in the chapter.

The Role of the α-Helix in DNA Recognition

How does a protein specifically recognize DNA? The size and shape of the α-helix (cylindrical; main chain is 4.6 Å, in diameter) are ideal for fitting into the major groove (helical diameter, 19 Å; major groove rise, 17 Å) and, not surprisingly, the vast majority of DNA-binding proteins have employed this strategy (Fig. 13.2E). The chemical diversity and flexibility of the amino acid side chains and the rotation of the helical axis endow the α-helix with a large number of potential recognition surfaces for binding a specific DNA sequence. Crystal studies also show that the disposition of this so-called recognition α-helix in the groove relative to the backbone axis of the DNA varies extensively among different regulatory proteins. However, because there is little evidence that an isolated α-helix is capable of independent recognition (Pabo and Sauer 1992), it has been proposed that interactions with the phosphate backbone, mediated by other protein elements, are essential for properly positioning the helix on its site. Thus, once on its site, the recognition helix is stabilized by a protein scaffold (including adjacent α-helices) and an intricate network of

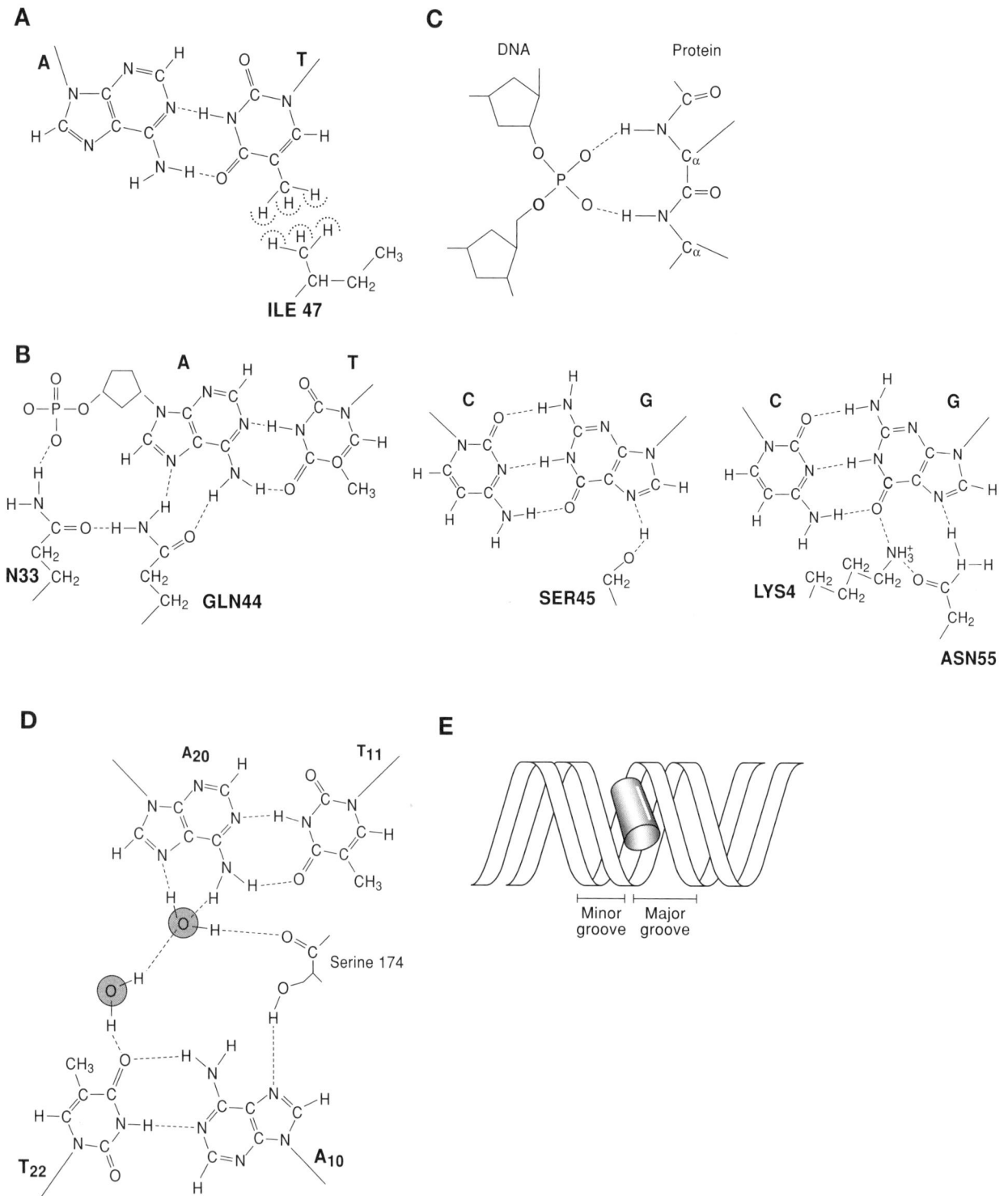

FIGURE 13.2. Chemical basis of interactions for DNA binding. (*A*, Adapted, with permission from Kissinger et al. 1990 [Copyright Cell Press] 1992.) (*B*, Adapted, with permission, from Branden and Tooze 1991 [Copyright 1991 Garland Publishing].) (*C*, Redrawn, with permission, from Jordan and Pabo 1988 [Copyright 1988 American Association for the Advancement of Science].) (*D*, Adapted, with permission, from Feng et al. 1994. [Copyright 1994 American Association for the Advancement of Science].) (*E*) An α-helix in major groove. (Adapted, with permission from Ptashne 1992 [Copyright Blackwell Science].)

nonspecific interactions with the DNA backbone. Despite the predilection of structures employing α-helices in the major groove, there are many examples of β-sheets used in DNA recognition, including the prokaryotic Met and Arc repressors (Somers and Phillips 1992; Raumann et al. 1994; Suzuki 1995) and the eukaryotic p53 protein (Cho et al. 1994). Furthermore, many proteins use, in addition to major groove interactions, flexible stretches of peptide that reach back and make contacts into the minor groove, thereby enhancing specificity.

Major and Minor Groove Specificity

We have described the energetics of DNA site recognition above, but we have not discussed what constitutes the complementary surface on DNA that permits it to dock with a protein surface in a sequence-specific fashion. That is, how do different DNA sequences mark themselves to be identified by a regulatory protein? The conformation or shape of the DNA, although displaying minor sequence-dependent changes in its dimensions and torsion angles, is nonetheless relatively uniform in structure, as are the phosphate backbone and sugar conformations. Instead, specificity is imparted by the sequence-dependent projection of chemical groups from the bases into the major and minor grooves.

When A and T, or C and G, hydrogen-bond to form the base pairs connecting the antiparallel strands of the DNA, all four bases expose chemical groups that are either multivalent or not engaged in pairing. These exposed groups could conceivably bond with amino acid side chains or the peptide backbone on the recognition surface of the regulatory protein. These chemical groups, however, could only impart specificity if the arrangement of accessible H-bond donors and acceptors in the major and minor groove changed dramatically with the DNA sequence. In 1976, Seeman and colleagues (Seeman et al. 1976) considered this issue by examining the arrangements of chemical groups in different base pairs. They concluded, for reasons that are described below, that each base pair displays a unique three-dimensional pattern of chemical groups in the major groove. A multibase recognition site displays a combinatorial and, hence, more elaborate array. Thus, the uniqueness of an array increases with the size of the site. This variation is most evident in the major groove and less evident in the minor groove.

How does the array vary? There are four combinations of base pairs—A:T, T:A, G:C, and C:G. Each base pair, in addition to containing H-bonded chemical groups that permit formation of the base pair, also displays certain unbonded chemical groups that fall into three categories: (1) the H-bond acceptor (ac); (2) the H-bond donor (do), or (3) the van der Waals contacts with a C5 hydrogen on cytosine (vdw-h) or the methyl group projected from the C5 of thymine (vdw-me) (Fig. 13.3). In the case of 5′AT3′, the major groove displays the spatial array ac/do/ac/vdw-me, whereas TA displays vdw-me/ac/do/ac. Similarly, a GC base pair displays ac/ac/do/vdw-H, whereas a CG base pair displays vdw-h/do/ac/ac. Thus, each base pair is unique with respect to the array of chemical groups. In contrast, the minor groove of a TA base pair displays ac/ac, whereas an AT displays ac/ac as well. A GC displays ac/do/ac, whereas a CG displays ac/do/ac as well. Thus, there is significant diversity in the major groove, and AT can be distinguished from TA, and GC from CG; the minor groove, on the other hand, is relatively bare and AT cannot be distinguished from TA or GC from CG, only AT/TA from GC/CG. For this reason, as well as the accessibility issue, the major groove appears to be the primary target for sequence specificity. This may be oversimplified because studies using substituted pyrroles revealed subtle differences in recognition of A-T versus T-A base pairs in the minor groove. These differences have not yet been observed in crystal structures of DNA–protein complexes (Kielkopf et al.

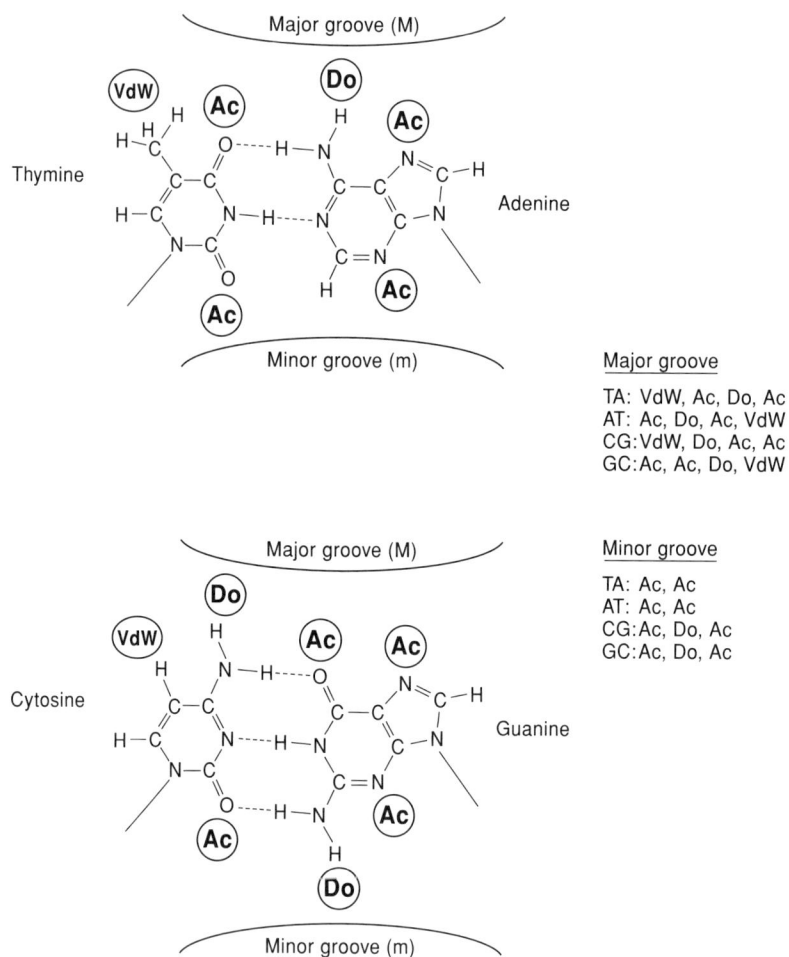

Major groove

TA: VdW, Ac, Do, Ac
AT: Ac, Do, Ac, VdW
CG:VdW, Do, Ac, Ac
GC:Ac, Ac, Do, VdW

Minor groove

TA: Ac, Ac
AT: Ac, Ac
CG:Ac, Do, Ac
GC:Ac, Do, Ac

FIGURE 13.3. H-bond donors and acceptors (Adapted, with permission, from Watson et al. 1987 [Copyright 1987 Benjamin Cummings Publishing].)

1998) but are likely to occur as the crystal and nuclear magnetic resonance (NMR) solutions of additional minor groove binding proteins are published.

Despite the predilection that most DNA-binding proteins have for the major groove, well-described cases of specific minor groove recognition do indeed exist. As described above, some proteins couple major and minor groove recognition, whereas others interact predominantly with the minor groove. Examples of the coupling approach include the Oct-1 POU and MATα-2 homeodomains, among others. These contact the major groove with the homeodomain helix-turn-helix or HTH motif and the minor groove with an extended amino-terminal arm (see, e.g., Wolberger et al. 1991; Klemm et al. 1994). In prokaryotes, the Hin recombinase also employs minor-groove as well as major-groove contacts (Feng et al. 1994). Some proteins recognize the minor groove exclusively. A more unusual mechanism of such minor groove specificity involves the binding of TBP to the TATA box (for a review, see Burley and Roeder 1996). TBP employs 10 anti-parallel β-sheets to form a concave undersurface that recognizes and binds in the minor groove of the TATA site, a point we cover later in this chapter. HMG domain proteins also employ extensive contacts with the minor groove, as discussed below and later in the chapter. LEF-1, an HMG-box protein, distorts and bends the DNA helix at the minor groove to facilitate the formation of an enhanceosome at the TCR-α enhancer. The HMG domain is an L-shaped molecule that molds the minor groove to accommodate the three α-helices (Grosschedl et al. 1994; Love et al. 1995). Both TBP and the HMG domain employ intercalating amino acids as part of their mechanism of specificity. Taken together, these examples show that minor groove binding is an important component for sequence recognition by several major classes of DNA-binding proteins.

Currently there is no universal code for the recognition of DNA sequences by proteins. The amino acid side chains or main-chain amido and carbonyl groups engage in sequence-specific interactions with a wide variety of bases or with the phosphate backbone. However, there may be a preference for forming hydrogen bonds with purines, and there is some degree of conservation in sequence recognition within families of DNA-binding proteins, as within the family of homeodomains or zinc fingers (for review, see Pabo and Sauer 1992). A considerable amount of research has been applied to devising zinc fingers with altered specificities (Rebar and Pabo 1994), and the results of these studies may soon reveal some general rules for amino acid–base pair interactions for at least one class of transcription factors (Pomerantz et al. 1995). Further research into DNA-binding interactions may yet reveal more discernible patterns for DNA sequence recognition.

Monomers and Dimers: Energetic and Regulatory Considerations

Repetition of a DNA recognition unit is one of the most widely employed strategies used by nature to design DNA-binding proteins (Fig. 13.4A). There are several ways in which this repetition is employed: (1) dimerization or formation of another higher-order oligomer (Fig. 13.4AII) and (2) multimerization of a DNA recognition unit (Fig. 13.4AIII). Because the affinity is exponentially related to the free energy of binding ($\Delta G = -RT\ln K_d$), doubling the binding energy by doubling the number of recognition units leads to an exponential increase in affinity of a dimer versus a monomer, or a monomer bearing tandem recognition units versus a monomer with a single unit. Dimers like the yeast GAL4 and GCN4 proteins bind to 17- and 7-bp sites, respectively, each site displaying pseudo-twofold rotational symmetry (Carey et al. 1989; Oliphant et al. 1989; Ellenberger et al. 1992). Crystal structures have revealed that the proteins bind DNA with each monomer recognizing one-half of the rotationally symmetric sites.

Many proteins increase their regulatory diversity by heterodimer formation, with each monomer recognizing one of the half-sites. This strategy has two purposes. It allows heterodimers to recognize a site bearing nonsymmetric half-sites, because each monomer has a different sequence preference. Alternatively, both partners in the heterodimer may have the same DNA-binding specificity but have unique regulatory properties. Excellent examples of the former include the binding of RXR/retinoic acid receptor (RAR) heterodimers to an asymmetric direct repeat (DR) on a promoter; RXR recognizes the 5′ half-site, and RAR (or another partner) recognizes the 3′ half-site. Depending on RXR's partner, the heterodimer complex can regulate transcription from several different DRs (for review, see Mangelsdorf and Evans 1995) and respond to different combinations of ligands. The Max protein product can also bind to a number of related proteins. Myc/Max and Mad/Max heterodimers bind a conserved 6-bp regulatory site, but the heterodimers have very different regulatory effects. The Myc/Max dimer activates cell-cycle-dependent genes, whereas Mad/Max dimers repress these same genes. Another example is the eukaryotic Jun protein, which can bind DNA either as a homodimer (Jun-Jun) or as a heterodimer (Jun-Fos, Jun-CREB). Jun homodimers bind weakly to the AP-1 promoter element, and Jun-Fos heterodimers bind tightly. Jun heterodimers with other family members also bind to a wide array of sites (for an older but insightful review, see Herschman 1991).

Multimerization of the DNA-binding motif within a single polypeptide is another powerful approach used to reinforce specificity. Multimeric proteins can contain either multiple identical recognition units or several units with distinct modular structures. The zinc-finger proteins like Zif 268 are a good example of the first case. In Zif 268, three C-C-H-H fingers recognize similar tandem sequence motifs on the double helix. The zinc fingers are well-designed to position the recognition helix of Zif 268 into the major groove by forming a "C-" shaped structure around the double helix (Pavletich and Pabo 1991). POU-domain homeobox proteins like Oct-1 employ the second method of binding-domain multimerization. Oct-1 contains two specific DNA recognition units, the POU homeodomain and the POU-specific domain. The POU homeodomain is a 3-helix subunit where helices 2 and 3 form the helix–turn–helix (HTH) (see Chapter 1) motif and minor-groove binding occurs via the amino-terminal arm. The POU-specific domain contains four α-helices, which employ extensive base and phosphate-backbone contacts with the double helix. These two domains together bind an 8-bp site, and both domains are required for efficient binding (Klemm et al. 1994).

Strategies for determining the oligomeric state. Determination of the oligomeric state of a protein, either in solution or when it binds to its site, is a relatively simple task. There are four general methods:

1. *Chemical and nuclease footprinting techniques.* These techniques reveal whether a dimeric protein binds to a symmetrical site. The approach can also indicate whether symmetrically opposed mutations in the dyad alter binding. We elaborate on these methods in the second section of this chapter.

2. *Heterodimer analysis.* This strategy requires that one have in hand a set of deletion derivatives that bind to the site and give rise to complexes with unique electrophoretic mobilities in electrophoretic mobility shift assay (EMSA) analyses. Two such derivatives are chosen (often they can be synthesized by in vitro transcription and translation) and assayed individually or in combination for the ability to bind the site. The association of monomers in an oligomer follows a binomial distribution $(a+b)^n$ where a is the concentration of one derivative, b is the concentration of the other, and n is the oligomeric state. If the protein were a dimer, then the distribution of EMSA complexes would be predicted by the equation, $a^2+2ab+b^2$. Thus, each derivative alone would generate one complex as shown in Figure 13.4B, and together these derivatives would form a new complex that migrated with intermediate mobility. If a and b were present at equimolar concentrations, the complexes would be present in a 1:2:1 distribution. Whatever the concentrations of a and b, they would be factored into the equation to predict the distribution. If the protein were a trimer, the distribution would be predicted by $(a+b)^3$.

There are two problems with this analysis. If the proteins contain a strong dimerization interface, the two monomers may not exchange upon mixing to form heterodimers, thereby confounding the analysis. One way to circumvent this problem is to use a variety of salt and buffer conditions, because some conditions may promote intersubunit exchange. Another more common option is to use in vitro transcription and translation to cosynthesize the two derivatives. Thus, the two synthetic RNAs can be combined and cotranslated in the same mixture, allowing the subunits the opportunity to associate as they are synthesized.

Figure 13.4C shows an example of an actual experiment using GAL4 derivatives. Two different GAL4 derivatives were synthesized individually, or cotranslated, incubated with a labeled 17-mer GAL4 site and fractionated on nondenaturing polyacrylamide gels. An autoradiograph of the gel is shown. Each derivative gave rise to a unique shifted com-

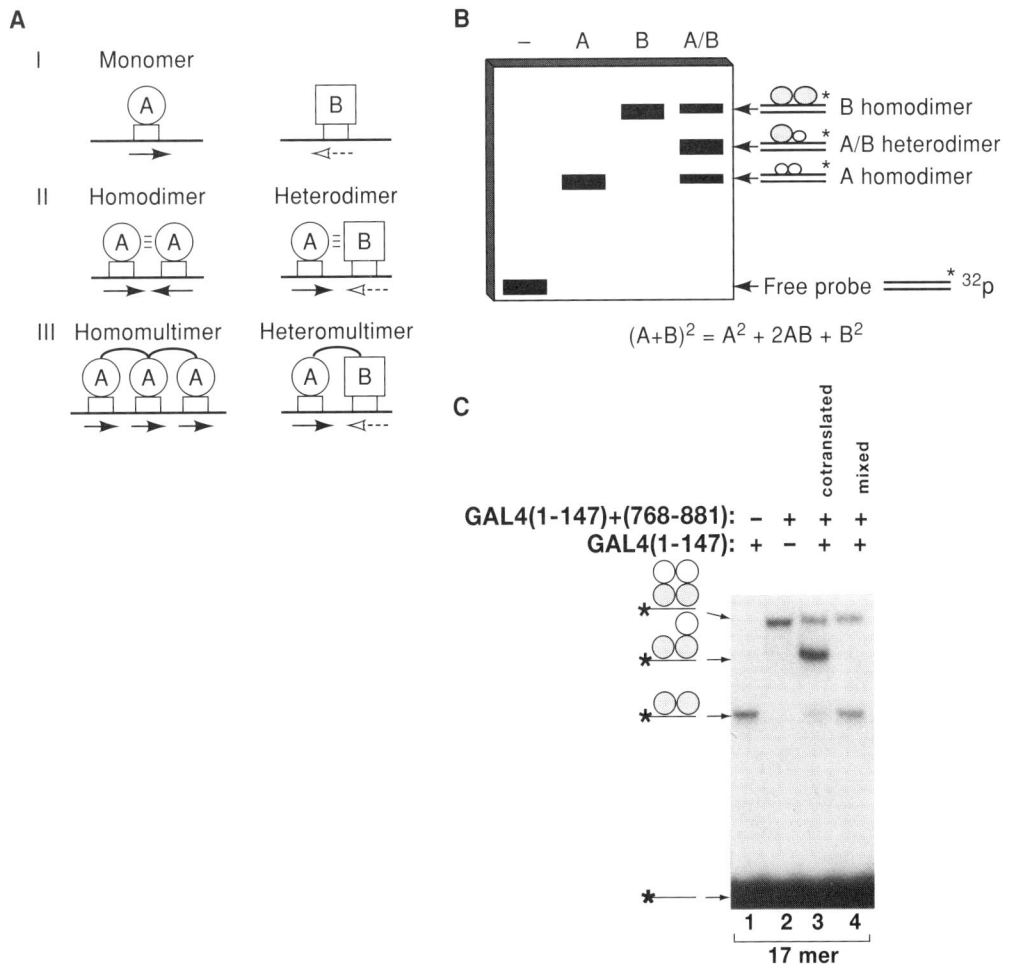

FIGURE 13.4. Heterodimer analysis. (*A*) Different strategies for repetition of a DNA-binding unit. (*B*) Theory. (*C*) GAL4-derivatives experiment. (Adapted, with permission, from Carey et al. 1989 [Copyright Academic Press Ltd.].)

plex (lanes 1 and 2), whereas the two derivatives together generated a new shifted complex with intermediate mobility (lane 3). The appearance of a single new complex was evidence for heterodimer formation and demonstrated that GAL4 bound its site as a dimer, as illustrated by the schematics next to the autoradiogram. However, mixing the two derivatives together did not generate the intermediate shifted complex, demonstrating that the GAL4 proteins had strong dimer interfaces and the subunits could not exchange once formed (lane 4).

3. *Chemical crosslinking followed by SDS-gel electrophoresis.* Glutaraldehyde is a common agent used in crosslinking experiments (it crosslinks primary amines). Different concentrations of recombinant regulatory factors are incubated with increasing amounts of glutaraldehyde and then, after quenching with primary amines (Tris, ethanolamine, etc.), the products are fractionated on SDS-polyacrylamide gels against untreated protein. The shift of the protein from its predicted molecular weight to a higher-molecular-weight species is generally indicative of higher-order oligomer formation; but be aware that glutaraldehyde is relatively nonspecific, thus controls like BSA must be used

in conjunction. Measuring dimer formation at concentrations of protein used for EMSA and in the presence and absence of DNA can also be helpful. Such information may help determine the mechanism of dimerization.

4. *Gel filtration resins.* Gel filtration chromatography is a relatively outdated method that can separate the monomeric and dimeric species on the basis of molecular mass. The approach is less reliable because the shape of the protein can dramatically influence the mobility of a protein in gel filtration columns. An accurate measure of molecular weight depends on a combination of gel filtration and sedimentation velocity experiments. The apparent molecular weight of the complex in such experiments can be compared with the molecular weight of the monomer to determine the oligomeric state.

Dissociation Constant Analysis

Because the biological activity (activation or repression) of a recognition site is often proportional to its affinity for a regulatory protein, affinity is a useful criterion for comparing proteins binding to related but distinct binding sites. The affinity of a protein for its site is generally expressed in terms of a dissociation constant or K_d. The K_d is an equilibrium constant that describes the ratio of reactants to products. K_d is expressed as a molar concentration. Typically, the K_d varies depending on the temperature, buffer, salt conditions, and method of measurement. The variability is unique to each protein and, hence, absolute measures of K_d for two different proteins are generally not comparable unless they were performed under the same conditions and preferably side-by-side. K_d values for DNA-binding proteins are measured by nitrocellulose filter binding, EMSA, or DNase I footprinting, although recent technological advances have resulted in machines such as the Pharmacia Biacore that use surface plasmon resonance (SPR) to measure K_d. SPR and other physical methods for K_d measurement, such as fluorescence anisotropy, are becoming more common in the gene expression field.

One common misconception is that the ratio of protein to DNA is a meaningful parameter in determining the K_d. The ratio is actually meaningless except under very special circumstances, which we will comment on. Instead, the absolute concentrations of protein and DNA, and the conditions under which binding is studied, are the essential parameters for the measurement. Because K_d measurements are becoming more common, we discuss some simple principles to consider when performing such an analysis. We begin with an analysis of a simple equilibrium between a DNA site (S) and an activator protein (P). The reaction is as follows: Activator [P] binds to its DNA site [S] and forms an activator:DNA complex [PS]. This is represented schematically in Equation 1, which by convention is most often illustrated as the dissociation reaction

$$PS \overset{K_d}{\leftrightarrow} P+S \tag{1}$$

The algebraic representation of the reaction is shown by Equation 2

$$K_d = \frac{[P][S]}{[PS]} \tag{2}$$

There are three standard scenarios that a researcher encounters when studying DNA–protein interactions:

1. *Subsaturating.* This condition occurs when there may be billions of molecules of S and P in a reaction mixture but the concentrations of both are well below the K_d. Under such conditions, even if equivalent numbers of molecules are present, the components are not at high enough concentrations to form an appreciable amount of complex (<1%). In an EMSA reaction, no shifted complex would be observed.

2. *One-component saturating.* When one component, P for example, is equal to or greater than the K_d while S is limiting, then P will begin to saturate S. In an EMSA, 50% of the probe would be saturated when P is equal to the K_d and, hence, 50% of the DNA would be shifted to the bound form. In such a case, while S was nearly saturated there could still be many molecules of unbound P.

3. *Stoichiometric interactions.* When both components are present at concentrations exceeding the K_d, they interact largely in a stoichiometric fashion. Thus, if P and S were present at equal concentrations, they would always exist in the form of a complex, in contrast to the subsaturating case.

Examples of the latter two scenarios are explained in Box 13.1.

Box 13.1.

Examples of K_d Scenarios

We examine the most common scenario, scenario 2, first and then scenario 3. Let us assume for a moment that the K_d of P for its site is 10^{-9} M. Let us now measure the occupancy of S when we set the P protein concentration equal to the K_d (10^{-9} M) and the S at the following two concentrations lower than the K_d: (A) 2×10^{-11} M and (B) 2×10^{-15} M. Because the K_d is a constant, the DNA and protein concentrations must be manipulated using approximations to come to a final tally equal to the K_d. The amounts of protein and DNA have been purposely arranged to minimize the manipulations. You would obtain the following result in case A (Eq. 2a): (Remember Eq. 2a is the same as Eq. 2 with the values filled in)

$$K_d = 10^{-9} = \frac{[10^{-9}][10^{-11}]}{[10^{-11}]} = \frac{[P][S]}{[PS]} \tag{2a}$$

Note that the amount of P that binds S (10^{-11}) is so negligible that it does not effectively change the concentration of free protein (10^{-9}M), and so an *approximation* of the concentration of free protein can be used in place of algebra (i.e., 10^{-9}). An exact concentration of the free protein can be calculated using algebra, but it would have very little effect on the overall result. Also note that we started with 2×10^{-11} M DNA, which is partitioned between S and PS.

$$\frac{[S]}{[PS]} = \frac{10^{-11}}{10^{-11}} = 1 = 50\% \text{ occupancy of DNA} \tag{3}$$

Because the amounts of bound DNA (PS) and free DNA (S) are equivalent (Eq. 3), then we say that the DNA is occupied 50% of the time. Hence, the operational definition of K_d—the concentration of free protein for which 50% of the DNA is bound when the DNA concentration is limiting (i.e., well below the K_d). The percentage of bound or complexed DNA is often referred to as the fractional saturation or occupancy.

We now examine case B. It is evident that if we adjust the DNA to 2×10^{-15} M, as shown below (Eq. 4), we observe the same fractional occupancy as when the DNA was 2×10^{-11} M. Because the concentration of protein that binds to the DNA is negligible, the ratio of the bound and unbound DNA remains 1, which is equivalent to 50% occupancy (Eq. 5).

$$10^{-9} = \frac{[10^{-9}][10^{-15}]}{[10^{-15}]} = \frac{[P][S]}{[PS]} \tag{4}$$

$$\frac{[S]}{[PS]} = \frac{10^{-15}}{10^{-15}} = 1 = 50\% \text{ occupancy of DNA} \tag{5}$$

The relevant points are: (1) The protein concentration can be thousands of fold greater concentration than the DNA, yet the DNA is only partially occupied. (2) When the DNA concentration is below the K_d, at any given protein concentration the same fractional occupancy of the sites will be observed.

Using this set of conditions, where the DNA concentration is below the K_d, one can plot either the S/PS (right axis) or the percent saturation of a probe (left axis) as a function of protein concentration expressed on a log scale. Figure 13.5 shows the parabolic increase in saturation as the protein concentration is raised. Note that in the simple bimolecular interactions described above, to change probe saturation from 10% to 90% or from S/PS of 9 to S/PS of 0.11 entails an 81-fold increase in protein concentration.

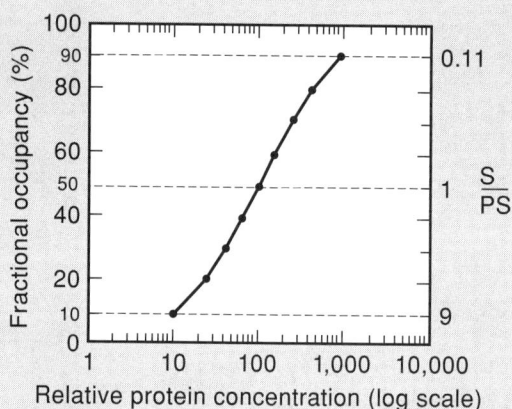

FIGURE 13.5. Binding isotherm for simple bimolecular reaction.

A different type of result from that described above is observed when the DNA concentration is near to or greater than the K_d (scenario 3). If the protein concentration added to the reaction is left at 10^{-9} and the DNA concentration is adjusted to 10^{-8} (Eq. 6), the following results are obtained using approximations. (You could also raise the protein instead. The K_d does not usually refer to any one component of the reaction.)

$$10^{-9} = \frac{[10^{-10}][10^{-8}]}{[10^{-9}]} = \frac{[P][S]}{[PS]} \tag{6}$$

The result is that almost all of the starting protein enters into a complex with the DNA. The overall free DNA concentration, however, does not change because it is much greater than the protein concentration.

Take another example. The starting protein and DNA concentrations added to the reaction are now raised well above the K_d to 10^{-7} for each. $[P]=10^{-7}$ and $[S] = 10^{-7}$

$$10^{-9} = \frac{[10^{-8}][10^{-8}]}{[10^{-7}]} = \frac{[P][S]}{[PS]} \tag{7}$$

To balance the amounts of P, S, and PS, almost all of the protein and DNA must interact in a stoichiometric fashion. Thus, almost all of the 10^{-7} M added DNA and protein form the PS complex. A smaller amount (10^{-8} M using approximations) remains in the form of free P and S. Although this latter scenario is commonly viewed to be occurring during standard DNA–protein interaction studies, in fact, usually one of the components is limiting and the situations described in Equations 2a and 4 apply. The scenario, however, is a common way to determine the activity of a DNA-binding protein as described in the text.

K_d *Determination*

There are several well-established methods for determining K_d, many of which employ EMSA or quantitative DNase I footprinting analysis. One of the most traditional approaches relies on employing the concept depicted in Equation 7: When the protein and DNA concentrations are raised above the K_d, the interaction becomes stoichiometric. Therefore, if the DNA concentration is known, the protein concentration can be determined empirically. The method relies on three conditions: (1) a protein that is sufficiently pure to accurately determine its absolute concentration by protein determination assays such as the Bradford, Lowry, or biuret methods; (2) the availability of ample amounts of an oligonucleotide bearing a high-affinity site; (3) an assay that can accurately measure binding of the protein in solution.

If EMSA is to be employed, preliminary experiments should be preformed to demonstrate that the EMSA and DNase I footprinting assays generate similar results. The EMSA is more quantitative because it is quite easy to measure binding by densitometric scanning of the unbound and bound DNAs. However, occasionally the EMSA conditions are not always optimal and can favor dissociation of the protein–DNA complexes during the running of the gel. An effect called caging, where the limited volume of the gel pores minimizes the dissociation or promotes rapid reassociation, may also influence the results.

The first step is to determine the apparent K_d of a protein by titrating the protein with a very low amount of ^{32}P-labeled oligonucleotide ($\sim 10^{-11}$ M). The concentration of protein that is required to generate 50% occupancy is known as the apparent K_d. The term "apparent K_d" is used because although the amount of protein is known, the amount of *active* protein is not. Many recombinant proteins, particularly those requiring special structures (e.g., zinc nucleated protein folds), are not entirely active due to partial denaturation or oxidation of key residues (e.g., cysteines) during purification.

To ensure that the DNA is indeed limiting (i.e., below the K_d) in the initial titration, it should be varied fivefold in either direction, and the fractional occupancy (50%) should remain the same. This concept is illustrated in Equations 2a and 5. Once the apparent K_d is known, the protein is raised 50- to 100-fold above the apparent K_d. This will lead to almost complete occupancy of the probe. At this stage the precise amount of active protein is unknown, but it is known that levels of the active protein have been raised well above the K_d into the range where it will interact stoichiometrically with the DNA. The next step is to gradually add unlabeled competitor DNA of known concentration. As the competitor oligonucleotide is raised, it begins to compete with the bound ^{32}P-labeled oligonucleotide for the protein. Because the DNA concentration exceeds the K_d, when the competitor oligonucleotide begins to compete (i.e., reduced occupancy in the EMSA), it is assumed that it must be interacting stoichiometrically with the protein. Eventually, all of the active protein in solution is quantitatively sequestered by the oligonucleotide, and the amount of oligonucleotide bound is identical to the amount of active protein.

Therefore, when the competitor begins to exceed the amount of active protein, unbound radiolabeled DNA is observed either by EMSA or footprinting. When 50% of the labeled oligonucleotide is competed to the unbound form, the oligonucleotide competitor would have exceeded the amount of active protein by twofold. The active protein concentration is then calculated to be equivalent to one-half of the oligonucleotide concentration. Because the molar concentration of oligonucleotide is known, the active protein concentration can be easily calculated. Once the active concentration is known, one can return to the original experiment and replace the apparent K_d with a true K_d. The method is facile and can be performed without any special algebraic manipulations. The advantage of this approach is that if the protein of interest is an oligomer, a dimer for example, the high concentrations of protein used in the stoichiometry method will favor the dimer species. Thus, one determines the active amount of dimer in the reaction mixture.

Note that because the free DNA is always related to the bound DNA by the K_d equation, oligonucleotide competition can be used to calculate the K_d at any concentration of protein. However, the algebraic manipulations are more complicated and, depending on the approach, make several assumptions regarding protein activity or the oligomerization status. One disadvantage of performing a measurement when the protein concentration is near or below the K_d is that, when the protein is an oligomer, unless the oligomerization constant is lower than the K_d, a large portion of the protein is in the monomeric form. Since only the dimer will be effectively competed by unlabeled oligonucleotide, this scenario could in principle result in a lower value for the K_d than that determined using the stoichiometry method.

Analysis and Modeling of DNA–Protein Interactions

Fortunately, due to the rapid advances made in solving the crystal or nuclear magnetic resonance (NMR) structures of DNA-binding proteins and their complexes with DNA, our knowledge of DNA recognition is quite advanced. Generally, knowing that a protein falls into a recognizable family for which a crystal structure is available can instantaneously provide a general overview of the sequence preference of a regulatory factor and provide an outline for understanding how the protein recognizes DNA.

Nuclease and chemical probing can provide invaluable information about how a protein recognizes its site. It can be used to refine a model deduced for a related family member, test the validity of a crystal or NMR structure, and explore the binding of previously uncharacterized regulatory protein families. In the following sections we discuss the information that can be obtained by such approaches. We first cover the approach necessary to identify a site if one has a DNA-binding protein but a high-affinity recognition site is not known.

Identification of a High-affinity DNA Recognition Site

In many cases, the investigator has a protein in hand because it was isolated from genetic screens to identify important regulatory proteins or perhaps scored as an oncogene. There may be some evidence that the protein binds DNA, but its sites of action are not known. Proteins such as Myc, MyoD, and orphan nuclear receptors are classic examples of this scenario. In such cases, how does one identify a site to initiate mechanistic studies on DNA binding?

In *Saccharomyces cerevisiae* and *Drosophila melanogaster*, organisms with facile genetics, many regulatory proteins have been cloned through genetic screens, and their promoter sites have been identified and confirmed by promoter mutagenesis of affected genes. Inactivating either the site or the factor genetically permits an assessment of whether the factor is acting through the site in vivo. Studies in mammalian systems, where there is not a facile genetic system, generally rely on a myriad of approaches that couple mutagenesis with either transfection or biochemical assays. The strategies listed below describe approaches for the identification of a high-affinity binding site necessary to initiate binding studies. The approaches do not describe how to determine whether that site is a physiological recognition sequence associated with a naturally responsive promoter. This issue is covered in detail in Chapter 9.

1. *Promoter scanning.* This is scanning the promoter of a responsive gene by DNase I footprinting and EMSA analysis using overlapping restriction fragments as probes.

2. *Selection techniques.* These approaches, which fall into several categories, are the preferred methods for identification of a high-affinity site. In one, the DNA-binding protein is coupled to a solid affinity matrix and a mixture of random DNAs (either oligonucleotides, random restriction fragments, or fragments encompassing a responsive promoter) is passed through. Fragments that are retained can be reselected and then subcloned for biochemical analysis (Kalionis and O'Farrell 1993). A variation of this technique is *s*elected *a*nd *a*mplified *b*inding site (SAAB) analysis (Blackwell 1995). The SAAB method employs a binding technique such as a solid affinity matrix or EMSA to identify DNA fragments or oligonucleotides that bind a protein. The DNA fragments are then purified and amplified by PCR and reselected. By varying the stringency of the binding reaction, high-affinity sites can be obtained for binding analyses. A variation of the method has also been called SELEX (systematic evolution of ligands by exponential enrichment).

3. *One-hybrid analysis.* An alternative genetic technique that has proven successful has been to use yeast as a tool to identify the binding site (described in Chapter 8). In these cases, the random DNA fragments or suspected binding sites from promoters or enhancers are cloned upstream of a yeast reporter gene comprising a core promoter linked to β-gal on a yeast vector, preferably integrated into the genome. The transcription factor is then introduced into yeast under control of a strong promoter (i.e., ADH). In some cases, a yeast activation domain is fused to the mammalian DNA-binding domain. This approach is the basis of the yeast one-hybrid assay for identifying promoter sites and DNA-binding domains from cDNA libraries (Wang and Reed 1993).

Basic Theory

All nucleolytic and chemical probing methods are based on the "nested set" theory used to describe DNA sequencing (see Maxam and Gilbert 1977). A nested set contains a unique end and a variable end (Fig. 13.7). When DNA molecules containing a unique end marked by an end label, usually ^{32}P, and a variable end represented by a modification or cleavage site, are fractionated by denaturing gel electrophoresis, the size of the labeled single-stranded DNA fragment, as measured on a gel, will be a precise measure of the distance to the variable end, i.e., its position relative to the unique end. The principle of sequencing and enzymatic and chemical probing is to produce DNA fragments containing a unique end and a variable end positioned at a specific base. The length of the radioactive single-stranded DNA molecules on a gel then specifies the precise position(s) of that base. In most footprinting methods, the unique end is generated by directly ^{32}P-end-labeling the DNA fragment.

General methods

In initiating any modeling analysis of a DNA–protein interaction, one begins with a simple approach. After the recombinant protein is purified, a DNA fragment bearing the site of interest is generated. It is best to use small DNA fragments whose banding pattern on a sequencing gel can be resolved to the base pair. Although synthetic oligonucleotides are generally adequate for measuring binding in gel retardation assays, they are, for reasons that will be elaborated on below, only suitable for studying binding of a small protein to a short site. Therefore, the binding site is generally first subcloned into an easily manipula-

Box 13.2

SAAB Analysis

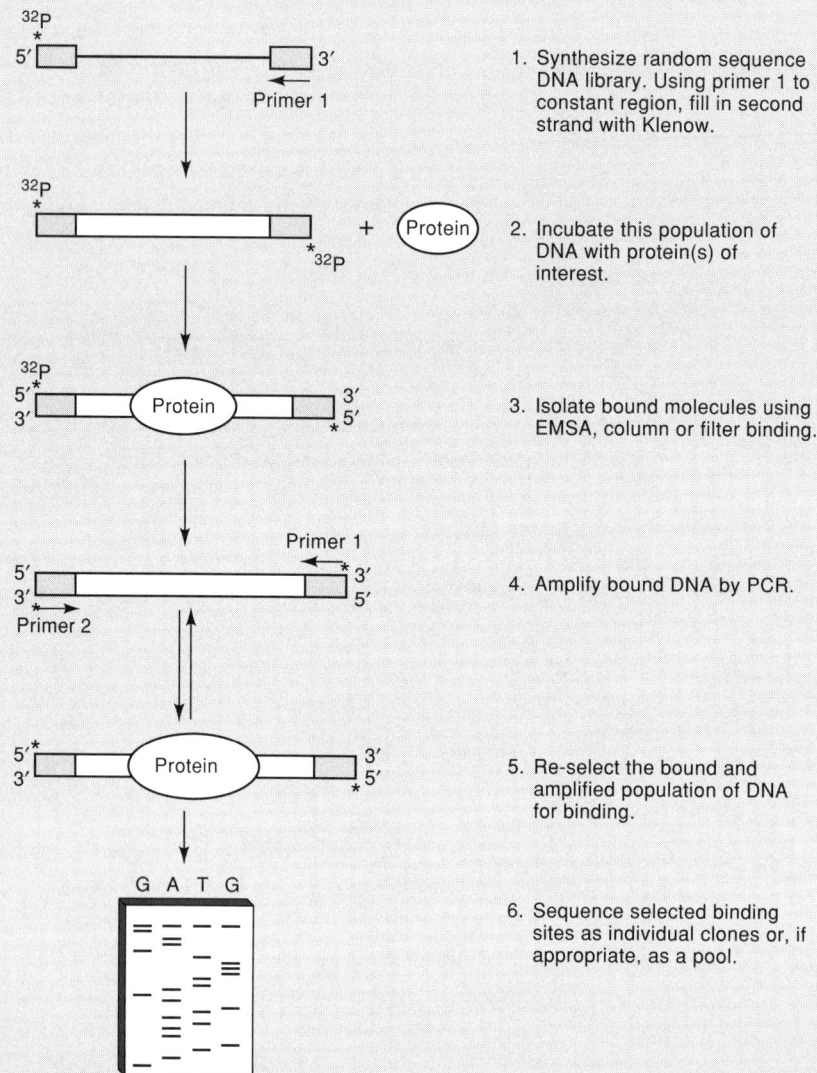

FIGURE 13.6. Flowchart for SAAB analysis (Adapted, with permission, from Blackwell 1995 [Copyright Academic Press].)

The steps shown in the figure are:

1. Synthesize random sequence DNA library. Using primer 1 to constant region, fill in second strand with Klenow.

2. Incubate this population of DNA with protein(s) of interest.

3. Isolate bound molecules using EMSA, column or filter binding.

4. Amplify bound DNA by PCR.

5. Re-select the bound and amplified population of DNA for binding.

6. Sequence selected binding sites as individual clones or, if appropriate, as a pool.

SAAB methodology (Fig. 13.6) was developed by Blackwell and Weintraub (Blackwell 1995) for comparing binding sites for homo- and heterodimers of the bHLH proteins MyoD and E2A. An independent but conceptually similar approach was developed by Ellington and Szostak (1990) to select RNA molecules with unique binding properties. The method, particularly when applied to RNA, has been termed SELEX in subsequent publications (Tuerk et al. 1992). The SAAB approach has since been used to identify the binding-site repertoire for dozens of regulatory proteins. The approach is imaginative and simple, employing PCR technology coupled with EMSA. An oligonucleotide that contains a random sequence flanked by constant PCR sites is constructed and ^{32}P end labeled. A second oligonucleotide complementary to the constant region is synthesized and annealed to the first oligonucleotide. The annealed oligonucleotide is then used as a primer for Klenow DNA polymerase, which is used to fill in the gap to generate a double-stranded molecule. The ^{32}P-labeled molecule is then incubated with saturating or subsaturating concentrations of the protein; the complexes are separated from unbound DNA by gel mobility analysis. The bound DNA containing the selected sequence is reamplified by PCR and subjected to one or more reselection steps. One advantage of this approach is the ability to manipulate the binding conditions and the number of selection cycles to distinguish high-affinity from low-affinity sites.

Note that the selection technique need not employ EMSA. Chromatography of the DNA over solid affinity matrices such as glutathione-S-transferase (GST)-transcription factor fusions attached to glutathione Sepharose or immunoprecipitation of the DNA–protein complexes with an antibody to the transcription factor are alternative methods of identifying the bound sites.

ble vector. Optimally, for sites in the range of 7 to 17 bp, the site is centered within a 50-bp region that can be either excised as a uniquely ^{32}P-end-labeled restriction fragment or prepared by PCR using a primer pair where only one primer is ^{32}P-labeled. After fragment purification and subsequent enzymatic or chemical analysis of the protein–DNA complex, the products are fractionated on a 7–10% polyacrylamide gel. This manipulation generates a cleavage ladder of sufficient detail for careful and quantitative analysis.

The two main approaches that give a broad indication of the binding of a protein to its site are DNase I (Galas and Schmitz 1978) and exonuclease III footprinting (Box 13.3). Due to the large size of the nucleases and steric blockage by the DNA-bound protein, the footprint is often considerably larger than the recognition site itself, and the precise edge of the binding site is difficult to locate. Nevertheless, these methods provide a necessary starting point for studying and modeling any DNA–protein interaction. There are several methods that employ small chemicals such as copper-phenanthroline or methidium-propyl EDTA that bind in the minor groove and cleave the DNA (Box 13.3). These reagents generally allow a more precise definition of the site borders (Sigman et al. 1991).

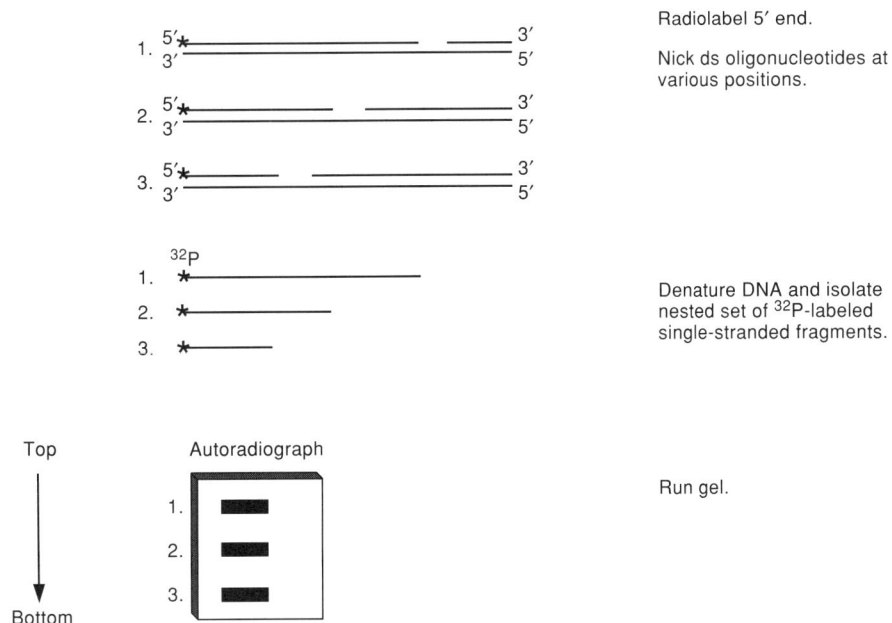

FIGURE 13.7. A nested set of oligonucleotides.

Box 13.3

DNase I and Exonuclease Footprinting

A

B

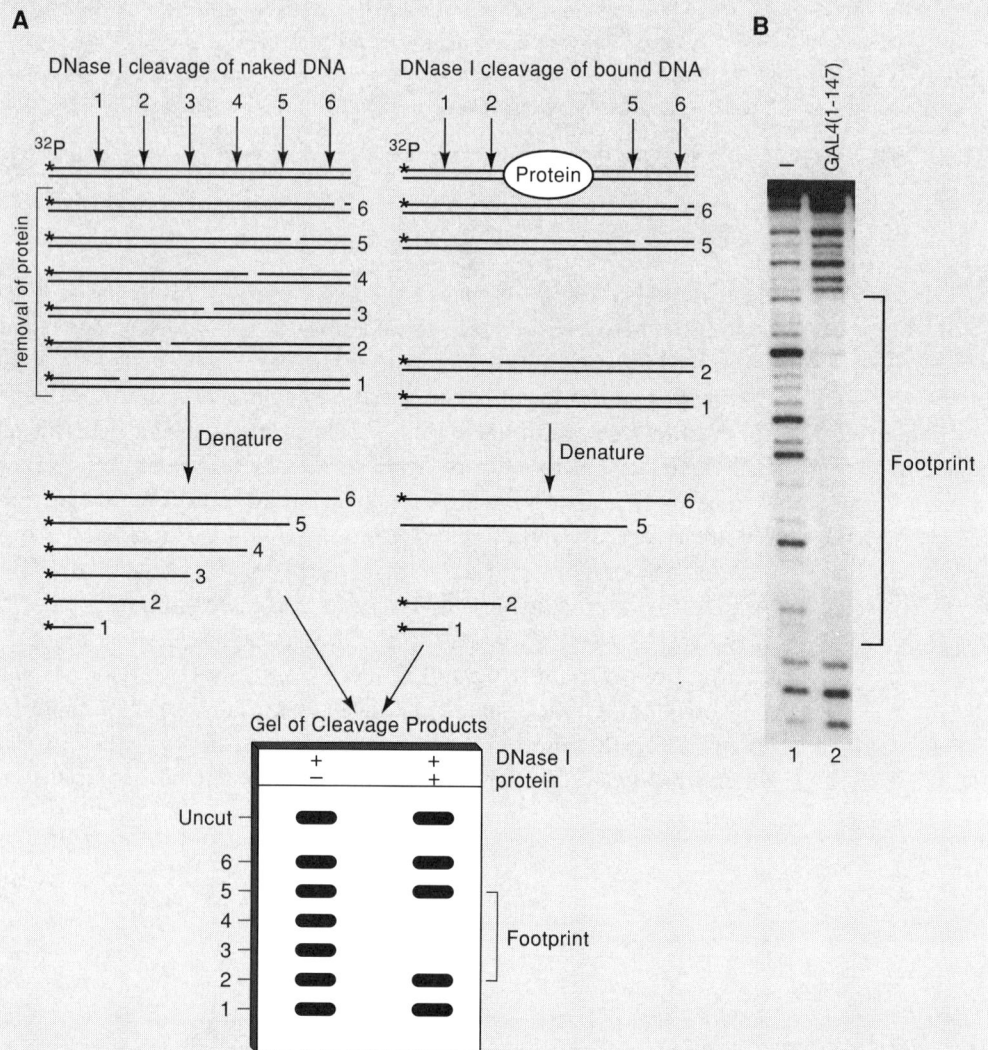

DNase I cleavage of naked DNA

DNase I cleavage of bound DNA

FIGURE 13.8. (*A*) DNase I footprinting theory. (*B*) Example of GAL4 DNA-binding domain (Adapted, with permission, from Carey et al. 1989 [Copyright Academic Press Ltd.].)

DNase I footprinting (Fig. 13.8) is one of the most widely used methods to study binding by a regulatory protein to its recognition sequence (see Protocol 13.1) (Galas and Schmitz 1978). DNase I interacts with the DNA in the minor groove. Although it binds and cleaves the minor groove in a relatively nonspecific fashion, its activity is dependent on certain sequence-based structural characteristics that the binding site adapts (Suck 1994). Local structural deformities within the site, like differences in propeller twist or local minor groove width, can significantly influence the binding and/or cleavage properties of DNase I.

In the presence of Ca^{++} and Mg^{++}, DNase I cleaves a single strand of the double-stranded DNA (manganese ions can lead to double-stranded cleavages) and generates a 5'-PO_4 and 3'-

OH products. Bound protein protects the DNA site from cleavage. For many proteins, the DNase I footprint is significantly larger than the recognition sequence (possibly because the DNase I protein itself is so bulky), and hence it is considered a low-resolution probe. The footprints are often staggered because DNase I is asymmetric in shape, and on a particular strand, will more closely approach the edge of the site on one side than the other. Figure 13.8A illustrates the footprinting theory, and Figure 13.8B shows an example using the GAL4 DNA-binding domain. It is relatively important in footprinting to cleave the DNA on average only once per molecule to observe a digestion ladder. This will entail only partial cleavage of the probe. DNase I will not cleave where protein is bound and the missing cleavage products form a "footprint on the gel."

Exonuclease III (*Exo*III) is a 28-kD nuclease isolated from *E. coli*. It preferentially binds to a free 3′ hydroxyl on double-stranded DNA and then cleaves inward in a semiprocessive fashion. It prefers a recessed or blunt 3′ hydroxyl group for binding, releases nucleoside monophosphates, and possesses a 3′ phosphatase activity. *Exo*III is employed by molecular biologists, in conjunction with single-stranded endonucleases, to create deletions of genes or their promoter regions (see Chapter 7). However, it can also be used to localize protein-binding sites (Siebenlist et al. 1980). The enzyme cleaves until it encounters the binding site, whereupon it halts (Fig. 13.9). Generally, given enough time, the nuclease can wait until the protein falls off its site and can pass through it, resulting in "read-through" on the autoradiograph. *Exo*III footprints are slightly smaller than DNase I footprints. For GAL4, *Exo*III generates a 21-bp footprint, whereas with DNase the footprint is 27 bp (Carey et al. 1989).

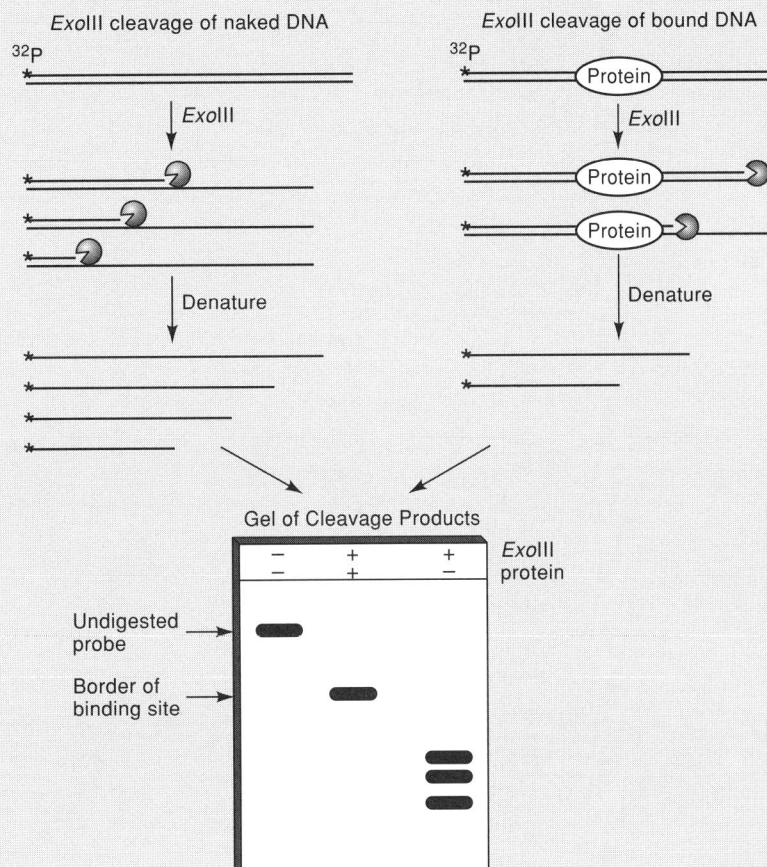

FIGURE 13.9. Exonuclease III footprinting.

Other general footprinting reagents

MPE and Cu-phenanthroline footprinting. Although DNase I and *Exo*III footprinting provide a broad indication of site binding, smaller chemical cleavage reagents can define the edges of the site more accurately, as well as provide important structural information on the interaction. Methidium propyl EDTA (MPE) and Cu-phenanthroline (OP-Cu) are two chemical cleavage reagents that can provide this type of information. Both reagents bind in the minor groove; MPE intercalates between the base pairs. MPE generates a diffusible hydroxyl radical that cleaves the DNA at the sugar/phosphate backbone. OP-Cu is believed to cause a more directed cleavage by attacking the C-1 hydrogen on the sugar. OP-Cu can be an especially useful reagent because the nicking assay can be carried out within an acrylamide gel on otherwise unstable protein–nucleic acid complexes (Sigman et al. 1991).

Minor Groove/DNA Backbone Probes

The positioning of a protein along the DNA backbone can provide important information about its mechanism of recognition. As described above, interaction of the protein with backbone phosphates serves an important role both in providing energy of binding and in allowing these interactions to position elements of the protein in the major groove. There are two main techniques for probing minor groove and backbone interactions: hydroxyl radical protection (Protocol 13.2 and Box 13.4) and ethylation interference (Protocol 13.3 and Box 13.4). Unlike DNase I and OP-Cu footprinting, which also bind the minor groove but give a larger picture of both major and minor groove interactions, hydroxyl-radical protection (because the radical is small and diffusible, unlike phenanthroline which must intercalate) and ethylation interference can provide a more detailed view of exclusively minor groove interactions. It is for this reason that these techniques have become so popular in footprinting assays. Hydroxyl radical footprinting identifies interactions near sugar residues, whereas ethylation interference identifies interactions with phosphates directly.

Box 13.4

Chemical Probes for Minor Groove Interactions

The hydroxyl radical attacks many residues of the deoxyribose sugar with a preference for the hydrogens on the deoxyribose (C5′ = C4′ > C3′ = C2′ = C1′) and abstracts the hydrogen atoms attached to those residues (see Dixon et al. 1991). It was previously thought to be a minor groove reagent, but it is more strictly stated as a probe for contacts near the sugar/phosphate backbone (Balasubramanian et al. 1998). The abstraction of the H atom initiates a series of electron-transfer reactions resulting ultimately in DNA strand scission. A hydroxyl radical molecule is as small as a water molecule and thus is not subject to the same steric restrictions as DNase I. Therefore, hydroxyl-radical footprinting reveals detailed information about primary contacts along the backbone and minor groove. Hydroxyl radical is somewhat insensitive to small perturbations in backbone structure except in the case of kinks and bends where the cleavage efficiencies can vary. Unfortunately, hydroxyl radical also attacks proteins. Some proteins are more sensitive than others to the radical or to the chemical used to generate it. Hydroxyl radicals are typically generated using the Fenton reaction, as shown below (Fig. 13.10A). Figure 13.10B shows a typical hydroxyl-radical footprinting reaction where increasing concentrations of GAL4 DNA-binding domain (lanes 3–5) were incubated with DNA and

A

$$[Fe\,(EDTA)]^{-2} + H_2O_2 \rightarrow [Fe\,(EDTA)]^{-1} + OH^- + OH^{\cdot}$$

Na-Ascorbate

B

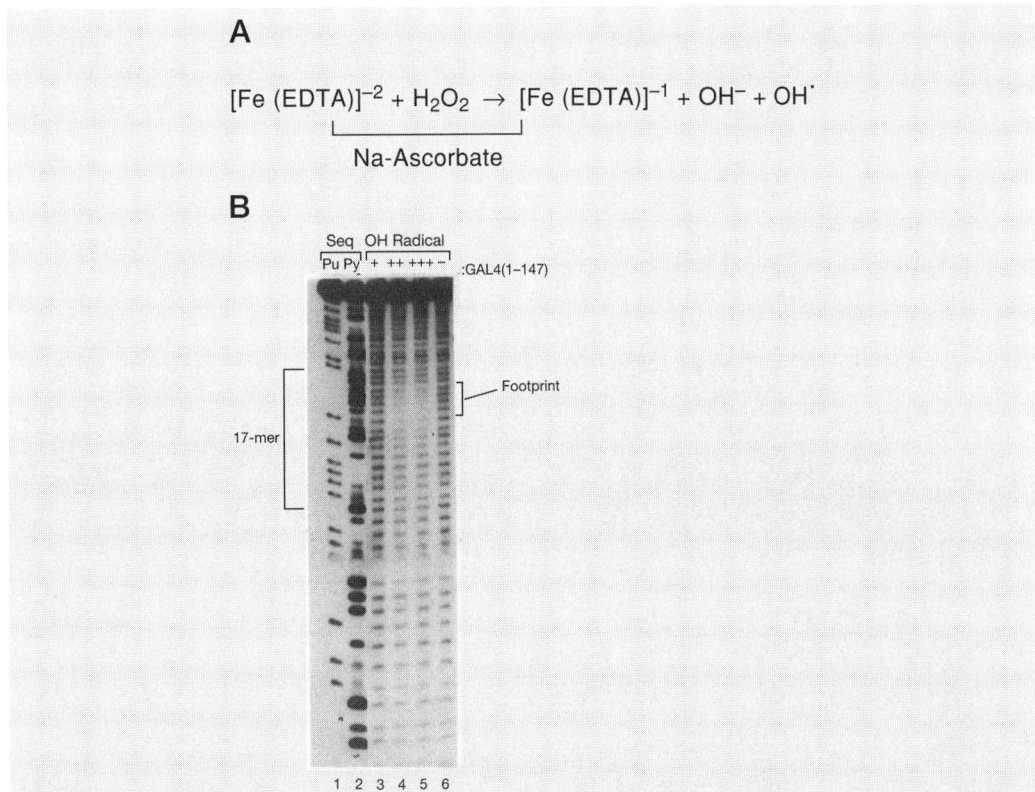

FIGURE 13.10. (*A*) Fenton reaction. (*B*) Hydroxyl-radical footprinting reaction of GAL4. (Adapted, with permission, from Carey et al. 1989 [Copyright Academic Press Ltd.].)

cleaved with radical. A small footprint located near the center of the 17mer was observed when compared with DNA alone (lane 6). The position of the footprint was determined by comparing the positions against a DNA sequencing ladder. Lanes 1 and 2 show Maxam-Gilbert purine and pyrimidine cleavage ladders. We will later employ this information in modeling GAL4-DNA interactions (see section on Modeling DNA-protein interactions).

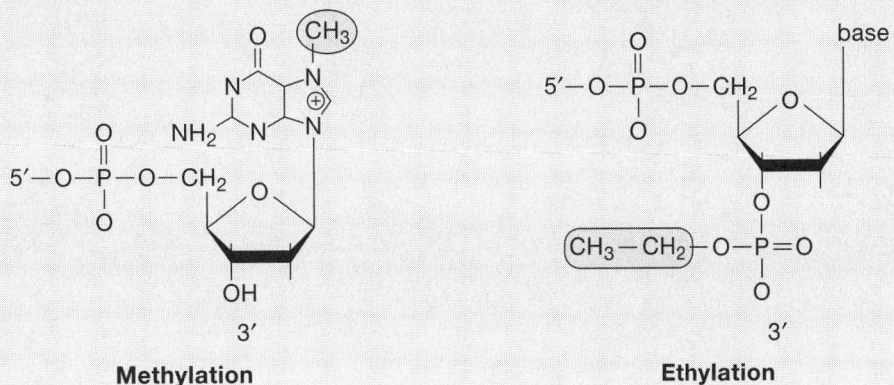

Methylation

Ethylation

FIGURE 13.11. Methylated Gs and ethylated phosphates.

Ethylation interference relies on the ability of ethylated phosphates to interfere with protein binding (Protocol 13.3) (Siebenlist et al. 1980; Manfield and Stockley 1994). The information content is similar but not identical to hydroxyl-radical footprinting, a point that we return to below. First, the DNA is ethylated with ethylnitrosourea, which modifies primarily phosphates along the backbone (Fig. 13.11). The modification is performed to ensure one ethylation per DNA molecule (Fig. 13.12). By taking into account the Poisson distribution of the process of modification, only about 10% of the molecules should be allowed to become ethylated before a significant amount of molecules are modified twice or more (Fig. 13.12). This can be determined by modifying DNA until approximately 10% of the starting probe is converted to a ladder of bands. It is critical to modify only once so that in the final analysis it is clear that the modification being detected— the one nearest to the ^{32}P-labeled end— is the one responsible for interference. The presence of too many modifications on the same molecule will confound the analysis. This same principle holds for other interference and protection assays.

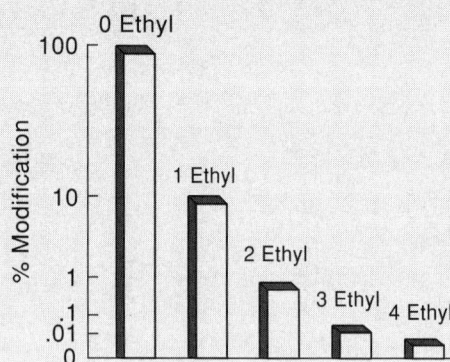

FIGURE 13.12. Optimal Poisson distribution of DNA molecules in different states of ethylation.

After modification, barely saturating amounts of the protein are bound to the DNA (Fig. 13.13A). Barely saturated amounts are used because although the modification decreases binding, it does not always abolish it. High concentrations of protein may overcome the deleterious effect of the modification. The unbound and bound DNA fractions are then separated by EMSA. The bound fraction of DNA is enriched in DNA molecules modified at positions that do not interfere with protein binding and, hence, are not in close proximity to the protein (steps 2 and 3). The unbound fraction is modified at positions that do interfere with protein binding. The points of modification can be identified by isolating the bound and unbound fractions (e.g., bands from an EMSA gel; step 3), treating the DNA with piperidine, which cleaves at the affected residues, and fractionating the cleaved fragments by denaturing gel electrophoresis (steps 4 and 5) alongside a sequencing ladder of the same fragment (Fig. 13.13A). The sequencing ladder can be generated by using the Maxam-Gilbert chemical method or by simply preparing a sequencing primer with the same 5′ end as the ^{32}P-label and then using the dideoxy sequencing method.

The data are usually clear in that most positions on a given stretch of DNA will not interfere and, hence, the bound fraction contains a ladder of bands spanning the length of the gel. A few bands will, however, be depleted in the bound fraction and enriched in the unbound fraction. These bands represent modifications that interfere with binding and, thus, positions that come into close proximity with bound protein. The data are generally summarized by superimposing

A

32P

5′ *————————————— 3′ C₂H₅   6
* ————————————— C₂H₅   5
* ————————————— C₂H₅   4
* ————————————— C₂H₅   3
* ————————————— C₂H₅   2
* ————————————— C₂H₅   1

1. Ethylate naked DNA. Only the labeled strand of the ds DNA is shown.

5′ *———————⬭——— 3′ C₂H₅   6
* ———————⬭——— C₂H₅   5
* ——————————— C₂H₅   4
* ——————————— C₂H₅   3
* ———————⬭——— C₂H₅   2
* —————⬭—————   1

2. Bind protein.

Ethyl groups interfere with binding at positions 3 and 4.

Molecules 1, 2, 5, 6 ◄— Bound protein

Molecules 3, 4 ◄— Unbound

3. Separate bound and unbound DNA molecules via EMSA. Isolate molecules from gel. Remove protein by phenol extraction.

5′ *——|——|——|——|——| 3′ 6
* ——|——|——|——| 5
* ——|——| 2
* ——| 1

} Bound

* ——|——|——|——| 4
* ——|——|——|——| 3

} Unbound

4. Cleave isolated DNA with piperidine.

Unbound Bound
6
5
4
3
2
1

5. Analyze cleavage products on sequencing gel.

B

Bottom strand

Seq

unbound | bound | A+G | G | C+T | C

17mer

GCGTTCTGAGAGGAGGC

1 2 3 4 5 6

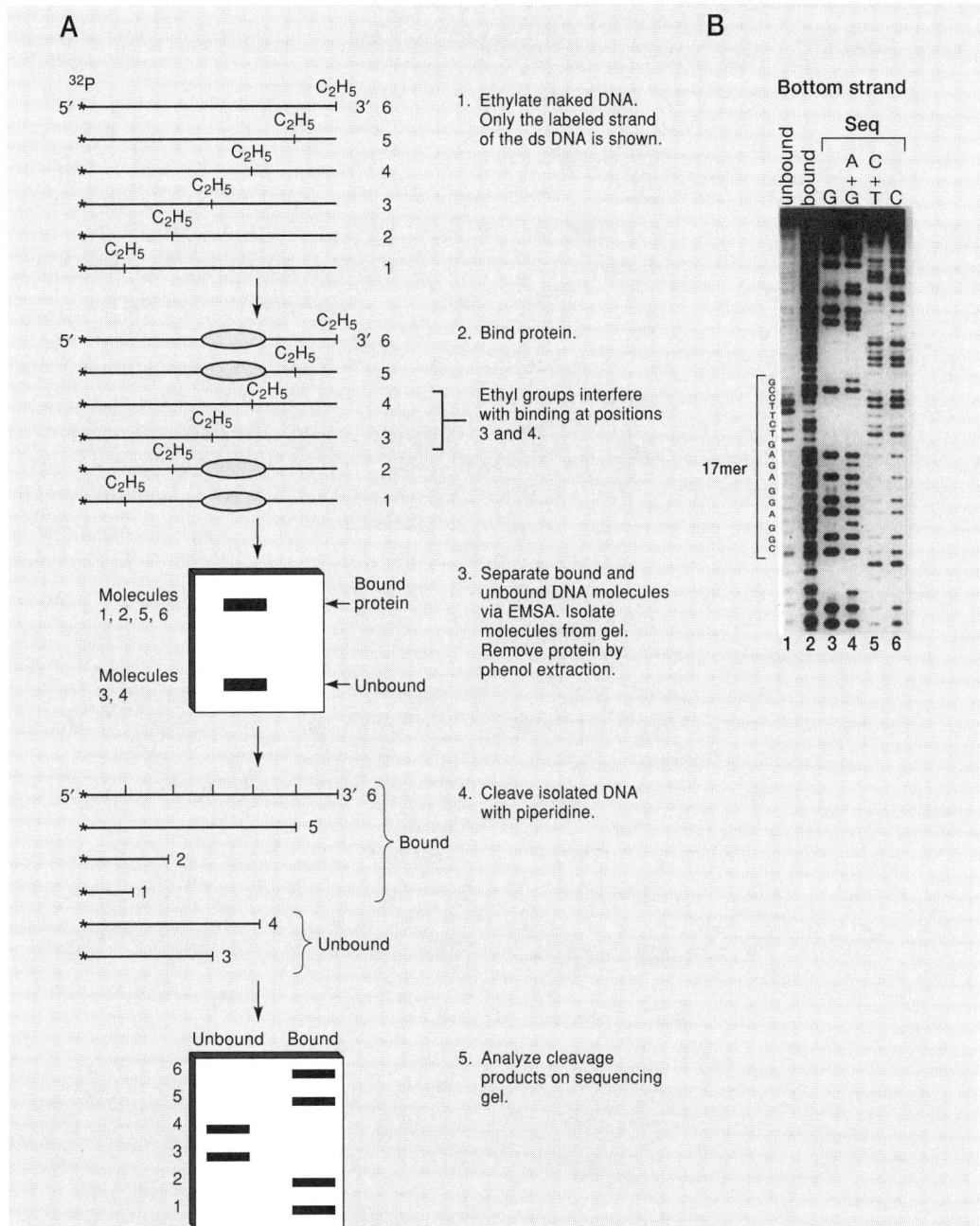

FIGURE 13.13. (*A*) Ethylation interference. (*B*) Experiment with GAL4 (Adapted, with permission, from Carey et al. 1989 [Copyright Academic Press Ltd.].)

the contacts onto a schematic B-DNA helix. If the protein falls into a family whose structure is known, the data can be modeled using that structure as a starting point.

Figure 13.13B illustrates a typical experiment using GAL4. The bands in the unbound lane represent phosphates that interfere with GAL4 binding, and the bands in the bound lane represent phosphates that do not interfere. Note the depletion of phosphates from the bound, and the enrichment in the unbound, lanes. Positions were determined by a Maxam-Gilbert sequencing ladder. We will later show how this information can be used for modeling GAL4's interaction with its site (see section on Modeling DNA-protein interactions).

Major Groove Probes

There are several approaches to understanding how a protein interacts in the major groove of the DNA. These include dimethyl sulfate (DMS) protection and interference (Siebenlist et al. 1980; Wissmann and Hillen 1991), the missing nucleoside approach (Hayes and Tullius 1989; Dixon et al. 1991), and standard site-directed mutagenesis (see Chapter 8).

DMS methods. The principle of DMS protection is similar to DNase I footprinting, but DMS is a very small reagent and hence can provide detailed information of an interaction. DMS methylates the ring nitrogens N7 of G (major groove) (Fig. 13.11) and N3 of A (minor groove). Unlike DNase I, which generates a protection over the entire binding site, only bases that come into very close proximity to a protein are protected from modification by DMS. Furthermore, bases protected from methylation can often correlate with bases directly involved in an interaction with protein, as deduced from crystal or NMR structures. However, this correlation does not always hold up because protein binding can inhibit methylation of uncontacted bases in a site. Therefore, the chemical probing analysis is often correlated with a mutational analysis of the protein and site. We discuss examples below to demonstrate the limits of these correlations. Some proteins are sensitive to conditions required for DMS protection or to DMS itself, which often modifies proteins, causing them to bind poorly (e.g., TFIIIA; Fairall et al. 1987). In such cases, an alternative is the DMS interference assay.

In DMS interference, the DNA is methylated in the absence of protein on average of once per molecule followed by isolation of the modified DNA. The modified DNA is then interacted with a protein, and the bound and unbound samples are isolated and analyzed. The methodology and DNA processing chemistry are much like ethylation interference. Many investigators proceed directly to DMS interference simply to avoid the potential complications of DMS protection.

Missing nucleoside and mutagenesis. Among the other approaches that can be employed to study major groove interactions, missing nucleoside and mutagenesis (Chapter 8) are the simplest. In mutagenesis, one or more bases in the site are replaced with other bases, and the change, usually a decrease, in affinity is measured. Certain bases play a major role in affinity. Altering these bases removes chemical groups critical to the interaction, which leads to a reduction in affinity.

The missing nucleoside approach is somewhat analogous to an interference experiment (Dixon et al. 1991). The hydroxyl radical is used to remove a nucleoside (base-sugar) prior to interacting with a protein. Afterward, the gapped DNA is incubated with saturating amounts of protein and the bound and unbound fractions are separated by EMSA. Again the bands are excised and electrophoresed on a gel alongside a sequencing ladder. The bound fraction is enriched in molecules missing nonessential bases, whereas the converse is true for the unbound fraction. The information content of this approach is probably greater than with the DMS and ethyl interference techniques, but again, the gapping can alter the structure of the DNA site, which may influence affinity.

Chemical substitution. Base analog or chemical substitution is an excellent, albeit technically difficult, method for determining the effect of substituting certain chemical groups on affinity. In this example, certain functional groups on the base are selectively replaced, keeping the remaining structure of the base intact. K_d analysis is used to measure the consequences. Base analog substitution can minimize the potential structural perturbations (see Lesser 1990). The substitution of inosine for guanosine is an example of using base analog substitution. Such an approach was used to study TBP binding. The thymines and adenines in the TATA box were exchanged with cytosines and inosines, respectively (Starr and Hawley

1991). These changes converted the major groove of TATAAAA to that of the sequence CGCGGGG but did not alter the display of chemical groups in the minor groove (which remained the same as with TATAAAA). The authors showed that the substitution had little effect on TBP binding, demonstrating that TBP was largely binding in the minor groove.

A primer on the energetics of major groove contacts. To understand the energetic consequences of altering an amino acid–base pair interaction via mutagenesis, recall that a typical hydrogen bond imparts 3–7 kcal of energy whereas a van der Waals bond imparts 1–2 kcal. Let's calculate the resulting decrease in affinity due to loss of one hydrogen bond (i.e., via a mutation). Using the Gibbs equation, we can calculate the decrease in affinity for the loss of 1 kcal of energy: if $\Delta G = -RT\ln K$, then at 25°C, $\Delta G = -(1.98)(273+25)2.3 \log K$, with each unit in the standard state. Each 10-fold decrease in affinity is therefore accompanied by a decrease of 1.36 kcal of free energy. Thus, for every loss of 1 kcal in binding energy, the dissociation constant decreases 7.35-fold. Therefore, the loss of one hydrogen bond (~3 kcal) between a base and an amino acid side chain results theoretically in a 22-fold decrease in affinity. One caveat of the mutagenesis approach is that changing the base can also have subtle effects on the overall structure of the site, i.e., different bend, twist, or roll angles which contribute to DNA recognition or deformability of the site (e.g., *Eco*RI). These caveats should be kept in mind when interpreting data, particularly with proteins that bend or kink the DNA.

Modeling DNA–Protein Interactions

To demonstrate how the chemical and nuclease probing technology can be employed to study a protein–DNA interaction, we have chosen two well-characterized examples in which the probing chemistry or "footprint phenotypes" (Yang and Carey 1995) can be compared directly with the protein–DNA structure. The examples discussed will be GAL4 and TBP. GAL4 binds in the major groove while TBP engages in predominantly minor groove contacts. In both cases, essential aspects of the DNA-binding mechanism were deduced from the results of chemical and nuclease protection and later confirmed by the crystal structure of the co-complexes (Carey et al. 1989; Lee et al. 1991; Starr and Hawley 1991; Kim et al. 1993a, b; Marmorstein et al. 1992).

GAL4. We attempt to show here that the crystal structure of GAL4 DNA-binding domain with its 17-bp site bears out many of the predictions made from modeling GAL4-DNA interactions using chemical and nuclease protection. *Figure 13.14 summarizes the domain* organization of GAL4 and essential elements of the crystal structure in a graphic form. *Fig. 13.15 then compares the contacts* made in the structure with those deduced from chemical probes.

The DNA-binding domain is located within the first 94 amino acids and can be subdivided into a DNA recognition domain between amino acids 1 and 65 and a dimerization domain from 66 to 94 (Fig. 13.14A). Embedded within the DNA recognition domain are 6 cysteines between amino acids 11 and 38 (boxed) that chelate two zinc ions to nucleate folding of the region into a structure that recognizes the GAL4-binding site. The cysteines and their corresponding protein fold are often called a binuclear cluster, and GAL4 was the founding member of a large fungal protein family with this conserved DNA recognition motif (Schwabe and Rhodes 1997; Todd and Andrianopoulos 1997). Although the DNA recognition domain is a monomer in solution, it forms a dimer on DNA. This dimer binds very weakly, and the natural dimerization region, located between amino acids 65 and 94, greatly increases affinity of GAL4 for its site (e.g., see Carey et al. 1989). Nevertheless, when attempting to crystallize GAL4, the fragment bearing 1–65 formed crystals with DNA and its structure was solved in 1992 (Marmorstein et al. 1992).

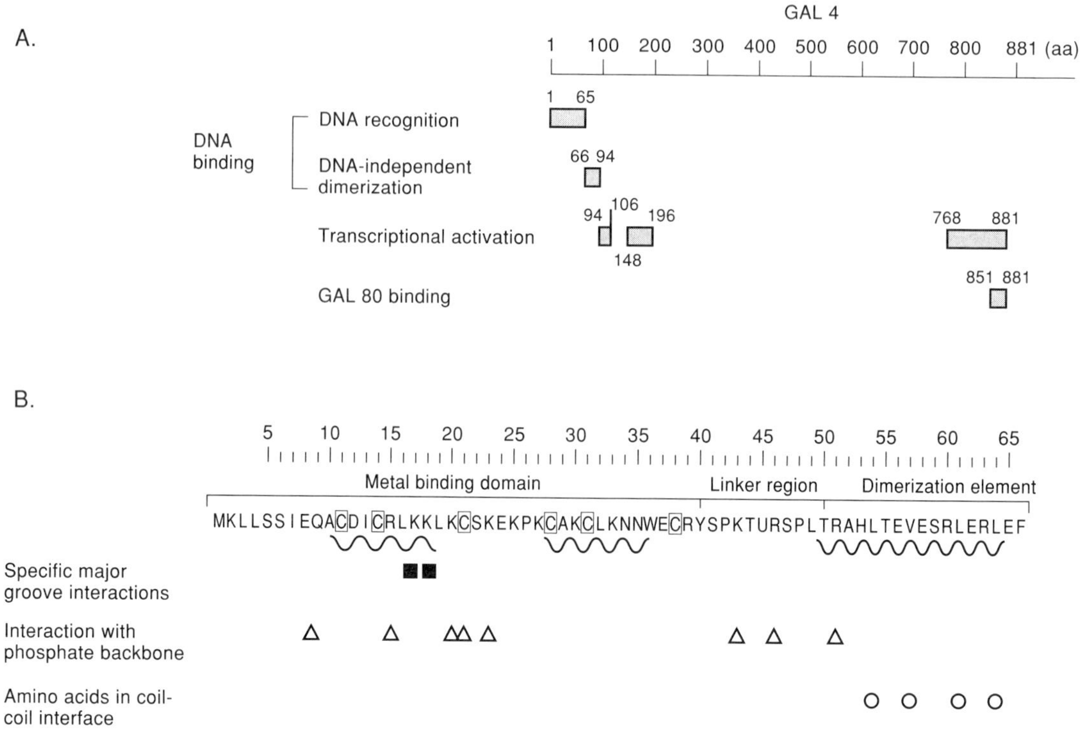

A.

GAL 4

1 100 200 300 400 500 600 700 800 881 (aa)

DNA binding

- DNA recognition — 1 65
- DNA-independent dimerization — 66 94

Transcriptional activation — 94 | 106 196 / 148 ; 768 881

GAL 80 binding — 851 881

B.

5   10   15   20   25   30   35   40   45   50   55   60   65

Metal binding domain | Linker region | Dimerization element

MKLLSSIEQACDICRLKKLKCSKEKPKCAKCLKNNWECRYSPKTURSPLTRAHLTEVESRLERLEF

Specific major groove interactions

Interaction with phosphate backbone

Amino acids in coil-coil interface

FIGURE 13.14. GAL4 and its DNA-binding site. (A) Domain organization of GAL4. (B) Crystal structure features in graphic form. (Modified, by permission from *Nature* [Marmorstein et al. 1992] Copyright 1992 Macmillan Magazine Ltd.)

Figure 13.15A summarizes chemical and nuclease footprinting results of GAL4 on DNA (Carey et al. 1989). The dyad site and symmetric nature of the GAL4 contacts shown in the figure were the original indication that GAL4 bound to its 17-bp site as a dimer, with each monomer contacting one-half of the dyad. However, when the interactions inferred from DMS, hydroxyl radical, and ethylation interference were displayed on a typical B-DNA helix, as illustrated in Figure 13.15B, left panels (two rotational views are shown), they revealed important information, which led to a more sophisticated model for how GAL4 bound its site. Most of the major groove DMS protections/interference were at G residues located at both ends of the dyad site (gray G residues), whereas the minor groove/backbone interactions extended from the Gs toward the center of the site. The affected phosphates implicated by ethylation interference are in black along the backbone and the affected sugar residues implicated by hydroxyl radical are indicated in boldface stick representations. On the basis of the heterodimer data shown in Figure 13.4 and chemical data in Fig. 13.15A, it was proposed in 1989 (Carey et al. 1989) that each monomer of GAL4 "makes sequence specific contacts in the major groove at the outer base pairs of the recognition site and then snakes along one strand of the phosphate backbone" with the "dimer contacts positioned over the center of the dyad." Remarkably, the structure solved in 1992 (Marmorstein et al. 1992) confirmed and significantly refined this simple model. The structure revealed that the interactions identified by chemical probing very closely match those interactions identified in the crystal structure.

This point is best illustrated by superimposing the crystal structure of GAL4 onto the DNA and indicating the contacts inferred from chemical probing. Again, two rotational views are shown in the right panels of Figure 13.15B. First, it is clear that GAL4 in the

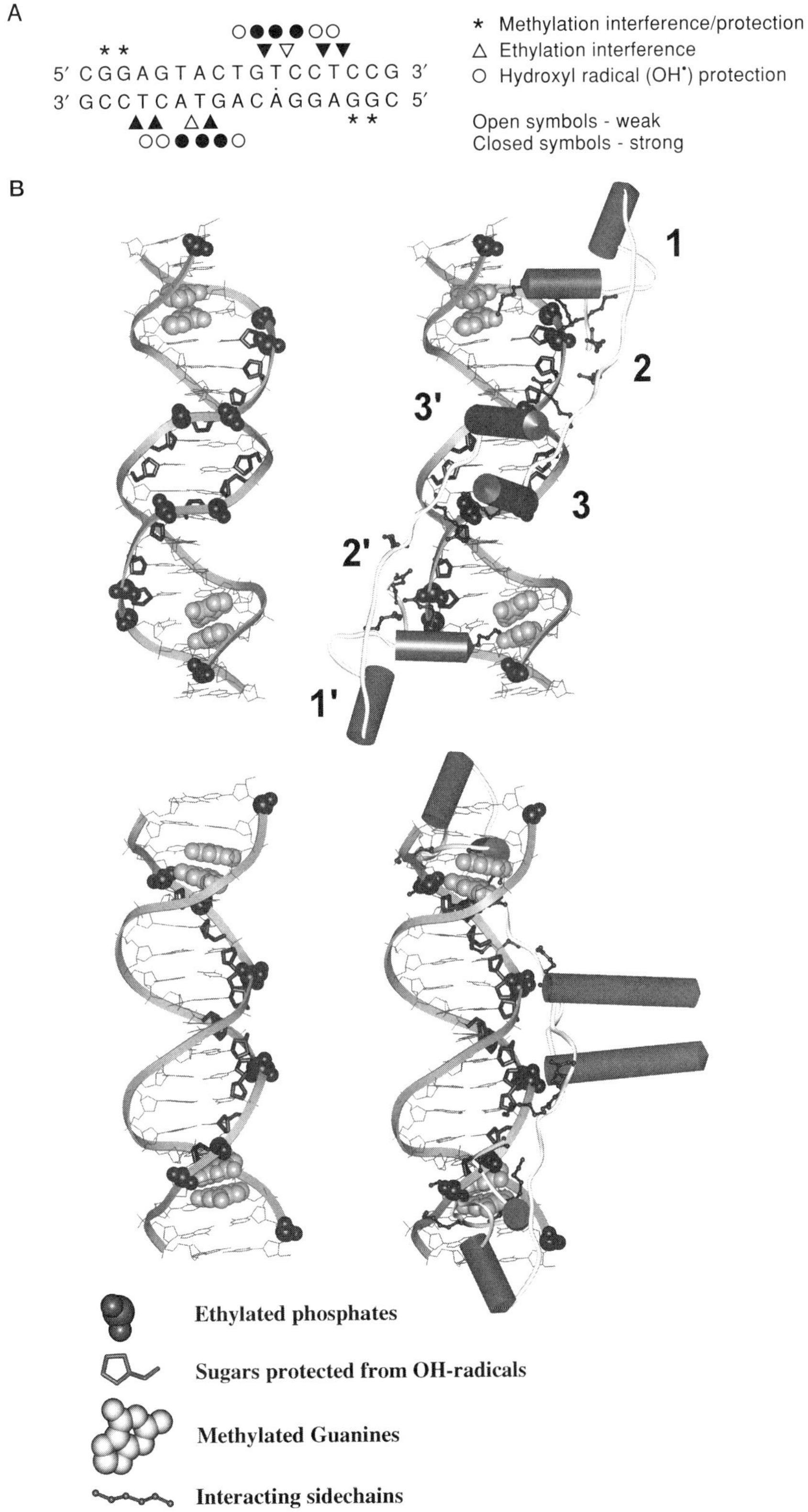

A

```
                O ●● ● OO
      * *       ▼ ▽   ▼▼
   5' C G G A G T A C T G T C C T C C G 3'
   3' G C C T C A T G A C A G G A G G C 5'
          ▲▲  △▲              * *
        OO ● ●● O
```

* Methylation interference/protection
△ Ethylation interference
O Hydroxyl radical (OH•) protection

Open symbols - weak
Closed symbols - strong

B

🔴 **Ethylated phosphates**

⬠ **Sugars protected from OH-radicals**

⚪ **Methylated Guanines**

〜 **Interacting sidechains**

FIGURE 13.15. (*A*) Chemical contact summary of GAL4 interaction with DNA. (Modified, with permission, from Carey et al. 1989 [Copyright Academic Press Ltd.].) (*B*) Crystal structure of interaction. (Crystal structures rendered by Michael Haykinson [UCLA] using the Molecular Graphics structure modeling computer program Insight II.)

structure interacts directly with the residues on DNA implicated in contact by chemical probing. More specifically, discrete structural features of the protein generate the contacts with DNA. The zinc-complexed DNA recognition module is located between amino acids 8 and 40 (see Fig. 13.14B). In this domain, the two atoms of zinc from each monomer complex with 6 cysteines to fold into a structure containing two perpendicularly disposed α-helices (Fig. 13.15B features 1 and 1′). One of the helices from each monomer inserts into the major groove perpendicular to the DNA axis and makes sequence-specific contacts with lysines 17 and 18 (black stick representations emerging from the cylindrical helix) and the exposed groups of the G residues at the end of the dyad (indicated as gray van der Waals representations). These Gs are those detected by DMS protection and interference in chemical probing experiments (Fig. 13.15A). Within the superfamily of fungal binuclear cluster transcription factors, most members contain conserved lysines or arginine at position 17 and lysine or histidine at position 18. For many of the known proteins, the contacts of these residues with a dyad CGG motif are a conserved feature of the mechanism of DNA recognition (Liang et al. 1996; Schwabe and Rhodes 1997; Swaminathan et al. 1997). Thus, the GAL4-DNA chemical probing data and the structure can be applied to this very large family of factors to help determine how they recognize their sites.

Immediately carboxy-terminal to the DNA-binding module in each monomer is a flexible unstructured linker (2 and 2′) that traces along the minor groove. Mainly lysines and arginines in this linker engage (see Fig. 13.15B) in nonspecific contacts with the phosphate backbone until it comes to the center of the dyad. The black stick representation shows the side chains projecting onto the DNA. These side chains interact with the residues implicated in ethylation interference and hydroxyl-radical protection (phosphates are in black van der Waals, and sugars are shown as sticks). The protein linker varies in length among members of the binuclear cluster family. The variation allows the different members to bind sites with different distances separating the 5′ and 3′ CGG motifs (Schwabe and Rhodes 1997).

The two monomers each form an α-helix in the center of the dyad (3 and 3′). The α-helices from each monomer associate to form a coiled-coil dimerization motif using leucines and a valine to form the hydrophobic interface (Fig. 13.15B). The base of this coiled-coil motif also engages in nonspecific contacts with the minor groove in the center of the dyad site. Again, the indicated ethylation interference and hydroxyl-radical contacts match those residues that interact with the base of the coiled coil in the crystal structure.

In conclusion, the model predicted from the chemical probing data (Carey et al. 1989) is in close agreement with the crystal structure (Marmorstein et al. 1992).

TBP. The TATA box binding protein, or TBP, is an example of a protein that binds exclusively in the minor groove and makes contacts with the base pairs and with the phosphate backbone. DNA sequence comparisons between the different species reveal a relatively conserved TBP core sequence of 180 amino acids required for TATA box binding and interaction with the general transcription factors. This core contains two 80-amino-acid direct repeats. The amino terminus of the TBP gene differs significantly among species, and it has been proposed that this region mediates species-specific interactions with components of the transcriptional machinery required for activation (Burley and Roeder 1996).

Prior to solving the crystal structure of TBP bound to a TATA box, researchers performed extensive chemical footprinting analysis to understand how TBP and TFIID docked with a TATA box. These analyses led to a model whose main elements were later confirmed by the crystal structure. Figure 13.16A illustrates the results of hydroxyl-radical footprinting, ethylation interference, and DMS protection and interference assays. Note the lack of major groove interactions (i.e., N7-methyl G residues) and the preponderance of putative minor groove contacts (i.e., N3-methyl A residues and ethylated phosphates) inferred from the chemical probes. These data were originally interpreted as support for a model whereby TBP makes primary contact with the minor groove (Lee et al. 1991; Starr and Hawley 1991).

The model was confirmed and extended by the crystal structure of the TBP:TATA co-complex solved independently by the laboratories of Burley and Sigler. Sigler and colleagues solved a yeast TBP co-crystal (Kim et al. 1993b), and the structure solved by Burley and colleagues was that of the *Arabidopsis* TBP (Kim et al. 1993a). As in the previous figure, the chemical contacts are superimposed on the DNA in the left panels (two rotational views) while the TBP structure is shown with DNA in the right panels of Figure 13.16B. The structure contains a 10-stranded antiparallel β-sheet (arrows) in a saddle-like conformation, with four α-helices (cylinders) crowning the upper surface of the saddle. The protein consists of two subdomains (Fig. 13.16B, 1 and 1′) corresponding to each direct repeat in the primary sequence. Each subdomain is organized into a sheet 1– helix 1 – sheet 2 –turn – sheet 3–sheet 4– sheet 5 –helix 2 configuration (Fig. 12.7; see Chapter 12). The β-strands constituting the underside or concave surface of the saddle form a large hydrophobic interface that interacts through van der Waals and hydrophobic contacts with the bases. A nominal number of specific hydrogen bonds (Sigler and colleagues identified 6 and Burley and colleagues identified 4) are formed within the 8-bp region in the minor groove. Binding induces a significant DNA deformation causing an 80° bend and unwinds the DNA 110°, resulting in the characteristic bent shape indicated in the figure. The deformation is caused by intercalation of two phenylalanines between the bases at both ends of the TATAAA. This intercalation buckles the bases, causing the overall site to bend. This can be visualized by focusing on the two upper A residues in the DNA, shown in gray van der Waals representation. The figure with TBP docked shows the intercalating phenylalanine inserting between the two As, causing them to buckle. The resulting bend and untwisting maximize the hydrophobic interface between the protein and the minor groove. Salt bridges between basic residues and the phosphates, water-mediated H-bonds, and van der Waals contacts with the sugars are the driving forces behind the interaction. Most importantly, Figure 13.16B shows that when the structure is docked with the DNA, the crystal interactions explain the hydroxyl-radical protections and almost all of the ethyl phosphates that interfere with TBP binding.

In conclusion, we have illustrated two examples, where the details of the chemical probing experiments allowed investigators to propose models for how the proteins interacted with DNA. Once the structures were solved, they confirmed and extended fundamental aspects of the original model. As more structures are solved, the interrelationship between the chemistry and the mechanism of DNA binding should provide ample information to allow an investigator to extrapolate the data to understand how DNA binding is mediated by unknown proteins or proteins that fit into defined families, one of whose members is of known structure.

Analysis of Promoter-specific Multicomponent Nucleoprotein Complexes

At the outset of this chapter, we described how clusters of proteins bound to promoters and enhancers (i.e., enhanceosomes) are probably best described as networks of protein–protein and protein–DNA interactions (see Chapter 1 and Carey 1998). The two driving forces behind the formation of enhanceosome complexes are cooperativity and DNA looping or bending. Cooperativity is a phenomenon whereby proteins assist one another binding to DNA so that together they bind more tightly than each one alone. Put another way, in the presence of its partner, a protein exhibits a lower K_d or higher affinity for its DNA site. DNA bending allows distal interactions to occur. DNA bending occurs when two distal proteins interact with the intervening DNA looping out. Within the persistence length of DNA (~140

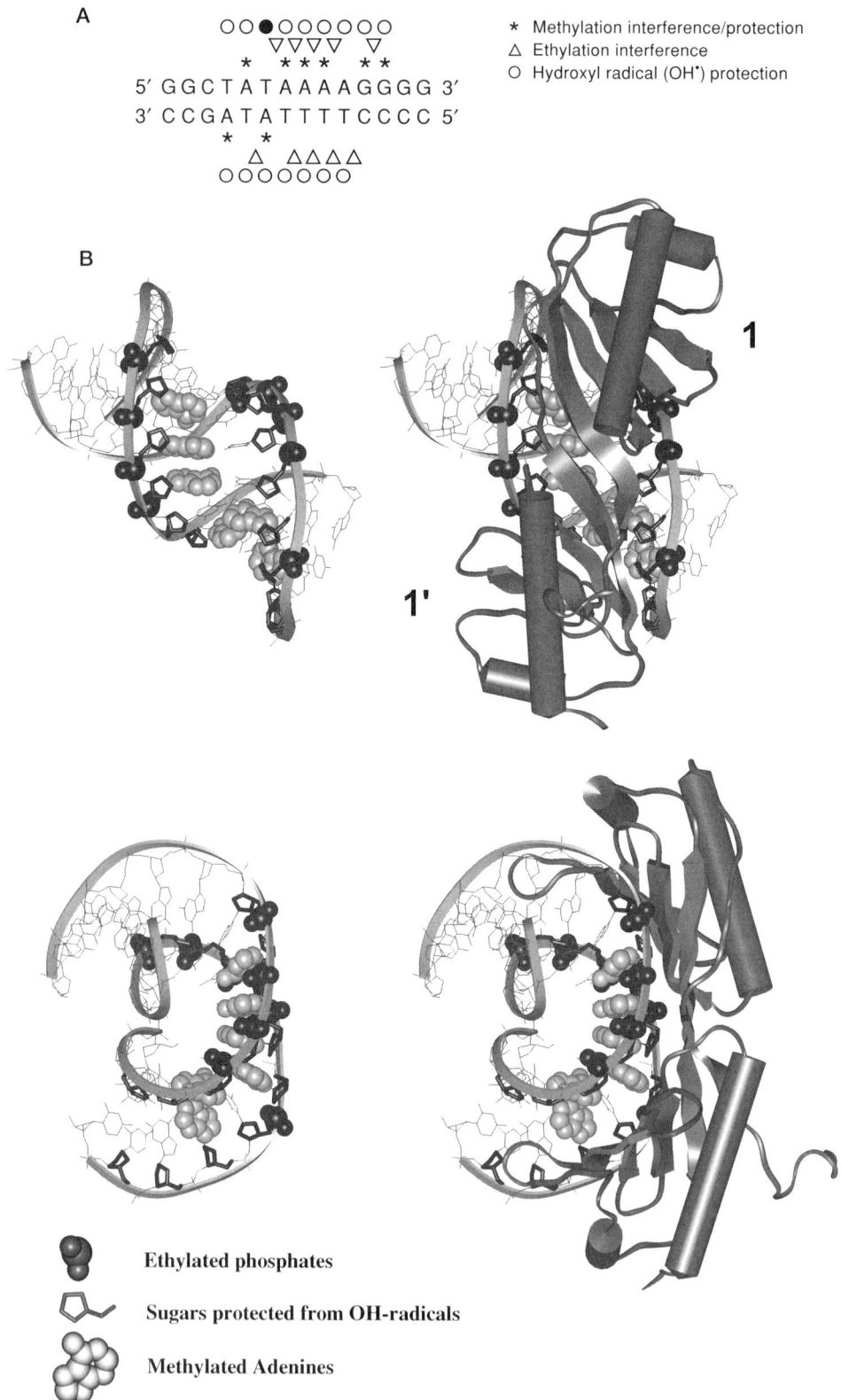

A

○○●○○○○○
　▽▽▽▽　▽
　*　***　**
5′ GGCTATAAAAGGGG 3′
3′ CCGATATTTTCCCC 5′
　*　*
　△　△△△△
○○○○○○

* Methylation interference/protection
△ Ethylation interference
○ Hydroxyl radical (OH•) protection

B

1

1′

● Ethylated phosphates

⬡ Sugars protected from OH-radicals

◉ Methylated Adenines

FIGURE 13.16. (*A*) Chemical contact summary of TBP–DNA interactions. (*B*) Crystal structure. (Rendered by Michael Haykinson [UCLA] using the Molecular Graphics structure modeling computer program Insight II.)

bp), DNA bending extracts an energetic penalty that must be paid by the strength of the protein–protein interaction or by a protein that stabilizes the bend. Beyond the persistence length, the energetic penalty is less substantial, but the probability of the protein–protein interaction decreases with distance. These effects have been modeled by biophysical chemists but are only beginning to be applied to transcription systems in a predictive fashion (Wang and Giaever 1988; Rippe et al. 1995).

DNA Binding Cooperativity

One of the key mechanisms for generating specificity is cooperative DNA binding. In light of the enhanceosome, this concept is being revisited in the eukaryotic gene expression field. We first discuss the theory and an example followed by a simple approach to studying the phenomenon. Of the many examples of cooperative binding by DNA-bound protein, the paradigm is λ repressor or cI (Johnson et al. 1981). Repressor binding has been analyzed using standard biochemistry and biophysical techniques, and the equilibrium has been analyzed mathematically and thus provides an excellent model for understanding and studying cooperativity (for review, see Hochschild 1991). In a bacteriophage lysogen, repressor maintains the lysogenic state by controlling transcription from two related rightward and leftward operators called O_R and O_L. For simplicity we describe the situation at O_R. At O_R, repressor dimers bind cooperatively to a high-affinity site called O_R1 and a low-affinity site called O_R2. The cooperativity is mediated by direct protein–protein interactions between the carboxy-terminal domains of the adjacent dimers. Repressor bound at O_R1 represses transcription from a promoter called P_R (promoter in the rightward direction), which activates the lytic cycle. Concurrently, repressor bound at O_R2 touches its target, RNA polymerase, and activates expression from P_{RM} (promoter for repressor maintenance). P_{RM} controls expression of the λ repressor and is a classic example of an autoregulated promoter.

The cooperative effect was initially measured by DNase I footprinting assays. By comparing the affinity of repressor for a ^{32}P-end-labeled DNA fragment containing O_R1 and O_R2 in their natural positions, or either O_R1 or O_R2 alone, Ptashne and colleagues showed that the repressor binds cooperatively. As described above, the DNA fragment was set at a concentration below the K_d (10^{-10} M). Then, the repressor was titrated in twofold steps until the DNA fragment was saturated as measured by the footprinting analysis. Recall that if the DNA concentration is below the K_d, then when PS=S (see Box 13.1), conditions where the DNA site is 50% occupied, the K_d =P. Put simply, the K_d is equivalent to the concentration of free protein required to generate 50% occupancy of the site. Using this approach, it was shown that the K_d values for the isolated O_R1 and O_R2 are 3.3 × 10^{-9} and 5 × 10^{-8}, respectively (Fig. 13.17A). However, in the presence of O_R1, the K_d for O_R2 decreases to approximately 10^{-9}. The amount of cooperativity can be expressed as the ratios of the O_R2 K_d measured in the presence and absence of O_R1.

Therefore, in evaluating whether proteins bind cooperatively, their affinities must be measured in isolation and together. In cases where different proteins are binding cooperatively, the same DNA template can be used to perform the analysis. However, in cases where a single protein is binding to multiple sites, the binding of protein to the wild-type template must be compared with templates where one or the other site is mutated. Generally, the cooperativity is easily visualized. Some investigators employ EMSA. In such cases, one protein will enhance the amount of probe shifted by the other in a greater than additive fashion. Often the data can be quantitated using a Hill plot of the bound and unbound DNAs. Although we do not discuss quantitative modeling of the phenomenon, such issues are covered in detailed reviews (Brenowitz et al. 1986; Hochschild 1991).

In a multicomponent complex, all protein–protein interactions should, in principle, have reciprocal effects on each other's stability. Obviously, in some cases, these effects will be dramatic, such as two proteins binding to weak sites and, in some cases, these will be less dramatic, such as the threefold effect λ repressor bound at O_R2 has on binding to O_R1 (Johnson et al. 1981).

In considering how an enhanceosome might work, one can predict that many of the interactions will be cooperative. As described in Chapter 1, this is indeed the case on both the T-cell receptor α (TCR-α) and interferon β (IFN-β) enhanceosomes (Giese et al. 1995; Thanos and Maniatis 1995). As an extension of this idea, it could be imagined that as the promoter-bound activators help to assemble the general machinery into a transcription complex, the complex should, in turn, help the activators to bind to their sites on DNA. This appears to be the case in the IFN-β enhanceosome (Kim and Maniatis 1997).

DNA Looping and Bending

DNA looping. The ability of two proteins to interact while both are bound to DNA forms the theoretical basis for DNA looping, long believed to be the mechanism by which distally bound activators interact with the general transcriptional machinery. The concept of DNA looping was first discussed in studies on the arabinose operon in *E. coli* and has been implicated in numerous biological processes. The phenomenon has been extensively investigated using the λ repressor system, and we discuss this as an example.

λ repressor as discussed above normally binds cooperatively to two 17-bp sites with a center-to-center distance (inclusive) of 25 bp or 2.4 helical turns. Insertion of 10 or 11 bp (called a helical increment because the periodicity of DNA is 10.5 bp per turn) has little effect on the cooperativity up to 8 turns of DNA away. However, insertion of nonhelical increments between sites separated by up to 8 turns away abolishes cooperativity. Figure 13.17 illustrates an example where insertion of 3.5 helical turns between the repressor binding sites prevents cooperativity, and where insertion of 4 helical turns restores it (Hochschild and Ptashne 1986). Electron microscopy and physical measurements have confirmed that the intervening DNA loops (Griffith et al. 1986). The looping within a plane requires only that the intervening DNA bend smoothly so that the adjacent proteins can interact. The inability of the proteins to interact with the introduction of nonintegral turns is due to the energetic penalty of twisting the DNA. At the distance stated, the cost would be 7 kcal or more. Presumably, the energy of the protein–protein interaction between adjacent dimers is not sufficiently large to maintain the interaction and absorb the cost of twisting. This assertion is supported by the observation that introduction of a short single-stranded gap between the sites, a gap that in principle allows free rotation around the single strand, restores the cooperativity even at nonintegral distances. Surprisingly, cooperativity can be observed up to 20 helical turns away, but the requirement for helical periodicity occurs only up to 8 turns. Presumably after 8 turns the energetic penalty for the twisting is less and the interaction is no longer proscribed (see Rippe et al. 1995 for a recent review on the energetics of looping and twisting). In analyzing a situation where two or more proteins are believed to be interacting, the effect of altering the helical phasing is generally used as evidence for an interaction.

Although looping can be inferred from protein–protein interactions that occur at a distance and generate cooperative binding of a protein to DNA, it is difficult to visualize looping directly. However, the DNase I cleavage pattern of the DNA between the sites can be informative. In the case of repressor bound to sites separated by 6 turns, the DNA between the sites exhibits a series of enhancements and protections with a 10-bp periodicity (Hochschild and Ptashne 1986). The explanation is that the DNA is smoothly bent, exposing the minor groove on the outside convex surface of the bend and protecting the minor groove on the inside concave surface. Such a pattern might be diagnostic of looping. Long-range interactions between activators in enhanceosomes might also display such periodicity.

There are several features of the λ repressor model system that will likely differ in other systems. Some proteins may be large and sufficiently flexible to interact without an energetic penalty, irrespective of their periodical relationship. Alternatively, one could imagine that the protein–protein interaction is strong enough to absorb the energetic cost. λ repressor indeed represents a novel protein in the sense that its protein–protein interaction, although unable to pay the cost of DNA twisting to bring the proteins into phase, can nevertheless absorb the penalty of in-phase DNA bending. There are many situations where this is not the case, including interactions among activators in the TCR-α and IFN-β enhanceosomes. These interactions require the action of sequence-specific DNA-bending proteins (Grosschedl 1995; Carey 1998).

DNA bending. Studies in prokaryotic and eukaryotic systems over the past several years have emphasized the important role DNA bending plays in DNA metabolic processes. In transcription, the role of DNA bending appears to be to align closely situated proteins into complexes that contact the general transcription machinery. Passive DNA bending within the persistence length of DNA (~140 bp) is constrained due to the rigidity of the DNA helix. Although certain protein–protein interactions can occur by passive bending or looping of the intervening DNA, these require a threshold amount of energy imparted by the protein–protein interaction. If this energy is not sufficient, the protein may only interact with the help of a third protein that has the ability to bend the DNA in a certain direction and to a specific angle, permitting the interaction. These bending proteins are called architectural proteins (Werner and Burley 1997).

A Cooperative binding

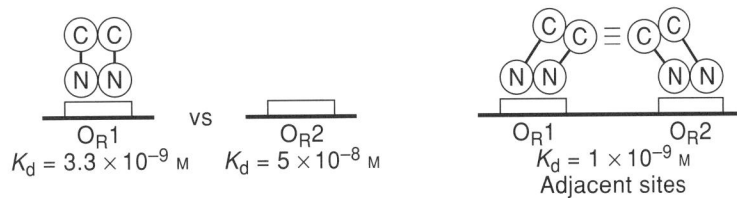

$K_d = 3.3 \times 10^{-9}$ M   vs $K_d = 5 \times 10^{-8}$ M

$K_d = 1 \times 10^{-9}$ M
Adjacent sites

B Cooperative binding and DNA looping

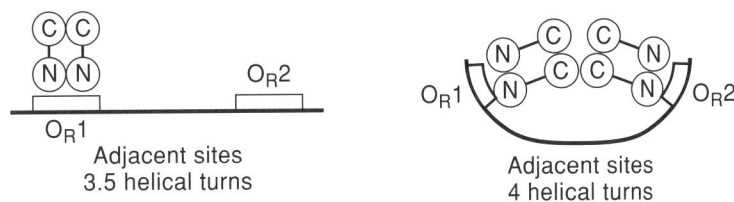

Adjacent sites
3.5 helical turns

Adjacent sites
4 helical turns

FIGURE 13.17. Cooperative DNA binding and DNA looping.

A prototypic example of a complex multiprotein structure involving cooperative protein–protein interactions, DNA bending, and looping is the intasome. The intasome and other prokaryotic recombination complexes are reviewed in Johnson (1995). The intasome is a protein–DNA complex involved in integrating and excising the 40-kbp bacteriophage λ genome to and from the *E. coli* chromosome. The excision reaction requires the phage-encoded integrase (int) and excision protein (Xis), and the *E. coli*-encoded proteins integration host factor (IHF) and Fis. Both IHF and Fis bend the DNA. The prophage genome is flanked by two attachment (att) sites called attL and attR (Fig. 13.18). Each attachment site contains phage and *E. coli* components; attL is composed of attP′ and attB, whereas attR is attP and attB′. All four proteins form a complex on each attachment site on super-coiled DNA (Fig. 13.18), the two attachment complexes interact to form the synaptic complex, and the integrase protein performs a concerted strand exchange and ligation reaction that circularizes the phage genome, excising it from *E. coli* and simultaneously religating the two free *E. coli* chromosome ends.

Within the complex, there are numerous examples of cooperativity between members of the complex, as illustrated in Figure 13.18. Integrase has two DNA-binding domains — one located on the carboxyl terminus, which binds the core sites flanking the recombination crossover point, and one on the amino terminus, which contacts the arms sites. The role of IHF is to bend the DNA to allow the amino- and carboxy-terminal domains to bind their respective sites cooperatively. Xis and Fis promote cooperative binding of Int molecules to P2. The role of IHF is strictly architectural. Thus, although IHF, a sequence-specific heterodimeric protein, can promote cooperative interactions by facilitating DNA looping, it can be replaced by the nonspecific bacterial HU protein or the eukaryotic HMG proteins at appropriate concentrations. It is envisaged that the sole role of IHF is to impart a directional bend. It does not directly contact other proteins or, if it does, the interaction is not essential. The intasome exemplifies how both DNA bending and cooperativity can function to assemble a complex nucleoprotein structure. The methods and approaches used for understanding intasome formation have served as a guide for understanding enhanceosome formation in higher eukaryotes.

IHF, in addition to its role in recombination, participates in gene activation on numerous bacterial promoters. One well-characterized example is activation of transcription by the *Klebsiella* NIF A activator protein, which binds upstream of the σ_{54} RNA polymerase holoenzyme and stimulates its transcription from responsive promoters (Hoover et al. 1990). IHF binding between the NIF A binding site and the nif H core promoter stimulates transcription presumably by bending the DNA and facilitating the ability of NIF A to contact the holoenzyme and induce open complex formation. EM and footprinting analyses have lent credence to this hypothesis.

Mechanisms of DNA Bending

In eukaryotes, DNA-bending proteins bind DNA either specifically or nonspecifically (see Werner and Burley 1997). Examples of specific proteins include: (1) TBP, discussed above, (2) LEF-1 involved in formation of the TCR-α enhanceosome, (3) the sex-determining factor SRY, and (4) HMG(I)Y, which is involved in formation of the IFN-β enhanceosome. LEF-1 contains a domain referred to as the HMG box. This domain was first identified in the eukaryotic HMG 1 and 2 proteins, which are good examples of nonspecific DNA-bending proteins. HMG box proteins bind to the minor groove. Crystal and NMR structures of the HMG box demonstrate an L-shaped structure with three α-helices (Fig. 13.19).

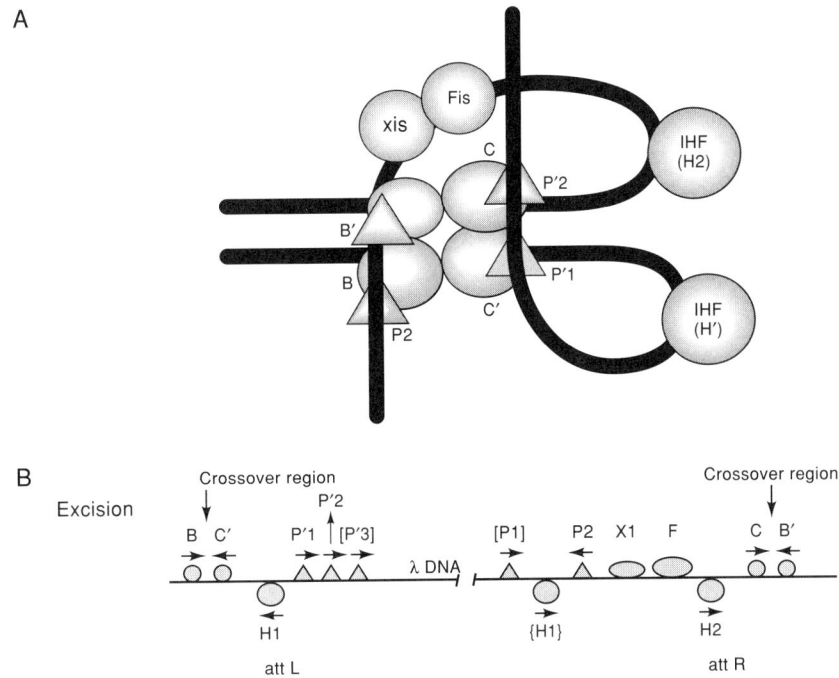

FIGURE 13.18. (*A*) Model of the intasome. (Redrawn, with permission, from Kim et al. 1990 [Copyright 1990 Cell Press] and Kim and Landy 1992 [Copyright 1992 American Association for the Advancement of Science].) (*B*) Binding sites for various proteins. (Adapted, with permission, from Johnson 1995 [Copyright Oxford University Press].)

Bending is facilitated by insertion of a more hydrophobic amino acid between the base pairs, resulting in unstacking of the bases, partial unwinding of the helix, and a bend toward the major groove.

Not all proteins that bend DNA bind in the minor groove. The best-characterized of these are prokaryotic regulatory proteins CAP and Fis (Pan et al. 1996). These dimeric proteins bend DNA by inserting recognition α-helices into adjacent major grooves. In the case of Fis, the center-to-center distance of the recognition helices is less than the center-to-center distance of the major grooves, and the protein must bend the DNA to bring the DNA's sites in register with the protein. In such cases the DNA may also wrap more extensively around the protein, further exaggerating the bend. Some eukaryotic proteins, including members of the bZIP family, also bend the DNA, apparently from the major groove.

Approaches for Studying Bending

There are three commonly used methods to measure DNA bending: EMSA, DNA cyclization, and electron microscopy (for reviews on theory and methodology, see Crothers et al. 1991, 1992). EMSA is one of the most tractable and informative methods, as it can provide information on the position, magnitude, and direction of the bend when compared with known standards.

When DNA migrates through a polyacrylamide or agarose gel, it is thought to slither through the gel pores. Although molecules of different length have the same charge density (due to the uniformity of the sugar/phosphate backbone) and, hence, experience the same electrical force, the smaller ones are able to move efficiently through the pores. Bent DNA slows the movement, with the effect being more pronounced when the bend is positioned in the center of a DNA molecule as opposed to the ends. The current view is that the mobility is inversely proportional to the square of the distance between the DNA ends. In a DNA molecule bent in the center, the distance between the ends is at a minimum and will therefore retard mobility to the greatest extent versus a linear molecule or a molecule bearing a bend positioned near the end of the molecule.

In a typical assay (see Fig. 13.20) the DNA site of interest is cloned into a vector containing two tandem polylinkers separated by a unique cloning site (see Zwieb and Adhya 1994). The DNA molecule is then cut with restriction endonucleases, which cleave once in each of the polylinkers. This generates a series of circularly permuted fragments, identical in sequence composition but differing in the position of the site relative to the ends of the polylinker. The DNA molecules are then labeled with ^{32}P and incubated with protein, and the complexes are fractionated on native polyacrylamide gels. The protein retards all of the fragments relative to unbound DNA. However, if the protein significantly bends the DNA helix, it will retard fragments bearing centrally positioned sites to a greater extent than fragments bearing a distally positioned site. The mobility of the fragments (vertical axis) measured as the distance from the end of the gel is then plotted against the position of the site in the DNA fragment. The curves display a minimum mobility at the bend center.

To determine the directionality of the bend, the binding site is placed on a DNA fragment bearing a phased A tract—stretches of 5 or more consecutive A residues that cause the DNA to bend in a known direction. A linker of various lengths is placed between the A tracts and protein-binding site. As the linker length is expanded or contracted, the protein-induced bend and the A-tract bend are rotated relative to one another. Under such condi-

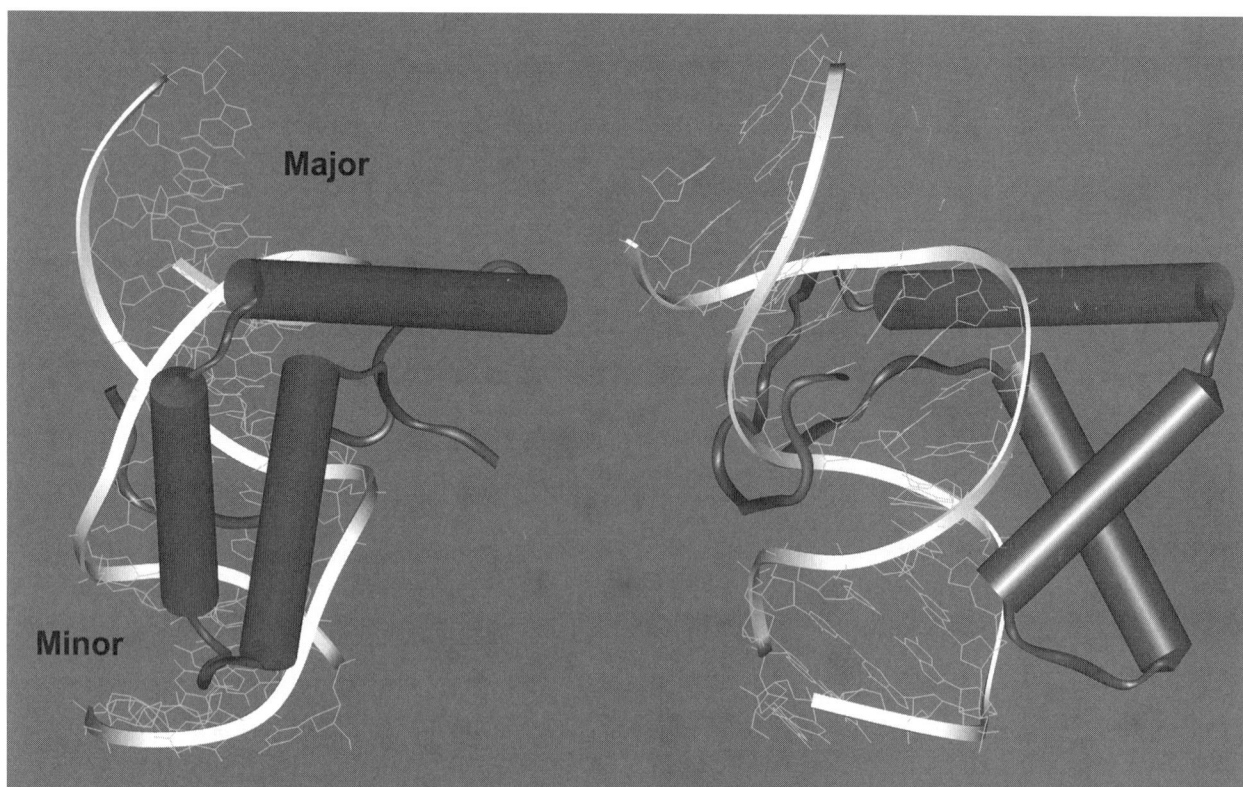

FIGURE 13.19. LEF-1 binding in minor groove and resulting DNA bend. (Crystal structures rendered by Michael Haykinson [UCLA] using the Molecular Graphics structure modeling program Insight II.)

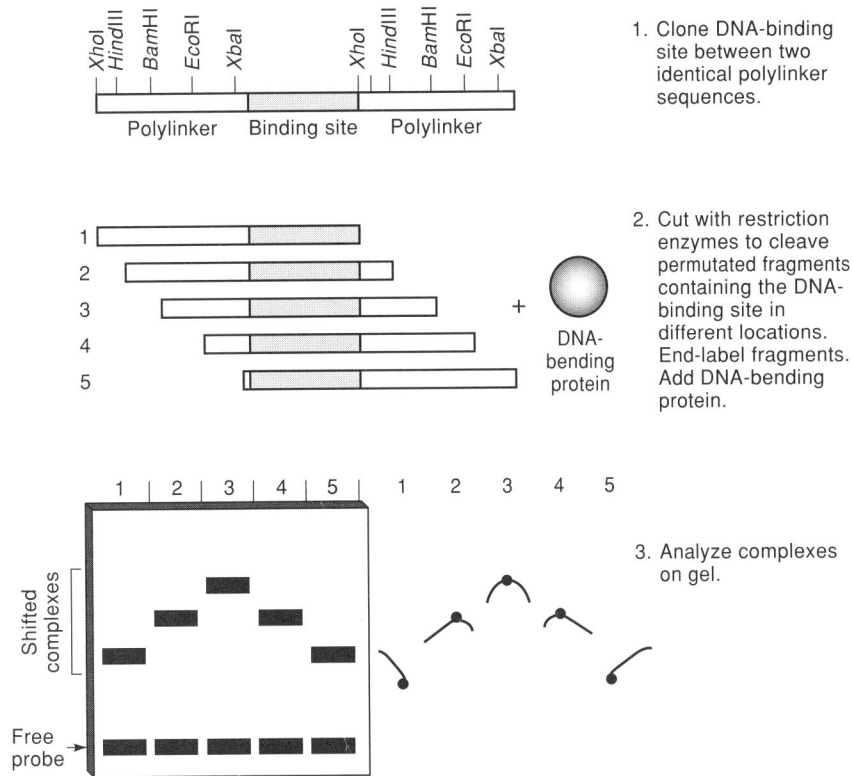

FIGURE 13.20. Analysis of DNA bending.

tions, the fragment will display a sinusoidal mobility pattern. When the bends are in the same direction, the mobility of the fragment reaches a minimum. At this point, one can be confident that the bend induced by the protein is in the same direction as the known bend induced by the A tracts. This bend can be assessed in the context of its natural site for its biological relevance.

The angle of A-tract bending has also been rigorously established and can therefore be used as a standard to determine the bend induced by the protein. In this experiment, the protein-binding site is again placed at the end of a DNA molecule containing phased A tracts. As the number of A tracts increases, so does the bend angle. Because the A tracts are separated by 10–11 bp, the direction of the bend for each A tract will be identical, and as additional A tracts are added, the bends add. The presence of a protein-binding site on the end allows for correction of the mobility. The protein–DNA complex with the bend site placed in the center is now compared with the A-tract standards (Crothers et al. 1991).

In summary, a concerted study of how eukaryotic activators bind DNA and assemble into enhanceosomes requires a significant effort to apply numerous methodologies to a highly sophisticated problem. The study of such complexes in the prokaryotic field took more than a decade and the commitment required was great. Nevertheless, the assembly and regulation of such nucleoprotein complexes almost certainly hold the key to the phenomenon of specificity during combinatorial control and gene regulation.

TECHNIQUES

PROTOCOL 13.1

DNase I Footprinting

The DNase I footprinting protocol was introduced to the research community in 1978 (Galas and Schmitz 1978). The method was based on the nested-set theory that formed the conceptual basis for DNA sequencing (Maxam and Gilbert 1977). Because of its simplicity, DNase I footprinting has found a wide following for both identifying and characterizing DNA–protein interactions.

The concept is that a partial digestion by DNase I of a uniquely ^{32}P-end-labeled fragment will generate a ladder of fragments, whose mobilities on a denaturing acrylamide gel and whose positions in a subsequent autoradiograph will represent the distance from the end label to the points of cleavage. Bound protein prevents binding of DNase I in and around its binding site and thus generates a "footprint" in the cleavage ladder (see Box 13.3 and Fig. 13.8). The distance from the end label to the edges of the footprint represents the position of the protein-binding site on the DNA fragment. The exact position of the site can be determined by electrophoresing a DNA sequencing ladder alongside the footprint.

DNase I cannot bind directly adjacent to a DNA-bound protein because of steric hindrance. Hence the footprint gives a broad indication of the binding site, generally 8–10 bp larger than the site itself. Furthermore, the crystal structure and extensive biochemical studies on DNase I (for review, see Suck 1994) show that it binds in the minor groove, contacts both strands of the sugar/phosphate backbone, and bends the DNA toward the major groove. The sequence-dependent width of the minor groove at a particular position will influence the cleavage efficiency. As a result, the cleavage ladders will not be uniform and certain gaps will exist. In some cases, proteins bind in these gaps and the footprints on such sites are not as dramatic as when the binding site is positioned over a region where the cleavage is more uniform and efficient. Footprinting can be performed with purified proteins or with crude extracts. The technical considerations for pure and crude systems are elaborated upon below and in Chapter 8.

TIME LINE AND ORGANIZATION

A protocol for performing a DNase I footprinting analysis is described below. Generally, one must prepare a uniquely end-labeled DNA fragment encompassing the protein-binding site. For small sites, double-stranded oligonucleotides that are labeled on a single strand are satisfactory. A highly active stock of DNase I must be obtained and prepared from a powder or obtained in a solution form. An entire analysis generally takes 3 days, because first the optimal concentrations of DNase must be determined before performing a dose–response curve with the protein of interest. Ideally the investigator has either a pure protein or an extract expressing the protein and preliminary data from mutagenesis studies, transfections, or EMSA indicating that a protein does indeed bind to the site.

Day 1: Prepare ^{32}P-end-labeled fragment (described in Protocol 13.6)
Day 2: Perform DNase I titration to optimize amounts and conditions for footprinting
Day 3: Use optimal amounts of DNase I to perform a dose–response curve with protein of interest

OUTLINE

DNase I titration (time commitment: 1 day)

Step 1: Pour gel from premade acrylamide/urea mix (30 minutes)

Step 2: Prepare buffers for DNase I footprinting (1 hour)

Step 3: Reactions: Prepare reactions and titrations (30 minutes), and incubate reaction mixtures at appropriate temperature (15 minutes)

Step 4: DNase I footprinting: Perform DNase I dilutions (15 minutes). Add DNase I and perform footprinting reactions (5 minutes). Terminate reactions and process products (1 hour)

Step 5: Gel electrophoresis (4–6 hours) and autoradiography (8–12 hours) or phosphorimager analysis

Step 6: Data analysis

Dose response curve (1 day)

Steps 1 and 2: As above

Step 3: Prepare serial dilutions of protein (30 minutes)

Step 4: DNase I footprinting as above

Step 5: As above

PROCEDURES

CAUTIONS: *Acrylamide, CaCl₂, Chloroform, DTT, Ethanol, Formamide, Glycerol, KCl, KOH, β-Mercaptoethanol, MgCl₂, Phenol, PMSF, Radioactive substances, SDS. See Appendix I.*

DNase I TITRATION

Step 1: Pour gel from premade acrylamide/urea mix

About 2 hours before beginning the experiment, pour an 8–12% polyacrylamide/urea gel depending on the fragment size. The gel will take about 1 hour to polymerize and 30 minutes to 1 hour to pre-run prior to sample loading. See Sambrook et al. (1989, pp. 13.45–13.57) for instructions on mixing and pouring acrylamide gels.

Step 2: Prepare buffers for DNase I footprinting

Four buffers are required for the DNase I footprinting protocol in addition to the protein. The amounts needed will have to be determined empirically depending on the scope of the study.

Buffer D: We often use as a DNA-binding buffer a buffer referred to as Buffer D from the Dignam and Roeder protocol for preparation of nuclear extract (Dignam et al. 1983). Nuclear extract is dialyzed against Buffer D containing PMSF and DTT. However, any buffer of similar composition is compatible with the assay. The key point is that the final salt and magnesium concentrations must be compatible with protein binding in vitro and with the catalytic activity of the DNase I. Buffer D alone has a shelf life of several months at room temperature but should ideally be filter-sterilized before long-term storage.

Buffer D (0.1 M KCl):

20 mM HEPES-KOH (pH 7.9)

20% glycerol

0.2 mM EDTA

0.1 M KCl

Protein diluent buffer: This is the buffer into which the DNA-binding protein is diluted. However, in crude extracts the protein is not diluted. In this model case the diluent is Buffer D containing a protein-stabilizing agent, such as nuclease-free BSA, an anti-aggregation agent such as 0.01–0.1% NP-40 or Triton X-100, and a reducing agent such as 0.1–1 mM DTT or 10–50 mM β-mercaptoethanol. These reducing agents are particularly important for protein with cysteines within the DNA-binding domain (i.e., zinc fingers or the basic region of many b-Zip proteins). This buffer should be prepared in small quantities and stored in aliquots in a freezer. Dilute solutions of reducing agents generally have short half-lives, so long-term storage of the solutions on ice or in the refrigerator is not recommended.

> 500 μl of Buffer D (0.1 M KCl)
> +1 μl of BSA (50 μg/μl)
> +1 μl of β-mercaptoethanol (14 M)

DNase I diluent buffer: This buffer is the same as the protein diluent and is designed to be compatible with it. $CaCl_2$ is added to enhance the activity of DNase I, which is essential when footprinting crude extracts or impure preparations of protein. This buffer should be prepared fresh and kept on ice or at 4°C.

> 500 μl of protein diluent buffer (0.1 M KCl)
> +20 μl of 1 M $CaCl_2$

Stop buffer: The Stop buffer terminates the DNase I reaction and prepares it for phenol extraction and ethanol precipitation. EDTA is added to chelate divalent cations necessary for DNase I activity; SDS is added to denature protein and strip them from DNA; sodium acetate is added to facilitate ethanol precipitation as is the carrier tRNA. This solution can be maintained as a stock for several months at room temperature.

> *Final concentrations*
> 400 mM sodium acetate
> 0.2% SDS
> 10 mM EDTA
> 50 μg/ml yeast tRNA

Immediately before using, aliquot the amount of Stop buffer needed for the experiment and add 1/1000 volume of Proteinase K solution (10 mg/ml in deionized, distilled water).

Formamide dye mix:
> 98% deionized formamide
> 10 mM EDTA (pH 8)
> 0.025% xylene cyanol FF
> 0.025% bromophenol blue

Step 3: Reactions

For the typical experiment using a recombinant DNA-binding protein, we initially prepare 12 reactions, which are used to titrate both the initial DNase I and DNA-binding protein concentrations. Higher concentrations of a DNA-binding protein, particularly with crude extracts, tend to inhibit the DNase I. Therefore, the concentration of DNase I used for naked DNA controls will be lower compared with the concentration used with high concentrations of a DNA-binding protein. Furthermore, the amount of DNase I required will be smaller with increasing fragment size. It is important in this initial step of the study to optimize the concentrations of DNase I needed for a variety of conditions.

1. Prepare a reaction mix containing the reagents described below.

   Note: All of the pipetting and reaction preparations are ideally done on ice. Prepare enough mix for two extra reactions, one for inevitable losses or inconsistencies during pipetting and one for a mock reaction without DNase I.
   Mix:

DNA template[a] (50 fmole/µl)	0.10 µl (1000–10,000 cpm)
poly dI:dC[b] (1 µg/µl)	0.20 µl (200 ng)
BSA (50 µg/ µl)	0.20 µl (10 µg)
β-mercaptoethanol (14 M)	0.10 µl
MgCl$_2$ (0.1 M)[c]	1.5 µl
Buffer D (0.1 M KCl)[d]	11.5 µl
H$_2$O	5.4 µl
Total volume	19.00 µl
14 Reaction batch	266 µl

   Recombinant activator protein is prepared at a concentration of 250 ng to 1 mg per ml. For the DNase I titration, 1 ml of neat, undiluted protein and a 1/40 dilution will be used in the initial DNase I titrations. For the dose-response curves (day 3), we perform a titration in threefold steps based on the results from day 2.

   [a] End-labeled activator-specific DNA probe (Protocol 13.6 for end-labeling DNA).

   [b] Amersham Pharmacia Biotech Inc. (#27-7880-01)

   [c] For crude extracts that may contain endogenous nucleases, binding reactions may work better if set up in the absence of MgCl$_2$, which can be added later in the DNase I diluent buffer.

   [d] For crude extracts, the Buffer D should be substituted by extract in Buffer D allowing at least 12 µl to be added. For less concentrated extracts, Buffer D and H$_2$O can both be substituted allowing 17 µl of extract to be added.

2. Aliquot the reaction mix into 13 0.5-ml siliconized eppendorf tubes on ice.

3. Thaw recombinant activator protein on ice immediately before titrating. Add 1 µl to 39 µl of protein diluent to make the low-concentration sample. Add protein and mix gently by tapping the tube with your finger (vigorous mixing using the pipetman is not recommended, as some proteins may be sensitive to the frothing often generated in this technique). Once the dilution is made, titrate 1 µl of neat activator protein into the first four tubes, 1 µl of the 1/40 dilution into the next four tubes, and 1 µl of protein diluent into the final four:

   Tubes 1–4: 1 µl of neat activator protein

   Tubes 5–8: 1 µl of a 1/40 dilution of activator protein

   Tubes 9–12: 1 µl of diluent alone

   For crude extracts, DNase I must be titrated with low and high amounts of extract. Note that crude extracts are highly inhibitory and large amounts of DNase must be added.

4. Place the tubes into a 30°C water bath for 20 minutes (or for crude extracts at room temperature or on ice). This is generally enough time for binding of an activator protein to reach equilibrium, although some proteins (e.g., TBP) bind more slowly.

Step 4: DNase I footprinting

The goal here is to determine roughly the amount of DNase I to employ for a given amount of DNA-binding protein. We therefore make two concentrations (high and low) of protein and, for each, add increasing concentrations of DNase I. We intend to identify concentrations of DNase I that generate an evenly distributed cleavage ladder, where ~50% of the DNA remains uncleaved. This amount of cleavage minimizes multiple cleavages within a single DNA molecule, which can affect the interpretation of data.

1. Choose four concentrations of DNase I for the initial titration.

   Note: The concentrations required will be highly dependent on the source of DNase I and the divalent cation concentrations in the mixture as well as the DNA-binding protein being employed. We typically use the Amersham Pharmacia Biotech Inc. FPLC-pure, RNase-free DNase I (cat. # 27-0514-01), which is delivered at a concentration of 5,000–10,000 units per ml.

2. Dilute the DNase I serially to 1/9, 1/27, 1/81, and 1/243 using DNase I diluent. The amount of DNase I added to a cleavage reaction will be 1 μl. Note that for fragments larger than that used here, smaller amounts of DNase I will be required.

3. After the 20-minute incubation in Step 3.4, perform sequential 1-minute DNase I digestions as follows:

   Use timer:

   > add 1 μl of DNase I to tube #1 at time n
   >
   > add 1 μl of DNase I to tube #2 at time $n+15$ seconds
   >
   > add 1 μl of DNase I to tube #3 at time $n+30$ seconds
   >
   > add 1 μl of DNase I to tube #4 at time $n+45$ seconds
   >
   > add 100 μl of stop buffer + Proteinase K to tube #1 at time $n+1$ minute
   >
   > add 100 μl of stop buffer + Proteinase K to tube #2 at time $n+1$ minute, 15 seconds
   >
   > add 100 μl of stop buffer + Proteinase K to tube #3 at time $n+1$ minute, 30 seconds
   >
   > add 100 μl of stop buffer + Proteinase K to tube #4 at time $n+1$ minute, 45 seconds

   Process one set of four reaction tubes per 2 minutes. It is convenient to place the four DNase I dilutions immediately behind the tubes into which you will be titrating. Remember to mix gently by tapping the tubes immediately after pipetting in the DNase I. Place a radioactive disposal vessel nearby, because the pipet tips will be mildly radioactive. For crude extracts, the DNase I digestions, like the binding reactions, may work better if performed on ice.

4. Repeat this entire cycle as soon as possible for the remaining two sets of tubes.

5. Perform a mock reaction alone without any DNase I gel so that in the final analysis the uncut probe can be compared to the cleavage ladders generated above.

6. Incubate all reaction tubes at 55°C for 15 minutes to allow Proteinase K digestion to occur. This is particularly important for cruder protein preparations. Meanwhile, assemble the polymerized gel into the gel apparatus. Pre-run the gel at 1000 V for 30 minutes.

7. Extract each reaction mix with an equal volume (100 μl) of phenol.

8. Centrifuge tubes for 2 minutes.

9. Transfer a constant volume (e.g., 100 μl) of the top (aqueous) phase to 1.5-ml non-siliconized eppendorf tubes.

10. Repeat steps 7–9 using an equal volume (100 μl) of phenol/chloroform.

11. Add 2x volumes of cold 95% ethanol to the 1.5-ml tubes to precipitate DNA. Mix well.

    Note: There is no need to add 3 M sodium acetate because the stop buffer contains high salt.

12. Incubate the tubes on dry ice, dry-ice ethanol, or in the –80°C freezer for 10 minutes.

13. Centrifuge the tubes at 14,000g in a microfuge for 10 minutes.

14. Decant ethanol from pellets using drawn-out, narrow-bore, pasteur pipets. The pellets are loose, so be careful!

    Note: Monitor tubes periodically with a Geiger counter to ensure that the pellet has not been accidentally drawn up into the pasteur pipet.

    Wash the pellets with 100 μl of 80% ethanol, centrifuge for 2 minutes, and remove the alcohol with drawn-out pipets. Allow the pellets to air-dry completely.

Step 5: Gel electrophoresis and autoradiography

1. Resuspend the pellets in a formamide dye mix.

2. Denature DNA by incubating samples for 2 minutes at 95°C.

3. Load the samples on a polyacrylamide-urea gel and perform gel electrophoresis.

4. Dry the gel under vacuum.

5. Perform autoradiography or phosphorimager analysis.

Step 6: Data analysis

The purpose of the DNase I titration is to optimize cleavage so that an evenly distributed cleavage ladder is obtained and at least 50% of the probe remains uncleaved. To assess the amount of cleavage accurately, the digestion ladders must be compared with a mock reaction containing probe alone but no DNase I. Also, with crude extracts, the mock reaction performed with extract reveals the presence of endogenous nucleases. Ideally, because the DNase I concentration is increased, there should be a concomitant decrease in the amount of intact probe and an increase in the amount of cleavage products. As the probe begins to disappear, there should be a change in the distribution of cleavage products from lower mobility to higher mobility bands as the DNase I begins to cleave the probe into smaller fragments. Ideally, the amount of DNase I that generates approximately 50% cleavage at each concentration of DNA-binding protein should be noted and used to calculate amounts for the dose–response curve.

In the next step, we normally perform a dose–response curve from 1 μl of DNA-binding protein to a 1/2000 dilution. Although it is unnecessary to adjust the DNase I for every protein concentration, it should be adjusted at concentrations at which the protein is likely to be inhibitory. We adjust the concentration based on the difference in inhibition between 1 μl of activator protein and the 1/40 dilution. We assume that the inhibition will be somewhat linear with respect to DNase I concentration and extrapolate the amounts accordingly.

PERFORMING A DOSE–RESPONSE CURVE

Step 1: Pour gel from premade acrylamide gel mix (Sambrook et al. 1989, pp. 13.45–13.57)

Step 2: Prepare buffers for DNase I footprinting

Step 3: Prepare serial dilutions of protein

1. Set up the reaction mixtures as described (pp. 473–474) except now add the recombinant activator protein in serial dilutions using the protein diluent buffer. Start with 1 μl of activator protein and perform threefold serial dilutions for about 8 points (1, 1/3, 1/9, 1/27, 1/81, 1/243, 1/729, 1/2187). For crude extracts, 1 to 17 μl should be titrated.

2. Prepare a batch for 10 reactions as in step 3.1, aliquot into 9 reaction tubes, and add 1 μl of the protein dilutions to each tube. Prepare one tube without protein for comparison.

3. Choose the optimal amount of DNase I determined on day 2. Add DNase I to each reaction. Perform digestion and process products as described above.

Step 4: Gel electrophoresis and autoradiography or phosphorimager analysis

Ideally, there should be a gradual increase in the appearance of the footprint as the DNA-binding protein concentration increases. The fractional occupancy or percent protection, two terms useful in affinity calculation, can be deduced by comparing bands from the cleavage ladder in the no-protein lanes with those in the footprint lanes using laser densitometry or phosphorimaging software. An unaffected band below and above the footprint can be used to normalize the DNase I cleavage efficiency. Most proteins will bind a DNA site gradually and follow normal Michaelis kinetics for a first-order reaction. However, many proteins bind as dimers and higher oligomers, and if the dimerization constant is near to the K_d for the site, protein binding may follow second-order or higher kinetics. If this technique is to be used quantitatively to model DNA-binding kinetics, read Koblan et al. (1992) for a more sophisticated overview of technical considerations.

ADDITIONAL CONSIDERATIONS

Generalizing the protocol

There is no steadfast rule for generalizing the protocol. The following factors should be considered:

- Smaller reaction sizes will favor less concentrated DNA binding proteins.
- Lower salt concentrations will favor weaker binding proteins.
- The final volume can be anywhere from 10 to 100 μl, depending on the reaction.
- Smaller reaction volumes can be used to conserve valuable materials.
- Different carrier DNAs can have different effects: Poly dI:dC, for example, is reported to compete for binding of TBP to a TATA box probe and is often replaced with poly dG:dC.

The following are some ranges that have been used successfully in our laboratories and the published work of others:

- 10–50 mM HEPES-KOH or Tris-HCl (pH 7.0–8.0). DNase I is compatible with a wide range of buffers.

- 0–10 mM $MgCl_2$. This divalent cation neutralizes the phosphates, and some DNA-binding proteins have been reported to bind it as a co-factor.

- 50–100 mM KCl, NaCl, potassium acetate, ammonium sulfate. Although most DNA-binding proteins are sensitive to unusually high concentrations of salt, salt optima can vary considerably depending on the protein. It is best to evaluate this empirically.

- 0–20% v/v glycerol. Glycerol is a stabilization reagent that lowers the water concentration in the reaction and mimics the in vivo environment. It is also a free-radical scavenger.

- 10–100 µg/ml BSA. BSA is another protein stabilizer, which can act as nonspecific carrier and prevent spurious attachment of dilute protein samples to surfaces (e.g., the walls of eppendorf tubes).

- 0.01–0.1% NP-40, Triton X-100. These nonionic detergents prevent nonspecific protein binding to surfaces but also act as anti-aggregation agents.

- 0.1–1 mM DTT or 10–50 mM β-mercaptoethanol. Absolutely essential reducing agents for many proteins.

- 0–1 µg carrier DNA, either calf thymus or synthetic co-polymers (dI:dC). Prevents nonspecific binding of contaminant during binding reactions and minimizes nonspecific binding of the protein being footprinted.

- Polyvinyl alcohol (PVA), polyethylene glycol (PEG), dimethyl sulfoxide (DMSO). PVA and PEG reagents can vary considerably in concentration. PVA and PEG are volume exclusion agents that increase the functional concentration of the protein in solution and decrease water concentration. DMSO is a denaturant that has some unusual stabilizing properties at low concentration, possibly by minimizing nonspecific binding of the protein or favoring conformational flexibility.

A detailed analysis of relevant parameters and how these influence DNA–protein interactions can be found in reviews by Record et al. 1991 and Koblan et al. 1992.

DNA fragment size

There are several important considerations for fragment size. Highly purified proteins in which the mechanism of DNA binding or affinity is being established should utilize small DNA fragments, such as sites cloned into the polylinkers of vectors such as pGEM or pUC (i.e., 50–100 bp in length). The small size results in better band resolution on polyacrylamide/urea sequencing gels and allows the same fragment to be compared in EMSA and footprinting studies. When attempting to locate a protein-binding site on a promoter, by necessity larger fragments must be employed (see Chapter 8). In crude extracts, where there is an abundance of nonspecific DNA-end-binding proteins like the Ku autoantigen, the sites of interest are ideally located 50 bp or so from the ends, so the footprint can be distinguished from the contaminating end-binders. Please note that for fragments larger than that used here, smaller amounts of DNase will be required.

Cruder extracts

Use of crude extracts in footprinting studies adds an additional complexity to the experimental design. Higher concentrations of DNase I must be employed to overcome the abundance of inhibitors. Some extracts are rich in nonspecific DNA-binding proteins and, consequently, low concentrations of extract (<10 μg) and a high concentration of nonspecific carrier DNA (1 μg or more) must be employed to minimize the inhibition. This must be balanced against the possible need for high concentration of extract to detect specific binding. Due to the presence of endogenous nucleases, it is prudent sometimes to perform the initial binding and DNase I reactions in the cold (4°C). However, higher concentrations of DNase I are necessary to cleave at lower temperatures, although these amounts must be determined empirically. Alternatively, $MgCl_2$ can be omitted from the preincubation to inhibit nuclease activity during the binding reaction. $MgCl_2$ is then added with the DNase I.

Time course

It is necessary occasionally to measure the kinetics of binding by a protein. Ideally, this means staggering the time that the protein is added to the binding mixture and the time that the mixture is placed at the proper incubation temperature. Simply adding the protein to a mixture on ice and assuming it will not bind to DNA until the mixture is incubated at 30°C is incorrect. Many binding studies are in fact performed on ice, as alluded to above. Thus, ideally the reaction mixture is brought to incubation temperature and then the DNA-binding protein is added.

Indeed, the rates at which some reactions reach equilibrium are very rapid (i.e., binding of GAL4-VP16 is complete within 2 minutes), whereas others are slow (binding of TBP can take an hour or more). In the former case, it is prudent to speed up the DNase I cleavage reaction considerably by increasing the concentration of DNase I and lowering the time of cleavage to about 15 seconds. Conversely, for slow reactions, the DNase I cleavage rate can be lowered by lowering the concentration of DNase I. Others have commented on the conditions needed for performing truly quantitative footprint for use in biophysical modeling (Brenowitz et al. 1986).

Using unlabeled DNA

It is possible to perform DNase I footprinting on circular DNA templates and to detect the cleavage points by indirect end-labeling or LM-PCR (see Chapter 10, Protocol 10.2 or Protocol 15.2 in Chapter 15). These methods are described in Gralla (1985) and Grange et al. (1997).

TROUBLESHOOTING

No cleavage

Possible cause: Sometimes nuclear extracts can be highly inhibitory at commercially available DNase I concentrations
Solution: If no cleavage by DNase I is observed, increase the DNase I, $MgCl_2$, or $CaCl_2$ concentrations and, most certainly, try a different batch of DNase I. Try lowering the extract concentration or further fractionating it by column chromatography. Additionally, purchase DNase I powder and dissolve it at higher concentrations.

No or weak footprint

Possible cause: This result could be due to low concentrations of protein, the absence of a physiological site in the fragment, or inhibitors of binding contaminating protein preparations.

Solution: Attempt to raise protein concentration, lower DNA concentration, or adjust binding parameters to optimize salt concentrations, pH, etc.

Smeary gels

Possible cause: Smeary gels are most often caused by: (1) not pre-running the gels to remove the residual ammonium persulfate in the acrylamide; (2) incomplete removal of protein during the phenol extraction step and subsequent co-precipitation with alcohol making the resulting pellet difficult to resuspend; or (3) the pellet contains too much salt.

Solution: In case (1), occasionally the bands become compressed at the salt front. Next time, pre-run the gel. In case (2), one observes black spots that smear down the gel from the wells. Try resuspending the sample more vigorously or phenol-extracting the sample one extra time before ethanol precipitation. Another option is to precipitate the sample with 2 M ammonium acetate, which tends to leave protein, small oligonucleotides, and nucleotides in solution while the larger DNA fragments precipitate. In case (3), often a reverse dovetail effect is observed where the banding pattern narrows as it approaches the lower part of the gel. Simply wash the pellet with 80% ethanol before drying and resuspending in formamide dye buffer.

PROTOCOL 13.2

Hydroxyl-radical footprinting

The hydroxyl-radical footprinting methodology was devised by Tullius and Dombroski in 1986 initially to study λ repressor and Cro binding to DNA. However, the method has wide applications to studying protein–DNA interactions, as well as structural perturbations (e.g., bending) that occur in DNA during protein binding. Hydroxyl radicals cleave DNA by abstracting a hydrogen from C4 of the sugar in the minor groove. Protein binding over the minor groove generally protects the sugar from cleavage (see Box 13.4 and Fig. 13.10). The method can give detailed information on protein binding to the minor groove. Although some of the information overlaps that obtained by ethylation interference, it is often complementary and can assist in modeling how a protein docks with its site. As described earlier, the radical is generated by Fe(II) EDTA, which cleaves hydrogen peroxide into a hydroxyl radical and a hydroxide ion. The radical then cleaves the sugar in a diffusion limited reaction. Ascorbic acid is included to regenerate the active Fe(II).

Hydroxyl radical is highly reactive but easily quenched by glycerol; therefore, glycerol, a component in most binding reactions, must be avoided. The radical reaction requires peroxide, which is a powerful oxidant and occasionally interferes with protein binding to DNA. The conditions can occasionally be adjusted to accommodate such scenarios. The radical also reacts with and cleaves proteins; long reaction times can lead to protein degradation.

TIME LINE AND ORGANIZATION

The reaction is set up almost identically to DNase I footprinting except for the absence of glycerol in the buffers. Often, the DNA plasmid quality is essential to the success of the method. Many preparations contain a low level of preexisting DNA cleavage. Because the hydroxyl radical generally decays quite rapidly, less than 5% of the DNA is usually cleaved in the process. Thus, low preexisting background cleavage can lead to signal-to-noise problems. We have found that it is important to use high-quality DNA purified twice over a CsCl density gradient. Furthermore, when using the alkali lysis method, the final DNA pellet after isopropyl alcohol-precipitation must be neutralized with Tris base prior to CsCl density centrifugation. Once high-quality DNA is in hand, an end-labeled DNA fragment is prepared. Often it is necessary to perform a DNase I footprint under the hydroxyl-radical reaction conditions to ensure that the protein of interest binds under those conditions. In fact, titration of the time of radical cleavage at different concentrations of protein may be useful in the event the protein has inhibitors like glycerol present. Once these preliminary steps are complete, the hydroxyl-radical footprint can be performed in a single day. The products can be analyzed via autoradiogram the following day.

OUTLINE

Step 1: Prepare buffers (30 minutes)
Step 2: Hydroxyl-radical protection (5–6 hours)
Step 3: Gel electrophoresis and autoradiography
Step 4: Data analysis

PROTOCOL: HYDROXYL-RADICAL FOOTPRINTING

Procedure

CAUTIONS: *Ammonium sulfate, Bromophenol blue, Ethanol, Formamide, H_2O_2, KCl, KOH, $MgCl_2$, Xylene. See Appendix I.*

Step 1: Prepare buffers

Binding buffer:

20 mM HEPES-KOH (pH 7.5)
50 mM KCl
5 mM $MgCl_2$
100 µg/ml BSA
5 µg/ml poly(dI-dC)
(use 50–100 µl per reaction)

Formamide dye mix:

98% deionized formamide
10 mM EDTA (pH 8)
0.025% xylene cyanol FF
0.025% bromophenol blue

Step 2: Hydroxyl-radical protection

1. Add 10–50 femtomoles of ^{32}P-labeled DNA fragment and saturating concentrations of recombinant DNA-binding protein to 50–100 µl of binding buffer.

2. Incubate for 15 minutes at room temperature or optimal binding temperature determined by DNase I footprinting, for example.

3. Add 10 µl of freshly diluted 100 µM $Fe(NH_4)_2SO_4$ and 10 µl of freshly diluted 200 µM EDTA (do not dilute into a glass container; use an eppendorf microfuge tube).

4. Add 10 µl of 10 mM ascorbic acid and 10 µl of 0.3% H_2O_2 solution, both freshly prepared from stocks. Incubate for 2 minutes at room temperature.

5. Terminate the reaction by adding 100 µl of:
   5 M ammonium acetate
   5 mM thiourea
   10 mM EDTA
   10 µg of tRNA
   The thiourea should be added fresh from a concentrated stock.

6. (Optional) Phenol-extract.

Step 3: Gel electrophoresis and autoradiography

1. Ethanol precipitate the DNA. Wash with 80% ethanol, dry, and resuspend in formamide dye mix.

2. Denature the DNA by incubation at 95°C for 2 minutes.

3. Load directly onto a 10% or 12% polyacrylamide/urea gel and electrophorese 4–6 hours, depending on fragment size.

4. Dry gel under vacuum.

5. Expose dried gel to film or phosphorimager.

ADDITIONAL CONSIDERATIONS

1. The hydroxyl radical should generate a ladder of DNA fragments approximately equal in intensity down the gel (except in stretches where DNA bending or twisting is occurring). The ladder should be dependent on the addition of peroxide, ascorbate, and Fe-EDTA. Appropriate controls lacking these chemicals must be performed in parallel.

   As described above, some proteins are not compatible with binding in peroxide. The relative balances of peroxide, ascorbate, and Fe-EDTA can easily be manipulated to accommodate alternative reaction conditions (i.e., lower peroxide but raise ascorbate). These issues are covered in detail by Dixon et al. (1991).

   To determine whether your protein is affected by the radical conditions, a DNase I footprinting reaction can be performed in the presence of the chemicals necessary for generating the radical. If certain chemicals or combinations thereof inhibit binding in a footprinting assay, the conditions must be adjusted accordingly.

2. The radical can also detect DNA alterations such as bending, because these are characterized by a series of alternating 5-bp regions of reduced and enhanced cleavage due to narrowing and expanding of the minor groove through the bend. The radical also detects other unusual DNA structures as discussed by Dixon and colleagues (1991).

3. There are other variations of this technology including the missing nucleoside experiment. DNA is first treated with hydroxyl radical to remove nucleosides randomly on average of one per molecule. The ability of a protein to bind to the gapped DNA is measured by EMSA. The bound and unbound DNAs are excised, purified, and fractionated on a polyacrylamide/urea gel. The bound fraction is enriched for molecules missing bases that are not important for protein binding, while the unbound fraction is enriched in molecules missing bases important for binding. The positions of the bases are identified on a polyacrylamide/urea gel, alongside a sequencing ladder of the fragment. The technique can be used to scan a DNA molecule at single-nucleotide resolution in a single experiment (Hayes and Tullius 1989).

TROUBLESHOOTING

No cleavage

Possible cause: Dilute solutions of ascorbic acid, peroxide, and EDTA can all become inactivated over short times. Even the dry reagents, particularly ascorbic acid, can go bad in moist climates.
Solution: If no cleavage ladder is observed, try using fresh chemicals. Also, ensure that neither the activator preparation nor buffers contain more than 1% glycerol. Even 1% glycerol, however, is inhibitory.

No specific cleavage

Possible cause: DNA not prepared carefully enough.
Solution: If a preexisting ladder exists in the absence if cleavage, use greater care in preparing the plasmid DNA and the DNA fragment. Try re-preparing it. Also, prepare fresh, deionized formamide loading dye mix.

PROTOCOL 13.3

Phosphate Ethylation Interference Assay

The ethylation interference assay was originally developed by Walter Gilbert and colleagues to study RNA polymerase binding to *E. coli* promoters (Siebenlist et al. 1980; Siebenlist and Gilbert 1980).

The technique (reviewed in Manfield and Stockley 1994) involves using ethylnitrosourea to ethylate phosphates within a ^{32}P-end-labeled DNA containing a protein recognition site. This modification neutralizes the phosphate charge and places a bulky ethyl group along the edge of the minor groove (Fig. 13.1). The ethylated phosphates interfere with proteins coming into close proximity to the minor groove or phosphate backbone of the binding site. Most modifications will have little effect because only a subset of phosphates within the site will interfere and most phosphates flanking the site will not interfere (see Fig. 13.13).

The bound and unbound DNAs are then separated by EMSA (or filter binding, immunoprecipitation, etc.). The gel is autoradiographed, and the bound and free DNA populations are excised from the gel and purified from the gel slices. The unbound fraction will be enriched in DNA molecules containing modifications at positions that interfere with protein binding. In contrast, the bound fraction will contain modifications that do not interfere with binding of the protein. Ethylation makes DNA susceptible to piperidine-induced cleavage. The cleavage products are then fractionated on a polyacrylamide/urea gel alongside a sequencing ladder.

Note that cleavage with alkali can leave the phosphate attached to either the 5′ or 3′ breakpoints. Thus, the cleavage products often migrate as closely spaced doublets on a gel.

TIME LINE AND ORGANIZATION

The ethylation interference assay can be performed in 2 days. Most of the organization involves first radiolabeling the fragment and determining by EMSA an amount of protein that is barely saturating (see Protocols 13.5 and 13.6). Higher amounts of DNA are used in the initial EMSA to ensure conditions necessary for adequate recovery of DNA during the actual procedure. The DNA fragment is then ethylated, purified, and subjected to another round of EMSA, whereupon the bound and unbound DNAs are excised and purified. The purified DNAs are then cleaved with piperidine and fractionated on a polyacrylamide urea sequencing gel alongside a sequencing ladder.

OUTLINE

Step 1: Prepare buffers (1 hour)
Step 2: Ethylation interference assay (10 hours)
Step 3: Cleave ethylated residues (1 hour)
Step 4: Fractionate DNA from bound and unbound fractions on polyacrylamide/urea gel (3 hours)

PROTOCOL: ETHYLATION INTERFERENCE ASSAY
PROCEDURE

CAUTIONS: *DTT, Ethanol, Ethylnitrosourea, Formamide, KCl, MgCl₂, NaOH, NaPO₄, Sodium cacodylate. See Appendix I.*

Step 1: Prepare buffers

Buffer A:

10 mM Tris-HCl (pH 8.0)
10 mM MgCl$_2$
100 mM KCl
1 mM DTT
0.1 mM EDTA

0.05 M Sodium cacodylate (pH 8.0)

Step 2: Ethylation interference assay

1. Add 100 µl of sodium cacodylate (pH 8.0) and 100 µl of 95% ethanol saturated with ethyl nitrosourea to 10 picomoles of ^{32}P-end-labeled DNA fragment.

2. Incubate for 1 hour at 50°C.

3. Precipitate DNA twice with 10 µl of 5 M ammonium acetate and 2.5 volumes of ethanol.

4. Resuspend the pellet in 100 µl of Buffer A.

5. On ice, add saturating concentrations of DNA-binding protein to 50,000 cpm of ethylated DNA fragment under standard binding conditions. Incubate at 30°C for 15 minutes.

6. Carefully load and electrophorese the protein–DNA mixtures through a pre-run 4.5% native polyacrylamide gel (Sambrook et al. 1989, pp. 13.45–13.57).

7. Expose gel to film; take care that the gel and the film are oriented correctly (see Chapter 4, Protocol 4.3, Step 3 for isolating a radiolabeled DNA fragment from a polyacrylamide gel).

8. Align gel with autoradiograph and excise bands corresponding to bound and unbound fractions with a razor blade.

9. Recover the DNA fragments from the gel by electroelution (see Protocol 13.6, Step 3).

10. Ethanol-precipitate and dry the DNA pellets.

Step 3: Cleave ethylated residues

1. Resuspend the pellets from the bound and unbound fractions in 10 mM NaPO$_4$ (pH 7.0), 1 mM EDTA.

2. Add 2.5 µl of 1 M NaOH. Incubate at 90°C for 30 minutes. During this step, the ethylated DNA backbone is cleaved with NaOH.

3. Add 1:1 formamide loading dye (see Protocol 13.1) to samples, and load onto a 10% polyacrylamide/urea gel alongside a sequencing ladder.

4. Dry gel under vacuum.

5. Expose the gel to film and perform autoradiography or phosphorimager analysis.

ADDITIONAL CONSIDERATIONS

1. The initial ethylation frequency must be determined empirically (see Fig. 13.2). Alter the amounts of ethylnitrosourea and/or the time of modification. Analyze the cleavage of naked DNA to ascertain conditions for less than one modified phosphate per DNA (see Box 13.4).

2. Remember that in interference techniques you need not obtain 100% binding to extract useful information. The bound lane, although enriched in ethylated phosphates, which do not interfere with binding, is depleted of ethyl phosphates, which do interfere. Simply comparing the cleaved DNAs from the bound lane with a standard ethylation ladder and determining the positions of missing bands in the bound lane can yield important information.

3. Ethylation conditions are not compatible with direct binding experiments and thus there is no ethylation protection procedure.

4. The data obtained from ethylation interference can often be useful in determining backbone contacts and can be compared and contrasted with hydroxyl radical data as discussed in Protocol 13.2.

TROUBLESHOOTING

The ethylated DNA does not gel shift when protein is added

Possible cause: This may indicate that the DNA is highly overethylated.
Solution: Try lower concentrations of ethylnitrosourea or shorter modification times.

Possible cause: The DNA may not have been purified adequately from the ethylnitrosourea.
Solution: Try re-precipitating the DNA once or twice with ethanol. Make sure that pH of resuspended DNA is near 7.

The bound and unbound DNA fractions from EMSA display the same band distribution when isolated and compared on a polyacrylamide/urea gel

Possible cause: This result is often indicative of preexisting banding or cleavage pattern, resulting from impure or old DNA.
Solution: Prepare a fresh plasmid prep and take the precautions outlined as per the hydroxyl radical footprinting procedure (see Protocol 13.2).

PROTOCOL 13.4

Methylation Interference Assay

The methylation interference and protection techniques were initially developed by Walter Gilbert and colleagues to study the interaction of *E. coli* RNA polymerase with the major groove of the DNA. The chemistry was an offshoot of that used in the chemical method of DNA sequencing invented by Maxam and Gilbert. Methylation is now widely employed as the major technique for studying interaction of proteins with major groove guanines, although it also detects adenines in the minor groove and cytosines in single-stranded DNA.

There are two versions of the method. In the protection technique, a protein is prebound to a ^{32}P-end-labeled DNA containing a recognition site and the protein–DNA complex is subjected to attack by DMS. Bound protein protects guanines (Gs) in the major and adenines (As) in the minor grooves from methylation and subsequent cleavage by piperidine. Although the protection technique works smoothly for many proteins, it has several drawbacks. Because DMS modifies certain buffers (e.g., Tris), the protection reaction was historically performed in cacodylate, which is rarely used as a standard protein-binding buffer these days. Additionally, some proteins are directly modified by DMS, and these modifications can disrupt their binding to DNA (e.g., TFIIIA). Furthermore, unlike interference techniques, protection techniques require nearly stoichiometric binding (~90%) to observe footprints and gain information on the binding site. For these reasons and others that are described below, the methylation interference technique has been widely employed in place of protection. It is certainly one of the highest-resolution methods for measuring the bases involved in sequence-specific recognition by proteins.

The technique involves first modifying a ^{32}P-end-labeled DNA containing a recognition site (on average once per DNA molecule; see Figure Box 13.4) and then binding the protein. Certain modifications prevent the protein from binding because they are at positions where the protein comes into close proximity with the DNA. Most modifications, however, have no effect.

The bound and unbound DNAs are then separated by EMSA (or filter binding, immunoprecipitation, etc.). The gel is autoradiographed and the bound and free DNA populations are excised from the gel and purified from the gel slices. The unbound fraction will be enriched in DNA molecules containing modifications at positions that interfere with protein binding. In contrast, the bound fraction will contain modifications that do not interfere with binding of the protein. Methylation weakens the nucleotide and makes it susceptible to piperidine-induced depurination. This leads, in turn, to scission of the DNA backbone. The cleaved fragments can then be fractionated on a sequencing gel alongside A+G and C+T chemical sequencing ladders to identify the affected positions.

TIME LINE AND ORGANIZATION

The DMS footprint can be performed in 2 days. On day 1, the DNA probe is methylated, subjected to EMSA with the protein of interest, isolated, and purified. On day 2, the methylated DNA probes are cleaved with piperidine, and the fragments are analyzed on a sequencing gel followed by autoradiography or phosphorimager analysis.

* Adapted from Ausubel et al. 1994, p. 12.3.1.

OUTLINE

Step 1: Prepare buffers (1 hour)
Step 2: Prepare methylated DNA probe (2 hours)
Step 3: Bind methylated probe to protein and isolate free and bound DNA by EMSA (5 hours)
Step 4: Cleave DNA with piperidine (3 hours)
Step 5: Analyze cleavage ladders on sequencing gels (4 hours)

PROTOCOL: METHYLATION INTERFERENCE ASSAY

Procedure

> **CAUTIONS:** *DMS, Ethanol, Formamide, β-Mercaptoethanol, MgCl₂, Piperidine, PMSF, Sodium cacodylate. See Appendix I.*

Step 1: Prepare Buffers

DMS reaction buffer:

50 mM sodium cacodylate (pH 8)
1 mM EDTA (pH 8) in dH$_2$O

Prepare 50 ml
Store at 4°C

DMS stop buffer:

1.5 M sodium acetate (pH 7)
1.0 M β-mercaptoethanol in dH$_2$O

Prepare 50 ml fresh

0.3 M Na acetate/ 0.1 mM EDTA:

Volume to use
5 ml of 3 M sodium acetate
10 μl of 0.5 M EDTA
pH solution to pH 5.2

dH$_2$O to 50 ml

Step 2: Prepare methylated DNA probe

1. Prepare a ^{32}P-end-labeled DNA probe (see Protocol 13.6 for a detailed protocol on how to end-label DNA).

2. Suspend approximately 10^6 cpm of probe in 5 to 10 μl of 10 mM Tris, pH 7.6, 1 mM EDTA.

3. Add 200 μl of DMS reaction buffer.

4. Add 1 μl of neat DMS in a fume hood.

5. Mix the tubes well by vortexing and incubate for 5 minutes at room temperature.

6. Add 40 μl of DMS stop buffer to the reaction.

7. Add 1 μl of 10 mg/ml tRNA and 600 μl of 95% ethanol. Mix and incubate for 10 minutes in a dry ice/ethanol bath. Centrifuge for 10 minutes at 14,000g in a microfuge at 4°C. Carefully remove the supernatant with a drawn-out pasteur pipet and dispose in liquid DMS waste.

8. Resuspend the pellet in 250 μl of 0.3 M sodium acetate/1 mM EDTA. Keep the tube on ice and add 750 μl of 95% ethanol. Mix and re-precipitate.

9. Repeat ethanol precipitation again exactly as in the previous step. Resuspend the pellet in 250 μl of 0.3 M sodium acetate/1 mM EDTA. Keep the tube on ice and add 750 μl of 95% ethanol, mix, and ethanol-precipitate.

10. Carefully remove the supernatant with a drawn-out pasteur pipet, wash the pellet with 80% ethanol, and microcentrifuge it for 10 minutes at 14,000 rpm. Again, carefully remove the supernatant with a drawn-out pasteur pipet and let the tube air-dry completely.

11. Measure the pellet for Cerenkov counts in a scintillation counter to determine cpm.

12. Resuspend the pellet in TE buffer at 20,000 Cerenkov cpm/ μl.

    Note: The DNA may be difficult to resuspend. Heat, vortex, and pipet the DNA up and down to solubilize.

Step 3: Bind methylated probe to protein and isolate probes

1. Prior to initiating this analysis, a study must be performed to optimize the amount of protein necessary to saturate an unmodified DNA fragment in an EMSA reaction. The DMS interference experiments will then utilize amounts of protein that are just barely saturating. Set up a series of three or four DNA-binding reactions in 0.5-ml siliconized eppendorf tubes (this procedure is the same one used for EMSA and DNase I footprinting). The reaction is scaled up slightly to increase the number of cpm, and multiple reactions are performed side by side to ensure uniformity in the result and also to again increase the number of cpm.

   For each reaction, mix:

DNA-binding protein	1.00 μl
^{32}P-end-labeled DNA probe (50 fmole/μl)	0.50 μl
poly dI:dC (1 μg/ μl)	0.20 μl
BSA (50 μg/ μl)	0.05 μl
β-mercaptoethanol (14 M)	0.05 μl
MgCl$_2$ (0.1 M)	0.75 μl
PMSF (0.1 M)	0.05 μl
dH$_2$O	3.80 μl
Buffer D$^+$ (0.1 M KCl)	6.60 μl
Total volume	13.0 μl

2. Incubate at 30°C for 1 hour.

3. Load binding reactions on a native (nondenaturing) 4.5% polyacrylamide gel as in the EMSA Protocol 13.5 (p. 493).

4. Autoradiograph the gel and cut out the bands corresponding to the protein–DNA complex (bound) and free probe. See Protocol 4. 2, Step 3, on how to isolate a radioactive fragment from a polyacrylamide gel.

5. Purify the DNA from each of the gel slices by electroelution (Protocol 13.6, Step 3).

Step 4: Cleave modified DNA with piperidine

1. Resuspend the pellets from bound and unbound samples in 100 μl of 1 M piperidine in a chemical hood.

2. Incubate the reaction mix at 90°C for 30 minutes.

   Note: Lock the tubes with lid locks, so that they do not pop open!

3. After incubation, place tubes on dry ice.

4. Make holes in the tops of the tubes with a large-gauge needle and dry the samples in a vacuum evaporator for 1 hour or until dry. Add 100 μl of distilled H_2O. Freeze and dry again. Repeat freezing and drying. Measure sample for Cerenkov counts to determine cpm.

Step 5: Analyze fragments on DNA sequencing gels

1. Based on the Cerenkov counts, add sufficient formamide sequencing gel-loading buffer to the pellet so that 1–2 μl will contain 3000 cpm of the sample to be loaded. Heat-denature the samples for 5 minutes at 90°C. Quickly chill on ice.

2. For an overnight exposure with intensifying screens, 3000 cpm/per lane is generally sufficient. It is critical to equalize the number of counts applied for the bound and free probe to allow accurate comparison between samples.

3. Load the samples from the free probe and the bound complex onto a 6–12% (depending on size of fragment) polyacrylamide/urea sequencing gel (see Sambrook et al. 1989, pp. 13.45–13.57). Electrophorese the samples as for a sequencing gel and expose gel to autoradiographic film or a phosphorimager screen overnight. A Maxam-Gilbert or dideoxy sequencing ladder should be electrophoresed alongside to distinguish the Gs from the As and to assign a precise position to each modification.

ADDITIONAL CONSIDERATIONS

1. Remember that in interference techniques you do not need to obtain 100% binding to obtain useful information. The bound lane, although enriched in methylated adenines and guanines, which do not interfere with binding, is depleted in methylated guanines, which interfere. Simply comparing the cleaved DNAs from the bound lane with a standard methylation ladder prepared without protein (and not separated by EMSA) and determining the positions of missing bands (in the bound lane) can yield important information.

2. Heavy overmethylation of DNA prior to the EMSA can lead to methylation at multiple positions in a single DNA molecule. This can confuse the analysis because it will not be clear which one of the methylated bases interfered with protein binding. Usually overmethylation is apparent based on band intensity; the uncleaved DNA on the sequencing gel is absent or present in low quantities relative to the cleavage products.

TROUBLESHOOTING

No difference between bound and unbound DNA

Possible cause: The methylation works well based on the cleavage patterns but the DNA from the bound and free samples generates the same cleavage pattern. It is plausible that there are no Gs in the region bound by the protein. In some instances, adding too much protein can overcome the negative influence of the methylated G, and most of the DNA will appear in the bound lane.

Solution: In this case, simply titrate down the protein concentration until the amount of unbound DNA begins to increase.

No cleavage of Gs

Possible cause: This can be because the DNA was undermethylated in the first place.
Solution: In such a case, a titration of the amount of DMS or time of treatment must be performed to optimize modification.

Possible cause: The piperidine may have gone bad.
Solution: Try again with fresh new reagents.

Pre-existing cleavage ladder

Possible cause: The plasmid DNA may have already been depurinated extensively during preparation.

Solution: Prepare fresh plasmid DNA. Take care to pH DNA prior to $CsCl_2$ gradient.

PROTOCOL 13.5

Electrophoretic Mobility Shift Assays

SUMMARY

In an electrophoretic mobility shift assay (EMSA) or simply "gel shift," a ^{32}P-labeled DNA fragment containing a specific DNA site is incubated with a cognate DNA-binding protein. The protein–DNA complexes are separated from free (unbound) DNA by electrophoresis through a nondenaturing polyacrylamide gel. The protein retards the mobility of the DNA fragments to which it binds. Thus, the free DNA will migrate faster than the DNA–protein complex. An image of the gel is used to reveal the positions of the free and bound ^{32}P-labeled DNA (Kerr 1995).

EMSA is one of the most sensitive methods for studying the binding properties of a protein for its site (see Chapter 8). It can be used to deduce the binding parameters and relative affinities of a protein for one or more sites or for comparing the affinities of different proteins for the same sites (Fried 1989). It can also be employed to study higher-order complexes containing multiple proteins. For example, a single protein binding to a single site would generate one predominant shifted complex. If another protein bound on top of that, it would generate an additional shift or "supershift." Finally, EMSA can be used to study protein- or sequence-dependent DNA bending (Crothers et al. 1991). The many uses of EMSA are described throughout the text. We have designed the protocol for a pure recombinant protein but describe modifications for cruder preparations in additional considerations.

A summary of the theory and other practical aspects of EMSA methodology are found in Fried and Bromberg (1997). Ideally, the DNA concentration in the reaction should be below the K_d to obtain accurate and physiological binding measurements. Because most eukaryotic transcription factors bind with K_d values of 10^{-9} to 10^{-10} M, the DNA should be below that. For a 10-μl reaction, 1 fmole of fragment is ~10^{-10} M DNA. For a 50-kD protein, 0.5 ng in a 10-μl reaction is equal to 10^{-9} M. Note that much DNA-binding activity can be lost during protein purification due to inactivating modifications (e.g., oxidation of key cysteines). Thus, 10^{-9} M protein added does not ensure 10^{-9} M active protein. For pure proteins, BSA or low concentrations of a nonionic detergent must be added to prevent binding of the protein to the plastic tube. Carrier DNA must be added because all proteins bind both specifically and nonspecifically to DNA and, in the absence of carrier DNA, the nonspecific binding occurs on the fragment and can obscure specific binding. DTT is added to prevent protein oxidation during incubation, particularly of cysteines, and KCl and $MgCl_2$ are added to minimize charge repulsion of the phosphates in the backbone and to help stabilize protein and DNA structure.

TIME LINE AND ORGANIZATION

The first step is to prepare a ^{32}P- end-labeled DNA fragment or double-stranded oligonucleotide containing the site of interest. Next a 4–4.5% native polyacrylamide gel is poured and pre-electrophoresed to remove ammonium persulfate (APS) from the gel.

Occasionally various buffer components are included in the gel running buffer to mimic the reaction buffer, although this is generally not necessary (see Chapter 8). Standard binding reactions are then performed and loaded gently into the wells. After a 4- to 6-hour run, depending on the DNA fragment size, the gel can be fixed in a methanol/acetic acid mixture, attached to Whatman 3MM paper, and dried under a vacuum. Ideally, 1000 cpm of labeled fragment is employed in the reaction. Because the radioactivity partitions between the bound and unbound DNA (two or so bands), this amount of radioactivity allows the resulting autoradiograph to expose and bands to be detected in a few hours to overnight.

OUTLINE

Step 1: Prepare buffers
Step 2: Electrophoretic mobility shift assay (EMSA)

PROTOCOL: ELECTROPHORETIC MOBILITY SHIFT

Procedure

CAUTIONS: *Acetic acid (concentrated), Acrylamide, Bromophenol blue, DTT, Glycerol, KCl, KOH, β-Mercaptoethanol, Methanol, MgCl$_2$, PMSF, Radioactive substances, TEMED. See Appendix I.*

Step 1: Prepare buffers

Gel fixing solution:

> 1 liter of fixative:
> 700 ml of dH$_2$O
> 200 ml of methanol
> 100 ml of acetic acid

Buffer D$^+$(0.1 M KCl):

> *Final concentrations*
> 20 mM HEPES-KOH (pH 7.9)
> 20% v/v glycerol
> 0.2 mM EDTA
> 0.1 M KCl
> 0.5 mM PMSF
> 1 mM DTT (add PMSF and DTT just before use)

Electrophoresis buffer:

> 0.5x TBE
> 1% glycerol

Step 2: Electrophoretic mobility shift assay (EMSA)

1. Pour a 40-ml 4.5% native acrylamide gel (1- to 1.5-mm spacers).

   final concentrations

   6.0 ml of acrylamide mix
   (30%, 29:1 acrylamide:bisacrylamide)   4.5% acrylamide mix
   4 ml of 5x TBE   0.5x TBE
   2 ml of glycerol (20% v/v)   1% glycerol
   28 ml of H$_2$O
   300 µl of APS (10%)   0.075% APS
   Add 30 µl of TEMED just before pouring.

   The gel must be pre-run at 10 mA for 2 hours.

2. Set up binding reactions in 0.5-ml siliconized eppendorf tubes.

Recombinant protein[a] (0.5–100 ng)	1.00 µl
^{32}P-DNA template[a] (ideally 1 fmole)	1.00 µl
poly (dI:dC)[a] (1µg/ µl)	0.20 µl
BSA (50 µg/µl)	0.25 µl
DTT (0.1 M)	0.10 µl
MgCl$_2$[b] (0.1 M)	0.75 µl
Buffer D$^+$(0.1 M KCl)	6.70 µl
Total volume	10.0 µl

   [a] These amounts must be titrated.

   [b] Usually omitted when using crude extracts.

3. Incubate at 30°C (room temperature or ice for crude extracts) for 1 hour.

4. Load the samples directly (no dye added) onto the pre-run 4.5% native polyacrylamide gel. Very carefully layer the mix onto the bottom of the well and watch the sclieren from the glycerol–buffer interface.

   Note: Add 5 µl of 10x DNA loading buffer with dyes to one lane that does not contain a reaction for use as a marker to determine how far the products have traveled in the gel.

5. Electrophorese at 10 mA for desired time (for a 30-bp fragment so that the bromophenol blue dye migrates about two-thirds of the way down the gel). When the gel run is complete, carefully pour out the buffer into the sink and remove the gel from the apparatus. Remove comb and split plates. Leave the gel attached to one plate.

6. *Optional:* Fix the gel in methanol/acetic acid for 15 minutes and then place on top of two sheets of Whatman 3MM paper (cat. #3030-917). Cover the top of the gel with Saran Wrap and dry on a gel dryer for 1 hour at 70°C.

7. Expose the dried gel to autoradiography film or phosphorimager overnight.

ADDITIONAL CONSIDERATIONS

1. For binding measurements on short DNA sites, perform these reactions with double-stranded oligonucleotides. When using restriction fragments, the effective limit is generally around 250 bp (see Chapter 8).

2. Large protein complexes greater than 1 Megadalton in mass are difficult to resolve and migrate quite slowly in the native polyacrylamide gels. For analyzing larger complexes, use either agarose or acrylamide-agarose composite gels.

3. Cruder preparations of protein, including nuclear extract, must be widely titrated to observe specific binding; a titration of additional carrier DNA is usually essential in such cases. When crude extracts are being used, often microgram amounts of extract or less are used. Many proteins in crude preparations bind nonspecifically to DNA and generate shifted complexes. To ensure specific binding, fragments bearing mutations in the site or competition with specific and nonspecific oligonucleotide competitors must be employed. See Chapter 8 for a discussion of the relevant parameters. Antibodies against the protein of interest can also be employed in supershift experiments to determine specificity. In such cases, however, a reaction containing the antibody itself should be performed and compared with pre-immune serum or an antibody prepared similarly.

4. Optimization of binding generally involves titrating the buffer pH, salt, and magnesium concentrations as well as the balance of carrier DNA and specific labeled fragment (see Fried 1989).

5. Some proteins simply do not bind well in gel-shift experiments and are more compatible with DNase I footprinting or another methodology (see Chapter 8). This can be due to high K_d values (weak binding) causing extensive dissociation during gel electrophoresis. Also recall that once the protein is in the gel, the buffer is TBE, which may not be compatible with binding. Simply testing binding in TBE can help determine whether this is the case. Other low-salt running buffers can be employed in place of TBE, including HEPES and TAE. Sometimes divalent ions can be added to the gel and buffers. In such cases, however, the conductivity of the gel must be kept to a minimum to prevent the gel from overheating, and sometimes if the buffering capacity of an electrophoresis buffer is low, the buffer should be recirculated among the top and bottom wells to maintain pH. Fried and Bromberg (1997) have discussed some relevant parameters that influence the stability of protein–DNA complexes during gel electrophoresis.

TROUBLESHOOTING

No binding

Possible cause: No binding generally indicates the protein is not concentrated, does not bind the site, or the buffer conditions are not compatible with protein binding during electrophoresis.
Solution: Try greater concentrations of protein, alter buffer pH and salt concentrations, try alternative electrophoresis running buffers, run gel at 4°C in cold room.

Smeary gels

Possible cause: Smeary gels may indicate protein dissociation during the electrophoresis, too much protein in the reaction, or electrophoresing the gel at too high a temperature.
Solution: Electrophorese the gel more slowly, reduce protein concentration, try running the gel in the cold room, alter the buffer conditions, or add more competitor DNA.

Too many bands

Possible cause: Too many bands are generally caused by nonspecific binding.
Solution: Try adding less protein and more carrier DNA.

PROTOCOL 13.6

Preparation of ^{32}P-end-labeled DNA Fragments

The generation of a uniquely ^{32}P-end-labeled DNA fragment is essential for performing DNA-binding experiments such as DNase I footprinting and ethylation interference. We briefly describe a protocol for end-labeling a restriction fragment (see also Chapter 4).

TIME LINE AND ORGANIZATION

For a plasmid DNA bearing a region containing the binding site of interest, the protocol involves cleaving with a single restriction endonuclease, which generates a 5′ overhang containing a phosphate. This is generally necessary for both common forms of fragment end-labeling using either phosphorylation with polynucleotide kinase or "filling in the end" with DNA polymerases such as Klenow fragment. We focus on the phosphorylation reaction. The phosphate is removed with calf intestinal phosphatase, bacterial alkaline phosphatase, and later the free 5′OH generated by this manipulation is phosphorylated with polynucleotide kinase and [γ-^{32}P]ATP. This in turn generates a plasmid now labeled at each end with γ-^{32}P. To generate a uniquely end-labeled DNA fragment, the labeled plasmid is heat-treated to inactivate any remaining kinase and then recleaved with the second endonuclease, releasing a short DNA fragment and a longer vector fragment. The DNA fragment is purified from the labeled vector on a 5–8% native polyacrylamide gel. The molar amount of plasmid DNA must be below the amount of ATP added to the reaction and the ATP must be of high specific activity to generate a fragment labeled to the extent necessary for many DNA-binding experiments.

In contrast, an oligonucleotide primer is generally unphosphorylated at the 5′ end and can be directly phosphorylated with kinase, bypassing the phosphatase step.

Preparation and labeling of DNA restriction fragments typically is performed in 1–2 days.

OUTLINE

^{32}P-end-labeling of a DNA fragment (2 days)

Step 1: Cleavage and labeling of the plasmid DNA (4 hours)
Step 2: Generating a uniquely labeled DNA fragment (5 hours)
Step 3: Electroelute the labeled DNA fragment (3 hours)

PROTOCOL: ^{32}P-END-LABELING OF A DNA FRAGMENT

PROCEDURE

CAUTIONS: *Acrylamide, Ammonium persulfate, Bromophenol blue, Chloroform, Ethanol, Phenol, Radioactive substances, TEMED, Xylene. See Appendix I.*

Step 1: Cleavage and labeling of the plasmid DNA

1. Digest 10 µg of plasmid DNA in a 1.5-ml microfuge tube with an enzyme generating a 5′-phosphate overhang. Typically, 1 µg of a 4-kb plasmid is equal to 380 fmoles of DNA. When cleaved once, it generates 760 fmoles of ends. Thus, 10 µg is equal to 7.6 pmoles of DNA ends. We label the DNA with100 µCi or 20 pmoles of high-specific-activity ATP (5000–7000 Ci/mmole).

DNA (10 µg)	20.0 µl
10x restriction enzyme buffer 1	5.0 µl
restriction enzyme 1	5.0 µl
calf intestinal phosphatase (10 units)	0.5 µl
dH$_2$O	19.5 µl
Total volume:	50.0 µl

2. Incubate at 37°C for 2 hours to cleave the plasmid and to phosphatase the DNA ends.

3. Add 1 µl of Proteinase K (stock at 10 mg/ml) and SDS to 0.1% and incubate at 55°C for 15 minutes to digest residual phosphatase and restriction enzyme.

4. Add 50 µl of TE (10 mM Tris pH, 7.9, 1 mM EDTA) and 100 µl of phenol/chloroform (1:1) and vortex gently. Centrifuge for 1 minute in a microfuge at 14,000g, then carefully remove 90 µl of the aqueous (top) phase and transfer to a new 1.5-ml eppendorf tube. Do not transfer any of the interphase; instead remove less of the aqueous phase. This manipulation effectively removes all of the phosphatase.

5. Add 10 µl of 3 M sodium acetate and 250 µl of 95% ethanol to the aqueous phase. Mix well and place on dry ice for 15 minutes. Collect the DNA precipitate as a pellet by centrifugation in a microfuge at 14,000g for 15 minutes. Carefully remove the ethanol using a drawn-out glass pasteur pipet.

6. Wash the pellet by adding 100 µl of 80% ethanol, centrifuge in microfuge for 1 minute, remove the ethanol with drawn-out pipet and air-dry completely.

7. Resuspend the dried DNA pellet in 16 µl of dH$_2$O by gently pipetting up and down several times. ^{32}P-end-label with polynucleotide kinase and [γ-^{32}P]ATP.

   Mix:
DNA	16.0 µl
10x One-Phor-All buffer (Amersham Pharmacia Biotech Inc. cat. # 27-0901-02)	3.0 µl
[γ-^{32}P]ATP 5-7000 Ci/mmole (10 mCi/ml aqueous)	10.0 µl
polynucleotide kinase (10 U/ µl)	1.0 µl
Total volume	30.0 µl

8. Mix very gently and then incubate at 37°C for 15 minutes. Exercise maximum precautions and shielding to prevent exposure to or spilling of ^{32}P. This reaction end-labels the free 5′OH ends generated by restriction endonuclease cleavage and phosphatase treatment (Steps 1–5).

9. After the end-labeling reaction, heat-inactivate the kinase by incubating at 65°C for 15 minutes.

10. Phenol/chloroform-extract, ethanol-precipitate, and wash the DNA pellet as in Steps 4–6. Keep a 50-ml conical tube nearby to store the radioactive drawn-out pipet.

Step 2: Create a uniquely labeled DNA

1. Resuspend the pellet in 10 μl of ddH$_2$O and digest with the second enzyme.

Mix:

DNA	10.0 μl
10x restriction enzyme 2 buffer	5.0 μl
restriction enzyme 2 (10 units)	5.0 μl
dH$_2$O	30.0 μl
Total volume	50.0 μl

2. Digest DNA at 37°C for 3 hours.

3. Add 5 μl of 10x DNA loading buffer and use sequencing pipet tips to load the mixture onto a native 8% polyacrylamide gel (for fragments in the 100-bp range) and electrophorese in 1x TBE buffer. On an 8% gel, the bromophenol blue dye migrates with the 45-bp marker, and the xylene cyanol dye runs with the 160-bp marker.

4. Prepare an 8% native acrylamide gel. Mix the following ingredients in a 50-ml conical disposable tube.

Mix:

40% acrylamide (29:1)	6 ml
5x TBE buffer	6 ml
dH$_2$O	18 ml
Total volume	30 ml

Add 300 μl of 10% APS

Add 30 μl of TEMED

Mix well and rapidly pour into gel plates. Insert comb and clamp the top of gel to hold the comb tightly. Allow the gel to polymerize for 1 hour, carefully remove the comb, and prerun the gel for 1 hour at 200 V to remove APS.

5. Load the sample slowly into gel wells by layering, and electrophorese until the bromophenol blue has migrated approximately halfway down the gel.

Step 3: Electroelute the labeled DNA fragment

1. Very carefully remove the gel from the vertical rig and pour radioactive buffer from lower chamber into radioactive waste container. Rinse the remaining gel apparatus with water in a sink. Rinse the plates with water and then carefully separate them using a spatula. Do this on bench-top paper behind a shield. Leave the gel on one plate and carefully cover the plate with plastic Saran Wrap. Bring the gel to the dark room and

expose it to XAR-5 photographic film (Kodak) for 3 minutes to locate position of radiolabeled fragment. Use fluorescent dye to mark plastic wrap around gel (dot in three places) or flash the gel three times with a camera flash apparatus. The latter generates an outline of the gel and plate on the film, and the former generates a series of black dots. Either method will be necessary to align the autoradiograph later against the gel to excise the band. Develop the film.

2. Align the film and gel using fluorescent markers or the flash outline. Excise the gel slice that corresponds to the probe using a razor blade and tweezers. Cut through the Saran Wrap.

3. Place the gel slice in dialysis tubing (10-mm flat diameter, 12,000–14,000 Dalton MWCO approx. 2 inches long) with 500 μl of 1x TBE and electroelute the DNA from the gel for 1 hour at 100 V. Place the dialysis tubing containing the gel slice into a horizontal agarose gel chamber containing 1x TBE. After electroeluting for 1 hour at 100 V, switch the direction of the electrodes and electroelute for 2 minutes. (This will elute any DNA off the tubing and back into the buffer.)

4. Use a 1-ml pipetman to remove buffer containing the eluted DNA fragment from the dialysis tubing and place in a 1.5-ml microfuge tube.

5. Extract the solution with an equal volume of phenol/chloroform.

6. Add 1/10th volume of 3 M sodium acetate and 2 volumes of 95% ethanol and precipitate DNA as above.

7. Resuspend the final pellet in TE at 10 K cpm/μl (as judged by a Geiger counter).

8. Use a scintillation counter to measure Cerenkov radioactivity (^{3}H channel) of 1 μl of probe dotted onto a 1-cm^2 piece of filter paper. Compare to radioactivity of 0.01 μCi of the starting ATP to determine incorporation and specific activity. Typically, 10,000 cpm is employed for a DNase I footprint, but much less is required for EMSA.

ADDITIONAL CONSIDERATIONS

1. Sometimes it is necessary to cleave restriction fragments with enzymes that do not generate unique sites. This is feasible as long as the ATP is in twofold or greater excess over the actual number of DNA ends. The ideal fragment for EMSA and footprinting reactions is 50–200 nucleotides in length.

2. An alternative method used by many researchers involves using PCR to generate the end-labeled fragment. In such a case, approximately 10 ng of plasmid DNA is incubated with 10 pmole each of two primers flanking the region of interest, one of which is end-labeled. After 20–30 cycles of PCR, an end-labeled fragment of high specific activity can be generated and recovered by purification on a native acrylamide gel as described in Steps 2 and 3 (above).

TROUBLESHOOTING

No or weak labeling of DNA fragments

Possible cause: DNA may not have been added.
Solution: Check concentration of DNA stocks by A$_{260}$ or by simply running a diagnostic agarose gel for plasmids to ensure that DNA was indeed added. If DNA was added, check kinase and buffers.

Possible cause: Old DTT can inactivate the kinase.

Solution: Fresh DTT is needed to maintain polynucleotide kinase activity.

Possible cause: Background labeling of contaminating nucleic acids. In plasmid preps contaminated with RNA the kinase will also phosphorylate RNA ends.

Solution: Removing RNA by CsCl EtBr gradients or simply rerunning a QIAGEN column can help to remove excess RNA. Do not add RNase; it creates a greater problem by increasing the number of ends.

Possible cause: Inactive kinase.

Solution: Use freshly purchased kinase.

REFERENCES

Ausubel F.M., Brent R.E., Kingston E., Moore D.D., Seidman J.G., Smith J.A., and Struhl K. 1994. *Current protocols in molecular biology*. John Wiley and Sons, New York.

Balasubramanium B., Pogozelski W.K., and Tullius T.D. 1998. DNA strand breaking by the hydroxyl radical is governed by the accessible surface areas of the hydrogen atoms of the DNA backbone. *Proc. Natl. Acad. Sci.* **95:** 9738–9743.

Blackwell T.K. 1995. Selection of protein binding sites from random nucleic acid sequences. *Methods Enzymol.* **254:** 604–618.

Branden C. and Tooze J. 1991. *Introduction to protein structure*. Garland Publishing, New York.

Brenowitz M., Senear D.F., Shea M.A., and Ackers G.K. 1986. Quantitative DNase footprint titration: A method for studying protein-DNA interactions. *Methods Enzymol.* **130:** 132–181.

Burley S.K. and Roeder R.G. 1996. Biochemistry and structural biology of Transcription Factor IID (TFIID). *Annu. Rev. Biochem.* **65:** 769–799.

Carey M. 1998.The enhanceosome and transcriptional synergy. *Cell* **92:** 5–8.

Carey M., Kakidani H., Leatherwood J., Mostashari F., and Ptashne M. 1989. An amino-terminal fragment of GAL4 binds DNA as a dimer. *J. Mol. Biol.* **209:** 423–432.

Cho Y., Gorina S., Jeffrey P.D., and Pavletich N.P. 1994. Crystal structure of a p53 tumor suppressor-DNA complex: Understanding tumorigenic mutations. *Science* **265:** 346–355.

Crothers D.M., Gartenberg M.R., and Shrader T.E. 1991. DNA bending in protein-DNA complexes. *Methods Enzymol.* **208:** 118–146.

Crothers D.M., Drak J., Kahn J.D., and Levene S.D. 1992. DNA bending, flexibility, and helical repeat by cyclization kinetics. *Methods Enzymol.* **212:** 3–29.

Dignam J.D., Lebovitz R.M., and Roeder R.G. 1983. Accurate transcription initiation by RNA polymerase II in a soluble extract from isolated mammalian nuclei. *Nucleic Acids Res.* **11:** 1475–1489.

Dill K.A. 1997. Additivity principles in biochemistry. *J. Biol. Chem.* **272:** 701–704.

Dixon W.J., Hayes J.J., Levin J.R., Weidner M.F., Dombroski B.A., and Tullius T.D. 1991. Hydroxyl radical footprinting. *Methods Enzymol.* **208:** 380–413.

Echols H. 1986. Multiple DNA-protein interactions governing high-precision DNA transactions. *Science* **233:** 1050–1056.

Ellenberger T.E., Brandl C.J., Struhl K., and Harrison S.C. 1992. The GCN4 basic region leucine zipper binds DNA as a dimer of uninterrupted α helices: Crystal structure of the protein-DNA complex. *Cell* **71:** 1223–1237.

Ellington A.D. and Szostak J.W. 1990. In vitro selection of RNA molecules that bind specific ligands. *Nature* **346:** 818–-822.

Fairall L., Rhodes D., and Klug A. 1986. Mapping of the sites of protection on a 5 S RNA gene by the *Xenopus* transcription factor IIIA. A model for the interaction [published erratum appears in *J. Mol. Biol.* 1987 Apr 5;194(3):581]. *J. Mol. Biol.* **192:** 577–591.

Feng J.-A., Johnson R.C., and Dickerson R.E. 1994. Hin recombinase bound to DNA: The origin of specificity in major and minor groove interactions. *Science* **263:** 348–355.

Fried M.G. 1989. Measurement of protein-DNA interaction parameters by electrophoresis mobility shift assay. *Electrophoresis* **10:** 366–376.

Fried M.G. and Bromberg J.L.1997. Factors that affect the stability of protein-DNA complexes during gel electrophoresis. *Electrophoresis* **18:** 6–11.

Galas D.J. and Schmitz A. 1978. DNase footprinting: A simple method for the detection of protein-DNA binding specificity. *Nucleic Acids Res.* **5:** 3157–3170.

Giese K., Kingsley C., Kirshner J.R., and Grosschedl R. 1995. Assembly and function of a TCR α enhancer complex is dependent on LEF-1-induced DNA bending and multiple protein-protein interactions. *Genes Dev.* **9:** 995–1008.

Gralla J.D. 1985. Rapid "footprinting" on supercoiled DNA. *Proc. Natl. Acad. Sci.* **2:** 3078–3081.

Grange T., Bertrand E., Espinas M.L., Fromont-Racine M., Rigaud G., Roux J., and Pictet R. 1997. In vivo footprinting of the interaction of proteins with DNA and RNA. *Methods* **11:** 151–163.

Griffith J., Hochschild A., and Ptashne M. 1986. DNA loops induced by cooperative binding of λ repressor. *Nature* **322:** 750–752.

Grosschedl R. 1995. Higher-order nucleoprotein complexes in transcription: Analogies with site-specific recombination. *Curr. Opin. Cell Biol.* **7:** 362–370.

Grosschedl R., Giese K., and Pagel J. 1994. HMG domain proteins: Architectural elements in the assembly of nucleoprotein structures. *Trends Genet.* **10:** 94–100.

Ha J.H., Spolar R.S., and Record M.T. Jr. 1989. Role of the hydrophobic effect in stability of site-specific protein-DNA complexes. *J. Mol. Biol.* **209:** 801–816.

Hayes J.J. and Tullius T.D. 1989. The missing nucleoside experiment: A new technique to study recognition of DNA by protein. *Biochemistry* **28:** 9521–9527.

Herschman H.R. 1991. Primary response genes induced by growth factors and tumor promoters. *Annual Review of Biochemistry* **60:** 281–319.

Hochschild A. 1991. Detecting cooperative protein-DNA interactions and DNA loop formation by footprinting. *Methods Enzymol.* **208:** 343–361.

Hochschild A. and Ptashne M. 1986. Cooperative binding of λ repressors to sites separated by integral turns of the DNA helix. *Cell* **44:** 681–687.

Hoover T.R., Santero E., Porter S., and Kustu S. 1990. The integration host factor stimulates interaction of RNA polymerase with NIFA, the transcriptional activator for nitrogen fixation operons. *Cell* **63:** 11–22.

Johnson A.D., Poteete A.R., Lauer G., Sauer R.T., Ackers G.K., and Ptashne M. 1981. λ Repressor and cro–Components of an efficient molecular switch. *Nature* **294:** 217–223.

Johnson R.C. 1995. Site specific recombinases and their interactions with DNA. In DNA-*protein: structural interactions* (ed. Lilleys D.M.J.), vol. 7, p. 141. Oxford University Press, New York and IRL Press at Oxford University Press, United Kingdom.

Jordan S.R. and Pabo C.O. 1988. Structure of the lambda complex at 2.5 A resolution: Details of the repressor-operator interactions. *Science* **242:** 893–899.

Kalionis B. and O'Farrell P.H. 1993. A universal target sequence is bound in vitro by diverse homeodomains. *Mech. Dev.* **43:** 57–70.

Kerr L.D., 1995. Electrophoretic mobility shift assay. *Methods Enzymol.* **254:** 619–632.

Kielkopf C.L., White S., Szewczyk J.W., Turner J.M., Baird E.E., Dervan P.B., and Rees D.C. 1998. A structural basis for recognition of A-T and T-A base pairs in the minor groove of B-DNA. *Science* **282:** 111–115.

Kim S. and Landy A. 1992. Lambda Int protein bridges between higher order complexes at two distant chromosomal loci, *attL* and *attR*. *Science* **256:** 198–203.

Kim S., Moitoso de Vargas L., Nunes-Duby S., and Landy A. 1990. Mapping of a higher order pretein-DNA complex: Two kinds of long-range interactions n Lambda, *attL*. *Cell* **63:** 773–781.

Kim T.K. and Maniatis T. 1997. The mechanism of transcriptional synergy of an in vitro assembled interferon-β enhancesome. *Mol. Cell* **1:** 119–130.

Kim J.L., Nikolov D.B., and Burley S.K. 1993a. Co-crystal structure of TBP recognizing the minor groove of a TATA element. *Nature* **365:** 520–527.

Kim Y., Geiger J.H., Hahn S., and Sigler P.B. 1993b. Crystal structure of a yeast TBP/TATA-box complex. *Nature* **365:** 512–520.

Kissinger C.R., Liu B., Martin-Bianco E., Kornberg T.B., and Pabo C.O. 1990. Crystal structure of an engrailed homeodomain-DNA complex at 2.8Å resolution: A framework for understanding homeodomain-DNA interactions. *Cell* **63:** 579–590.

Klemm J.D., Rould M.A., Aurora R., Herr W., and Pabo C.O. 1994. Crystal structure of the Oct-1 POU domain bound to an octamer site: DNA recognition with tethered DNA-binding modules. *Cell* **77:** 21–32.

Koblan K.S., Bain D.L., Beckett D., Shea M.A., and Ackers G.K. 1992. Analysis of site-specific interaction parameters in protein-DNA complexes. *Methods Enzymol.* **210:** 405–425.

Lee D.K., Horikoshi M., and Roeder R.G. 1991. Interaction of TFIID in the minor groove of the TATA element. *Cell* **67:** 1241–1250.

Lesser D.R., Kurpiewski M.R., and Jen-Jacobson L. 1990. The energetic basis of specificity in the Eco RI endonuclease–DNA interaction. *Science* **250:** 776–786.

Liang S.D., Marmorstein R., Harrison S.C., and Ptashine M. 1996. DNA sequence preferences of GAL4 and PPR1: How a subset of Zn2 Cys6 binuclear cluster proteins recognizes DNA. *Mol. Cell. Biol.* **16:** 3773–3780.

Love J.J., Li X., Case D.A., Giese K., Grosschedl R., and Wright P.E. 1995. Structural basis for DNA bending by the architectural transcription factor LEF-1. *Nature* **376:** 791–795.

Luisi B. 1995. DNA-protein interaction at high resolution. In *DNA-Protein: Structural interactions* (ed. Lilley D.M.J.). Oxford University Press, New York and IRL Press at Oxford University Press, United Kingdom.

Manfield I. and Stockley P.G. 1994. Ethylation interference. *Methods Mol. Biol.* **30:** 125–139.

Mangelsdorf D.J. and Evans R.M. 1995. The RXR heterodimers and orphan receptors. *Cell* **83:** 841–850.

Marmorstein R., Carey M., Ptashne M., and Harrison S.C. 1992. DNA recognition by GAL4: Structure of a protein-DNA complex. *Nature* **356:** 408–414.

Maxam A.M. and Gilbert W. 1977. A new method for sequencing DNA. *Proc. Natl. Acad. Sci.* **74:** 560–564.

Narayana N., Ginell S.L., Russu I.M., and Berman H.M. 1991. Crystal and molecular structure of a DNA fragment: d(CGTGAATTCACG). *Biochemistry* **30:** 4449–4455.

Oliphant A.R., Brandl C.J., and Struhl K. 1989. Defining the sequence specificity of DNA-binding proteins by selecting binding sites from random-sequence oligonucleotides: Analysis of yeast GCN4 protein. *Mol. Cell. Biol.* **9:** 2944–2949.

Otwinowski Z., Schevitz R.W., Zhang R.G., Lawson C.L., Joachimiak A., Marmorstein R.Q., Luisi B.F., and Sigler P.B. 1988. Crystal structure of trp repressor/operator complex at atomic resolution [published erratum appears in Nature 1988 Oct 27;335(6193):837]. *Nature* **335:** 321–329.

Pabo C.O. and Sauer R.T. 1992. Transcription factors: Structural families and principles of DNA recognition. *Annu. Rev. Biochem.* **61:** 1053–1095.

Pan C.Q., Finkel S.E., Cramton S.E., Feng J.A., Sigman D.S., and Johnson R.C. 1996. Variable structures of Fis-DNA complexes determined by flanking DNA-protein contacts. *J. Mol. Biol.* **264:** 675–695.

Pavletich N.P. and Pabo C.O. 1991. Zinc finger-DNA recognition: Crystal structure of a Zif268-DNA complex at 2.1Å. *Science* **252:** 809–817.

Pomerantz J.L., Sharp P.A., and Pabo C.O. 1995. Structure-based design of transcription factors. *Science* **267:** 93–96.

Ptashne M. 1992. *A genetic switch: Phage λ and higher organisms*, 2nd edition. Cell Press and Blackwell Scientific Publications, Cambridge, Massachusetts.

Ptashne M. and Gann A. 1998. Imposing specificity by localization: Mechanism and evolvability [published erratum appears in *Curr. Biol.* 1998 Dec. 3;8(24):R897] *Curr. Biol.* **8:** R812–822.

Raumann B.E., Rould M.A., Pabo C.O., and Sauer R.T. 1994. DNA recognition by β-sheets in the Arc repressor-operator crystal structure. *Nature* **367:** 754–757.

Rebar E.J. and Pabo C.O. 1994. Zinc finger phage: Affinity selection of fingers with new DNA-binding specificities. *Science* **263:** 671–673.

Record M.T. Jr., Ha J.H., and Fisher M.A. 1991. Analysis of equilibrium and kinetic measurements to determine thermodynamic origins of stability and specificity and mechanism of formation of site-specific complexes between proteins and helical DNA. *Methods Enzymol.* **208:** 291–343.

Rippe K., von Hippel P.H., and Langowski J. 1995. Action at a distance: DNA-looping and initiation of transcription. *Trends Biochem. Sciences* **20:** 500–506.

Sambrook J., Fritsch E.F., and Maniatis T. 1989. *Molecular cloning: A laboratory manual*, 2nd edition. Cold Spring Harbor Laboratory, Cold Spring Harbor, New York.

Schwabe J.W. and Rhodes D. 1997. Linkers made to measure. *Nat. Struct. Biol.* **4:** 680–683.

Seeman N.C., Rosenberg J.M., and Rich A. 1976. Sequence-specific recognition of double helical nucleic acids by proteins. *Proc. Natl. Acad. Sci.* **73:** 804–808.

Shakked Z., Guzikevich-Guerstein G., Frolow F., Rabinovich D., Joachimiak A., and Sigler P.B. 1994. Determinants of repressor/operator recognition from the structure of the trp operator binding site. *Nature* **368:** 469–473.

Siebenlist U. and Gilbert W. 1980. Contacts between *Escherichia coli* RNA polymerase and an early promoter of phage T7. *Proc. Natl. Acad. Sci.* **77:** 122–126.

Siebenlist U., Simpson R.B., and Gilbert W. 1980. *E. coli* RNA polymerase interacts homologously with two different promoters. *Cell* **20:** 269–281.

Sigman D.S., Kuwabara M.D., Chen C.H., and Bruice T.W. 1991. Nuclease activity of 1,10-phenanthroline-copper in study of protein-DNA interactions. *Methods Enzymol.* **208:** 414–433.

Somers W.S. and Phillips S.E. 1992. Crystal structure of the met repressor-operator complex at 2.8 Å resolution reveals DNA recognition by β-strands. *Nature* **359:** 387–393.

Starr D.B. and Hawley D.K. 1991. TFIID binds in the minor groove of the TATA box. *Cell* **67:** 1231–1240.

Suck D. 1994. DNA recognition by DNase I. *J. Mol. Recognit.* **7:** 65–70.

Suzuki M. 1995. DNA recognition by a β-sheet. *Protein Eng.* **8:** 1–4.

Swaminathan K., Flynn P., Reece R.J., and Marmorstein R. 1997. Crystal structure of a PUT3-DNA complex reveals a novel mechanism for DNA recognition by a protein containing a Zn2Cys6 binuclear cluster. *Nat. Struct. Biol.* **4:** 751–759.

Thanos D. and Maniatis T. 1995. Virus induction of human IFN β gene expression requires the assembly of an enhanceosome. *Cell* **83:** 1091–1100.

Todd R.B. and Andrianopoulos A. 1997. Evolution of a fungal regulatory gene family: The Zn(II)2Cys6 binuclear cluster DNA binding motif. *Fungal Genet. Biol.* **21:** 388–405.

Tuerk C., MacDougal S., and Gold L. 1992. RNA pseudoknots that inhibit human immunodeficiency virus type 1 reverse transcriptase. *Proc. Natl. Acad. Sci.* **89:** 6988–6992.

Tullius T.D. and Dombroski B.A. 1986. Hydroxyl radical "footprinting": High-resolution information about DNA-protein contacts and application to λ repressor and Cro protein. *Proc. Natl. Acad. Sci.* **83:** 5469–5473.

Wang J.C. and Giaever G.N. 1988. Action at a distance along a DNA. *Science* **240:** 300–304.

Wang M.M. and Reed R.R. 1993. Molecular cloning of the olfactory neuronal transcription factor Olf-1 by genetic selection in yeast. *Nature* **364:** 121–126.

Watson J.D., Weiner A.M., and Hopkins N.H. 1987. *Molecular biology of the gene*, 4th edition. Benjamin/Cummings Publishing, Menlo Park, California.

Werner M.H. and Burley S.K. 1997. Architectural transcription factors: Proteins that remodel DNA. *Cell* **88:** 733–736.

Wissmann A. and Hillen W. 1991. DNA contacts probed by modification protection and interference studies. *Methods Enzymol.* **208:** 365–379.

Wolberger C., Vershon A.K., Liu B., Johnson A.D., and Pabo C.O. 1991. Crystal structure of a MATα-2 homeodomain-operator complex suggests a general model for homeodomain-DNA interactions. *Cell* **67:** 517–528.

Yang J. and Carey J. 1995. Footprint phenotypes: Structural models of DNA-binding proteins from chemical modification analysis of DNA. *Methods Enzymol.* **259:** 452–468.

Zwieb C. and Adhya S. 1994. Improved plasmid vectors for the analysis of protein-induced DNA bending. *Methods Mol. Biol.* **30:** 281–294.

CHAPTER 14

Crude and Fractionated Systems for In Vitro Transcription

Important issues

- *Can and should specific transcription of a gene be reproduced in vitro?*
- *The extract system, template, and an appropriate assay are critical choices when attempting to reproduce specific transcription patterns in vitro.*
- *Several approaches can be taken to achieve proper regulation in vitro.*
- *Pure or fractionated systems are necessary for detailed mechanistic studies.*

INTRODUCTION

In 1979 the laboratory of Robert Roeder reproduced the accurate transcription of an mRNA-encoding gene in a test tube (Weil et al. 1979a). This landmark discovery represented a significant step toward elucidating the pathways governing gene regulation by making it possible to understand the mechanism of promoter action at the molecular level. Although the initial biochemical studies employed viral promoters, recent variations of the approach more accurately recapitulate the complex expression patterns of cellular genes. Because of the expense and time commitment necessary to initiate a biochemical study, one must evaluate whether development of an in vitro transcription system is practical for the gene under study.

Clearly, the in vitro system is not an appropriate substitute for in vivo aproaches for analyzing a gene's regulation. However, if in vivo studies (see Chapters 2–9) have indicated that the gene is regulated in an interesting fashion, an investigator may wish to reproduce the transcription of the gene in vitro. The in vitro system is an essential starting point for deducing the biochemical mechanism of regulation.

If there is a compelling rationale for proceeding, there are several considerations when attempting to develop the system:

- Can the system be set up from a tissue or cell line where the factor or promoter of interest is active? The availability of cell lines that properly regulate a gene or the abundance of an appropriate tissue from an animal source are key logistical issues in the decision to set up an in vitro system.

- What type of DNA template will be employed for the analysis? For example, the template can include a natural promoter, an enhancer, or can be an artificial template bearing either portions of a functional enhancer or a multimerized activator binding site. The choice of template will ultimately depend on the needs of the investigator and the questions being asked.

- Finally, will regulation of the gene require recreation of a special nucleoprotein structure such as an enhanceosome or chromatin? Recent studies have emphasized the role of ATP-dependent chromatin remodeling machines and histone acetylases and deacetylases in activation and repression of transcription. The accurate regulation of a gene may therefore require recreating chromatin templates to observe the effects of co-activators and co-repressors (see, e.g., Sheridan et al. 1995; Kraus and Kadonaga 1998).

Many of the early studies on the biochemistry of eukaryotic gene regulation were performed on viruses, including SV40 and adenovirus. Viruses were attractive as experimental organisms because they contain strong, constitutive promoters/enhancers. Additionally, a rudimentary knowledge of the genetics existed that could be coupled with biochemistry and molecular biology to produce insights into gene control. In contrast, most cellular genes are expressed at low-to-moderate levels and employ special, tissue-specific activators and coactivators that may not be present at high concentrations in extracts. Hence, although the viral promoters can serve as important models, a key challenge is recreating the transcription of important regulated cellular genes in a tissue-specific fashion. In this chapter we discuss strategies for pursuing this goal and alert the reader to cumbersome technical problems and experimental artifacts encountered in such an analysis. We refer to primary references for details of the technology but attempt to summarize the major decision-making processes here.

As indicated above, an in vitro transcription assay may be extremely valuable if the investigator plans to proceed to more advanced studies of the factors responsible for transcriptional regulation. To develop an in vitro transcription assay, the requirements are as follows:

- A source of cells that can be obtained in large enough quantities to prepare a concentrated protein extract
- Knowledge of the transcription start site for the gene of interest
- An assay for measuring specific promoter activity
- Ideally, when analyzing a complex promoter, the availability of one or more regulatory proteins in recombinant form will simplify the analysis because the low concentrations of such factors in crude extracts present a problem for advanced studies.

CONCEPTS AND STRATEGIES

Preparation of Extracts

Cell Choice

In vitro transcription assays can be carried out with extracts from either tissues or cell lines. Because of the large number of cells typically used for extract preparation, some tissues, such as rat liver, have proven to be an inexpensive and plentiful source of cells. Extracts from solid tissues, however, are more problematic due to heterogeneity in cell type, proteolysis problems, difficulties in cell lysis, and so on. Given the large number of specific and general transcription factors needed for transcription, these problems can hinder the development of an efficient in vitro assay. For this reason, transformed cell lines are more commonly used during extract preparation for in vitro transcription assays. We therefore focus largely on cultured cells, although we present a protocol for preparation of whole tissue extracts.

It is difficult to recommend a precise number of cells for extract preparation as the needs vary depending on the problem being addressed. In general, however, extracts obtained from fewer than 10^9 cells can be difficult to prepare. The difficulty lies in the fact that the protein extract must remain concentrated during preparation. The use of a small number of cells results in less concentrated extracts because the lower volumes of buffers and the small size of the cell pellet can lead to greater losses than with larger volumes. Additionally, certain manipulations, including dounce homogenization, tube transfers, and dialysis, are also more difficult to perform with smaller volumes. There are, however, some methods that employ alternative lysis techniques and minimize the number of manipulations. These techniques have been successful for production of mini extracts (see Protocols 14.1, 14.2, and 14.3), although most laboratories that have achieved success with in vitro assays employ extracts prepared from 10^9–10^{10} or more cells.

To prepare extracts from tissue-culture cells, it is important for cells to be healthy and growing exponentially. HeLa cells, for example, grow at 5×10^8 cells per liter in spinner culture. HeLa cells, unlike many other lines, can be grown outside of a CO_2 incubator, and the growth medium can be autoclaved and supplemented with inexpensive serums from bovine or horse. Typically, 2–3 ml of extract can be prepared from 2 liters of HeLa cells grown in 4-liter flat-bottomed Florence flasks at 37°C in a temperature-controlled cabinet or a standard warm room. For labs making a long-term commitment, it is recommended that the facilities necessary to grow 32–50 liters of cells a week be set up. Smaller volumes of cells can be grown to higher densities if they are placed in special growth chambers, or cell culture fermentors, which maintain a constantly buffered and nutritionally supplemented environment. However, such equipment is expensive and requires constant maintenance.

Many adherent tissue-culture cells (e.g., fibroblasts), unlike HeLa cells and some lymphocytes, are not easily adapted to spinner culture and generate typically 10^6 to 10^7 cells in a 10-cm dish. At the upper end of this scale, 100 dishes are necessary to equal the output of 1–2 liters of HeLa cells. Alternatively, roller bottles can be used to increase the scale, particularly with lymphoctes that do not grow well in spinner cultures. Nevertheless, large-scale growth is still more challenging than using cells that can be adapted to spinner culture. It is therefore considerably more time-consuming and difficult to prepare an extract from cells that do not grow in spinner culture. For this reason, the investigator should assess whether outcome from the in vitro studies will merit the input, time, and cost. To gauge the logistical problems, imagine that an investigator wishes to prepare an extract from 100 plates of cells of a standard adherent cell line that grows to 5×10^6 cells per plate. The investigator begins with one plate of cells and, on the basis of the 24- to 48-hour doubling times characteristic of cultured cells, the investigator can expect that it will take 2 weeks or more to scale up to 100 plates. The harvested cell pellet prior to extract preparation will be approximately 1 ml in volume, which will generate typically 2 ml of extract using a standard nuclear extract procedure. A typical in vitro transcription experiment uses approximately 20 μl (~ 100–200 μg) of extract. Therefore, on average, 2 ml of extract generates enough material for 100 reactions. Because a typical experiment involves 20 reactions, 2 weeks of work will yield five in vitro transcription experiments. Furthermore, for extracts prepared at the scale discussed, the failure rate can be high for novice experimenters.

Thus, recreating the tissue-specific expression of a gene in vitro is one of the most difficult approaches in the transcription field and entails the commitment of significant time and resources. Due to the rich variety of cell-specific activators and coactivators, proper regulation in vitro may only be achieved in extracts generated from tissues or cell lines where the gene is normally active. One major advantage is gained when all or many of the transcriptional activators are available in recombinant form and can be added exogenously. In such a case, heterologous tissues such as HeLa cells can be used as an easy abundant source of general transcription factors, which are then supplemented with exogenous activators.

Extract Preparation Method

The Dignam and Roeder procedure (Dignam et al. 1983a) for preparation of nuclear extracts from tissue-culture cells has proven successful in numerous cell lines ranging from HeLa cells to B cells (BJAB/Namalwa) to prostate cancer cell lines (LNCaP) (see Protocol 14.1). The approach involves collecting a cell pellet and resuspending it in an excess of hypotonic buffer. This treatment causes the cells to swell, and the distended cells can be disrupted by standard hand dounce homogenization. Careful douncing fractures the cells but leaves the smaller nuclei intact such that these can be separated from the remaining cellular and cytoplasmic debris by low-speed centrifugation. The nuclei are then resuspended in a moderate salt buffer (0.38–0.42 M NaCl), which extracts the transcriptional components out of the nucleus. The resulting extracts have been shown to be rich in histones, although efforts to recreate chromatin in the Dignam extracts have not been successful.

It is worth noting that the concentration of KCl (0.42 M) used in the Dignam/Roeder protocol was determined empirically as the optimal concentration for transcription of adenovirus promoters (Dignam 1983a). In other words, that concentration of KCl resulted in the extraction of the appropriate balance of positively acting transcription factors needed for transcription and negatively acting proteins that might interfere with transcription. Presumably, lower salt concentrations would have been less efficient in extracting the positive transcription factors; higher concentrations presumably resulted in the extraction of additional inhibitory proteins (i.e., such as histone H1, which typically is released at 0.6 M KCl) that reduce the promoter activity. With this in mind, it is advisable to optimize the extracts for a new promoter by preparing them with variable concentrations of KCl.

After extraction of proteins from the nuclei, the extracts are either directly dialyzed or precipitated with ammonium sulfate. For many cell lines and genes, in vitro transcription activity can be enhanced by ammonium sulfate precipitation of the extract. Ammonium sulfate is often used in protein purification procedures as an early step to achieve a few-fold purification because some proteins do not precipitate with the concentrations needed for precipitation of all the relevant factors (see Scopes 1994). The protein pellet can be dissolved and dialyzed. Ammonium sulfate precipitation is most useful for concentrating extracts. The unprecipitated extracts, particularly from a small cell volume, are often 2–5 mg protein/ml. The extracts can be concentrated to 10–20 mg/ml by efficiently precipitating the proteins and resolubilizing in a smaller volume. This allows an increased amount of protein to be added to the typical in vitro transcription reaction. After precipitation and resuspension, the extract is dialyzed to reduce the salt concentration.

Although the Dignam method has been widely successful for extract preparation from tissue-culture cells, it has not been applied to solid tissues because it is difficult to generate the single-cell suspension necessary to efficiently distend the cells for manual douncing. However, there are many alternative methods for generating solid tissue extracts (Gorski et al. 1986; Conaway et al. 1996; Stuempfle et al. 1996). One general approach has been to disrupt the tissue first by mincing it with a scalpel and scissors in cold isotonic buffer (see Protocol 14.2) followed by an incubation in hypotonic buffer. The minced and swollen tissue is then subjected to homogenization with a mechanical dounce. Hand douncing is also possible if the tissue is minced finely enough. The resulting extract is layered onto a sucrose step gradient and the nuclei are pelleted and separated from cellular debris. These nuclei can be extracted using either the Dignam method or a variation in which the nuclei are lysed, the chromatin is precipitated, and the transcription components are extracted (see Protocol 14.2).

The primary rationale for pursuing a solid tissue extract is that the gene of interest is expressed there and the tissue is a good source of gene-specific regulatory proteins. However, if the solid tissue extract does not support activation on its own, one strategy is to supplement HeLa extracts with a purified regulatory factor or a chromatographic fraction from the solid tissue extract. The HeLa extract is a good source of general transcription factors and ubiquitous cellular activators like Sp1. Supplementation of the extract with exogenous tissue-specific activators is sufficient to recreate certain layers of gene regulation in vitro. This latter approach has proven successful with numerous activators provided the activator is not abundant in the HeLa extracts. GAL4-VP16 is an example of an activator that can be added exogenously, because there are no proteins in HeLa extracts that recognize the GAL4 site (Carey et al. 1990a,b). Indeed, if an endogenous activator is present and it is essential to recreate the activator activity, it may have to be removed by immunoprecipitation or affinity chromatography. Maniatis and colleagues have removed the interferon-β (IFN-β) enhancer-binding proteins from mammalian extracts by DNA and immunoaffinity chromatography (Kim and Maniatis 1997). If these latter approaches fail, fractionation of the extract can sometimes separate the activator from the general transcription factors.

If specific promoter activity is not achieved with crude nuclear extracts or ammonium sulfate-precipitated extracts, several other variations can be attempted. One alternative is to attempt a whole-cell extract preparation (Manley et al. 1980, 1983). The additional proteins present in that extract might be more active for some promoters.

Another strategy that has proven useful is to perform a crude fractionation procedure with a heparin-agarose resin. With this resin, all of the basal transcription factors and many DNA-binding proteins elute between 0.2 and 0.5 M KCl, whereas many other proteins that may inhibit transcription do not bind to the column (see Dynan and Tjian 1983). To perform heparin-agarose chromatography, the crude extract or ammonium sulfate-precipitated extract should be dialyzed into Buffer D (recipe given below, p. 529). If possible, it is useful to ensure that the salt concentration has been sufficiently reduced by performing a conductivity measurement. The dialyzed extract can then be applied to a heparin agarose column of at least 1–2 ml. In principle, the extract is added such that more than 10 and up to 40 mg of protein is added per milliliter of resin. Although this latter amount of protein vastly exceeds the stated capacity of the column, it ensures that the eluted protein is fairly concentrated. However, it may not be practical when working with small amounts of extract. Small to moderate potential losses in extract amounts due to flowthrough should be weighed against the advantages of having a more concentrated extract. For small (1–2 ml) columns (which can be run in a Bio-Rad Econo-Column, cat. #737-0511), the protein can be passed through the column with gravity two to three times to obtain efficient protein binding. For larger columns, the protein should be passed through the column at a standard rate of 1–4 column volumes/hour. After binding, the column is washed with loading buffer and then with loading buffer containing 0.1 or 0.2 M KCl. The proteins are then eluted with loading buffer containing 0.5 M KCl. The eluate is dialyzed and tested for activity.

Transcription Assays

General Considerations

To establish an in vitro transcription assay for a new gene, it is necessary first to confirm that the extract is functional using a strong promoter. This is generally accomplished during the initial period when the extract, buffer conditions, and promoters are being optimized. As a standard template, promoters are used that rely only on basal transcription factors or on ubiquitous upstream activators. The adenovirus major late promoter (AdMLP) is a commonly used promoter that directs strong transcription in vitro. The AdMLP contains both TATA and Inr elements that direct significant transcription in vitro in the absence of upstream activators, and it also contains binding sites for the USF protein and other activators that enhance transcription. If this promoter does not initially function in extracts from the cell line of interest, it may be useful to purchase or prepare an extract from HeLa cells (Promega sells HeLa nuclear extract, cat. #3091). The HeLa extracts can be used to troubleshoot basic aspects of the in vitro transcription assay.

With the control promoter, it is first necessary to titrate both template and extract concentrations, which can both influence promoter activity. The optimal concentrations of extract and template are likely to be similar but not identical for the control promoter and the promoter of interest, but titrations should be performed with each promoter. If strong activity is detected with the control promoter, but not with the experimental promoter, it is probably best to begin systematically altering conditions to obtain promoter-specific activity. We discuss some common options in a later section.

Several assays can be used to measure promoter activity. They fall into two groups. In one, transcription reactions are performed with unlabeled NTPs and the resulting mRNA is measured using primer extension, S1 nuclease, or RNase protection assays. These assays are covered in Chapter 4. Primer extension is traditionally the most effective assay, particularly with the crude RNA products obtained from in vitro transcription reactions. If this assay is used, it is important to use a primer that will not bind to endogenous cellular mRNA contaminating crude extracts. The second group of assays involves the direct labeling of transcripts with ^{32}P-labeled NTPs. There are two types of assays in this group: runoff transcription and the G-less cassette assay (see Box 14.1). There are two significant logistical considerations in choosing which method to employ: (1) the time it takes to process the transcription products (i.e., from test tube to autoradiograph or phosphorimager) and (2) the specific activity of the products (i.e., the time it takes to visualize a signal). In this regard, the primer extension method is slower but is generally more sensitive than direct labeling.

Box 14.1

Methods for Measuring Transcription In Vitro

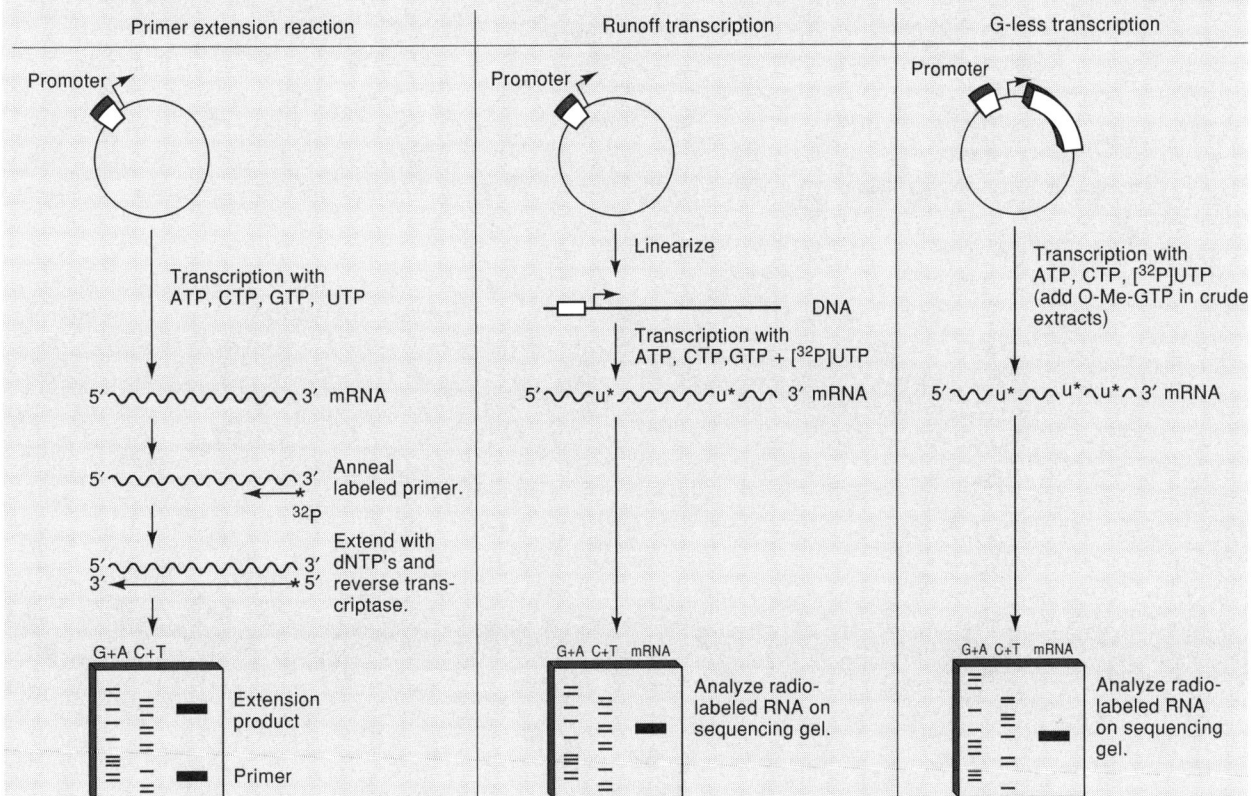

FIGURE 14.1. Primer extension, runoff, and G-less transcription.

Primer extension

As described in Chapter 4, a popular method for measuring transcription is primer extension (Fig. 14.1). Among methods for start site mapping in vivo, primer extension is the one most easily adapted for in vitro transcription. In this method, the mRNA is isolated and hybridized to a ^{32}P-end-labeled primer positioned about 50–150 nucleotides from the start site (see Protocol 4.1). The primer is extended by reverse transcriptase, and the labeled cDNA products are fractionated on an 8–10% polyacrylamide/urea gel. Generally, the addition of 0.1 pmole of primer to the extension reactions is sufficient because it will be in vast excess over the amount of mRNA, facilitating quantitative primer hybridization and extension. Because it is theoretically possible to label the primer quantitatively, the sensitivity of primer extension can, in principle, be many times more than that of the G-less assay or runoff transcription (because each mRNA can be extended by a ^{32}P-labeled primer). In practice, however, the labeling efficiency of the primer and the hybridization efficiency and extension by reverse transcriptase are not quantitative. Thus, the sensitivity of primer extension is lower than stated but, nevertheless, more efficient than internal labeling with [α-^{32}P]-NTPs (see below). One additional advantage of primer extension is that the method only measures a short fragment of RNA, so a small degree of cleavage of the mRNA by endogenous RNases in the extract or during sample manipulation does not hinder the analysis. This is particularly important in vitro where it is difficult to inhibit RNase during the manipulations necessary to generate an in vitro product. However, as described in Chapter 4, one must be wary of the secondary structure of the transcript because this can limit the efficiency of extension by reverse transcriptase. Addition of actinomycin D to the primer extension reaction can prevent spurious snap-back synthesis, which can generate primer extension products that do not represent the natural start site. Additionally, the primer extension reaction has a narrow pH optimum, particularly with the avian reverse transcriptase, and must be performed in a carefully buffered reaction. (See Chapter 4 and the accompanying protocols for other considerations in primer design and experimental setup.)

One potential problem when performing the transcription reactions on natural templates using extracts prepared from a natural tissue source (where the gene is normally expressed) is that the extracts are frequently contaminated with mRNA from the gene of interest. These contaminants may present a formidable background problem when one is attempting to measure an in vitro transcription signal. For this reason, heterologous sequences or primer tags are often cloned 90 bp downstream of the promoter and used in place of a primer complementary to natural downstream sequences. A common approach is to subclone the promoter into a multiple cloning site and to use the flanking sequencing primers for primer extension.

The major disadvantage of primer extension in vitro is the amount of time it takes to proceed through both the in vitro transcription reactions and primer extension reactions versus G-less cassette and runoff methods, where the internally labeled RNAs can be directly loaded onto gels. However, when studying a new promoter, the enhanced sensitivity of primer extension and the ability to authenticate the start site in vitro represent significant advantages over the other methods.

Runoff Assays

To perform a runoff assay (Fig. 14.1), a typical template is linearized 200 or so base pairs downstream of the promoter and transcription is measured in the presence of α-^{32}P-radiolabeled NTPs. Often the reactions contain three unlabeled nucleotides at a concentration of 200 μM and one labeled nucleotide at a concentration of 10–20 μM. Due to the high specific activity of radionuclides and the limits to the amount of radioactivity that can be added per reaction, a small amount of the labeled nucleotide is diluted with a low concentration of the same nucleotide unlabeled. Thus, for example, 1–10 μCi of [α-^{32}P]UTP would be added to a reaction containing 20 μM unlabeled UTP. The UTP concentration is close to the minimum amount necessary to observe efficient transcription because of the high NTP K_m for RNA polymerase II. In the presence of nucleotides, Pol II will initiate and elongate until it runs off the end of the DNA template generating a radiolabeled mRNA.

Although runoff transcription is simple, the labeling efficiency of the transcripts is not high. For example, in a 50 μl reaction with 20 μM UTP, the addition of 10 μCi of [α-^{32}P]UTP (given a theoretical maximum of ~9000 Ci/mmole) would translate to one pmole of labeled UTP for every nmole of unlabeled UTP. Thus, only 1 in 1000 U residues would be radioactively labeled. Given an average of 50 uracils (Us) in a 200-nucleotide-long transcript, only 1 in 20 transcripts would contain a radioactive U per reaction. Lowering the unlabeled UTP concentration to 5 or 10 μM or raising the amount of labeled nucleotide could improve efficiency, although one can never increase the specific activity beyond that of the preparation of α-^{32}P-labeled nucleotide, and there are practical limits to the amount of radioactivity that can be added to a reaction. Once the reaction is complete, the products are fractionated on polyacrylamide/urea gels. Although 40-cm sequencing gels are adequate, shorter gels take less time to run. The actual resolution is not necessarily important unless one is comparing multiple start sites or if there is a contaminant RNA migrating nearby.

Linearization of the template close to the promoter can lead to an inhibition of transcription because of the preponderance of proteins in crude extracts that bind to DNA ends. Therefore, it is prudent, if not essential, to linearize the template as far away as possible from the promoter in a fashion that still allows one to visualize the RNA product.

One potential problem when using radiolabeled NTPs is that background transcription is quite high in extracts. Nonspecific RNA polymerase III transcription can occur on many plasmid template DNAs. Also, Pol I, II, and III all carry out a reaction called end-to-end transcription, where the polymerase binds to the ends of the templates and transcribes inward. The different products can be distinguished by addition of the elongation inhibitor α–amanitin. In a mammalian system, 2 μg/ml amanitin inhibits Pol II, 200 μg/ml inhibits Pol III, and Pol I is insensitive to all concentrations but can be inhibited by actinomycin D; actinomycin D intercalates into double-stranded DNA and is a general inhibitor of many DNA-enzyme reactions. Finally, mammalian extracts are rich in enzymes that will add nucleotides onto the 3′ end of contaminating RNA in the extract. As the most abundant contaminants are generally 5S and tRNAs, they tend to become ^{32}P-end-labeled and generate 65–120-base bands on the gel. The lack of inhibition by α-amanitin and the appearance of the labeled RNAs in the absence of template DNA are the most common ways for identifying the products of such reactions.

In general, runoff assays are less efficient than assays employing the G-less cassette, mainly because the G-less reaction uses circular supercoiled DNAs, which are better templates than linearized DNA. However, runoff transcription does have one major advantage. Sometimes, insertion of a G-less cassette into a promoter region will disrupt promoter sequences downstream of the transcription start site. If this is the case, it may be wise to employ the runoff transcription assay.

The G-less cassette

The G-less cassette designed by Sawadogo and Roeder (1985) is one of the simplest and fastest methods for measuring transcription (Fig. 14.1). It is often a 360-bp sequence devoid of G residues on the noncoding strand. The resulting message is G-less. In general, the 360-nucleotide G-less cassette is introduced into the template by subcloning the sequence at the start site of transcription. The cloning generally can be done using PCR.

Typically, the template is incubated in the extract for 1 hour with ATP, CTP, and [α-^{32}P]UTP supplemented with unlabeled UTP. GTP is omitted to minimize background transcription. Because GTP is absent from the reaction, transcription should in principle terminate at the end of the G-less cassette, and the synthesis of small nonspecific RNAs should be minimized because there is no GTP. After purification of the products, the mixtures are loaded onto 6% polyacrylamide/urea gels. As detailed above, only 1 in 1000 U residues will be radiolabeled.

In crude extracts, despite extensive dialysis, there are often large enough quantities of contaminating GTP to allow the polymerase to proceed past the end of the cassette. In such cases, a distinct G-less transcript is not observed and more background transcription is observed. One solution is to treat the reaction products with RNase T1, which cleaves at guanines. This will eliminate the background transcripts and trim the G-less cassette messages down to a distinct 360-nucleotide RNA, which can be visualized on a gel. However, there will be random transcripts, initiating upstream of the G-less cassette, that read through it. Thus, treatment with RNase cannot distinguish between properly and randomly initiated mRNAs that pass through the cassette. To minimize such readthrough transcription, a GTP analog called 3′-O-methyl-GTP can be added in excess to the reaction. 3′-O-Me-GTP lacks a 3′ OH and acts as a chain-terminating analog. In the presence of 3′-O-Me-GTP, the polymerase will stall when it attempts to incorporate a G into the elongating transcript. Inclusion of 3′-O-Me-GTP reagent has been shown to dramatically enhance the specificity of a reaction in crude extracts contaminated with GTP.

The G-less cassette is available in many different sizes, and a smaller version of the G-less cassette can be fused to an internal control template for normalizing the efficiency of different reactions. Finally, the effect of replacing downstream sequences with the G-less cassette should be evaluated to ensure that critical regulatory sequences are not eliminated inadvertently. Recent studies by many labs have shown that the sequences immediately downstream of the start site (i.e., the DPE) can contribute to core promoter activity (see Chapter 1).

Choice of Template

The proper choice of template for an in vitro transcription assay is an important aspect of any study (Fig. 14.2A). Most in vitro systems employing naked DNA templates do not faithfully recapitulate the action of promoter regulatory elements located more than a few hundred base pairs from the start site of transcription. However, in some cases, special chromatin templates allow such long-distance effects to be observed. Chromatin templates are also preferred for the study of some regulated promoters.

The choice of template depends on the analysis. If the goal is to understand the cooperative and synergistic effects of multiple regulatory proteins in a natural context, a natural promoter should be employed (Fig. 14.2A). Because distal regulatory elements may not be accurately regulated, these can sometimes be studied by generating a template in which they are positioned close to the core TATA box (i.e., by deleting the intervening DNA or placing the enhancer upstream of a well-characterized core promoter). If the goal is simply to analyze the properties of a single regulatory protein, an artificial template bearing multimerized tandem copies of a responsive element can be constructed (Fig. 14.2B). In all cases, however, it is best to employ promoters that have been analyzed by transfection analyses. Transfection results can then be compared with the in vitro results to determine whether the biochemical system is faithfully recreating appropriate levels of gene expression seen normally in vivo. Figure 14.2C displays the results of a typical analysis where artificial templates bearing different numbers of multimerized GAL4 sites were incubated in HeLa extracts with GAL4 (1–147), GAL4-AH, and GAL4-VP16. Transcription was measured by primer extension. The results demonstrate the dynamic range of the HeLa system where transcription levels respond to the number of sites and strength of the activation domain.

A. Natural template

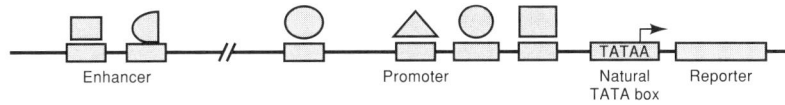

B. Model templates using various regulatory elements and reporter genes

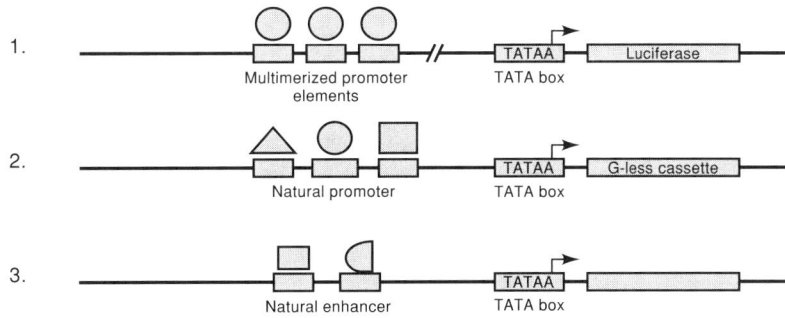

C.

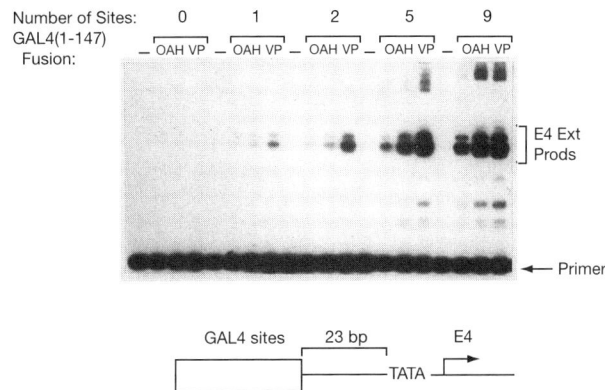

FIGURE 14.2. Templates used for in vitro transcription. (Reprinted by permission from *Nature* [Carey et al. 1990. Copyright 1990 Macmillan Magazines Ltd.].)

Often, when attempting to optimize expression of a promoter in vitro, it is difficult to distinguish the transcription being driven from the upstream promoter elements (activated transcription) and the transcription deriving solely from the core promoter (basal transcription). As the template concentration is raised, basal transcription will begin to predominate over activated transcription because activators become diluted out and distributed among different template molecules. Furthermore, high concentrations of template will also drive the binding of the basal factors, thereby circumventing the need for activator recruitment. Under such conditions, the core promoter elements alone may be driving expression. To demonstrate that the transcription being observed is regulated, a template bearing the upstream promoter of interest should always be compared with a template bearing the core promoter alone to demonstrate that the conditions are appropriate for regulated transcription.

In situations where exogenous activator is being added, DNA template titrations should be performed in both the presence and absence of activator. The optimal amount of activator can be determined by titrating in the recombinant protein and measuring transcription at each template concentration. Note that high DNA template concentrations can lead to a situation where very high concentrations of activator need to be added for complete filling of the template sites. Under such conditions the activator concentration may cause inhibition by squelching (Carey et al. 1990a). In such a case, lowering the template should permit lower concentrations of activator to be employed.

The inability to observe activation on naked DNA does not necessarily imply that the activator is not directly stimulating preinitiation complex assembly. High basal levels of transcription in the extract or transcription system may obscure the effects of activators. These effects may be more apparent when transcription complex assembly is competing with chromatin. The activities of several mammalian promoters and enhancers have been reproduced on chromatin templates including activation from the TCR-α, HIV-1, and β - globin genes (Sheridan et al. 1995, 1997; Barton and Emerson 1996; Mayall et al. 1997).

Chromatin Systems

There are several cell-free systems that will assemble plasmids into chromatin templates, including *Drosophila* embryo extracts, or mixtures of purified histones or nucleosomes incubated under the proper conditions with the DNA template (see Table 14.1) (Becker et al. 1994; Pazin et al. 1994, 1996; Bulger et al. 1995; Barton and Emerson 1996; Steger et al. 1997). The systems vary in purity, and some contain numerous chromatin modification activities such as the SWI2/SNF2 ATP-dependent remodeling complex, several related complexes bearing SWI2 homologs (e.g., ISWI), or histone acetylases.

There are different approaches to using chromatin templates. One approach is to determine whether the activator facilitates transcription during the process of chromatin assembly. For example, the activator alone or in the presence of the general machinery would be incubated with a DNA template in an embryo extract capable of assembling the template into chromatin. If activation is observed, one might imagine two mechanisms. In one, the activator could be blocking access of the promoter to chromatin, thereby passively facilitating assembly of a basal transcription complex. Alternatively, the activator could be directly stimulating complex assembly by interactions with the transcriptional machinery.

In an alternative approach, the chromatin is preassembled onto the template and the activator and general machinery are added later. If the activator stimulates transcription, one might imagine two possible mechanisms. The activator might gain access to promoters preassembled into chromatin and actively remove the chromatin by recruiting chromatin remodeling enzymes. This remodeling would allow the general machinery to bind passively. Alternatively, the activator might facilitate chromatin remodeling and also actively promote transcription complex assembly.

Some activators are known to recruit components of the chromatin remodeling and general transcription machinery actively and independently. For example, GAL4-VP16 directly recruits both the SWI/SNF complex and histone acetylases to DNA, while also interacting directly with TAFs, other coactivators, and general transcription factors to stimulate transcription (Kingston et al. 1996; Utley et al. 1998). There is also a growing body of literature suggesting that the chromatin remodeling machinery may be a component of the holoenzyme and that recruitment of the general machinery and chromatin remodeling occur in a concerted fashion (e.g., see Wilson et al. 1996).

Currently, it is difficult to assess which of the mechanisms discussed above is operational, although defined criteria will become available as the field matures and as more investigators use chromatin systems in their studies. Several recent studies of chromatin

TABLE 14.1. *Chromatin reconstitution systems*

Method	Summary	Applications	Comments	Chromatin analysis assays	Technical reference
Drosophila S150 or S190	Plasmid DNAs mixed with histones into *D. melanogaster* S150 embryo extract	Assembly of nucleosomal arrays for in vitro transcription	Key components appear to be histone chaperones and energy-dependent remodeling activities (i.e., NURF, CHRAC, SWI/SNF). Pure systems currently in development.	1. Micrococcal nuclease digestion (see Chapter 10) 2. Supercoil relaxation assay 3. Electron microscopy	1. Becker et al. 1994 2. Robinson and Kadonaga 1998
Urea/salt or salt dialysis of histones	^{32}P-labeled restriction fragments or plasmids mixed with urea and/or high salt with purified histones from chicken erythrocytes or HeLa nuclei	Assembly of mononucleosomes, dinucleosomes, and arrays for studying DNA binding of transcription factors to chromatin and chromatin remodeling; in vitro transcription applications	Can assemble arrays using this technique but need to add chromatin-remodeling components to generate regularly spaced arrays. Also need to carefully match histone–DNA ratios. Can also space arrays by adding specific 5S gene fragments, which contain positioning sequences.	1. Arrays – see 1–3 above. 2. Native gel electrophoresis of mono, dinucleosomes. 3. DNase I digestion of mononucleosomes and dinucleosomes (identify 10-bp cleavage periodicity). 4. Can use restriction enzyme accessibility to map nucleosome positions (see Chapter 10).	1. Studitsky et al. 1996 2. Ura and Wolffe 1996
Octamer exchange	Micrococcal nuclease digestion of chromatin, isolate mononucleosomes by gel filtration, mix with ^{32}P-labeled DNA in high salt followed by dilution or dialysis	Assembly of mononucleosomes and dinucleosomes for studying DNA binding of transcription factors to chromatin and chromatin remodeling	Quick and easy system for preparing nucleosomes, but contains contaminating genomic DNA from chromatin.	1. See 2–4 above	1. Utley et al. 1996

TABLE 14.2. *Select biochemical studies on transcription factor action on chromatin*

1. Enhanceosome assembly and function on chromatin templates in vitro
Mayall et al. 1997
Summary: The TCR α enhancer activity was tested in vitro on chromatin templates using purified T-cell transcription factors (LEF-1, AML1, and Ets-1) and the cAMP-responsive transcription factor CREB.

2. Energy-dependent chromatin remodeling
A. Transcription factor binding to mononucleosomes
Utley et al. 1997
Summary: A recent article summarizing how transcription factors with different DNA-binding domains compare regarding their ability to bind nucleosomes in the presence of SWI/SNF. The degree of SWI/SNF stimulation of factor binding by a factor was inversely related to the extent that binding is inhibited by the histone octamer. Cooperative binding to a nucleosome reduced the degree of stimulation by SWI/SNF. The authors conclude that cooperative binding of factors to nucleosomes is partially redundant with the function of the SWI/SNF complex.

B. In vitro transcription
Mechanics of ATP machines
Mizuguchi et al. 1997
Summary: The ATP-dependent *Drosophila* nucleosome remodeling factor (NURF) was shown on a preassembled chromatin template to facilitate transcription mediated by a GAL4-derived activator. The authors conclude that once nucleosome remodeling is completed, NURF is not continuously required for recruitment of the general transcriptional machinery and transcription.

Ito et al. 1997
Summary: The ability of ACF, an ISWI protein distinct from NURF, was tested in nucleosome remodeling and shown to mediate promoter-specific nucleosome reconfiguration by Gal4-VP16 in an ATP-dependent manner.

Activator-specific regulation
Armstrong et al. 1998
Summary: The erythroid Kruppel-like factor (EKLF), necessary for stage-specific expression of the human β-globin gene, was shown to require a SWI/SNF-related chromatin remodeling complex, EKLF coactivator-remodeling complex 1 (E-RC1), to generate a DNase I hypersensitive, transcriptionally active β-globin promoter on chromatin templates in vitro.

Mechanics of initiation and elongation in GTF-reconstituted systems
LeRoy et al. 1998
Summary: This study employed purified general factors, the energy-dependent chromatin remodeling factor RSF, and the chromatin elongation factor FACT to define the minimal factor requirements for activator-dependent transcription on chromatin templates in vitro.

3. Systems for studying the effect of histone acetylation in vitro
A. Transcription factor binding
Vettese-Dadey et al. 1996
Summary: The authors assembled nucleosomes from acetylated and unacetylated histones and showed in binding experiments that cores containing the most highly acetylated forms of histone H4 have the highest affinity for transcription factors.

B. In vitro transcription
Steger et al. 1998
Summary: Four histone acetyltransferase (HAT) complexes from yeast were employed to test whether acetylation regulates HIV-1 transcription in vitro. HAT activities acetylating either histone H3 (SAGA, Ada, and NuA3) or H4 (NuA4) stimulate HIV-1 transcription from preassembled nucleosomal templates in an acetyl CoA-dependent manner. HIV-1 transcription from histone-free DNA was not affected by the HATs. Restriction enzyme cleavage of chromatin was enhanced by the HATs, suggesting that histone acetylation leads to nucleosome remodeling.

Nightingale et al. 1998
Summary: Chromatin, containing either control or acetylated histones, was reconstituted to comparable nucleosome densities and characterized by electron microscopy after psoralen cross-linking, as well as by in vitro transcription. Acetylation of histones, and particularly of histone H4, affected transcription at the level of initiation. Acetylation stimulated heat shock factor binding to chromatin. The experiments demonstrated that histone acetylation can enhance activator access to their target sites in chromatin.

systems that can be employed as guides for studying transcription factor binding, ATP-dependent remodeling, and the effects of acetylation are listed in Table 14.2.

Optimization of Conditions

The initial step in carrying out an in vitro transcription study is to optimize the assay conditions to recapitulate in vitro the expression patterns observed in vivo or in transfection studies. There are several parameters to consider when optimizing a reaction, including extract concentration, template concentration, buffer conditions, and the necessity for special supplements.

Extract optimization generally involves titrating 20–250 µg of crude extract (measured by a Lowry or Bradford assay) in a 25- to 50-µl reaction mixture. When titrating the extract, the template should be kept constant— somewhere between 50 and 500 ng of DNA for a plasmid-based circular template and less for linear fragments. Conversely, when titrating the template, the extract should be kept constant. In most cases, the extracts are rich in nonspecific DNA-binding proteins and addition of variable amounts of carrier DNA (i.e., plasmid vector, poly(dI:dC), poly(dG:dC), or calf thymus DNA) is necessary to prevent inhibition of transcription.

The buffer and incubation conditions are also critical. Typically, the in vitro transcription reaction is optimally active with 50–100 mM monovalent salt (NaCl, KCl, ammonium acetate, potassium acetate), 5–10 mM $MgCl_2$, and 20 mM HEPES or 50 mM Tris (pH 7.9) as a buffer. Standard time courses have revealed that the reactions do not synthesize products for more than 1 hour at 30°C, presumably because some component in the extract has become inactivated due to covalent modification, thermal instability, or endogenous proteases. The salt, magnesium ion, buffer conditions, time course, and temperature should also be optimized for each extract and template. The effects of volume exclusion reagents such as polyethylene glycol or polyvinyl alcohol, or special supplements such as spermidine, can also be tested because they often enhance biochemical reactions involving nucleic acid metabolism.

Some researchers find it useful to include a reference plasmid in the system. This manipulation allows one to visualize a second transcript that serves as a control for product recovery or to determine whether one set of buffer conditions is optimal for multiple templates. However, the reproducibility of an effect is generally a better strategy because the inclusion of exogenous templates can lead to competition between the control and test templates. If, however, one feels compelled to include an internal control, it must be shown that the control is unaffected by the presence of the test template and vice versa.

Crude extracts often contain abundant RNA-binding proteins that can sequester the product. Therefore, after completing the transcription reaction, the products are often treated with Proteinase K for 15 minutes at 55°C in a buffer containing 0.2% SDS to degrade such proteins prior to phenol extraction and ethanol precipitation. Extraction by phenol alone followed by phenol chloroform is highly recommended to extract proteins completely. Phenol-chloroform alone does not efficiently extract the protein and can lead to significant difficulties later in the processing of the product (e.g., smeary bands on gels, insoluble pellets after ethanol precipitation).

Fractionated Systems

It has been almost two decades since the Roeder and Sharp laboratories (Weil et al. 1979a; Manley et al. 1980) demonstrated that crude homogenates of HeLa cells, but not Pol II alone, would mediate accurate initiation of cloned adenovirus genes. The studies demonstrated that extracts contained accessory factors that controlled transcription by Pol II. These factors were eventually called the general transcription machinery because they were used by numerous promoters bearing only a core promoter.

Roeder's lab was the first to show that these factors could be separated into distinct fractions by chromatography over phosphocellulose columns (Matsui et al. 1980). By loading the extract onto a phosphocellulose column at 0.1 M KCl and then eluting with successive steps of higher salt concentrations, Matsui and colleagues were able to isolate several fractions that were inactive on their own but together recreated specific transcription. These fractions were called TFIIA (the 0.1 M KCl breakthrough), TFIIB (0.35 M KCl eluate), TFIIC (0.5 M KCl eluate), and TFIID (1 M KCl eluate). The TFII designation is derived from *transcription factor for Pol II*. TFIIC was subsequently shown to be a protein (poly ADP-ribose polymerase) that suppressed nonspecific transcription by binding to nicks on the DNA template, thereby increasing the amount of polymerase available for specific transcription. Thus, this protein is not a true general transcription factor.

Reinberg and Roeder subsequently subdivided the fractions into several other factors using additional chromatography resins such as DEAE-cellulose and heparin-Sepharose (see, e.g., Reinberg and Roeder 1987). In addition to Pol II, there have been at least six distinct general factors discovered (TFIIA, TFIIB, TFIID, TFIIE, TFIIF, TFIIH) that are required for transcription. These factors in combination with coactivators from what is called the upstream stimulatory activity (USA) coactivator fraction (Orphanides et al. 1996) are sufficient to support basal and activated transcription in vitro. Many of the polypeptides encoding the various factors have now been purified to homogeneity and their genes have been cloned (see Box 14.2 and Protocols 14.6–14.10 for details and references on purification). There are many sources for obtaining vectors and cell lines. A partial list is given in Table 14.3. More details on this topic are reviewed in Chapter 1.

Holoenzyme

A recently described form of the transcriptional machinery is the holoenzyme. The holoenzyme was originally discovered in yeast by analyzing suppressors of the cold-sensitive phenotype of CTD deletions in Pol II (Koleske and Young 1994). These suppressors encoded proteins, called suppressors of RNA polymerase B (II), and were therefore called SRBs. Analysis of large multiprotein complexes containing the SRBs revealed that they contained Pol II, TFIIB, TFIIF, and TFIIH and could be complemented for basal and activated transcription by addition of TFIID and TFIIE. Subsequent studies in mammalian cells revealed several holoenzymes, many of which have been either shown by immunoblotting to contain mammalian SRB homolog (the yeast SRB7, 10 or 11 proteins are called SRB7, CDK8 and cyclin C in mammalian cells) or are likely to contain them based on related studies (Ossipow et al. 1995; Chao et al. 1996; Gold et al. 1996; Pan et al. 1997; Neish et al. 1998).

Several studies have isolated intact holoenzymes containing almost all of the general factors except TFIIA. In one study, the holoenzyme was isolated by immunopurification with an anti-MO15 antibody. MO15 is CDK7, the kinase component of TFIIH (Ossipow et al. 1995). In another study, an affinity column displaying the conserved domain of the elongation factor TFIIS was used to affinity-purify a holoenzyme. In the latter case, the holoenzyme responded in vitro to GAL4-VP16 (Pan et al. 1997). Other holoenzymes have also been reported, including one containing BRCA1, p300, the BRG1 subunit of SWI/SNF, and several general factors, but not TFIID (Neish et al. 1998). The composition and properties of current versions of the mammalian holoenzyme have been reviewed recently (Parvin and Young 1998).

TABLE 14.3. *A partial list of expression systems for human transcription factor purification*

Transcription factor	Expression system	Literature reference
TFIIA: 1. large subunit 2. small subunit	bacterial	Dejong and Roeder 1993 Dejong et al. 1995
TFIIB	bacterial	Ha et al. 1991
TFIID	human cell lines:	
	HeLa FLAG-tag TBP	Chiang et al. 1993
	HeLa HA-tag TBP	Zhou et al. 1992
TFIIE 1. large subunit 2. small subunit	bacterial	Chalut et al. 1994
TFIIF 1. large subunit 2. small subunit	bacterial	Wang et al. 1993
PC4 (USA component)	bacterial	Ge and Roeder 1994
TFIIH complex	human cell lines:	
	HeLa and fibroblast containing HA-tagged XPB	Winkler et al. 1998

TFIIH subunits	baculovirus expressing
	    recombinant subunits; can be reconstituted
	    into complex
		Tirode et al. 1999

p300	baculovirus	Schiltz et al. 1999
PCAF	baculovirus	Yang et al. 1996
SWI/SNF complex	human cell lines:	
	Jurkat cell line HA-tagged BRG1	Wang et al. 1996
	HeLa cell FLAG-tagged Ini1	Schnitzler et al. 1998
SWI/SNF subunits	baculovirus expressing recombinant subunits; can be reconstituted into complex	Phelan et al. 1999
Human mediator complex	human cell lines: HeLa cell	
	FLAG-tagged SRB10 (CDK8)	Gu et al. 1999

Many of the bacterial and baculovirus cell lines are available as His$_6$-tagged protein that can be purifed by Ni-NTA affinity chromatography (Qiagen) or as FLAG-tagged proteins. The HA-monoclonal antibody resins and peptides are available from BAbCO. The FLAG-monoclonal M1 and M2 monoclonal antibody resins and peptides are available from BAbCo and SIGMA. MonoFLAG and polyclonal antibodies for many of the transcription factors are available from Santa Cruz Biotechnology. Pol II monoclonal antibodies are available from BAbCO and QED. The cDNAs are available from the literature sources cited and from the ATCC.

Mediator Subcomplexes

In yeast, the holoenzyme can be subdivided into a mediator complex that includes Pol II but lacks the general factors (reviewed in Myer and Young 1998). Both conventional chromatography and immunoaffinity chromatography against CDK8 (human SRB10) were employed to purify a mammalian mediator complex lacking the general factors but containing human SRBs, Pol II, the SWI/SNF complex, and the acetyltransferases p300 and PCAF (Cho et al. 1998). More recent studies using FLAG-tagged CDK8 have identified further subcomplexes containing largely SRB and MED proteins (Gu et al. 1999) (see Chapter 1). In some in vitro assays, the complexes act positively (Gu et al. 1999) whereas in others they act negatively (Sun et al. 1998).

Partially Fractionated Systems

The ability to study transcription in fractionated systems has allowed experiments to be performed on the precise mechanism by which activators stimulate transcription. In model systems with artificial templates, the fractionated systems were used to show that various complexes containing TFIIA, TFIID, and TFIIB were subject to regulation by different activators (Lin and Green 1991; Wang et al. 1992b). Furthermore, Maniatis and col-

Box 14.2

Purified Transcription Factors

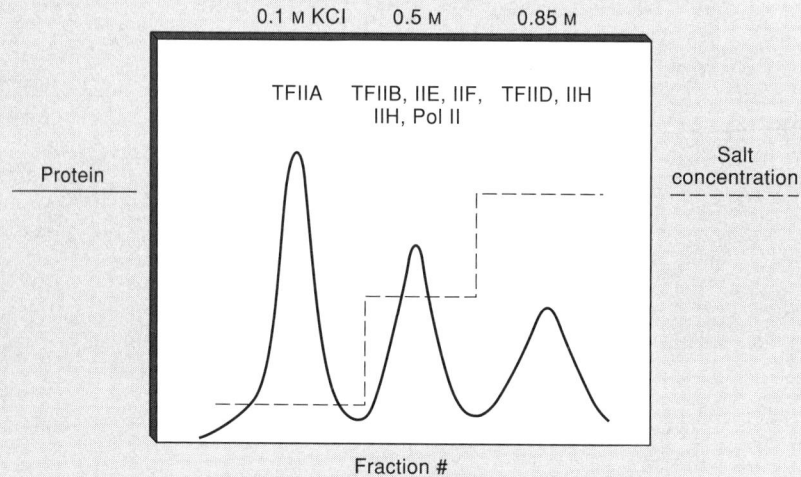

FIGURE 14.3. (*A*) Fractionation of GTF components by column chromatography.

TFIID

TATA-binding protein (TBP) was the first general factor cloned (Hahn et al. 1989). Although TBP purifies as a single polypeptide from yeast, in mammalian cells it is stably associated with the TBP-associated factors (TAF$_{II}$s). In HeLa extracts 10 or more TAF$_{II}$s co-immunoprecipitate with TBP antibodies. Although TFIID can be purified conventionally, two labs have generated HeLa cell lines bearing either hemagglutinin (HA)- or FLAG-tagged versions of TBP. The epitope is placed on or near the amino terminus of TBP, where it has been shown to have little inhibitory effect on TFIID's biochemical activities (Zhou et al. 1992; Chiang et al. 1993). The recombinant HeLa cell lines are grown, in spinner culture, and extracts from these cell lines are passed over phosphocellulose columns. The 0.5–0.85 M eluate (Fig. 14.3) is incubated with an immunoaffinity resin containing the HA or FLAG antibody coupled to protein A-Sepharose. The unbound protein is washed away with high-salt buffers and the TFIID is eluted from the resin with 1 mg/ml of HA or FLAG peptide to compete the TFIID from the resin. The TFIID generated in this fashion is highly purified and can be used for in vitro transcription, DNase I footprinting, EMSA, and crosslinking assays. Heat treatment of HeLa extracts at 45°C for 15 minutes selectively inactivates TFIID, providing a convenient system to measure recombinant TFIID activity (Nakajima 1988). This is, however, a somewhat simple system for measuring TFIID, and some other activities are also inactivated to varying extents.

TFIIB

After TBP, TFIIB was the next general factor cloned. It was purified to homogeneity by Reinberg's lab (Ha et al. 1991) and subjected to microsequencing analysis, and the cDNA was cloned from a library. TFIIB is a 316-amino-acid, 33-kD protein. It has been inserted into a T7 expression vector and is easily overexpressed and purified to homogeneity from *E. coli*. Usually the bacterial extracts are subjected to preliminary polyethylenimine and ammonium sulfate steps to remove nucleic acid and contaminant proteins, respectively. The TFIIB is then purified over a single phosphocellulose column. TFIIB strongly stabilizes TBP binding to TATA oligonucleotides in native polyacrylamide gel or magnesium agarose EMSAs. Some activator recruitment studies can also be performed in these systems because several activators directly contact TFIIB, and one functional consequence is recruitment. The crystal structure of TFIIB has been solved (Nikolov et al. 1995).

TFIIA

TFIIA was first isolated and cloned from *Saccharomyces cerevisiae* (Ranish et al. 1992). It consists of two subunits called TOA1 and TOA2. Genetic studies and tetrad analysis showed that both subunits were essential for yeast survival. Using the yeast genes, TFIIA was then cloned from humans and *Drosophila* (DeJong and Roeder 1993; Ma et al. 1993; Yokomori et al. 1993, 1994; Bernstein et al. 1994; Ozer et al. 1994; Sun et al. 1994; DeJong et al. 1995). TFIIA in humans is a complex of three subunits (α, β, γ) of 37, 19, and 13 kD and is found associated with a subpopulation of TFIID. The two largest subunits of the three-subunit TFIIA proteins are derived from a larger precursor homologous to the yeast TOA1 gene. This precursor is cleaved proteolytically to generate the mature subunits.

Footprinting studies (Lieberman 1994; Lieberman and Berk 1994) and the crystal structure of the TFIIA-TBP-TATA ternary complex reveal that TFIIA stabilizes the binding of TFIID to a TATA box by contacting TBP directly using one subunit and by contacting DNA with the other. Biochemical studies have shown that the α-β precursor will function with the γ subunit in biochemical assays. Recombinant TFIIA is fully functional in TBP and TFIID binding assays and in in vitro transcription assays.

TFIIF

TFIIF contains two subunits termed RAP30 and RAP74 because they were originally isolated as RNA polymerase-associated proteins in affinity chromatography experiments. RAP30 was cloned by Jack Greenblatt and RAP74 by Zack Burton (Sopta et al. 1989; Aso et al. 1992; Finkelstein et al. 1992). The genes are available as bacterial expression vectors, and the proteins can be overproduced and purified from *E. coli* (Wang et al. 1993, 1994). The recombinant subunits can be mixed and retain all the known functional properties of TFIIF. The subunits associate in the form of a heterotetramer.

TFIIE

TFIIE contains two subunits of 34 and 56 kD. These subunits are, like TFIIF, present as a heterotetramer. Both subunits were cloned as a collaborative effort between Robert Tjian's and Danny Reinberg's labs (Ohkuma et al. 1990; Inostroza et al. 1991; Peterson et al. 1991). Both subunits can be expressed from T7 expression vectors, and the subunits are functional in in vitro transcription assays (see Chalut et al. 1994).

TFIIH

TFIIH, because of its wide variety of subunits, its many biochemical activities, and its dual roles in DNA repair and transcription, has become one of the most studied general factors. The protein has been extensively characterized in both mammalian cells and in yeast (where it is called **factor b**). Experiments in Reinberg's lab and in J-M. Egly's lab have shown that TFIIH consists of at least eight subunits and possesses several biochemical activities (for review, see Drapkin and Reinberg 1994; Seroz et al. 1995; Svejstrup et al. 1996). Highly purified TFIIH acts as an ATP-dependent DNA helicase. TFIIH also has a kinase activity that in the presence of ATP will phosphorylate the carboxy-terminal domain of Pol II. The phosphorylation causes a shift in mobility of the largest subunit in an SDS gel and converts the Pol II from what is called the IIA form to the IIO form. The effect does not require DNA or other components of the transcriptional machinery. TFIIH can be purified by conventional chromatography using antibodies against various subunits and immunoblots to monitor the chromatographic profiles. Be aware that, despite popular belief, there are many chromatographic complexes of TFIIH with different subsets of the subunits. TFIIH has also been subjected to purification by immunoaffinity chromatography (Winkler et al. 1998). Egly and colleagues have also recently succeeded in generating intact TFIIH from subunits expressed in baculovirus (Tirode et al. 1999).

Pol II

Mammalian RNA polymerase II contains 12 subunits. They have all been cloned, but Pol II has not yet been reconstituted from the recombinant subunits. Instead Pol II can be purified to homogeneity either by standard chromatographic procedures or by immunoaffinity chromatography using a monoclonal antibody against the carboxy-terminal domain (CTD) (Thompson et al. 1990; Thompson and Burgess 1996). Basically, extracts are prepared from the nuclei left over after preparation of the Dignam nuclear extract and used as a source of the Pol II. We will briefly cover the purification of the components below but the reader is advised to consult more extensive sources.

leagues used fractionated systems to understand how the natural IFN-β enhanceosome assembled transcription complexes (Kim and Maniatis 1997). Thus, fractionated systems have great utility in understanding regulation at the molecular level.

We discuss two fractionated systems for recreating gene activation in vitro —partially fractionated and purified systems. Partially fractionated mammalian systems that retain much of their ability to support gene activation can be prepared from standard nuclear extracts. These systems, although lacking in purity, nevertheless have provided important information on key steps in the activation process. Although homogeneous systems are more difficult, they have several advantages when analyzing specific steps in the transcription process, including the ability to perform DNase I footprinting and EMSA analyses.

A partially fractionated system with high activity can be prepared from the Dignam HeLa cell nuclear extract by chromatographing it over a phosphocellulose P-11 column pre-equilibrated in 0.1 M KCl (see Protocol 14.6). The column is then sequentially developed with steps of 0.3, 0.5 (TFIIB, TFIIE, TFIIF, TFIIH, Pol II), and 0.85 M KCl (TFIIH, TFIID, USA). This system is ideal for visualizing kinetic stages in assembling a transcription complex and can be employed for some rudimentary transcription assays, including permanganate probing and sarkosyl challenge of various steps (see Chapter 15).

The phosphocellulose flowthrough fraction contains TFIIA activity and can be further purified over DEAE-cellulose (DE-52) to generate a highly active fraction with TFIIA activity. The 0.5 M P-11 fraction is dialyzed and applied to a DE-52 column. TFIIB activity flows through the column, and Pol II, TFIIE, TFIIF, and a small amount of TFIIH are eluted.

The 0.85 M P-11 fraction containing TFIID activity is dialyzed and applied to a DE-52 column. After washing extensively with pre-equilibration buffer, TFIID is eluted and the eluate is applied directly to a pre-equilibrated heparin-Sepharose column. The column is washed with the pre-equilibration buffer and TFIID activity is eluted. (A detailed protocol is included below; see Protocol 14.10.)

Recombinant TFIIB is overexpressed in *E. coli* and purified by phosphocellulose chromatography (see Wang et al. 1992a). RNA polymerase II can be purified from the nuclear pellet of the Dignam and Roeder extract by a procedure involving conventional and affinity chromatography (see Protocol 14.7).

Pol II, recombinant TFIIB, and the crude fractions containing IIA, another containing IID, and the last one containing IIE, IIF, IIH, and Pol II, are all then dialyzed against the transcription buffer and used directly in cell-free transcription experiments. Typically, the fractions are mixed in combination and individually to demonstrate that they are enzymatically pure (e.g., that the TFIID fraction does not contain TFIIA or the TFIIA fraction does not contain TFIIE, TFIIF, TFIIH, and Pol II by complementation analysis) and that the entire set is required for basal and activated levels of transcription from a responsive reporter template. The in vitro reactions are designed to contain either a complete set or subsets in which one or more factors are omitted to prove that they are complementary. These partially pure systems tend to display higher activity than highly pure systems but are incompatible with DNase I footprinting or EMSA assays. Nevertheless, because TFIIA, TFIID, and TFIIB have been found to be major targets of regulation, the partially fractionated systems have great utility.

Factor-depleted Systems

If the goal of the study is to measure the effect of specific activators on a single general factor, there are some rapid and facile methods for quantitatively removing that putative target factor from a HeLa extract. These systems tend to be highly active because they contain many of the coactivators necessary to achieve activated transcription, and, in each case, a single factor is depleted selectively thus maintaining the proper balance of the remaining factors, which is necessary to confer high transcriptional activity on the nuclear extracts.

A useful system is one depleted of TFIID, where TFIID can be added back to complement transcription. Heat treatment of HeLa extracts at 45°C for 15 minutes is sufficient to inactivate TFIID completely, while having only a moderate effect on the other factors (Nakajima et al. 1988). This is the most convenient system to set up because it requires no special reagents or chromatography equipment. A second method relies on slowly passing a nuclear extract pre-equilibrated in 0.5 M KCl over a phosphocellulose column. The resulting breakthrough is devoid of TFIID and can be supplemented with the TFIID fraction eluted from the same column at 0.85 M KCl. Note that this fraction also contains coactivators (USA) (see Ge et al. 1996). In some cases, the TFIID is further fractionated over DE-52 and/or heparin-Sepharose to purify and concentrate it further.

Removal of TFIIA is also easily achieved by passing a nuclear extract slowly over a column of an equal volume (or no more than twofold excess) of nickel (NTA)-Sepharose. The large subunit of TFIIA has a string of histidines, causing it to bind tightly to nickel. The resulting extract can be complemented with recombinant TFIIA, or the natural TFIIA can be eluted with imidazole and then employed (Ozer et al. 1994).

Removal of TFIIB can be accomplished using affinity-purified high-titer TFIIB antibodies, which are available from several academic and commercial sources (e.g., Santa Cruz Biotechnology Inc.). Preincubation of a resin comprising TFIIB antibody coupled to protein A-Sepharose with HeLa extract is sufficient to deplete the protein quantitatively from the extract. The extract then serves as a system to study TFIIB activity. TFIIH, Pol II, TFIIE, TFIIF, and various coactivators can all be removed in a similar fashion. Because some of these factors are associated with one another in the extract, a subset in the form of a holoenzyme, then some immunoprecipitations will result in a decrease in the overall activity of the remaining components.

Highly Fractionated Systems

Systems composed of all of the general factors individually, many in recombinant form, can be generated in vitro and will support low to moderate levels of activation (Ge et al. 1996; Maldonado et al. 1996; Tantin et al. 1996). The setup of such systems involves purification of recombinant human TFIIA, TFIIB, TFIIE, and TFIIF from *E. coli*, immunopurification of TFIID and Pol II from HeLa extracts (Thompson et al. 1990; Zhou et al., 1992), conventional or immunopurification of TFIIH from HeLa extracts, and purification of USA fraction from HeLa extracts. As our knowledge of the activation process evolves, these systems will increase in purity. It is important to consult the literature, particularly the work coming from the Roeder, Reinberg, Tjian, and Egly laboratories to fully evaluate the current status of the purified systems.

TECHNIQUES

Preparation of Nuclear and Whole Cell Extracts: An Overview (Protocols 14.1–14.3)

Cell-free extracts serve as a source of regulatory factors for DNA binding, in vitro transcription, and protein purification experiments (Ge et al. 1996; Maldonado et al. 1996; see Roeder 1996). Historically, three classes of extracts have been successfully employed to recreate transcription by Pol II in vitro. The first cell-free system was a cytoplasmic HeLa cell S100 extract (Weil et al. 1979a), which when supplemented with Pol II led to transcription of plasmid templates in vitro. This was followed shortly by the whole-cell extract (Manley et al. 1980), which was easy to prepare and did not require the addition of pure Pol II. Finally, the nuclear extract was developed initially by Dignam and Roeder for use with HeLa cells (Dignam et al. 1983a). Later, additional types of nuclear extracts were developed that differed in their method of preparation and were more suited for use with solid tissues (Gorski et al. 1986; Stuempfle et al. 1996). Because of their transcriptional efficiency and purity, the Dignam et al. (1983a) nuclear extracts have gained the widest use.

Traditional nuclear extract preparations have relied on a large number of cells, 10^{10} or more, although efficient mini-extract protocols for use with as few at 3×10^7 cells have been developed (Lee et al. 1988; Osborn et al. 1989). There are several methods for preparing nuclear extracts. The methods often vary with the cell type, and it is easier to employ tissue-culture cells from dishes or spinner or roller bottles than to employ solid tissues. Nevertheless, the choice of method depends largely on the application and the availability of tissue or cell lines for the extract. If the goal of the study is to obtain a system for addition of exogenous tissue-specific regulatory proteins or studies on the general machinery, then HeLa nuclear extracts prepared by the method of Dignam et al. (1983a) are a good source of general transcription factors and are highly active on a specific activity basis (transcription per mg of extract). Once the facilities are in place, HeLa cells can be generated in spinner culture at a cost of ~$5/liter excluding labor ($15–$20/liter commercially).

HeLa nuclear extracts are relatively easy to prepare, although the time (doubling time 24 hours) and facilities necessary to grow the cells are a disadvantage. In the Dignam and Roeder method, the cells are swollen in a hypotonic buffer and lysed, and the nuclei are separated from the cytoplasmic extract by centrifugation. The resulting nuclear pellet is then incubated in a medium salt buffer (0.42 M NaCl) to extract the transcriptional components. There is minimal proteolysis during the procedure and 1 liter of culture generates 1 or 2 grams of cells (depending on cell density at harvest) and 1–2 ml of extract at 6–10 mg/ml. This is enough extract for ~50–100 in vitro transcription reactions, depending on the scale. Other cell lines, including some lymphocytes, also grow in large-scale spinner culture or in roller bottles, but sometimes these cells require CO_2 for growth (unlike HeLa cells, which use a bicarbonate buffer system and autoclavable medium). Nevertheless, the extracts produced from these cells (e.g., Namalwa; see Foulds and Hawley 1997) are of high quality.

If the goal is to begin studying transcription of a gene in its native cell type and a tissue-culture source is unavailable, many investigators might wish to obtain nuclear extracts from solid tissues. Solid tissues are generally more difficult to manipulate because they contain many cell types, including connective tissues. Much effort is expended both in the preparation of the nuclei from the tissue and in preventing proteolysis. For example, a single rat contains 10 g of liver tissue that eventually only generates 6–7 mg of high-quality nuclear extract, less than that obtained from 1 liter of HeLa cells. Solid tissues also must first be minced with a homogenizer and then lysed with a mechanical Dounce. The nuclei are subsequently purified by centrifugation over sucrose step gradient and finally extracted. The presence of contaminants, including nucleases and proteases, during nuclear extract preparation can severely affect the activity of the extract. Certain tissues like liver and intestine are rich in such enzymes. For this reason, some tissues or cell lines yield more active extracts than others. Optimization of the extraction conditions is always necessary for obtaining the most favorable balance of general factors and inhibitors. This is determined by trial and error (discussed in Dignam et al. 1983a).

We describe two protocols for large-scale preparation of nuclear extracts. The first is the Dignam and Roeder extract (Dignam et al. 1983a), which has proven reliable, easy, and applicable to a wide variety of tissue-culture lines. We use this protocol in our laboratories and have included extensive notes on additional considerations and troubleshooting. We also describe a rat liver protocol, provided by Joan and Ron Conaway and derived from Gorski and colleagues (1986), that can be used for solid tissues. Stuempfle and colleagues recently described a modification of the procedure, which is claimed to be easier and more broadly applicable (Stuempfle et al. 1996). One protocol for whole-cell extracts is also included in this section because this is a reasonable approach to attempt when nuclear extracts fail (Manley et al. 1983).

PROTOCOL 14.1

The Dignam and Roeder Nuclear Extract

For the Dignam and Roeder nuclear extract preparation (Dignam et al. 1983a), the cells are harvested by centrifugation, washed, and resuspended in a hypotonic buffer, which causes the cells to swell. Subsequently the cells are lysed by homogenization with a hand-held Dounce. The nuclei are pelleted by centrifugation, the cytoplasmic supernatant is decanted, and the nuclear pellet is resuspended by Douncing in a moderate salt buffer. After stirring the suspension for 30 minutes to allow extraction of transcription factors, the nuclei are centrifuged again and the resulting supernatant or extract is dialyzed against Buffer D for use in transcription experiments.

TIME LINE AND ORGANIZATION

The Dignam/Roeder nuclear extract preparation takes 1 day after harvest of the cells. Although it is imperative to use fresh cells, the protocol can be stopped after preparation of nuclei and the nuclear pellet can be frozen for later usage. Alternately, the protocol can be stopped prior to dialysis. All manipulations have to be carried out on ice or in a 4°C cold box. Chill all buffers, centrifuge tubes, rotors, and the Dounce homogenizer prior to use. The extracts after preparation have a half-life of 12 hours at 4°C; therefore, minimizing dialysis time is essential.

OUTLINE

Preparation of nuclear extracts from HeLa cells (1 day)

Step 1: Prepare buffers and chill (1 hour)
Step 2: Harvest and lyse cells (4 hours or more, depending on amount)
Step 3: Prepare nuclear extract (8 hours)

PROCEDURE

CAUTIONS: *DTT, Glycerol, KCl, KH$_2$PO$_4$, KOH, MgCl$_2$, PMSF. See Appendix I.*

Step 1: Prepare buffers and chill

Buffer A$^+$ (0.01 M KCl):

10 mM HEPES-KOH (pH 7.9 at 4°C)
1.5 mM MgCl$_2$
10 mM KCl
0.5 mM DTT (add just before use)

Buffer C:

20 mM HEPES-KOH (pH 7.9)
25% v/v glycerol
0.42 M NaCl
1.5 mM MgCl$_2$
0.2 mM EDTA
0.5 mM PMSF
0.5 mM DTT (add PMSF and DTT just before use)

Buffer D⁺ (0.1 м KCl):
> 20 mм HEPES-KOH (pH 7.9)
> 20%. v/v glycerol
> 0.1 м KCl
> 0.2 mм EDTA
> 0.5 mм PMSF
> 1 mм DTT (add PMSF and DTT just before use)

Phosphate-buffered saline (PBS):
> Make from 10 x PBS stock: per liter:

KCl	2.0 g
KH_2PO_4	2.0 g
NaCl	80.0 g
$HPO_4 \times 7H_2O$	21.6 g
H_2O to 1 liter	

Step 2: Harvest and lyse cells

1. Grow the HeLa cells at 37°C in spinner flasks. Cells may also be obtained from commercial sources (National Cell Culture Center, Minneapolis, MN 55433, tel.: 800-325-1112). Count the cells using a hemacytometer prior to harvest, and then harvest at $0.5–1 \times 10^6$ cells/ml by centrifugation at 3000 rpm for 10 minutes. Ideally, for cultures larger than 2 liters in size, harvesting is done in 1-liter bottles using a Sorvall RC-3B centrifuge.

2. Wash the pellet twice gently with 10 ml of PBS per liter of cells; centrifuge between washes. Transfer the suspension into conical graduated centrifugation tubes before the last centrifugation.

3. All subsequent steps must be performed at 4°C. Determine the volume of the resultant cell pellet using the graduations on the centrifugation tube. Ideally, 1 liter of cells yields 1–2 ml packed cell volume (PCV). Resuspend the pellet in 5 PCVs of Buffer A.

4. Lyse the cells with 10 strokes in a Dounce homogenizer fitted with a B-type pestle. This step has to be done slowly and gently, especially for the down strokes, to prevent nuclear lysis. Stop douncing if the suspension becomes viscous.

5. Check the lysate under a microscope for cell lysis. The nuclei can also be stained with Trypan blue to monitor lysis.

Step 3: Prepare nuclear extract

1. Centrifuge the nuclear pellet obtained in Step 2 for 20 minutes at 10,000 rpm in a Sorvall SS-34 rotor. Decant the supernatant containing the cytoplasmic material. Carefully remove any lipids near the lip of the centrifuge tube with a pipet or Kimwipe. Designate the resultant pellet as crude nuclei.

2. Resuspend the crude nuclei in 3 ml of Buffer C per 10^9 cells by gentle dounce homogenization (10 strokes with a B-type pestle).

3. Stir the resulting suspension very gently with a magnetic stirring bar for 30 minutes to 1 hour at 4°C and then centrifuge for 30 minutes at 10,000 rpm (Sorvall SS-34 rotor). This step extracts the transcription components from the nuclei. Save the pellet for Pol II purification (see Protocol 14.9).

4. Transfer the resulting clear supernatant to a dialysis bag and dialyze against 50 volumes of Buffer D$^+$ for 5 hours at 4°C. The dialysis bag must be spinning vigorously to achieve complete dialysis in the time stated. Longer dialysis times lead to a decrease in transcription activity. Measure conductivity to ensure dialysis is complete.

5. Often, a small amount of precipitate forms during dialysis. If this occurs, centrifuge the dialysate in a Sorvall SS-34 rotor at 4°C, 10,000 rpm for 20 minutes.

6. Discard the precipitate. Freeze the supernatant, designated as nuclear extract, as 100-µl to 1-ml aliquots in liquid nitrogen or on dry ice.

7. Analyze an aliquot for protein concentration by the method of Bradford using BSA as a standard.

8. Store the extract at –80°C. Typical yields are about 6 – 10 mg/ml, or 15 –20 mg of protein from 10^9 cells.

ADDITIONAL CONSIDERATIONS

1. Extraction step. The salt concentration in the extraction step is optimized for HeLa extracts. Higher salt concentrations begin to extract inhibitors, whereas lower salt concentrations do not efficiently extract the general transcription components. Other cell types may exhibit different optimal extraction conditions, and these should be determined empirically as described by Dignam et al. (1983a).

2. Extract concentration. The key to an active extract is high protein concentration. Extracts below 5 mg/ml often are marginally active. Some investigators couple the Dignam and Roeder procedure with an ammonium sulfate concentration step where solid ammonium sulfate (0.3–0.4 g/ml of supernatant) is added to the final extract (after Step 3.4) while stirring gently. This concentration of ammonium sulfate is sufficient to salt out or precipitate most of the protein in the extract. After 30 minutes, the precipitate is collected by centrifugation, resuspended in transcription buffer (two- to fourfold more concentrated than the starting volume), and dialyzed to remove excess ammonium sulfate.

3. Dialysis time. There are two considerations when dialyzing the extract. First, long dialysis times can lead to excess protein precipitation, which decreases the protein concentration and activity of the extract. It is generally impossible to resuspend the protein precipitated in this fashion. Second, the extract is subject to random proteolysis and protein inactivation. Therefore, inclusion of fresh protease inhibitors and a reducing agent are necessary to maintain the activity of the extract. In both cases, it is important to minimize dialysis time by vigorous stirring of the dialysate, use of large pore tubing (~10,000 MW cutoff), and optimizing the surface area (i.e., by using long narrow tubing versus short wide tubing). Always monitor the conductivity of the dialysate prior to storage to ensure that it has equilibrated with the buffer. If the following guidelines are employed, the extract can usually be dialyzed to equilibration in 5 hours or less.

4. Miniextracts. For some experiments, or for optimizing extraction conditions, it may be necessary to generate extracts from smaller numbers of cells. There are two techniques that work with as few as 3×10^7 cells (i.e., the equivalent of 60 ml of HeLa cells). In one the cells are lysed by passage through a syringe needle and the nuclei are centrifuged and extracted in a microfuge tube (Lee et al. 1988), and in the other the cells are lysed with a hypotonic buffer and NP-40 (Osborn et al. 1989). The extract pro-

duced by these techniques can be employed in DNase I footprinting, EMSA, and in vitro transcription experiments.

TROUBLESHOOTING

No transcriptional activity

Possible cause: Problems with extract.

Solution: First check the technique using a control extract. Promega and other manufacturers market transcriptionally active HeLa extracts. If the Promega extract is functioning with your template, consider that the protein concentration in your extract may be too low or the salt concentration too high. Check the protein concentration by Bradford and salt concentrations by conductivity and determine whether they are within acceptable ranges (i.e., see in vitro transcription, Protocol 14.4).

Possible cause: The extract contains inhibitors.

Solution: Perform a mixing experiment. Mix your extract in with a control extract and determine whether it inhibits.

Possible cause: Occasionally the extract preparation fails because of the quality of the HeLa cells or other unknown causes.

Solution: Such extracts occasionally can be supplemented with a limiting factor to resuscitate them, but the best advice is to start over with fresh buffers and cells.

PROTOCOL 14.2

Preparation of Nuclear Extracts from Rat Liver

This protocol was provided to us by Joan and Ron Conaway and is a modification of Gorski et al. (1986). It is a method for preparation of nuclear extracts from rat liver and serves as a model for nuclear extracts from other solid tissues. The livers are first dissected and then minced prior to cell disruption in a Teflon homogenizer. The nuclei are purified over a sucrose cushion and the resulting nuclear pellet is extracted. Unlike the Dignam and Roeder procedure (Dignam et al. 1983a), the nuclei are lysed with ammonium sulfate and the chromatin is centrifuged out. The resulting extract is precipitated by addition of solid ammonium sulfate and the pellet is resuspended and dialyzed to generate the final extract. Because liver represents a highly heterogeneous tissue sample with numerous protease activities, there is an emphasis on significant amounts of protease inhibitors, maintaining samples at 0–4°C, and speed of preparation. Note that this procedure is significantly more involved and time-consuming than the procedure that uses tissue-culture cells.

OUTLINE

Preparation of nuclear extracts from rat liver (2 days)

Step 1: Prepare buffers and equipment (1 hour)
Step 2: Prepare tissue and lyse cells (3–4 hours)
Step 3: Prepare nuclear extracts (2 hour, 30 minutes)
Step 4: Dialyze extract and determine protein concentration (5 hours)

PROCEDURE

CAUTIONS: *Animal treatment, Antipain, Aprotinin, DTT, Glycerol, KCl, KOH, Leupeptin, MgCl$_2$, NaOH, (NH$_4$)$_2$SO$_4$, Pepstatin A, PMSF, SDS. See Appendix I.*

Step 1: Prepare buffers and equipment

- This protocol is designed for nuclear extracts from 20 rats. Keep everything on ice at all times. Before starting this protocol, be sure all homogenization equipment and flasks are cold.

  Have ready in the cold room:
  five 1.0-liter glass beakers
  three 0.5-liter glass beakers
  Teflon-steel homogenizer
  two 250-ml glass cylinders
  two 500-ml glass cylinders
  two 1-liter glass bowls
  centrifuge tubes
  centrifuge rotors

- All buffers should be ice-cold before use.

Phosphate-buffered saline (PBS):

For 10x PBS stock: per liter:

KCl	2.0 g
KH_2PO_4	2.0 g
NaCl	80.0 g
$HPO_4 \times 7H_2O$	21.6 g

Homogenization buffer:

Make 500 ml fresh for 10 rats, 1.0 liter fresh for 20 rats.

For 0.5 l:	stock solution	final conc.	volume
sucrose	2.5 M	2.0 M	400 ml
HEPES-KOH (pH 7.6)	1.0 M	10 mM	5.0 ml
KCl	2.0 M	25 mM	6.25 ml
EDTA	0.5 M	1 mM	1.0 ml
glycerol	100%	10%	50.0 ml
ddH_2O			33.25 ml
ADD BEFORE USE:			
spermine	0.5 M	0.15 M	0.15 ml
spermidine	0.2 M	0.5 mM	1.25 ml
PMSF	10 mg/ml	10 µg/ml	500 µl
benzamidine	100 mM	100 µM	500 µl
antipain	1 mg/ml	1 µg/ml	500 µl
leupeptin	5 mg/ml	1 µg/ml	100 µl
soybean trypsin inhibitor	1 mg/ml	1 µg/ml	500 µl
DTT	1.0 M	1.0 mM	500 µl
pepstatin A	1 mg/ml	1 µg/ml	500 µl

Nuclear lysis buffer:

Need 20 ml for 10 rats, 40 ml for 20 rats. Make a 1-liter stock and store at 4°C.

For 1.0 l:	stock solution	final conc.	volume
HEPES-KOH (pH 7.6)	1.0 M	10 mM	10.0 ml
KCl	2.0 M	100 mM	50.0 ml
$MgCl_2$	1.0 M	3 mM	3.0 ml
EDTA	0.5 M	0.1 mM	0.2 ml
glycerol	100%	10%	100.0 ml
dd H_2O			831.1 ml
ADD BEFORE USE:			
PMSF	1.0 M	0.5 mM	500 µl
benzamidine	100 mM	100 µM	1.0 ml
antipain	1.0 mg/ml	1 µg/ml	1.0 ml
leupeptin	5 mg/ml	1.0 µg/ml	200 µl
soybean trypsin inhibitor	1.0 mg/ml	1.0 µg/ml	1.0 ml
DTT	1.0 M	1.0 mM	1.0 ml
pepstatin A	1.0 mg/ml	1.0 µg/ml	1.0 ml

Dialysis buffer:

Make a fresh 2-liter preparation. Store at 4°C.

for 2.0 l:	stock solution	final conc.	volume
HEPES-KOH (pH 7.6)	1.0 M	25 mM	50.0 ml
KCl	2.0 M	40 mM	40.0 ml
EDTA	0.5 M	0.1 mM	0.4 ml
glycerol	100%	10%	200.0 ml
SDS	20%	0.5%	50.0 ml
ddH$_2$O			1657.4 ml
ADD BEFORE USE:			
PMSF	1.0 M	0.1 mM	0.2 ml
DTT	1.0 M	1 mM	2.0 ml

Dialysis buffer for pellet resuspension:

To 40 ml of dialysis buffer above, add:

for 40 ml:	stock solution	final conc.	volume
spermine	0.5 M	0.15 mM	12.0 µl
leupeptin	5 mg/ml	1 µg/ml	8.0 µl
spermidine	0.2 M	0.5 mM	100 µl
PMSF	10 mg/ml	10 µg/ml	40 µl
antipain	1 mg/ml	1 µg/ml	40 µl
soybean trypsin inhibitor	1 mg/ml	1 µg/ml	40 µl
DTT	1.0 M	1.0 mM	40 µl
pepstatin A	1 mg/ml	1 µg/ml	40 µl
benzamidine	100 mM	100 µM	40 µl

4 M (NH$_4$)$_2$SO$_4$:

Saturated solution; keep at 4°C.
Per liter: Add 532 g of (NH$_4$)$_2$SO$_4$
Adjust to pH 7.9 with NaOH.

Step 2: Prepare tissue and lyse cells

1. Sacrifice 20 Sprague-Dawley male rats according to accepted animal-use protocols. Prepare, in advance, three 0.5-liter beakers— two with 300 ml of ice-cold PBS, and one with 300 ml of ice-cold homogenization buffer + protease inhibitors and DTT.

2. Excise the livers and immerse them twice in PBS and once in homogenization buffer (measure the weight of the homogenization buffer beaker before and after adding the livers). Record the net weight (about 200 g) of the livers.

3. Transfer the livers to a 150-mm petri dish. Mince them into 3 x 3-mm pieces with scalpels and scissors and then transfer to a beaker.

4. Add homogenization buffer to a final volume of 500 ml and homogenize on a larger-scale Teflon-steel homogenizer. Transfer the homogenate to a glass cylinder, and add homogenization buffer to 648 ml. Mix by inversion.

5. Have 2 x 12 40-ml centrifuge tubes (use open Beckman ultraclear tubes) ready with 10 ml of homogenization buffer in each. Add 27-ml aliquots of homogenized livers on top of the homogenization buffer in the centrifuge tubes. Balance and then centrifuge using a prechilled SW28 rotor at 25,000 rpm for 60 minutes at 0°C.

6. Remove the top lipid layer with a pasteur pipet, drain the supernatant, cut the centrifuge tube in half, and wipe the tube wall with a Kimwipe.

Step 3: Prepare nuclear extracts

1. Resuspend each of the nuclear pellets in 1 ml of nuclear lysis buffer. Wash the tubes with a few milliliters of additional buffer. Transfer each nuclear pellet to a 15-ml Dounce homogenizer. Dounce 10 times (using a tight A-type pestle) on ice. Transfer to a precooled graduated glass cylinder. Measure and record the volume.

2. Measure the DNA concentration by diluting 15 μl into 285 μl of 0.5% SDS solution and read the absorbance at 260 nm. Multiply the value by 20.

3. Add 4 M $(NH_4)_2SO_4$ to 0.33 M (= 1/12 volume) by mixing gently with a pipet. Chromatin will appear after a short while. Continue to mix for 30 minutes (use a slow spinning oval stir bar, or a bowl on a rotary shaker at 4°C). The chromatin extraction is now complete. Record the appearance of the solution.

4. Transfer the chromatin to Ti45 centrifuge tubes (65 ml polycarbonate) and centrifuge at 45,000 rpm for 60 minutes to pellet the chromatin, at 4°C (capacity of a Ti45 rotor = 390 ml).

5. Collect the supernatant into a cylinder using a 25-ml pipet. Be careful not to take up any of the viscous solution. Record the volume collected.

6. Transfer the pellet to a glass beaker and slowly add 0.35 g/ml of solid $(NH_4)_2SO_4$. Stir for 15 minutes and then let sit for an additional 15 minutes on ice after it has dissolved.

7. Centrifuge in GA-14 centrifuge tubes (Beckman, 250 ml polycarbonate) at 12,000 rpm for 45 minutes.

8. Remove the supernatant and centrifuge again for 10 minutes at 12,000 rpm to tighten the pellet.

9. Remove the supernatant with a pasteur pipet. The pellet can be kept on ice overnight without losing significant activity.

Step 4: Dialyze extract and determine protein concentration

1. Resuspend the pellet (using a curved-tip pasteur pipet) in 2x 3.0 ml of dialysis buffer containing protease inhibitors (per 20 rats worth of preparation). (Keep some buffer frozen for controls.) Wash with additional 2x 1 ml dialysis buffer.

2. Transfer to a dialysis bag and dialyze against 1 liter of dialysis buffer for 4 hours, changing buffer once.

3. Transfer the nuclear protein into microfuge tubes and centrifuge for 5 minutes. Record the total volume.

4. Aliquot the supernatant and freeze it in liquid nitrogen (52 μl for transcription assays, 22 μl for EMSA assays, 1 ml for batch storage). Store at –70°C.

5. The protein concentration should be about 6–10 mg/ml. Determine the protein concentration using a Bradford Bio-Rad assay (cat. # 500-0001; dilute 5 μl into 200 μl of ddH$_2$O, take 20 μl into 1 ml of Bradford reagent). Read the absorbance at 595 nm. Record OD = μg/0.5 μl. Record the total protein. Expect 6–7 mg of protein from one adult rat liver, which should be enough for 50–100 transcription reactions (50–150 μg/reaction).

PROTOCOL 14.3

Preparation of Whole-cell Extract

Manley and colleagues developed the protocol for preparing whole-cell extracts (Manley et al. 1980). HeLa cells are incubated in a hypotonic buffer and lysed by Dounce homogenization. In the same mixture, the nuclei are lysed by addition of a lysis buffer and saturated ammonium sulfate. The chromatin is removed by high-speed centrifugation and the supernatant/extract containing the soluble proteins is decanted and precipitated with solid ammonium sulfate. The pellet is resuspended and excess salt is subsequently removed by dialysis into a buffer compatible with transcription. The extract can be aliquoted and frozen for long-term storage.

OUTLINE

Preparation of whole-cell extract (1 day)

Step 1: Prepare buffers and chill (1 hour)
Step 2: Harvest and lyse cells
Step 3: Prepare whole-cell extract (5–6 hours), dialyze extract (6 hours)

PROCEDURE

CAUTIONS: $(NH4)_2SO_4$, DTT, Glycerol, KCl, KH_2PO_4, KOH, $MgCl_2$, NaOH, PMSF. See Appendix I.

Step 1: Prepare buffers and chill

All buffers should be ice-cold before use.

PBS (make from 10x stock):

Per liter:

KCl	2.0g
KH_2PO_4	2.0g
NaCl	80.0g
$HPO_4 \times 7H_2O$	21.6g
H_2O to	1 liter

Hypotonic buffer:

10 mM Tris-HCl (pH 7.9)
1 mM EDTA
5 mM DTT
0.5 mM PMSF (add just before use)

Lysis buffer:

> 50 mM Tris-HCl (pH 7.9)
> 10 mM MgCl$_2$
> 2 mM DTT
> 25% sucrose
> 50% v/v glycerol
> 0.5 mM PMSF (add just before use)

4 M (NH$_4$)$_2$SO$_4$:

> Saturated solution; keep at 4°C
> Per liter: Add 532 g of (NH$_4$)$_2$SO$_4$
> Adjust to pH 7.9 with NaOH.

1 N NaOH solution

Resuspension buffer:

> 25 mM HEPES-KOH (pH 7.9)
> 100 mM KCl
> 12 mM MgCl$_2$
> 0.5 mM EDTA
> 2 mM DTT
> 17% v/v glycerol
> 0.5 mM PMSF (add just before use)

Step 2: Prepare whole-cell extract

1. The Manley protocol highly recommends preparing whole-cell extracts from HeLa cells, although the procedure has been done using other tissue-culture lines. Grow HeLa cells at 37°C in spinner flasks. Cells may also be obtained from commercial sources (National Cell Culture Center, Tel: 800-325-1112). Count the cells in a hemocytometer prior to harvest.

2. Harvest cells at 0.5×10^6 cells/ml by centrifugation at 3000 rpm for 10 minutes. Ideally for cultures larger than 2 liters in size, harvesting is done in 1-liter bottles using a small Sorvall RC3B centrifuge.

3. Wash the pellet twice with PBS, centrifuging between washes. Transfer the suspension into conical graduated centrifugation tubes before the last centrifugation.

4. All subsequent steps must be performed at 4°C. Determine the volume of the resultant cell pellet using the graduation on the centrifugation tube. Ideally, 1 liter of cells yields 1–2 ml packed cell volume (PCV). Resuspend pellet in four PCVs of hypotonic buffer. The cells will visibly swell in the hypotonic buffer. Incubate on ice for 20 minutes.

5. Lyse the cells by homogenization in a Dounce homogenizer using eight strokes with a B pestle. Dounce slowly and gently, especially for the down-strokes. This step lyses the cells.

6. Add four PCVs of lysis buffer. Gently mix the suspension in a beaker stirring slowly. Add one PCV of saturated (NH$_4$)$_2$SO$_4$ dropwise, with continued gentle stirring. After this addition, gently stir the highly viscous lysate for an additional 30 minutes. Stirring

must be very gentle to prevent shearing of the DNA, which would interfere with its removal in the next step. Nuclear lysis can be detected by increased viscosity after approximately half the $(NH_4)_2SO_4$ has been added. Occasionally, lysates appear clumpy and only slightly viscous, rather than extremely viscous and uniform as usually observed. This is acceptable.

7. Carefully pour the extract into polycarbonate tubes and centrifuge at 45,000 rpm in a Beckman SW 50.2 rotor for 3 hours.

8. Transfer the supernatant so as not to disrupt the pellet (the last 1 or 2 ml will be left behind). Determine the volume of the high-speed supernatant. Precipitate the protein and nucleic acid by addition of solid $(NH_4)_2SO_4$ (0.33 g/ml of solution) to the supernatant. After the $(NH_4)_2SO_4$ is dissolved, add 1 N NaOH [0.1 ml/10 g of solid $(NH_4)_2SO_4$], and stir the suspension for an additional 30 minutes. This step neutralizes the acidity of the $(NH_4)_2SO_4$.

9. Collect the precipitate by centrifugation at 10,000g for 20 minutes (completely decant the supernatant), and dissolve it in resuspension buffer to 5% of the original volume of the high-speed supernatant.

10. Dialyze the extract against two changes of 50–100 extract volumes of the resuspension buffer for a total of 8 hours. The volume of the solution increases 30–50% during dialysis. Determine the conductivity of 1/100 dilution (in H_2O) of the dialysate to ensure it is the same as the dialysis buffer. Dialysis against buffer containing lower salt concentrations results in less active lysates, again as a result of increased protein precipitation.

Step 3: Prepare extract for storage and make aliquots

Centrifuge the dialysate in a Sorvall centrifuge at 10,000g for 10 minutes to remove insoluble material. The supernatant is then divided into small aliquots (0.2–0.5 ml). The extract can be thawed and quick-frozen several times without loss of activity, and will retain full activity at –80°C for at least a year. Lysates contain between 15 and 30 mg of protein per milliliter and up to 2 mg of nucleic acid per milliliter. One liter of cells should yield about 2 ml of whole-cell extract, or enough for 100–400 assays. More concentrated extracts are desirable because with these, the same optimal protein concentration can be obtained in reaction mixtures with a smaller volume of lysate. In this manner, the salt concentration in the in vitro reaction mixture can be lowered (high salt severely inhibits transcription).

IN VITRO TRANSCRIPTION ASSAYS (PROTOCOLS 14.4 AND 14.5)

PROTOCOL 14.4

In Vitro Transcription Using HeLa Cell Extracts and Primer Extension

The first studies on recreating transcription of exogenous genes in cell-free systems were for Pol III. RNA polymerase III-transcribed genes (i.e., 5S and tRNAs) were generally small (50–150 bp), and extracts from *Xenopus* oocytes (Birkenmeier et al. 1978) or HeLa cells (Weil et al. 1979b) would synthesize full-length transcripts of these genes upon addition of nucleotides and exogenous plasmid DNA templates. Many of the methods for studying Pol II transcription derived from the early Pol III studies. Although Pol II systems were developed that accurately transcribed exogenous templates (Weil et al. 1979b; Manley et al. 1980), these systems were inefficient relative to Pol III, and for the most part still are incapable of producing a full-length mRNA transcript. Nevertheless, these systems served as an important starting point for the subsequent fractionation and characterization of the transcriptional machinery (Matsui et al. 1980; Segall et al. 1980). The initial studies employed runoff transcription of linearized plasmid templates. Mapping of the 5′ end by RNase fingerprinting was an essential aspect of early studies to show that the in vitro systems were generating RNAs with the natural in vivo 5′ ends. The earliest studies employed adenovirus and SV40 genes as DNA templates because these viruses were known to contain strong constitutive promoters, and a rudimentary viral genetics existed that could be used to help elucidate the mechanism.

Although the general approach has not changed over the years, assays such as primer extension and G-less cassette have replaced runoff transcription and fingerprinting to measure transcription and map the 5′ ends of the mRNA. The nuclear extract has become the extract of choice in many in vitro studies due to its enhanced efficiency (Dignam et al. 1983a). Natural cellular templates have begun gradually to replace the viral genes as the emphasis in the field shifts from understanding basic aspects of mechanism (i.e., general transcription factor function) to understanding regulation.

TIME LINE AND ORGANIZATION

Typically, DNA templates bearing activator binding sites are co-incubated with nuclear extracts, which provide the general transcription factors, and the corresponding activators to synthesize mRNA. In some cases, the activator is present in the extract (e.g., Sp1) and templates lacking or containing sites are compared. The mRNA is then quantitated by primer extension assays in which a short, labeled DNA oligonucleotide is first annealed to a complementary region in the mRNA, then extended by a reverse transcriptase to the 5′ end of the mRNA. The resulting labeled cDNA is then fractionated and detected on a polyacrylamide/urea gel. The amount of the product is a measure of the transcriptional activation by the activators. Alternately, a runoff transcription or G-less cassette assay can substitute. The template is usually naked DNA, but chromatin reconstitution systems are becoming more popular and many investigators are employing chromatin templates.

The initial goal of any study is generating extracts of sufficient quality from the appropriate cell types and optimizing the transcription of the template by varying multiple reaction parameters. The initial in vitro transcription reaction and the analysis of the products can be accomplished in 2 days with the extract, templates, and reagents in hand. A detailed protocol for extract preparation can be found in Protocol 14.1. The primer extension reaction on day 2 requires ^{32}P-end-labeled primer. Chapter 4 (Protocol 4.1) contains a protocol for end-labeling via polynucleotide kinase.

Day 1: In vitro transcription and primer labeling
Day 2: Reverse transcription and PAGE analysis

OUTLINE

In vitro transcription (4 hours)

Step 1: Prepare buffers (30 minutes)
Step 2: Reaction– In vitro transcription and isolation of products (2 hours 30 minutes)

Primer hybridization, extension, and gel electrophoresis (8 hours)

Step 1: Prepare buffers (30 minutes)
Step 2: Prepare 10% urea gel from premade acrylamide/urea mix (30 minutes)
Step 3: Reaction–Primer hybridization and extension/reverse transcription (4 hours)
Step 4: PAGE analysis of reaction products (4 hours)

PROCEDURE

In vitro transcription

CAUTIONS: *Chloroform, DTT, Glycerol, KCl, KOH, MgCl$_2$, Phenol, PMSF, SDS. See Appendix I.*

Step 1: Prepare buffers

Buffer D$^+$ (0.1 M KCl):
    20 mM HEPES-KOH (pH 7.9)
    20% v/v glycerol
    0.1 M KCl
    0.2 mM EDTA
    0.5 mM PMSF
    1 mM DTT (add PMSF and DTT just before use)

Stop buffer:
    0.3 M sodium acetate
    0.2% SDS
    10 mM EDTA
    50 µg/ml yeast tRNA

Step 2: Transcription reaction

1. In an eppendorf tube, mix together on ice:

HeLa cell nuclear extract	15 μl
Buffer D⁺ (0.1 M KCl)	10 μl[a]
25 m M NTPs (NOT dNTPs!)	0.8 μl
DNA template	50 ng [b]
carrier DNA (50 ng/μl)	200 ng
0.1 M MgCl$_2$	3.0 μl
activator protein (200 ng/μl)	1.0 μl [c]
H$_2$O	to 40 μl

   [a] HeLa extract and Buffer D can be varied but should total 25 μl.

   [b] The specific and nonspecific DNA should be titrated but should total 250 ng. Nonspecific DNA is ideally the plasmid vector used in cloning of the promoter being assayed (i.e., GEM3, pUC18).

   [c] If necessary, perform serial dilutions of activator to determine the optimal concentration.

2. Incubate the reaction mix at 30°C for 1 hour.

3. Add 100 μl of Stop buffer plus 1 μl of 10 mg/ml Proteinase K. Incubate at 55°C for 15 minutes.

4. Add 140 μl of phenol, vortex well, and centrifuge for 2 minutes. Repeat once with 1:1 phenol:chloroform.

5. Transfer the supernatant to a fresh tube. To avoid the organic phase, take only 130 μl of the supernatant. Add 260 μl of ethanol, vortex well, and incubate on dry ice for 10 minutes.

6. Centrifuge at 14,000 rpm in a microfuge for 15 minutes. The pellet should be visible. The pellet contains mRNA. Carefully aspirate out the supernatant with a pasteur pipet (draw out the tip of the pipet by flaming it first).

7. Store the pellet at –20°C or use the pellet for primer extension.

Primer hybridization and extension

CAUTIONS: *Acrylamide, Actinomycin D, Bromophenol blue, DTT, Ethanol, Formamide, KCl, MgCl$_2$, Radioactive substances, SDS. See Appendix I.*

Step 1: Prepare buffers

Hybridization buffer (2✕):
   600 mM NaCl
   20 mM Tris-HCl (pH 7.6)
   2 mM EDTA
   0.2% SDS (add before use)

10x Reverse transcriptase buffer:

> 0.5 M Tris-HCl (pH 8.6) (25°C)
> 0.75 M KCl
> 0.1 M MgCl$_2$

Primer extension buffer:

> 2 µl of 10x reverse transcriptase buffer (Promega #1701)
> 2 µl of 1 mg/ml actinomycin D
> 0.8 µl of 25 mM dNTPs
> 0.2 µl of DTT (1 M)
> 0.3 µl of reverse transcriptase (200 units/µl, Promega #M1701)
> 0.3 µl of RNasin
> add ddH$_2$O to total of 10 µl
> Prepare as a batch mix for multiple reactions.

Step 2: Prepare 10% urea gel from premade acrylamide/urea mix

About 2 hours before beginning the experiment, pour a 10% polyacrylamide/urea gel. The gel will take about 30 minutes to polymerize, and 30 minutes to 1 hour to pre-run prior to sample loading. See Sambrook et al. 1989 (pp. 18.49–18.54) for instructions on mixing and pouring polyacrylamide gels.

Step 3: Reaction– Primer hybridization and extension/reverse transcription

1. Resuspend the pellet obtained from the in vitro transcription reaction in 9 µl of ddH$_2$O. Add 11 µl of a mix containing 10 µl of 2x hybridization buffer and 1 µl of end-labeled primer (0.1 pmole).

2. Incubate at 37°C for 2 hours behind a radioactive shield or at an empirically determined temperature (discussed in Protocol 4.1). At this point, the primer is annealed to the mRNA. The following steps purify the mRNA:DNA hybrids for primer extension.

3. Add 200 µl of 1 M ammonium acetate and 200 µl of isopropyl alcohol.

4. Vortex and let stand at room temperature for 10 minutes. Centrifuge at 14,000 rpm in a microfuge for 15 minutes. The pellet should be visible, but it is small and loose so it is important to monitor the pellet with a Geiger counter from this step onward. Remove the supernatant with a drawn-out pasteur pipet.

5. Add 400 µl of cold 70% ethanol, wash, and briefly centrifuge. This step washes away the residual amount of ammonium acetate.

6. Aspirate the supernatant. Let the pellet air-dry briefly. Residual alcohol will inhibit the primer extension reaction.

7. Resuspend the pellet in 10 µl of 10 mM Tris-HCl (pH 8.3). On ice, add 10 µl of primer extension buffer and mix well.

8. Quickly transfer the tubes to a 45°C water bath and incubate for 1 hour.

9. Chill on ice.

Step 4: PAGE analysis of reaction products

1. Add 20 μl of formamide dye mix. Heat to 90°C for 3 minutes and immediately chill on ice.

   CAUTION: *Formamide. See Appendix I.*

2. Load 10 μl onto a 10% polyacrylamide/urea sequencing gel.

3. Run the gel until the bromophenol blue dye has migrated two-thirds of the way down the gel.

4. Transfer gel to Whatman 3MM paper. Dry gel under vacuum.

   Note: The gel can be fixed at this point, although it is not essential (see Protocol 14.4, Step 3.4).

5. Expose gel to film or phosphorimager cassette.

ADDITIONAL CONSIDERATIONS

1. The protocol can be adapted for runoff transcription by linearizing the DNA template with a restriction endonuclease and adding a radiolabeled NTP (see Protocol 14.5). Generally, the template is linearized 300 nucleotides downstream of the start site. Linearization closer is often inhibitory due to DNA-end-binding proteins contaminating the extract, and linearization farther away reduces the likelihood of observing a full-length transcript because of premature stalling by Pol II.

2. For primer extension, the primer should have a $T_{\rm m}$ of 75°C or above and should ideally be GC-rich, lack any obvious dyads (to prevent hairpin formation), and be positioned 60–90 nucleotides from the presumed start site.

3. Typically, one begins an in vitro study by titrating the amount of extract and the amount of DNA in the reaction. Less important are the buffer pH and the KCl and $MgCl_2$ concentrations, although these should fall within acceptable ranges. Note that organic salts such as potassium acetate and glutamate are popular in the yeast in vitro studies, and some investigators are now adding these and volume exclusion agents such as PVP to the mammalian extracts to enhance their efficiency.

4. It is rare to observe transcription regulated over long distances (i.e., by enhancers) in the current mammalian in vitro systems except when strong activators or chromatin templates are involved.

5. Control reactions with 2 μg of α-amanitin should always be performed at the earliest phase of the study to ensure that the observed transcription is due to Pol II.

TROUBLESHOOTING

No transcription

Possible cause: The extract may not be transcriptionally active.
Solution: Control extracts and templates can be purchased commercially (e.g., from Promega) to ensure that the methodology and technique are adequate. HeLa cells can also be purchased to ensure that the technique for extract preparation is working. Large quan-

tities of cells can be obtained from Cellex Biosciences in Minnesota (1-800-325-1112) or from the NIH-sponsored National Tissue Culture Center.

Weak signal

Possible cause: Suboptimal conditions.
Solution: Optimize by varying the DNA and extract. Check the primer extension reaction, the specific activity of the primer, etc.

Possible cause: Secondary structure in the RNA: Is a series of bands observed of lower molecular weight than expected? Sometimes this is due to snap-back synthesis or premature pausing. Also, buffers may have gone bad.
Solution: To prevent the former, keep the hybridization and extension temperatures as high as possible. Sometimes simply change the position of the primer. For the latter, note that the avian reverse transcriptase is somewhat pH sensitive. Replace the buffers and try fresh or new dNTPs. If the signal is simply weak, check whether the primer is labeled to a high specific activity. In addition, do not expect a high signal from a gene that is transcribed at a low level in vivo. Viral promoters such as the AdMLP are 100-fold more active than many natural cellular promoters.

Smeary gels

Possible cause: This may be due to inefficient removal of protein from the transcription reaction, which carries through to primer extension.
Solution: Fresh proteinase K should be added to deproteinize the reactions. A phenol extraction step performed without chloroform is necessary to extract protein efficiently from the products of the transcription reaction.

PROTOCOL 14.5

G-less Cassette In Vitro Transcription Using HeLa Cell Nuclear Extracts

INTRODUCTION

The G-less cassette assay was developed to simplify the measurements of RNA polymerase II (Pol II) transcription on circular DNA templates (Sawadogo and Roeder 1985). The G-less cassette is a 365-nucleotide segment of DNA lacking guanine (G) residues on the nontemplate strand. In principle, a full-length transcript can be generated in an in vitro reaction lacking GTP, an omission that leads to suppression of most random, nonspecific transcription throughout the plasmid. This method, like runoff transcription, generates radiolabeled RNA products, directly bypassing the necessity and extra time required to perform primer extension or other indirect mRNA product measurements. Unlike runoff transcription, which requires a cleaved end, the G-less assay can be performed on a circular, supercoiled plasmid, which in many systems is a more efficient template. In practice most crude systems, like HeLa nuclear extracts, contain low amounts of contaminating GTP which lead to small amounts of background transcription and occasionally can cause random upstream transcription to read through the G-less cassette.

To minimize these artifacts, the reaction generally contains 3′-O-Me-GTP, a chain terminating analog of GTP that causes transcription to cease when it is incorporated into a growing transcript, much like the dideoxy analogs used in DNA sequencing. The reaction products are cleaved with T1 RNase, which cleaves RNAs at G residues, further reducing background transcription; the G-less mRNA remains intact while small random RNAs are digested.

TIME LINE AND ORGANIZATION

In contrast to the previous protocol, using a G-less cassette as a template for in vitro transcription bypasses the need for a primer extension step. The nascent transcripts are directly labeled with radioactive ATP, CTP, or UTP. The reaction products are immediately analyzed via PAGE and autoradiography. The reactions can be performed in a single day and the data can be examined the following day. The original G-less plasmid generates a 365-nucleotide product. The main delay is in cloning the G-less cassette immediately downstream of the natural transcription start site. With this template in hand, the protocol takes a single day and involves incubating DNA template, nuclear extract, and [^{32}P]UTP, CTP, and ATP. The reaction is terminated, and the products are isolated and fractionated via PAGE. In general, the gel is autoradiographed overnight.

OUTLINE

G-less cassette in vitro transcription (6–7 hours)

Step 1: Prepare buffers (30 minutes)
Step 2: Prepare 6% polyacrylamide/urea gel (30 minutes)
Step 3: In vitro transcription reaction (2 hours)
Step 4: Gel electrophoresis and product analysis (3-4 hours)

PROCEDURE

CAUTIONS: *Acetic Acid (see also General Cautions for Acids and Bases), Acrylamide, Bromophenol blue, Chloroform, DTT, Ethanol, Formamide, Glycerol, Methanol, KCl, KOH, Phenol, PMSF, Radioactive substances, SDS. See Appendix I.*

Step 1: Prepare buffers

Buffer D+(0.1 M KCl):

20 mM HEPES-KOH (pH 7.9)
20% v/v glycerol
0.1 M KCl
0.2 mM EDTA
0.5 mM PMSF
1 mM DTT (add PMSF and DTT just before use)

Stop buffer:

0.3 M sodium acetate
0.2% SDS
10 mM EDTA
50 ng/µl tRNA
plus 10 µg/ml of Proteinase K (add just before use)

Step 2: Prepare 6% polyacrylamide/urea gel

About 2 hours before beginning the experiment, pour a 6% polyacrylamide/urea gel. The gel will take about 30 minutes to polymerize, and 30 minutes to 1 hour to pre-run, prior to sample loading. See Sambrook et al. 1989 (pp. 18.49–18.54) for instructions on mixing and pouring acrylamide gels.

Step 3: In vitro transcription reaction

1. In an eppendorf tube, mix together:

HeLa cell nuclear extract	15 µl
Buffer D$^+$(0.1 M KCl)	10 µl[a]
250 µM ATP, CTP	
10 µM UTP	
T1 RNase (Life Technologies[GIBCO BRL]cat. # 18030-015; dilute a 2000 units/µl stock (1/100)	0.5 µl
[α-^{32}P]UTP (4 µCi)	0.4 µl
DNA template	50 ng
carrier DNA (50 ng/µl)	200 ng
0.1 M MgCl$_2$	3.0 µl
activator protein (200 ng/µl)	1.0 µl[b]
H$_2$O	to 40 µl

[a] HeLa extract and Buffer D can be varied but should total 25 µl.

[b] Perform serial dilutions of activator if necessary to determine the optimal concentration.

2. Incubate reaction mix at 30°C for 1 hour.

3. Add 100 µl of Stop buffer plus 1 µl of 10 mg/ml Proteinase K. Incubate at 55°C for 15 minutes.

4. Transfer to room temperature and add 140 µl of phenol, vortex well, and centrifuge for 2 minutes. Repeat once with 1:1 chloroform.

5. Transfer the supernatant to a fresh tube. To avoid contamination by the organic phase, remove only 130 µl of the supernatant. Add 260 µl of ethanol, vortex well, and incubate on dry ice for 10 minutes.

6. Centrifuge at 14,000 rpm in a microfuge for 15 minutes. The pellet, which should be visible, contains mRNA. Carefully aspirate out the supernatant with a pasteur pipet (draw out the tip of the pipet by flaming it first).

Step 3: Gel electrophoresis and product analysis

1. Air-dry the pellet and add 20 µl of formamide loading buffer. Heat samples at 90°C for 2 minutes. Quickly chill on ice.

2. Equilibrate the samples to room temperature and load half of the sample on to a 6% polyacrylamide-urea gel.

3. Electrophorese until the bromophenol blue dye has migrated two-thirds of the way down the gel.

4. Carefully transfer the gel to 3MM Whatman paper (cat. #2300-916).

    Note: Some investigators cut off the lower portion of gel containing unincorporated [^{32}P]UTP. Some fix the gel while it is still attached to the plate in a bath of 10% methanol/ 5% acetic acid.

5. Dry the gel under vacuum.

6. Expose the gel to film or phosphorimager cassette.

ADDITIONAL CONSIDERATIONS

1. The G-less cassette must be cloned at the precise start site of transcription and replaces the natural downstream sequences. However, if downstream elements such as the recently discovered downstream promoter element are important for promoter activity, their replacement by the G-less cassette may decrease transcription levels (Burke and Kadonaga 1997).

2. The G-less signal is dependent on the specific activity of the ^{32}P-labeled nucleotide. Even when high-specific-activity nucleotides are used, they still must be diluted with unlabeled nucleotides up to the limiting K_m so that the polymerase can initiate and elongate transcription. For example, 500 Ci/mmole ^{32}P is already diluted approximately 20-fold with unlabeled UTP. Further dilution by addition of more cold UTP can lead to even lower specific activities. In the reaction described above, 4 μCi contains ~ 4.4×10^{-10} mmoles of UTP added to a reaction containing 4×10^{-7} mmoles of unlabeled UTP. Thus, approximately 1 of 1000 molecules of UTP is radioactive. Assuming equal amounts (~122) of adenines, cytosines, and uracils in the G-less cassette, only 1 in 8 transcripts actually contains a radiolabeled nucleotide. Note also that crude extracts are sometimes contaminated with micromolar quantities of NTPs, and this can further limit the sensitivity of the signal.

3. The size of the G-less cassette varies in some constructs, and a control template can be constructed with a smaller G-less cassette, whose product can be measured simultaneously as an internal control.

TROUBLESHOOTING

Weak signal

Possible cause: Reagents may need adjustment.
Solution: Check the specific activity of the nucleotides and adjust accordingly. Sometimes the maximal amount of ^{32}P-labeled nucleotides and the lowest amount of unlabeled carrier must be adjusted by titration to optimize the amount of signal.

No signal

Possible causes: Extracts may be inactive.
Solutions: Is the extract transcriptionally active for the control template? Does primer extension work better as a method of RNA analysis?

A signal is observed but there is no activation

Possible cause: Occasionally in crude systems a large fraction of the signal is due to readthrough transcription caused by a strong random upstream promoter and the presence of high contaminating GTP in the extract.
Solution: By leaving out RNase T1, one can distinguish the normal sized 365-nucleotide transcript from an upstream readthrough product. Unfortunately, a correctly initiated transcript may not terminate properly for the same reason. Therefore, the best solution is

to minimize contaminating GTP by dialyzing the extract further to remove contaminating GTP and adding 3′-OMe-GTP. In addition, again try a different measurement method where activation is known to occur.

TRANSCRIPTION FACTOR PURIFICATION (PROTOCOLS 14.6–14.10)

Roeder and colleagues were the first to show that cell-free systems capable of transcribing mRNA templates could be subdivided chromatographically into fractions that were inactive on their own but would biochemically complement one another for transcription when mixed together (Matsui et al. 1980). The major advantage of the fractionated in vitro transcription system is that it permits an investigator to determine the rate-limiting steps in transcription from a particular promoter and determine how activators influence transcription complex assembly.

There are four types of systems: factor-depleted systems (Nakajima et al. 1988; see Ozer et al. 1994), partially fractionated systems (see Wang et al. 1992b), homogeneous systems (see Dignam et al. 1983b; Ge et al. 1996; Maldonado et al. 1996; Tantin et al., 1996), and holoenzymes (see Cho et al. 1997; Cujec et al. 1997; Pan et al. 1997). In factor-depleted systems one or more factors are removed from the extract or simply inactivated (i.e., removed chromatographically, by immuno- or affinity-depletion, or by certain treatments, which selectively inactivate a factor). The advantage of such systems is that they are easily generated, highly active, and focus on a single factor whose depletion and subsequent addition back to the extract can be carefully controlled. The disadvantage is the crude nature of the extract and the limited ability to manipulate only a single component.

In the partially fractionated systems, the HeLa extract is subdivided into fractions containing different subsets of the general factors. In these systems, all of the fractions can be crude or, alternately, some can be crude and others replaced by purified natural or recombinant factors. This latter approach produces transcriptionally efficient and biochemically manipulable systems which still contain some crude fractions that can hamper interpretation.

The homogeneous systems contain all pure general factors (TFIIA, TFIIB, TFIID, TFIIE, TFIIF, and TFIIH) and are active for basal transcription but are largely unresponsive to activators unless supplemented with crude coactivator fractions such as the USA fraction. Even so, these systems do not fully recapitulate the high level of activation obtained with crude extracts.

Finally, recent studies suggest that the general factors are assembled into a holoenzyme that can be purified using traditional or affinity chromatography. The advantage is that such systems are reasonably pure, but the physiological functions of the holoenzyme are still unclear.

Protocol 14.6 describes the development of a system containing a mixture of partially fractionated and pure transcription factors. The protocol is designed for a lab just beginning a biochemical investigation into a gene's regulatory mechanism. Protocols for purifying certain key factors to homogeneity (Protocols 14.7, 14.8, 14.9, 14.10) are also included, however, so that these factors can be employed in advanced assays to study how activators influence transcription complex assembly, including Mg-agarose EMSA, permanganate probing, DNase I footprinting, and so on (see Chapters 13 and 15). The system first contains three fractions generated from partial fractionation of HeLa extracts: (1) TFIIA; (2) TFIID; and (3) a fraction containing TFIIE, IIF, IIH, and Pol II, called E/F/H/pol (which may even be a subform of a holoenzyme because several investigators have purified these as a complex through multiple chromatographic steps), and one recombinant factor (TFIIB). The TFIIA and TFIID fractions can be replaced by recombinant TFIIA, immunopurified TFIID, and the USA coactivator fraction. USA is normally part of the crude TFIID fraction but can be separated from it chromatographically. The efficiency of transcription in this system increases upon addition of purified Pol II. The subunits of TFIIE and TFIIF, as well as some TFIIH subunits, can be purified in recombinant form from various expression systems; we do not describe their purification here. However, protocols for purification of TFIIE, TFIIF, TFIIH, and other factors to homogeneity (including alternatives to the protocols presented here) are available in the literature (see, for example, Ge et al. 1996; Maldonado et al. 1996). Purified coactivators such as SMCC, ARC, DRIP, and human mediator can also be purified and added as necessary (see Chapter 1; also Table 14.3).

The flowchart below (Fig. 14.4) summarizes the procedure; a detailed protocol follows.

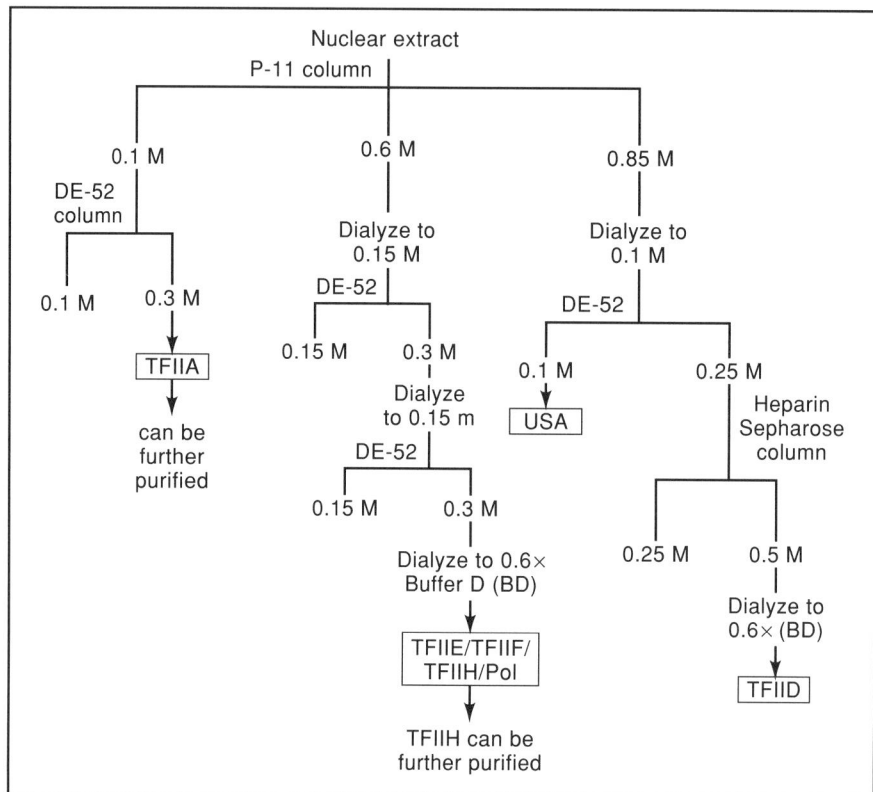

FIGURE 14.4. Flowchart of transcription factor purification.

PROTOCOL 14.6

Preparation of a Crude Fractionated System

This protocol describes preparation of a crude, partially fractionated system based on an elaboration of the protocol of Matsui et al. (1980). More details regarding current protocols can be found in Ge et al. (1996). The partially fractionated system is generally prepared from HeLa nuclear extract (see Protocol 14.1). The general transcription factors are separated into distinct fractions by chromatography over phosphocellulose, DE-52, and heparin-Sepharose columns. Although these systems are impure, they have led to important discoveries in activation of transcription and the identification of new transcriptional coactivators. The final system contains a crude TFIIA fraction, a crude TFIID fraction, and the E/F/H Pol II fraction. Addition of recombinant TFIIB to this system completes the requirements for transcription. However, ideally, pure Pol II is added to the system to increase its efficiency. TFIIA, TFIID, and other factors can be purified to homogeneity and substituted for the cruder fractions described here. The crude system and its applications were employed in Wang et al. (1992b) to study rate-limiting steps in pre-initiation complex assembly.

TIME LINE AND ORGANIZATION

Following the flowchart, it is important to plan the steps carefully. Save and label all eluates as well as the flowthrough fractions. All reactions have to be carried out at 4°C. Keep eluates on ice.

Day 1: Prepare columns, make buffers

Day 2: Initial fractionation of HeLa extract over phosphocellulose and purification of TFIIA

Days 3–4: Purify the E/F/H/Pol fraction

Days 5–6: Purify TFIID and USA fractions

OUTLINE

Step 1: Prepare buffers and all columns (1 day)

Note that several steps can and should be done concurrently.

Step 2: Initial fractionation of extract and purification of TFIIA (1 day)

Load extract and elute P-11 column with KCl step-gradient.
Purify TFIIA from P-11 flowthrough using a DE 52 column.

Step 3: Purification of a fraction containing TFIIB, Pol II, TFIIE, TFIIF, and TFIID and TFIIH (1–2 days)

Partially purify the E/F/H/Pol fraction from the P-11 (0.6 M KCl) column over a DE-52 column.

Step 4: Purification of the TFIID-USA fraction (1–2 days)

Chromatograph the P-11 0.85 M pool over DE-52.
Chromatograph the DE-52 0.25 M pool over heparin-Sepharose.

PROCEDURE

CAUTIONS: *DTT, Glycerol, KCl, KOH, PMSF. See Appendix I.*

Step 1: Prepare buffers and all columns

Prepare buffers

Buffer D⁺(0.1 M KCl):
> 20 mM HEPES-KOH (pH 7.9)
> 20% v/v glycerol
> 100 mM KCl
> 0.1 mM EDTA
> 1 mM DTT
> 1 mM PMSF
> Add DTT and PMSF just before use.

Prepare buffer D⁺ containing 0.05 M KCl, 0.1 M KCl, 0.15 M KCl, 0.2 M KCl, 0.3 M KCl, 0.5 M KCl, 0.6 M KCl, 0.85 M KCl, and 0.6× Buffer D⁺ (0.1 M KCl), respectively (i.e., prepare Buffer D [0.1 M KCl] without PMSF and DTT and make a 0.6× solution).

Prepare columns

Pour and equilibrate all columns. Prior to the final equilibration step, wash the columns with the highest salt buffer called for in the purification procedure. During this wash step, monitor pH and conductivities of column eluate. Phosphocellulose (P-11) needs to be extensively pre-equilibrated. The pH will change during the high-salt elution phases if the column is not properly equilibrated. Scopes (1994) provides general guidelines for handling column resins, as does *Current Protocols in Protein Science* (Coligan 1996).

Prepare a:
30-ml P-11 (Whatman #4071-050) column; equilibrate with Buffer D⁺ (0.1 M KCl)
10-ml DE-52 (Whatman #4057-050) column; equilibrate with Buffer D⁺ (0.1 M KCl)
10-ml DE-52 column; equilibrate with Buffer D⁺ (0.15 M KCl)
8-ml DE-52 column; equilibrate with Buffer D⁺ (0.15 M KCl)
3-ml heparin-Sepharose column (Amersham Pharmacia Biotech #17-0407-1); equilibrate with Buffer D⁺ (0.3 M KCl)

Step 2: Initial fractionation and purification of TFIIA (day 1)

Load extract and elute P-11 column with KCl step-gradient

1. Apply 25 ml (about 250 mg of protein) of HeLa cell nuclear extract (see Protocol 14.1 for HeLa nuclear extract preparation) to a 30-ml P-11 column pre-equilibrated in

Buffer D⁺ (0.1 M KCl). Ideally, the flow rate is 1 ml/minute during the load and elution of the primary protein peaks. The wash and late elution phases can be 2–3 ml per minute. Note that if there is significant precipitation of nuclear extract during thawing, the extract should be centrifuged in a Sorvall SS-34 rotor at 10,000 rpm for 10 minutes at 4°C to remove the precipitate. Otherwise the precipitated material will clog the column and reduce the flow rate to unacceptable levels.

2. Wash the column with 90 ml of Buffer D⁺ (0.1 M KCl).

3. Sequentially elute with 90-ml steps of Buffer D⁺ containing 0.3, 0.6*, and 0.85 M KCl. Collect 5-ml fractions. The 0.1 M KCl flowthrough contains TFIIA; the 0.6 M KCl fraction contains TFIIB, TFIIE, TFIIF, some TFIIH and Pol II; and the 0.85 M KCl fraction contains TFIID, the USA coactivators, and some TFIIH.

   *Note: 0.6 M is used here rather than 0.5 M because it leads to better activity of the E/F/H/Pol fraction in transcription assays.

4. A Bradford protein assay or a UV monitor should be used to collect and pool only fractions containing concentrated protein. A maximum of 30 ml for the TFIIA and TFIIE/F/H/Pol II fractions should be collected and only 15–20 ml of the TFIID fractions. This keeps the protein concentrated and helps to maintain activity. The 30-ml pool containing the TFIIA fraction can be loaded directly onto the next column.

Purify TFIIA from P-11 flowthrough using a DE-52 column

1. Apply the P-11 flowthrough pool containing TFIIA directly to a 10-ml DE-52 column pre-equilibrated in Buffer D⁺ (0.1 M KCl). The flow rate during the load is 0.5 ml per minute and 1 ml per minute during the elution and wash phases.

2. Wash the column with 30 ml of Buffer D⁺ (0.1 M KCl).

3. Elute TFIIA activity with 30 ml of Buffer D⁺ (0.3 M KCl). Collect 2-ml fractions. Use Bradford assay to determine protein concentration and pool only 5–10 ml of the fractions containing the most concentrated protein.

4. Dialyze the TFIIA eluate to 0.6x Buffer D⁺ (0.1 M KCl).

Step 3: Purification of a fraction containing Pol II, TFIIE, TFIIF, and TFIIH (days 2–3)

Partially purify the E/F/H/Pol fraction from the P-11 (0.6 M KCl) column over a DE-52 column

1. Dialyze the 0.6 M P-11 fraction for 4–6 hours against 2 liters of Buffer D⁺ (0.15 M KCl).

2. Monitor the conductivity; it should be within 10% of the conductivity of the starting dialysis buffer. Then centrifuge the dialysate at 10,000g for 10 minutes in a Sorvall centrifuge to remove the material that precipitated. Again, removal of precipitate is essential to maintain the proper flow rates.

3. Apply the supernatant to a 10-ml DE-52 column pre-equilibrated with Buffer D⁺ (0.15 M KCl). TFIIB activity flows through the column.

4. Wash the column with 30 ml of Buffer D⁺ (0.15 M KCl).

5. Pol II, TFIIE, and TFIIF (referred to as E/F/H/Pol in the text) are eluted with 30 ml of Buffer D⁺ (0.3 M KCl). The peak fractions are collected.

6. Dialyze the pooled 10 ml of eluate against Buffer D$^+$ (0.15) for 4 hours, determine the conductivity, and reapply to DE-52. Elute again with Buffer D$^+$ (0.3 M KCl). This removes the residual TFIIB and much, but not all, of Pol II.

7. Dialyze the eluate to 0.6x Buffer D$^+$ (0.1 M KCl).

Step 4: Purification of the TFIID-USA Fraction

Chromatograph the P-11 0.85 M pool over DE-52

1. Dialyze the pooled 0.85 M P-11 fraction (20 ml maximum) containing TFIID activity against 2 liters of Buffer D$^+$ (0.1 M KCl).

2. Check the conductivity; to remove any precipitated material, centrifuge the dialysate at 10,000 rpm for 10 minutes in an SS-34 (or equivalent) rotor in a Sorvall centrifuge.

3. Apply the eluate to an 8-ml DE-52 column pre-equilibrated with Buffer D$^+$ (0.1 M KCl). The flow rate should be 1 ml/minute during the load.

4. Wash the column at 1 ml/minute with 24 ml of pre-equilibration buffer.

5. Elute TFIID with Buffer D$^+$ (0.25 M KCl). The column flow rate should be 0.5 ml/minute.

6. Collect 1.5-ml fractions. Use the Bradford assay to pool the peak 6–7 ml of eluate.

Chromatograph the DE-52 0.25 M pool over heparin-Sepharose

1. Apply the DE-52 eluate directly at a flow rate of 0.25 ml/minute to a 3-ml heparin-Sepharose column pre-equilibrated in Buffer D$^+$ (0.1 M KCl).

2. Wash the column with 9 ml of Buffer D$^+$ (0.3 M KCl) and elute TFIID activity with Buffer D$^+$ (0.5 M KCl). Use the Bradford assay to monitor protein concentration, and pool 2–3 ml of the fractions containing the most concentrated protein.

ADDITIONAL CONSIDERATIONS

1. The relevance of purity. It is our experience that the more fractionated a system becomes, the less responsive it becomes to activators. Thus, a system composed of only basal factors including TFIID does not respond well to activators, whereas a crude extract or the crude fractionated system described here does. Although we favor purifying proteins to homogeneity to study their biochemical activities, it is really a judgment call dependent on the particular problem. The key issue is whether significant mechanistic information can be gained by using cruder systems that are highly responsive to activators versus the pure systems that, even when supplemented with coactivator fractions like USA, are less active. When using pure systems to study certain steps (e.g., DA complex assembly, recruitment of TFIIB), however, it is incumbent upon the investigator to show that those steps affected by activators in isolated biochemical reactions are limiting steps in the context of an intact activator-responsive transcription system. Thus, transcription complex assembly experiments must always be accompanied by, for example, kinetics of transcription experiments.

2. Activity. Nuclear extract rapidly loses activity even at 4°C. Thus, incubation of a nuclear extract for 12 hours at 4°C leads to a 50% reduction in activity, probably due to proteolysis and protein oxidation. Maintaining all buffers and fractions in the cold, speed in purification, and constant addition of protease inhibitors and DTT have significant stabilizing effects on the resulting fractionated system.

3. Elution conditions. The elution conditions for the fractionation have been optimized to generate the highest activity fractions with the lowest protein concentration. The elution conditions can be adjusted empirically to maximize activity. However, lowering the salt concentrations used for loading and eluting the columns can lead to cross-contamination. For example, lowering the salt concentration for loading of the E/F/H/Pol fraction onto DE-52 can lead to contamination by TFIIB and excess Pol II, which we hope to add separately as purified proteins.

4. Reconstituting transcription. After adjusting all of the fractions to 0.6x Buffer D, the different fractions can be mixed together in various ratios within 40-μl transcription reactions with 200 ng of DNA template, 200 μM NTPs, and 7 mM $MgCl_2$ to optimize activity. Typically, the fractions are mixed together to optimize activity, and then reactions lacking individual fractions are performed to ensure that the resulting fractionated system is dependent on each factor or fraction.

TROUBLESHOOTING

No activity when mixing resulting fractions

Possible cause: The relative ratios of the factors may not be correct.
Solution: Try altering the relative ratios of factors. Refer to Wang et al. (1992b).

Possible cause: The extract or factors may not be active.
Solution: Test the starting extract to determine if it was active. Consider also that one of the factors became inactivated during purification. Refer to additional considerations and attempt the procedure again. It is very important to save fractions from all the early steps and to show that they are all active in complementation assays. Then, if the TFIID fraction from heparin-Sepharose, for example, is inactive when assayed, but the DE-52 TFIID fraction is active, this allows the investigator to narrow the source of the problem. Similar approaches can be taken for other factors.

Cross-contamination of fractions

Possible cause: This is generally caused by using an incorrect salt concentration during elution: The concentration used for elution may vary depending on the pH, source of resin, or for other spurious reasons. This is apparent when addition of an incomplete set of factors supports transcription.
Solution: If the amount of contamination is low, this problem is often easily remedied. It may be necessary to optimize the elution conditions or re-run a particular column. As described above, it is important from the beginning to isolate enzymatically pure fractions. The complementation capabilities should be measured as early as the P-11 column to ensure that the procedure is working.

Weak transcriptional activity

Possible cause: This has many causes. The fractions may be too dilute.
Solution: Re-running them through the last column in the purification, one-half to one-third the original size, and repeating the elution might help. Alternatively, concentrate the fractions using approaches such as the Centricon concentrators.

PROTOCOL 14.7

Purification of Recombinant TFIIB from *E. coli*

TFIIB is the only purified recombinant factor necessary for developing the partially fractionated system described above. Recombinant TFIIB is relatively easy to purify and fully substitutes for the crude TFIIB fraction generated in the flowthrough fraction of the DE-52 phase of the E/F/H/Pol II fractionation. TFIIB was originally cloned and overexpressed in *E. coli* by Reinberg and colleagues (Ha et al. 1991). The overexpression employs an IPTG-inducible T7 expression system. The expression vector is freshly transformed into the appropriate T7 host strain. The colonies are grown overnight and inoculated into a large-scale broth which is grown to an A_{600} of 0.6 and induced for several hours. The cell pellet can be stored in the –80°C freezer overnight prior to extract preparation.

TIME LINE AND ORGANIZATION

The purification of recombinant TFIIB requires several days from harvest of the induced cells. The phosphocellulose column and buffers should be prepared in advance, because it often takes several hours to properly equilibrate the resin. The polyethylenimine (PEI) must also be prepared in advance and adjusted to the correct pH. The idea is to prepare an extract from induced cells and perform preliminary fractionation steps including PEI and ammonium sulfate cuts to remove nucleic acid and much contaminating *E. coli* protein. The ammonium sulfate pellet is then resuspended and fractionated over phosphocellulose. The procedure can and should be stopped after the ammonium sulfate precipitation, before column chromatography. The pellet can be stored in the –80°C freezer while the investigator monitors the early stages of the fractionation by SDS-PAGE.

OUTLINE

Purification of rTFIIB (3 days)

Step 1: Prepare column and buffers; overexpress TFIIB in *E. coli* (day 1)
Step 2: Prepare homogenate (day 2)
Step 3: Purify rTFIIB over P-11 column (day 3)

PROCEDURE

CAUTIONS: *DTT, Glycerol, IPTG, KCl, KOH, Leupeptin, β-Mercaptoethanol, $(NH_4)_2SO_4$, Pepstatin A, PMSF. See Appendix I.*

Step 1: Prepare column and buffers

Buffer D+ (0.1 M KCl):
> 20 mM HEPES-KOH (pH 7.9)
> 20% v/v glycerol
> 100 mM KCl
> 0.1 mM EDTA
> 1 mM DTT
> 1 mM PMSF
> The protease inhibitors and DTT are added immediately prior to use.

Buffer A:

> 20 mM HEPES-KOH (pH 7.9 at 25°C)
>
> 200 mM NaCl

Buffer C:

> 20 mM HEPES-KOH (pH 7.9)
>
> 1 mM EDTA
>
> 0.5 mM PMSF
>
> 10 μg/ml leupeptin
>
> 10 μg/ml pepstatin A
>
> 20 mM β-mercaptoethanol

Also prepare Buffer C with 0.2 M and 0.6 M NaCl. The protease inhibitors and β-mercaptoethanol are added immediately prior to use.

Prepare a 3-ml P-11 column, equilibrate with Buffer C (0.2 M KCl).

Step 2: Prepare homogenate

TFIIB can be overexpressed in and purified from *E. coli* using an expression vector provided by Danny Reinberg's laboratory (Ha et al. 1991). All manipulations should be performed at 4°C. All buffers should be chilled to 4°C.

Note: Throughout Step 2, take small samples and analyze induction and initial steps on a mini SDS-PAGE. To monitor induction of protein, resuspend pellets from a 1-ml culture in 100 μl of dH$_2$O and 100 μl of 2× SDS-sample dye. Sonicate for 20 seconds. Heat the samples at 95°C for 2 minutes and load 20 μl onto a 10% mini-SDS gel. TFIIB is a 33-kD protein and should be easily visible. An antibody can be purchased from Santa Cruz Biotechnology Inc. (polyclonal: sc-225) or Berkeley Antibody Co. (BAbCO) (monoclonal: MMS-273R-500) if the identification of TFIIB is uncertain.

1. Grow 6 liters of *E. coli* [BL21(DE3)] harboring the TFIIB T7-based expression vector to an A$_{600}$ of 0.6. Remove a 1-ml sample for gel analysis (before induction).

2. Induce TFIIB expression by addition of IPTG to 1 mM.

3. After 3 hours, remove a 1-ml sample for gel analysis (after induction), then harvest the remaining cells by centrifugation for 10 minutes at 5000 rpm in a Sorvall GS-3 rotor. Wash the pellets in a total of 1 liter of Buffer A (0.2 M NaCl). Recentrifuge and resuspend in 75 ml of Buffer A (0.2 M NaCl) containing 20 mM β-mercaptoethanol and 1 mM PMSF. Resuspended cells can be frozen at –80°C overnight.

4. Lyse the cells by sonication.

5. Remove the insoluble debris by centrifugation at 10,000 rpm for 10 minutes and 4°C in a SS-34 rotor. Transfer the supernatant to a 200-ml glass beaker.

6. While stirring gently on ice or in a 4°C cold room, add 10% PEI (pH 7.9) to the supernatant, dropwise, to a final concentration of 0.1%. This step precipitates the nucleic acid and some acidic proteins.

7. After 30 minutes, remove the fine white precipitate containing nucleic acid by centrifugation at 10,000 rpm in a Sorvall SS-34 rotor for 10 minutes.

8. Transfer the supernatant to a clean 200-ml beaker on ice and, while stirring gently, slowly add 0.2 g of solid (NH$_4$)$_2$SO$_4$ per milliliter of supernatant. The supernatant at this stage contains 60% of the original TFIIB as some precipitates with PEI.

9. After 30 minutes at 0°C, remove the precipitate by centrifugation for 10 minutes at 10,000 rpm. Transfer the supernatant to a clean beaker and, while stirring, add an additional 0.05 g of $(NH_4)_2SO_4$ per milliliter of supernatant.

10. After 30 minutes, collect the precipitate containing about 40% of the starting TFIIB by centrifugation as above. These $(NH_4)_2SO_4$ cuts are very effective in removing contaminating protein. Monitor TFIIB fractionation by rapid minigel to ensure that TFIIB expression was induced and that it is indeed present in the second $(NH_4)_2SO_4$ pellet. TFIIB will occasionally precipitate at the lower $(NH_4)_2SO_4$ concentration.

11. Resuspend the second $(NH_4)_2SO_4$ pellet by Dounce homogenization in Buffer C to a conductivity equal to Buffer C (0.2 M NaCl). If the salt concentration is too low, add additional NaCl from a 3 M stock solution. If the salt concentration is too high, dialyze briefly against Buffer C (0.2 M NaCl).

12. Remove insoluble debris by centrifugation at 10,000 rpm in a Sorvall for 10 minutes at 4°C.

Step 3: Purify rTFIIB over P-11

1. Apply the supernatant now containing about 30% of the original TFIIB at a flow rate of 0.25 ml per minute to a 3-ml P-11 column pre-equilibrated in Buffer C (0.2 M NaCl).

2. Wash the column with 9 ml of Buffer C (0.2 M NaCl).

3. Elute TFIIB with 9 ml of Buffer C (0.6 M NaCl). Monitor the peak fractions by Bradford assay and by a mini-SDS gel.

4. Often the resulting TFIIB is so concentrated (>1 mg/ml) that it must be diluted more than 10-fold prior to use in the transcription assay. Therefore, it is not essential to dialyze it against 0.6x Buffer D^+(0.1 M KCl) like the other factors because it can be diluted into that buffer during the assay. The pooled factor preparation should, however, be subdivided into 10-μl and 50-μl aliquots, each used only a few times before disposal, and stored at –80°C.

ADDITIONAL CONSIDERATIONS

1. Standard precautions. Recombinant TFIIB is highly active in reconstituted transcription systems. The procedure is not difficult to perform but involves all of the standard precautions employed in protein purification, including use of protease inhibitors, speed of purification, minimizing extract frothing and keeping the temperature at or below 4°C for all manipulations.

2. Monitoring purification. All steps must be carefully monitored by SDS-PAGE minigels. The gels should ideally be electrophoresed and stained immediately to monitor purification during preparation of the extract. Do not proceed until it is clear that the preliminary steps have worked. The use of minigels and the rapid pace of the purifications should allow at least two gels to be run in a single day between steps of the purification.

3. Ammonium sulfate cuts. Regarding the preliminary steps, the concentrations of ammonium sulfate and PEI required to achieve purification are dependent on the concentration of the extract. Differences in the length of induction, the OD at which

the cells were induced, the amount of lysis by sonication or otherwise might lead to protein extracts varying in protein concentration. For this reason, the SDS-PAGE gel must be run before column chromatography to ensure that TFIIB is present in the last ammonium sulfate cut. Higher protein concentrations favor precipitation of TFIIB at lower concentrations of ammonium sulfate.

TROUBLESHOOTING

TFIIB passes through P-11 column

Possible cause: Salt concentration is too high or column pH is incorrect.
Solution: Check salt concentration and lower it next time. Check pH of column. Higher pH values favor weaker binding to phosphocellulose.

No induction during extract preparation

Possible cause: Transformed *E. coli* or IPTG has gone bad.
Solution: Re-transform plasmid into appropriate cells and induce fresh. Ensure that correct strain is being used by using another inducible vector as a positive control. Prepare fresh IPTG.

No protein in final AS pellet

Possible cause: TFIIB was not induced.
Solution: Check induction, and make sure protein is soluble. Then check earlier fractions.

Protein precipitates during PEI precipitation

Possible cause: PEI was added too quickly, causing clumping and coprecipitation of TFIIB.
Solution: Very slowly add to the lowest concentration that causes formation of a fine white precipitate. If TFIIB is still precipitating at this point, raise the salt concentration of the buffer and repeat.

PROTOCOL 14.8

Purification of Recombinant TFIIA

TFIIA has many roles in transcription. It antagonizes the action of inhibitors that block TBP's binding to either DNA or to other general factors. It increases affinity of TBP for DNA. Finally, it acts as a coactivator, facilitating the effects of activators on the transcriptional machinery. Human TFIIA isolated from HeLa nuclear extracts exists as a heterotrimer composed of α (37 kD), β (19 kD), and γ (13 kD). The α and β subunits are encoded by a single gene, whose polypeptide is cleaved proteolytically in vivo. The genes encoding α–β and γ are remarkably well-conserved from yeast to human. *E. coli* expression vectors for the recombinant subunits can be obtained from several academic laboratories (DeJong and Roeder 1993; Ma et al. 1993; Ozer et al. 1994; Sun et al. 1994). There are different methods for reconstituting the α-β and γ subunits into TFIIA that is functional for in vitro transcription. Processing of α-β is not necessary in the in vitro systems. Note that although we are presenting a protocol for recombinant TFIIA, authentic HeLa TFIIA is not difficult to purify because the large subunit encodes a fortuitous internal stretch of histidines, allowing the protein to adhere to an NTA-Sepharose column. Despite this natural modification, the recombinant subunits have been His$_6$-tagged to further facilitate purification of the recombinant protein. Both groups studying human TFIIA purified the recombinant protein using QIAGEN NTA-Sepharose columns as directed by the manufacturer (Ozer et al. 1994; Sun et al. 1994). Because our protocol is exactly as described in these publications, we do not present it in detail but summarize the important features.

TIME LINE AND ORGANIZATION

The entire purification of both TFIIA subunits can take 1 day. An overnight culture of the *E. coli* is induced for 3 hours and harvested. The two *E. coli* pellets, one containing the α-β subunit and the other containing the γ subunit, are resuspended in binding buffer as per QIAGEN instructions and loaded directly onto NTA-Sepharose minicolumns. After eluting with a pH gradient, there is an SDS-PAGE step to identify peak fractions, followed by a 12-hour dialysis to renature the subunits.

OUTLINE

Purification of TFIIA subunits (1 day)

Step 1: Culture *E. coli* (3 hours)
Step 2: Harvest and resuspend *E. coli* pellets
Step 3: Load on NTA-Sepharose minicolumns
Step 4: Dialysis (12 hours)

PROCEDURE

CAUTIONS: *DTT, HCl (see also General Cautions for Acids and Bases), IPTG, NaH₂PO₄, PMSF. See Appendix I.*

Step 1: Grow and induce *E. coli*

Grow 2 liters of *E. coli* harboring either the α-β or γ subunit expression vectors to an A_{600} of 0.7 and induce with 0.5 mM IPTG.

Step 2: Harvest and resuspend *E. coli* pellet

Harvest the *E. coli* and resuspend in QIAGEN lysis/binding buffer containing 6 M guanidine HCl, 0.1 M NaH₂PO₄, 0.1 M Tris-HCl (pH 8.0). In the initial steps, denature and purify the subunits separately. In the final step, the subunits will be mixed and renatured by dialysis.

Step 3: Load on NTA-Sepharose minicolumns

1. Load the *E. coli* lysates onto small 1.5-ml Ni-NTA-Sepharose columns at room temperature. Elute in stepwise fashion with a buffer containing 1 mM DTT, 1 mM PMSF, 8 M urea, 0.1 M NaH₂PO₄, 0.01 M Tris, adjusted to pH 4.5, 5.3, and 6.3 using 0.1 N HCl.
2. Collect 0.75-ml fractions. Monitor elution of the α-β and γ subunits by SDS-PAGE.

Step 4: Dialysis

1. The subunits generally elute at pH 4.5 and pH 5.3.
2. Mix pooled peak fractions of the two subunits in equimolar amounts and dialyze in a stepwise fashion against 1 liter of Buffer D (0.1) containing 2 M urea, Buffer D (0.1) containing 0.5 M urea, and finally Buffer D (0.1). All buffers should contain PMSF and DTT. Each dialysis step should take 4 hours at 4°C.
3. After dialysis, centrifuge out the insoluble TFIIA at 10,000 rpm in a Sorvall SS-34 rotor at 4°C. Store the supernatant in aliquots at –80°C.

   Note: Sometimes it is necessary to concentrate the dialysate by Centricon.

PROTOCOL 14.9

Affinity Purification of RNA Pol II

The transcriptional activity of the fractionated system described above is stimulated considerably by additional Pol II (see Wang et al. 1992b). Pol II was originally discovered and purified by Roeder and Rutter (1969). The initial methodologies employed extensive column chromatography. The three nuclear RNA polymerases could be separated on DEAE-Sephadex A-25 columns into three peaks (Pol I, II, and III). The activities were assayed using a nonspecific template incorporation assay, where different nonspecific templates [calf thymus DNA or poly(dA:dT)] were incubated with radiolabeled NTPs and the resulting small synthetic RNAs were precipitated from free unincorporated NTPs using TCA precipitation. Pol I, II, and III could also be distinguished by sensitivity to α-amanitin (Lindell et al. 1970) and further purified via multiple chromatographic steps over a period of 1 or more weeks. However, the development of monoclonal antibodies has greatly facilitated purification, and it can now be accomplished in a matter of days. The following protocol describes a general immunoaffinity method developed by R. Burgess for purifying Pol II to homogeneity (Thompson and Burgess 1996). We have combined the protocol with elements of a protocol developed by Reinberg and colleagues for purification of Pol II from the pellets remaining after the HeLa nuclear extract preparation. Generally, these protocols are performed with substantial amounts of materials to optimize recovery; however, the protocol works when scaled down.

TIME LINE AND ORGANIZATION

The entire procedure of affinity purification of Pol II can be performed in 3 days starting with HeLa nuclei. The most convenient place to stop is after the ammonium sulfate precipitation step (Step 2.6). Simply decant the supernatant away from the pellet and freeze until the next day. The supernatant after the DEAE cellulose column (Step 3.3) can also be frozen and stored at −80°C overnight while checking for the presence of Pol II by immunoblotting. Buffers (without PMSF and DTT) and columns, particularly the immunoaffinity resin, should be prepared in advance. Always save sample aliquots throughout the preparation that can be later assayed for activity or by immunoblotting to monitor the efficiency of different steps.

OUTLINE

Purification of RNA polymerase II (3 days)

Step 1: Prepare buffers and DEAE cellulose column (day 1)
Step 2: Homogenize nuclei and prepare extract (day 1)
Step 3: Purify Pol II via DEAE cellulose (day 2)
Step 4: Immunopurify Pol II (day 3)

PROCEDURE

CAUTIONS: *DEAE, DTT, Glycerol, KCl, KOH, MgCl$_2$, (NH$_4$)$_2$SO$_4$, PMSF. See Appendix I.*

Step 1: Prepare buffers and DEAE cellulose column

Buffer D$^+$ (0.1 M KCl):
> 20 mM HEPES-KOH (pH 7.9)
> 20% v/v glycerol
> 100 mM KCl
> 0.1 mM EDTA
> Add protease inhibitors and DTT just before use:
> 1 mM DTT
> 1 mM PMSF

Buffer B$^+$:
> 200 ml of:
> 50 mM Tris-HCl (pH 7.9)
> 25% v/v glycerol
> 5 mM MgCl$_2$
> 2 mM DTT
> 0.5 mM PMSF (add just before use)

Binding buffer:
> 10 mM Tris-HCl (pH 7.6)
> 1 mM EDTA
> 200 mM (NH$_4$)$_2$SO$_4$

Ammonium sulfate solutions:
> 3.8 M (NH$_4$)$_2$SO$_4$

Elution buffer:
> Binding buffer + 40% PEG 8000 or ethylene glycol

Prepare a 60-ml DEAE cellulose column (DE-52 from Whatman # 4057-050). Wash the column once with Buffer D$^+$, containing 0.5 M (NH$_4$)$_2$SO$_4$, in place of KCl and equilibrate the column with Buffer D$^+$ containing 145 mM (NH$_4$)$_2$SO$_4$ in place of KCl.

Step 2: Homogenize nuclei and prepare extract

Keep all reagents, and perform all procedures, on ice or in a 4°C cold room.

1. Prepare nuclear pellets from HeLa cells according to Protocol 14.1, Steps 1–3.4. Thaw three nuclear pellets (from an equivalent of 100 liters of HeLa cells) on ice and mix together. Add 36 ml of Buffer B per 10^{10} cells (from original extract) to yield about 170 ml.

2. Using an A-type pestle, Dounce-homogenize several strokes to resuspend the thawed nuclei. The pellet will swell in volume and become viscous.

3. Transfer the suspension to a plastic beaker for sonication. Sonicate at 5 × 20-sec intervals, using a Heat Systems sonicator (or equivalent) at power setting 9. This step shears

the chromatin and reduces viscosity. Sonicate on ice or in the cold room. Cool down to 4°C between intervals.

4. While stirring, slowly add 3.8 M $(NH_4)_2SO_4$ to a final concentration of 0.3 M. The solution will become viscous again.

5. Centrifuge at 40,000 rpm for 1 hour at 4°C in a Ti-45 rotor. Record the volume of the supernatant (about 193 ml).

6. Transfer the supernatant to a 500-ml glass beaker while avoiding the loose viscous pellet and gradually add 0.42 g/ml solid $(NH_4)_2SO_4$ over 30 minutes, stirring slowly on ice. After this addition, allow to stir for 30 additional minutes. Centrifuge at 40,000 rpm in a Beckman Ti-45 rotor for 1 hour at 4°C. Try to remove all of the supernatant and wipe the sides of the tube dry with a Kimwipe to remove as much salt solution as possible. The Pol II is in the pellet.

7. Resuspend the pellet in 20 ml of Buffer D$^+$. Check the conductivity against a 145 mM solution of $(NH_4)_2SO_4$. Add a sufficient volume of Buffer D$^+$ to the pellet to adjust the conductivity of the resuspended pellet to the conductivity equal to Buffer D$^+$ containing 145 mM $(NH_4)_2SO_4$. You will probably need to dilute the sample about 10-fold. An alternative at this stage is to resuspend in 100 ml of Buffer D and dialyze against Buffer D$^+$ containing 145 mM $(NH_4)_2SO_4$ for 4–6 hours.

8. Centrifuge the resuspended pellet or dialysate at 40,000 rpm for 1 hour in the Beckman ultracentrifuge.

Step 3: Purify Pol II via DEAE cellulose

1. Load the supernatant onto the DE-52 column. Wash with 180 ml of Buffer D$^+$ containing 145 mM $(NH_4)_2SO_4$ and develop the column with 180 ml of Buffer D containing 0.5 M $(NH_4)_2SO_4$. Collect 10-ml fractions. Use the Bradford assay to identify peak protein fractions. Pool approximately 30–45 ml of the most concentrated fractions. The polymerase activity can be detected using a nonspecific template assay (see Roeder and Rutter 1969) or using the 8WG16 monoclonal antibody (Thompson and Burgess 1996).

2. Dialyze the peak fraction for 6 hours into 2 × 5 liters of binding buffer. A precipitate will form during the dialysis.

3. Centrifuge the dialysate at 8000 rpm for 15 minutes at 4°C in a SS-34 rotor to remove the precipitate. Pol II will be in the supernatant.

Step 4: Immunopurify Pol II

1. Prepare the immunoaffinity resin by covalently cross-linking 4 ml (or 4 mg) of pure 8WG16 (available from QED Bioscience and BAbCO) anti-Pol II antibody or ascites fluid (Thompson and Burgess 1996) to 2 ml of protein-A–Sepharose beads. See Protocol 14.10 for preparation of immunoaffinity columns.

2. Rinse beads in 20 ml of binding buffer, plus 1 mM DTT and 0.5 mM PMSF. Rotate for 20 minutes at room temperature. Rinse twice more in binding buffer (20 ml the first time, 10 ml the second time).

3. Add half of the DEAE-peak (15–30 ml) to the immunoaffinity resin (freeze the other half in 50-ml conical, capped disposable plastic test tubes). Rock, end-over-end, for 3.5 hours at 4°C to allow binding of Pol II to the antibody.

4. Centrifuge the beads down gently in a clinical microfuge at 2000 rpm for 3 minutes.

5. Remove the supernatant.

6. Wash the beads three times in binding buffer (20 ml for first rinse; 10 ml each for second and third rinses).

7. Add 5 ml of elution buffer. Incubate 20 minutes at room temperature.

8. Centrifuge the beads down (IEC or equivalent clinical centrifuge, 2000 rpm, 3 minutes), and transfer supernatant to a 15-ml conical tube. Centrifuge the supernatant again to remove residual beads that did not pellet the first time. Transfer the supernatant to a new conical tube. The supernatant contains the purified Pol II preparation.

9. Repeat the elution step five times and keep the eluates separate.

10. The Pol II preparation is dialyzed against 0.6x Buffer D$^+$ (0.1 M KCl) and divided into small 100-µl aliquots and stored at –80°C.

ADDITIONAL CONSIDERATIONS

1. DE-52 Pol II. The elution must be performed as described. One of the essential elements is keeping the DE-52 Pol II fractions as concentrated as possible. This also facilitates binding to the immunoaffinity resin. When scaling down the preparation, often it is wise to overload the DE-52 column mildly so as to generate an eluate with highly concentrated Pol II.

2. Elutions. Occasionally, subsequent elutions tend to yield distinct forms of Pol II in various CTD phosphorylation states. We have observed that the first elution produces a Pol II fraction in which the CTD phosphorylation state is the highest; subsequent elutions yield Pol II with increasingly lower CTD phosphorylation states. The amount of Pol II and composition can be quantitated by immunoblotting and by analysis of the fractions on silver stained 6–8% polyacrylamide SDS gels (see Sambrook et al. 1989, pp. 18.49–18.57).

TROUBLESHOOTING

The Pol II passed through the DE-52 column

Possible cause: Lower pH and higher salt tend to weaken binding of Pol II to DE-52.
Solution: Check salt and pH of extract.

No binding to antibody beads

Possible cause: The fact that the 8WG16 antibody is functional in an immunoblot does not necessarily mean it is present at high enough concentration to be useful in immunopurification.

Solution: Check antibody in an ELISA assay against a peptide or GST-CTD fusions. Often it is advisable to compare the titer of your antibody with an antibody obtained from an investigator who has used it successfully in immunopurification.

Pol II is not functional

Possible cause: Perhaps the protein has been inactivated.

Solution: Try again, and use greater care in purification, avoiding warm temperatures and frothing. Use fresh DTT and protease inhibitors to maintain stability.

PROTOCOL 14.10

Purification of Epitope-tagged TFIID

TFIID is one of the most critical factors in transcription complex assembly because it recognizes a core promoter and interacts with chromatin and activator proteins. Roeder and colleagues developed methods for traditional purification of TFIID from HeLa extracts (Nakajima et al. 1988). Although such methods resulted in relatively pure TFIID that could be employed in DNase I footprinting and in vitro transcription assays, the methods required large quantities of starting nuclear extract, and there were considerable activity losses during chromatography. The cloning of human TBP permitted Berk and colleagues to employ an approach called epitope tagging to modify TFIID so it could be purified to homogeneity using immunoaffinity chromatography in a simple two-step procedure with limited loss of activity. Berk fused a short peptide containing the influenza virus hemagglutinin (HA) tag onto the amino terminus of TBP and used a retroviral transfer system to generate a HeLa cell line stably expressing HA-tagged TBP (Zhou et al. 1992). Extracts from the cell line contained TFIID, which stably incorporated the epitope-tagged TBP. TFIID could be partially purified from these extracts using phosphocellulose chromatography and then immunopurified using a resin containing protein A-Sepharose beads crosslinked to the 12CA5 or HA.11 (Berkeley Antibody Company [BAbCO], #MMS-101P) monoclonal antibody against the influenza epitope. The TFIID could be eluted from the washed immunoaffinity resin in pure form using an HA peptide. The resulting TFIID contained a complete complement of TAFs and could be employed in both transcription, EMSA, and footprinting assays, and its purity is well suited for many other studies. Roeder and colleagues developed a similar procedure for adding a FLAG tag onto TBP (Chiang et al. 1993). Note that, unlike the cruder preparations of TFIID, the affinity-purified TFIID did not contain the USA coactivators. Thus, although the pure TFIID can be employed in the fractionated system described in this section, the USA coactivator fraction must be added to achieve high levels of activated transcription; TFIID alone generates only basal levels.

TIME LINE AND ORGANIZATION

Day 1: Prepare buffers, P-11, and immunoaffinity resin
Day 2: Fractionate nuclear extract over P-11
Day 3: Immunoblot P-11 fractions and bind peak fractions to immunoaffinity resin overnight
Day 4: Elute TFIID from immunoaffinity resin

OUTLINE

Step 1: Prepare nuclear extract from cell line expressing HA-TBP (see Protocol 14.1)

Step 2: Prepare P-11 columns, buffers, and immunoaffinity resin (1 day)

Prepare buffers (1 hour)
Prepare P-11 column (2 hours)
Prepare HA-antibody resin (4–5 hours)

The entire procedure requires 4 days after a nuclear extract containing HA-TBP has been prepared. Cell growth and extract preparation are described in Zhou et al. (1992). Cells expressing HA-TBP can be obtained from the Arnold Berk laboratory (UCLA).

Step 3: Purify epitope-tagged TFIID (3 days)

Fractionate TFIID from nuclear extract
Bind TFIID to antibody affinity resin and elute TFIID

Step 4: Analyze purified TFIID via SDS-PAGE and immunoblotting

PROCEDURE

CAUTIONS: *DMP, DTT, Ethanolamine, Glycerol, KCl, KOH, PMSF. See Appendix I.*

Step 1: Prepare nuclear extract

See Protocol 14.1 for nuclear extract preparation. Prepare from HA-tagged cell line (Zhou et al. 1992).

Step 2: Prepare P-11 column, buffers, and immunoaffinity resin

Prepare buffers

Buffer D+:

20 mM HEPES-KOH (pH 7.9)
20% v/v glycerol
0.1 mM EDTA
Add protease inhibitor and DTT just before use
1 mM DTT
1 mM PMSF
Prepare Buffer D with no salt, 0.1 M KCl, 0.2 M KCl, 0.25 M KCl, 0.5 M KCl, and 1 M KCl; add protease inhibitors if indicated (Buffer D+).

0.2 M Sodium borate (pH 9.0)

0.2 M Ethanolamine (pH 7.9)

Prepare P-11 column

1. Prepare a 30-ml phosphocellulose column (P-11, Whatman #4071-050) in Buffer D. A 30-ml column is used for 50 ml (~500 mg) of nuclear extract prepared from 32 liters of HeLa cells. After equilibration according to the manufacturer's instructions, wash the column with 120 ml of Buffer D (1 M KCl) and pre-equilibrate with 60 ml of Buffer D+ (0.5 M KCl).

2. Monitor the conductivity of Buffer D. Monitor the conductivity of HeLa nuclear extract sample. Add 3 M KCl to adjust the conductivity of the extract to equal the conductivity of Buffer D$^+$ (0.5 M KCl).

Prepare HA-antibody resin

1. Couple HA.11 or 12CA5 antibody to protein-A–Sepharose beads. This protocol uses 0.5 g of beads and yields approximately 1.0 ml, and can be scaled accordingly. Measure 0.5 g of protein-A–Sepharose (Amersham Pharmacia Biotech #17-0780-01) and resuspend in 5.0 ml of Buffer D containing 0.25 M KCl + 1 mM PMSF (elution buffer). Rock at room temperature for 3 minutes.

2. Centrifuge at 1500g (maximum speed in an IEC clinical centrifuge) for 5 minutes. Wash 2x with 20 ml of elution buffer.

3. Add 1–2 ml of HA.11 or 12CA5 antibody (1 mg/ml). BAbCO markets the HA.11 antibody peptide, although the hybridoma line expressing the 12CA5 antibody against the HA epitope is available from academic sources. Rock for 1 hour at room temperature.

4. Centrifuge at 1500g for 5 minutes.

5. Wash the beads 2x with 10 volumes (~ 20 ml) of 0.2 M sodium borate (pH 9.0).

   Save a 10 μl bead (100 μl suspension) aliquot for gel. Centrifuge at 3000 rpm for 2 minutes in a microfuge. Remove the supernatant.

6. Resuspend the beads in 10 volumes of 0.2 M sodium borate.

7. Add DMP, (Sigma #D8388) up to 20 mM (in 20 ml, 103.6 mg DMP). Rock for 30 minutes at room temperature.

8. Wash with 10 volumes of 0.2 M ethanolamine (pH 7.9).

9. Incubate for 2 hours at room temperature with 10 volumes of 0.2 M ethanolamine (pH 7.9).

10. Wash the beads 2x with 15 ml of Buffer D (0.25 M KCl) + PMSF.

11. Resuspend in 1 ml of Buffer D (0.25 M KCl) + 0.5 mM PMSF. Save a 20 μl aliquot for a gel.

12. Boil, load, run, and stain SDS-PAGE (12%) to verify cross-linking.

    Load 2 μl of beads before cross-linking; load 20 μl of beads after cross-linking. Monitor large and small antibody chains (see Harlow and Lane, 1999).

Step 3: Purify epitope-tagged TFIID

Fractionate TFIID from nuclear extract

1. Load the extract prepared from the special HeLa strain bearing the epitope-tagged TFIID onto the 30-ml phosphocellulose (P-11) column.

2. Wash the column with 90 ml of Buffer D$^+$ (0.5 M KCl).

3. Elute slowly (at about 1 ml/minute) with two column volumes of Buffer D (1 M KCl). Collect 3-ml fractions with a fraction collector.

4. Determine the protein content of the fractions with a Bradford protein assay. Pool 15 ml of the peak fractions. It is essential to pool only the most concentrated samples.

5. Dialyze the eluate against Buffer D (0.25 M KCl) for 2 hours.

Bind TFIID to antibody affinity resin and elute TFIID

1. For each milliliter of dialysate, add 100 μl of protein-A-Sepharose beads coupled to antibody.

   Note: Use siliconized 1.5-ml eppendorf tubes to avoid protein loss. Bind the protein to the antibody-coupled beads 4 hours to overnight on a rocker at 4°C.

2. Centrifuge the beads down for about 15 seconds at full speed (14,000 rpm) in a microfuge.

3. Wash the beads three times with Buffer D⁺ (0.25 M KCl). To wash, pipet 1 ml of buffer onto the protein/bead slurry, vortex the tube very gently, and incubate on ice for 5 minutes. Centrifuge the beads down for 15 seconds in a microfuge, and then remove the supernatant.

4. Dilute the hemagglutinin (HA) peptide to 10 mg/ml in Buffer D (0.1 M KCl) for long-term storage. Then, dilute the peptide to 1.5 mg/ml in Buffer D (0.25 M KCl) just before use.

5. Mix peptide solution with beads in a 1:1 ratio (i.e., 125 μl of beads elute with 125 μl of HA peptide mix in Buffer D) containing 0.25 M KCl, 1 mM PMSF, 1.5 mg/ml HA peptide, 0.05% NP-40, and 0.2 mg/ml BSA.

6. Incubate with rocking at room temperature (25°C) for 20 minutes and at 30°C for 20 minutes in a water bath. Wipe off the outside of the tube with ethanol. Punch a very small hole with a small-gauge needle into the bottom of the tube. The goal here is to allow the liquid to flow through without the beads, so just barely break the surface of the tube with the needle.

7. Place the 1.5-ml eppendorf tube containing the beads into another 1.5-ml tube and tape or Parafilm the two together. Place both in a 50-ml falcon tube at 4°C and centrifuge in a clinical centrifuge for 4 minutes or until beads are dry. To remove any residual beads that broke through, recentrifuge the eppendorf tube bearing the TFIID and then carefully transfer the supernatant to a fresh siliconized tube, avoiding any beads at the bottom of the eppendorf tube.

8. Aliquot the eluted TFIID into 20-μl samples.

Step 4: Analyze purifed TFIID via SDS-PAGE and immunoblotting

1. Titrate the eluted TFIID and fractionate on a 10% SDS gel against a titration of purified marker protein.

2. Analyze the protein via Western blot. Use antibody HA.11 (1:500 diluted) against the HA tag as a primary antibody in protein detection.

ADDITIONAL CONSIDERATIONS

1. Binding to immunoaffinity column. It is important to monitor the extract before and after binding of the P-11 eluate to the immunoaffinity resin. At least one-half to two-thirds of the material should bind. This is apparent when comparing starting and depleted extracts. Similarly, when eluting from the immunoaffinity resin, only one-half to two-thirds of the TFIID is expected to elute under the conditions shown. Higher salt concentrations (i.e., 0.5–1 M) favor more quantitative elution but are not

immediately compatible with DNA-binding experiments.

2. TFIID concentration. Typically, 200 ng or so of immunopure TFIID is employed in DNase I footprinting or in vitro transcription experiments. This is equivalent to 1–3 μl of affinity column eluate. A typical yield from 300 mg of nuclear extract is a maximum of 20 μg and usually less. If the TFIID is eluting at more dilute concentrations, it may be impossible to obtain clear DNase I footprints although the protein is still usable for in vitro transcription and EMSA. Often the molarity of the TFIID can be determined by preparing recombinant TBP from *E. coli*. Known molar mounts of TBP can be compared to TFIID by immunoblotting with a TBP antibody (Santa Cruz Biotechnology, Inc.). Often when we are attempting to preserve TFIID we employ this approach rather than micro-Bradfords or silver-stained gels for estimating concentration.

3. Composition. Occasionally a silver-stained SDS gel is employed to confirm the subunit composition of TFIID. Additionally, an immunoblot can be performed with antibodies against TBP and various TAFs available from Santa Cruz Biotechnology, Inc.

4. Contamination with Protein A beads. It is imperative to remove all of the protein-A beads from the TFIID preparation. Contamination with beads can cause problems in the footprinting and EMSA reactions employing the TFIID. Typically the beads can be seen by flicking the tube. In such a case, centrifuge the pure TFIID in a microfuge and carefully transfer the supernatant, but avoid the bead pellet, to another tube.

TROUBLESHOOTING

The TFIID is eluting at concentrations too dilute for biochemical use

Possible cause: Low concentration starting material or low titer antibody are the two most common causes where the procedure fails.
Solution: Make sure that the P-11 TFIID peak material is as concentrated as possible. Try binding only the most concentrated fractions from P-11 column.
Solution: Test titer of the HA.11 or 12CA5 antibody and compare the result to either a commercial sample or a sample provided by another investigator.

TFIID does not elute from resin

Possible cause: This is a common problem and usually involves a peptide that has gone bad, or is present at too low concentrations, or use of low salt concentrations during elution.
Solution: Try new peptide and slightly higher (by increments of 100 mM) salt concentrations.

REFERENCES

Armstrong J.A., Bieker J.J., and Emerson B.M. 1998. A SWI/SNF-related chromatin remodeling complex, E-RC1, is required for tissue-specific transcriptional regulation by EKLF in vitro. *Cell* **95:** 93–104.

Aso T., Vasavada H.A., Kawaguchi T., Germino F.J., Ganguly S., Kitajima S., Weissman S.M., and Yasukochi Y. 1992. Characterization of cDNA for the large subunit of the transcription initiation factor TFIIF. *Nature* **355:** 461.

Barton M.C. and Emerson B.M. 1996. Regulated gene expression in reconstituted chromatin and synthetic nuclei. *Methods Enzymol.* **274:** 299–312.

Becker P.B., Tsukiyama T., and Wu C. 1994. Chromatin assembly extracts from *Drosophila* embryos. *Methods Cell. Biol.* **44:** 207–223.

Bernstein R., DeJong J., and Roeder R.G. 1994. Characterization of the highly conserved TFIIA small subunit from *Drosophila melanogaster. J. Biol. Chem.* **269:** 24361.

Birkenmeier E.H., Brown D.D., and Jordan E. 1978. A nuclear extract of *Xenopus laevis* oocytes that accurately transcribes 5SRNA genes. *Cell* **15:** 1077–1086.

Bulger M., Ito T., Kamakaka R.T., and Kadonaga J.T. 1995. Assembly of regularly spaced nucleosome arrays by *Drosophila* chromatin assembly factor 1 and a 56-kDA histone-binding protein. *Proc. Natl. Acad. Sci.* **92:** 11726–11730.

Burke T.W. and Kadonaga J.T. 1997. The downstream core promoter element, DPE, is conserved from *Drosophila* to humans and is recognized by TAFII60 of *Drosophila. Genes Dev.* **11:** 3020–3031.

Carey M., Leatherwood J., and Ptashne M. 1990a. A potent GAL4 derivative activates transcription at a distance in vitro. *Science* **247:** 710.

Carey M., Lin Y.S., Green M.R., and Ptashne M. 1990b. A mechanism for synergistic activation of a mammalian gene by GAL4 derivatives. *Nature* **345:** 361.

Chalut C., Lang C., and Egly J.M. 1994. Expression in *Escherichia coli*: Production and purification of both subunits of the human general transcription factor TFIIE. *Protein Expr. Purif.* **5:** 458–467.

Chao D.M., Gadbois E.L., Murray P.J., Anderson S.F., Sonu M.S., Parvin J.D., and Young RA. 1996. A mammalian SRB protein associated with an RNA polymerase II holoenzyme. *Nature* **380:** 82–85.

Chiang C.M., Ge H., Wang Z., Hoffmann A., and Roeder R.G. 1993. Unique TATA-binding protein-containing complexes and cofactors involved in transcription by RNA polymerases II and III. *EMBO J.* **12:** 2749–2762.

Cho H., Maldonado E., and Reinberg D. 1997. Affinity purification of a human RNA polymerase II complex using monoclonal antibodies against transcription factor IIF. *J. Biol. Chem.* **272:** 11495–11502.

Cho H., Orphanides G., Sun X., Yang X.J., Ogryzko V., Lees E., Nakatani Y., and Reinberg D. 1998. A human RNA polymerase II complex containing factors that modify chromatin structure. *Mol. Cell. Biol.* **18:** 5355–5363.

Coligan J.E., Dunn B.M., Ploegh H.L, Speicher D.W., and Wingfield P.T., eds. 1996. *Current protocols in protein science* (series ed. Benson Chanda V.). John Wiley and Sons, New York.

Conaway R.C., Reines D., Garrett K.P., Powell W., and Conaway J.W. 1996. Purification of RNA polymerase II general transcription factors from rat liver. *Methods Enzymol.* **276:** 194–207.

Cujec T.P., Cho H., Maldonado E., Meyer J., Reinberg D., and Peterlin B.M. 1997. The human immunodeficiency virus transactivator Tat interacts with the RNA polymerase II holoenzyme. *Mol. Cell. Biol.* **17:** 1817–1823.

DeJong J. and Roeder R.G. 1993. A single cDNA, hTFIIA/α, encodes both the p35 and p19 subunits of human TFIIA. *Genes Dev.* **7:** 2220–2234.

DeJong J., Bernstein R., and Roeder R.G. 1995. Human general transcription factor TFIIA: Characterization of a cDNA encoding the small subunit and requirement for basal and activated transcription. *Proc. Natl. Acad. Sci.* **92:** 3313.

Dignam J.D., Lebovitz R.M., and Roeder R.G. 1983a. Accurate transcription initiation by RNA polymerase II in a soluble extract from isolated mammalian nuclei. *Nucleic Acids Res.* **11:** 1475–1489.

Dignam J.D., Martin P.L., Shastry B.S., and Roeder R.G. 1983b. Eukaryotic gene transcription with purified components. *Methods Enzymol.* **101:** 582.

Drapkin R. and Reinberg D. 1994. The multifunctional TFIIH complex and transcriptional control. *Trends Biochem. Sci.* **19:** 504.

Dynan W.S. and Tjian R. 1983. Isolation of transcription factors that discriminate between different promoters recognized by RNA polymerase II. *Cell* **32:** 669–680.

Finkelstein A., Kostrub C.F., Li J., Chavez D.P., Wang B.Q., Fang S.M., Greenblatt J., and Burton Z.F. 1992. A cDNA encoding RAP74, a general initiation factor for transcription by RNA polymerase II. *Nature* **355:** 464.

Foulds C.E. and Hawley D.K. 1997. Analysis of the human TATA binding protein promoter and identification of an ets site critical for activity. *Nucleic Acids Res.* **25:** 2485–2494.

Ge H. and Roeder R.G. 1994. Purification, cloning, and characterization of a human coactivator, PC4, that mediates transcriptional activation of class II genes. *Cell* **78:** 513–523.

Ge H., Martinez E., Chiang C.M., and Roeder R.G. 1996. Activator-dependent transcription by mammalian RNA polymerase II: In vitro reconstitution with general transcription factors and cofactors. *Methods Enzymol.* **274:** 57–71.

Gold M.O., Tassan J.P., Nigg E.A., Rice A.P., and Herrmann C.H. 1996. Viral transactivators E1A and VP16 interact with a large complex that is associated with CTD kinase activity and contains CDK8. *Nucleic Acids Res.* **24:** 3771–3777.

Gorski K., Carneiro M., and Schibler U. 1986. Tissue-specific in vitro transcription from the mouse albumin promoter. *Cell* **47:** 767–776.

Gu W., Malik S., Ito M., Yuan C.X., Fondell J.D., Zhang X., Martinez E., Qin J., and Roeder R.G. 1999. A novel human SRB/MED-containing cofactor complex, SMCC, involved in transcription regulation. *Mol. Cell* **3:** 97–108.

Ha I., Lane W.S., and Reinberg D. 1991. Cloning of a human gene encoding the general transcription initiation factor IIB. *Nature* **352:** 689–695.

Hahn S., Buratowski S., Sharp P.A., and Guarente L. 1989. Isolation of the gene encoding the yeast TATA binding protein TFIID: A gene identical to the SPT15 suppressor of Ty element insertions. *Cell* **58:** 1173–1181.

Harlow E. and Lane D. 1999. *Using antibodies: A laboratory manual.* Cold Spring Harbor Laboratory Press, Cold Spring Harbor, NY.

Inostroza J., Flores O., and Reinberg D. 1991. Factors involved in specific transcription by mammalian RNA polymerase II. Purification and functional analysis of general transcription factor IIE. *J. Biol. Chem.* **266:** 9304.

Ito T., Bulger M., Pazin M.J., Kobayashi R., and Kadonaga J.T. 1997. ACF, an ISWI-containing and ATP-utilizing chromatin assembly and remodeling factor. *Cell* **90:** 145–155.

Kim T.K. and Maniatis T. 1997. The mechanism of transcriptional synergy of an in vitro assembled interferon-β enhanceosome. *Mol. Cell* **1:** 119–129.

Kingston R.E., Bunker C.A., and Imbalzano A.N. 1996. Repression and activation by multiprotein complexes that alter chromatin structure. *Genes Dev.* **10:** 905–920.

Koleske A.J. and Young R.A. 1994. An RNA polymerase II holoenzyme responsive to activators. *Nature* **368:** 466-469.

Kraus W.L. and Kadonaga J.T. 1998. P300 and estrogen receptor cooperatively activate transcription via differential enhancement of initiation and reinitiation. *Genes Dev.* **12:** 331–342.

Lee K.A., Bindereif A., and Green M.R. 1988. A small-scale procedure for preparation of nuclear extracts that support efficient transcription and pre-mRNA splicing. *Gene Anal. Tech.* **5:** 22–31.

LeRoy G., Orphanides G., Lane W.S., and Reinberg D. 1998. Requirement of RSF and FACT for transcription of chromatin templates in vitro. *Science* **282:** 1900–1904.

Lieberman P. 1994. Identification of functional targets of the Zta transcriptional activator by formation of stable preinitiation complex intermediates. *Mol. Cell. Biol.* **14:** 8365.

Lieberman P.M. and Berk A.J. 1994. A mechanism for TAFs in transcriptional activation: Activation domain enhancement of TFIID-TFIIA–promoter DNA complex formation. *Genes Dev.* **8:** 995.

Lin Y.S. and Green M.R. 1991. Mechanism of action of an acidic transcriptional activator in vitro. *Cell* **64:** 971–981.

Lindell T.J., Weinberg F., Morris P.W., Roeder R.G., and Rutter W.J. 1970. Specific inhibition of nuclear RNA polymerase II by α-amanitin. *Science* **170:** 447–449.

Ma D., Watanabe H., Mermelstein F., Admon A., Oguri K., Sun X., Wada T., Imai T., Shiroya T., Reinberg D., and et al. 1993. Isolation of a cDNA encoding the largest subunit of TFIIA reveals functions important for activated transcription. *Genes Dev.* **7:** 2246–2257.

Maldonado E., Drapkin R., and Reinberg D. 1996. Purification of human RNA polymerase II and general transcription factors. *Methods Enzymol.* **274:** 72–100.

Manley J.L., Fire A., Samuels M., and Sharp P.A. 1983. In vitro transcription: Whole cell extract. *Methods Enzymol.* **101:** 568–582.

Manley J.L., Fire A., Cano A., Sharp P.A., and Gefter M.L. 1980. DNA-dependent transcription of adenovirus genes in a soluble whole-cell extract. *Proc. Natl. Acad. Sci.* **77:** 3855–3859.

Matsui T., Segall J., Weil P.A., and Roeder R.G. 1980. Multiple factors required for accurate initiation of transcription by purified RNA polymerase II. *J. Biol. Chem.* **255:** 11992–11996.

Mayall T.P., Sheridan P.L., Montminy M.R., and Jones K.A. 1997. Distinct roles for P-CREB and LEF-1 in TCR α enhancer assembly and activation on chromatin templates in vitro. *Genes Dev.* **11:** 887–899.

Mizuguchi G., Tsukiyama T., Wisniewski J., and Wu C. 1997. Role of nucleosome remodeling factor NURF in transcriptional activation of chromatin. *Mol. Cell* **1:** 141–150.

Myer V.E. and Young R.A. 1998. RNA polymerase II holoenzymes and subcomplexes. *J. Biol. Chem.* **273:** 27757–27760.

Nakajima N., Horikoshi M., and Roeder R.G. 1988. Factors involved in specific transcription by mammalian RNA polymerase II: Purification, genetic specificity, and TATA box-promoter interactions of TFIID. *Mol. Cell. Biol.* **8:** 4028–4040.

Neish A.S., Anderson S.F., Schlegel B.P., Wei W., and Parvin J.D. 1998. Factors associated with the mammalian RNA polymerase II holoenzyme. *Nucleic Acids Res.* **26:** 847–853.

Nightingale K.P., Wellinger R.E., Sogo J.M., and Becker P.B. 1998. Histone acetylation facilitates RNA polymerase II transcription of the *Drosophila* hsp26 gene in chromatin. *EMBO J.* **17:** 2865–2876.

Nikolov D.B., Chen H., Halay E.D., Usheva A.A., Hisatake K., Lee D.K., Roeder R.G., and Burley S.K. 1995. Crystal structure of a TFIIB-TBP-TATA-element ternary complex. *Nature* **377:** 119.

Ohkuma Y., Sumimoto H., Horikoshi M., and Roeder R.G. 1990. Factors involved in specific transcription by mammalian RNA polymerase II: Purification and characterization of general transcription factor TFIIE. *Proc. Natl. Acad. Sci.* **87:** 9163.

Orphanides G., Lagrange T., and Reinberg D. 1996. The general transcription factors of RNA polymerase II. *Genes Dev.* **10:** 2657.

Osborn L., Kunkel S., and Nabel G.J. 1989. Tumor necrosis factor α and interleukin 1 stimulate the human immunodeficiency virus enhancer by activation of the nuclear factor γ B. *Proc. Natl. Acad. Sci.* **86:** 2336–2340.

Ossipow V., Tassan J.P., Nigg E.A., and Schibler U. 1995. A mammalian RNA polymerase II holoenzyme containing all components required for promoter-specific transcription initiation. *Cell* **83:** 137–146.

Ozer J., Moore P.A., Bolden A.H., Lee A., Rosen C.A., and Lieberman P.M. 1994. Molecular cloning of the small (gamma) subunit of human TFIIA reveals functions critical for activated transcription. *Genes Dev.* **8:** 2324–2335.

Pan G., Aso T., and Greenblatt J. 1997. Interaction of elongation factors TFIIS and elongin A with a human RNA polymerase II holoenzyme capable of promoter-specific initiation and responsive to transcriptional activators. *J. Biol. Chem.* **272:** 24563–24571.

Parvin J.D. and Young R.A. 1998. Regulatory targets in the RNA polymerase II holoenzyme. *Curr. Opin. Genet. Dev.* **8:** 565–570.

Pazin M.J., Kamakaka R.T., and Kadonaga J.T. 1994. ATP-dependent nucleosome reconfiguration and transcriptional activation from preassembled chromatin templates. *Science* **266**: 2007–2011.

Pazin M.J., Sheridan P.L., Cannon K., Cao Z., Keck J.G., Kadonaga J.T., and Jones K.A. 1996. NF-κ B-mediated chromatin reconfiguration and transcriptional activation of the HIV-1 enhancer in vitro. *Genes Dev.* **10**: 37–49.

Peterson M.G., Inostroza J., Maxon M.E., Flores O., Admon A., Reinberg D., and Tjian R. 1991. Structure and functional properties of human general transcription factor IIE. *Nature* **354**: 369.

Phelan M.L., Sif S., Narlikar G.J., and Kingston R.E. 1999. Reconstitution of a core chromatin remodeling complex from SWI/SNF subunits. *Mol. Cell* **3**: 247–253.

Ranish, J. A., Lane, W. S., and Hahn, S. 1992. Isolation of two genes that encode subunits of the yeast transcription factor IIA. *Science* **255**: 1127.

Reinberg D. and Roeder R.G. 1987. Factors involved in specific transcription by mammalian RNA polymerase II. Purification and functional analysis of initiation factors IIB and IIE. *J. Biol. Chem.* **262**: 3310.

Robinson K.M. and Kadonaga J.T. The use of chromatin templates to recreate transcriptional regulatory phenomena in vitro. *Biochemica et Biophysica Acta*, 1998 Aug. 19. 1378 (1):M1-6.

Roeder R.G. 1996. Nuclear RNA polymerases: Role of general initiation factors and cofactors in eukaryotic transcription. *Methods Enzymol.* **273**: 165–171.

Roeder R.G. and Rutter W.J. 1969. Multiple forms of DNA-dependent RNA polymerase in eukaryotic organisms. *Nature* **224**: 234–237.

Sambrook J., Fritsch E.F., and Maniatis T. 1989. *Molecular cloning: A laboratory manual.* Cold Spring Harbor Laboratory, Cold Spring Harbor, New York.

Sawadogo M. and Roeder R.G. 1985. Factors involved in specific transcription by human RNA polymerase II: Analysis by a rapid and quantitative in vitro assay. *Proc. Natl. Acad. Sci.* **82**: 4394–4398.

Schiltz R.L., Mizzen C.A., Vassilev A., Cook R.G., Allis C.D., and Nakatani Y. 1999. Overlapping but distinct patterns of histone acetylation by the human coactivators p300 and PCAF within nucleosomal substrates. *J. Biol. Chem.* **274**: 1189–1192.

Schnitzler G., Sif S., and Kingston R.E. 1998. Human SWI/SNF interconverts a nucleosome between its base state and a stable remodeled state. *Cell* **94**: 17–27.

Scopes R.K. 1994. *Protein purification: Principles and practice*, 3rd ed. Springer-Verlag, New York.

Segall J., Matsui T., and Roeder R.G. 1980. Multiple factors are required for the accurate transcription of purified genes by RNA polymerase III. *J. Biol. Chem.* **255**: 11986–41991.

Seroz T., Hwang J. R., Moncollin V., and Egly J.M. 1995. TFIIH: A link between transcription, DNA repair and cell cycle regulation. *Curr. Opin. Genet. Dev.* **5**: 217.

Sheridan P.L., Mayall T.P., Verdin E., and Jones K.A. 1997. Histone acetyltransferases regulate HIV-1 enhancer activity in vitro. *Genes Dev.* **11**: 3327–3340.

Sheridan P.L., Sheline C.T., Cannon K., Voz M.L., Pazin M.J., Kadonaga J.T., and Jones K.A. 1995. Activation of the HIV–1 enhancer by the LEF–1 HMG protein on nucleosome–assembled DNA in vitro. *Genes Dev.* **9**: 2090–2104.

Sopta M., Burton Z.F., and Greenblatt J. 1989. Structure and associated DNA–helicase activity of a general transcription initiation factor that binds to RNA polymerase II. *Nature* **341**: 410.

Steger D.J., Owen–Hughes T., John S., and Workman J.L. 1997. Analysis of transcription factor-mediated remodeling of nucleosomal arrays in a purified system. *Methods* **12**: 276–285.

Steger D.J., Eberharter A., John S., Grant P.A., and Workman J.L. 1998. Purified histone acetyltransferase complexes stimulate HIV-1 transcription from preassembled nucleosomal arrays. *Proc. Na. Acad. Sci.* **95**: 12924–12929.

Studitsky V.M., Clark D.J., and Felsenfeld G. 1996. Preparation of nucleosomal templates for transcription in vitro. *Methods Enzymol.* **274**: 246–256.

Stuempfle K.J., Koptides M., Karinch A.M., and Floros J. 1996. Preparation of transcriptionally active nuclear extracts from mammalian tissues. *Biotechniques* **21**: 48–50, 52.

Sun X.Q., Ma D., Sheldon M., Yeung K., and Reinberg D. 1994. Reconstitution of human TFIIA activity from recombinant polypeptides: A role in TFIID mediated transcription. *Genes Dev.* **8**: 2336–2348.

Sun X.Q., Zhang, Y., Cho H., Rickert P., Lees E., Lane W., and Remberg D. 1998. NAT, a human complex containing Srb polypeptides that functions as a negative regulator of activated transcription. *Mol. Cell* **2:** 213–222.

Svejstrup J.Q., Vichi P., and Egly J.M. 1996. The multiple roles of transcription/repair factor TFIIH. *Trends Biochem. Sci.* **21:** 346.

Tantin D., Chi T., Hori R., Pyo S., and Carey M. 1996. Biochemical mechanism of transcriptional activation by GAL4–VP16. *Methods Enzymol.* **274:** 133–149.

Thompson N.E. and Burgess R.R. 1996. Immunoaffinity purification of RNA polymerase II and transcription factors using polyol–responsive monoclonal antibodies. *Methods Enzymol.* **274:** 513–526.

Thompson N.E., Aronson D.B., and Burgess R.R. 1990. Purification of eukaryotic RNA Polymerase II by immunoaffinity chromatography. *J Biol. Chem.* **265:** 7069.

Tirode F., Busso D., Coin F., and Egly J.M. 1999. Reconstitution of the transcription factor TFIIH: Assignment of functions for the three enzymatic subunits, XPB, XPD, and cdk7. *Mol. Cell* **3:** 87–95.

Ura K. and Wolffe A.P. 1996. Reconstruction of transcriptional active and silent chromatin. *Methods Enzymol.* **274:** 257–271.

Utley R.T., Cote J. Owen-Hughes T., and Workman J.L. 1997. SWI/SNF stimulates the formation of disparate activator-nucleosome complexes but is partially redundant with cooperative binding. *J. Biol. Chem.* **272:** 12642–12649.

Utley R.T., Owen-Hughes T.A., Juan L.J., Cote J., Adams C.C., and Workman J.L. 1996. In vitro analysis of transcription factor binding to nucleosomes and nucleosome disruption/displacement. *Methods Enzymol.* **274:** 276–291.

Utley R.T., Ikeda K., Grant P.A., Cote J., Steger D.J., Eberharter A., John S., and Workman J.L. 1998. Transcriptional activators direct histone acetyltransferase complexes to nucleosomes. *Nature* **394:** 498–502.

Vettese-Dadey M., Grant P.A., Hebbes T.R., Crane-Robinson C., Allis C.D., and Workman J.L. 1996. Acetylation of histone H4 plays a primary role in enhancing transcription factor binding to nucleosomal DNA in vitro. *EMBO J.* **15:** 2508–2518.

Wang B.Q., Lei L., and Burton Z.F. 1994. Importance of codon preference for production of human RAP74 and reconstitution of the RAP30/74 complex. *Protein Expr. Purif.* **5:** 476–485.

Wang B.Q., Kostrub C.F., Finkelstein A., and Burton Z.F. 1993. Production of human RAP30 and RAP74 in bacterial cells. *Protein Expr. Purif.* **4:** 207–214.

Wang W., Carey M., and Gralla J.D. 1992a. Polymerase II promoter activation: Closed complex formation and ATP-driven start site opening. *Science* **255:** 450.

Wang W., Gralla J.D., and Carey M. 1992b. The acidic activator GAL4-AH can stimulate polymerase II transcription by promoting assembly of a closed complex requiring TFIID and TFIIA. *Genes Dev.* **6:** 1716–1727.

Wang W., Xue Y., Zhou S., Kuo A., Cairns B.R., and Crabtree G.R. 1996. Diversity and specialization of mammalian SWI/SNF complexes. *Genes Dev.* **10:** 2117–2130.

Weil P.A., Luse D.S., Segall J., and Roeder R.G. 1979a. Selective and accurate initiation of transcription at the Ad2 major late promoter in a soluble system dependent on purified RNA polymerase II and DNA. *Cell* **18:** 469–484.

Weil P.A., Segall J., Harris B., Ng S.Y., and Roeder R.G. 1979b. Faithful transcription of eukaryotic genes by RNA polymerase III in systems reconstituted with purified DNA templates. *J. Biol. Chem.* **254:** 6163–6173.

Wilson C.J., Chao D.M., Imbalzano A.N., Schnitzler G.R., Kingston R.E., and Young R.A. 1996. RNA polymerase II holoenzyme contains SWI/SNF regulators involved in chromatin remodeling. *Cell* **84:** 235–244.

Winkler G.S., Vermeulen W., Coin F., Egly J.M., Hoeijmakers J.H., and Weeda G. 1998. Affinity purification of human DNA repair/transcription factor TFIIH using epitope-tagged xeroderma pigmentosum B protein. *J. Biol. Chem.* **273:** 1092–1098.

Yang X.J., Ogryzko V.V., Nishikawa J., Howard B.H., and Nakatani Y. 1996. A p300/CBP-associated factor that competes with the adenoviral oncoprotein E1A. *Nature* **382:** 319–324.

Yankulov K. and Bentley D. 1998. Transcriptional control: Tat cofactors and transcriptional elongation. *Curr. Biol.* **8:** R447–449.

Yokomori K., Admon A., Goodrich J.A., Chen J.L., and Tjian R. 1993. *Drosophila* TFIIA-L is processed into two subunits that are associated with the TBP/TAF complex. *Genes Dev.* **7:** 2235.

Yokomori K., Zeidler M.P., Chen J.L., Verrijzer C.P., Mlodzik M., and Tjian R. 1994. *Drosophila* TFIIA directs cooperative DNA binding with TBP and mediates transcriptional activation. *Genes Dev.* **8:** 2313.

Zhou Q., Lieberman P.M., Boyer T.G., and Berk A.J. 1992. Holo-TFIID supports transcriptional stimulation by diverse activators and from a TATA-less promoter. *Genes Dev.* **6:** 1964–1974.

CHAPTER 15

Approaches for Studying Transcription Complex Assembly

Important issues

- *Assays for measuring intact transcription complexes can be used to determine how an activator stimulates transcription.*

- *There are many assays for studying individual steps in gene activation, particularly the early steps (e.g., TFIID binding to DNA).*

- *A reconstituted transcription system should be set up to understand the mechanism of activation at a biochemical level.*

INTRODUCTION

One of the most challenging problems in the field of mammalian gene regulation is understanding the biochemical mechanism by which an activator or enhanceosome activates transcription. The current view is that activators stimulate transcription largely by promoting assembly of a large multicomponent transcription complex at the core promoter (see Chapter 1; for review, see Ptashne and Gann 1997). This complex comprises RNA polymerase II (Pol II), the general factors, coactivators, and possibly chromatin-remodeling components. By analogy with its prokaryotic counterpart, the complex can be termed the preinitiation complex or PIC. ATP hydrolysis isomerizes the complex from the so-called "closed complex," in which the start site remains intact, to an "open complex," in which the DNA encompassing the start site becomes melted. Finally, Pol II initiates transcription by forming the first phosphodiester bond in the mRNA, escapes from the promoter, and then elongates through the gene (for review, see Orphanides et al. 1996). In many cases, the polymerase reinitiates transcription so that multiple mRNAs can be produced. Each step can be subdivided even further and, on the basis of studies in prokaryotic systems, each step is theoretically subject to regulation.

For these reasons, biochemical studies have focused on many different steps in the transcription process. In the past, most of the emphasis was placed on the early stages in PIC assembly because these are important for establishing transcription of the gene (for review, see Hori and Carey 1994; see Pugh 1996). However, more recent studies point to effects of some activators on late steps, including the binding or activity of TFIIH, an effect that may influence subsequent elongation by Pol II (for review, see Reines et al. 1997). Other studies have supported the idea that a major effect of activators is to stimulate reinitiation (Sandaltzopoulos and Becker 1998). Reinitiation is a logical point of regulation, because after elongation certain general factors (i.e., TFIID) are thought to remain attached to the promoter, while others fall off and diffuse away. Therefore, because the complex is already partially assembled, the effects of activators on recruitment of subsequent molecules of factors and polymerase may be fundamentally different from formation of the initial transcription complex. This difference may be important in regulating the actual rate of transcription. Finally, transcription in vivo occurs in the context of chromatin. Activators must therefore overcome the negative influence of chromatin before they can recruit the general machinery (for review, see Orphanides et al. 1996; Struhl 1998). When initiating a biochemical study, the investigator must consider each of these steps as potentially regulable.

The study of activation mechanisms involves choosing an appropriate system. Systems based both on model templates and on natural promoters have been widely employed in the field. Typically, model systems are composed of a single activator protein and a reporter bearing multimerized sites. Studies in these systems can provide general biochemical insights into how a particular activator functions at the mechanistic level. However, recent experiments have led to the view that a sophisticated nucleoprotein structure called the enhanceosome forms at a promoter/enhancer (see Kim and Maniatis 1997; Carey 1998). It is the enhanceosome that ultimately interacts with chromatin remodeling machines and with the general transcription machinery. Therefore, a key question the researcher must ask when initiating an analysis is whether the mechanism of an individual activator renders important information or whether it is necessary to address the issue of regulation from a natural promoter with all of its inherent complexities. To a certain extent, this problem involves two major considerations: (1) the availability of an in vitro transcription system to couple transcription (see Chapter 14) with binding studies and (2) a consideration of how much is known about the system already from in vivo studies or from previous biochemical analyses.

Once the system has been chosen and the reagents are available, one should ask the following questions: Does an activator assemble a functional transcription complex or does it act after the complex has formed or both? If the activator stimulates complex assembly, which steps are affected? What is the effect of chromatin on transcription complex assembly? We discuss approaches to these questions by summarizing how certain methods were originally employed to generate our current biochemical view of mammalian transcription and gene activation. It will become clear that certain aspects of transcription complex assembly have been studied in great depth whereas studies on other aspects are just beginning to appear in the literature. We take somewhat of a chronological approach when describing the experiments. This is not to say the early methods are outdated; indeed, many are still widely employed. We begin by describing functional studies where the readout is transcription. However, the main emphasis is on assays to measure mechanistic steps in the transcription process. We have not focused on chromatin because few studies have physically analyzed complex assembly on chromatin templates (i.e., by footprinting or crosslinking). Furthermore, coactivator/mediator complexes and holoenzymes have only recently begun to reach the stage of purity appropriate for study by physical analysis (see Chapter 14). Nevertheless, many of the approaches described here will be appropriate for analyzing problems involving such complexes as knowledge of their properties evolves.

We discuss both basal and activated transcription. We focus on the basal systems because they reveal the order in which factors build into a complex in the absence of regulatory factors. Moreover, the information gained from such analyses can point to important regulatory checkpoints. Nevertheless, it is important to consider that basal transcription is carried out by a small set of factors including Pol II, TBP, TFIIB, TFIIE, TFIIF, and TFIIH. Many additional polypeptides, including TATA-binding protein associated factors (TAFs), mediators, and chromatin remodeling machines, are essential for activators to function in vivo. At this moment, it is not entirely clear whether these additional proteins, necessary for activation, alter fundamental mechanistic aspects of the transcription process or simply alter the amount of transcription complex formed.

It should be apparent to the reader at the outset that the various methods all have certain limitations and that the experimental design will undoubtedly influence the outcome of a study. The concentration of various factors in a mixture and the methodology being employed can impose a particular rate-limiting step on the reaction that otherwise (i.e., in vivo) might not be limiting (for discussion, see Kingston and Green 1994). In the absence of genetics, however, it is often difficult to evaluate the validity of biochemical studies. Nevertheless, the high degree of conservation between the general transcription factors from yeast to humans has made the yeast system an invaluable aid in deciphering the mechanism of eukaryotic gene activation (for review, see Stargell and Struhl 1996; Hampsey 1998). Therefore, experiments in yeast help to reveal some of the limitations to the approaches used in assessing the validity of the experimental results in mammalian systems.

CONCEPTS AND STRATEGIES

Formation of the Basal Preinitiation Complex

The concept of a transcription complex originally derived from the observation that the purified eukaryotic core polymerases could not recognize and transcribe a promoter on their own, whereas crude mammalian extracts could. This observation inspired the idea that the extracts contain ancillary factors eventually called "the general transcription machinery" (see Chapter 14). Subsequent fractionation and characterization of the extract revealed the presence and identities of such factors. In this section, we describe selected studies aimed at deciphering how the factors assemble into complexes.

Kinetic Studies

The original approach to studying the basal transcription complex and the first credible biochemical arguments that a complex was indeed being assembled were based on kinetic studies in crude cell-free systems (Hawley and Roeder 1985, 1987). When a HeLa extract was incubated with the adenovirus major late promoter (AdMLP) and nucleotides, there was a lag period before transcription reached a peak rate of synthesis. In contrast, if the template was preincubated with the extract for a period of time (~1 hour) in the absence of nucleotides, transcription reached peak rates instantaneously upon nucleotide addition. The implication from these studies was that there was a rate-limiting step in transcription that was likely due to the time it took for a transcription complex to assemble. The complex assembled during the preincubation was initially termed the "rapid start complex." Simple time-course studies like these, where the readout is transcription of an mRNA, can be performed in the presence and absence of activators, and they are an easy starting point for the investigator interested in understanding the mechanism of transcription of a promoter (Fig. 15.1) (see, e.g., Johnson and Krasnow 1992). Such studies when coupled with methods such as sarkosyl challenge can yield important information on activator function.

Sarkosyl Probing

The sarkosyl approach is based on many results in the literature showing that prokaryotic transcription complexes in different stages of assembly or function display differential sensitivity to agents that disrupt protein–protein and protein–DNA interactions (e.g., heparin). Hawley and Roeder (1985, 1987) found that the ionic detergent sarkosyl could be used in eukaryotic cell-free systems to delimit different steps in Pol II transcription, namely, the difference between initiation, elongation, and reinitiation. The protocol was to preincubate the template with the nuclear extract and then add various concentrations of sarkosyl at either the beginning or the end of the preincubation, or after addition of nucleotides. The idea was that the earlier steps in transcription would be sensitive to lower concentrations of detergent because the complex becomes more stable as additional components are added and the process of transcription begins to occur (Fig. 15.2).

Low concentrations of sarkosyl (0.015%) were found to prevent transcription if added with the DNA template (time a, Fig. 15.2). In contrast, if the template was added to the extract and allowed time to assemble into complex before sarkosyl addition, transcription became resistant to low concentrations of the detergent. However, if higher concentrations of sarkosyl were added (0.1%), they inhibited initiation even after complex assembly (time b, Fig. 15.2). Finally, however, if polymerase was allowed to initiate in the presence of

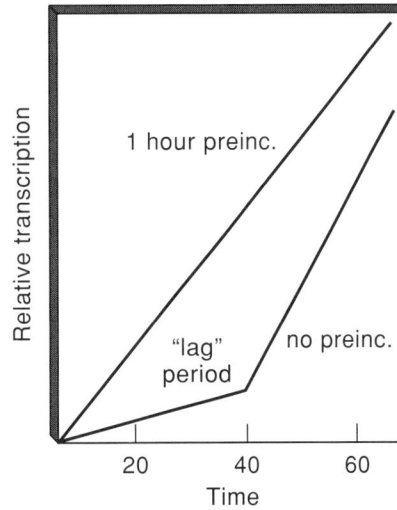

FIGURE 15.1. Kinetics of transcription.

nucleotides and begin elongation, transcription then became resistant to 0.1% sarkosyl. These experiments demonstrate that sarkosyl can be used to distinguish several steps in the transcription process. Sarkosyl challenge studies have been widely useful in distinguishing different stages in transcription and have been used both in crude systems (Zenzie-Gregory et al. 1992) and in reconstituted systems (see Chapter 14) to distinguish functionally important complexes containing different subsets of pure factors (Lieberman 1994; Kim and Maniatis 1997; see below). Approaches other than sarkosyl challenge have also been employed to probe the differential sensitivity of basal transcription complexes, including the use of increased temperature and salt concentrations (see, e.g., Cai and Luse 1987a).

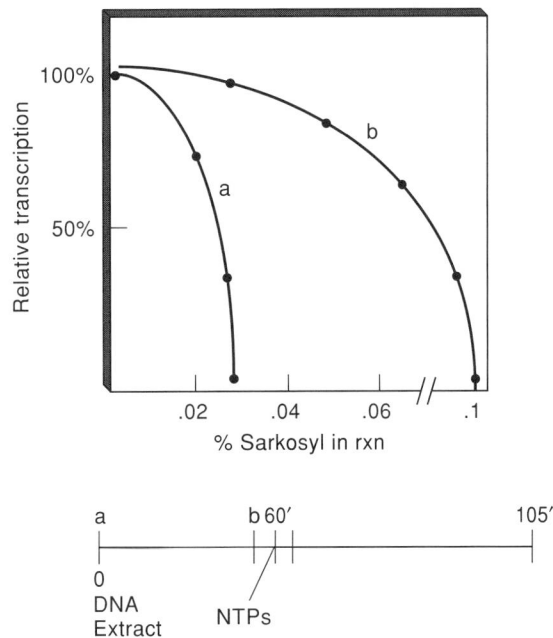

FIGURE 15.2. Sarkosyl inhibition.

In crude systems, sarkosyl challenge is often used to restrict transcription to a single round (i.e., to allow initiation but prevent reinitiation). Indeed, single-round sarkosyl challenge experiments can be used to study whether an activator is stimulating transcription complex assembly. If greater levels of transcription are observed in a single-round assay in the presence of activator, one can conclude that the activator is stimulating complex assembly.

Template Commitment Experiment

Another approach used to study transcription complex assembly is called template commitment. This technique can be employed to complement physical analyses of complex assembly (i.e., chemical and nuclease footprinting and crosslinking) and provide important functional information on the stability of the complex during the process of transcription. Either intact extracts or various fractions from a reconstituted system are preincubated for a period of time with one DNA template and then challenged with a competitor DNA. If a factor binds DNA tightly, then it will bind during the preincubation and commit DNA template #1 to transcription. Put simply, template #1 will be transcribed preferentially upon addition of nucleotides, the remaining factors, and a competitor template. If, however, both the preincubated and competitor templates are transcribed after the preincubation, it can be concluded that the template was not committed and the step was not limiting. This general approach was used to establish that TFIID binding committed a gene to transcription (Van Dyke et al. 1989) and provided functional evidence that TFIID binding, previously studied by DNase I footprinting, was indeed a crucial step in establishing assembly of an active transcription complex.

DNase I Footprinting and EMSA Studies of Transcription Complex Assembly

Physical information on binding and conformation of the basal transcription complex was first achieved by nuclease and chemical footprinting techniques. DNase I and methidium propyl EDTA footprinting experiments by Roeder's lab demonstrated that highly fractionated but not pure TFIID bound to the TATA box and nucleated binding of the other general factors. These studies also revealed that TFIID binding to and conformation on the TATA box could be influenced by activators and provided the framework for future studies on activator recruitment of complexes containing TFIID (see below) (Sawadogo and Roeder 1985; Horikoshi et al. 1988; Nakajima et al. 1988; Van Dyke et al. 1988).

The cloning of the TBP subunit of TFIID and further purification of the remaining general factors permitted a more refined analysis of the basal complex. The small size and availability of pure TBP permitted the use of electrophoretic mobility shift assay (EMSA) and DNase I footprinting methodologies to study basal transcription complexes. Using this methodology, Buratowski and Sharp (Buratowski et al. 1989), and subsequently Reinberg and colleagues (for review, see Orphanides et al. 1996), established the stepwise pathway for basal transcription complex assembly. This contribution provided an important and crucial framework for later studies on activated transcription.

In the initial experiments, Sharp and colleagues (Buratowski et al. 1989) employed a ^{32}P-labeled restriction fragment bearing the AdMLP. The radiolabeled fragment was incubated with TBP alone or with various combinations of fractionated general factors. The complexes were resolved by native polyacrylamide gel electrophoresis, i.e., EMSA. To illustrate the approach, consider binding of TBP and TFIIA. TBP alone bound poorly and generated a rapidly migrating complex with DNA. Addition of TFIIA, which did not bind DNA on its own, strongly stimulated TBP binding to form what is referred to as the TA subcomplex. The TA complex migrated with slower mobility than the T complex, demonstrating that TFIIA had indeed joined the complex. By varying the combinations of factors and examining the complexes formed, the stepwise assembly pathway could be conclusively established.

Low-resolution information on the location of factor binding within the promoter was deduced by combining the EMSA assay with DNase I footprinting (Fig. 15.3). Once formed, the complexes can be subjected to a light digestion with DNase I prior to gel electrophoresis. Inhibition of further cleavage with EDTA left the DNA nicked on only a single strand of the ^{32}P-labeled probe, but otherwise intact. The complexes were then resolved by EMSA. Because the DNA was nicked along a single strand, it did not dissociate from the protein in the native gel (Buratowski et al. 1989). After autoradiography, the nicked DNA was excised from the gel, purified away from the protein, and then resolved on a polyacrylamide/urea sequencing gel to visualize the DNase I footprints of the individual complexes. This method has many applications for activated transcription and should be kept in mind as we discuss this issue below.

The early studies by Buratowski and Sharp established the methodology and the initial steps, but the factor preparations used were somewhat crude and could not be employed to assign late steps in the pathway unambiguously. Subsequent studies by Reinberg's lab, using highly purified or recombinant general factors, eventually established the final pathway (Fig. 15.4) (for review, see Orphanides et al. 1996). In the first step, TBP binds to the TATA, a reaction that is aided and enhanced by TFIIA (or TFIIB in some experiments). The TA subcomplex serves to recruit TFIIB (TAB complex), which in turn recruits RNA polymerase II bound to TFIIF (TABpolF). Once polymerase is bound, it recruits TFIIE, which then recruits TFIIH to form the TABpolEFH complex. The IIA form of Pol II, lacking a phosphorylated CTD of the largest subunit, binds the complex. In the presence of ATP, TFIIH phosphorylates IIA to generate the IIO form. The presence of ATP also induces DNA melting in a reaction requiring TFIIH, a topic we return to in later sections. One minor caveat is that although the aforementioned scheme is the most likely pathway from a thermodynamic standpoint, it is possible that alternative pathways might exist under conditions different from those used in the Reinberg lab experiments (i.e., unusually high concentrations of certain factors).

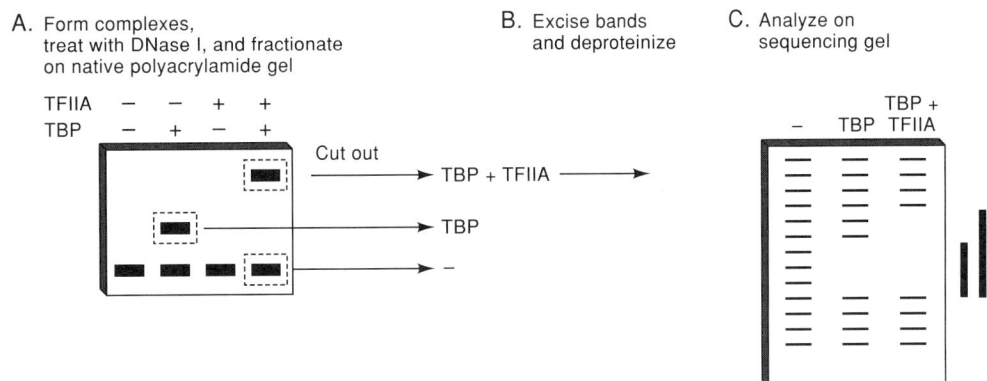

FIGURE 15.3. Schematic of EMSA complexes and footprint.

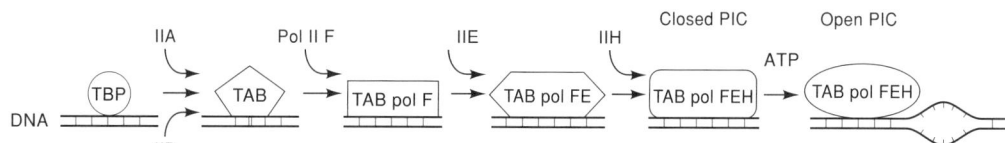

FIGURE 15.4. Schematic of the steps in transcription complex assembly.

Photocrosslinking

Although DNase I footprinting yields information on the position of factors on the DNA, the resolution of the footprints is somewhat low, and it is not always clear what proteins are contacting which portions of the DNA fragment. For example, as a complex is being assembled from free factors, new footprints may appear as new factors are added. The new footprints are not necessarily caused by direct binding of the newly added factor. Instead, the new factor may induce a conformational change in prebound factors that alters their binding. Therefore, the use of photoactivatable crosslinking reagents is essential to provide a detailed picture of the direct protein–DNA interactions occurring within the basal transcription complex, particularly for probing conformational changes in the complex.

There have been two general crosslinking methods applied to Pol II complexes. One method employs 5-[*N*-(*p*-azidobenzoyl)-3-aminoallyl]-dUTP or N_3RdUTP. N_3RdUTP crosslinking was first applied to transcription complex assembly in the Pol III system by Geiduschek and colleagues (see Fig. 15.5A) (Bartholomew et al. 1991, 1993). N_3RdUTP positions a photoreactive azide group 9–10 Å from the uracil pyrimidine ring. The reagent probes protein interactions with the major groove of the DNA. The second method involves attaching a phenyl azide group to a phosphorothioate-derivatized phosphate (see Fig. 15.5B) (Lagrange et al. 1996; Kim et al. 1997). This method positions the photoactivatable group 9.7 Å from the phosphate backbone and, in principle, probes interactions in both the major and minor grooves. The crosslinking groups are added onto oligonucleotides that are ^{32}P-end-labeled. The oligonucleotides are hybridized to a single-stranded DNA template and primer-extended using DNA polymerase to create double-stranded molecules. The labeled, derivatized promoter fragment is removed by restriction endonuclease digestion and gel-isolated.

Once the derivatized fragment is prepared, different combinations of recombinant factors are bound to it, and the resulting complexes are subjected to UV irradiation (312–350 nm). The UV treatment activates the azide group, generating a nitrene, a highly reactive group that crosslinks the DNA to the protein. The protein–DNA complexes are then subjected to extensive digestion with DNase I, which cleaves all but a few nucleotides of the crosslinked DNA away from the protein. The ^{32}P label is generally positioned close to the crosslinking group; thus, the small amount of DNA that remains attached to the protein after DNase I treatment contains a ^{32}P label. The radiolabel allows the protein to be detected on an autoradiograph of an SDS-polyacrylamide gel. Because the basal factors are recombinant or highly pure, the sizes of the subunits are known and can be used to ascertain the identity of the labeled protein.

Three recent studies illustrate the techniques' utility in studying basal complex assembly. In one study, Ebright and colleagues employed photocrosslinking to demonstrate that the largest and second largest subunits of Pol II form an extended 220-Å channel that interacts with AdMLP DNA upstream and downstream of the start site (Lagrange et al.

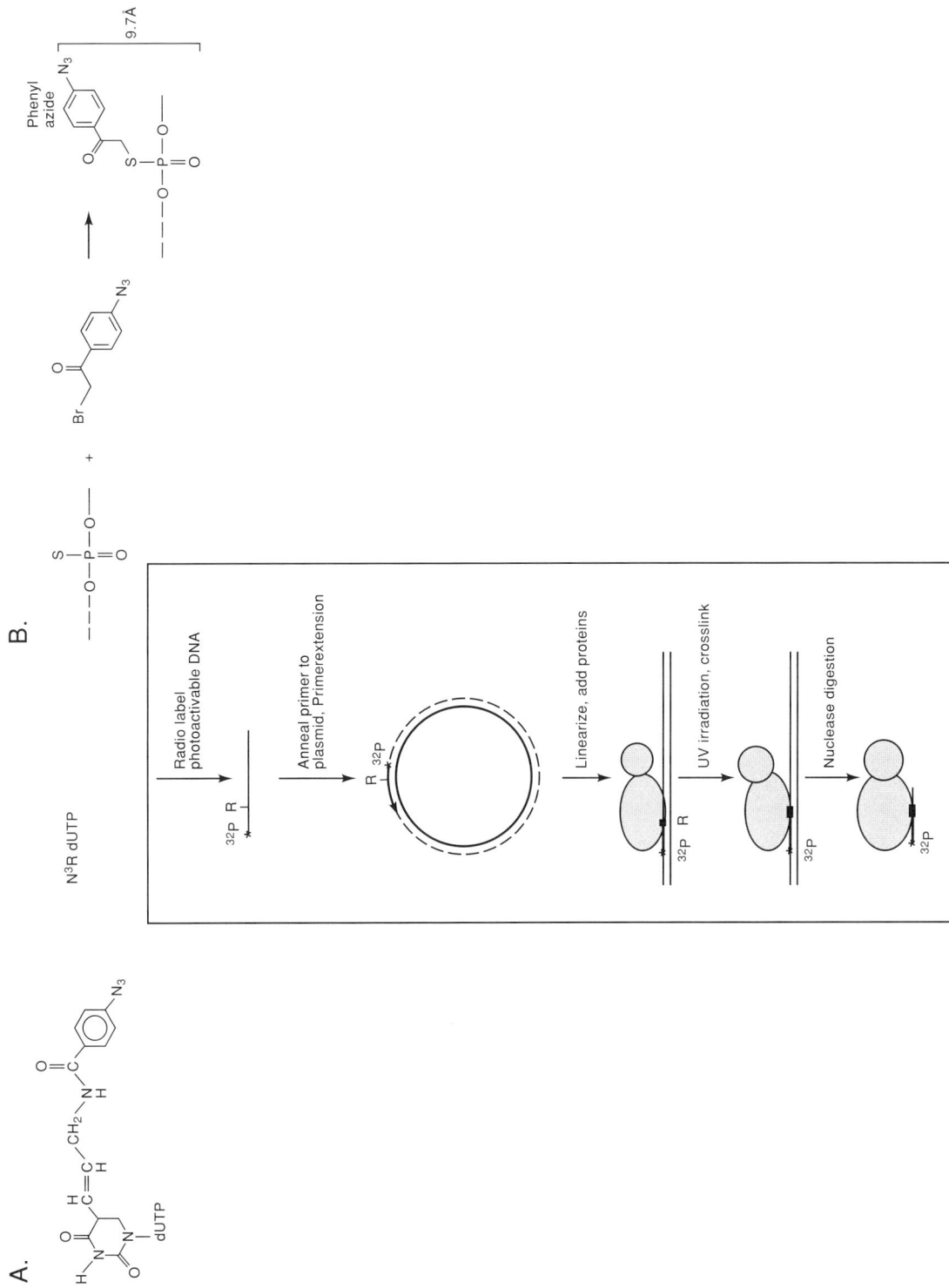

FIGURE 15.5. Crosslinking of proteins to DNA. (*A*, Adapted, with permission, from Bartholomew et al. 1991 [Copyright American Society for Microbiology; Lagrange et al. 1996 [Copyright 1996 National Academy of Sciences, U.S.A.].) (*B*, Adapted, with permission, from Lagrange et al. 1996 [Copyright 1996 National Academy of Sciences, U.S.A.].)

1996). Roeder and colleagues used the technique to study TFIID binding and the positioning of TAF$_{II}$s along the AdMLP. The authors showed that TFIIA induced a critical conformational change in the binding of TFIID to DNA, an effect that may hint at the mechanism of TFIIA in activated transcription (Oelgeschlager et al. 1996). Finally, to examine how crosslinking data are utilized in modeling complex protein–DNA interactions, consider the data generated by Coulombe and colleagues (Robert et al. 1998). Figure 15.6A summa-

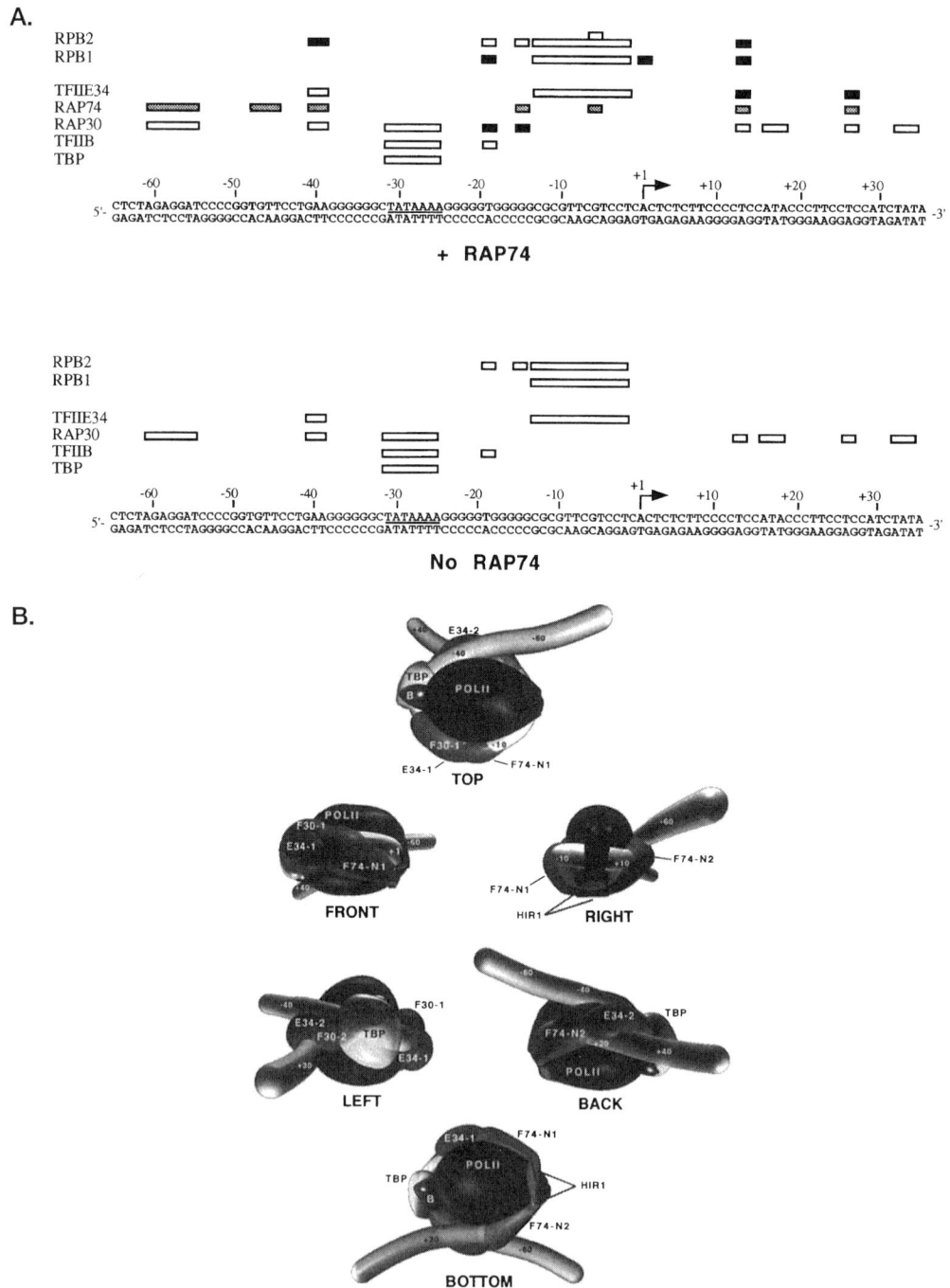

FIGURE 15.6. (A, Reprinted, with permission, from Robert et al. 1998 [Copyright 1998 Cell Press].) (B, Reprinted, with permission, from Robert et al.1998 [Copyright 1998 Cell Press].)

rizes the positions of weak (open boxes) and strong (closed boxes) crosslinking of the various general factors to the AdMLP in the presence and absence of RAP74. On the basis of these data, the authors of the study concluded that RAP74 induces a conformational change in the basal complex leading to wrapping of the DNA around Pol II as shown in Figure 15.6B (Robert et al. 1998). These changes are manifested by the more extensive crosslinking to the RPD1 and 2 subunits of Pol II in the presence versus the absence of RAP74.

Structure–Function Analyses of the General Machinery

The analysis of the general machinery by mutagenesis and crystallography has provided additional mechanistic details of the eukaryotic transcriptional apparatus. The most comprehensive approaches have combined the assays described above with mutagenesis to reveal the surfaces of each of the factors participating in an interaction. To date, TFIIA, TFIIB, TBP, and several TAFs have been crystallized and their structures solved. In many cases, the mutagenesis and the crystal structures agree perfectly, although in other cases discrepancies have arisen. The clearest understanding results from a compendium of the structural, mutagenesis, and, most importantly, functional analyses of the mutants in reconstituted transcription systems. This issue is covered in Chapter 12 and is not described any further in this chapter.

Open Complex Formation, Initiation, and Promoter Escape

ATP Analogs and an Energy-dependent Step

Transcription in both prokaryotes and eukaryotes follows conceptually similar paths. The polymerase and its associated factors form a closed complex on promoter DNA, which is followed by formation of an open complex in which the DNA encompassing the start site is melted. Unlike the common bacterial σ70 systems, the Pol II (but not Pol I or Pol III) machinery requires ATP to form the open complex. The polymerase then initiates transcription and eventually escapes from the promoter and elongates. There is good evidence that Pol II, like its bacterial counterpart, can undergo iterative cycles of abortive initiation where the polymerase initiates but fails to clear the promoter.

Identification of the important energetic role for ATP hydrolysis was one of the first applications of the in vitro transcription systems developed by Roeder and colleagues in the early 1980s (Bunick et al. 1982). Roberto Weinmann's lab incubated the Pol II-transcribed AdMLP and E4 promoters and the Pol III-transcribed VA gene in an extract and assayed whether transcription would occur in the presence of a nonhydrolyzable ATP analog called AMP-PNP. AMP-PNP contains an imido group between the β and γ phosphorus atoms. Recall that transcription requires hydrolysis of the α–β bond of nucleotides. AMP-PNP, which contains an intact α–β bond, had been shown to support transcription initiation and elongation by bacterial RNA polymerases. Although transcription by Pol III was unaffected by the AMP-PNP, Pol II transcription was completely abolished. Other β–γ imido derivatives, including GMP-PNP, had no effect on transcription and fully supported Pol II elongation when ATP was present. If dATP was added to the mixtures in place of ATP, transcription by Pol II occurred in the presence of AMP-PNP. This observation led to the conclusion that dATP was providing a hydrolyzable β–γ bond that satisfied the energy-requiring step. The energy allowed Pol II to initiate transcription, and elongation occurred by utilization of the intact α–β bond of AMP-PNP. This important discovery initiated almost 15 years of research to elucidate the energetic role of ATP in Pol II transcription.

Several subsequent kinetic studies by the laboratories of Roeder and Luse examined the role of ATP β–γ bond hydrolysis (Sawadogo and Roeder 1984; Luse and Jacob 1987).

Collectively, their data pointed toward the Pol II open complex as target of the β–γ bond requirement. More direct support of this view eventually came from a study by Timmers, which showed that supercoiled but not linear AdMLP could bypass the requirements for ATP β–γ bond hydrolysis and for the general factor TFIIH in a reconstituted transcription system (Timmers 1994). The fact that negative supercoiling was known to promote DNA melting suggested that formation of the open complex was indeed the step affected by ATP hydrolysis. Furthermore, the observation that TFIIH was bypassed suggested that it was the factor carrying out the melting step. TFIIH was known to contain DNA helicase and kinase subunits that both utilized ATP (for review, see Orphanides et al. 1996), and it was speculated that one of these activities was responsible for DNA melting. A preponderance of evidence strongly supports the role of the TFIIH ATP-dependent helicase in DNA melting, although no one has formally demonstrated a direct interaction of open complex with TFIIH (i.e., by crosslinking).

Permanganate Probing

Permanganate sensitivity is an excellent method for studying open complex formation directly. The method was first used to measure open complex formation by *E. coli* RNA polymerase and was subsequently applied to open complexes formed by RNA polymerase III (Kassavetis et al. 1992). Gralla and colleagues later employed the method to study open complex formation by Pol II (Wang et al. 1992a). Potassium permanganate works by selectively modifying thymidines in single- but not double-stranded DNA (Fig. 15.7). When the start site melts, the thymidine residues become susceptible to modification by permanganate. The modified thymidines block elongation by *Taq* DNA polymerase in a primer extension reaction (see below) and are also sensitive to cleavage with piperidine. The stall or cleavage products can be resolved on polyacrylamide/urea gels and serve as an indicator of open complex formation. Using HeLa nuclear extracts, Gralla's lab showed that melting of the adenovirus E4 start site to form Pol II open complexes, as measured by permanganate sensitivity, required the ATP β–γ bond. This technique also has great utility in studying activated transcription and is covered below in more detail (also see Protocol 15.1).

Premelted Templates

The studies of Roeder, Luse, Gralla, and others led to the hypothesis that ATP hydrolysis was driving DNA melting during open-complex formation. Furthermore, Timmers' study suggested that in the absence of ATP, the free energy of supercoiling could substitute for β–γ bond hydrolysis. One prediction of this hypothesis was that a premelted DNA template should also bypass the requirement for ATP and the factors that melt the start site. One of our labs created an adenovirus E4 DNA template bearing a 10-bp mismatch over the start site (Tantin and Carey 1994) and showed that this template indeed bypassed the requirement for the β–γ phosphoanhydride bond of ATP during activated transcription. Timmers' lab found that the same premelted E4 template would bypass the requirement for TFIIE and TFIIH in a basal transcription system with purified factors (Holstege et al. 1995), a conclusion reached independently by Greenblatt's group using a similar approach (Pan and Greenblatt 1994). One remaining issue was whether activators regulated the melting step. Studies in highly pure reconstituted systems showed that the activator GAL4-VP16 was still required for activated transcription on the premelted templates. However, in the activated system, the premelted templates only bypassed the requirement for TFIIH but not TFIIE (Tantin et al. 1996), suggesting that other proteins necessary for activated transcription (TAFs or coactivators) impose a requirement for TFIIE on activated transcription.

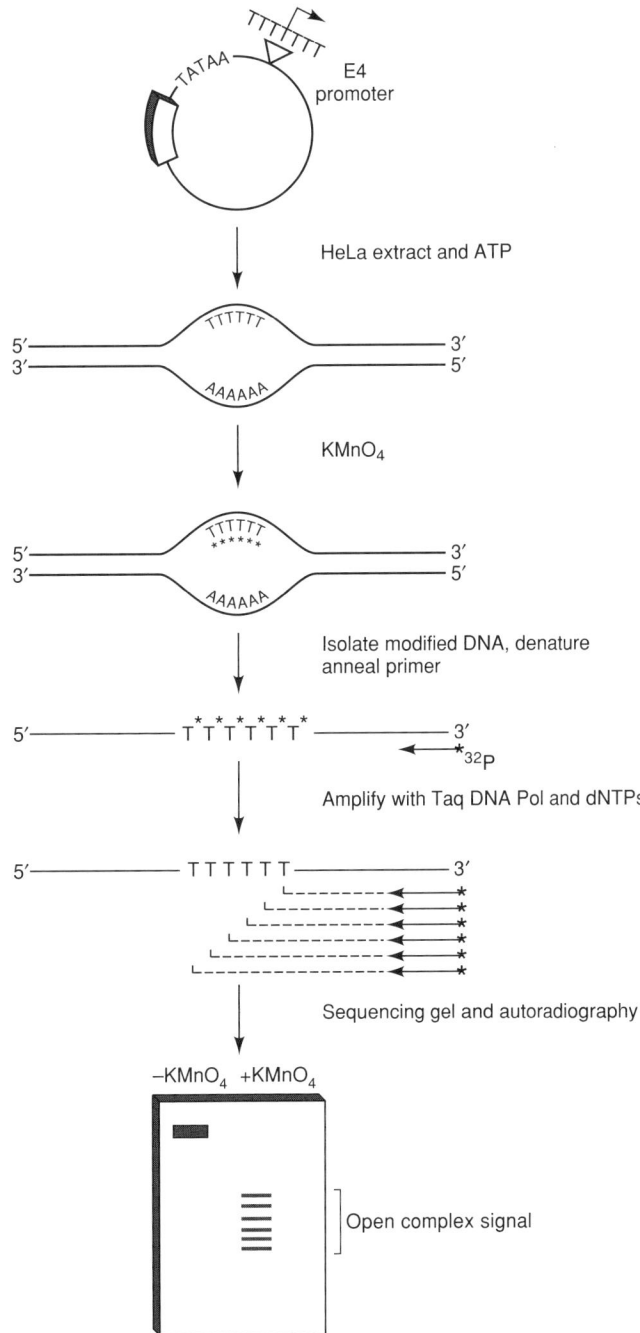

FIGURE 15.7. Permanganate probing.

The Transition to Elongation

The transition from the open complex to productive elongation is a critical step in transcription of a gene. After initiation, the polymerase has several choices: It can elongate a short distance (i.e., 10 nucleotides or so) but not release from the general transcription machinery (abortive initiation), it can elongate further (i.e., within 100 bp of the start site) and pause (promoter-proximal pausing; see Chapter 3), or it can continue productive elongation. Studies have suggested that all of these steps are subject to regulation by activators (for review, see Bentley 1995). Indeed, activators like GAL4-VP16 have been observed both to stimulate promoter escape and to prevent promoter-proximal pausing in model systems (Kumar et al. 1998). This is not an artifact of the model systems, because some natural activators like the HIV Tat protein have been found to prevent promoter-proximal pausing in vivo (for review, see Yankulov and Bentley 1998).

Currently, the details regarding the specific steps involved in the transition to elongation are murky, although TFIIH and the TAK complex have been implicated in the process (for review, see Jones 1997; see Chapter 3). We do not focus here on the transition from a promoter-proximal pause to elongation, because this is a complex issue beyond the scope of this chapter. Instead, we emphasize the role of TFIIH in promoter clearance and choose a few select studies to illustrate how this problem was studied. The main point, as we discuss at the end of this section, is that activators do indeed intervene in the transition from initiation to promoter escape. Biochemical assays allow the investigator to probe the mechanism.

We first describe the use of the immobilized template approach to study the activity and composition of basal transcription initiation and elongation complexes. Then we discuss applications of the permanganate probing and premelted template approaches toward studying promoter escape. Finally, we describe a biochemical study showing that activators directly influence promoter escape.

Reinberg's lab used an immobilized template assay in an elegant experiment to study the composition of the transcription complex as it begins productive elongation (Zawel et al. 1995). The immobilized template approach was developed initially to purify RNA polymerase III (Pol III) transcription complexes (Kasher et al. 1986) and subsequently applied to analyze Pol II complexes (Arias and Dynan 1989). A biotinylated promoter fragment was immobilized on streptavidin beads and basal factors were added (Fig. 15.8). After washing to remove unbound factors, various combinations of nucleotides were added to control the extent of elongation. The resulting elongation complexes were washed again to remove factors that dissociated during elongation. A restriction site between the core promoter allowed the investigators to release the elongation complexes and examine by immunoblotting their composition relative to what was bound at the promoter. It was found that TFIID (TBP) remained behind after initiation and elongation while TFIIB, TFIIE, TFIIF, and TFIIH were released. TFIIE was released before formation of the tenth phosphodiester bond and TFIIH was released after transcription to +30. TFIIF displayed the ability to remain associated at least transiently with Pol II during elongation and, in fact, could dissociate and rejoin a paused polymerase.

Timmers' lab further investigated the elongation transition at a structural level. Permanganate sensitivity was used to probe structural changes in the open complex as Pol II began elongation. The transition was controlled using OMeGTP, a chain-terminating analog of GTP that stalled Pol II when it became incorporated into the growing mRNA chain on templates bearing G residues at various positions downstream of the initiation site (Holstege et al. 1997). The authors found that formation of a stable, productive open complex required ATP hydrolysis and the general factors including TFIIE and TFIIH. The initial open complex (from –9 to +1) could be reversed with ATPγS, a competitive but non-hydrolyzable analog of ATP. After a 4-nucleotide mRNA was produced, the complex became insensitive to inhibition by ATPγS, suggesting that the complex had undergone a critical transition during initiation. The authors then found that the transcription bubble expanded continuously until +11, when the start site closed behind the bubble. It was speculated that this stage represented the transition from initiation to promoter escape.

The mechanism of promoter escape and the factors influencing it were examined in several important studies. Experiments by Goodrich and Tjian (1994) were the first to sug-

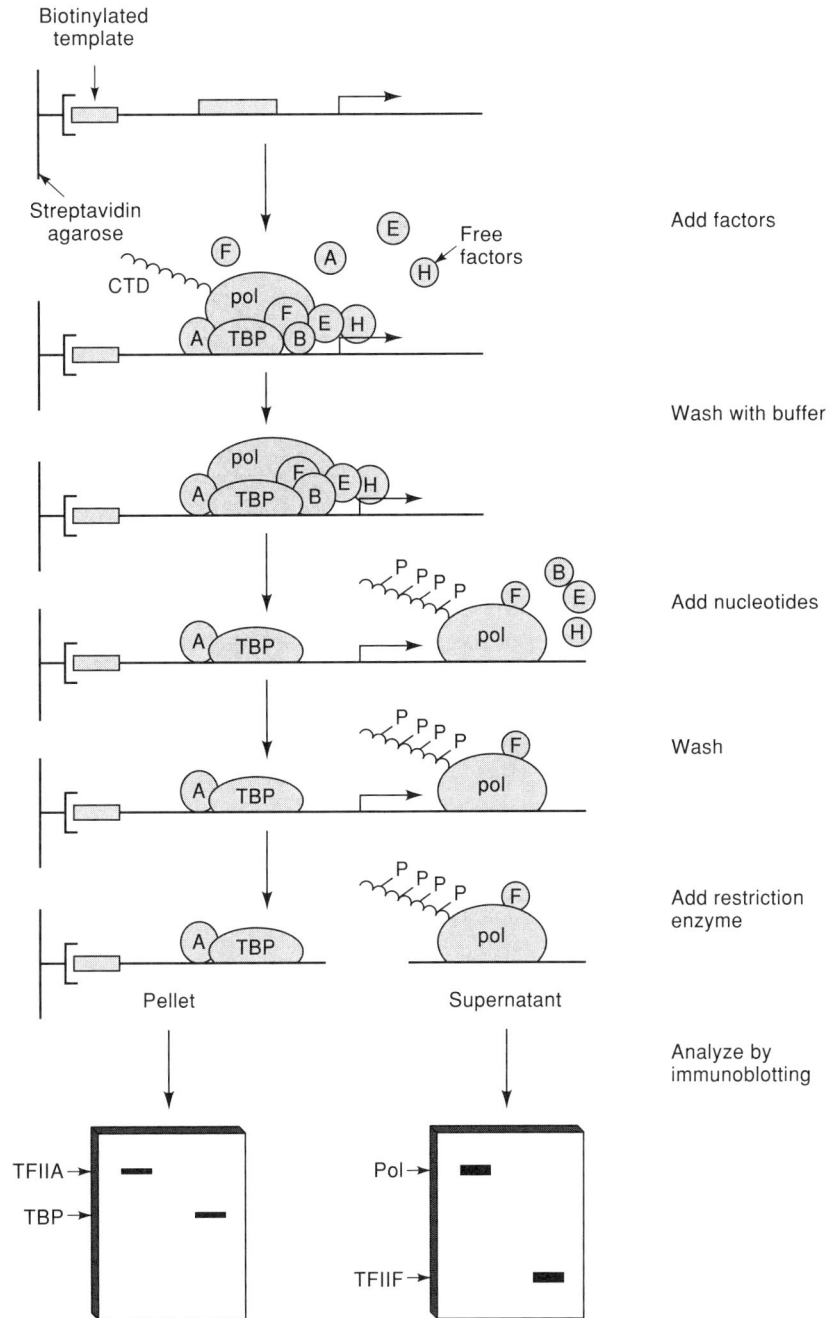

FIGURE 15.8. Immobilized template (procedure).

gest that TFIIH participates in promoter escape. This initial observation was confirmed and extended in later studies by the Conaways and colleagues that employed AdMLP templates premelted at the start site. Because such templates bypass the TFIIH requirement for initiation (see above), TFIIH could be added to the reaction after initiation to probe its effects on promoter escape (Dvir et al. 1997a). To study initiation on the premelted templates, the authors employed a dinucleotide priming assay. In this assay, a dinucleotide complementary to the start site was used as a primer for polymerase II (Samuels et al.

1984). Pol II adds a single radiolabeled nucleotide onto the primer, and the efficiency of this reaction is considered a measure of initiation. Although TFIIH had little effect on the efficiency of dinucleotide priming from the premelted template, it substantially enhanced the production of full-length transcripts when a complete set of nucleotides was added. The conclusion was that TFIIH, in addition to its role in start site melting, is required for efficient release of Pol II from the promoter. The lack of sensitivity of the elongation reaction to a CTD kinase inhibitor suggested that the DNA helicase activity of TFIIH, rather than its kinase activity, was required for promoter escape.

The idea that promoter escape could be differentially regulated by activators was raised in several biochemical studies. In a more recent example, Reinberg's lab found that in the absence of TFIIH, short abortive dinucleotide-primed RNAs were formed and Pol II elongation stalled at positions +12 to +17 downstream of the start site. Addition of TFIIH and ATP stimulated production of full-length transcripts, but the short RNAs were still observed. However, addition of the activator GAL4-VP16 led to a reduction of short transcripts and an increase in the production of full-length transcripts, suggesting that activator was indeed stimulating promoter escape (Kumar et al. 1998). Supporting in vivo results had previously been reported by Bentley and colleagues (see, e.g., Bentley 1995). However, the issue is complicated because the formal mechanistic difference between promoter clearance and promoter-proximal pausing has not been established conclusively in eukaryotic systems.

Assembly of Activated Complexes at a Promoter

To this point, we have focused largely on the mechanics and structure of the basal complex, although we have touched on various aspects of activation. The ensuing sections cover several assays that have proven informative in understanding the biochemical action of activators.

The Immobilized Template Approach

Michael Green's laboratory employed the immobilized template approach to identify the steps in transcription complex assembly that were affected by activators (Lin and Green 1991). The approach is an excellent way to study activator mechanism because it allows functional studies to be performed on complexes assembled in crude extracts and permits the identification by immunoblotting of components recruited into the complex by activators (Fig. 15.9). The initial experiments distinguished two main hypotheses: (model 1) that an activator helps to assemble a complex from free components and (model 2) that an activator isomerizes a preassembled complex from an inactive to an active form.

An immobilized template was incubated in the presence (+) and absence (−) of the activator GAL4-AH with HeLa nuclear extract and washed with an excess of buffer. The activator was then added back to reactions that lack it in the preincubation. If the role of activator was to assemble the transcription complex, addition of activator back to the mixture lacking it in the preincubation would have no effect because the complex had not formed and all of the free factors had been washed away (model 1). In contrast, if the complex had formed in the absence of activator, addition of activator back to the reaction would lead to enhanced levels of transcription (model 2). It was found that addition of the activator back to the preincubation mixture after the wash step had no effect on transcription. In contrast, inclusion of the activator in the preincubation stimulated formation of active complexes, which synthesized mRNAs after the wash step. The result implied that the activator stimulated the assembly of a complex rather than its isomerization (model 1). This is not to imply that the

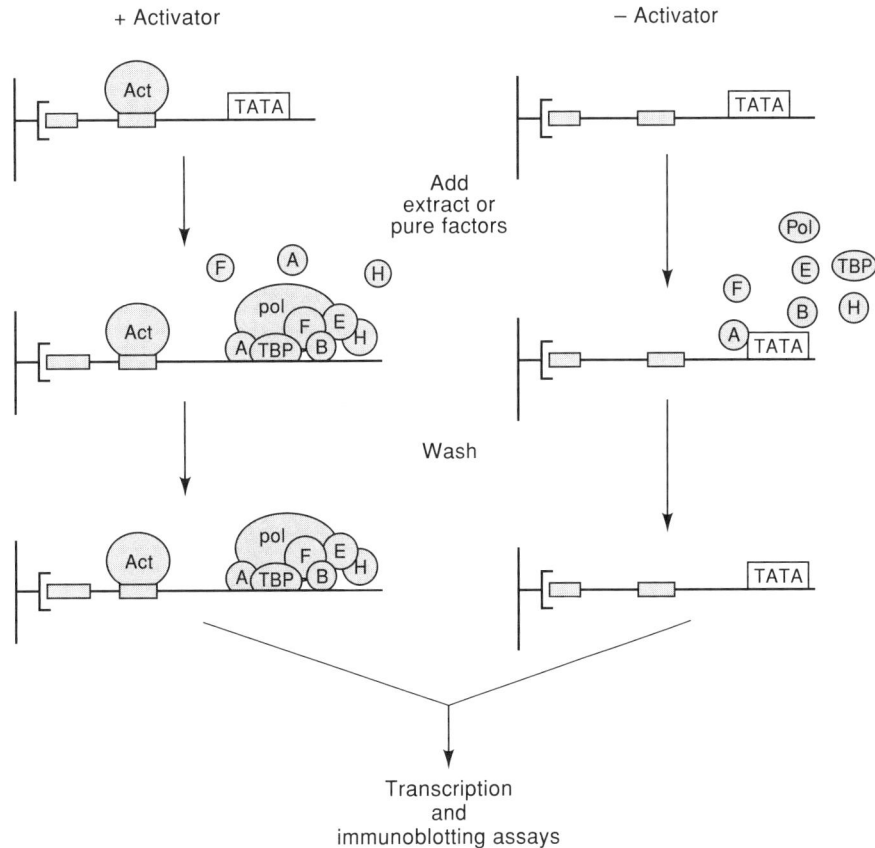

FIGURE 15.9. Immobilized template (+ activator, – activator).

activator cannot also alter the conformation of the complex. Indeed, activators apparently do both in prokaryotes, often through similar types of protein–protein contacts. However, the activator must first recruit the complex to DNA for any isomerization events to occur.

This same approach can be applied to deduce the stage affected by activator binding. Various combinations of the factors are first preincubated in the presence and absence of activator, followed by a washing step to remove proteins that did not assemble into complexes. The immobilized DNA–protein complexes are supplemented with the missing factors. If the activators stimulated binding of the factors in the preincubation, addition of the remaining factors and nucleotides after the wash will lead to transcription. Using this approach, Green's group showed that TFIID bound to the template in either the presence or absence of activator but that TFIIB bound to TFIID only when both the activator and TFIID were preincubated with TFIIB. If TFIID alone was present, TFIIB failed to bind and was washed away.

The experiments of Hahn and colleagues are a particularly elegant example of the immobilized template approach for studying the mechanism and dynamics of transcription complex assembly (Ranish et al. 1999). The authors assembled complexes from yeast extracts prepared from wild-type cells or cells containing conditional mutations in TBP, TFIIA, TFIIB, Pol II, or several SRB proteins. The effect of GAL4-AH on complex assembly was measured by analyzing the transcriptional activity of the wild-type and mutant complexes, and the factor composition by immunoblotting. Recall that in yeast systems the holoenzyme lacks TFIID; this component must be added to holoenzyme preparations to observe transcriptional activity. The authors were able to show that the DA complex (see Chapter 1) bound in one step and the holoenzyme in a subsequent step. Mutations in any components that bind after TFIID and TFIIA binding (i.e., TFIIB, SRBs, and Pol II) affected binding of all subsequent factors such as TFIIE, TFIIF, and TFIIH. This experiment represents perhaps the most compelling argument to date for the recruitment of the holoenzyme by an activator in vitro.

One drawback of the approach is that the background binding of crude general factor preparations to the DNA and the resin can generate problems when analyzing the complex either functionally or by immunoblotting. Therefore, careful controls for specificity, including but not restricted to promoter and activator mutants, must be included in the study. Furthermore, a systematic analysis of wash and buffer conditions may be necessary to optimize the binding effects. Many of these issues are covered in the Hahn study (Ranish et al. 1999), including a discussion of the specific activity of the complexes. The Hahn lab website can be consulted for additional details of the experiments (http://www.fhcrc.org/science/basic/labs/hahn).

Gel Filtration

The gel filtration approach is conceptually similar to the immobilized template and can be used to study different stages of transcription complex assembly. It was first applied to study factor addition in Pol III transcription complexes (Carey et al. 1986). The basic scheme is that complexes are assembled from nuclear extract or fractionated factors on a plasmid DNA template and then applied to a gel filtration column. The column separates the plasmid DNA and any attached factors away from free, unbound factors by molecular sieving. The excluded (void) volume containing the plasmid DNA and bound factors can then be assayed by immunoblotting or transcription complementation assays.

The columns are ideal for identifying early steps in transcription complex assembly and have been successfully applied to studying activator-mediated binding of TFIIB to TFIID. In one example, TFIID and TFIIB were incubated in the presence and absence of GAL4-VP16, and the excluded volume was complemented by addition of the missing factors and nucleotides. It was shown that TFIIB bound to TFIID only in the presence of activator. Surprisingly, the activator promoted binding of TFIIB, but not the other general factors, even when TBP was substituted for TFIID (Choy and Green 1993; Roberts et al. 1995).

Permanganate Probing to Study Activation

We have discussed the utility of employing the permanganate technique to study the open complex. However, because open complexes measure complete preinitiation complex formation, the technique can be employed in both kinetic and thermodynamic experiments to quantitate the amount of transcription complex stimulated by an activator (Wang et al. 1992a).

In a typical in vitro reaction, plasmid template DNA is incubated in a nuclear extract in the presence and absence of activator. ATP is added to allow DNA melting. After a preincubation period during which activated complexes are allowed to assemble, potassium permanganate is added. The modification reaction is terminated and the modified DNA is subjected to primer extension using the *Taq* DNA polymerase linear amplification reaction in a thermocycler. *Taq* stalls at the modified thymines. The open complex signals of activated and basal reactions are then compared on polyacrylamide/urea gels.

The most immediate application is to ascertain whether the activator stimulates formation of an open complex and whether the relative levels of open complex parallel the levels of transcription (Wang et al. 1992a). The results of this analysis can be quite informative when determining the effects of the activator on initiation versus reinitiation. Because the open complexes are a measure of what will effectively become the first round of transcription, the open complexes should first be compared to the transcription product generated in a single-round assay. The most efficient way to restrict transcription of the template to a single round is a sarkosyl challenge experiment after formation of the ATP-dependent open complex. As discussed above, sarkosyl permits elongation by the preinitiated polymerase but effectively prevents reinitiation. If the fold-stimulation of open complexes by activator equals the fold-stimulation of transcription in a single-round assay (i.e., by densitometry or phosphorimager quantitation), it can be tentatively concluded that the activator acts at the step of preinitiation complex formation. If the amount of open complex was similar in the presence and absence of activator but the single-round transcription assay demonstrated stimulation, it would be concluded that the activator was affecting a step after initiation (i.e., promoter escape, elongation). If the rate of reinitiation and initiation are proportional, the open complex signal should still be proportional to the transcription signal in a multiround assay. If, however, the open complex was stimulated in the presence of activator but the fold-stimulation was less than that observed in a multiround transcription assay, it could be tentatively concluded that the activator is stimulating reinitiation at a greater rate than initiation. Such an observation might warrant further investigation of the mechanism of reinitiation (see below).

Another application would be to compare the relative open complex and transcriptional signals on different templates for the purpose of studying regulatory phenomena like synergy. It is instructive to examine such an application to understand how the technology is employed and interpreted. Figures 15.10 and 15.11 compare open complex formation with transcription to study activation by the Epstein-Barr lytic cycle *trans*-activator ZEBRA (Chi and Carey 1993). Model reporter templates were generated bearing different numbers of tandem ZEBRA-binding sites positioned upstream of the adenovirus E4 core promoter. The region encompassing the E4 start site contains 6 thymidines and is an ideal substrate for the permanganate reaction.

When the reporter template was incubated in a HeLa nuclear extract in the presence of ZEBRA and dATP, the thymidines encompassing the start site became sensitive to permanganate (bracketed region, Fig. 15.10). ATP was required for complex formation because omitting dATP abolished the effect. The open complexes were transcriptionally active and due to Pol II, because if the nucleotides were added, the permanganate sensitivity disap-

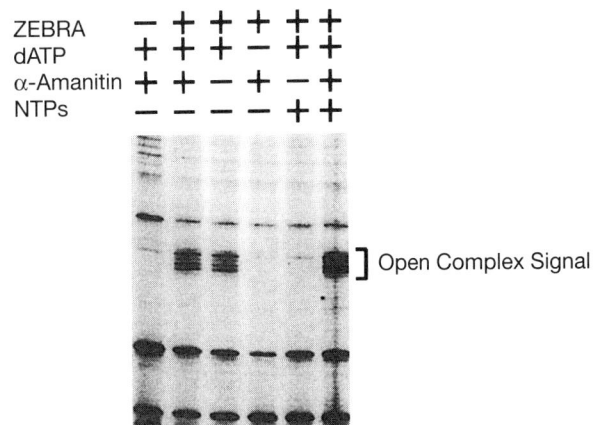

FIGURE 15.10. Open complex formation compared with transcription (see text for details). (Reprinted, with permission, from Ellwood et al.1998 [Copyright Cold Spring Harbor Laboratory Press].)

peared. This observation is consistent with elongation of Pol II away from the promoter and concomitant reclosing of the DNA around the start site. If, however, nucleotides were added with the elongation inhibitor α-amanitin, the complexes remained in place, confirming that they were indeed transcriptionally competent Pol II open complexes.

Figure 15.11 shows that the site-dependent synergistic effect of ZEBRA measured in transcription assays (left panel) paralleled the relative amount of open complex formed on these same templates (right panel). Thus, the synergistic effects of ZEBRA on transcription in a multiple-round assay are first manifested as synergistic increases in transcription complex assembly. This example illustrates that transcription is somewhat proportional to transcription complex assembly. Had the transcription results diverged from the open complex results, ZEBRA's effects on elongation would have been investigated.

The permanganate assay can also be used to study reinitiation. If the preinitiation complex is allowed to assemble and nucleoside triphosphates are then added, the polymerase will initiate and elongate transcription. Addition of the elongation inhibitor α-amanitin after nucleotide addition stalls polymerase elongation downstream but does not inhibit reinitiation. The time it takes for the open complex to reform can then be compared with the time it takes to form the initial open complex. The methodology was used by Gralla and colleagues to show that the rate of reinitiation in the presence of activator is faster than that of initiation (Jiang and Gralla 1993).

The permanganate assay can also be employed in reconstituted systems to study rate-limiting steps in transcription complex assembly (Wang et al. 1992b). Kinetic experiments with ZEBRA and GAL4-AH have shown that it takes 30 minutes to 1 hour to assemble a full level of open complex in a HeLa nuclear extract. The binding of activator to the DNA and the ATP-dependent isomerization are rapid events based on independent measurements, so it was assumed, on the basis of other studies in the field, that the 30-minute lag was required for the activator to assemble the closed complex. The step affected by activator could be ascertained by preincubating combinations of the general transcription factors with activator for 30 minutes and then adding the remaining factors and performing

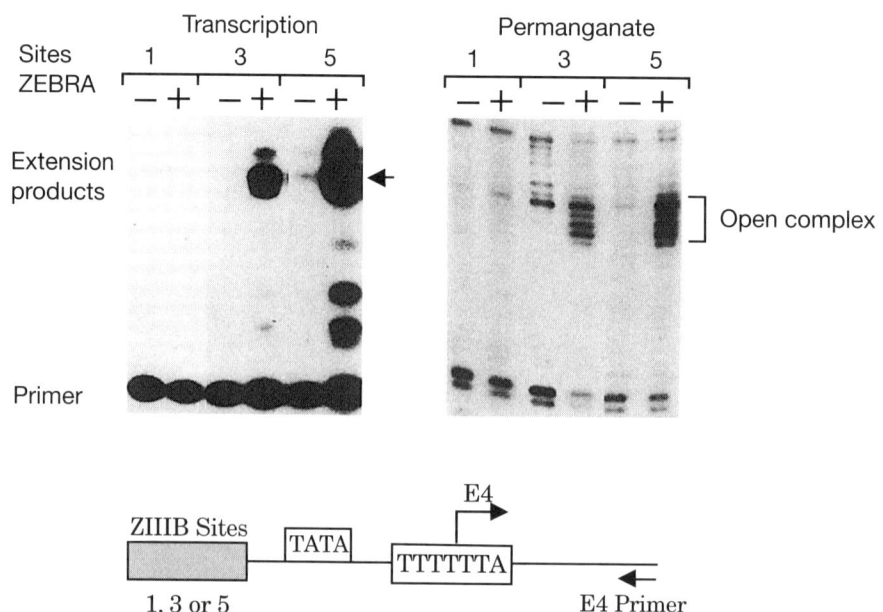

FIGURE 15.11. Open complex formation, compared with transcription (see text for details). (Reprinted, with permission, from Ellwood et al.1998 [Copyright Cold Spring Harbor Laboratory Press].)

a kinetic analysis of open complex assembly. If the preincubation had bypassed a slow activator-responsive step, open complexes would form almost instantaneously upon addition of the missing components and ATP. If the preincubation had not bypassed such a step, the complexes would take 30 minutes to assemble. Indeed, preincubation of either ZEBRA or GAL4-AH with crude TFIID and TFIIA fractions bypassed a slow step. This assay was in fact the first to show that the activator functioned during formation of the TFIIA–TFIID complex, an issue we return to below. One disadvantage of the kinetic approach is that the complexes cannot be analyzed for their composition due to the presence of unbound factors in the HeLa extracts. If the goal is to physically analyze binding of the transcription complex and to analyze its composition, other approaches must be employed.

EMSA and DNase I Footprinting Analyses of the TFIID–TFIIA Complex

Perhaps the most straightforward methods for analyzing assembly of activated transcription complexes are EMSA and DNase I footprinting. The availability of pure transcription factors has made this analysis somewhat simpler. Roeder and colleagues had shown that the activator USF enhanced the stability of the TFIID footprint on the DNA (Sawadogo and Roeder 1985). However, it was never formally shown that TFIID was recruited. Experiments by Berk and colleagues using purified TFIID and TFIIA and employing EMSA and DNase I footprinting analyses were the first direct functional evidence for recruitment (Lieberman and Berk 1994). The approach is important because many activators have been shown to intervene at the TFIID–TFIIA-binding step, and biochemical and genetic experiments in yeast have confirmed the global importance of this step in gene regulation. It is likely that many activators will ultimately be found to affect this step (for review, see Hampsey 1998).

In the EMSA studies, magnesium-agarose gels, rather than polyacrylamide, were employed because of the large size of the TFIID–TFIIA complexes (Fig. 15.12). A DNA template bearing the E4 start site and upstream ZEBRA sites was incubated with recombinant ZEBRA, purified HA-tagged TFIID (D) (see Chapter 14) and purified TFIIA (DA). The complexes were then fractionated on magnesium-agarose gels. D or DA bound poorly and little shifted complex was observed. ZEBRA alone shifted the probe marginally in these gels but strongly stimulated an EMSA complex containing TFIID and TFIIA (DAZ). ZEBRA had little effect on TBP–TFIIA complex assembly, demonstrating that the increased recruitment was due to TAFs (not shown). The precise role of the $TAF_{II}s$ is unknown. They could either directly interact with ZEBRA or stabilize a conformation of the complex that allows productive interaction with TBP and TFIIA.

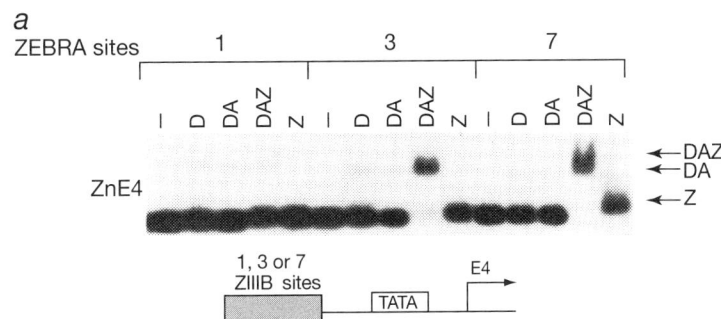

FIGURE 15.12. ZEBRA promotion of synergistic DA complex formation (see text for details). (Reprinted, by permission, from *Nature*, Chi et al. 1995 [Copyright 1995 Macmillan Magazines Ltd.].)

The strong ZEBRA-mediated recruitment observed in EMSAs can also be observed in DNase I footprinting, although the EMSAs are easier to quantitate. Chi and colleagues used the same assay to show that multiple ZEBRA molecules synergistically recruited the DA complex, suggesting that ZEBRA was contacting different subunits of the DA complex (Chi et al. 1995). The ability of TFIIB to interact with the DA complex requires activator, and DNase I footprinting suggested that ZEBRA can induce a conformational change in the complex and that this change correlates with the ability of TFIIB to bind (Chi and Carey 1996). Because TFIIB binds well to a TA but not to a DA complex in the absence of activators, it is assumed that the TAFs have a negative effect on transcription by blocking access of TFIIB to TBP; it was implied that the activator-induced conformational change neutralizes the inhibition.

In summary, the magnesium-agarose EMSA method can be applied to study the assembly of complexes containing TFIIA, TFIID (or TBP), and TFIIB. The method is simple and easily quantifiable. The presence of individual components can be assayed by immunodepletion of the components from the reaction mixture or by antibody supershifts.

Although these binding assays show that the DA and DAB complexes are key targets, other approaches show that the complexes can be further stabilized by addition of coactivators. Lieberman used similar model templates and ZEBRA to demonstrate that preincubation of ZEBRA with DAB and the USA coactivator fraction (see Chapter 14) stabilized the complex such that it survived challenge with sarkosyl and allowed subsequent transcription upon addition of the remaining factors (Lieberman 1994). Furthermore, the formation of the ZDABUSA complex stabilized the complex to challenge with ZEBRA-binding-site oligonucleotides. Similar approaches were later applied to the IFN-β enhanceosome to show that it also stimulated assembly of a complex containing DAB and the USA fraction (Kim et al. 1997).

Recent studies have shown that activators, in addition to interacting with TFIIA, TFIID, and TFIIB, also interact tightly with large coactivator complexes. Because these complexes are purified, it is likely that the assays described in this chapter will be applied to understand how such complexes build onto TFIID and TFIIA and how they influence initiation, reinitiation, and transcription on chromatin templates.

Assembly and Analysis of TFIID Subcomplexes

One of the strengths of the biochemical approach is the ability to manipulate the protein components involved in a reaction. This in turn allows specific hypotheses to be posed and tested. Whereas several of the GTFs are monomeric or dimeric, larger GTFs like TFIID and TFIIH pose a problem— the subunits do not readily assemble during the course of a reaction. Much effort is required to assemble and analyze these complexes. One of the most elegant approaches in this regard is that of Tjian and colleagues, who used purified TAF$_{II}$ subcomplexes to determine a direct relationship between activator–TAF$_{II}$ interactions and activated transcription (Chen et al. 1994). The premise was simple: to correlate the interaction between a TAF$_{II}$ and an activator with a functional assay. Although Tjian and colleagues have applied this to many activator–TAF$_{II}$ interactions, we will choose one that exemplifies the experimental approach.

The chimeric activator GAL4-NTF-1 was used in affinity chromatography experiments to identify which of the *Drosophila* TAF$_{II}$s bound. It was found that only TAF$_{II}$150 bound tightly. TAF$_{II}$ subcomplexes were then constructed by binding an epitope-tagged version of TAF$_{II}$250 to an antibody affinity resin; the HA monoclonal antibody linked to protein A-Sepharose. A 10-fold excess of TBP and other relevant TAF$_{II}$s was then incubated with the TAF$_{II}$250 resin; subcomplexes were allowed to form and then eluted from the resin with the

HA peptide. The TAF$_{II}$s and TBP are assembled into TFIID in a very specific fashion (Fig. 15.13A). TAF$_{II}$250 binds TBP first. Most of the other TAF$_{II}$s are then thought to build around the TAF$_{II}$250–TBP complex. TAF$_{II}$150 binds to the TAF$_{II}$250–TBP forming a TAF$_{II}$150–250–TBP subcomplex. The TAF$_{II}$ subcomplexes can then be tested in in vitro transcription reactions for the ability to support GAL4-NTF-1 transcriptional activation (Fig. 15.13B). GAL4-NTF-1 efficiently activated in the presence of the TBP–TAF$_{II}$250– TAF$_{II}$150 subcomplex but not in the presence of TBP or TBP–TAF$_{II}$250. Thus, there was a direct correlation between the ability of the activator to interact with a specific TAF$_{II}$ and to stimulate transcription in the subcomplex.

Although we have emphasized the interaction of TAF$_{II}$s and activators, the use of subcomplexes to probe mechanism is becoming increasingly common in the gene expression field. Subunits from numerous macromolecular complexes have been cloned, purified from baculovirus and other expression systems (see Chapter 11), and reconstituted into various subcomplexes for use in mechanistic studies. Recent examples include TFIIH (Tirode et al. 1999) and SWI/SNF (Phelan et al. 1999). However, as different subunits of the human mediator complex are cloned (see Chapter 1), it is likely that the reconstitution and affinity approaches will again become extremely useful for understanding how different activators and enhanceosome complexes interact with the mediator and how the mediator interprets those interactions during regulated transcription.

Future Directions

The state of the art in the transcription field is improving at a rapid pace. Intact holoenzymes and subcomplexes containing mammalian mediator and SRB proteins have been isolated from several sources. The next stage will be to purify such complexes to homogeneity in quantities large enough to perform recruitment and mechanistic experiments. Methods for reconstituting the complexes from individual subunits will also be required for detailed structure–function studies. Although the role of TAF$_{II}$s in gene activation may be controversial, the approaches employed to study the interaction of TAF$_{II}$ subcomplexes with activators are widely applicable.

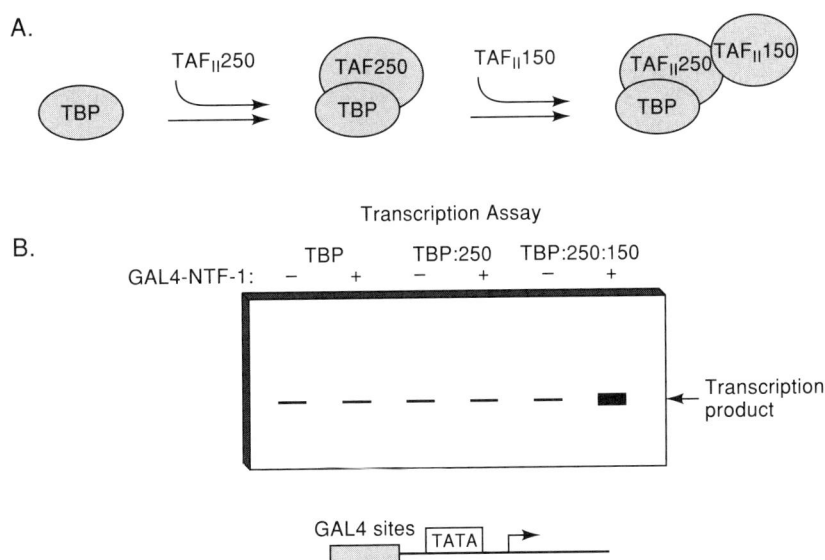

FIGURE 15.13. Schematic of TAF reconstitution experiment.

Recent studies of the yeast Pol II holoenzyme have allowed some simple tests of function. For example, one study by Ptashne and colleagues showed that excess holoenzyme could bypass the requirement for activators in an in vitro system. This result is consistent with the idea that activators function simply by recruiting holoenyme to a promoter (Gaudreau et al. 1998). Assays such as DNase I footprints or photocrosslinking, however, have yet to be achieved on the holoenzyme complexes. Such experiments will be necessary to understand the detailed workings of the complex and the roles of its myriad components.

Many studies have already been completed with systems composed of model activators like GAL4-VP16 or ZEBRA and idealized templates. Future challenges will be to understand how more sophisticated enhanceosome complexes function to interact with the general machinery. In this respect, the sarkosyl and oligonucleotide challenge experiments of Kim and Maniatis and the EMSA experiments of Thanos and colleagues showing interaction of the enhanceosome with CBP/P300 are noteworthy because they represent direct biochemical demonstration of the recruitment of a coactivator in a functional in vitro system (Kim and Maniatis 1997; Merika et al. 1998). Such assays allow one to probe the role of individual activator and activator mutants on cooperative recruitment and synergistic gene activation.

The preponderance of histone acetylase and ATP-dependent remodeling machines implicated as interacting with activators suggests that chromatin removal is a key step in complex assembly (for review, see Struhl 1998; Varga-Weisz and Becker 1998). Whether this step is mechanistically linked to transcription complex assembly is unknown, although it clearly must precede it. There have been many reports showing the effect of chromatin remodeling activities on transcription of chromatin templates but few that directly examine binding of the general machinery by footprinting or crosslinking assays. Some early studies probed binding of TBP and activators in the presence of SWI/SNF, but these did not examine the transcription complex assembly directly. More recently, Workman's significant study showing activator-targeted histone acetylation and Emerson's study demonstrating activator-specific recruitment of an ATP-dependent remodeling machine are advances in this area (Armstrong et al. 1998; Utley et al. 1998). Future studies must go further to probe complex assembly on chromatin structurally, how it is influenced by activators, and how chromatin remodeling and transcription complex assembly are linked.

PROTOCOL 15.1

Potassium Permanganate Probing of Pol II Open Complexes

SUMMARY

One measure of an activator's ability to stimulate preinitiation complex assembly is stimulation of the Pol II open complex. The reagent most commonly employed to visualize open complexes is potassium permanganate ($KMnO_4$). The method was developed by Gralla and colleagues in 1989 (Sasse-Dwight and Gralla 1989) to study prokaryotic RNA polymerase open complexes and was subsequently applied to Pol II and Pol III complexes (Kassavetis et al. 1990; Wang et al. 1992a).

Potassium permanganate modifies thymine residues in single- but not in double-stranded DNA. Permanganate has been used to identify the Pol II open complexes on promoters in nuclear extracts (Wang et al. 1992a) or in vivo in several experimental organisms (Giardina and Lis 1993). DNA melting by Pol II in vitro is energy dependent, so ATP is added to the reaction mix. dATP is commonly used to avoid the possibility that the ATP will be accidentally incorporated into a transcript. α-Amanitin prevents elongation that can be facilitated by residual amounts of NTPs present in the nuclear extract. The modified thymines are detected by primer extension via linear PCR. In the PCR, *Taq* polymerase stalls at the modified thymines on the template. The resulting products fractionate as a series of bands on a polyacrylamide urea sequencing gel. Permanganate will generate a low background modification of thymines in double-stranded DNA. If the open complexes form on only a small fraction of DNA templates, it may be difficult to detect the signal against the background. Thus, the initial experiment is performed using a wide range of template concentrations. Promoters with an abundance of T residues near the start site are preferable because they enhance the potential signal.

It is critical that the start site be mapped by primer extension or some other technique to ensure that the start site region is known. This requirement is important in mammalian cell extracts; however, in yeast sometimes the open complex and start sites show some differences in location. The reactions are performed at 30°C and contain DNA template, nuclear extract, and activator, much like the in vitro transcription experiment. Typically, the reaction contains 10–50 ng of a template present on a 4- to 5-kbp plasmid. The reaction also contains nonspecific carrier DNA. After a brief 15-minute incubation to permit assembly of the preinitiation complex, permanganate is added and the DNA is modified for a period of 1 or more minutes. The DNA is then purified and subjected to primer extension analysis; the products are fractionated on polyacrylamide urea sequencing gels. Reactions lacking and containing activator are compared, as are reactions lacking and containing ATP. Titrations of the DNA template, activator, and a time course are recommended. Often the amount of open complex correlates well with transcription. In the case of GAL4-VP16 on templates bearing different numbers of sites, the amount of open complex paralleled the amount of transcription (Wang et al. 1992a). A similar phenomenon was observed by comparing the transcriptional potency of VP16 mutants versus their ability to form open complexes (Jiang et al. 1994).

TIME LINE AND ORGANIZATION

The method is simple and requires the standard components for an in vitro transcription reaction in addition to a ^{32}P-end-labeled DNA primer positioned 60–90 nucleotides downstream of the transcription start site complementary to the top strand. The reaction itself and primer extension can be completed in 1 day. The result can be detected on the second day. It may be necessary to optimize the reaction; this process could take a week or more. Transcriptionally "weak" extracts or promoters make it more difficult to detect signals. The permanganate is a strong oxidant and destroys protein–DNA interactions over long incubation times. Typically, one compares the open complex on templates bearing and lacking activator binding sites or reactions containing and lacking activator.

Day 1: Potassium permanganate probing
Day 2: Data analysis

OUTLINE

Potassium permanganate probing (1 day)

Step 1: Prepare buffers and reagents (20 minutes)
Step 2: Prepare reactions and perform footprint (1 hour, 30 minutes)
Step 3: Perform primer extension reaction (2–3 hours)
Step 4: Fractionate primer extension products on gel (3–4 hours)

Data analysis

PROCEDURE

CAUTION: *KMnO$_4$, see Appendix I.*

Step 1. Prepare buffers and reagents

KMnO$_4$ stock:

0.37 M stored in lightproof bottle
Prepare working solution by diluting 5 μl of the stock into 32 μl of H$_2$O immediately prior to use.

Stop buffer:

0.4 M sodium acetate
0.1% SDS
10 mM EDTA
50 μg/ml yeast tRNA

Step 2: Prepare reactions and perform footprint

1. Prepare reaction mixture.
   Mix:

nuclear extract (10 mg/ml)	25 μl[a]
template (10 ng/μl)	1.0 μl[b]
carrier DNA (200 ng/ μl)	1.0 μl
$MgCl_2$ (100 mM)	3.3 μl
dATP (20 mM)	1.0 μl
α-amanitin (80 ng/ μl)	1.0 μl
H_2O	To 40 μl

   [a] HeLa extract and buffer D (see Protocol 15.2) can be varied but should total 25 μl.

   [b] DNA template concentration can be varied but total DNA should be 200 ng of bulk DNA.

2. Incubate for 30 minutes at 30°C.

3. Add 5 μl of $KMnO_4$ working solution. Incubate at 30°C for 4 minutes.

4. Add 3 μl of β-mercaptoethanol to quench $KMnO_4$ reaction. Add 100 μl of stop buffer, plus 1 μl of Proteinase K (10 mg/ml).

5. Incubate for 15 minutes at 37°C. Phenol-extract. Do not phenol/chloroform-extract until phenol-extracted at least once.

6. Pellet in a microfuge for 10 minutes at 14,000 rpm. Wash the pellet with 80% ethanol. Air-dry.

7. Resuspend the pellet in 60 μl of TE.

Step 3: Perform primer extension reaction

1. Mix:

DNA product from Step 2	20 μl
10x *Taq* polymerase buffer	5.0 μl
dNTPs (25 mM)	0.5 μl
labeled primer (0.3 pmoles / μl)	1.0 μl
Taq polymerase (1–5 units)	0.25 μl
$MgCl_2$ (100 mM)	1.5 μl[a]
H_2O	to 50 μl

   [a] As in many Taq-dependent primer extension reactions, the magnesium concentration must be determined empirically.

   Overlay with mineral oil unless using insulated chamber thermocycler.

2. Set the thermocycler as follows: 90°C, 1 minute; 55°C, 2 minutes; 72°C, 2 minutes. Repeat for 30 cycles.

3. Remove the supernatant and add 33 μl of 5 M ammonium acetate. Add 250 μl of ethanol. Mix well. Incubate on dry ice for 10 minutes.

4. Pellet in a microfuge for 10 minutes at 14,000 rpm. Wash the pellet with 80% ethanol and air-dry.

5. Resuspend the pellet in 4 μl of formamide sequencing dye. Fractionate products on an 8% sequencing gel alongside a dideoxy sequencing ladder generated with the same primer used in the primer extension. The ^{32}P-labeled primer can be used if the sequencing is done with *Taq* polymerase in an amplification reaction. Otherwise, use unlabeled primer and ^{35}S or ^{32}P nucleotide labeling mixes.

Step 4: Data analysis

Use the markers to identify the presumed position of the open complex. The open complex should form at the start site. The gel should show a clear difference between the activator and the nonactivated lanes. Sometimes an open complex is observed in the absence of ATP due to endogenous nucleotide in the mammalian nuclear extract.

ADDITIONAL CONSIDERATIONS

1. The method can be employed with an end-labeled DNA instead of using primer extension to detect the modified thymines (Ts). In such a case, the reaction products are treated first with piperidine as in Maxam-Gilbert sequencing, or, often, strand scission occurs without treatment and the products can be loaded directly onto a gel.

2. When using primer extension to detect the products, it is important to have a mock-treated DNA, or DNA incubated in the extract alone, as a control. The primer extension is sometimes sensitive to natural pause sequences in untreated DNA.

3. This method has also been used to study early events in elongation and promoter escape.

TROUBLESHOOTING

No specific signal

A permanganate-dependent banding pattern is observed but no open complex signal is seen in the presence versus absence of activator or ATP.
Possible cause: Inactive nuclear extracts, or the template does not contain accessible Ts near the start site.
Solution: Assay transcriptional activation in the extract to ensure activity. Often lowering the template concentration improves signal to noise. Add more Ts to the template via mutagenesis to enhance sensitivity.

No modification signal

Possible cause: Excess nucleotide in extract can cause elongation and inability to observe a signal.
Solution: Ensure that α-amanitin is added.
No permanganate-dependent modification is observed.

Possible cause: The permanganate may have gone bad.
Solution: Try fresh permanganate at higher concentrations or slightly longer incubation times.

PROTOCOL 15.2

Magnesium-agarose EMSA of TFIID Binding to DNA

The general transcription factor TFIID is a key target for regulation because its binding to a core promoter is the nucleating step in transcription complex assembly. Many eukaryotic activators stimulate recruitment of the TFIID when its concentration is made limiting at a promoter in vitro. Activator-stimulated recruitment of TFIID has traditionally been studied by DNase I footprinting. However, recent studies have shown that magnesium-agarose gels can separate large complexes containing TFIID, TFIIA (the DA complex), and TFIIB (the DAB complex) and permit a quantitative measurement of how activators stimulate assembly of such complexes. The advantage of the EMSA is that the reactions can be performed under subsaturating conditions where a TFIID footprint might not be observed (Lieberman and Berk 1994; Zerby and Lieberman 1997). Typically, the activator is incubated with a ^{32}P-labeled DNA template, recombinant TFIIA purified from *E. coli*, and immunopurified TFIID; TFIID is the HA epitope-tagged version from a special strain constructed by Berk and colleagues (Zhou et al. 1992), but the FLAG-tagged versions of Roeder and colleagues (Chiang et al. 1993) should also work.

The DNA–protein complexes containing TFIID migrate with reduced mobility on magnesium-agarose gels for two reasons: (1) the TBP subunit induces a sharp bend in the DNA causing altered mobility; and (2) the large size of the TBP–TAF$_{II}$ complex. By comparing the binding of TFIID over a wide concentration range, with and without activator, one can assess whether the activator interacts with TBP or with one of the TAF$_{II}$s. Additional factors such as TFIIA and TFIIB can be added subsequently to quantify their contributions to assembly of the transcription complex.

The method is sensitive and informative, but as with most biochemical assays, the results should be combined with in vitro transcription studies (i.e., kinetic analyses) to assess the physiological relevance of the DA complex during transcription. We have found DA complex assembly has been shown to correlate well with the level of gene activation and formation of activator stimulated open complexes (Chi and Carey 1993; Chi et al. 1995).

TIME LINE AND ORGANIZATION

The protocol is relatively simple to perform but requires special reagents that may not be readily available, including the general factors TFIIA, TFIIB, and TFIID. The bulk of the time will be spent purifying and characterizing these reagents prior to successfully employing the technique. TFIIA and TFIIB are available in recombinant form (Promega cat. #E3790). However, TFIID is difficult to obtain, and facilities must be available for large-scale growth of the strain bearing the epitope-tagged TFIID (see Chapter 14). A method that has been steadily increasing in popularity is to create recombinant TBP–TAF$_{II}$ subcomplexes using TBP from *E. coli* and TAF$_{II}$s overexpressed in baculovirus systems (Chen and Tjian 1996).

Ideally, the binding reactions contain 10 ng of ~200-bp ^{32}P-labeled DNA template in addition to the proteins. Occasionally, 0.1% Tween 20 is added to prevent aggregation. Typically, after 1 hour at 30°C, the mixtures are loaded onto horizontal, 10-cm long, 1.4% slab agarose gels (~100 ml) prepared in 45 mM Tris base, 45 mM boric acid, and 5 mM magnesium acetate; the gels are electrophoresed for 3 hours at 50 V in the same buffer, dried onto Whatman DE81 paper (cat. # 3658-917), and autoradiographed by exposure to XAR-5 film or phosphorimager analysis (Kodak). The agarose should be of the low EEO variety for consistent results.

Day 1: ^{32}P-Labeled promoter fragment (see Protocol 13.6)
Day 2: Magnesium agarose gel electrophoresis
Day 3: Data analysis

OUTLINE

Label promoter fragment with [γ-^{32}P]ATP (see Protocol 13.6)

Magnesium agarose gel electrophoresis (1 day)

Step 1: Prepare buffers (30 minutes)
Step 2: Pour Mg-agarose gel (30 minutes)
Step 3: Set up binding reactions (1 hour)
Step 4: Electrophoresis (3 hours)
Step 5: Autoradiography (1 hour)

Data analysis

PROCEDURE

CAUTION: *Boric acid, Glycerol, KCl, KOH. See Appendix I.*

MAGNESIUM AGAROSE GEL ELECTROPHORESIS

Step 1: Prepare buffers

10x gel buffer:

54 g of Tris base
27.5 g of boric acid
10.7 g of magnesium acetate
Add dH$_2$O to 1.0 liter

Buffer D (0.1 M KCl):

20 mM HEPES-KOH (pH 7.9)
20% v/v glycerol
0.2 mM EDTA
0.1 M KCl
Use 0.6x Buffer D (0.1 M KCl) for binding reactions

Step 2: Pour Mg-agarose gel

Prepare a 1.4% Mg-agarose gel using low EEO agarose (Fisher cat. # BP160-500) in 1x gel buffer. Let the mixture cool briefly and then pour the gel, avoiding clumps. The gels are generally quite small in size and should be run for short periods of time, occasionally in the cold room to minimize dissociation.

Step 3: Set up binding reactions

The typical reaction will contain activator and TFIID supplemented with either TFIIA or TFIIB. The conditions described incorporate all three general factors. Also, the amounts are based on recombinant proteins in our labs and ideally should be titrated over a broad range in this type of analysis. All of the factors employed are generally dialyzed against buffer D containing 0.1 M KCl.

1. The mixture should be prepared on ice and a batch should be made to accommodate multiple reactions. Note that the final KCl concentration needs to be 60 mM and final Buffer D concentration needs to be 0.6x. Therefore, if the volumes of any of the transcription factors added are changed, adjust the reaction by adding a diluent with Buffer D + 0.1 M KCl. The reactions should be prepared in 0.5-ml siliconized microfuge tubes.

   Mix:

immunopure, epitope-tagged TFIID[a] (200 ng/µl)	2.0 µl
recombinant TFIIA[a] (20 ng/µl)	1.0 µl
recombinant TFIIB[a] (160 ng/µl)	1.0 µl
activator	1.0 µl
γ-^{32}P-end-labeled probe (10,000 cpm/µl)	0.1 µl
dG:dC (1.0 mg/ml)	0.2 µl
BSA (50 mg/ml)	0.1 µl
β-mercaptoethanol (14 M)	0.05 µl
$MgCl_2$ (0.1 M)	0.75 µl
PMSF (100 mM)	0.05 µl
H_2O	3.35 µl
Buffer D (0.1 M KCl)	3.0 µl
Total	13.0 µl

   [a] Binding can be analyzed with TFIID alone or with TFIIA and TFIIB.

2. Incubate the mixtures at 30°C for 15–60 minutes. If space and reagents permit, two time points are recommended.

Step 4: Electrophoresis

The gels are ideally 0.5-cm-thick mini-gels (i.e., 10 cm long, 8 cm wide). The mixtures should be gently loaded into the wells; focus on layering the mixture. The glycerol from the Buffer D should be enough to permit the sample to layer. Load a standard gel dye mix on a side lane. The gel should be electrophoresed at 50 V for 3 hours at room temperature. The bromophenol blue should remain on the gel. Use shorter electrophoresis times for shorter gels.

Step 5: Autoradiography or phosphorimager analysis

Cut a sheet of Whatman DE81 paper to the size of the gel slab. Place the gel on top of this and cover the gel with Saran Wrap. Place two pieces of Whatman filter paper underneath, and vacuum-dry the gel at 65°C for 1 hour or until completely dry. Transfer the dried gel to a film or phosphorimager cassette for autoradiography.

Data analysis

Ideally, the free probe will migrate immediately above the bromophenol blue dye front. Addition of activator generally induces only a small change in mobility in an agarose gel (unless the activator alters the shape of the DNA). The DA complex alone should, however, give rise to a shift of much greater mobility than activator alone. The activator–DA complex may not migrate significantly higher, but the relative ratio of DA–activator complex to the activator alone should be greater than when the DA complex alone is compared with free probe. Often we use the ratios of the complexes as determined by a phosphorimager to carefully quantitate the effect of the activator.

ADDITIONAL CONSIDERATIONS

1. The TFIID must be titrated over a broad range to observe an effect of activator. At low concentrations of TFIID, little effect is seen on recruitment by ZEBRA (Chi et al. 1995), for example, while at high concentrations the DNA becomes saturated without added activator. It is critical to perform a dose–response titration to identify the relevant window of concentration.

2. TFIIA, TFIIB, and almost certainly other factors will influence the ability of an activator to recruit the complex. Because all of these transcription complex assembly reactions are somewhat cooperative, an activator interacting strongly with TFIIB might recruit TFIID indirectly by virtue of the activator binding TFIIB while TFIIB is binding to TFIID. Again, it is important to quantify these additional contributions through the use of titrations.

3. For reasons that are unclear, TBP-containing complexes sometimes migrate in these gels with mobilities very similar to TFIID. Although the TBP complex is generally unaffected by activator, it can be used to study the contributions of activator–TFIIA, and activator–TFIIB, interactions to TBP recruitment in the absence of TAF$_{II}$s.

4. The gel system is amenable to many different conditions. Sometimes it is difficult to resolve one complex from another (i.e., the DA from the DAB complex). Titrating magnesium acetate concentration sometimes helps to distinguish between the two complexes. Furthermore, the gels can be electrophoresed at a variety of temperatures, which change the relative mobilities of the complexes. The binding reactions themselves can also tolerate changes in salt concentration.

TROUBLESHOOTING

No DA complex or activator binding

Possible cause: It is plausible that under some assay conditions certain activators will not bind stably to promoter DNA; however, it is unlikely that DA will not bind.
Solution: If no TFIID binding is observed, add more TFIID fraction. The affinity of TFIID for various core promoters varies widely and, as mentioned above, the method tends to work best over narrow windows of TFIID concentration. Thus, it is advisable simply to increase TFIID if binding is not observed.

Smeary gels

Possible cause: Sometimes the preparations of immunopurified TFIID still have some protein A-Sepharose beads, which can retain the complexes in the wells.

Solutions: To avoid this, spin the TFIID preparation in a microfuge before use and carefully pipet the supernatant into a new siliconized tube. Low-purity TFIID can also cause the complexes to be retained in the wells. Additional wash steps during the TFIID immunopurification procedure can remove contaminants and alleviate the problem. In addition, check to ensure that low EEO agarose is being used. Check purity of proteins. Add 0.05% NP-40.

No stimulation of DA by activator

If over a wide range of conditions the DA complex is observed and accumulates in a dose-responsive manner yet the activator is not stimulating its formation, one might conclude that the activator is not directly interacting with TFIID. For example, a GAL4 derivative containing the intact VP16 activation domain stimulates DA complex formation, whereas a similar derivative containing only the amino-terminal half of the VP16 activation domain fails to stimulate. Yet both activate transcription very well in HeLa in vitro systems. Data such as these have pointed to other interactions between activators and the general machinery as being more important for in vitro gene activation.

REFERENCES

Arias J.A. and Dynan W.S. 1989. Promoter-dependent transcription by RNA polymerase II using immobilized enzyme complexes. *J. Biol. Chem.* **264:** 3223–3229.

Armstrong J.A., Bieker J.J., and Emerson B.M. 1998. A SWI/SNF-related chromatin remodeling complex, E-RC1, is required for tissue-specific transcriptional regulation by EKLF in vitro. *Cell* **95:** 93–104.

Bartholomew B., Kassavetis G.A., and Geiduschek E.P. 1991. Two components of *Saccharomyces cerevisiae* transcription factor IIIB (TFIIB) are stereospecifically located upstream of a tRNA gene and interact with the second-largest subunit of TFIIIC. *Mol. Cell Biol.* **11:** 5181–5189.

Bartholomew B., Durkovich D., Kassavetis G.A., and Geiduschek E.P. 1993. Orientation and topography of RNA polymerase III in transcription complexes. *Mol. Cell. Biol.* **13:** 942–952.

Bentley D.L. 1995. Regulation of transcriptional elongation by RNA polymerase II. *Curr. Opin. Genet. Dev.* **5:** 210–216.

Brown S.A. and Kingston R.E. 1997. Disruption of downstream chromatin directed by a transcriptional activator. *Genes Dev.* **11:** 3116–3121.

Bunick D., Zandomeni R., Ackerman S., and Weinmann R. 1982. Mechanism of RNA polymerase II–specific initiation of transcription in vitro: ATP requirement and uncapped runoff transcripts. *Cell* **29:** 877–886.

Buratowski S., Hahn S., Guarente L., and Sharp P.A. 1989. Five intermediate complexes in transcription initiation by RNA polymerase II. *Cell* **56:** 549–561.

Cai H. and Luse D.S. 1987a. Transcription initiation by RNA polymerase II in vitro. Properties of preinitiation, initiation, and elongation complexes. *J. Biol. Chem.* **262:** 298–304.

———. 1987b. Variations in template protection by the RNA polymerase II transcription complex during the initiation process. *Mol. Cell. Biol.* **7:** 3371–3379.

Carey M. 1998. The enhanceosome and transcriptional synergy. *Cell* **92:** 5–8.

Carey M.F., Gerrard S.P., and Cozzarelli N.R. 1986. Analysis of RNA polymerase III transcription complexes by gel filtration. *J. Biol. Chem.* **261:** 4309–4317.

Chen J.L. and Tjian R. 1996. Reconstitution of TATA-binding protein-associated factor/TATA-binding protein complexes for in vitro transcription. *Methods Enzymol.* **273:** 208–217.

Chen J.L., Attardi L.D., Verrijzer C.P., Yokomori K., and Tjian R. 1994. Assembly of recombinant TFIID reveals differential coactivator requirements for distinct transcriptional activators. *Cell* **79:** 93–105.

Chi T., Lieberman P., Ellwood K., and Carey M. 1995. A general mechanism for transcriptional synergy by eukaryotic activators. *Nature* **377:** 254–257.

Chi T. and Carey M. 1993. The ZEBRA activation domain: Modular organization and mechanism of action. *Mol. Cell. Biol.* **13:** 7045–7055.

———. 1996. Assembly of the isomerized TFIIA–TFIID–TATA ternary complex is necessary and sufficient for gene activation. *Genes Dev.* **10:** 2540–2550.

Chiang C.M., Ge H., Wang Z., Hoffmann A., and Roeder R.G. 1993. Unique TATA-binding protein-containing complexes and cofactors involved in transcription by RNA polymerases II and III. *EMBO J.* **12:** 2749–2762.

Choy B. and Green M.R. 1993. Eukaryotic activators function during multiple steps of preinitiation complex assembly. *Nature* **366:** 531–536.

Dvir A., Conaway R.C., and Conaway J.W. 1997a. A role for TFIIH in controlling the activity of early RNA polymerase II elongation complexes. *Proc. Natl. Acad. Sci.* **94:** 9006–9010.

Dvir A., Tan S., Conaway J.W., and Conaway R.C. 1997b. Promoter escape by RNA polymerase II. Formation of an escape-competent transcriptional intermediate is a prerequisite for exit of polymerase from the promoter. *J. Biol. Chem.* **272:** 28175–28178.

Ellwood K., Chi T., Huang W., Mitsouras K., and Carey M. 1998. Cooperative assembly of RNA polymerase II transcription complexes. *Cold Spring Harbor Symp. Quant. Biol.* **63:** 253–261.

Gaudreau L., Adam M., and Ptashne M. 1998. Activation of transcription in vitro by recruitment of the yeast RNA polymerase II holoenzyme. *Mol. Cell* **1:** 913–916.

Giardina C. and Lis J.T. 1993. DNA melting on yeast RNA polymerase II promoters. *Science* **261:** 759–762.

Goodrich J.A. and Tjian R. 1994. Transcription factors IIE and IIH and ATP hydrolysis direct promoter clearance by RNA polymerase II. *Cell* **77:** 145–156.

Gralla J.D. 1996. Global steps during initiation by RNA polymerase II. *Methods Enzymol.* **273:** 99–110.

Hampsey M. 1998. Molecular genetics of the RNA polymerase II general transcriptional machinery. *Microbiol. Mol. Biol. Rev.* **62:** 465–503.

Hawley D.K. and Roeder R.G. 1985. Separation and partial characterization of three functional steps in transcription initiation by human RNA polymerase II. *J. Biol. Chem.* **260:** 8163–8172.

———. 1987. Functional steps in transcription initiation and reinitiation from the major late promoter in a HeLa nuclear extract. *J. Biol. Chem.* **262:** 3452–3461.

Holstege F.C., Fiedler U., and Timmers H.T. 1997. Three transitions in the RNA polymerase II transcription complex during initiation. *EMBO J.* **16:** 7468–7480.

Holstege F.C., Tantin D., Carey M., van der Vliet P.C., and Timmers H.T. 1995. The requirement for the basal transcription factor IIE is determined by the helical stability of promoter DNA. *EMBO J.* **14:** 810–819.

Hori R. and Carey M. 1994. The role of activators in assembly of RNA polymerase II transcription complexes. *Curr. Opin. Genet. Dev.* **4:** 236–244.

Horikoshi M., Hai T., Lin Y.S., Green M.R., and Roeder R.G. 1988. Transcription factor ATF interacts with the TATA factor to facilitate establishment of a preinitiation complex. *Cell* **54:** 1033–1042.

Jacob G.A., Kitzmiller J.A., and Luse D.S. 1994. RNA polymerase II promoter strength in vitro may be reduced by defects at initiation or promoter clearance. *J. Biol. Chem.* **269:** 3655–3663.

Jiang Y. and Gralla J.D. 1993. Uncoupling of initiation and reinitiation rates during HeLa RNA polymerase II transcription in vitro. *Mol. Cell. Biol.* **13:** 4572–4577.

Jiang Y., Yan M., and Gralla J.D. 1996. A three-step pathway of transcription initiation leading to promoter clearance at an activation RNA polymerase II promoter. *Mol. Cell. Biol.* **16:** 1614–1621.

Jiang Y., Triezenberg S.J., and Gralla J.D. 1994. Defective transcriptional activation by diverse VP16 mutants associated with a common inability to form open promoter complexes. *J. Biol. Chem.* **269:** 5505–5508.

Johnson F.B. and Krasnow M.A. 1992. Differential regulation of transcription preinitiation complex assembly by activator and repressor homeo domain proteins. *Genes Dev.* **6:** 2177–2189.

Jones KA. 1997. Taking a new TAK on tat transactivation. *Genes Dev.* **11:** 2593–2599.

Kasher M.S., Pintel D., and Ward D.C. 1986. Rapid enrichment of HeLa transcription factors IIIB and IIIC by using affinity chromatography based on avidin-biotin interactions. *Mol. Cell. Biol.* **6:** 3117–3127.

Kassavetis G.A., Blanco J.A., Johnson T.E., and Geiduschek E.P. 1992. Formation of open and elongating transcription complexes by RNA polymerase III. *J. Mol. Biol.* **226:** 47–58.

Kassavetis G.A., Braun B.R., Nguyen L.H., and Geiduschek E.P. 1990. *S. cerevisiae* TFIIIB is the transcription initiation factor proper of RNA polymerase III, while TFIIIA and TFIIIC are assembly factors. *Cell* **60:** 235–245.

Kim T.K. and Maniatis T. 1997. The mechanism of transcriptional synergy of an in vitro assembled interferon-β enhanceosome. *Mol. Cell* **1:** 119–129.

Kim T.K., Lagrange T., Wang Y.H., Griffith J.D., Reinberg D., and Ebright R.H. 1997. Trajectory of DNA in the RNA polymerase II transcription preinitiation complex. *Proc. Natl. Acad. Sci.* **94:** 12268–12273.

Kingston R.E. and Green M.R. 1994. Modeling eukaryotic transcriptional activation. *Curr. Biol.* **4:** 325–332.

Kumar K.P., Akoulitchev S., and Reinberg D. 1998. Promoter-proximal stalling results from the inability to recruit transcription factor IIH to the transcription complex and is a regulated event. *Proc. Natl. Acad. Sci.* **95:** 9767–9772.

Lagrange T., Kim T.K., Orphanides G., Ebright Y.W., Ebright R.H., and Reinberg D. 1996. High-resolution mapping of nucleoprotein complexes by site-specific protein-DNA photocrosslinking: Organization of the human TBP-TFIIB-DNA quarternary complex. *Proc. Natl. Acad. Sci.* **93:** 10620–10626.

Lieberman P. 1994. Identification of functional targets of the Zta transcriptional activator by formation of stable preinitiation complex intermediates. *Mol. Cell. Biol.* **14:** 8365–8375.

Lieberman P.M. and Berk A.J. 1994. A mechanism for TAF$_{II}$s in transcriptional activation: Activation domain enhancement of TFIID-TFIIA-promoter DNA complex formation. *Genes Dev.* **8:** 995–1006.

Lin Y.S. and Green M.R. 1991. Mechanism of action of an acidic transcriptional activator in vitro. *Cell* **64:** 971–981.

Luse D.S. and Jacob G.A. 1987. Abortive initiation by RNA polymerase II in vitro at the adenovirus 2 major late promoter. *J. Biol. Chem.* **262:** 14990–14997.

Merika M., Williams A.J., Chen G., Collins T., and Thanos D. 1998. Recruitment of CBP/p300 by the IFN β enhanceosome is required for synergistic activation of transcription. *Mol. Cell* **1:** 277–287.

Nakajima N., Horikoshi M., and Roeder R.G. 1988. Factors involved in specific transcription by mammalian RNA polymerase II: Purification, genetic specificity, and TATA box-promoter interactions of TFIID. *Mol. Cell. Biol.* **8:** 4028–4040.

Oelgeschlager T., Chiang C.M., and Roeder R.G. 1996. Topology and reorganization of a human TFIID-promoter complex. *Nature* **382:** 735–738.

Orphanides G., Lagrange T., and Reinberg D. 1996. The general transcription factors of RNA polymerase II. *Genes Dev.* **10:** 2657–2683.

Pan G. and Greenblatt J. 1994. Initiation of transcription by RNA polymerase II is limited by melting of the promoter DNA in the region immediately upstream of the initiation site. *J. Biol. Chem.* **269:** 30101–30104.

Phelan M.L., Sif S., Narlikar G.J., and Kingston R.E. 1999. Reconstitution of a core chromatin remodeling complex from SWI/SNF subunits. *Mol. Cell* **3:** 247–253.

Ptashne M. and Gann A. 1997. Transcriptional activation by recruitment. *Nature* **386:** 569–577.

Pugh B.F. 1996. Mechanisms of transcription complex assembly. *Curr. Opin. Cell Biol.* **8:** 303–311.

Ranish J.A., Yudkovsky N., and Hahn S. 1999. Intermediates in formation and activity of the RNA polymerase II preinitiation complex: Holoenzyme recruitment and a postrecruitment role for the TATA box and TFIIB. *Genes Dev.* **13:** 49–63.

Reines D., Dvir A., Conaway J.W., and Conaway R.C. 1997. Assays for investigating transcription by RNA polymerase II in vitro. *Methods* **12:** 192–202.

Robert F., Douziech M., Forget D., Egly J.M., Greenblatt J., Burton Z.F., and Coulombe B. 1998. Wrapping of promoter DNA around the RNA polymerase II initiation complex induced by TFIIF. *Mol. Cell* **2:** 341–351.

Roberts S.G., Choy B., Walker S.S., Lin Y.S., and Green M.R. 1995. A role for activator-mediated TFIIB recruitment in diverse aspects of transcriptional regulation. *Curr. Biol.* **5:** 508–516.

Samuels M., Fire A., and Sharp P.A. 1984. Dinucleotide priming of transcription mediated by RNA polymerase II. *J. Biol. Chem.* **259:** 2517–2525.

Sandaltzopoulos R. and Becker P.B. 1998. Heat shock factor increases the reinitiation rate from potentiated chromatin templates. *Mol. Cell. Biol.* **18:** 361–367.

Sasse-Dwight S. and Gralla J.D. 1989. KMnO$_4$ as a probe for lac promoter DNA melting and mechanism in vivo. *J. Biol. Chem.* **264:** 8074–8081.

Sawadogo M. and Roeder R.G. 1984. Energy requirement for specific transcription initiation by the human RNA polymerase II system. *J. Biol. Chem.* **259:** 5321–5326.

———. 1985. Interaction of a gene-specific transcription factor with the adenovirus major late promoter upstream of the TATA box region. *Cell* **43:** 165–175.

Stargell L.A. and Struhl K. 1996. Mechanisms of transcriptional activation in vivo: Two steps forward. *Trends Genet.* **12:** 311–315.

Struhl K. 1998. Histone acetylation and transcriptional regulatory mechanisms. *Genes Dev.* **12:** 599–606.

Szentirmay M.N. and Sawadogo M. 1994. Sarkosyl block of transcription reinitiation by RNA polymerase II as visualized by the colliding polymerases reinitiation assay. *Nucleic Acids Res.* **22:** 5341–5346.

Tantin D. and Carey M. 1994. A heteroduplex template circumvents the energetic requirement for ATP during activated transcription by RNA polymerase II. *J. Biol. Chem.* **269:** 17397–17400.

Tantin D., Chi T., Hori R., Pyo S., and Carey M. 1996. Biochemical mechanism of transcriptional activation by GAL4-VP16. *Methods Enzymol.* **274:** 133–149.

Timmers H.T. 1994. Transcription initiation by RNA polymerase II does not require hydrolysis of the beta-gamma phosphoanhydride bond of ATP. *EMBO J.* **13:** 391–399.

Tirode F., Busso D., Coin F., and Egly J.M. 1999. Reconstitution of the transcription factor TFIIH: Assignment of functions for the three enzymatic subunits, XPB, XPD, and cdk7. *Mol. Cell* **3:** 87–95.

Utley R.T., Ikeda K., Grant P.A., Cote J., Steger D.J., Eberharter A., John S., and Workman J.L. 1998. Transcriptional activators direct histone acetyltransferase complexes to nucleosomes. *Nature* **394:** 498–502.

Van Dyke M.W., Roeder R.G., and Sawadogo M. 1988. Physical analysis of transcription preinitiation complex assembly on a class II gene promoter. *Science* **241:** 1335–1338.

Van Dyke M.W., Sawadogo M., and Roeder R.G. 1989. Stability of transcription complexes on class II genes. *Mol. Cell. Biol.* **9:** 342–344.

Varga-Weisz P.D. and Becker P.B. 1998. Chromatin-remodeling factors: Machines that regulate? *Curr. Opin. Cell Biol.* **10:** 346–353.

Wang W., Carey M., and Gralla J.D. 1992a. Polymerase II promoter activation: Closed complex formation and ATP-driven start site opening. *Science* **255:** 450–453.

Wang W., Gralla J.D., and Carey M. 1992b. The acidic activator GAL4-AH can stimulate polymerase II transcription by promoting assembly of a closed complex requiring TFIID and TFIIA. *Genes Dev.* **6:** 1716–1727.

Yankulov K. and Bentley D. 1998. Transcriptional control: Tat cofactors and transcriptional elongation. *Curr. Biol.* **8:** R447–449.

Zawel L., Kumar K.P., and Reinberg D. 1995. Recycling of the general transcription factors during RNA polymerase II transcription. *Genes Dev.* **9:** 1479–1490.

Zenzie-Gregory B., O'Shea-Greenfield A., and Smale S.T. 1992. Similar mechanisms for transcription initiation mediated through a TATA box or an initiator element. *J. Biol. Chem.* **2674:** 2823–2830.

Zerby D. and Lieberman P.M. 1997. Functional analysis of TFIID-activator interaction by magnesium-agarose gel electrophoresis. *Methods* **12:** 217–223.

Zhou Q., Lieberman P.M., Boyer T.G., and Berk A.J. 1992. Holo-TFIID supports transcriptional stimulation by diverse activators and from a TATA-less promoter. *Genes Dev.* **6:** 1964–1974.

Cautions

General Cautions

The following general cautions should always be observed.

- **The absence of a warning** does not necessarily mean that the material is safe, since information may not always be complete or available.

- **Proper disposal procedures** must be used for all chemical, biological, and radioactive waste.

- Consult your local safety office for specific guidelines on **appropriate gloves**.

- **Acids and bases** that are concentrated should be handled with great care. Wear goggles and appropriate gloves. A face shield should be worn when handling large quantities. Strong acids should not be mixed with organic solvents as they may react. Especially, sulfuric acid and nitric acid may react highly exothermically and cause fires and explosions. Strong bases should not be mixed with halogenated solvent as they may form reactive carbenes which can lead to explosions. For proper disposal of strong acids and bases, dilute them by placing the acid or base onto ice and neutralize them. **Do not** pour water into them. If the solution does not contain any other toxic compound, the salts can be flushed down the drain.

- Never **pipet** solutions using mouth suction. This method is not sterile and can be dangerous. Always use a pipet aid or bulb.

- **Halogenated and nonhalogenated** solvents should be kept separate (e.g., mixing chloroform and acetone can cause unexpected reactions in the presence of bases).

- **Photographic fixatives and developers** also contain chemicals that can be harmful. Handle them with care and follow manufacturer's directions.

- **Power supplies and electrophoresis equipment** pose serious fire hazard and electrical shock hazards if not used properly.

- The use of **microwave ovens and autoclaves** in the lab requires certain precautions. Accidents have occurred involving their use (e.g., to melt agar or bactoagar stored in bottles or to sterilize). Often the screw top is not completely removed and there is not enough space for the steam to vent. When the containers are removed from the microwave or autoclave, they can explode and cause severe injury. Always completely remove bottle caps before microwaving or autoclaving. An alternative method for routine agarose gels that do not require sterile agar is to weigh out the agar and place the solution in a flask.

- Procedures for the **humane treatment of animals** must be observed at all times. Consult your local animal facility for guidelines.

Acetic acid (concentrated) must be handled with great care. It is harmful by inhalation, ingestion, or skin absorption. Wear appropriate gloves and goggles and use in a chemical fume hood.

Acrylamide (unpolymerized) is a potent neurotoxin and is absorbed through the skin (the effects are cumulative). Avoid breathing the dust. Wear appropriate gloves and a face mask when weighing powdered acrylamide and methylene-bisacrylamide. Use in a chemical fume hood. Polyacrylamide is considered to be nontoxic, but it should be handled with care because it might contain small quantities of unpolymerized acrylamide.

Actinomycin D is a teratogen and a carcinogen. It is highly toxic and may be fatal if inhaled, ingested, or absorbed through the skin. It may also cause irritation. Avoid breathing vapors. Wear appropriate gloves and safety glasses, and always use in a chemical fume hood. Solutions of actinomycin D are light-sensitive.

α-Amanitin is highly toxic and may be fatal by inhalation, ingestion, or skin absorption. Symptoms may be delayed for as long as 6–24 hours. Wear appropriate gloves and safety glasses and always work in a chemical fume hood.

Ammonium acetate may be harmful by inhalation, ingestion, or skin absorption. Wear appropriate gloves and safety glasses. Use in a chemical fume hood.

Ammonium persulfate is extremely destructive to tissue of the mucous membranes and upper respiratory tract, eyes, and skin. Inhalation may be fatal. Wear appropriate gloves, safety glasses, and protective clothing. Use only in a chemical fume hood. Wash thoroughly after handling.

Ammonium sulfate ($[NH_4]_2SO_4$) may be harmful by inhalation, ingestion, or skin absorption. Wear appropriate gloves and safety glasses.

Antipain may be harmful by inhalation, ingestion, or skin absorption. Wear appropriate gloves and safety glasses. Use in a chemical fume hood.

Aprotinin may be harmful by ingestion, inhalation, or skin absorption. It may also cause allergic reactions. Exposure may cause gastrointestinal effects, muscle pain, blood pressure changes, or bronchospasm. Wear appropriate gloves and safety glasses. Do not breathe the dust. Use only in a chemical fume hood.

Boric acid (H_3BO_3) may be harmful by inhalation, ingestion, or skin absorption. Wear appropriate gloves and goggles. Use in a chemical fume hood.

Bromophenol blue may be harmful by inhalation, ingestion, or skin absorption. Wear appropriate gloves and safety glasses. Use in a chemical fume hood.

Calcium chloride ($CaCl_2$) is harmful by inhalation, ingestion, or skin absorption. Wear appropriate gloves and safety glasses. Use in a chemical fume hood.

Chloramphenicol is harmful by inhalation, ingestion, or skin absorption and is a carcinogen. Wear appropriate gloves and safety glasses. Use in a chemical fume hood.

Chloroform is irritating to the skin, eyes, mucous membranes, and respiratory tract. It is a carcinogen and may damage the liver and kidneys. Wear appropriate gloves and safety glasses and always use in a chemical fume hood.

Chloroquine may be harmful by inhalation, ingestion, or skin absorption. Prolonged exposure can lead to permanent eye damage. Wear appropriate gloves and safety goggles.

Diethylaminoethyl (DEAE) is harmful by inhalation, ingestion, or skin absorption. Wear appropriate gloves and safety glasses and use in a chemical fume hood.

Diethyl pyrocarbonate (DEPC) is a potent protein denaturant and is a suspected carcinogen. Aim bottle away from you when opening it; internal pressure can lead to splattering. Wear appropriate gloves and lab coat, and use in a chemical fume hood.

Dimethyl pimelimidate (DMP) is irritating to the eyes, skin, mucous membranes, and upper respiratory tract. It can exert harmful effects by inhalation, ingestion, or skin absorption. Avoid breathing the vapors. Wear appropriate gloves, face mask, and safety glasses and do not inhale.

Dimethyl sulfate (DMS) is extremely toxic and is a carcinogen. Avoid breathing the vapors. Wear appropriate gloves and safety glasses. Use only in a chemical fume hood. Dispose of solutions containing dimethyl sulfate in accordance with MSDS recommendations.

Dimethyl sulfoxide (DMSO) is harmful by inhalation or skin absorption. Wear appropriate gloves and safety glasses. Use in a chemical fume hood. DMSO is also combustible. Store in a tightly closed container.

Dithiothreitol (DTT) is a strong reducing agent that emits a foul odor. Wear lab coat and safety glasses and use in a chemical fume hood when working with the solid form or highly concentrated stocks.

Ethanol may be harmful by inhalation, ingestion, or skin absorption. Wear appropriate gloves and safety glasses.

Ethanolamine is toxic and harmful by inhalation, ingestion, or skin absorption. Handle with care and avoid any contact with the skin. Wear appropriate gloves and goggles and use in a chemical fume hood. Ethanolamine is highly corrosive and reacts violently with acids.

Ethidium bromide is a powerful mutagen and is moderately toxic. Consult the local institutional safety officer for specific handling and disposal procedures. Avoid breathing the vapors or dust. Wear appropriate gloves when working with solutions that contain this dye.

Ethylnitrosourea, see *N-Nitroso-N-ethylurea*

Formamide is teratogenic. The vapor is irritating to the eyes, skin, mucous membranes, and upper respiratory tract. It may be harmful by inhalation, ingestion, or skin absorption. Wear appropriate gloves and safety glasses. Always use a chemical fume hood when working with concentrated solutions of formamide. Keep working solutions covered as much as possible.

Glycerol may be harmful by inhalation, ingestion, or skin absorption. Wear appropriate gloves and safety glasses. Use in a chemical fume hood.

Hydrochloric acid (HCl) is volatile and may be fatal if inhaled, ingested, or absorbed through the skin. It is extremely destructive to mucous membranes, upper respiratory tract, eyes, and skin. Wear appropriate gloves and safety glasses and use with great care in a chemical fume hood. Wear goggles when handling large quantities.

Hydrogen peroxide is corrosive, toxic, and extremely damaging to the skin. It is harmful by inhalation, ingestion, and skin absorption. Wear appropriate gloves and safety glasses and use only in a chemical fume hood.

Isopropyl-β-D-galactopyranoside (IPTG) is harmful by inhalation, ingestion, or skin absorption. Wear appropriate gloves and safety glasses.

$K_2HPO_4 \cdot 3H_2O$ **(dibasic)** and KH_2PO_4 **(monobasic)**, see **Potassium phosphate**

KCl, see **Potassium chloride**

KMnO$_4$, see **Potassium permanganate**

KOH and **KOH/methanol,** see **Potassium hydroxide**

Leupeptin (or its **hemisulfate**) may be harmful by inhalation, ingestion, or skin absorption. Wear appropriate gloves and safety glasses. Use in a chemical fume hood.

Magnesium chloride (MgCl$_2$) is harmful by inhalation, ingestion, or skin absorption. Wear appropriate gloves and safety glasses, and use only in a chemical fume hood.

Magnesium sulfate (MgSO$_4$) may be harmful by inhalation, ingestion, or skin absorption. Wear appropriate gloves and safety glasses. Use in a chemical fume hood.

β-**Mercaptoethanol (2-Mercaptoethanol)** may be fatal if inhaled or absorbed through the skin and is harmful if ingested. High concentrations are extremely destructive to the mucous membranes, upper respiratory tract, skin, and eyes. Wear appropriate gloves and safety glasses. Always use in a chemical fume hood.

Methanol is poisonous and can cause blindness. It is harmful by inhalation, ingestion, or skin absorption. Adequate ventilation is necessary to limit exposure to vapors. Avoid inhaling these vapors. Wear appropriate gloves and goggles. Use only in a chemical fume hood.

MgCl$_2$, see **Magnesium chloride**

3-(N-Morpholino)-propanesulfonic acid (MOPS) may be harmful by inhalation, ingestion, or skin absorption. It is irritating to mucous membranes and upper respiratory tract. Wear appropriate gloves and safety glasses and use in a chemical fume hood.

(NH$_4$)$_2$SO$_4$, see **Ammonium sulfate**

Na$_2$HPO$_4$, see **Sodium hydrogen phosphate**

NaH$_2$PO$_4$, see **Sodium dihydrogen phosphate**

NaOH, see **Sodium hydroxide**

NaPO$_4$, see **Sodium phosphate**

N-**Nitroso-***N*-**ethylurea** may be harmful by inhalation, ingestion, or skin absorption. Wear appropriate gloves and safety glasses. Do not breathe the dust.

Pepstatin A may be harmful by inhalation, ingestion, or skin absorption. Wear appropriate gloves and safety glasses. Use in a chemical fume hood.

Phenol is highly corrosive and can cause severe burns. Wear appropriate gloves, goggles, and protective clothing. Always use in a chemical fume hood. Rinse any areas of skin that come in contact with phenol with a large volume of water and wash with soap and water; do not use ethanol!

Phenylmethylsulfonyl fluoride (PMSF) is a highly toxic cholinesterase inhibitor. It is extremely destructive to the mucous membranes of the respiratory tract, eyes, and skin. It may be fatal if inhaled, ingested, or absorbed through the skin. Wear appropriate gloves and safety glasses and always use in a chemical fume hood. In case of contact, immediately flush eyes or skin with copious amounts of water and discard contaminated clothing.

Piperidine is highly toxic and is corrosive to the eyes, skin, respiratory tract, and gastrointestinal tract. It reacts violently with acids and oxidizing agents. Do not breathe the vapors. Keep away from heat, sparks, and open flame. Wear appropriate gloves and safety glasses and use in a chemical fume hood.

PMSF, see **Phenylmethylsulfonyl fluoride**

Polyvinylpyrrolidone may be harmful by inhalation, ingestion, or skin absorption. Wear appropriate gloves and safety glasses. Use in a chemical fume hood.

Potassium chloride (KCl) may be harmful by inhalation, ingestion, or skin absorption. Wear appropriate gloves and safety glasses.

Potassium hydroxide (KOH and **KOH/methanol)** solutions should be handled with great care. Wear appropriate gloves.

Potassium permanganate (KMnO$_4$) is an irritant and is explosive. Use all solutions in a chemical fume hood.

Potassium phosphate (K$_2$HPO$_4$·3H$_2$O [dibasic]; KH$_2$PO$_4$ [monobasic]) may be harmful by inhalation, ingestion, or skin absorption. Wear appropriate gloves and safety glasses. Do not breathe the dust.

Radioactive substances: Wear appropriate gloves when handling. Consult the local safety office for further guidance in the appropriate use and disposal of radioactive materials. Always monitor thoroughly after using radioisotopes.

SDS, see **Sodium dodecyl sulfate**

Sodium cacodylate may be carcinogenic. It is highly toxic and may be fatal by inhalation and ingestion, or skin absorption. It also may cause harm to the unborn child. Effects of contact or inhalation may be delayed. Do not breathe the dust. Wear appropriate gloves and safety goggles and use only in a chemical fume hood.

Sodium dihydrogen phosphate (NaH$_2$PO$_4$) (sodium phosphate, monobasic) may be harmful by inhalation, ingestion, or skin absorption. Wear appropriate gloves and safety glasses. Use in a chemical fume hood.

Sodium dodecyl sulfate (SDS) is harmful if inhaled. Wear a face mask when weighing SDS.

Sodium hydrogen phosphate (Na$_2$HPO$_4$) may be harmful by inhalation, ingestion, or skin absorption. Wear appropriate gloves and safety glasses. Use in a chemical fume hood.

Sodium hydroxide (NaOH) and **solutions containing NaOH** are caustic and should be handled with great care. Wear appropriate gloves and a face mask. Concentrated bases should be handled in a similar manner.

Sodium phosphate (NaPO$_4$) is an irritant to the eyes and skin. It may be harmful by inhalation, ingestion, or skin absorption. Wear appropriate gloves and safety goggles. Do not breathe the dust.

N,N,N′,N′-**Tetramethylethylenediamine (TEMED)** is extremely destructive to tissue of the mucous membranes and upper respiratory tract, eyes, and skin. Inhalation may be fatal. Prolonged contact can cause severe irritation or burns. Wear appropriate gloves, safety glasses, and other protective clothing and use in a chemical fume hood. Wash thoroughly after handling. Flammable: Vapor may travel a considerable distance to source of ignition and flash back. Keep away from heat, sparks, and open flame.

Thiourea may be carcinogenic and is harmful by inhalation, ingestion, or skin absorption. Wear appropriate gloves and safety glasses and use in a chemical fume hood.

UV light (see **UV radiation**) can damage the retina of the eyes. Never look at an unshielded UV light source with naked eyes. View only through a filter or safety glasses that absorb harmful wavelengths.

UV radiation (see **UV light**) is dangerous, particularly to the eyes. UV radiation is also mutagenic and carcinogenic. To minimize exposure, make sure that the UV light source is adequately shielded. Wear protective safety glasses that efficiently block UV light. Wear protective appropriate gloves when holding materials under the UV light source.

Xylene must always be used in a chemical fume hood. It is flammable and may be narcotic at high concentrations.

Suppliers

Commercial sources and products have been included in the text for the user's convenience and should not necessarily be construed as an endorsement by the authors. With the exception of those suppliers listed in the text with their addresses, all suppliers mentioned in this manual can be found in the BioSupplyNet Source Book and on the Web site at:

http://www.biosupplynet.com

If a copy of BioSupplyNet Source Book was not included with this manual, a free copy can be ordered by using any of the following methods:

- Complete the Free Source Book Request Form found at the "Get the Source Book" link on the Web site:

  http://www.biosupplynet.com
- E-mail a request to info@biosupplynet.com
- Fax a request to 516-349-5598

Trademarks

The following trademarks and registered trademarks are accurate to the best of our knowledge at the time of printing. Please consult individual manufacturers and other resources for specific information.

5′/3′ RACE Kit	Boehringer Mannheim Corp.
Affi-Gel	Bio-Rad Laboratories, Inc.
Altered Sites	Promega Corp.
Bac-N-Blue	Invitrogen Corp.
Bac to Bac	Life Technologies, Inc.
Bluescript	Stratagene
Capture-Tec	Invitrogen Corp.
Centricon	Amicon, Inc.
CLONfectin	CLONTECH Laboratories, Inc.
Coomassie	Imperial Chemical Industries, Ltd.
CytoGem	Packard Bioscience
Dual-Luciferase	Promega Corp.
Econo-Column	Bio-Rad Laboratories, Inc.
Erase-a-Base	Promega Corp.
Exo-Size	New England Biolabs, Inc.
ExSite	Stratagene
FLASH CAT	Stratagene
FastTrack Kit	Invitrogen Corp.
Gene Pulser	Bio-Rad Laboratories, Inc.
GeneRunner Sequence Analysis Program	Hastings Software, Inc.
GenomeWalker	CLONTECH Laboratories, Inc.
Kimwipe	Kimberly-Clark Corp.
Kodak	Eastman Kodak
Lasergene Biocomputing Software	DNASTAR, Inc.
LipoTAXI	Genetic Applications, LLC
Luer-Lok	Becton Dickinson
Macintosh	Apple Computers, Inc.
Marathon cDNA Amplification Kit	CLONTECH Laboratories, Inc.
Muta-Gene	Bio-Rad Laboratories, Inc.
NucTrap Probe Purification Column	Stratagene
pBlueBac	Invitrogen Corp.
pCAT	Promega Corp..
pCAT3	Promega Corp.
pCITE	Novagen, Inc.
PerFect Lipid	Invitrogen Corp.
pET	Novagen, Inc.

pGEM	Promega Corp.
pHook	Invitrogen Corp.
Pipetman	Gilson Medical Electronics
pRetro-Off	CLONTECH Laboratories, Inc.
pRetro-On	CLONTECH Laboratories, Inc.
ProFection	Promega Corp.
PrimerSelect	DNASTAR, Inc.
pT7Blue	Novagen, Inc.
QuikChange	Stratagene
Retro-Xpress	CLONTECH Laboratories, Inc.
Saran Wrap	S.C. Johnson & Sons, Inc.
Sephadex	Amersham Pharmacia Biotech
Sepharose	Amersham Pharmacia Biotech
Single Tube Protein	Novagen, Inc.
SpeedVac	Savant Instruments, Inc.
SuperFect	QIAGEN, Inc.
SUPERSCRIPT	Life Technologies, Inc.
Teflon	E.I. DuPont deNemours and Co.
Talon	CLONTECH Laboratories, Inc.
Tfx Transfection Reagents	Promega Corp.
TNT	Promega Corp.
Transfectam	Promega Corp.
Whatman	Whatman International Ltd.
ZWITTERGENT	Calbiochem-Novabiochem International, Inc.

Index

DNA bending, 468

functions, 6, 11, 584–585

minor groove specificity, 440–441, 458–459, 462–463

site-directed mutagenesis, structure–function analysis, 419–421

substitution for TFIID in assays, 11

TBP. *See* TATA-binding protein

T-cell receptor α (TCRα), gene disruption experiments of protein binding, 305

TCRα. *See* T-cell receptor α

TdT. *See* Terminal transferase

Template commitment assay, basal preinitiation complex studies, 584

Terminal transferase (TdT)

D′-binding protein analysis

abundance of in vitro complexes, 294–295

correlation between DNA-binding sequence and control element sequence, 296–297

gene disruption experiments, 304–305

relative binding affinity comparisons, 303–304

transcription initiation analysis

primer extension assay, 104–105

RNase protection assay, 108

S1 nuclease assay, 111–112

transient transfection assay, 60, 144

TFIIA

complex with TFIID, analysis, 599–600

conformational changes during transcription complex assembly, 11

depleted transcription systems, 525

DNA recognition, 9

functions, 6, 10, 585, 596

purification

fractionation from HeLa cell nuclear extracts, 552–553

overview, 523–525

recombinant protein expressed in *Escherichia coli*

cell induction and lysis, 561

dialysis, 561

NTA-Sepharose chromatography, 561

overview, 560

reconstitution, 560–561

time line and organization, 560

recombinant protein expression systems, 521

TFIIB

altered specificity analysis of protein–protein interactions, 418, 420

conformational changes during transcription complex assembly, 11

depleted transcription systems, 525

functions, 6, 10, 595–596

purification

affinity chromatography, 413

overview, 522–523, 525

recombinant protein expressed in *Escherichia coli*

ammonium sulfate precipitation, 557–559

buffer and column preparation, 556–557

homogenate preparation, 557–558

monitoring, 558

overview, 556

phosphocellulose chromatography, 558

time line and organization, 556

troubleshooting, 559

recognition element (BRE), 9

recombinant protein expression systems, 521

TFIID

assembly and analysis of subcomplexes, 600–601

complex with TFIIA, analysis, 599–600

components and functions, 8, 10, 14, 584, 595–596, 607

conformational changes during transcription complex assembly, 11

core promoter binding, 9

crosslinking of complexes, 588–589

depleted transcription systems, 525

magnesium-agarose electrophoretic mobility shift assay

binding reactions, 609

buffer preparation, 608

data analysis, 610

gel

autoradiography, 609

casting, 608

electrophoresis, 609

overview, 607

protein titrations, 610

time line and organization, 607–608

troubleshooting, 610–611

purification

epitope-tagged protein

buffer and column preparation, 568–569

concentration, 571

contamination with protein A beads, 571

immunoaffinity chromatography, 570–571

nuclear extract preparation, 568

overview, 567

phosphocellulose chromatography, 569

time line and organization, 567–568

troubleshooting, 571

Western blot analysis, 570–571

fractionation from HeLa cell nuclear extracts with USA, 554

overview, 522, 524

recombinant protein

assembly of complexes, 390–391

expression systems, 521

TFIIE

affinity chromatography, 413

conformational changes during transcription complex assembly, 11

functions, 6, 10, 590, 592

purification

overview, 523

partial fractionation from HeLa cell nuclear extracts, 553–554

recombinant protein expression systems, 521

TFIIF

conformational changes during transcription complex assembly, 11–12

DNA recognition, 9

elongation regulation, 81

functions, 6, 10, 585

purification

overview, 523

partial fractionation from HeLa cell nuclear extracts, 553–554

recombinant protein expression systems, 521

TFIIH

ATP-dependence, 590